# Handbook of Industrial Oil and Fat Products

## Volume 4

Rendering, Manufacturing Processes for Emulsifiers, Recovery and Refining of Glycerine, Antioxidants, Fats and Oils in Soap, Surfactants, Paints, Varnishes, Leather Processing, Textile Processing, Pharmaceuticals, Natural Waxes, Cosmetics. Quality Assurance, Storage, Stability and Transport of Fats and Oils

**Handbook of Industrial Oil and Fat Products** in 4 Volumes

**Volume 1**

Chemistry of Fats and Oils, Chemical and Physical Changes in Edible Fats, Fatty Acids and their Derivatives, Toxicity and Safety of Fats and Oils

**Volume 2**

Preparation, Properties and Uses of Various Oils such as Coconut Oil, Canola Oil, Corn Oil, Cottonseed Oil, Olive Oil, Palm Oil, Castor Oil, Rice Bran Oil, Rapeseed Oil, Sesame Oil, Soyabean Oil, Sunflower Oil, Safflower Oil, Peanut Oil

**Volume 3**

Butter, Cheese, Casein, Margarine and Spread, Shortening, Cooking Oils, Salad Oils, Marine Oils, and Frying Technology

**Volume 4**

Rendering, Manufacturing Processes for Emulsifiers, Recovery and Refining of Glycerine, Antioxidants, Fats and Oils in Soap, Surfactants, Paints, Varnishes, Leather Processing, Textile Processing, Pharmaceuticals, Natural Waxes, Cosmetics. Quality Assurance, Storage, Stability and Transport of Fats and Oils

# Handbook of Industrial Oil and Fat Products

**Volume 4**

Rendering, Manufacturing Processes for Emulsifiers, Recovery and Refining of Glycerine, Antioxidants, Fats and Oils in Soap, Surfactants, Paints, Varnishes, Leather Processing, Textile Processing, Pharmaceuticals, Natural Waxes, Cosmetics. Quality Assurance, Storage, Stability and Transport of Fats and Oils

**G. S. BRECK** MSc, PhD
**S. C. BHATIA** BE (Chemical), MBA

CBS

CBS PUBLISHERS & DISTRIBUTORS

New Delhi • Bangalore

**Handbook of Industrial Oil and Fat Products** •Volume **4**



First Edition 2008

ISBN : 978-81-239-1514-2



*Published by :*
Satish Kumar Jain for CBS Publishers & Distributors,
4819/XI Prahlad Street, 24 Ansari Road, Ansari Road,
Daryaganj, New Delhi 110002, India.
E-mail : cbspubs@vsnl.com
Website : www.cbspd.com

*Production Director :* Vinod K. Jain

*Branch Office :*
Seema House, 2975, 17th Cross, K.R. Road,
Bansankari 2nd Stage, Bangalore - 560070
Fax : 080-26771680 • E-mail : cbsbng@vsnl.net

*Printed at :*
India Binding House, Noida, UP (India)

# Preface

This volume covering rendering, manufacturing processes for emulsifiers, recovery and refining of glycerine, and antioxidants is the fourth of the four-volume ***Handbook of Industrial Oil and Fat Products***. It also discusses fats and oils in soap, surfactants, paints, varnishes, leather processing, textile processing, pharmaceuticals, natural waxes, cosmetics. Quality assurance, storage, stability and transport of fats and oils by-product utilisation, analysis of oils and fats, and environmental aspects of animal and vegetable oil processing are important chapters of this volume.

The *Handbook* makes an essential reading for all students and teachers, professionals, researchers and industrialists, company executives and managers dealing with fats and oils. It is meant as an effective management tool for them and as a handy aid for their project responsibilities. It is also a valuable source of information for those preparing for or already associated with the oil technology, chemical engineering, food processing and foodservice industries, etc. To use these products efficiently, it is necessary to have a good understanding of the properties and behaviour of oils and fats.

The book contains valuable and necessary information on preparation, properties, applications and analysis of various oils and fats. Oils and fats constitute one of the three major classes of food products. The other two are proteins and carbohydrates. Oils and fats are essential nutrients of the human diet and are concentrated source of energy. In any study of human nutritional requirements, the contribution of each of these major food classes must be understood to set up proper diets for individuals. Knowledge of vitamins and minerals as well as the role of oxygen, water, fibre and trace elements is also essential.

Chapter 1 focuses on rendering which is the process of cooking raw material to remove the moisture and fat from the solid protein material. Emulsifiers are one of the unique ingredients in the food industry and are used in baking, margarine, formulated drinks, chocolate manufacturing, dough modification and many other applications. Keeping this in mind, Chapter 2 concentrates on manufacturing processes for emulsifiers. Chapter 3 is devoted to fats and oils in soap. Surfactants are compounds that reduce surface tension when dissolved in water or water solutions, or which reduce interfacial tensions between two liquids, or between a liquid and a solid. Keeping this in mind, Chapter 4 deals with fats and oils in surfactants. Chapter 5 focuses on fats and oils in paints, varnishes and related products. Chapter 6 deals with fats and their derivatives in leather industry.

Chapter 7 concentrates on fats and oils in textile industry. Chapter 8 is devoted to fats and oils in pharmaceuticals. Chapter 9 explains fats and oils in natural waxes such as beeswax. Various natural waxes of mineral and animal origin and vegetable waxes are discussed in detail. Cosmetics owe their origin to the use of fats because fats constitute the base of the cosmetics industry. Keeping this in mind, Chapter 10 focuses on fats and oils in cosmetics.

Chapter 11 is confined to the recovery and refining of glycerine. This chapter explains in detail the processing principles, refining and recovery of glycerine. Chapter 12 concentrates on oil processing and quality assurance, all aspects of production from unit operations, quality assurance and process control are discussed in detail. Chapter 13 deals with storage, stability, packaging and transport of fats and oils.

Chapter 14 focuses on utilisation of by-products obtained from various seed meals and oils and are used in various industries and in human foods. Antioxidants are compounds used in fats and food to prevent deterioration, rancidity, or discolouration caused by oxidation, and can prevent or delay oxidation of a molecule. Keeping this in mind, Chapter 15 is devoted to antioxidants and their importance in fats and oils. Chapter 16 concentrates on analysis of oils and fats. Chapter 17 is devoted to environmental aspects of animal and vegetable oil processing.

The text in this book is well supplemented with diagrams and tables. The treatment of all topics is in a cogent and lucid style aimed at enabling the reader to grasp the information quickly and easily. The references and index are also given at the end of the book.

While painstaking care has been taken into producing a useful and exhaustive reference/textbook, the authors would welcome any constructive criticism and creative feedback from the students, teachers and professionals in the industry. E-mail communications may please be sent to us at cbspubs@vsnl.com.

We wish to thank all those colleagues who provided us figures and photographs and offered us important advice during the preparation of this handbook.

**G. S. BRECK**
**S.C. BHATIA**

# Contents

## Contents in Detail

## Chapter 4 Surfactants 60-103

## Chapter 6 Fats and their Derivatives in the Leather Industry *135-146*



## Chapter 7 Fats and Oils in Textile Industry *147-151*



## Chapter 8 Fats and Oils in Pharmaceuticals *152-168*

## Chapter 12 Oil Processing and Quality Assurance *254-266*



## Chapter 13 Storage, Stability, Packaging and Transport of Fats and Oils *267-285*

## Chapter 14 By-products Utilisation *286-304*



## Chapter 15 Antioxidants and their Importance in Fats and Oils *305-327*

## Chapter 16 Analysis of Oils and Fats *328-499*

Chapter 1

# Rendering

## INTRODUCTION

The production of animal by-products into useful items is not a recent event. For example, early Eskimos and Native Americans would utilise the parts of the animals that could not be eaten. Hides and skins were used for clothing and shelter, bones provided utensils, and crude soap was made from melted fat and wood ash. However, it took the development of other industries (i.e. soap and candle making) before the rendering industry became an established industry.

The production of soap played a large part in the early development of rendering. The first record of rendering was the melting down of animal fat to obtain tallow.

Candle making also played a major role in the development of the rendering industry. The method of making candles using tallow goes back to the days of the Romans. They used tallow as the combustible material surrounding a flax or cotton wick. As the candles became available to the 'common' people, the need for tallow increased. Rendering was a process developed over the better part of 1800 years as a direct off-shoot of the candle making and soap industries. During this period, tallow was derived primarily from the fat of sheep and goats. This was because recovered fat from goat and sheep was harder and made better candle fat.

As cattle slaughter made the transition from the farm and ranch to the city, so did the rendering operation. Until this time, rendering existed in concert with soap manufacturing. Once the rendering companies discovered it was more profitable to produce tallow and sell it to the soap manufacturer than making it into soap themselves, the rendering industry, as it is known today, came into being.

The nutritional benefits of the rendered by-products were not fully understood until the nutritional needs of the domesticated animals produced for food was known. Until the nutritional benefits were understood, little fat was used in animal rations, and the protein meal was applied to the land as a fertiliser. Once the breakthrough in using rendered animal proteins and fats in feed occurred, the contribution of rendered products to the total value of the slaughtered animals was recognised. Methods of rendering also changed to improve the quality of the by-products made through rendering. Optimisation of by-product use reduces the ultimate cost to the consumer of the primary product, meat. The rendering industry is often referred to as the 'invisible industry.' Few people know or understand the industry's purpose and the important role it plays in the recycling of products from the slaughter of animals into useful products. This chapter is designed to give an overview and history of the rendering industry.

Rendering is the process of cooking raw material to remove the moisture and fat from the solid protein material. The raw material used in the rendering process includes animal offal, bones and trimming from meat processors, wholesalers and retailers and used kitchen and restaurant grease. The fat captured from the rendering process can be converted into edible or inedible fats. Edible fats are used commercially for the production of baked goods, margarine and frying oil. Inedible fats are used as an energy source for livestock feeds or used as industrial fats and oils for production of a variety of products such as soaps, lubricants and biodiesel fuel. The solid material is high in protein and can be processed into meat meal, meat and bone meal, and poultry meal to be used in livestock feed. The edible rendered protein produced with edible fats has human food application as a high quality protein additive.

## MODERN-DAY RENDERING

The rendering industry represents a major force in maintaining a clean environment by the disposal of perishable waste material generated by the beef, pork and poultry industries. The rendering process provides these same industries with high energy and high quality protein ingredients to supplement the feed grains that go to produce the beef, veal, pork, poultry and milk. The industry also supplies a large portion of its products to the pet food industry. Most significant is the fact that tallow, grease and animal protein are renewable resources.

## BY-PRODUCTS

The rendering industry performs the important job of recycling inedible tissues of animals into protein meals and fats and oils that are valuable animal feed and industrial ingredients.

### Meals

Animal by-product meals help fulfill protein and mineral needs in animal feeds. These meals are incorporated into rations produced for cattle, swine, poultry and aquaculture. List of by-product meals followed by the product description is given below:

1. Blood meal: This product is from clean, fresh animal blood, exclusive of all extraneous material such as hair, stomach belching, and urine, except as might occur unavoidably in good manufacturing processes; there is a guaranteed crude protein of 85 per cent.
2. Hydrolysed feather meal: This product results from the treatment, under pressure, of clean, undecomposed feathers from slaughtered poultry; there is a guaranteed crude protein of 80 per cent.
3. Meat and bone meal: This product is produced from mammalian tissues, including bone but exclusive of any added blood, hair, hoof, horn, hide trimmings, manure, stomach and rumen contents, except in such amounts as may occur unavoidably in good processing practices; this product ranges in crude protein from 45 to 54 per cent.
4. Meat meal: The product is produced from mammalian tissues inclusive of products listed in meat and bone meal, except it does not contain any substantial amount of bone; this product ranges in crude protein from 52 to 60 per cent.
5. Poultry by-product meal: This product consists of ground, rendered, clean parts of the carcass of slaughtered poultry, such as necks, feet, undeveloped eggs and intestines, exclusive of feathers, except in such amounts as might occur unavoidably in good processing practices; the range of crude protein is 58–62 per cent.

6. Fish meal: This product is from clean, dried ground tissue of undecomposed whole fish or fish cuttings, either or both, with or without the extraction of part of the oil; the range of crude protein is 60–65 per cent.

As with any dry feed ingredient, if stored in the proper environment, by-product meals can be kept for long periods of time. The rendering process kills bacteria. However, the meals can be re-contaminated through various environmental sources. Since, animal proteins provide a favourable climate for the growth of bacteria, extra care must be taken during storing and handling to prevent bacterial growth. The biggest factor is moisture; therefore, meals should be kept in a dry environment. The presence of moisture will also cause the fat in the meal to become rancid. Antioxidants can be added to the meal, either during the cooking process or during the blending of various meals to increase the shelf-life of the meals.

## Oils

The fats and oils produced from the rendering process provide energy in the most concentrated form available for use in animal feeds. Addition of fat in feed also reduces dustiness of the feed and makes it more palatable. Fats and oils also provide important ingredients for further processing for the making of many products such as soap, fatty acids and glycerol. Edible fats produced in the rendering industry may be used for commercial baked goods, fast-food restaurants, margarine production, or medicinal applications such as omega-3 fatty acids. The following is a list and description of liquid products resulting from the rendering process.

1. Feed-grade animal fat: derived primarily from the offal of beef and pork.
2. Poultry fat: derived from poultry offal.
3. Choice white grease: derived primarily from the rendering of pork offal.
4. Tallow: derived from the rendering of beef offal.
5. Yellow grease: derived primarily from restaurant grease.
6. Blended animal-vegetable fat: can include blends of all types of animal fat, vegetable oil, acidulated vegetable oil, soapstock and/or restaurant grease.
7. Fish oil (marine oil): derived from the rendering of whole fish or cannery waste.

As the soap-making industry grew and became more refined, the quality standards and specifications for animal fat became more precise to meet the soap industry's need. The specifications and terminology, from the soap industry, form the basis for the quality standards and characteristics used today. Some of these specifications include titer, iodine value, free fatty acid, refined and bleached colour, moisture, impurities and unsaponifiables.

Titer and iodine values (IV), both being a measurement of fat hardness, are important because the hardness of fat varies with different animal species. Hard fats make firm soap, while soft fat makes soft soap. The harder fats resist oxidation more than the softer fats, so that the firmer soaps do not go rancid as fast as the softer soaps. Free fatty acid (FFA) is a measure of the amount of hydrolysis that has taken place within the fat molecule. Time, temperature and the presence of moisture all favour the hydrolysis of fat into free fatty acids and glycerol. Refined and bleached colour (R & B) is a factor that is determined by the degree of damage done to the fat. A low R & B colour indicates the fat was rendered from high quality raw materials using good rendering techniques. This is important for finished soap quality. Moisture, impurities and unsaponifiables (MIU) are important determinants of fat quality. Excess MIU will cause deterioration of the fat during storage; therefore, the lower the MIU, the better the quality.

Animal fats are not difficult to handle and if kept in good condition, the fat can be kept for long periods of time. Several factors will determine how long animal fats can be kept. Fats should be stored and handled in steel or iron tanks and piping, avoiding any contact with brass or copper.

Even short exposure to brass and copper will cause the fat to go rancid. Dry, clean fat will not be damaged by moderate heat; however, excess MIU with moderate heat will increase the FFA and instability of the fat while increasing the probability of rancidity. Peroxide value is a measure of stability. When fats are handled properly and stabilised with a proper amount and type of stabiliser such as an antioxidant, the fats will remain stable with a low peroxide value for long periods.

The fats and oils go through many processes before they are made into their finished form. These processes are described in more detail below.

The major use of edible tallow and lard is for cooking and frying. Until recently, edible tallow was used quite extensively by the fast-food industry for frying foods such as french fries. However, due to the perceived health concerns regarding cholesterol, tallow has been replaced by vegetable oils and human consumption of tallow has declined. Nonetheless, the use of edible tallow in other markets has expanded, including the manufacturing of foods such as bakery products and margarine.

The majority of the rendered fat produced is inedible tallow and grease. A large portion of the market for inedible tallow and grease is accounted for by animal feeds. However, there are non-feed uses for inedible fats also.

## RENDERING PROCESSES

### Meals

#### *Blood meal*

Blood meal may be prepared by several different processes. Cooker dried blood is produced in a conventional batch cooker. Spray dried blood is a two-step procedure by which moisture is first removed by a low temperature evaporator under vacuum until it contains approximately 30 per cent solids and then is spray dried. Flash dried blood meal is also a two-step process and is shown in Fig. 1.1.

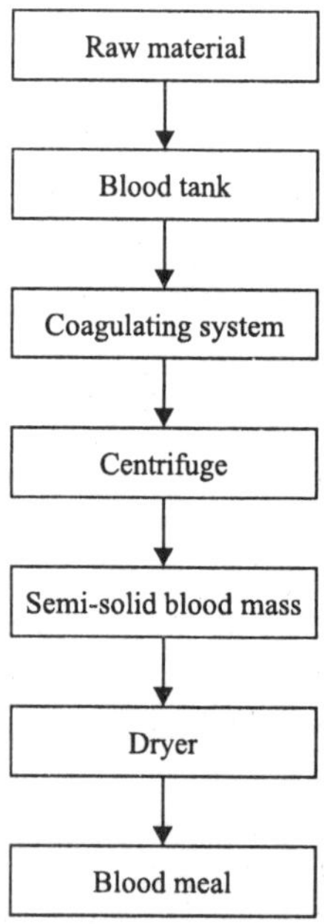


**Fig. 1.1.** Flow diagram of a typical blood system.

The blood is first put into a coagulating system. The raw blood is pumped through a heated pipe (live steam injected into a pipe), which causes coagulation. The coagulated blood is then transferred to a dewatering device (such as a centrifuge), and a large portion of the moisture is removed, yielding a semi-solid protein. The coagulated blood may also have the water removed by cooking in a condenser to a semi-solid state. The semi-solid blood mass is then transferred to a rapid drying facility, such as a flash dryer, where the more tightly bound water is rapidly removed. Ring dried blood is similar to flash dried blood meal, except that the rapid drying facility is a ring dryer. Ring drying the blood produces a better quality product than the conventional method, because it uses warm air to dry the blood mass and not warm metal. The blood is not exposed to a direct heat surface and consequently the amino acids (e.g. lysine) are not destroyed.

Care must be taken in the processing and/or purchasing of blood meal. These different processing methods produce differnt types of blood meal, which vary in crude protein and amount and bioavailability of lysine and other amino acids. Table 1.1 shows a summary of different processing methods and the resulting lysine availability.

**Table 1.1.** Total and biological availability of lysine with various methods of processing.

| *Method* | *Total lysine (g/16 g N)* | *Biological availability (% total)* |
|---|---|---|
| Conventional | 7.6 | 49 |
| Ring dried, low temperature | 9.5 | 78 |
| Ring dried, high temperature | 8.3 | 55 |
| Flash dried | 9.2 | 94 |
| Spray dried | 8.7 | 83 |

### *Feather meal*

Feathers are rendered either using a batch hydrolyser or a continuous hydrolyser. Poultry feathers consist of the protein keratin, which is essentially indigestible. To make the protein digestible and available for the domesticated animal, the feathers must be hydrolysed. This is done in a hydrolyser under pressure (approximately 30–50 psi for 45–60 min) to break the bonds that hold the protein together (such as amino acid bonds, sulphur-sulphur bonds of cystine and hydrogen bonds between non-adjacent amino acids). Figure 1.2 shows the basic flow diagram for rendering feathers. The raw material may first go through a spreader to cut up any large particles (such as heads or feet) that may have entered the system. The raw feathers are then fed into the hydrolyser and the protein bonds are broken down. The hydrolysed feather is then fed into a dryer to remove excess moisture to yield feather meal. At this point the meal can then be ground to produce a finer ground feather meal.

### *Meat and bone meal, meat meal and poultry by-product meal*

Understanding that beef, pork and poultry processing procedures are similar, for simplification, only the rendering process of ruminant meat and bone meal will be discussed.

Raw material collection is the first and an important step in the rendering process. The timely and efficient collection of raw materials is a must to the industry.

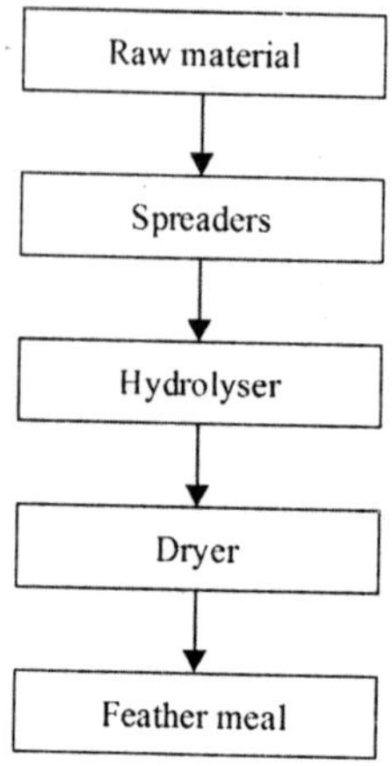


**Fig. 1.2.** Flow diagram of a typical feather system.

The raw material needs to be clean and fresh to yield the highest quality products. Dirty and decayed offal will lead to a darker colour fat, which is undesirable. Figure 1.3 shows a typical flow diagram of the production of meat and bone meal and tallow. The raw material is first collected by the trucks and then discharged into a receiving bin. The material is then screw conveyed to a crusher, pre-breaker, or similar device for size reduction: 2.5 to 5.0 cm (1–2 inches). It is important to reduce the size of material before processing to increase production rate, decrease energy costs, and lead to uniform cooking, which enhances the quality of the finished product.

At this point, the raw material then goes into either a batch cooking or continuous cooking system. Within the cooking system, the heating and grinding action breaks down the fat cells of animal tissue so the fat is released. The rendering process consists essentially of two steps. First, the raw material is heated or 'cooked' to evaporate the moisture. The second step is the removal of fat.

The batch cooking system is the process by which batches of raw material are cooked according to a repetitive cycle. There are two methods of batch cooking: wet rendering and dry rendering. For the wet rendering method, the material is cooked by steam under pressure in large closed tanks. The material is boiled with open steam under pressure (25–75 psi internal pressure) from the bottom of the cooker. After approximately 1.5–2.5 hr., the pressure is then reversed and applied from the top. This causes the water and sludge to settle to the bottom of the cooker, where it is drained off. After the water is drained off, the fat is removed from the bottom and diverted to another tank for further processing. The tankage then goes to the press for further fat removal.

The dry rendering process is the newer and more efficient method of cooking. All the material is cooked in its own grease by dry heat in open steam-jacketed drums until the moisture has evaporated. Efficient cookers usually range in temperature from 115 to 120°C and take approximately 1.5–4 hr. Agitators in the cooker reduces the cooking time. However, care must be taken because increased agitation speed usually increases the amount of fines or particulate matter in the fat. After the cooking process is complete, the material goes across a screen that allows the free fat to run off. The tankage is then conveyed to a press where the residual fat is removed, resulting in a product that contains 6–10 per cent fat. The continuous cooking system is similar to the dry rendering batch method, except the raw material is fed into the system in a semi-continuous fashion. This is the most modern and efficient cooking method. A continuous system normally consists of a single cooker, while the batch system consists of multiple cooker units.

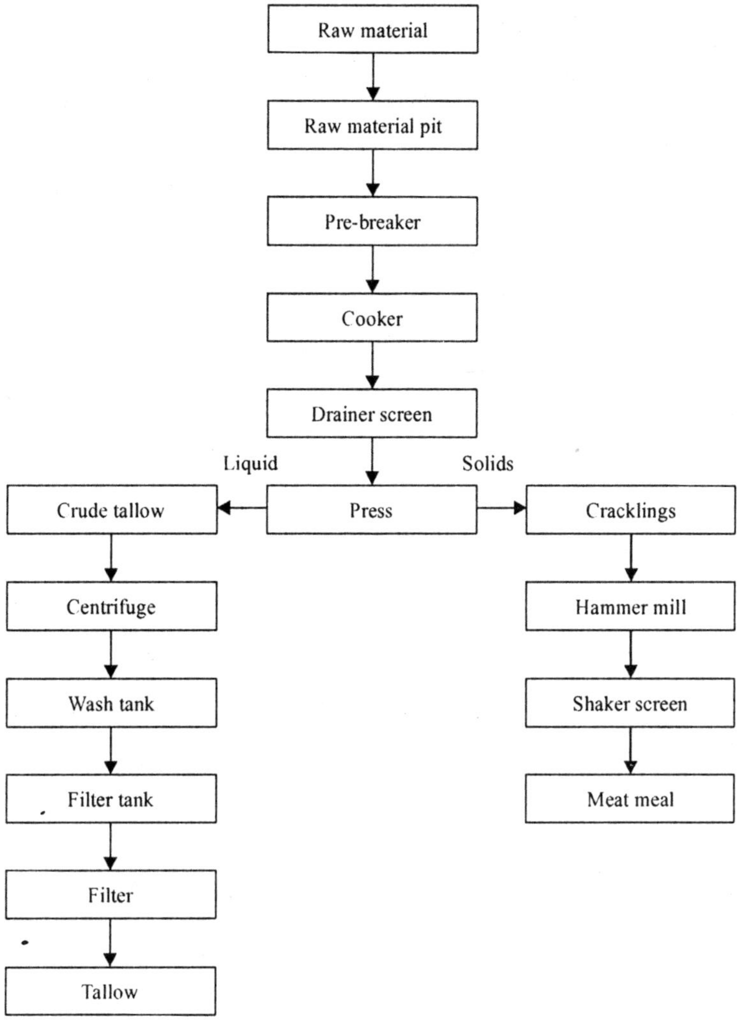


**Fig. 1.3.** Flow diagram of a typical beef system.

Because of this, continuous cooking systems offer many advantages over batch cooking methods, including reduced labour requirements, less floor space required (one cooker versus several), lower maintenance and operating costs (less time, temperature and fuel consumption with one cooker), and easier control of air and water pollution (more enclosed environment with the one cooker). The continuous system produces a more consistent product than the batch system. Since the flow is essentially constant, process variables are more easily controlled. However, care must be taken because some raw material could be short circuited without sufficient contact time to destroy the bacteria adequately in the raw material. Temperature of the continuous process usually falls into two categories: low temperature (100–115°C) and high temperature (130–150°C). The time the material stays in the cooker ranges from 30 to 60 minutes (depending on temperature: the lower the temperature, the longer it stays in the cooker) and the material is continuously discharged. The product is usually 4–6 per cent moisture.

After the cooking process is completed, the cooked material is drained and pressed to remove the fat using either a hydraulic or screw press (Fig. 1.3). The solid protein (cracklings) is then normally ground with a hammer mill. After the hammer mill, the meal passes over a shaker screen to remove any coarse material and to produce a uniform protein meal by-product. The fat discharged from the press normally contains fine solid particles, which are removed by centrifugation or filtration.

### *Fish meal*

There are two general methods of manufacturing fish meal: the dry process and the wet process. In the dry process, the oil produced during cooking is not removed. This process can be used when the starting material is fish with low oil content. In the wet process, a portion of the fish oil in the product is removed during processing. Fish meal made from fish with a high oil content, such as menhaden and anchovy, must have the oil removed to produce a stable product that can be easily handled. Most fish meal is produced using the wet method, which is usually a continuous cooker process similar to meat and bone meal production. The raw material is ground and fed through heat exchangers into a cooker. The cooked material is then hydraulically pressed against a screen to remove water, soluble protein and oil. What is left is called presscake. The presscake is then dried and ground to produce fish meal.

## Oils

### *Greases*

There are several steps taken to process the liquid once it is obtained from the presses during rendering or from street collection of restaurant grease. A typical flow diagram of a grease processing system is shown in Fig. 1.4.

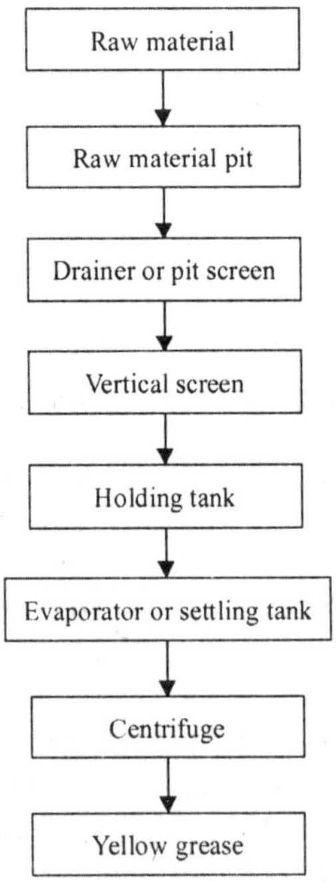


**Fig. 1.4.** Flow diagram of a typical grease system.

The grease is brought into the rendering plant in either 50-gal drums or bulk tankers. The raw grease is dumped into a receiving pit and then is screwed conveyed over a drainer screen and then a vertical screen to remove any foreign material such as rags, plastics, etc. The grease can then go to a vertical settling tank (holding tank) and is held while the water and more impurities are settled out or the grease goes to an evaporator system, which evaporates the water from the grease. Once the water is removed, the grease then goes through a centrifuge to remove more impurities. Once the moisture and impurities are at an acceptable level, the grease is pumped into a storage tank.

### *Tallow*

A flow diagram of a typical tallow system is shown in Fig 1.3. After the press, the crude tallow is pumped to a centrifuge, where small proteinacious material is removed and then to a holding tank washing and/or filtering to remove any impurities. After filtering, tallow can undergo further processing, including refining, bleaching, deodourising, fatty acid splitting and hydrogenation.

The term refining refers to any purifying treatment to remove any impurities in the tallow. These impurities may include free fatty acids, phosphatides, or other material found undesirable in the fat. The terms exclude bleaching and deodourising. Bleaching refers to the treatment of the tallow solely to reduce the colour of the oil. Deodourisation is the term used for the removal of fractions in the fat (e.g. free fatty acids and peroxides) that may cause flavours and odours.

Individual fatty acids are produced by 'splitting' the triglyceride to form fatty acids and glycerol. The glycerol is then separated and sold as a by-product. Distillation is used to separate and purify the fatty acids for a wide range of industrial applications, including surfactants, soaps, plastics, resins, rubber, paints, lubricants, textiles and cosmetics.

Hydrogenation is the addition of hydrogen to the carbon-carbon double bonds in the presence of a metal catalyst. This results in the conversion of liquid oils to hard or plastic fats and soft fats to firmer products. It also results in improved resistance of the fats and oils to deterioration through oxidation.

### *Fish oil*

The liquid removed from the fish presscake is separated by centrifugation into an oil fraction and a water soluble protein fraction called solubles. The solubles are often concentrated by evaporation and either added back to the presscake before drying, sold separately as 'fish solubles', or discard. Fish meal plants operating onboard fish processing ships usually discard the solubles.

The fish oil produced after the solubles have been removed can be further processed and utilised in various applications, such as a component of paint, a feed additive for aquaculture diets, edible oil used in preparing a variety of human foods and pharmaceutical uses. Currently, the primary pharmaceutical application is in the use of fish oil for its high omega-3 fatty acids. The omega-3 fatty acids are being used as oral supplements in connection with reducing certain cardiovascular disorders.

## ENVIRONMENTAL PROBLEMS

Environmental issues are important to the industry. As stated earlier, the renderers are the original recyclers. Without rendering, by-products from the meat-packing industry would fill up the landfills. Many modern rendering plants are utilising new techniques to help conserve energy and water. Heat recovery systems are available to recover waste heat from processing equipment. This heat can then be used to run other equipment. Water vapour that is released during the rendering process (cookers and evaporators) may be condensed and the water then used for other in-plant operations. Although the rendering industry can be viewed positively for its role in recycling organic material, it has the potential to create environmental problems in the air, water and on the land.

### Air

Air pollution is typically in the form of odours. During the rendering process, numerous breakdown products are formed or released from the raw materials. These compounds travel with the water being evaporated in the cooking process. If released directly to the air, though not hazardous, they would be quite objectionable. For this reason, condensers are placed after the cooking unit to cool the gas flow. This condenses the steam back into water as well as condensing some of the low boiling-point organics. Many of the odorous compounds are soluble and remain with the condensed liquid, which is sent to a waste-water treatment plant. Any materials that remain in the gas phase from the condenser are either incinerated (using a steam boiler or a designated burner) or scrubbed by passing them through an oxidising media.

Since, rendering processes are not tightly sealed, some odorous vapours will escape into the surrounding room. This room air is also scrubbed through oxidising media. Large volumes of air are drawn through the scrubber to keep the building at a negative pressure so that any leaks in the structure will be drawing fresh air in.

### Water

Water is produced in large quantities during the rendering process. The two main areas of waste-water production are from condensers and plant wash up. Wash-water is high in fat, while condenser water can have a $BOD_5$ (a measure of organic material) of 10,000 parts per million (ppm) and ammonia of 750 ppm. For comparison, domestic sewage has a $BOD_5$ of about 200 ppm and ammonia of 20 ppm. For this reason, rendering wastewater cannot be discharged directly to a stream and most municipal plants will require some sort of treatment before discharge to their facilities.

A typical rendering waste-water plant will contain a skimmer or dissolved air flotation (DAF) unit, which floats fatty and lightweight solids for recovery. This is followed by an extended aeration biological treatment system. In this system, bacteria are used to digest organic material ($BOD_5$ and ammonia, as well as others) that remains in the water. These bacteria are settled out in clarifiers so that only clear water continues through the process. High quality streams may require phosphorus removal, filtration and/or chlorination before discharge.

### Land

Excess bacteria generated in the waste-water treatment plant as well as other solids generated in various areas of the rendering process are often applied to the land either as a disposal option or as a beneficial soil conditioner. Care must be taken to ensure that nutrient loadings do not exceed vegetative demand, or they will migrate off site and create potential pollution problems (including elevated nitrates in ground water).

Dissolved solids or salts in land applied material can also create problems. They chemically collapse the clay structure, which normally allows soil to hold water. Plants will exhibit droughtlike symptoms even with adequate rainfall. Applying sludge in liquid form greatly increases the potential for salt overload.

### Environmental Issues

Environmental considerations, particularly in terms of water and air pollution control, will continue to have an impact on the economics of the business. As Federal regulations in odour control and water pollution get stricter, production costs will increase. As with other industries, smaller operations that do not have the financial ability to make substantial capital investments may be forced to close.

Another major environmental concern will be of dwindling energy sources. A lot of energy is used during the rendering process and from the fleet of trucks needed to haul the raw and finished material to the steam generators that supply the heat for the cooking process. As energy sources dwindle and become more expensive, the renderer will be forced to find more efficient and economical rendering equipment. The renderer will also be forced to supply fuel to the truck fleet that will be cleaner burning and more abundant. One consideration is that of biodiesel. This involves the formation of methyl esters from plant oils and animal fats to be used in modified diesel engines. The biodiesel is cleaner buring than today's diesel and is made from a renewable source, fats and oils.

### Government Regulations

As with any business, the renderer will have to abide by all governmental regulations. This will include following Occupational Safety and Health Administration (OSHA) regulations, health care regulations and other employment acts.

### Raw Material

Changes in consumer preference of meat cuts will change the face of the meatpacking industry. Due to consumer demand, meatpackers are having to trim the fat on the meat cuts closer, leaving less to be trimmed when the meat is prepared for retail sale at the local supermarket level. This will lead to more fat and bone staying in the packing plant and less at the supermarket. This change will increase the need for the rendering plant to be located next to the packing plant and produce a shortage of material in the areas served by metropolitan renderers. These urban renders will have to rely more on profits from processing restaurant grease and less from rendering fat and bone from butcher shops and supermarkets.

## FUTURE OUTLOOK

When looking at the future outlook of the type of raw material available to the renderer, DAF, sludge and trap grease must be discussed. DAF, as stated earlier, is involved with removing fatty substances from the water stream (earlier, is involved with removing fatty substances from the water stream). Sludge is the solids left from the waste-water treatment plant. Currently, some renderers are able to take the DAF and sludge and use it for land application. The problem is the amount of metals (such as iron, zinc and aluminium) found in DAF. As environmental regulations increase, the land application of the DAF and sludge will diminish. The problem then arises of how to dispose of this DAF and sludge. If it is to be rendered, the decision must be made as to which rendering process this material should be applied. Trap grease, or catch basin skimmings, is also a problem. A catch basin is a trap that separates fat-bearing material from the water before it goes to the sewer. The catch basins have a series of baffles so that the fat-bearing material will rise to the top and can be easily skimmed off. As the popularity of fast-food restaurants increases, the need for cleaning out the grease traps will also increase. This grease is high in moisture, particulate matter and free fatty acids, making it more costly to process and yielding lower quality finished fat. The problems become determining the feeding value of this grease product and finding a market.

Thus, with the increasing population of human there will be a geographical shift in the animal population. Animals will become more efficient in utilising feed and have increased production (e.g. milk, meat, eggs) either through generic manipulation and/or proper diet formulation. This will increase the need for high quality feed ingredients (such as animal by-products). As the nutrient requirements of animals are better defined, the marketing of the protein products will change from pricing on a crude protein basis to individual amino acid levels. This will shift a product such as meat and bone meal, or a blend of different animal proteins, from a commodity to a value-added product.

Considering the previously mentioned variables, the future of the rendering industry appears optimistic. Nevertheless, as in many industries, the most efficient producer with the lowest operating costs will ultimately survive and prosper.

Chapter 2

# Manufacturing Processes for Emulsifiers

## INTRODUCTION

A stable mixture of two or more immiscible liquids held in suspension by small percentages of substances are called emulsifiers. These are of two types—proteins or carbohydrate polymers, which act by coating the surfaces of the dispersed fat or oil particles, thus preventing them from coalescing, these are sometimes called protective colloids. Other types are long-chain alcohols and fatty acids, which are able to reduce the surface tension at the interface of the suspended particles because of the solubility properties of their molecules. Soaps behave in this manner; they exert cleaning action by emulsifying the oily components of soils. All such substances, both natural and synthetic, are known collectively as detergents.

Polymerisation reactions are often carried out in emulsion form; a wide variety of food and industrial products are emulsions of one kind or another, e.g. floor and glass waxes, drugs, paints, shortenings, textile and leather dressings, etc.

All emulsions are comprised of a continuous phase and a disperse phase; in an oil-in-water (o/w) emulsion (milk), water is the continuous phase and butter fat (oil) the disperse phase; in a water-in-oil (w/o) emulsion (butter) free fat (from crushed fat globules) is the continuous phase and unbroken fat globules plus water droplets are the disperse phase.

Emulsifiers are one of the unique ingredients in the food industry. They are used in baking, mayonnaise, margarine, formulated drinks, chocolate manufacturing, dough modification and many other applications. In a typical oil processing plant, emulsifiers such as monoglycerides and diglycerides, propylene glycol esters, lecithin, sorbitan esters, polyglycerol esters and lacilated and ethoxylated monoglycerides could be used to formulate different products.

However, the most common emulsifiers manufactured in an oil processing plant are monoglycerides and diglycerides and propylene glycol esters. Few oil plants manufacture lacilated monoglycerides. Other speciality type esters are manufactured by companies that specialised in emulsifiers. This chapter discusses the manufacturing processes for monoglycerides, diglycerides and propylene glycol monoesters. Critical process variables, finished product quality and troubleshooting of the processes are also discussed in detail.

Monoglycerides are formed when one of the three hydroxyl groups in a glycerol molecule is replaced with a fatty acid radical. Although the fatty acid can be located at any of the three locations, the end hydroxyl groups are primarily reacted. The middle hydroxyl group is more difficult to esterify.

## METHODS FOR MAKING MONOGLYCERIDES

Monoglyceride can be made via one of the following methods:

1. Direct esterification of glycerol and fatty acids.
2. Reaction of glycerol halohydrins with fatty acid salts of alkali metals or silver.
3. Esterification of glycerol with triglycerides in presence of a catalyst, which produces mixed glycerides and is generally followed by industrial manufacturers of monoglycerides.

Direct esterification of equimolar mixture of glycerol and fatty acids would yield monoglycerides, diglycerides and triglycerides. It was found that the yield of monoglycerides could be increased if the reaction was carried out in presence of a solvent in which both fatty acids and glycerol were soluble. Similar results were obtained by using dioxane as a solvent.

## DIRECT ESTERIFICATION OF GLYCEROL AND FATTY ACIDS

Direct esterification of glycerol and fatty acids produce monoglycerides, diglycerides and triglycerides at various levels. Composition of the end product depends on the ratio of glycerol: fatty acids, type of fatty acids and the reaction conditions applied in the process. In an equimolar mixture of glycerol and palmitic acid, predominately diglycerides and triglycerides were formed during initial reaction. This may indicate that diglycerides and triglycerides are the intermediate products formed in the process of making monoglycerides. A later study showed that direct esterification of glycerol with palmitic, oleic, and stearic acids, produced monoglycerides, diglycerides and triglycerides. It was demonstrated that triglycerides and diglycerides formed first and broke down to monoglycerides as the reaction progressed.

It was found that lauric, stearic, or oleic acid would completely esterify with glycerol when the mixture was heated at 230–400°C (446–464°F) for 3 hr. At 170–180°C (338–356°F), this mixture required prolonged heating for the same reaction. Lauric and oleic acids produced primarily monoesters and diesters, while stearic acid produced predominately diesters and triesters. It is interesting to note how the type of fatty acids used in the reaction determines what type of ester linkage is produced in glycerolysis. Reaction with equimolar proportion of fatty acids and glycerol at 180°C (356°F) produced primarily diglycerides and triglycerides. However, when the initial reaction step was followed by a second reaction step at 240–245°C (464–473°F), mostly monoglycerides and diglycerides were produced with virtual elimination of triglycerides as seen at the lower reaction temperature.

Direct esterification can be carried out with or without the presence of a catalyst. Non-catalytic reaction process needs to be carried out at high tempertures. This requires a long reaction time and tends to produce darker coloured product.

In direct esterification process, 1-monoglycerides and 1,3-diglycerides are found as the predominant reaction product. Synthesis of 2-monoglyceride or 1,2-diglyceride is difficult. It has been demonstrated that positional migration occurs from the 2 position to positions 1 or 3. This results in the formation of 1-monoglycerides and 1,3-diglycerides.

## REACTION WITH HALOHYDRINS

During the beginning of the nineteenth century, researchers made glyceryl esters of fatty acids by reacting glycerol halohydrins with alkali or silver salts of fatty acids. This method was considered reliable, since it was believed that the carboxyl group would take the position originally occupied by the chloride ion. However, later research showed that the end product is not predictable because of the migration of the carboxyl group from the middle position to the end ones. It was shown that the ester linkage at the middle position does not migrate when the acyl group is a cyclic fatty acid radical instead of being an aliphatic acid radical.

## REACTION WITH TRIGLYCERIDES

Reaction between triglycerides in presence of excess glycerol and a catalyst, produces a mixture of monoglycerides and diglycerides plus some unreacted triglyceride and catalyst residue. Mixed glycerides were obtained by reacting a 1:1 mixture of triglycerides and glycerol in presence of trisodium phosphate as a catalyst.

*Elder* and *Richardson* patented a process for making monoglycerides and diglycerides. They prepared monoglycerides by reacting glycerol with triglyceride, using an alcoholate catalyst at 140–290°C (284–554°F). They also used glyceroxide as a catalyst, used mechanical agitation and excess glycerol to produce a higher yield of monoglycerides. In another patent, they described a process for making monoglycerides by using alcoholates and alkali metal soap as catalyst and removing the excess water by blowing a stream of inert gas through the reaction mixture. They also conducted the same reaction under vacuum to remove the excess water. The same workers described their process of making propylene glycol monoesters by reacting triglyceride in excess of propylene glycol using metal soap of $Ca^{++}$, $Zn^{++}$ or monovalent alkali metals as catalyst, in another patent.

*Christensen* reported a method for making monoesters and diesters of glycerol. He reacted metal glyceroxides with fatty acids at 100–200°C (212–392°F). He claimed that by reacting 1-glyceroxide with fatty acids he obtained 1-monoglyceride. Similarly, by reacting, 1,2-diglyceroxide with fatty acids, 1,2 or 1,3 diglycerides would be produced.

*Richardson* and *Eckey* used dioxane, as solvent along with glycerol and triglyceride, in the reaction mixture and glyceroxide as catalyst. The reaction was carried out under reflux at 107°C (225°F). Reaction was judged to be complete when the mixture turned clear. At this point, the reaction mixture was neutralised with hydrochloric acid and the dioxane was removed by stripping the mixture with steam or nitrogen at 90–100°C (194–212°F). The residual dioxane was removed at 120°C (248°F). The resultant product from and ether extract showed a yield of 82 per cent of the theoretical value.

*Reiser* and *Isbell* described a process of obtaining a 90 per cent yield of monoglycerides. Triglyceride and isopropylidene glycerol were reacted in presence of an acidic or a alkaline catalyst at 100–200°C (212–392°F). This produced monoester of isopropylidene glycerol. This was then hydrolysed, to produce glycerol monoester and acetone. Acetone was distilled to obtain pure emulsifier.

## COMMERCIAL PRODUCTION OF MONOGLYCERIDES

In most manufacturing plants, monoglycerides are made in a batch process by reacting triglyceride with excess glycerol in presence of a catalyst. The finished product generally contains 35–50 per cent monoglycerides. The rest of it consists mostly of diglycerides, some unreacted triglycerides (c. 10 per cent), residual glycerol (3–4 per cent) and free fatty acids (1–3 per cent). It would be pointed out that the free fatty acids as measured is not true fatty acid, but titratable acid. Phosphoric acid, added in slight excess at the neutralisation step leaves some residue in the product along with the acid salts and are titrated as free fatty acid.

Higher concentration of monoglycerides can be obtained in a standard batch reactor by using a higher proportion of glycerol and higher amount of catalyst. This drastically reduces net pounds of monoglycerides produced per batch of reactants because a lower proportion of fat is used, making the process less economical. Higher concentration of monoglycerides is mostly made through molecular distillation process. There are some patented processes that can allow one to produce higher monoglycerides content in the product (80–90 per cent monoglycerides or higher). These are specialised processes and are not commonly used.

## Equipment for Batch Process

All equipment for a batch glycerolysis process (except deodourised oil storage and caustic storage tanks) is recommended to be made of stainless steel. Either 304 or 316 stainless steel is suitable. However, for the phosphoric acid storage and handling system, 316L stainless steel is recommended. A simplified flow diagram for batch glycerolysis process is shown in Fig. 2.1.

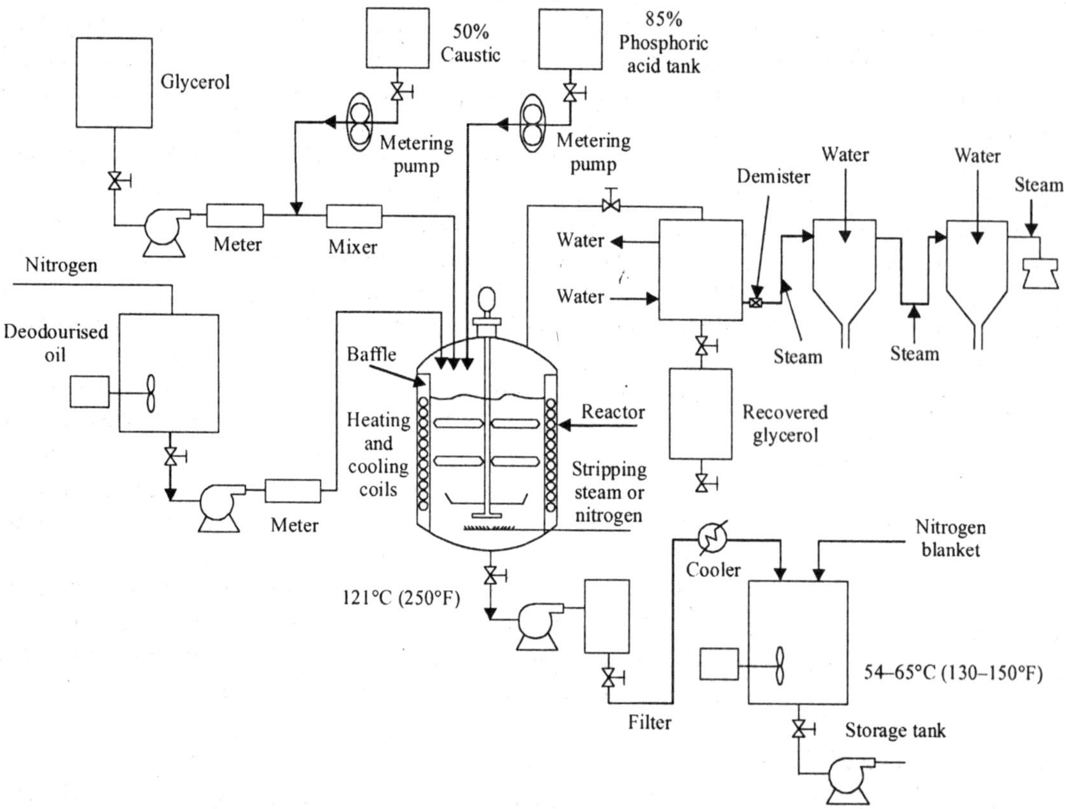


**Fig. 2.1.** Typical flow sheet for the batch glycerolysis process.

Deodourised oil, at about 55–60°C (130–150°F), is stored in the storage tank under nitrogen and is metered into the reactor through a batch meter, as needed. Glycerol from the storage tank is pumped through another batch meter. On the downstream side of this meter, a 50 per cent NaOH solution is metered in, either by a metering pump or through a batch meter. The two liquid streams are then mixed intimately in an in-line mixer before it is charged into the reactor. Phosphoric acid (85 per cent concentration) is stored in a 316L stainless steel tank and is metered into the reactor at the end of reaction to neutralise the alkali catalyst. Reactants are mixed intimately in the reactor by using a top entering agitator with specially designed impellers to achieve efficient mixing. Some of the installations use steam or nitrogen sparging for agitation. The reactor has high pressure steam heating and water cooling coils. There are baffles along the wall of the reactor built to prevent vortex created by the top entering agitator. The reactor is a pressure vessel, capable of withstanding full vacuum as well as positive pressure, which is typically 75 psig. A three-stage vacuum ejector system with an in-line surface condenser and a condensed receiver are provided. This pulls glycerol vapour through the condenser, where it is condensed and then collected in the receiver below. Glycerol-free finished product is then filtered to remove phosphate salt, cooled further and stored under nitrogen in a 304 stainless-steel tank. External cooler is recommended. The product could also be cooled down in the reactor and then filtered. However, this increases the total cycle time for a batch. Free fatty acids are separated from the glycerol condensate by gravity. Glycerol is re-used. Fatty acids are sold, usually at a premium. There are few continuous process plants for manufacturing monoglycerides which will be discussed later.

### Processing Steps for Glycerolysis: Batch Process

Typical steps for making commercial monoglycerides are listed below:

1. Deodourised oil is charged into the reactor at 82.2–85°C (180–185°F) under 25–50 mm Hg absolute pressure; this deaerates the oil as the reactor is being loaded.
2. Required amount of glycerol, mixed with 0.1–0.3 per cent sodium hydroxide (dry basis) is added to the reactor; sodium hydroxide solution is either pre-mixed with the glycerol or the two streams are blended through an in-line mixer, while charging the reactor; caustic solution must be added slowly into the glycerol stream to ensure uniform mixing of the two before pumping into the reactor.
3. Reaction mixture is mixed using the mechanical mixer; mixing can also be achieved by using steam or nitrogen sparging or external recirculation.
4. Reactants are then heated to 115.5–126.6°C (240–260°F) with agitation and gentle steam sparging to remove excess water from the sodium hydroxide solution; the reactor is maintained at an absolute pressure of 50–100 mm Hg; a small amount of glycerol escapes with the stripping steam, which is condensed in the surface condenser and returned to the reactor at the end of the drying operation.
5. After reactants have been dried for 15–30 minute at 121.1–126.6°C (250–260°F), the temperature is raised to 165.5–204.4°C (330–400°F) with no vacuum and no steam or nitrogen sparging. Mechanical mixing with either agitator or external recirculation is continued.
6. Reaction mixture is agitated at 165.5–190.5°C (330–375°F) for 30–60 minute, depending on the reaction mixture and end product desired.
7. At end of this period, the reaction mixture is neutralised with 85 per cent phosphoric acid.
8. Amount of acid added should be sufficient to produce some acid salts of sodium with phosphoric acid; the pH of reaction mixture should be acidic after neutralisation, which generally results in the formation of some dihydrosodium and monohydrosodium salts in addition to trisodium salt of phosphoric acid.
9. Unreacted glycerol is then removed from the reactor by vacuum distillation at about 176.6°C (350°F), with steam or inert gas stripping, using an absolute reactor pressure of less than 0–3 mm Hg.
10. Reaction product is then cooled to 54.4–65.5°C (130–150°F); the product is mixed with filter aid and filtered or it is centrifuged to remove residual glycerol and phosphate salt, without using filter aid.

11. Product is then stored in 304 or 316 (preferred) stainless-steel storage tank with nitrogen protection.

### Critical Reaction Variables

The primary object in glycerolysis process is to obtain maximum yield of monoglycerides at the lowest possible reaction time. This is achieved by manipulating a number of reaction variables, such as the ratio of glycerol:fat (or fatty acid), reaction temperature and catalyst concentration. Reaction time becomes important when the reaction temperature or catalyst concentration is low because, under these conditions, the reaction reaches equilibrium at a much slower rate.

Numerous studies can be cited describing and characterising reaction variables. It seems they all lead to the following reaction variables as being critical.

1. Ratio of glycerol:fat (or fatty acid).
2. Reaction temperature.
3. Type and concentration of catalyst.
4. Reaction time.

There are two other variables that are also important. These are:

1. Degree of mixing.
2. Reaction pressure.

#### *Ratio of glycerol to triglycerides*

*Feuge* and *Bailey* showed that when reaction equilibrium is reached at about 200°C (392°F), the reaction mixture becomes homogeneous. The proportions of monoglycerids, diglycerides, triglycerides, and unreacted glycerol conform to a pattern of random distribution with respect to the OH groups in the esterified glycerol.

#### *Reaction temperature*

As reaction temperature increases, rate of formation of monoglycerides increases. Therefore, during initial reaction, proportion of monoglycerides in the reaction mixture is low and it increases as the reaction temperature increases. This phenomenon is less pronounced when a high concentration of catalyst is used. This is true when glycerolysis is carried out with either triglycerides or with fatty acids. As mentioned earlier, it was found that an equimolar mixture of glycerol and fatty acid produced mostly triglycerides and diglycerides when reacted at 180°C (356°F). When the same reaction was conducted at 240°C (464°F), the reaction product contained mostly monoglycerides and diglycerides. It has been mentioned earlier that a higher yield of monoglycerides can be obtained at temperature of 170–180°C (336–356°F) by either carrying out the reaction for a longer time, with a higher proportion of glycerol and a higher level of catalyst or by using a solvent as described earlier.

To get miscibility of glycerol in the range of 20–30 per cent, reaction temperature must be around 250°C (482°F). Products obtained at this high temperature showed some abnormal outcome (Table 2.1). Based on the random distribution principle, an estimated monoglyceride content of 66 per cent would have been expected (with 20–21 per cent of the OH group esterified); actual yield was 57–59 per cent. In addition, calculated values of triglycerides appeared to be negative.

This means that after appropriate amount of total combined glycerol was accounted for in the monoglycerides, there was a surplus of glycerol in the product, even when the balance of reaction product was assumed to be only diglycerides. This could be feasible only if some reaction products

**Table 2.1.** Products prepared by reaction of fat and glycerol at 250°C: operational and analytical data.

| Run number | Reaction time (h:min) | Glycerol (% of fat basis) | Per cent OH esterified: On basis of glycerol present originally | Per cent OH esterified: On basis of glycerol in final product | Total glycerol in reaction product (%)[a] | Total glycerol in fatty portion (%) | Mono-glycerides in fatty portion (%) | Calculated composition of fatty portion (% by weight): Mono-glycerides | Di-glycerides | Tri-glycerides | Hydroxyl value, fatty portion |
|---|---|---|---|---|---|---|---|---|---|---|---|
| 21 | 4:00 | 31.46 | 25 | 26.1 | 31.05 | 21.08 | 53.20 | 53.2 | 48.2 | –1.4 | 212.7 |
| 22 | 4:00 | 41.95 | 20 | 20.4 | 36.49 | 21.95 | 57.53 | 57.5 | 46.6 | –4.1 | 224.8 |
| 25 | 1:00 | 41.95 | 20 | 20.7 | 36.45 | 21.41 | 57.80 | 57.8 | 42.5 | –0.3 | – |
| 31 | 0:20 | 41.95 | 20 | – | – | – | 59.30 | – | – | – | – |
| 29 | 5:00[b] | 41.95 | 20 | – | – | – | 58.90 | – | – | – | – |

[a] Corrected for soap formed by NaOH used as catalyst; calculated on basis of soap free product.

[b] Reacted at 250°C for 1 hr.; phenol added to make the material homogeneous at a lower temperature and reaction continued to an additional 5 hr. at 180°C.

contained higher number of OH groups than possible in glycerides. This led to the conclusion that there were some polymerisation products formed at this high temperature.

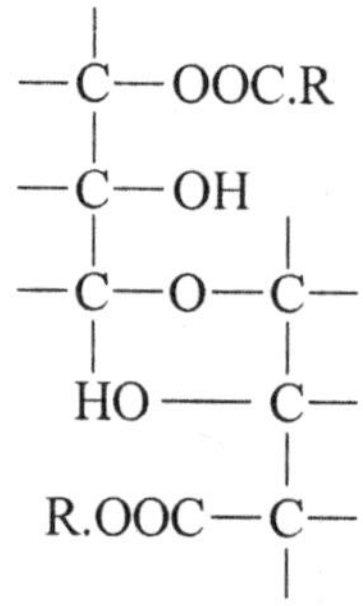


Possible polymers formed in the reaction at 250°C.

It was also concluded that most optimum temperature for glycerolysis is about 200°C (392°F). In reality, most glycerolysis reactions are carried out around 176.6°C (350°F) for batch process and about 232.2–260°C (450–500°F) for continuous process.

*Rheineck* and co-workers showed that in a glycerolysis process at 225°C (437°F) using linseed oil, the monoglycerides content started to decrease after it peaked. This could also be due to polymerisation, as described above.

### *Catalyst type and concentration*

The most common catalyst used for glycerolysis is NaOH. However, several researchers have shown that there are some differences in the performance between various catalysts. It has been noted that trivalent metals are not effective as catalysts for glycerolysis and that bivalent metal ions are less effective than monovalent metals. It has been indicated that $Sn^{++}$ and $Zn^{++}$ are effective catalysts. *Christensen* reported a higher yield of monoglycerides using glyceroxide as instead of alkali metal soap as catalyst.

*Rheineck* and co-workers showed that sodium glyceroxide was a more effective catalyst than sodium hydroxide in glycerolysis reaction between linseed oil and glycerol at 200°C (392°F). *Rheineck* also showed that at the optimum reaction temperature, reaction proceeds more rapidly when higher catalyst level is used. *Rheineck* showed that the glycerolysis reaction for linseed oil does not reach equilibrium at 150°C (302°F) even after 150 minute of reaction, using 0.18 per cent NaOH as catalyst. Equilibrium was reached in 2 hr. and 1 hr., when reaction temperature was 200°C (392°F) and 225°C (437°F), respectively, using 0.18 per cent NaOH as catalyst in both cases. At 0.09 per cent NaOH and 200°C (392°F), equilibrium was reached in 135 minute, while the equilibrium was reached in 75 minute when 0.11 per cent sodium glyceroxide was used as catalyst.

*Rheineck* proposed a reaction mechanism as shown in Fig. 2.2, which implies the true catalyst in the reaction is sodium glyceroxide and not sodium hydroxide. It also implies the following:

1. Sodium glyceroxide molecule ionises, forming glyceroxide ion.
2. Glyceroxide ion thus formed reacts with a triglyceride molecule, forming a complex.
3. The complex breaks down into monoglyceride and an intermediate product containing two fatty acid moieties.
4. Intermediate product can undergo acyl migration, forming 1,3- or 1,2-diglyceride ion.
5. It can react with glyceroxide ion forming 1,3- or 1,2-diglyceride.
6. The reaction continues until an equilibrium is reached, based on the composition of the starting reaction mixture.

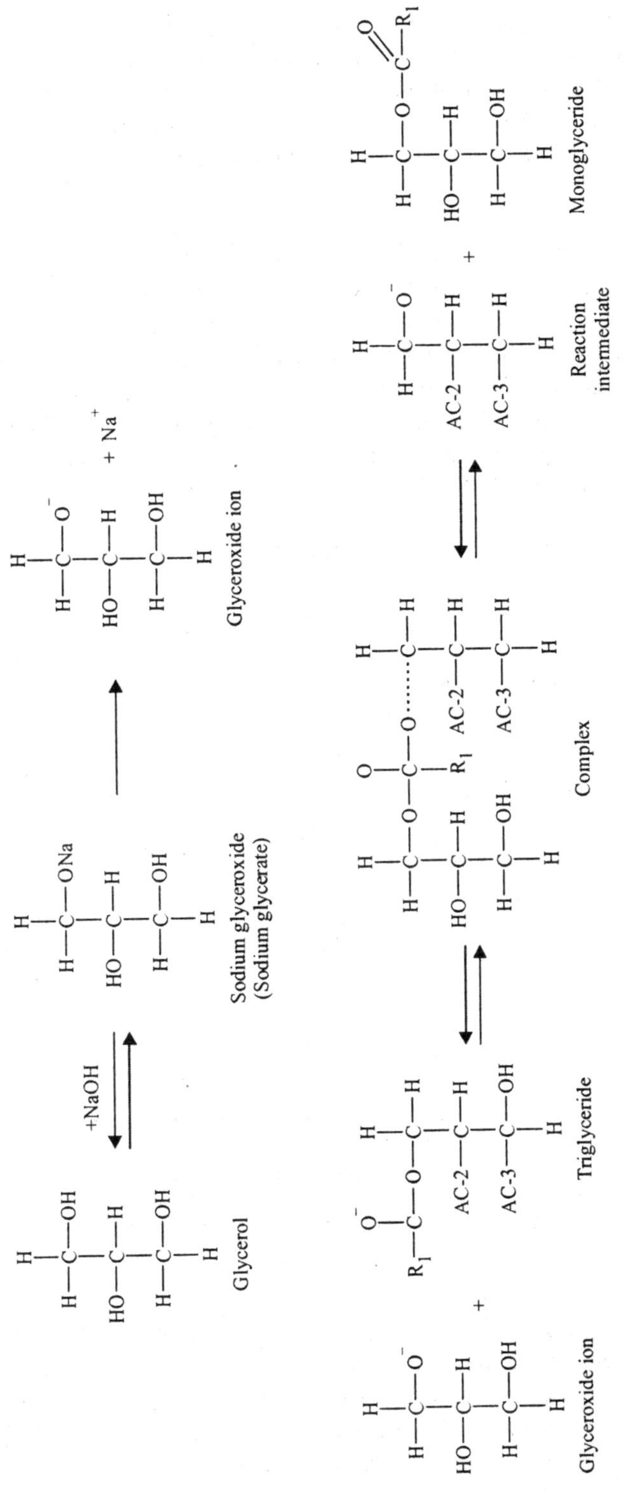


**Fig. 2.2.** Proposed reaction mechanism in glycerolysis. AC, fatty acid moiety (R—C(=O)—O).

It is conceivable that in glycerolysis reaction, sodium or metal glyceroxides are formed *in situ* and the reaction proceeds. Rate of reaction is dictated by the rate of glyceroxide formation. This could be an explanation for the fact that *Rheineck* and co-workers found sodium glyceroxide to be a more effective catalyst than sodium hydroxide. Also, *Christensen* obtained a higher percentage monoglyceride conversion when glyceroxide was used instead of alkali metal soap.

### Reaction time

*Feuge* and *Bailey* showed that the reaction equilibrium was reached after 3.5 hr. when fat was reacted at 200°C (392°F) with 16.63 per cent of its own weight of glycerol containing 0.1 per cent sodium hydroxide. The reaction mixture became clear and homogeneous soon after the reaction equilibrium was reached and the combined glycerol at the point was about 18.4 per cent of the fat (Fig. 2.3). Slower reaction rates have been noted at lower temperatures. It was shown that at lower temperature, the reaction equilibrium leaned toward triglycerides and diglycerides formation. Therefore, reaction time can be important when the reaction temperature is low or catalyst concentration is low, i.e. a longer reaction time must be provided to reach the final steady state condition.

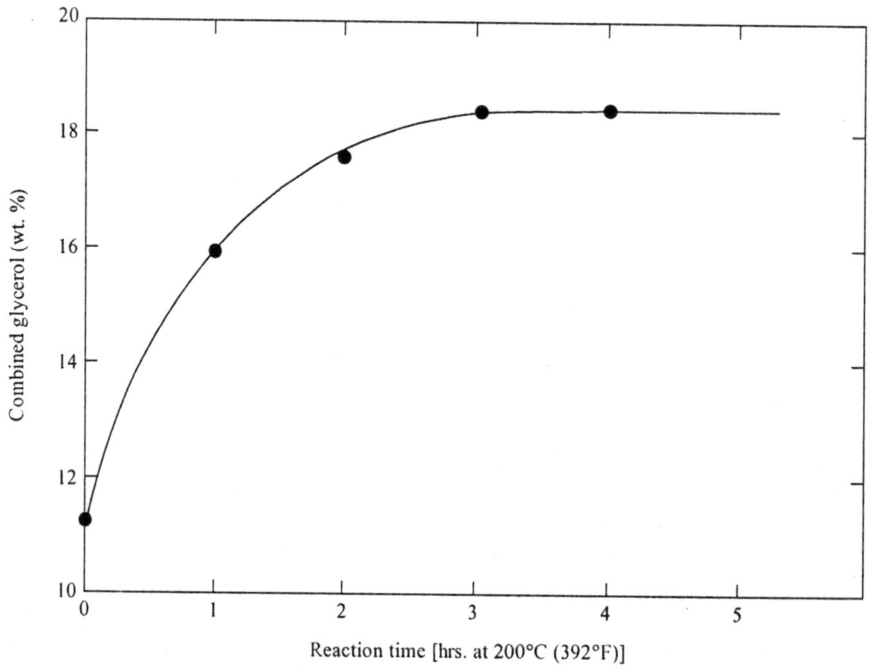


**Fig. 2.3.** Combined glycerol content versus reaction time.

### Degree of mixing

*Hilditch* and *Rigg* observed that miscibility of glycerol and triglyceride is of considerable difficulty in the glycerolysis process. Reaction occurs in the fat phase and any glycerol remaining as a discrete phase does not take part in the reaction. Figure 2.4 shows per cent glycerol in fat at temperatures of 175–250°C (347–482°F). The amount of miscible glycerol in fat over the entire temperature range is based on the amount of glycerol reacted and not on the basis of original glycerol present in the fat.

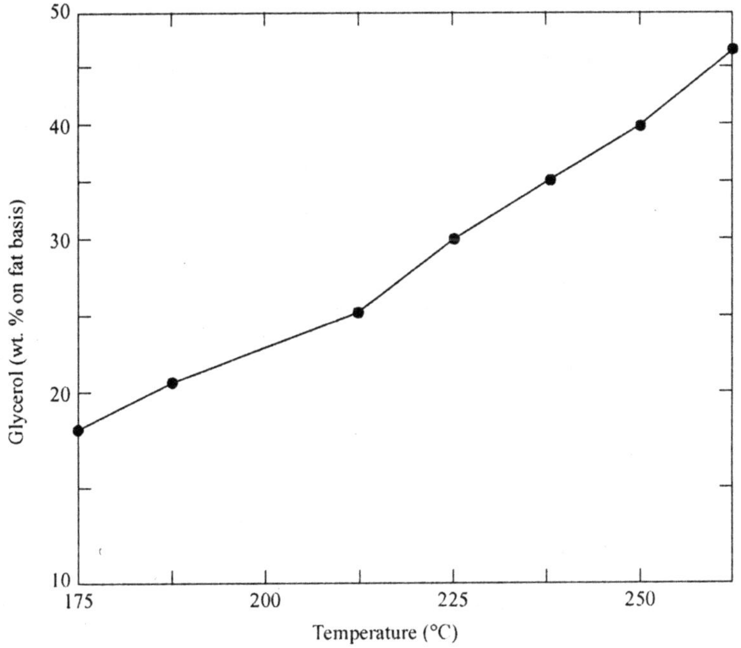


**Fig. 2.4.** Maximum amount of glycerol reactable and miscible with fat at different temperatures.

It is quite obvious that solubilising glycerol into fat is a difficult proposition. This is why several researchers have used solvents to improve solubility and reaction rate. None of the patents or papers published on preparation or production of monoglycerides addresses the importance of mixing, except some of the patents on continuous process of glycerolysis. Mechanical mixing was mentioned in a patent for a continuous reactor.

In glycerolysis, the reaction mixture requires intimate mixing simply because of the immiscible nature of the reactants. The system requires radial mixing to disperse glycerol and fat into a near perfectly dispersed phase. At the same time, it needs to turnover the reaction mixture from the top of the reactor to the bottom as rapidly as possible via axial mixing. The basic concept of the proposed mixing is shown in Fig. 2.5.

To give an illustration, an external pump in recirculation system with a capacity of 500 gal/minute of liquid flow can turnover a 3000 gal reaction mixture in 6 minutes, or 10 times/hr. A mechanical mixer, with a capacity of 7000–8000 gal of pumping capacity can turn over the reaction mixture in less than 30 seconds, which is 120 times/hr., which is 12 times more rapid than external recirculation. In addition, with an external recirculation loop, the liquid withdrawn and returned to the reactor does not experience any shear action, except for the fraction of a second it is in the pump housing. A properly designed mechanical mixer, as described in Fig. 2.5 produces continuous shear action on the entire volume of the reaction mixture. This is why a mechanical mixer is more effective and can produce much improved results.

There are several commercial mixers that are suitable for this purpose. The reactor vessel must have baffle arrangement to avoid the vortex formation. Proper selection of motor horsepower and mixer rpm is critical to ensure adequate mixing, especially at lower temperatures, when the viscosities of the reactant liquids are high. In addition, 85 per cent phosphoric acid is used for neutralising the metal soap at the end of reaction. Phosphoric acid has much higher density and viscosity than the reaction mixture.

A standard turbine type liquid mixer or inert gas or steam sparging is not enough to disperse the phosphoric acid uniformly in the reaction mixture effectively. This can result in insufficient neutralisation and reversal of the reaction during distillation of unreacted glycerol.

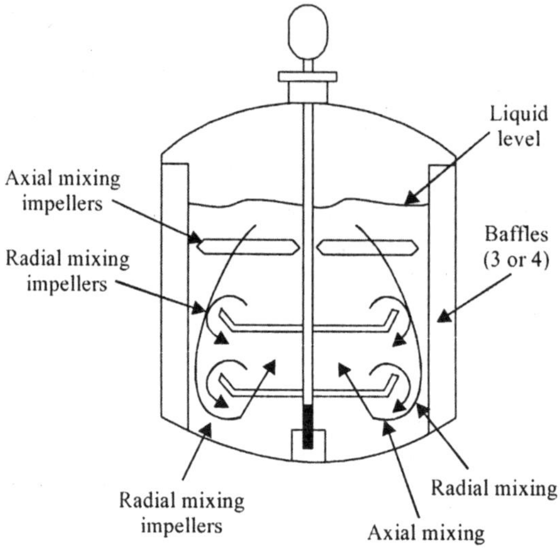


**Fig. 2.5.** Desired mixer configuration for glycerolysis reaction.

Various researchers have emphasised the importance of neutralisation to prevent reversal of the reaction during distillation of glycerol. It has been observed that during distillation of excess glycerol, there is always a drop in the monoglyceride content. This reversal of reaction is more dramatic when the reaction products are either not neutralised or partially neutralised.

Formation of sodium salt of proper acidity is also important during neutralisation. As mentioned earlier, the pH of reaction mixture must be between 5 and 7 after neutralisation, but pH should preferably be 5 or less to ensure proper stability of the product during glycerol removal by distillation and also in storage thereafter.

### *Operating pressure*

Since, glycerolysis is a liquid-phase reaction, effect of pressure on the reaction is not expected to be significant. However, reactor pressure becomes important when distillation of excess glycerol is conducted or during reaction to prevent any glycerol loss. It will be noted later that controlling the proper reaction pressure is more important during glycolysis, where propylene glycol monoesters are produced. This is because propylene glycol boils at a much lower temperature than glycerol. This is also true when other lower boiling alcohols are involved in the process.

The use of positive pressure can be beneficial in some instances where a small amount of water in glycerol is used to catalyse the process of esterification. Castor oil was reacted at 250°C (482°F) with glycerol containing 4–5 per cent water and at a reaction pressure of 200 psig produced by $CO_2$ gas in superincumbent space. The presence of water increased monoglyceride yield. *Kochar* claimed high yield of monoglycerides by using $CO_2$ pressure of 100 psig at a temperature of 240°C (464°F) from a reaction mixture, containing 2000 parts of glycerol as a catalytic agent. Under similar conditions, coconut oil or peanut oil has also been reported to produce 73–74 per cent α-monoglycerides. The pressure appeared beneficial in facilitating the reaction. It is possible that in both of these examples, $CO_2$ and water assisted hydrolysis of fat and accelerated the reaction process.

## Some Finer Points on Glycerolysis

To obtain monoglycerides with consistent quality, a number of process areas must be fully understood and carefully controlled, as outlined below:

1. Ratio of glycerol: triglyceride must be maintained constant, since the reaction equilibrium depends on it.
2. Catalyst concentration must be adequate. Sodium hydroxide solution can be weaker than the usual 50° Bé or the NaOH in the solution could have partially reacted with $CO_2$ in the air and formed $Na_2CO_3$, which is not as effective as NaOH for glycerolysis reaction. If the 50° Bé caustic is ever diluted with non-softened water, NaOH can react with the $Ca^{++}$ and $Mg^{++}$ carbonate or bicarbonate or chloride-sulphate and lose its effectiveness as a catalyst. Therefore, the NaOH solution should be checked for per cent $Na_2O$. For 50° Bé caustic, this value should be 32 per cent. A proportional upward adjustment should be made in the amount of caustic added, if per cent $Na_2O$ is lower. If this is done, an upward adjustment for the phosphoric acid addition should also be made to ensure proper neutralisation.
3. Monoglyceride yield can be lower than expected, even if glycerol and catalyst levels are satisfactory. This can happen if the reaction temperature is low or the reaction time is shortened. In addition, insufficient neutralisation of the catalyst will reverse the reaction during distillation of excess glycerol.

   A reaction mixture containing a lower proportion of glycerol yields a lower monoglyceride level in the product. Assuming the catalyst is not fully neutralised, the reaction mixture continuously loses glycerol during distillation and thus the ratio of glycerol:fatty ingredients is reduced. This would correspond to a reaction mixture with a lower level of excess glycerol and would produce lower monoglyceride content in the product.

   Neutralisation of the catalyst is important, but it is also equally important to control the type of phosphate salt produced. Trisodium phosphate is more desirable because it is crystalline and can be easily removed by filtration. However, neutralisation is done with excess phosphoric acid to obtain an acidic pH. This produces acid salts: $NaH_2PO_4$ and some $Na_2HPO_4$. These salts are not crystalline and remain dispersed in the product. These cannot be removed from the product by simple filtration. It has been shown that in the molecular distillation process, monoglyceride yield dropped from 55.7 to 7.4 per cent, when residue from molecular distillation process was recycled as a starting raw material along with fresh triglycerides. The results are shown in Table 2.2.

**Table 2.2.** Reduction in monoglyceride conversion, using molecular still residue.

| *Esterification* | *Residue (lb)* | *Fresh lard (lb)* | *α-Monoglyceride content (%)* |
|---|---|---|---|
| First | 0 | 2100 | 55.7 |
| Second | 840 | 1260 | 53.6 |
| Third | 1190 | 910 | 45.9 |
| Fourth | 1030 | 1070 | 12.2 |
| Fifth | 1415 | 685 | 9.9 |
| Sixth | 1500 | 600 | 7.4 |

It was found that $NaH_2PO_4$ formed during neutralisation cannot be removed from the emulsifier in filtration. It remains in the product and causes reversal of reaction or actually reacts with the NaOH, forming $Na_2HPO_4$ or $Na_3PO_4$ during the reaction, causing a deficit in the catalyst supply. The following remedy for this problem was proposed.

Mix molecular still residue with diatomaceous earth to remove acid phosphate salts by filtration and/or centrifuge the residue to separate the acid salt, which is removed in the heavy glycerol phase. The residue, free of sodium acid phosphate salt, is satisfactory as a starting material and provides good monoglyceride yield during subsequent uses (Table 2.3).

**Table 2.3.** Recycle of molecular still residue after centrifugation.

| *Esterification* | *α-Monoglyceride (%)* | *Acid value of truncated residue used in the run* |
|---|---|---|
| Initial | 59.2 | – |
| Second | 61.8 | 1.1 |
| Third | 47.2 | 3.17 |
| Fourth | 62.1 | 0.53 |

4. Lack of intimate contact between phosphoric acid and the reaction mixture during neutralisation can lead to reversal of reaction. Phosphoric acid must be added slowly to allow for its dispersion. If agitation in the reactor is done with steam or nitrogen, rapid addition of phosphoric acid could make it drop directly to the bottom of the liquid and below the sparge ring in the reactor. At this point, agitation by steam or nitrogen stripping becomes ineffective.
   If steam or nitrogen agitation is applied, the sparging should be vigorous and the vent or the vacuum system must be left open during sparging. This will strip some of the glycerol and may cause partial reversal of reaction. Therefore, a mechanical agitator is considered more desirable for the process.
5. Control of pH is important from the neutralisation standpoint. It is important to maintain a pH of $< 5$ for the monoglycerides for its colour stability in storage.

## Troubleshooting

To obtain the desired end product in glycerolysis, it is important to observe certain operating conditions and techniques carefully. However, during production, deviations from the normal outcome of the process may be experienced.

It is important to understand the cause-and-effect relationship in such situations. In addition, the operator should know what corrective action must be taken to correct the deviations (Table 2.4). While the illustration in Table 2.4 may not cover all probable situations that can occur in glycerolysis process, it covers the most common ones. The Table also provides the general concept of troubleshooting and can be expanded on in a specific operating plant.

## Obtaining High Monoglyceride Content

In a typical operation, monoglycerides content in the final product is 35–50 per cent, with few cases ranging up to 55–60 per cent. These products also contain diglycerides and unreacted triglycerides. At 35 per cent monoglycerides, the diglycerides content is expected to be around 46–48 per cent and the remainder is made of predominantly triglycerides. On the other hand at 50 per cent monoglyceride content in the emulsifier, diglycerides and triglycerides content would be 40–41 per cent and 8–10 per cent respectively.

**Table 2.4.** Troubleshooting glycerolysis process.

| *Symptoms* | *Assignable cause* | *Corrective action needed* |
|---|---|---|
| Per cent monoglyceride is low | Insufficient amount of glycerol | |
| | This can be caused by | |
| | Not enough glycerol added | Check and correct glycerol addition |
| | Glycerine flashed off or physically got lost due to entrainment during drying | Control glycerol pumping rate: reduce it if needed; reduce vacuum if necessary; reduce stripping steam |
| | Not enough catalyst | Add more catalyst |
| | Catalyst may be < 50° Bé | Check degree Bé |
| | Per cent $Na_2O$ is < 32% | Check per cent $Na_2O$ and make the necessary increase in catalyst addition |
| | Catalyst is either not neutralised or incompletely neutralised before distilling excess glycerol. This causes reversal of reaction. | Ensure proper neutralisation by checking the correct amount of phosphoric acid, maintaining excellent agitation in the reactor and/or adding the acid slowly for proper dispersion |
| Product looks too red | The base stock is too red | Use lighter colour oil |
| | Incomplete neutralisation | Add sufficient phosphoric acid and ensure proper neutralisation; check and make certain the pH < 5 in the finished product |
| High free glycerol in product | Inadequate distillation due to | Check for |
| | Low temperature | Proper distillation temperature |
| | Poor vacuum | Ensure proper vacuum |
| | Insufficient stripping steam | Use correct volume of stripping steam |
| | Short distillation time | Increase distillation time |

Commercial monoglycerides produced by the standard process are mostly made from partially or fully hydrogenated fat. Thus in commercially produced emulsifiers, unreacted fat could be all high melting trisaturated glycerides. In certain applications, such as cake or icing, the mouth melt of the product is critical.

Using a commercially prepared monoglyceride emulsifier could impact texture or mouthfeel of the product. In those cases, it is advantageous to use an emulsifier that has a monoglyceride content of 80–90 per cent, or higher. A smaller percentage of these emulsifiers in the shortening formula would provide the required level of monoglyceride needed in the product, without causing any negative impact on performance of the end product.

Present-day high monoglyceride containing emulsifiers (> 90 per cent α-monoglycerides) are manufactured in a two-step process. At first, standard emulsifier containing about 50 percent monoglycerides is produced. The product is cleaned to remove the neutralised catalyst residue and then distilled in a centrifugal molecular still at about 148.8°C (300°F) at an absolute pressure of 0.001 mm

Hg. Figure 2.6 explains the basic internal construction of a molecular still. Commercial-grade monoglyceride after proper cleaning is pumped at the top of the still on a spinning disc, which distributes the liquid in a thin film on the inner surface of the still. The still is heated from the outside along the vertical shell and a vacuum of < 0.01 mm Hg is maintained.

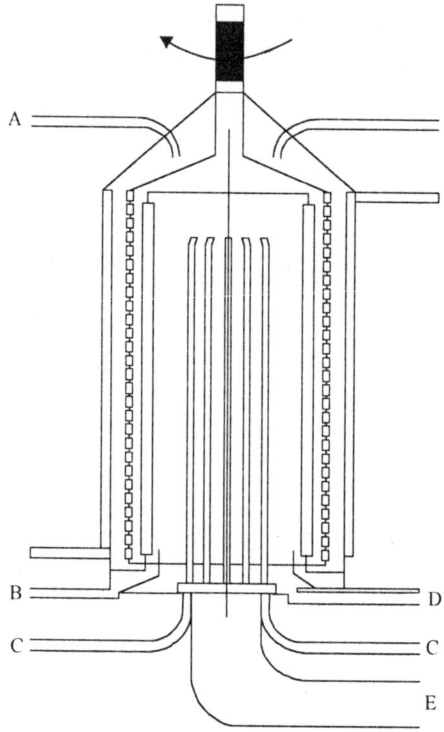


**Fig. 2.6.** Cut-away view of a molecular still.

The liquid runs down the inside of the shell while a scraper device continuously scrapes the liquid on the wall, creating a turbulent downward flow in the film. Monoglycerides evaporate under low pressure and high temperature and more toward the centre of the still where the vapour comes in contact with surface condenser tubes. The vapour condenses and is removed from the still through outlet D. Distillation residue is removed continuously through port B and the vacuum is maintained in the still through port C. Distilled monoglyceride thus obtained is then cooled and stored in 304 stainless-steel tanks under nitrogen.

The residue contains diglycerides and triglycerides and acid salts from neutralisation. The residue is cleaned and the heavy phase, which was about 1 per cent of the total volume of residue, contained practically all of the sodium salts from the neutralisation.

*Kuhrt* and co-workers, fractionated commercial monoglyceride in a molecular still whose operation was described. They took 41 per cent monoglyceride and fractionated it at various temperatures (Table 2.5). Temperature and pressure conditions are shown for the various fractions. The first two fractions, each constituting approximately 5 per cent of the total feed were taken out at lower temperatures to remove lower molecular weight components. The next three fractions, each about 10 per cent of the total, were collected and had > 90 per cent monoglycerides. The residue fraction contained a low level of monoglycerides.

**Table 2.5.** Operation data for the separation of monoesters by molecular distillation of a partially hydrogenated vegetable oil monoglyceride mixture.

| *Fraction* | *Temperature (°C)* | *Pressure (mm Hg)* | *Per cent cut* | *Per cent monoester* |
|---|---|---|---|---|
| Charge | – | – | (100) | 41 |
| 1 | 80–140 | 0.018 | 5.2 | 67 |
| 2 | 140–142 | 0.016 | 4.8 | 84 |
| 3 | 142–148 | 0.011 | 9.2 | 93 |
| 4 | 148–153 | 0.04 | 10.0 | 97 |
| 5 | 153–168 | 0.04 | 9.9 | 94 |
| 6 | 168–200 | 0.04 | 10.0 | 23 |
| 7 | 200–206 | 0.04 | 9.9 | 12 |
| Residue | – | – | 39.6 | 1 |

Table 2.6 lists the results after fractionation of several commercial monoglycerides. In every case, distilled product contained 2.3–2.6 times the monoglyceride content of the feed. Significant reduction of colour and free fatty acids were attained in the distilled product.

**Table 2.6.** Analytical data on products containing saturated acid monoesters of glycerol.

| *Monoglyceride products* | *Per cent monoester* | *Per cent glycerol* | *Per cent FFA* | *Melting point (°C)* | *Colour 5.25 inches lovibond* |
|---|---|---|---|---|---|
| Monolaurate | | | | | |
| Commercial mixture | 35 | 9.5 | 1.9 | 47 | 35 Y/4.0R |
| Distilled monoester | 92 | 0.7 | 1.3 | 57 | 20 Y/2.4 R |
| Monopalmitate | | | | | |
| Laboratory mixture | 38 | 3.6 | 2.7 | 61 | 60 Y/13.5 R |
| Distilled mixture | 93 | 0 | 1.3 | 72 | 8 Y/1.2 R |
| Monostearate | | | | | |
| Commercial mixture | 40 | 3.9 | 2.3 | 59 | 12 Y/1.9 R |
| Distilled monoester | 94 | 0.5 | 1.2 | 71 | 10 Y/1.5 R |

A minimum of 10°C rise in the melting point was observed in the distillate, compared with the feed. There was also a drastic reduction in catalyst residue content in the distilled product. For example, commercial product may contain 300–800 ppm of ash. Distilled product had < 2 ppm of ash.

There are many advantages of the molecular distillation process:

1. Monoglyceride content is > 90 per cent. Therefore, a smaller percentage of this emulsifier in shortening (or other products) can provide desired functionality.
2. Product is much lighter in colour; in some cases, distilled monoglycerides made from saturated fatty acids may be as clear as water. This could be valuable in products where lighter colour is a desirable attribute.
3. Free fatty acid content is low and roughly 50 per cent of that found in commercial grade monoglycerides; in some baked products, excess free fatty acids from the emulsifier can be undesirable. Distilled product definitely offers an advantage in such applications.

4. Distilled product contains < 2 ppm of ash compared with 300–800 ppm found in commercial product; this means the catalyst residue level is lower, which provides better stability in distilled monoglyceride as well as in the products made with it.
5. Absence of high level of triglycerides can be beneficial in some products, in terms of texture and/or mouthfeel.

### Smoke Point of Superglycerolated Product

Commercially produced superglycerolated product emits excessive amount of smoke in the fryer, when heated to frying temperature 176.6°C (350°F). It was shown that the smoke point of superglycerinated shortening could be increased to 171.1°C (340°F) or better, by controlling free glycerol content to < 0.02 per cent. *Monick* indicated that based on 6 per cent use of emulsifier in the shortening, free glycerol in the emulsifier must not exceed 0.3 per cent. At a level higher than this, the shortening will have a smoke point less than 171°C (340°F). Figure 2.7 shows the reduction of smoke point in shortening with an increase in free glycerol content. Smoke point of the shortening drops sharply with slight increase of free glycerol up to 0.12 per cent and then it remains steady at about 93.3°C (200°F).

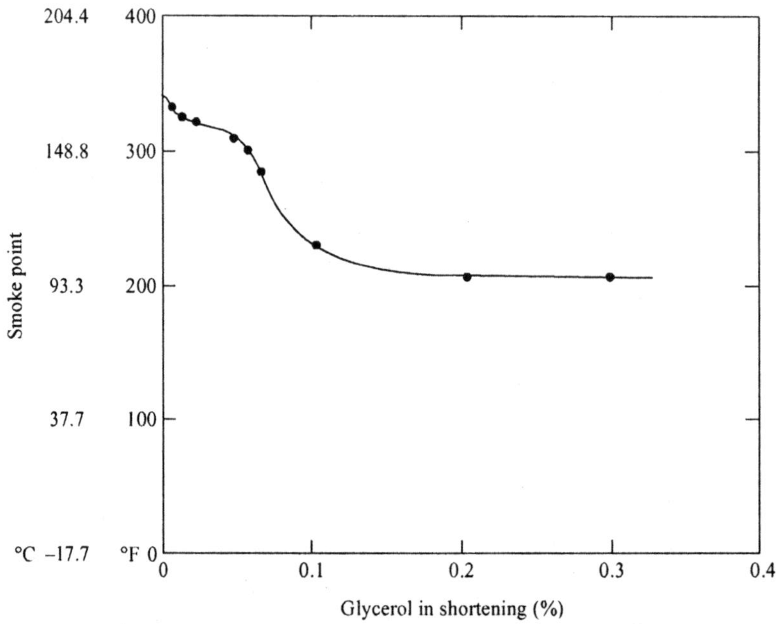


**Fig. 2.7.** Effect of glycerol in shortening smoke point.

*Monick* demonstrated that by a two-stage extraction of commercial grade emulsifier with 5 per cent NaCl solution, he was able to reduce the free glycerol content from 5.51 to 0.14 per cent. He also developed a continuous process where he stripped monoglycerides and glycerol in a thin film evaporator. The temperature was 171°C (340°F) and 6 mm Hg absolute pressure was used. Stripping steam was injected at the bottom of the evaporator. The fatty layer, separated from the aqueous part in the condenser, contained monoglycerides with little free glycerol (< 0.3 per cent).

*Richardson* and *Eckey* patented a process by which the excess glycerol was removed from the emulsifier in two steps. First, majority of the free glycerol was steam distilled after neutralisation. Then, the emulsifier was added at 4.5 per cent level of a batch of shortening toward the tail end of its own deodourisation cycle. Thus the emulsifier was distilled twice, without exhibiting any reversal of reaction.

## Continuous Process

Several continuous processes have been patented to manufacture monoglycerides. There are three patented processes that are unique in their approach.

*Chang* and *Wiederman* patented a countercurrent reactor. They used a column packed with glass helices. The column was 91.4 cm (36 inch) long and 1.91 cm (0.75 inch) in diameter. It was heated to 250°C (482°F) from the outside.

Glycerol with 0.2 per cent NaOH was fed from the top at 3 gram/minute. Conttonseed oil with 45 per cent glycerol and 0.7 per cent NaOH were homogenised and pumped through the bottom at 2 gram/minute. The outlet at the bottom was adjusted to allow 3 gram of glycerol to be taken out per minute. The fatty reactant materials went through an outlet at the top and brought in contact with a bed of Permutite Q, an acid-carbon exchange bed. The ion exchange resin bed was maintained at 60°C (140°F). The finished product had 70 per cent monoglycerides, after separating excess glycerol and phosphate salt of sodium. *Wiederman* emphasised the need for intimate mixing between the reactants. He also stressed the importance of catalyst neutralisation along with rapid cooling, applied to the reaction products.

*Allen* and *Campbell* patented a continuous process for manufacturing superglycerolated or superglycolated emulsifiers. In this process, the reaction mixture, containing 69 per cent soyabean oil and 31 per cent glycerol (pre-mixed with 0.29 per cent NaOH), was mixed in an in-line mixer and fed at the bottom of a vertical reactor. The reactor column had several horizontal perforated plates through which the reacting liquids rose. In each compartment, liquid was mixed with a paddle mixer, driven by a common vertical shaft. The reactor was maintained at 40 psig pressure and at a reaction temperature of 248.8°C (480°F) by providing heat through an outer jacket, using Dowtherm vapour. Reaction mixture left the reactor at the top through a back pressure control valve and was fed at the top of a thin film evaporator. The evaporator was maintained at 215.5°C (420°F) by heating through an external jacket. The pressure in the evaporator was maintained at 3 mm Hg.

Glycerol evaporated, was condensed in a condenser and was collected for re-use. The residue leaving the evaporator was cooled down to 168.3°C (335°F) within 2 minutes after it left the reactor. It was then neutralised with 85 per cent phosphoric acid. The mixture was filtered to remove phosphate salts. The final product had 58.2 per cent monoglycerides and 100 ppm ash, much lower than the usual 300–800 ppm ash found in commercial monoglycerides. The process is simple. *Allen* and *Campbell* emphasised the need for intimate mixing and rapid cooling prior to neutralisation.

*Birnbaum* patented a continuous process for making superglycerinated emulsifier. Anhydrous glycerol was heated with 1 per cent anhydrous sodium hydroxide. Oil was deaerated and then heated to 232.2°C (450°F). About 68.3 per cent deaerated fat at 232.2°C (450°F) was mixed with 31.7 per cent glycerol. The glycerol contained 1 per cent NaOH, pre-dissolved in it at 300°F. A continuous stream of the emulsifier, after neutralisation, was added to the reaction mixture at the rate of 10 per cent. The three streams of liquid were mixed in a mixing nozzle and fed into a Votator scraped wall heater-reactor, maintained at 182.5°C (483°F).

Reactants left the reactor after 8 minute and were held in a heated holding tank for 30 minutes at 263.8°C (507°F). At the holding tank outlet, the reaction mixture was neutralised with 85 per cent phosphoric acid. The product was cooled to 148.8°C (300°F) in less than 1 minute, in a Votator cooler, using water as a coolant. As mentioned earlier, a portion of the neutralised emulsifier was blended back into the reactants. The neutralised mixture was then pumped into a thin film evaporator, maintained at about 148.8°C (300°F) and operating under vacuum. Glycerine evaporated, was condensed and recovered. The residue containing salt from neutralisation was collected in an open tank where it was mixed with filter aid.

The mixture was then filtered, cooled and stored. The filtered product showed the following analyses when 10 per cent monoglycerides mixture was blended as one of the reactants:

| | |
|---|---|
| α-Monoglycerides | 60.7 per cent |
| Free glycerol | 8.46 per cent |
| Diglycerides and triglycerides | Balance of the mixture |

The above material was vacuum distilled and the resultant product had an α-monoglycerides content of 66.2 per cent. When the reaction mixture contained only 5 per cent recycled monoglyceride, the filtered product contained 60.47 per cent α-monoglycerides and 7.95 per cent free glycerol. The vacuum distilled product contained 65.75 per cent α-monoglycerides. Results indicate a nominal reduction in the yield of α-monoglycerides when recycled monoglycerides level was reduced to 5 per cent.

This process is a bit complicated due to the fact that a steam of monoglycerides needs to be recycled. Total time of reaction is also extended because of dependence on the holding tank. Although *Birnbaum* did not indicate any polymerisation during reaction, an extended residence time at such high temperature could be conducive to the formation of thermal polymers. Treatment of the neutralised product with filter aid in an open tank at 148.8°C (300°F) would be considered undesirable in the operation of a modern fats and oils processing plant.

A continuous system appears to use more excess glycerol than a batch process. *Chang* and *Weiderman* used 262 parts glycerol for 100 parts oil in the reaction mix, with a monoglyceride yield of 70 per cent. *Allen* and *Campbell* used 44.9 parts glycerol for 100 parts oil, and *Birnbaum* used 46.4 parts glycerol for 100 parts oil in the reaction mixture. Referring to theoretical yield calculations of *Feuge* and *Bailery* both *Allen* and *Birnbaum* obtained better yield of per cent monoglycerides than *Chang* and *Weiderman*. This could be explained by the fact that *Allen* used pre-mixed reactants and used a well-stirred reactor. *Birnbaum* used a holding tank, which extended the reaction time to compensate for lack of mixing as in *Allen's* process. *Chang's* process did not have the degree of mixing of *Allen's* process, nor did it provide the extra residence time as in *Birnbaum's* process. However, none of the processes had a yield of monoglycerides close to theoretical value.

It should be pointed out that any operation system discussed in this chapter can be used to produce both monoglycerides and propylene glycol monoesters. Propylene glycol is a lower boiling alcohol than glycerol. Therefore, precautions must be taken to prevent loss of propylene glycol during process.

In addition, when the reaction mixture is exposed to low absolute pressure during the distillation step, there is a sudden surge of polyhydric alcohol vapour leaving the reaction mixture. This is not so critical in a continuous process, because the rate of reactant flow is carefully regulated. However, in a batch process, the amount of vapour released during the first few minutes during distillation can be as much as 5–10 times the average rate. Therefore, the surface condenser must have sufficient cooling capacity. Otherwise, a lot of vapour will not condense and will be lost through the vacuum ejector system.

### Manufacturing Propylene Glycol Monoesters

Propylene glycol is a dihydroxy alcohol, while glycerol contains three hydroxyl groups. Thus propylene glycol ester loses its emulsifying property, if the emulsifier contains high level of diglycerides. This would be similar to having a high level of triglycerides in the superglycolated emulsifier.

Propylene glycol monoesters of saturated fatty acids, such as palmitic and stearic acids, are excellent aerators in cake batter. Monoesters of stearic acid have higher melting points than palmitic acid has. Both types of monoesters are found useful in the baking process, while propylene glycol monopalmitate is recognised as a better aerating agent.

As mentioned earlier, propylene glycerol monoesters and glycerol monoesters can be made using the same equipment. If it is a batch process, the surface condenser must be considerably larger for the propylene glycerol process. Thus, in a system where both types of emulsifiers are to be made, the condenser must be sized for propylene glycol vapour load.

## Processing Steps for Glycolysis: Batch Process

1. Processing steps for charging the reactants to the reactor and drying (if 50 per cent NaOH solution is used instead of dry glycerol and NaOH mixture) are identical to the process for making monoglycerides.
2. Reaction is normally carried out at about 190.5–218.3°C (375–425°F). Because of low boiling point for propylene glycol, the reaction pressure must be at least 15–20 psig to prevent loss of propylene glycol. Superincumbent pressure in the reactor is created by sparging nitrogen through sparger. Sparging is stopped when the desired pressure is reached in the reactor.
3. Reaction mixture is also cooled down to about 148.8°C (300°F) for neutralisation. At higher temperature, a brown colour tends to form in the product, which is believed to be the reaction product between propylene glycol and phosphoric acid. This step is not so critical in glycerolysis, since it is carried out at a much lower temperature.
4. After neutralisation, temperature is reduced to 121.1–126.6°C (250–260°F). Absolute pressure in the reactor is raised to about 4 mm Hg. A large portion of the propylene glycol flashes off the reactor and is condensed in the surface condenser. As mentioned earlier, the condenser must be designed to handle the huge vapour load created at the onset of flash evaporation. Recovered propylene glycol is sent to its storage tank.
5. After the vapourisation step, temperature in the reactor shows a slight increase. As this point, steam distillation is started and residual propylene glycol, mixed with glycerol, is recovered. This fraction must be analysed to determine the level of glycerol. During glycolysis, glycerol from triglycerides is released. This carries on a parallel reaction, producing mono and di-glycerides.
6. Neutralised and glycol-free product from the reactor is then treated with filter aid and filtered (or cleaned by centrifugation as in the glycerolysis reaction product).

All reaction variables discussed under glycerolysis apply in glycerolysis process. In a typical commercial operation, a high ratio of propylene glycol to fat is used because the emulsifier must have high monoester (>65 per cent) to be functional in baking. Therefore, in a typical commercial oil processing plant, the monoester content would be 65–75 per cent in propylene glycol esters while the monoglycerides made at the same location may have only 35–50 per cent monoester.

Commercial monoesters of propylene glycol can be concentrated further in a molecular distillation process, like monoglycerides.

## Troubleshooting

The troubleshooting processes for glycolysis and glycerolysis are similar. It must be remembered that propylene glycol has low boiling point and therefore, the reaction must be carried out under a positive pressure of 15–20 psig. Glycerolysis can be conducted at much lower pressure. Low yield of propylene glycol monoesters can occur for the same reasons as outlined in Table 2.4. However, there is one additional factor, as had already been mentioned, which is due to the loss of glycol due to lower pressure in the vector during reaction and neutralisation.

Glycerolysis proceeds simultaneously in the glycolysis process. The amount of monoglycerides produced depends largely on the amount of free glycerol available. The free glycerol level in the steam stripped portion of the propylene glycol increases with the number of its re-use. Therefore, when the recovered propylene glycol is used without checking its glycerol content, the finished emulsifier may contain high level of monoglycerides. This may not be desirable in certain applications. Therefore, free glycerol in the distillate must be checked and its re-use must be stopped at a pre-determined level of glycerol. The mixture can be taken through fractional distillation process to separate the two components and re-used. This can be done by a manufacturer of emulsifiers. However, for a standard fats and oils plant, where emulsifiers are produced for internal consumption, an additional still for separating propylene glycol from glycerol might not be economically feasible. This will require an economic justification, based on the production volume and the amount of mixed polyhydric alcohol generated.

Chapter 3

# Fats and Oils in Soap

## INTRODUCTION

Soap is the water-soluble product of the reaction between a fatty acid ester and an alkali (usually sodium hydroxide), with glycerol as the by-product. A soap is actually a specific type of salt in which the hydrogen of the fatty acid is replaced by a metal, which in common soaps is usually sodium. Upon contact with water, soap lowers the surface tension of water and thus permits the emulsification of fat-bearing soil particles. A typical commercial cleansing soap is made by reacting sodium hydroxide with a fatty acid. The lower the hydrogen content of the acid, the thinner the soap. The by-product of the reaction is glycerol. Many different carboxyl-containing substances are used, including rosin, tall oil, vegetable and animal oils and fats (stearic, palmitic and oleic acids). Olive oil is used for castile soaps; transparent soaps are made from decolourised fats. The specific gravity of soaps is slightly more than 1.0; inclusion of air gives a floating product. Water solutions of sodium soaps in the form of bar, chip or powder form are universally used as mild emulsifying detergents for washing textiles, skin, paint, etc. Medically, soap is used as an antidote for poisoning by ingestion of mineral acids or heavy metals. Liquid green soap is made with potassium hydroxide and a vegetable oil.

Heavy-metal soaps (loosely called metallic soaps) are those formed by metals heavier than sodium (aluminium, calcium, cobalt, lead, zinc). These soaps are not water soluble; specific types of these soaps are used in lubricating greases, gel thickeners and in paints as driers and flatening agents.

## STRONG BASE SOAPS

### Specific Characteristics of Raw Materials Used by the Soap-Making Industry

Non-fatty raw materials going in to the making of soaps are generally of industrial quality. One should, however, pay attention to the following points:

Water should be free from metal ions and alkaline-earth. Softened water or condensate from boiler would suit perfectly. Soda, potash and salt are of common quality. In certain continuous processes (such as Alfa Laval) the percentage of salt in the soda should be perfectly controlled.

Descriptions of fats used in soap-making industry have already been given in the preceding chapters. Here we will restrict the discussion to give important points for the soap industry. By their structure, length of their chain–different fats, give to the soap its lather qualities and hardness whereas their physical qualities—colour and odour—partly define the quality of a soap.

## Main Fats Used in the Soap-Making Industry

### Oils and fats in $C_{16}$–$C_{18}$

Tallow is the fat of this category that is used mostly in Europe. It gives a relatively soft soap with a fine stable foam that develops slowly. Cattle tallow is preferred to horse fat, for the latter contains a lot of polysaturated chains and is thus, more easily oxidisable. It is also preferred to sheep because of the odour a sheep tallow transmits to the soap. Two per cent industrial tallow is generally used by soap manufacturers not because of its percentage of acidity but because of the quality it represents and because it corresponds to a standardised quality. In the mixture with tallow, one can introduce a little lard but one should take care of not lowering the titer too much, for the soap would then become too soft.

Palm oil is also used worldwide. It gives to the soap the texture and lather characteristics identical to those given by the tallow. As the palm oil is used particularly in cooking, for soap-making one is using more the palm solid or palm stearin which comes from winterisation of the oil and which is therefore richer in saturated acids. For this reason, this stearin is used only in a mixture with tallow or with palm oil in order not to have too high a level of saturated fatty acids and therefore a tendency to crack. Percentages in the range of 40 per cent are frequently used.

### Lauric oils

Coprah and palm kernel oils are the oils used mostly in soap-making. Soaps obtained with these oils have the particularity of foaming rapidly even in cold water, but this lather is not very stable. When mixed with tallow or palm in the range of 10–20 per cent, they increase the level of soap suds. Sometimes higher proportions of these oils are used: 40–50 per cent. In this case, in order to tone down the aggressiveness of the skin of the soaps in $C_{12}$, one adds super fatteners that give a greasy foam.

### Other oils and raw materials

Many other oils can be used in soap-making, but the purpose in doing that is to drain the by-products of oil refining for manufacturing household soaps. Raw materials such as the tall oil or rosin were used earlier, but today they are not in use. On the other hand, fatty acids and methyl esters are being used more and more.

### Physical criteria of quality of raw materials

In order to obtain a good quality soap with regard to colour and odour, it is indispensable to first use raw materials of well defined quality in a standard manufacturing process. The object of this chapter is to define these criteria for principal raw materials.

### Tallow

At the time of saponification, the tallow should generally have been bleached to a certain level of colour, always the same. It is therefore necessary to have a valid measurement of this colour. Colour measurements of tallow for the soap industry were for a long time done as for other uses (FAC colour, Lovibond, etc.), but it has been noticed that there was no correlation with the colour of the soap. The method of colour measurement was developed by the ITERG. It measures the colour of a potash soap solution obtained from the tallow considered. This method, described in the standard NFT 60–216, is now very often used and has often been improved by soap manufacturers as, for the colour lower than 0.5 commonly achieved in toilet soap industry, they found it less than precise. This method has permitted the soap manufacturers and the smelters of tallow to define purchase specifications. But between its receipt and utilisation, the tallow can degrade and if, in order to obtain the prescribed iodine value the supplier has whitened his tallow, it will then be practically impossible to rewhiten it. To avoid this problem and to have a good colour on saponification, the soap manufacturer prefers to buy crude tallow and bleach it himself. In order to be certain of obtaining the iodine value that was fixed after whitening, the soap manufacturer can develop with his supplier a whitening test with the iodine value as the colour measurement method. The concentration of cutting, the methods of extraction, the work between the smelters and soap manufacturers have permitted to improve very substantially the quality of the tallow. One can thus see that wet melting gave the tallow more easily bleachable than the tallow obtained by dry melting. This is due to the fact that with this technique the tallow is found washed and thus freed of proteins and other matters that are an obstacle to the action of bleaching earths.

Another impurity present in the industrial tallow is the polyethylene coming from fat collection bags. It is particularly soluble in the tallow and remains in the molten soap. At the drying stage, it is deposited in the exchangers of the dryers to eventually give black specks in the soap. It should be noted that the smelters of tallow have to a great extent reduced its content, which is only to the order of 50 ppm.

### *Palm oil*

Palm oil contains a high concentration of carotenoids, which should be eliminated before saponification. Whitening of palm oil calls upon the same techniques as for other oils. It is possible to obtain soaps of very acceptable colour, but it is with this oil that one has the maximum number of problems of colour reversion.

### *Coconut and palm kernel oil*

These oils are much less sensitive to oxidations than the ones we have already discussed. Therefore coconut and palm kernel oils are less prone to colour reversion. There will therefore be the possibility of buying them raw in order to bleach them just before use or to buy them already bleached.

### *Taking delivery and bleaching*

In order to take delivery and store crude oils, ordinary steel is used even for quite long duration. In this latter case, one allows to cool down the fat in order to limit oxidation and one heats it again only at the time of use.

Bleaching of fats is done in a very traditional manner—by the action of bleaching earth, in some cases added with active carbon. It is recommended to do the bleaching in vacuum in order to only minimally adulterate the fat, the bleaching temperature being between 110–115°C. In order to limit oxidation, the temperature of the bleached oil is lowered to 50–60°C just after filtration. When one uses tallow containing polyethylene it is recommended to cool down again before filtration so as to retain maximum polyethylene, on the filter. In this case, the filtration is more difficult. Different matters can be bleached separately or together. The advantage of separate bleaching is that one can choose a bleaching process that suits the bleaching conditions for the concerned raw material.

Bleaching is generally the only treatment before saponification. However, in the case of use of fats of very mediocre quality and in order to improve the odour of the finished soap, a more or less thorough deodourisation can be done. The quality of the soap will depend both on the process and the quality of the raw materials. The present trend is to improve to the maximum the quality of raw materials in order to have a simplified saponification process.

The bleached products should be stored in good conditions, that is, in stainless steel vats, whose temperature is maintained by circulation of hot water or vapour at low pressure with slow agitation in order to avoid local overheatings which degrade the colour. An additional precaution is the introduction of nitrogen just after bleaching at the time of cooling in order to avoid any oxidation.

### *Fatty acids and methyl esters*

#### *Fatty acids*

They are traditionally employed in manufacturing laundry soaps for which they have always been recovered as the by-products of oil refining. Today it is possible to have fatty acids of very good quality, permitting manufacture of toilet soaps. With respect to triglycerides, they have the following advantages and disadvantages:

1. They permit us to select more precisely the cut of fatty acids that one desires and to eliminate the undesirable ones like those with very short chains.
2. They permit us to have a very simple equipment at the level of saponification and not the recovery of glycerine.
3. The storage conditions will be more delicate: stainless steel, short storage time, under nitrogen.
4. No recovery of glycerine, thus economic balance sheet to be prepared.
5. Limiting of sources of procurement.

The conclusion is that fatty acids are generally used either for very small tonnages or by industrial manufacturers of fatty acids.

#### *Methyl esters*

Like the fatty acids, they have advantages and disadvantages with respect to triglycerines:

1. They are stable and liquid at standard temperature, which permits us to transport them easily and without deterioration.
2. Methylation permits us to improve the quality of the original fat and therefore to use the fat of average quality in order to obtain a soap of good quality.
3. The saponification equipment is a little more simpler than that for the triglycerides, but the presence of methanol makes it mandatory to have explosion-proof installations. One should also foresee the recovery of methanol.
4. Elimination of methanol from soap is an important point to be considered, for this alcohol is more dangerous than glycerol.

Methyl esters, though having certain advantages, are employed only in special economic conditions.

## Physicochemical Aspect of Water–Soap Mixtures

Soaps and particularly sodium soaps, could be considered as ordinary substances. But on the contrary, they constitute one of the more complex cases of physicochemistry of relations between phases such as the phases in metal alloys. For a long time this has been a problem in rationaling their production because the soap industry has relied, practically up to the second World War on the empiricism of skilled soap makers.

### *Pure soaps*

This results essentially from the dualistic aspect of the properties of the soap molecule, which is the prototype of an amphiphile. In fact, a soap is composed of a fatty chain in the normal sense of the word with a carboxylate group its end.

A fatty chain is characterised by its insolubility in water (though it may have a minor affinity for water) and a lack of polar bonds, which would be susceptible to interaction with dipoles of water. On the other hand, the carboxylate group is essentially compatible with water because of its polarity and its ionisation. There is thus a duality of two essential properties (amphiphilic).

In solid, anhydrous or less hydrated state, hydrocarbon chains associate among themselves by the dispersion forces of *London* (instantaneous dipoles due to electronic circulation) acting at a small distance van der walls force proportion to $1/r^7$. On the other hand, carboxylate groups associate among themselves through dipolar interactions, and it follows that the solids obtained are of smectic type, molecules being arranged in double layers with opposite directions. One can distinguish in these solids two plans of terminal groups (methyl groups and carboxylate groups) and paraffinic strata.

From this scheme, the actions of water and temperature are best expressed as coordinates of a phase diagram with temperature and water concentration as coordinates.

In the case of a pure soap and water with the pressure being constant (that of coordinates atmospheric pressure) the phase rule of Gibbs is written as:

$$V = C + 1 - \varphi$$

where,

V = variance

C = number of independent constituents = 2

$\varphi$ = number of phases in equilibrium

That is, $V = 3 - \varphi$, which means that one can encounter in the diagram only three phases in equilibrium, the invariant being a point for a binary mixture. Elsewhere, there will be an equilibrium with two phases or one phase only.

By action of temperature, a fatty chain can offer a very high number of conformational states, as a result of a possible rotation around each C–C bond, which leads to 3 conformers, *trans*, $g^+$ and $g^-$. The number of these conformers increases considerably with the number of carbon atoms (n); and in fact, as $3^{(n-3)}$.

Carboxylate groups are, on the contrary, in very strong interactions and it has been recognised that they remain for a long time 'crystallised' as the temperature increases. They become separated (anhydrous soap) at very high temperatures (sodium laurate, transition subneat $\rightleftarrows$ neat at 241°C.

At these temperatures, the chains get liquefied well before this event. But this liquefaction does not bring about total molecular disorder. If there are disorders at short distances due to liquid chains, there subsists a long-distance order because of the non-broken interactions between carboxylate groups.

On account of this, the resulting paraffinic liquid is a constraint state.

It should be added that even if one achieves decrystallisation of carboxylate groups, the long-distance order is not destroyed (lamellar neat phase) and one should further increase the temperature in order to arrive at a micellar dispersion.

On the other hand, if a crystalline soap is submitted to the action of water, then this water naturally acts at the carboxyl group level, which causes swelling and then separation of these groups. Various macrostructures may succeed each other as the concentration of water increases and the curvature of strata grows.

From these properties entering into competition, there results finally a richness of phases which, crystalline polymorphs apart, are all mesophases. Initially it was *Mcbain* who established the phase diagrams of water/pure sodium and potassium soaps, of saturated fatty acids and oleic acid and for the first time threw light on the difficult physicochemistry of soaps and relations between their phases.

In dilute solutions, sodium soap molecules are dispersed only for very low concentrations (critical micellar concentration) and generally remain associated in the micelles whose form varies depending both on the nature of the soap and the concentration of the solution.

*McBain's* phase diagrams, mainly established with the help of a polarisation microscope and with the *McBain–Bakr* balance for poorly concentrated water phases, quite often continue to be considered. However, more recently *Perron* have reviewed these diagrams with the help, in particular, of differential thermal microanalysis. From these studies, covering sodium soaps of even saturated fatty acids from C12 to C18 on the one hand and sodium oleate on the other, diagrams in which different phases are correctly linked and which differ quite notably from those of *McBain*, have been established.

### *Saturated sodium soaps*

In reality, anhydrous sodium laurate in conjuction with myristate presents nine organised phases, a transition in the crystalline state (perhaps comparable to *McBain*, Curd $\rightleftarrows$ Supercurd transition) could be detected only for myristate. These phases are, in the order of increasing temperature: crystal, subwaxy I and II, waxy I and II, superwaxy, discs phase, subneat, neat and isotropic liquid. Waxy and subneat phases are mesomorphic phases, with ribbons structure, ribbons becoming narrower as the temperature increases. The neat phase the unstable lamellar phase for which the carboxylate groups are disorganised. The discs phase consists of flat discs (Fig. 3.1).

For sodium palmitate and stearate, there exists only one subwaxy phase whereas the superwaxy phase is transformed directly into the subneat phases when the temperature is increased. A transition in the crystalline state was observed in the two cases.

For anhydrous soap, it has been shown that the liquefaction of chains occurs at the transition: crystal $\rightleftarrows$ subwaxy. Decrystallisation of carboxylate groups occurs at the subneat $\rightleftarrows$ neat transition where one passes from a ribbon phase to the lamellar neat phase.

For sodium laurate, the subneat ribbon phase is preceded in the temperature gradient by the formation of a curious transitory discs phase starting from the superwaxy ribbon phase. Myristate also offers this phase, but in a very narrow temperature range (about 1°C).

These so-called 'anhydrous phases' admit very low quantities of water and are connected with each other and with aqueous neat phases by a succession of peritectic reactions at high temperatures as long as the hemihydrate does not exist. At much lower temperatures, one observes eutectic reactions between anhydrous crystalline varieties and at least one of the polymorphs ($C_3$) of the said hemihydrate.

With regard to this compound, three crystalline forms have been recognised with non-congruent fusion for $C_3$.

In the presence of more water than hemihydrate, all the phases that are poor in water as well as the polymorphs $C_1$ and $C_2$ are in equilibrium with the lamellar neat phase, $C_1$ with in equilibrium either with the middle soap, with the intermediate phases or the isotropic liquid.

The formation of the suppemeat phase, which *McBain* believed to have observed, does not take place. It and is replaced by an 'imminent' monotectic reaction (curve $T_i$ at high temperatures and low water contents). The monotectic reaction corresponds to a certain incompatibility (with critical temperature) between two types of micellar isotropic solutions of which one is probably inverse (spherical or better cylindrical) and the other is direct. This phenomenon is accentuated while passing from laurate to stearate. Variety $C_1$ of the hemihydrate is in equilibrium with the lamellar neat phase, but also with the middle phase, then with the phases called intermediary $I_1$ and $I_2$, already detected by X-ray diffraction by *Luzzati*, and finally with a very dilute isotropic liquid.

Let us recall that the middle phase is composed of cylinders arranged according to a centered hexagonal symmetry. These different equilibriums are expressed in the phase diagrams by a series of invariant lines in this region, confirming the existence of intermediate phases.

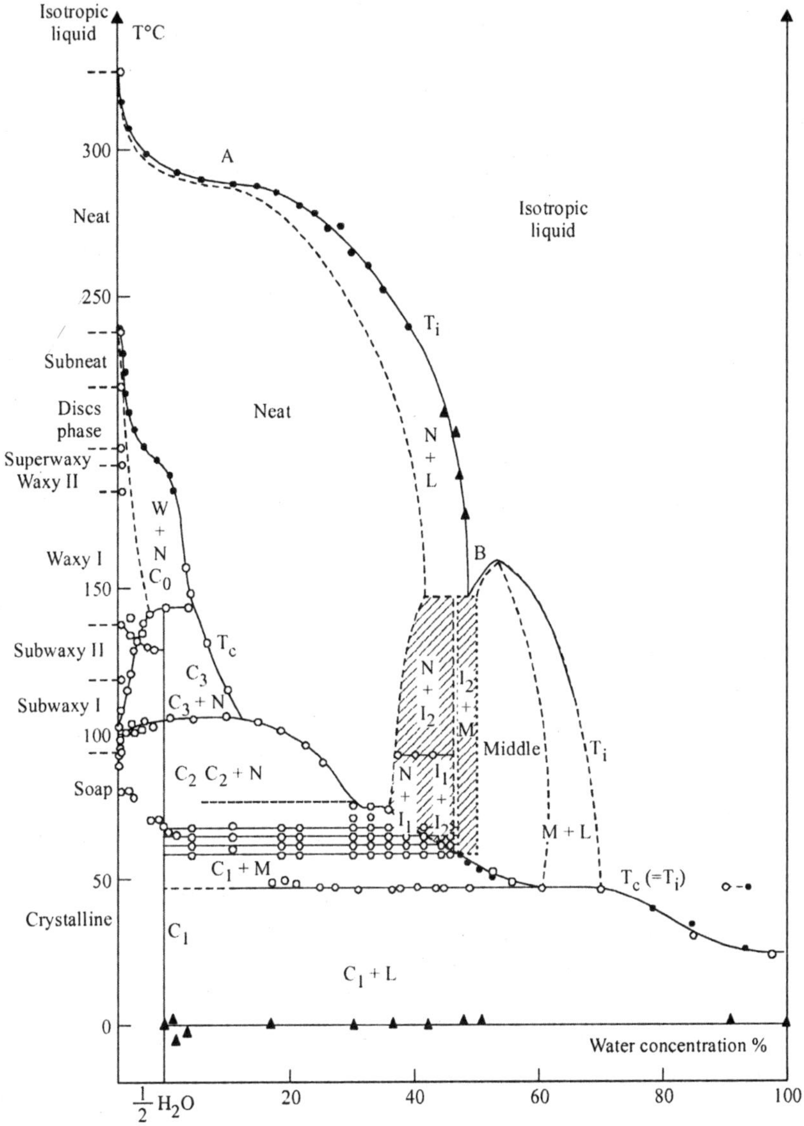


**Fig. 3.1.** Phase diagram of sodium-water laurate system.

An important precision, with regard to equilibria between hemihydrate and water, is that the sodium soaps of pure saturated fatty acids do not show any gel phase. Temperatures of invariant reactions, when passing from coagel to mesomorphous phases, are higher and higher and are distributed in a gradually narrower interval as the fatty chain of the soap is lengthened.

The accurate and coherent determination of different invariant reactions, especially for low water contents, has permitted us to define the Tc curve in this region, which starts, in fact, at the subneat ⇄ neat transition (whereas *McBain* could not trace it beyond 130°C, where the non-congruent fusion of crystalline species $C_3$ takes place).

It is therefore no longer a question of comparing this curve to the various transformations from crystalline states to mesomorphous phase with disordered chains. In reality and on the contrary, the $T_c$ curve represents essentially the locus of temperatures at which the carboxylate polar heads 'decrystallise' as a function of the water content. Moreover, the liquefaction of fatty chains can take place either simultaneously with decrystallisation (diluted solution of soap and relation with the Krafft point and the critical micellar concentration), or preceding this decrystallisation, the gap between the two phenomena increasing as water content decreases.

As regards the crystallisation of water, its study has revealed that as long as the water content is lower than that of hemihydrate, this water is totally bounded. Beyond, two crystalline curves of water for each soap were obtained, to which corresponds a unique melting point at 0°C (Fig. 3.2).

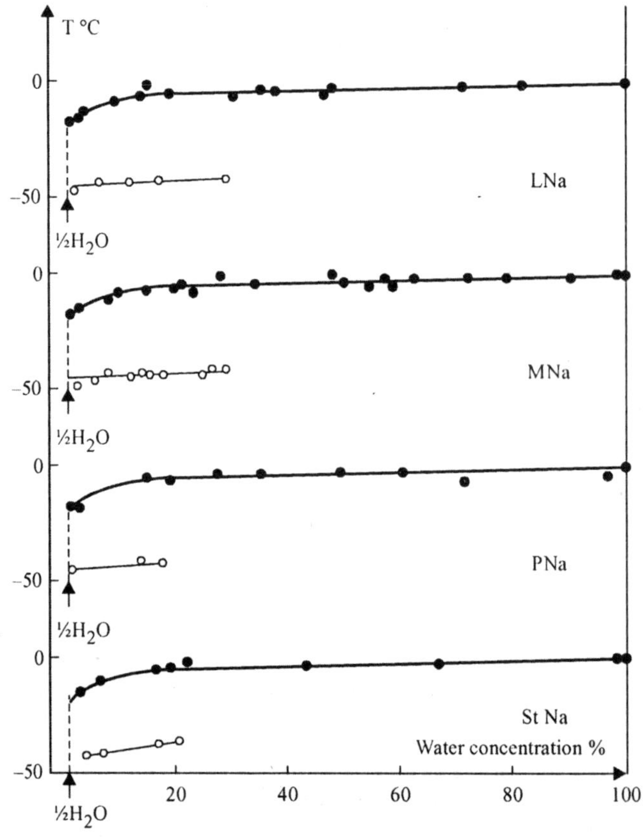


**Fig. 3.2.** Crystallisation of water in sodium laurate–myristate–palmitate–and stearate–water systems.

This corresponds to the existence of two waters–one more or less free, which starts as soon as the water content for hemihydrate is surpassed, for which crystallisation takes place from −15 to −25°C depending on the length of hydrocarbon chain, to reach 0°C at large dilutions; the other is more bounded water, of 'intermediary' type, which crystallises from hemihydrate between −45 and −35°C, to disappear, or to be no more detectable, between 20 and 30 per cent of water. These two types of water could correspond, one to the water surrounding crystallites, the other to an internal water in these crystallites.

A similar study (ATD, X-ray diffraction and IR spectrometry) was carried out on pure sodium oleate by *Curat*. For anhydrous soap it was found, contrary to earlier estimations for which the oleate seemed to present the same phases as the saturated soaps, the following series (increasing temperatures): three crystalline phases, whose disorder increases with temperature, two face-centered cubic phases, one centered rectangular phase called 'ribbon phase' and two lamellar neat phases, of which one is comparable to the normal neat phase and finally the isotropic liquid (Fig. 3.3).

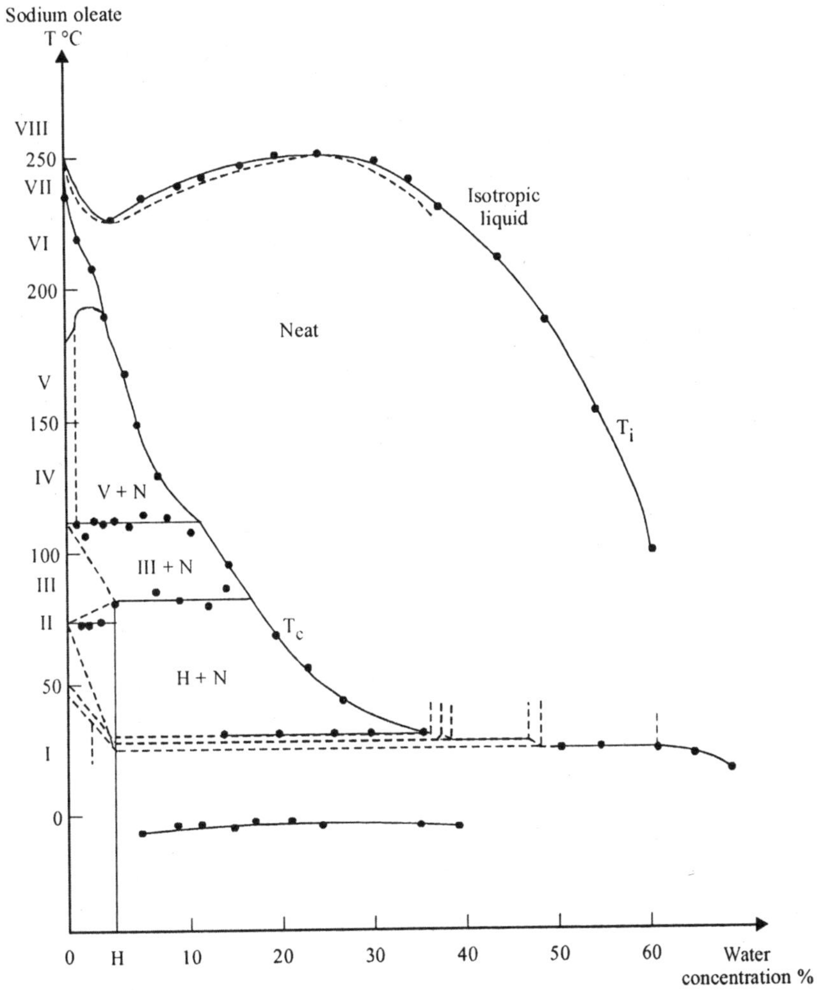


**Fig. 3.3.** Phase diagram of the sodium oleate–water system.

In the presence of water and contrary to saturated soaps, sodium oleate exists in the form of monohydrate, without polymorphism. This monohydrate is found in the diagram in equilibrium with the neat phase, or the middle phases, or the micellar isotropic liquid and at least with one intermediary phase.

The thermal effects in relation with the transitions are weaker than for saturated soaps and this occurs as the water content increases. This posed a problem in establishing the diagram for water content higher than 40 per cent by weight.

The passage from the coagel (also no gel is observed) to mesomorphous phases takes place in a narrow temperature interval. Equilibria between the phases derived from the anhydrous soap and the neat phase are analogous to those of saturated soaps (peritectics). The eutexy also appears with the existence of the non-congruent melting of the monohydrate.

For the crystallisation of water, only a single curve could be constructed when starting from the monohydrate, indicating a unique type of water with restricted degree of liberty.

### *Commercial soaps*

In practice, when one has to deal with a mixture of saturated and unsaturated fatty chains, coming from the fatty materials chosen before saponification, the principles referred to earlier and particularly the rule of phases, seem *a priori* impossible to consider.

Fortunately, this is not ruled out as in fact, what is distinct and independent in the sense of the rule of phases, it is essentially, 'the soap' on one hand and water on the other, each of very different constitution and with contrasting properties. From this point of view, nuances concerning the chain lengths and unsaturation are secondary with respect to this basic duality. This makes it possible to obtain a certain phase diagram for industrial soaps, at least in order to recognise the limits of the two main phases—neat and middle—and their relations and also the coagel (and gel in this complex case), the other phases and their connections being 'jumbled up' by the real complexity of the system.

Soap makers are, on the other hand, required to salt out the soap by brine and sodium chloride then intervenes as a third constituent of a ternary diagram, at low salt concentrations, whereas a 'grained' phase occurs, as well as the domain of coacervation.

### *Ternary diagrams*

Soap makers have always produced soap in presence of salt. This empirical practice was improved following the works carried out by *McBain*, particularly on the physicochemical study of the ternary mixtures soap/water/salt. These works are now well known. We have limited ourselves to the reproduction of a typical ternary diagram in Fig. 3.4.

*Prévot* on the use of diagrams of phases during manufacturing of soap has shown the importance of this diagram for a soap maker using either the batch process or the continuous process.

## Production of Neat Soap with 63 per cent Fatty Acids

### *Saponification reaction*

Saponification is a reaction between a triglyceride, or more generally an ester and a strong base, which produces soap and an alcohol. The reaction between a fatty acid and a base, which also produces soap is neutralisation. The saponification reaction can be described as follows:

$$\text{Triglyceride} + \text{strong base} = \text{soap} + \text{glycerol}$$

In this reaction, the triglyceride is characterised by its saponification value, which is the number of milligrams of potash required to saponify 1 gram of triglyceride. In the case of fatty acids, there is a neutralisation value with the same definition. In actual practice, soap manufacturers often use the saponification value expressed as $Na_2O$. The whole saponification theory, which the reader will find in 'Soap manufacture', will not be discussed here. It is important, however, to recall Lascaray's curve representing the speed of the reaction (Fig. 3.5).

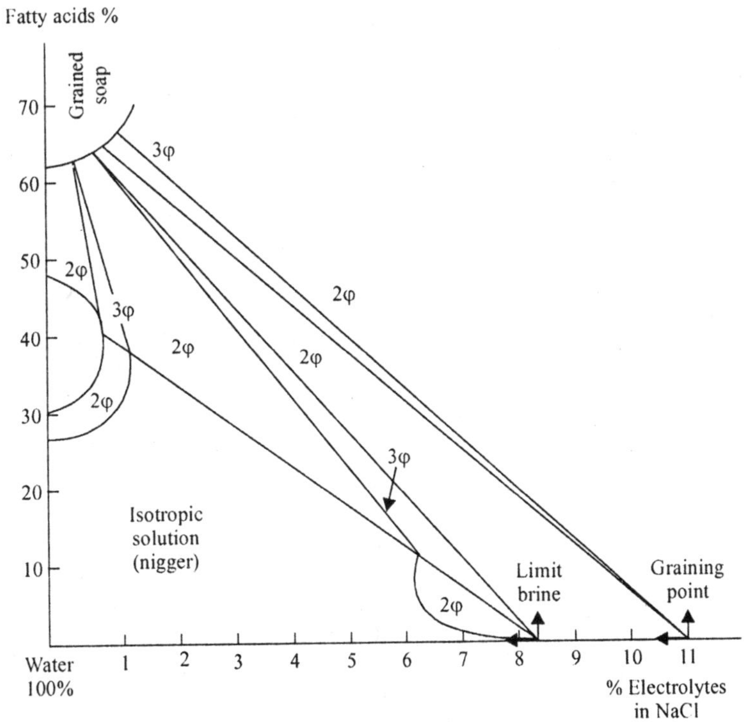


**Fig. 3.4.** Ternary diagram.

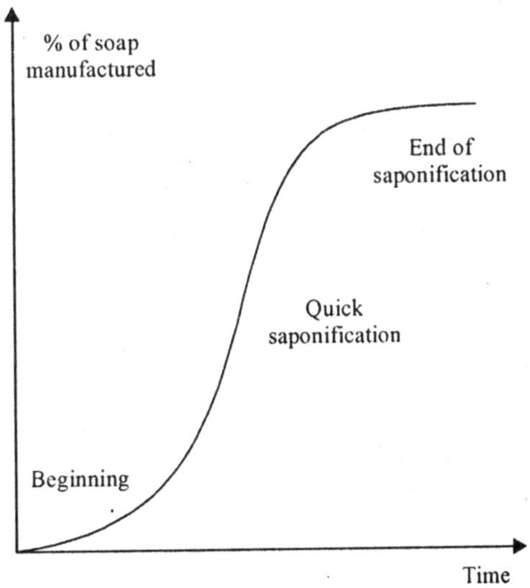


**Fig. 3.5.** Saponification reaction rate.

This curve shows the three stages of the reaction. In the first stage, the speed is slow but increases gradually. In the second stage, it accelerates rapidly and in the third stage, it decreases gradually until saponification is complete. What we have here is an autocatalytic reaction. In actual practice, this graph is very important; in batch saponification, we must make sure that the reaction has started before introducing the reactants gradually. Otherwise, the kettle may boil over due to the heat of the reaction leading to an accident. In continuous processes, the reactants are introduced at the same time in the circulating mass of soap, during quick saponification stage and the reaction occurs immediately.

Saponification may be obtained by different processes: both discontinuous and continuous. However, these processes are still commonly used in small factories. We are giving here only the characteristics of the major processes still in use, knowing fully well that they are not the only ones since big companies have developed their own processes.

### *Batch process*

It is the oldest and a well known process. This process consists of three traditional phases—saponification, washing and fitting. *Prévot* has shown the composition of soap during these different phases in a ternary diagram. This process is also known under other names such as kettle process or Marseilles process.

#### *Equipment*

All the operations are carried out in large insulated kettles having heating coils and an open coil for introducing steam directly, thus ensuring that the contents are stirred. These kettles are also provided with delivery pipes for water, oils, soda and brine, a valve at the bottom for draining out the residues and a swing-pipe to draw off the contents to the desired level.

#### *Saponification*

It consists of two operations: initial saponification, also known as the 'Killing Change' and boiling. The killing change is meant to obtain a 95 per cent saponification so that the glycerine lye at the bottom of the batch is not too alkaline. This operation is generally carried out with soap mass from a previous boil so that the reaction starts faster as we have already seen. After settling and clarifying the glycerine lye, which is sent for treatment, the saponification is completed by the addition of soda until the alkalinity becomes constant. This operation was used earlier for graining out soap and because of this, it took a long time; it has now been improved by working in more homogeneous phases to produce a neat soap.

In the course of saponification and washing, the soap and lye move countercurrent. It is thus possible to recover the maximum amount of glycerol and use minimum quantity of water and salt.

### *Washing*

Washing is meant to remove glycerine and impurities that appear during saponification. But they are removed only partly and washing by itself is not enough to produce a clean soap if the basic raw material is of a poor quality. Generally two or three washes are enough for household soap but for toilet soap four or five washes are required. In the case of countercurrent washes, it is possible to obtain a lye with a relatively high glycerine content: normally an average of 12 per cent.

#### *Fitting*

This operation is an essential part of the discontinuous process used in manufacturing soaps. It has a threefold purpose:

1. Reducing the percentage of electrolytes in neat soap.
2. Bringing about a practically fixed concentration of 63 per cent of fatty acids.
3. Purifying the soap.

At the time of fitting or finishing, the concentration of the mass should be in the two-phase region of the neat-isotropic solution. At the end of the washing operation, the composition of the mass in the kettle is such that in the diagram its point is on the right of the constant triangle, at a level of 55 per cent fatty acids. To reach the fitting stage, it is enough to add water slowly until the fitting point is reached, so as to move to the other side of the constant triangle. At this state, practical knowledge and experience of the soap-maker will ensure success of the fitting operation. However, the most reliable method of finding out if the concentration is right is to centrifuge a sample, taking care to maintain a constant temperature. To perform separation at a very high speed, it is necessary to remain as far to the right as possible of the two-phase region; this leads to the most favourable soap/niger ratio. In case it is necessary to purify the soap further, more water should be added to increase the niger. Neat/niger separation is quite slow because the two phases have very similar densities and their viscosities are relatively high: this operation generally requires 36–48 hr. Neat soap can be pumped out by placing the swing-pipe at the desired height after sampling and analyses.

There are many different methods of recovering fatty acids from niger but the most rational solution, which also produces soap of a uniform quality, consists of:

1. Collecting niger in a batch and graining it out by adding salt; the lye so obtained is recycled.
2. When enough soap is present in the niger kettle, more water is added to obtain both neat and niger from niger.
3. Neat is proportionally recycled in batches to start the reaction.
4. Niger from niger may be treated in the same manner or it can be recycled to produce an inferior quality soap.

The batch process offers the advantages of being flexible, of needing very simple equipment and not very skilled labour—except for the master soap-maker. But it also has the disadvantages of consuming too much energy in the form of steam, of taking up too much space and of requiring large quantities of raw material. Finally, it produces lye with a relatively low glycerine content.

### *Continuous processes*

The main processes commonly used in the soap industry is discussed below.

#### *Sharples process*

In this process (Fig. 3.6) the decantation operations are replaced by centrifugation. Further, there is no fitting stage. However, if the need arises, it has to be done in a kettle. This process consists of four stages:

Step 1: Primary saponification and separation of the glycerine-containing lye. This stage corresponds to the strong change in the kettle process where there is 98 per cent saponification and the glycerine-containing lye is slightly alkaline with a 17 per cent glycerine content.

Step 2: Finishing saponification. Soap from the first stage is mixed with lye from the third stage containing an excess of caustic soda and the mixture is centrifuged.

Step 3: Washing, where soap obtained from the previous stage is mixed with lye from the fourth step (after adjusting the balance of caustic soda and electrolytes) before it is centrifuged.

Step 4: Finishing of curd soap. Soap from the third stage is mixed with a lye having an exact composition so that the mixture is at a point just to the right of the constant triangle in the *McBain* diagram where neat is separated from lye. This separation occurs during the last centrifugation. Soap thus obtained contains about 0.5 per cent glycerine. This process is more advantageous than the kettle process because it needs a small quantity of raw material, consumes very little steam, does not take up much space and results in a lye with a high glycerine content.

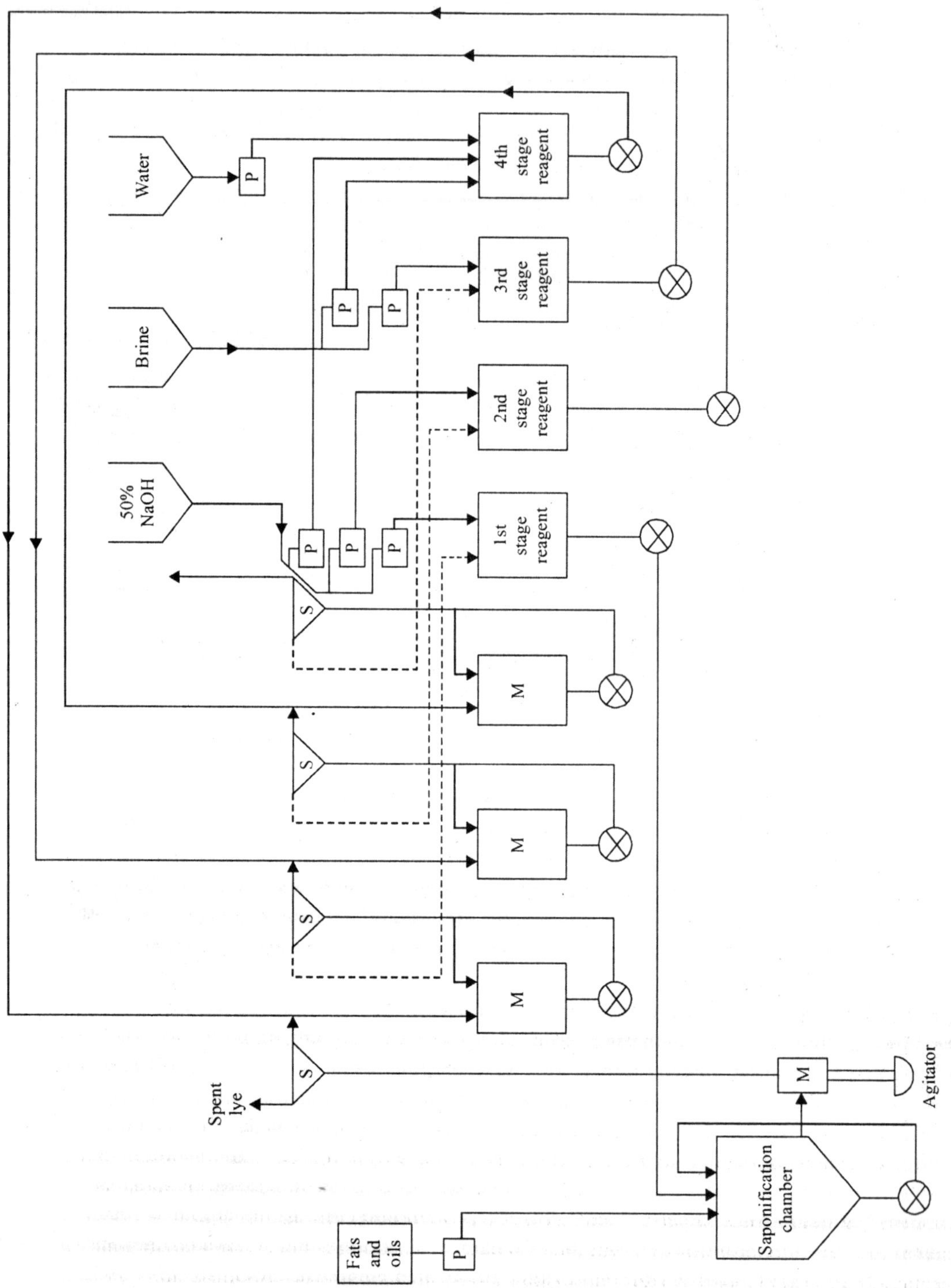


**Fig. 3.6.** Sharples process. P = Controlled volume pump; M = Mixer; S = Separator.

*Monsavon process*

This process was developed in France on the basis of a theoretical study of soap-making and the ternary soap/water/salt diagram. It consists of three stages: saponification, washing and fitting. Only neat/niger separation is done in traditional kettles.

### Saponification

Fats and caustic soda are mixed thoroughly into a fine emulsion at a high speed (3000 rpm) in a homogeniser. The emulsion leaving the 'colloidal mill' passes through a reaction tube and saponification is completed very quickly. Soap is stored in a kettle at the base of the reaction tube and then fed into a washing tower. It has a 63 per cent concentration at this stage which means that its viscosity is at its lowest, making deaeration of the soap possible.

### Washing

Washing is performed in a tower of tiered washing sections. Each section has its own mixing compartment where soap from the lower section is mixed with the lye coming from the section above. The mixture finally reaches the decantation tank where soap and lye are separated, the soap moves up and is fed into the upper section and the lye is pumped into the lower section. In this washing system, it is possible to recover about 16 per cent of glycerine from sweet water (Fig. 3.7).

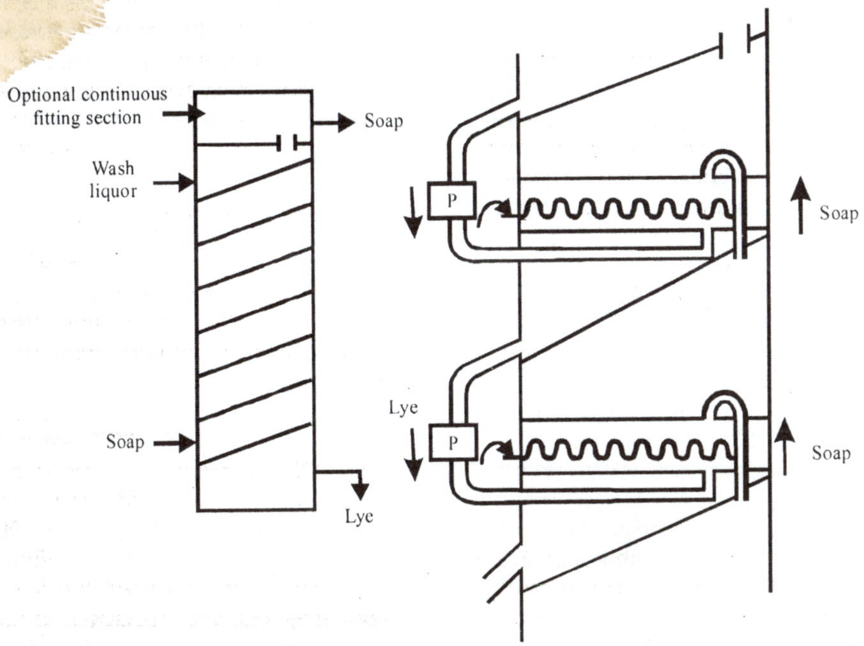


**Fig. 3.7.** Monsavon washing tower.

### Fitting

At the top of the tower, there is a special compartment where it is possible to add a special fitting solution to the soap after the last wash in order to obtain the desired concentration; the soap is then withdrawn into traditional kettles for settling.

This process has the advantage of being time-saving with a low consumption of steam and electricity and offers a high glycerine recovery. Unfortunately, a continuous settling process has not yet been developed and it is necessary to use traditional kettles for settling.

## Alfa-Laval Process

### *Saponification*

It takes place in an enclosed system or a recirculation loop where fats and caustic soda are continuously fed in by automatic proportioning pumps. Temperature inside the system is relatively high (135°C) and is maintained at a constant level by the heat of the reaction. Because of this, the reaction is almost immediate even though the mixture remains in the reactor for a short time (a few minutes). A special feature of this process is the automatic soda dosage control by constant measurement of soap viscosity, which varies according to the level of electrolytes. Another special feature is that it is possible to saponify at 63 or 73 per cent fatty acid content. Viscosity curves for 63 and 73 per cent fatty acids are given in Figs 3.8 and 3.9.

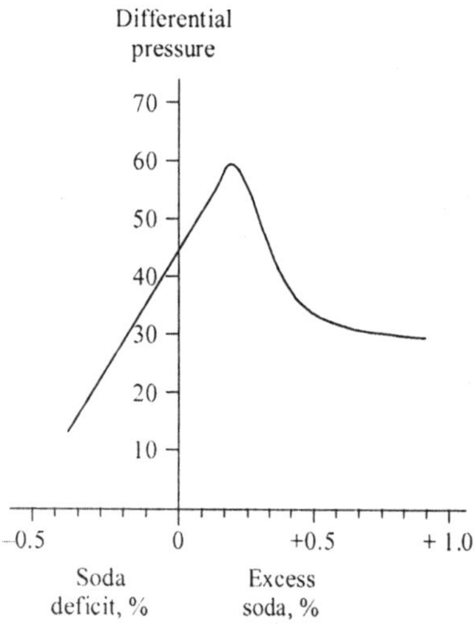


**Fig. 3.8.** Viscosity of soap with 63 per cent fatty acids.

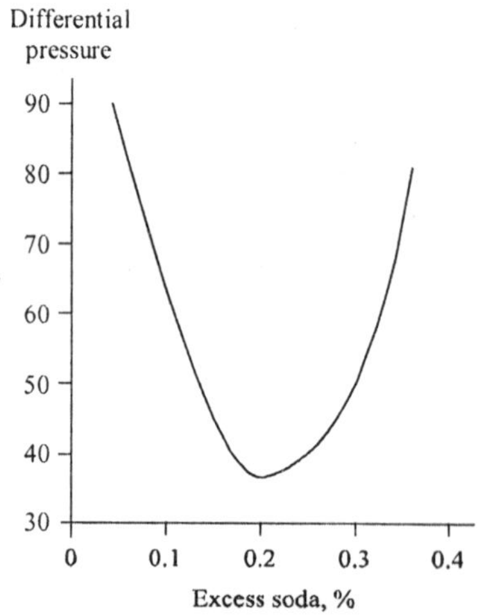


**Fig. 3.9.** Viscosity of soap with 73 per cent fatty acids.

Viscosity of the mass is measured by taking into account the difference in pressure at the time of entry into and exit from the recirculation pump. The speed of the caustic soda pump is regulated by selecting a particular working point to maintain uniform viscosity and percentage of caustic soda. Figure 3.10 shows a diagram of the saponification stage in an Alfa-Laval unit.

### *Washing*

Different solutions are proposed, which are given below:

1. The oldest solution consists of a series of washing stages with a mixing and a centrifuging section where soap and lye move at countercurrent.
2. Use of extractors so that mixing and separation can be carried out by the same equipment; but its operation is rather difficult and manufacturers have suggested a third solution.

3. Use of a washing tower with revolving plates. This solution seems to have been adopted now by all manufacturers in preference to the traditional process.

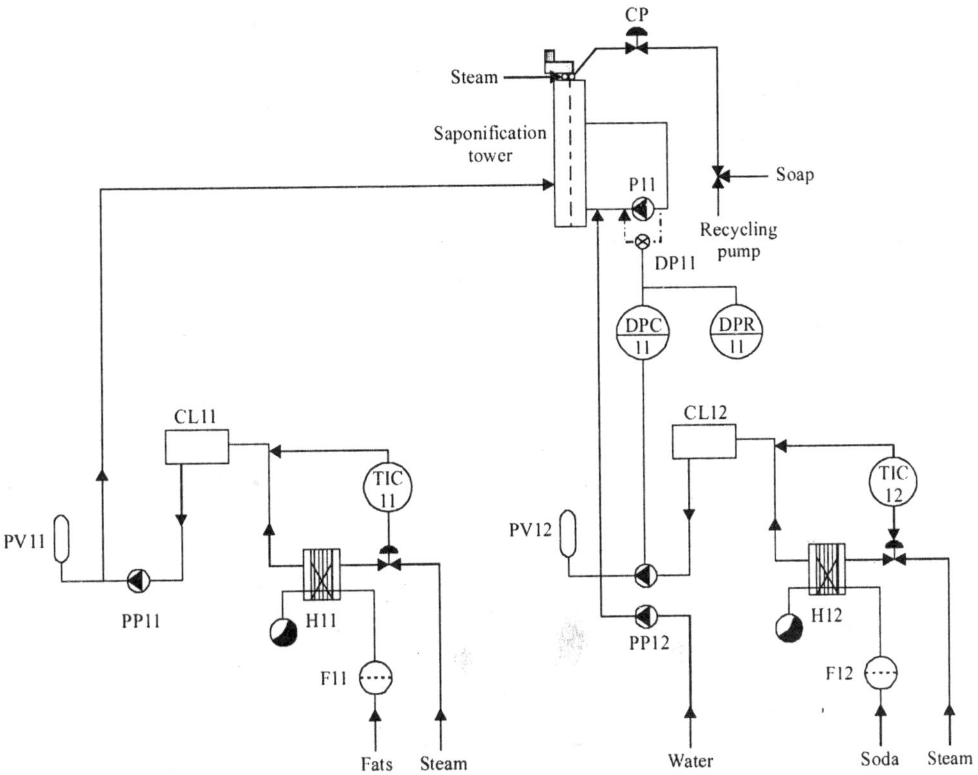


**Fig. 3.10.** Saponification stage in an Alfa-Laval unit.

But irrespective of the process used, the soap that leaves the washing stage is properly centrifuged. The concentration of the lye is within the normal limits and the resultant curd soap can be used as it is after adjusting its alkalinity or it can be subjected to a fitting operation.

### *Fitting*

The fitting equipment is identical to the saponification equipment with the addition of one or more centrifuges for separating curd soap form niger. The viscosity curve, which depends on the electrolyte concentration, is thc same as for 63 per cent saponification of fatty acids. Composition of fitting lye is adjusted automatically to maintain the right viscosity. Liquid soap is pumped into the centrifuges to separate it from niger, which is partially recycled during the washing stage. Advantages of this process are the automatic adjustment of addition of soda or electrolytes, low volume of material, high glycerine recovery and concentration of glycerine in glycerine lyes (25–30 per cent of glycerol).

### *Mazzoni process*

This process and others based on similar ideas are those that are most commonly used today. A complete diagram of the Mazzoni plant is given in Fig. 3.11. It shows the tower version of the plant for washing and neutralising curd soap by fatty acids.

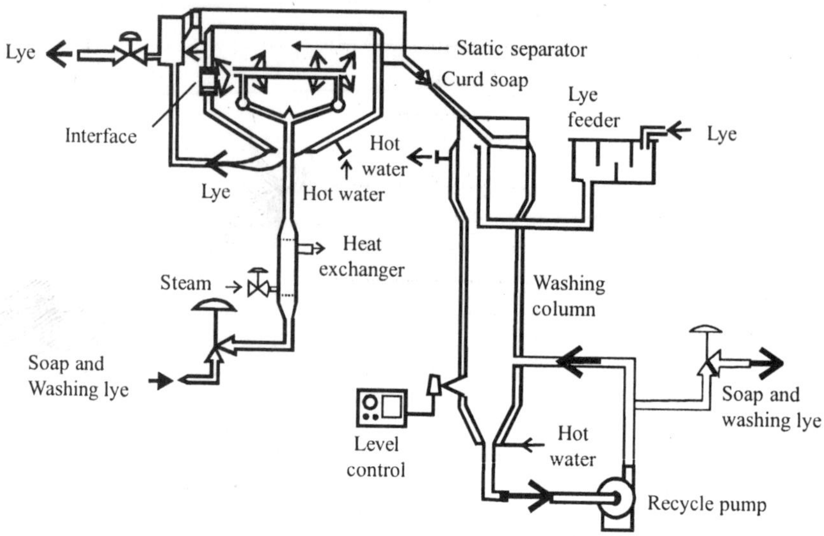


**Fig. 3.11.** Main features of a Mazzoni washing stage.

## Saponification

Saponification is carried out in an autoclave provided with a recycling pump. A temperature of 120°C and the presence of a soap mass permit very rapid saponification. Due to an excess of 0.5 per cent of soda, very little proportion of the unsaponifiable matter is left after saponification.

A cooler placed behind the autoclave reduces the soap temperature to 88°C so as to simultaneously release excess residual saponified and unsaponified alkali. The mixture is then pumped into a static decanter to separate the soap from the glycerine lye.

## Washing cycles

This step follows the traditional pattern described in many manuals. It is done in several stages with the soap-mass passing countercurrent to the lye. Each stage consists of a mixer with a constant level recycling placed above a heat exchanger, which maintains the temperature at the desired level and a static separator. The plant generally has three washing stages.

For some time now, some manufacturers have proposed the use of a washing tower with revolving disks for proper washing and for obtaining a lye with a higher concentration of glycerine. A concentration of 25–34 per cent glycerine is reported for 0.2–0.6 per cent soap, depending on the kind of fat used and the proportion between spent lye and fats.

The washing section in both options ends with centrifuging to separate the neat from the lye before the final finishing stage.

In the Mazzoni process, finishing can be done in two different ways—either by neutralisation or by fitting. In the case of neutralisation, the excess soda contained in the soap at the end of the washings stage is neutralised by the addition of fatty acids as shown in the general diagram of the plant. In case of fitting, a fitting solution is added to the liquid soap when it leaves the centrifuge and enters the mixer. It is then centrifuged once or several times to separate curd from niger.

This process has the advantage of being flexible in the choice of raw materials, saving energy and labour. It is also quite easy to maintain and is not very costly. The washes are effective and the lyes have a high content of glycerine (25–30 per cent glycerol).

### *Latest developments in processes*

The typical fitting operation is not commonly used in modern soap-making processes for the following reasons:

1. It is a very delicate operation.
2. It involves a series of procedures for recycling fatty acids contained in the niger.
3. There has been a definite improvement in the quality of fats and particularly of tallow and because of the new improved bleaching techniques, it is possible to saponify fats with a very low iodine index. Some prefer elaborate processing of raw material, practically a refining process, to avoid the fitting stage.
4. Adjustment of electrolytes and especially of free caustic soda, is obtained by the controlled addition of distilled fatty acids.

In continuous processes, the washing operation is now carried out in towers. It is therefore more efficient and the spent lyes have a higher glycerine content. The only disadvantage of washing towers is that, if several qualities of soap with different compositions are to be produced, it is necessary to change the composition of electrolytes in the lye present in the towers. In fact, concentration of this lye being almost the same as that of spent lye, as regards the mixture of fats, any change in the latter is bound to bring about a change in the former.

### *Other processes*

#### *Soap from fatty acids*

Making soap from fatty acids is a simple neutralisation operation for directly obtaining 63 per cent curd soap. This neutralisation may be effected either by plain soda or by 80 per cent sodium carbonate. Finishing is also done with soda. In Mazzoni SCC neutralisation plant, sodium carbonate is used. When plain soda is used, there is no deaeration of soap. Percentage of free soda is automatically controlled by electrode potential (mV).

#### *Methyl ester soap*

However, this process has not been commonly used in the soap manufacturing industry, except in Japan where *Lion* has designed a plant along these lines. In principle, saponification of a methyl ester is not different from the saponification of a triglyceride, but the process is different because the methyl alcohol released is volatile and highly inflammable; it is therefore necessary to eliminate methyl alcohols as much as possible and that is not a simple operation.

## Drying and Finishing of Soap

Traditionally, many methods have been employed to cool and dry liquid soap such as running, setting, cooling presses, belt dryers, atmospheric flash dryers, etc. but only vacuum dryers are in use today. The process is automated and better regulated. A distinction must be made between drying up to 70 per cent FA used in manufacturing household soaps (or similar types) and drying up to more or less 80 per cent FA for toilet soaps. The drying equipment is almost the same for both types of soaps, the only difference is that up to 70 per cent, the soap is very sticky and cannot be stored in the form of chips contrary to chips of toilet soaps.

#### *Drying of household soaps*

Figure 3.12 gives a diagram showing the different stages of finishing household soaps. At the drying stage, it is possible to introduce a scent so that the bar of soap has a pleasant smell; this is all the more important because today we find more and more toilet soaps in small packs manufactured by this method.

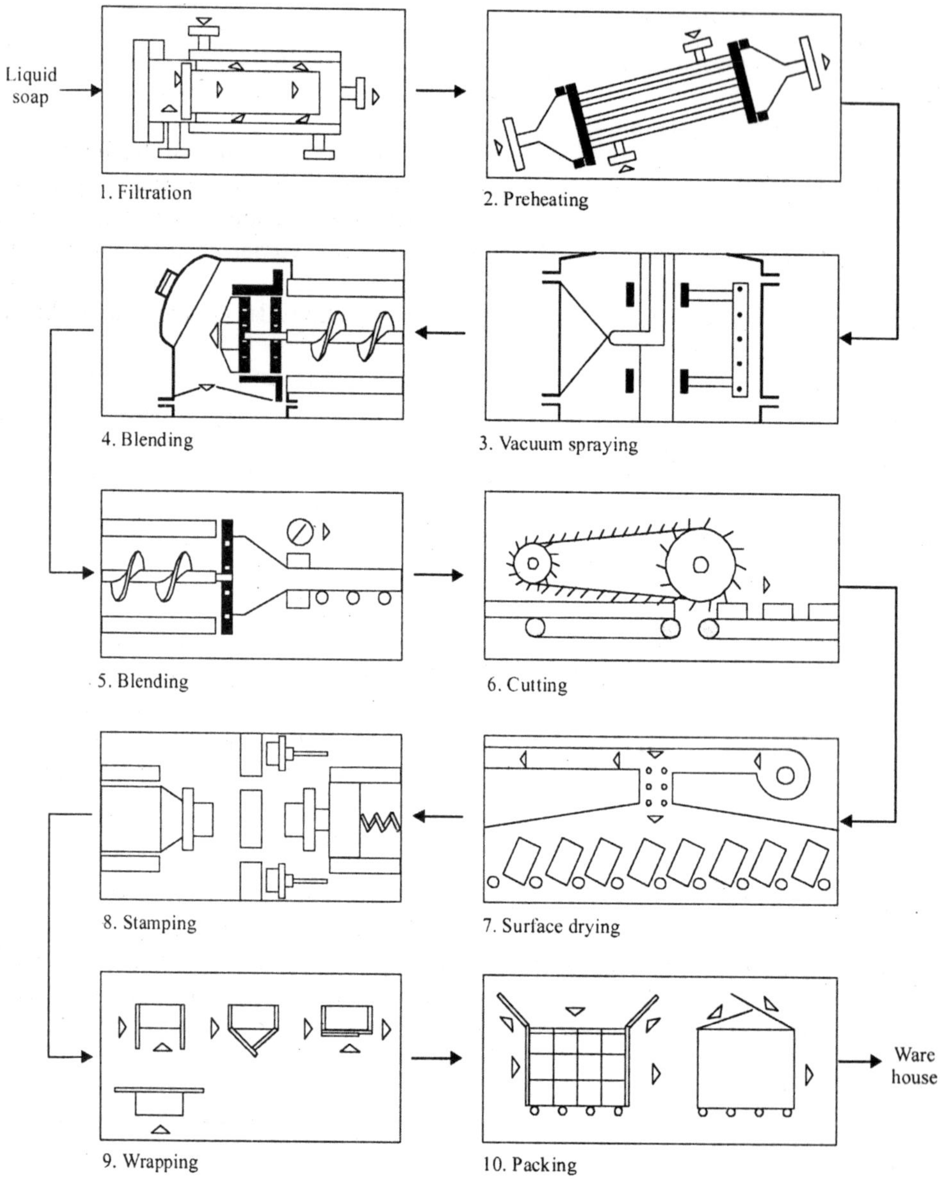


**Fig. 3.12.** Finishing process for household soap.

The bar of soap is then cut into uniform pieces by an automatic cutting machine. The small cakes then move into 'surface dryers' where their surface is hardened before they are moulded and finally wrapped in cellophane or other types of wrappers normally used for toilet soaps.

*Drying of toilet soap*

In case of toilet soap, the wet soap is dried up to a level of 80 per cent FA in a vacuum dryer. Although the soap is shaped like little rolls of the same size as the dryer's nozzle, these rolls are called chips out of sheer habit. Either before or in the course of drying, ingredients such as anti-rancid and sequestering agents are added to preserve the chips, together with optical dyes and superfattening agents, to make the end product more homogeneous.

Traditionally, toilet soaps are dried to a moisture content of about 13 per cent and a fatty acids content of 79 per cent. Superfatted soaps are drier with 9–10 per cent moisture for 82–83 per cent fatty acids. On leaving the dryer, the chips are stored in silos for a short period, for a day or two. This rest period seems to improve the soap's behaviour during the finishing stage.

### *Finishing of toilet soap*

Finishing of toilet soap is generally a continuous process. Figure 3.13 shows the layout of the finishing unit.

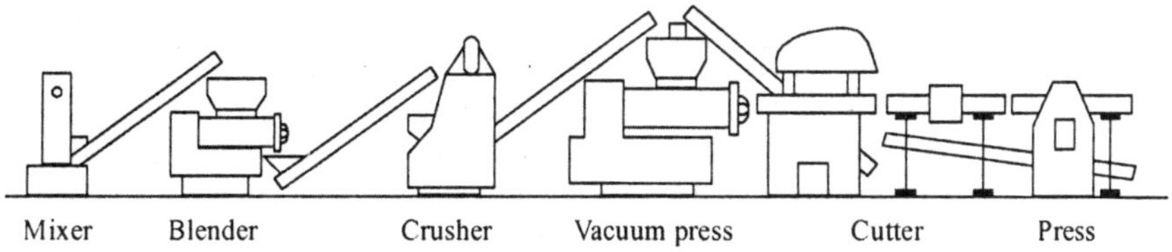


**Fig. 3.13.** Machine layout for a toilet soap finishing unit.

### Mixing

Ingredients normally found in soaps such as scent and colour, are added to the soap chips. Mixing can be done either in discontinuous mixers, which are more or less automatic, or in continuous mixers. Both processes have their own advantages: the discontinuous process is more flexible as it is possible to control the dosage of materials while the continuous process is simpler.

### Refining

After the initial mixing, the soap has to be refined and homogenised. Refining is done in roll mills and in preplodders or in refiners where the soap is pushed through fine meshes. In the latter case, both mixing and refining are done in the same machine called a mixer-refiner.

### Pressing

The soap is plodded further and shaped into a long, continuous bar. To avoid the formation of bubbles, this operation is carried out in vacuum. This stage is very important if the soap is to maintain its shape during use. The soap may crack if it is not properly compressed.

### Cutting

A continuous cutter cuts the bar into pieces of a uniform length. To avoid variations in weight, it is essential that the machine's feed movement should correspond to the movement of the soap-bar because the cutter's motor serves only to remove friction.

### *Pressing*

In recent years, there has been a lot of technological progress in this field. At first there were horizontal presses that could only make banded soaps. They were followed by vertical presses specially designed to produce bandless soap. Manufacturers then came up with horizontal presses capable of extruding both banded and bandless soaps. In order to increase the production speed, presses have been provided with four moulds instead of two and it is possible to make 300 cakes in a minute. Machinery manufacturers now offer a new generation of presses that run at a high speed (can make more than 500 soaps a minute). They make both banded and bandless soaps, which have an excellent appearance. These are the USN presses manufactured by Binacchi & Co. and the STUR press manufactured by Mazzoni.

To ensure that the soap does not stick to the mould and the surface remains perfectly smooth, all modern presses are provided with a cooling system for the moulds. Due to condensation of a thin film of water on the mould surface, it is possible to avoid using brine; use of brine tends to spoil the soap's appearance.

#### *Wrapping and packaging*

The procedure used for packing household soaps is quite simple. They are generally wrapped in cellophane paper or packed in cardboard boxes. It is now almost impossible to find soaps without any packaging. Toilet soap is generally wrapped in paper and sometimes packed in cardboard boxes. This packaging prevents evaporation of scent and moisture, protects the soap from scratches and makes it more attractive to the consumer. Earlier, soap was wrapped in wax paper to prevent evaporation but is now rarely used since an evaporation barrier is provided by coating the outer wrapper with a thermosealing material. Soap is protected from damage by a light, corrugated cardboard wrapper, which also serves as an attractive packing. The outermost packing gives the soap a commercial look. It is generally made from glossy printed paper. The introduction of mixed paper–polyester sheets and the latest improvements in printing technology provide very attractive packing material, which is also effective against evaporation.

## Glycerine Recovery

From glycerine lye coming out of soap factories, four stages are required to obtain commercial glycerine, which can be used in the pharmaceuticals, cosmetics or paint industry.

### *Lye processing*

Glycerine lye, which flows out at the washing stage of a saponification installation, has a variable composition according to the procedure used and the saponified fat mixture. Besides glycerine and salt, lyes contain free caustic soda and a certain number of impurities coming from raw materials and saponification reaction. The objective of the treatment is to eliminate these impurities. Quality of the final glycerine depends on the quality of this operation. Better the treatment, easier will the following operations be, particularly bleaching and deodourisation.

The treatment is simple: excess soda is neutralised by hydrochloric acid and then iron or aluminium salt (chloride or sulphate) is stirred, which, by hydrolysing itself, lowers the pH to about 4.5, a value for which hydroxide is insoluble and which traps impurities while flaking. Metallic soaps are also formed with fatty acids which are more or less oxidised, contained in the form of soaps in lye. Generally, aluminium salts are preferred to ferric salts because they give an easy filtrable precipitation and a non-sticking oil-meal cake. On the other hand, ferric chloride is highly corrosive. For this reason, aluminium sulphate is widely used but has the disadvantage of carrying sulphates that do not have the same salting out power as chloride and it is necessary to put them in solution at the time of use. Today, we find a very good compromise with aluminium chloride solution in hydrochloric acid. Thus, we have the advantage of having only one anion in the whole procedure, an easily filtrable precipitation and a reaction in liquid form, which is well adapted to continuous procedures. In many of the soap factories, treatment is done in a discontinuous manner, with skimming of precipitation and filtering on press filter. Plastic is largely used as construction material to avoid corrosion, filter plates, for example, are now made of polypropylene. Some soap factories have put up continuous installations that help to obtain a more regular and better quality treatment. Suppliers of soap factory equipment also propose the use of continuous installations that are more or less automated such as the 'DCA' installation of Mazzoni.

Before sending lye to the following stage, it is necessary to increase the pH to a value close to neutrality in order to avoid corrosion of the evaporators. During pH adjustment, a slight reprecipitation of hydroxide may take place; it is recommended to filter it before concentration.

### *Concentration*

The object of this operation is to increase the concentration of lye from an initial 10 to 30 per cent (according to procedure) to about 80 per cent. This concentration is achieved by evaporation of water under vacuum in appliances having a simple, double or triple effect. These equipment have the advantage of economising on energy; vapour consumption decreases by 30 per cent between a single and a triple effect. During operation, the concentration in salt increases and thus we have crystallisation of this salt, which is recovered and recycled after wash. Many equipment manufacturers propose the use of continuous installations and Kestner evaporation unit is an example of this. This is a double-action unit with a finishing stage and continuous production of crude glycerine with an optimum thermal yield. Crude glycerine contains about 80 per cent glycerol, 9 per cent salt and impurities, which bring ash content to 0.5 per cent above the salt percentage. Crude glycerine is also characterised by its percentage of organic matter, which is proportionate to the processing. Many soap manufacturers stop recovery of glycerine at this stage and resell the crude form to companies having distillation facilities. It should also be pointed out that in certain countries, concentration operation is not carried out, lye is directly thrown away as it is highly diluted and recovery is not profitable any longer.

### *Distillation*

Glycerine distillation operation has to be carried out under vacuum because glycerine decomposes when distilled under atmospheric pressure. Distillation appliances are still functioning under a vacuum of 6–12 mm mercury and a distillation temperature of 150–170°C. Quick introduction of vapour produces stirring of the mass and a steaming effect.

Distillation units are laid out for continuous or batch production and a certain number of manufacturers propose these installations. Glycerine distillation consumes large quantities of vapour and consumption can reach 2.5–4 tonnes vapour/tonnes distilled glycerol, depending on the type of distillation.

During distillation, it is very important to avoid movement of liquid particles, which in turn may pollute glycerine. Distillation should be carried out in an alkaline medium in order to avoid the presence of fatty acids. Distilled glycerine at a concentration greater than 99 per cent often has a yellowish colour and a typical burnt smell. Therefore bleaching and deodourising are required.

### *Bleaching*

Bleaching of glycerine is done either in batches or in a continuous column. The bleaching and deodourising agent is carbon black of a superior so-called 'glycerine' quality which avoids, among other things, use of chlorides whose maximum permitted percentage is very low by Codex quality standards. Bleaching is done at a temperature of about 80°C in order to avoid the return of colour or odour.

## METALLIC SOAPS

'Metallic soap' is the term used for all kinds of soaps whose cation is a metal, other than sodium or potassium.

All these soaps have one thing in common, which is their insolubility in water (property on which one of the manufacturing procedures depends) and their aptitude to produce gels in organic solvent solutions. Apart from monovalent metals, each metal can produce different soaps with the same anion: neutral soap and basic soap.

For example, theoretically aluminium can produce three soaps; a neutral soap and two basic soaps:

$$(R\ COO)_3\ Al, \quad (R\ COO)_2\ AlOH, \quad R\ COO\ Al\ (OH)_2$$

The presence of aluminium tristearate and monostearate has been much discussed.

## Manufacturing Process

There are two traditional methods of making metallic soaps, one is based on their insolubility in water and the other on the property of neutralising fatty acids or saponifying glycerides.

### *Precipitation process (or double decomposition)*

If, to an aqueous solution of sodium soap (or potassium), one adds an aqueous solution of a metallic salt, for example calcium chloride, a precipitation of calcium soap can be obtained immediately, according to the reaction.

$$2\ (R\ COO^-\ Na^+) + (Cl^-)_2\ Ca^{++} \longrightarrow \underbrace{(R\ COO)_2\ Ca}_{\downarrow} + 2\ (Cl^-Na^+)$$

This theoretical reaction roughly represents all complex reactions which give rise to variable proportions (depending on pH and concentration) of basic soaps, calcium hydroxide or free fatty acid.

After precipitation, soap is purified by washing with water or by dissolution in an organic solvent. Water eliminates any possible excess of sodium soaps or metallic salts. Organic solvents allow elimination of excess metallic salt or metalloid hydroxide that may have precipitated.

Neither of the two techniques allows elimination of free fatty acids.

A variant of this process for the preparation of metallic soaps of the zinc group consists in adding a salt solution of the desired metal to an ammonium soap solution containing excess ammonia. Metallic soap precipitates when ammonia is eliminated by heating.

One could also consider that the European Unilever patent, concerning triglyceride saponification by soda in aqueous solution and precipitation of metallic soap by a metal salt in presence of acetone, is a variant of the precipitation process.

### *Melting process*

By making fatty acids react to heat (generally at temperature around 200°C) on oxides, hydroxides or volatile salts (carbonates, formates, acetates) of the metal in question, the metallic soap is directly obtained according to the reaction:

$$2\ RCOOH + PbO \longrightarrow (R\ COO)_2\ Pb + H_2O \uparrow$$

$$2RCOOH + CO_3\ Pb \longrightarrow (R\ COO)_2\ Pb + CO_2 + H_2O \uparrow$$

Generally an oxide or metallic salt in powder form is poured into a hot fatty acid in excess in order to avoid any non-combined metals in fatty acids. A continuous process may be used.

Numerous variants have been proposed, out of which some introduce steam or water, while others suggest direct action of fatty acids on finely divided metal, in the presence of steam and air, or action of fatty acids on metallic alcoolates, or to get metallic hydroxide to react on fatty acids in the presence of acetone and finally use of glycerine instead of fatty acids.

This procedure is in harmony with the ancient practice or preparing 'drying oils' for paints by action of litharge on linseed oil under heat. This process continues to be used with other metallic oxides and allows preparing metallic soaps *in situ*, i.e. directly in the fats in which one wants to incorporate them.

Metallic soaps prepared *in situ* are less expensive but also less pure; their performance is inferior to soaps prepared separately.

### Choice of Cation and Fatty Acids

1. Only metals possessing several degrees of oxidation can play the role of oxidation catalyst and can be used as 'drying agents'. This is the case of lead, manganese and cobalt.
   The nature of fatty acids does not modify their catalysts properties. But to ensure that the drying oil remains stable during storing, preceding its use, one should avoid using polyunsaturated acids. Presently, only short-chain saturated acids or naphtenic acids are used.
2. Saturated fatty acids give with lithium, alkaline-earth metals, aluminium, soaps that easily form gels in fats, mineral oils and organic solvents.

Melting points of these soaps (and consequently characteristics of gel prepared with them) depend very little on length of chain, but, are very different depending on the nature of cation and type of soap (neutral, basic soap) as seen from the values given below:

| | |
|---|---|
| Lithium stearate | 220°C |
| Magnesium stearate | 145°C |
| Aluminium tristearate | 120°C |
| Aluminium distearate | 160°C |

On the contrary, unsaturation of fatty acids has a considerable influence on the melting point of soaps; e.g. magnesium oleate has a melting point of 80°C.

Water solubility of saturated fatty acid metallic soaps is always lower than 0.02 per cent, irrespective of the fatty acid or metal, except for lithium soaps whose short-chain acid salts are more soluble: 0.2 per cent for laurate.

Chapter 4

# Surfactants

## INTRODUCTION

Surfactants are a group of materials that have evolved with the growing and ever-changing needs of society. They pervade the whole spectrum of the industrial horizon, and, as such, deserve a close look as regards their present global status and future potential. Substances containing in the same molecule groups which are strongly hydrophobic (water repellent) and hydrophilic (water loving) orient themselves in different ways when dissolved or dispersed in water or nonaqueous solvents. Such orientation, which could also be in the form of aggregates of oriented molecules, termed micelles, is known as 'Surface Activity' and materials exhibiting this are termed 'Surface Active Agents', often abbreviated as 'Surfactants'. Figure 4.1 indicates the orientation of surfactants in several systems.

Practically almost every product in commercial use today has come in contact, at some stage or other of it's production process, with surfactants, which are classified by the nature of their hydrophilic moiety. The surfactant mosaic is made up of anionics, cationics, nonionics and amphoterics in broad classification. However, surfactants grouped together as polymeric surfactants, biosurfactants etc. depending on their end-use importance or functional aspects are gaining prominence.

The hydrophilic moiety carries a negative ionic charge in anionics, a positive ionic charge in cationics and is neutral in nonionics. The amphoterics contain, or have the potential to form, both positive and negative functional groups under specified conditions. The surfactants, in general, are governed by several criteria such as:

1. Solubility in water, non-aqueous media and two-phase systems.
2. Hydrophile-lipophile balance (HLB).
3. Surface tension.
4. Foaming characteristics.
5. Cloud point.
6. Orientation at interfaces.
7. Substantivity such as adsorption on solid surfaces.
8. Micelle formation.
9. Functional properties such as detergency, wetting, emulsifying, solubilising and dispersing.
10. Reactivity such as complex forming ability and antimicrobial properties.

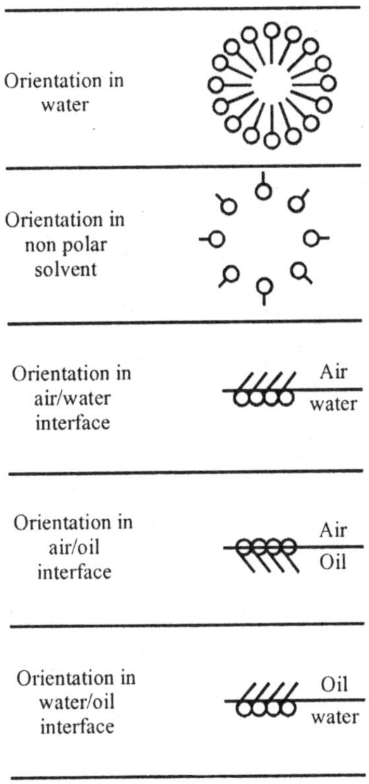


**Fig. 4.1** Orientation of surfactant in several systems.

Synthetic surfactants emerged on a large scale only with the use of petroleum derived fractions in the 1950s. Over the last 50 years there has been an enormous number of new surfactants produced and many new applications. Apart from the natural and synthetic surfactants, biosurfactants have emerged on the horizon. These are a collective class of biosurface-active materials, which are basically obtained by manipulating fermentation processes with very specific microorganisms. Several biosurfactants have surface-active and emulsification properties far superior to either synthetic or natural surfactants. In addition, they have the green attributes of bio-compatibility, complete bio-degradability and superior eco-toxic characteristics.

The usual picture of a surfactant as a simple rod-like hydrophobe with a hydrophilic head has changed in recent times. Many new surfactants are now being produced of more complex structure with more than one hydrophilic head. Such products form micro-emulsions more readily.

Polymeric materials, especially high molecular weight ones, have been used to give colloidal stability since almost the dawn of civilisation. In particular, naturally occurring polymeric materials, such as egg white, gum arabic and animal glues have been used to make water-based inks and paints pigmented with carbon black or metal oxide pigments. In recent years polymeric surfactants have been developed for wide applications in surface coatings. Surfactant technology thus is continuing to grow aided by a large well-established base industrial potential.

The number of surfactants available commercially today are several thousands. As a result many chemists, engineers, biologists and others, having the need to employ surfactants, feel daunted by the prospect of ploughing through the literature to find the best material for their purpose. Even if inclined to do so, they may face the investment of inordinate amounts of time and effort without guarantee of success. The breadth and complexity of the science and technology of surfactants makes it difficult to compress all the information available in a small volume. However in this chapter attempt has been made to give sufficient insight into the various classes of surfactants, major raw materials, key physico-chemical aspects, applications, typical formulations and other allied matters of relevance to the surfactants industry.

## TYPES OF SURFACTANTS

### Anionic Surfactants

By far the largest class of surfactants in general use today are the anionics (including soaps). For thousands of years in the history of the application of surface activity to the needs of mankind, the soaps, derived from natural animal and vegetable fats and oils, dominated the field. The major historical and economical advantage of the fatty acid soaps has always been the ready availability of the raw materials from natural, renewable sources. The weakness of the classical soaps, due to the sensitivity of the carboxylate salts which mainly constitute them, in many processes requiring surfactant action, led to the development of synthetic surfactants. Classically, anionic surfactants are divided into four basic categories, namely:

1. Carboxylates, of which common soap is the most important.
2. Sulphonates, of which linear alkyl benzene sulphonate (LABS) is the most significant and is the largest used synthetic surfactant worldwide.
3. Sulphuric acid esters such as alcohol sulphates or sulphated oils, the earliest of synthetic surfactants.
4. Other category includes the nitrogen-bearing taurates and sarcosinates, phosphorus-bearing chemicals such as the phosphate esters and others and industrial surfactants of the alkylnaphthalene-formaldehyde condensate type.

The four basic categories of anionic surfactants fall under two large classes of products—'true surfactants' and 'quasi-surfactants'. The 'true surfactants' are those which exhibit the important properties of lowering the surface tension of water and the interfacial tension between water and oils, of wetting and foaming actions, of emulsification etc. The most important of these surfactants are soap, alkylbenzene sulphonates, alcohol ether sulphates and alcohol sulphates. The 'quasi-surfactants' have feeble powers in terms of reducing the surface tension or wetting action, but are vastly superior in properties such as those of dispersion, thickening, spreading, penetrating, defoaming, demulsifying etc. Typical of these are lignin sulphonates and alkylnaphthanlene sulphonates. Most of these quasi-surfactants are used in the production of industrial commodities.

Surfactants are seldom used directly in the application sectors and are always components of formulations containing both surface-active and non-surface active chemicals. The market segments for anionic surfactants can be categorised as:

1. The household products sector.
2. The personal care products sector.
3. Industrial and institutional cleaners.
4. Other industrial applications.

In the household sector there are three important segments :

1. Laundry detergents.
2. Dishwashing detergents.
3. Household cleaners.

'True' anionics are consumed to a large extent in these applications. The usage is dominated by Linear alkylbenzene sulphonate (LAS), Alcohol ethoxy sulphate (AES), Sodium alpha sulphomethyl laurate (SAS), Alkyl sulphates (AS), Alpha olefin sulphonate (AOS) new anionics such as methyl ester sulphonates and phosphate esters. In all the segments of the household sector the anionics are used alongwith other surfactant types, mainly nonionics.

The laundry detergent segment is by far the most important in terms of the volume of surfactants used and the value contribution.

Whereas in the global context the laundry detergent segment is the predominant one, in the Industrial Triad the other segments are significant contributors. The three principle types of end products in the household laundry products segment are :

1. 'Conventional' or 'universal' powders.
2. Liquid detergents.
3. Superconcentrates.

Momentous changes have taken place in recent years in these products and multifunctional detergents incorporating bleach, fabric softeners etc. have landed in the market.

The personal care products sector also comprises of three segments, namely :

1. Hair care.
2. Skin care.
3. Oral hygiene products.

Anionic surfactants are mainly used in these segments, but they have to compete with other types such as amphoterics and cationics. At the global level, the skin care segment dominates with consumption of large volumes of soaps. In the Industrial Triad, however, both soap and other surfactants usage in this segment is limited. The hair-care sector is dominated by shampoos, which consume large volumes of anionics such as AES, AS and soap. The oral-hygiene segment is a small user of surfactants, with the anionics as the major type.

Industrial and Institutional Cleaners are products used in:

1. Food processing and commercial dishwashing.
2. Transportation.
3. Commercial laundry and dry cleaning.
4. Metals industry.

Anionics figure in large volume in these applications, LAS being the major anionic. Nonionic alcohol ethoxylates are also used in this sector.

Amongst industrial uses, anionics have substantial roles in textiles, petroleum processing, emulsion polymerisation of plastics and lattices, leather and pesticidal formulations. Other application areas include cement and concrete, paints and inks, froth flotation of ores, and laying of asphalt in road making.

### Cationic Surfactants

Cationic surfactants first became important when the commercial potential of their bacteriostatic properties was recognised in early fifties. Since than a wide class of cationics with literally hundreds of molecular structures have been introduced as commercial products, although they do not approach the importance of the anionic materials in sheer quantity or value. In most of these products advantage is taken of the 'substantivity' of the cation and it's semipermanent adsorption on surfaces which are normally found in nature to be negatively charged. The adsorption of the cation further makes the substrate surface oleophilic.

Until the availability of straight-chain petroleum-based surfactants, the sole sources of raw materials for cationic surfactants were vegetable derivatives of fatty amines of 1, 2 or 3 alkyl chains bonded directly or indirectly to a cationic nitrogen group. The most important classes of these cationics are the amine salts, quaternary ammonium compounds and the amine oxides. The cationic nature of the fatty amines and derivatives has given rise to a wide variety of functions in many applications. These functions may be recognised as being derived from four basic functional properties, namely, (i) surface active; (ii) substantive; (iii) reactive; and (iv) antimicrobial as represented in Fig. 4.2. These properties are exploited for the following surfactant functions in various applications as indicated in Table 4.1.

**Table 4.1.** Basic properties of fatty amines and their derivatives and their surfactant functions in various applications.

| *Basic property* | *Surfactant functions* |
|---|---|
| Surface active | Emulsifying, wetting, thickening, foaming |
| Substantivity | Softening, antistatic properties, lubrication, corrosion inhibition, adhesion, hydrophobation |
| Reactive | Flocculation, decolourisation, ion exchange |
| Antimicrobial | Algicidal, fungicidal, bactericidal |

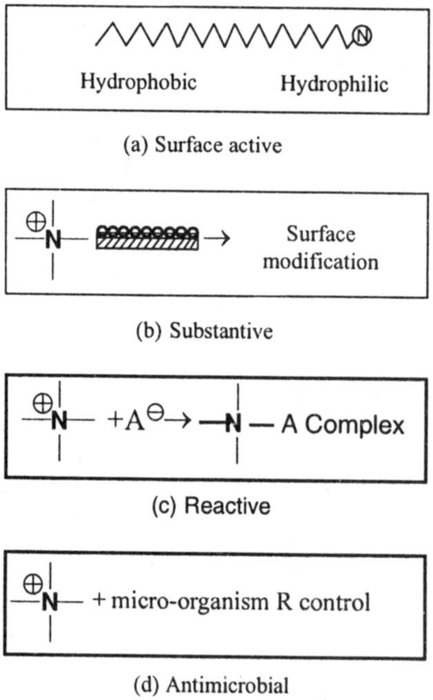


**Fig. 4.2.** Four basic functional properties of cationic surfactants.

Amongst the cationic surfactants there are two important categories which differ mainly in the nature of the nitrogen containing group. The first consists of the alkyl nitrogen compounds such as simple ammonium salts containing at least one long-chain alkyl group and one or more amine hydrogens, and quaternary ammonium compounds in which all amine hydrogens have been replaced by organic radical substitutions. The amine substitutions may be either long or short chain substitutions. The counterion may be a halide, sulphate, acetate etc. The second category contains heterocyclic materials typified by the pyridinium, morpholinium and imidazolinium derivatives.

The economic importance of the cationic surfactants has increased greatly because of their unique properties. Their substantive property is exploited in fabric softeners, bitumen emulsions, oil-muds used in the petroleum industry, hair-care products, corrosion inhibition in oilfield and water treatment segments. The cationics are also important in the textile industry as waterproofing and dye-fixing agents. Other application areas are in flotation processing of mineral ores, lubrication, as surface modifiers in electrostatic charge control. Their biological activity seems to be characteristic of almost all-cationic surfactant species, irrespective of minor structural changes, and is utilised in killing or inhibiting the growth of many micro-organisms.

Nonionic surfactants, which fit into a broader spectrum, do not carry any discrete charge when dissolved in water, unlike anionics or cationics. Some of the important advantages of this electrical neutrality are:

1. Lower sensitivity to the presence of electrolytes in the system.
2. A lessened effect of solution pH.
3. The synthetic flexibility of being able to design the required degree of solubility into the molecule by the careful control of the size of the hydrophilic group.

Several nonionics are currently in commercial use. The most numerous and most important technically amongst these are the polyoxyethylenes (POE) of the general formula:

$$RX\ (CH_2CH_2O)_nH$$

where R is usually a typical surfactant hydrophobic group, but may also be a polyether such as polyoxypropylene and X is O, N or another functionality capable of linking the polyoxyethylene chain to the hydrophobe. In most cases the value of 'n', the average number of oxyethylene units in the hydrophilic group must be greater than 5 or 6 to impart sufficient water solubility to make the materials useful. The most common pathway to the POE surfactants is by the reaction of ethylene oxide (EO) with alcohols, acids, amides, mercaptans etc. If primary amines are used double chain surfactants of the form:

$$RN\ [(CH_2CH_2O)_nH]_2$$

can be formed with different molecular complexities. Hydrophilicity, in the case of nonionics is, provided by hydrogen bonding with water molecules. Oxygen atoms and hydroxyl groups readily form strong hydrogen bonds whereas ester and amide groups do so less readily. Hydrogen bonding provides solubilisation in neutral and alkaline media. Under highly acidic conditions oxygen atoms are protonated giving a quasi-cationic character. As each atom of oxygen can make only small contribution to water solubility, dissolution of nonionics in aqueous media necessitates the presence of several oxygen atoms in the molecule. Polyoxyethylene solubilisation has thus been the key to the substantial and growing potential of the nonionics. The compatibility of the nonionics with both ionic and amphoteric surfactants gives a boost to this class as it gives rise to synergistic systems. Nonionics like alcohol ethoxylates and alkylphenol ethoxylates serve as building blocks for deriving anionic surfactants by sulphation. A new dimension to the nonionics resulted on account of the polyoxyethylene-polyoxypropylene block copolymers, wherein the hydrophobic tendency of the propylene oxide units is taken advantage of in tailoring the products.

Apart from ethoxylated products, alkanolamides, carboxylic amides, glycerol esters, sorbitan esters, amine oxides, 'Surfynols' based or acetylenic diols, 'Silwet' surfactants comprising of siloxane back-bone with organic polyalkylene pendants, alkyl polyglucosides, fluorosurfactants, triesters of phosphoric acid and carbohydrate derived products figure in the nonionics market.

Amongst the varied applications of nonionic surfactants, emulsification, dispersion, detergency and foaming are important phenomena. Emulsions are used in a large number of applications such as foods, cosmetics, medicinals, agriculture, paints and numerous industrial processes. Their ability to control the stability of small particles is important in detergents. An important aspects of this is that the surfactant amount necessary in a composition is relatively low as compared with anionics. The nonionics serve as both foam boosters and foam inhibitors. Anionic derivatives of nonionics are particularly good high foaming detergents, whereas other derivatives serve as low foamers. Large volume uses of nonionics associated with their foaming properties is in various types of cleaners, latex paints, insecticidal and herbicidal wettable powders and emulsifiable concentrate, aerosol foam products, spray metal cleaning, foam drilling of oil and gas wells, and emulsion and suspension polymerisation. Other applications are as demulsifiers of crude oils, lubricants, antistatic agents, thickeners and emollients, dyeing assistants and pigment dispersants.

## Amphoteric Surfactants

The true amphoteric surfactant carries both cationic and anionic groupings within the molecule, acting as anionic materials in alkaline pH and cationic at acid pH. This structure makes it not only possible to combine them with the other categories of surfactants without the problem of property inhibition, but also with the choice of selecting the ionic character. Although the amphoteric surfactants represent the lowest volume in the total global surfactants market, the unique properties they can impart to a formulation are improving their market position. Amphoteric surfactants have been available for many years. Early amphoterics were glycinates based on imidazoline chemistry and were the basis of many toiletry products such as mild shampoos. Betaines were also developed and used increasingly in personal care products due to their skin friendly, foaming and free rinsing properties. Although a large group of organic functionalities hold the potential for producing amphoteric surfactants, the materials mostly encountered in this category are:

1. Imidazoline derivatives.
2. Betaines and sulpho-betaines.
3. Amino acid derivatives.
4. Lecithin and related phosphatides
5. Protein based products.
6. Mono and dipropionates.

Most commercially important imidazoline-derived amphoteric surfactants can be described as fatty acid—amino-ethylethanolamine condensates of the general structure:

$$RCONHCH_2NR'R''$$

where R is the fatty acid residue, and R' as well as R" are different functionalities. Their mildness and toxicity led to their use in shampoos and body care products. Their amphoteric nature also makes them useful in a wide range of water types, ranging from hard to soft water and high to low pH values. Such flexibility makes them useful in cleaning formulations under a variety of conditions.

Betaine surfactants can formally be considered to be special members of the ring-opened, imidazoline-derived surfactants. However, they do not exhibit many of the characteristics of other amphoterics with regard to their solubility and electrical nature in alkaline solution. They are compatible with anionics at all pH values and do not have any problems of complex formation. The carboxyl-containing betaines have been found to form external salts in very strong acids (such as hydrochlorides) while the sulphobetaines do not do so. This class is generally insensitive to the presence of electrolytes and usually perform well in hard water. The carboxyl betaines have found a number of commercial applications as levelling and wetting agents in textile processing, detergents, scrubbing compounds, antistatic agents, as fabric softeners, as lime soap dispersants and in personal care products. The sulphobetaines are useful in special areas such as the control of static charge in photographic films.

$NC_{10}–C_{12}$ and N $C_{12}–C_{18}$ fatty acid amidoethyl-N-(2-hydroxy ethyl) glycinates are the types of amino acid derivatives used for the production of strongly alkaline industrial cleaning agents, baby shampoos and other mild surfactant shampoos.

Lecithins are commercially popular phosphatidyl surfactants and are high volume products. Because of their natural raw material sources, the lecithin surfactants are normally found as rather complex mixtures represented by:

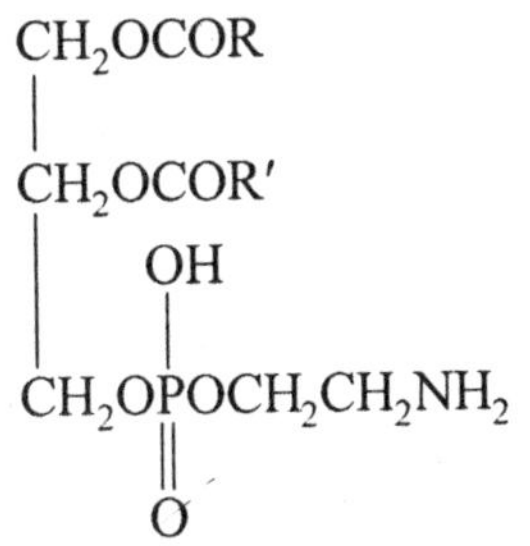

The natural lecithins have very limited solubility in water. Having good oil solubility, they have found extensive commercial application as non-aqueous emulsifiers, dispersants and wetting agents in diverse areas such as marine paints, inks, foods and cosmetics. The largest use of lecithins is in the processing of food and animal feeds and pet foods as emulsifiers or wetting agents. In the food industry they are used in baking, cake mixes, chocolates and ice creams. In cosmetics, they are used in creams, oils, soaps and hair care products. They simulate the phospholipids that are natural moisturisers in the skin and render the skin elastic. In hair care products they prevent the extraction of lipids. In soap their amphoteric nature represses the alkalinity and improves the stability of lather. Other applications of lecithins are as dispersants for dyes and pigments in the paint, ink and textile industries, in formulations of oil penetrants, in fat liquoring of leather, and as buffering agents in lubricating and other oils. Natural sources of lecithin range from egg yolks to many seed oils such as cottonseed, sunflower and soyabean. The most common source however is soyabean due to its relatively lower cost and greater availability.

Protein hydrolysed by acids and bases leads to the formation of polypeptides and amino acid residues, the proportion determined by the temperature and duration of hydrolysis. Protein sources can be vegetable protein byproducts, leather scraps and fish scales. Protein-based surfactants are produced from these hydrolysates by reaction with fatty acid chlorides, condensation with fatty acids and amines and reaction with polyamines. They find application in detergents soap and toilet bar formulations.

The need for salt-free amphoterics by industries to overcome the problems associated with high electrolyte levels and corrosion resulted in the growth of amphoteric mono- and dipropionates. Some of the many other advantages of these products are excellent dirt lifting properties, excellent hydrotropic properties, excellent free rinsing properties they impart to detergent systems, stability over a wide pH range, low toxicity, biostatic properties and lubricity. Currently, these propionate amphoterics are recommended for formulating general cleaning products, environmentally friendly vehicle cleaning systems, personal care products such as hand cleaners, shower gels, hair shampoos and foam baths, microemulsions of terpene based products and aqueous slat/chain lubricants.

## Polymeric Surfactants

Colloidal stabilisation by polymeric materials in both aqueous and non-aqueous systems is described as 'Steric stabilisation'. A limited range of polymers fit this role. The polyisobutylene chain is frequently used for stabilisation in very nonpolar organic solvents like aliphatic hydrocarbons. In solvents of higher polarities such as aromatic hydrocarbons, esters and ketones, various polyesters are often used. These polyesters may be based on hydroxycarboxylic acids, with the molecular weight of the hydroxy-carboxylic acid governing the range of solvents in which the surfactant is most effective. Alternatively, a condensation polyester from a mixture of diacid, diol and a chain terminating acid or alcohol may be used. A polyethylene oxide chain may be incorporated into polyesters of this type to increase compatibility with more polar solvents. Polymethylmethacrylate is also used to provide steric stabilisation to solvents of higher polarities. The polyalkylene oxide chains can be used to give steric stabilisation in the most polar organic solvents like alcohols, glycols and glycol ethers.

Polymeric dispersants have come into prominence to solve the complex rheological problems associated with spreading, levelling, relatively capital and energy expensive dispersion processes in the ink and paint industries.

Although surface coatings are the main outlets for sophisticated polymeric surfactants, a variety of other applications have started to develop in the last few years. These include their use in oil-based drilling muds, lubricating oil additives such as ashless dispersants, processing of electroceramics, magnetic tapes and discs and inverse emulsion polymerisation processes.

## Biosurfactants

Biosurfactants, derived from fermentation processes with very specific micro-organisms, depending on their nature or state typically display strong features in some respects while at the same time lacking in the versatility of conventional surfactants. Several biosurfactants have surface-active and emulsification properties far superior to either synthetic or natural surfactants. Based on their technical competence and social dictates they have the capability to penetrate all current fields of surfactant use. In particular, they have high potential for use in household detergents and cleaners and personal care products.

'Emulsan', a fully commercialised biosurface-active material is one of the best discovered emulsifiers and is classified as a 'Bioemulsifier'. 'Surfactin', another biosurfactant, has the ability to lower the surface tension of 0.1 N sodium bicarbonate solution from 72 dynes/cm to 27.9 by the addition of as little as 0.005 per cent. Other products such as rhamnose-lipids and sophorose-lipids obtained with Torulopsis bombicola KSM35 also show excellent surfactant properties. Biological products of genetic- and protein-engineering such as Lipolase or Cellulase impart restoration and repair characteristics to fabrics and are viewed as bio-detergents. In the personal care sector, biosurfactants termed biocosmetics are expected to take long strides. Sophorose- lipids are already in use in this sector as high value skin moisturisers. In the oilfield industry, microbial enhanced oil recovery (EOR), cleaning up of offshore spills, cleaning of oil tankers, pumping of crudes using bioemulsifiers, demulsification of crudes from the enriched oil recovery, extraction of bitumen from tar sands and viscosity reduction of heavy crudes are potential application areas for these surfactants. Lecithins, described earlier under amphotenics, are also bio-surface active materials used largely as food emulsifiers. Recent projections by some agencies show significant penetration of the biosurfactants into the conventional surfactant markets in the next two decades.

Widely used in consumer and industrial applications such as soaps and cleaners, laundry detergents, cosmetics and toiletries, non-consumer applications for surfactants can also be found in almost every industry. Except for cleaning uses, most of them consume minute quantities of the surfactants as processing aids. Industries where they are used include rubber, plastics, textiles, construction, petroleum-drilling, ore processing, food processing, pharmaceuticals, herbicides, machinery and metals. As new regulations and changing consumer demands arise, the surfactants industry is evolving rapidly. However, many of the surfactants initially alleged as being potential pollutants due to lot of pressure from the environmentalists are now being cleared by environmental agencies. Simultaneous with the development of new products for new or changing industrial applications, surfactants are beginning to attract attention in energy and environmental protection applications, as they are being tried out in pollution-control systems, paper processing and ultra-thin films. Main driving forces influencing changes in the surfactant industry are:

1. The relative state of the economy.
2. Consumer demand for 'green' products that are milder, safer and cheaper.
3. High expenses due to regulatory compliance.
4. Insurance and environmental costs.
5. Technical obsolescence.
6. Erosion of commodity/specialty segment differentiation.
7. General public perception of the chemical industry.

## RAW MATERIALS OF SURFACTANTS

The surfactant industry has undergone major changes over the last few years. Surfactant companies generally have been rationalising their product ranges in order to concentrate on their core surfactant business. Surfactant consumers are now looking for ingredients with increasing number of properties at the same prices and products of consistent quality. The surfactant industry is also turning towards raw materials from renewable sources and is trying to be less dependent on petrochemical inputs. New molecules like the alkyl polyglycosides (APG) and N-alkyl glucosamide (AGA) are now being increasingly produced.

Amongst the various aspects related to the surfactants industry, key factors are :

1. Surfactant raw materials.
2. Major surfactants.
3. Important physico-chemical characteristics.
4. Technologies.

There are two major materials from which the organic component of surfactant chemicals is predominantly obtained, namely :

1. Hydrocarbons, from petroleum and gas drilling operations.
2. Fats and oils from vegetative, animal and marine sources.

The paper industry is an ancillary source and supplies two rather important products—tall oils from which fatty acids can be obtained and lignosulphonates.

Petroleum is a unique source for benzenoid and naphthenoid surfactant chemicals such as alkylbenzene sulphonates, alkylphenols and alkylnaphthalene sulphonates. It is also the source of the most important petrochemical intermediate, ethylene and many straight chain chemical intermediates such as normal paraffins and alpha olefins from which several surfactants are derived.

Fats and oils are unique sources for straight chain, saturated and unsaturated compounds that serve as alternatives to petroleum intermediates for surfactants. An important example is that of detergent alcohols. Other intermediates derived from fats and oils are methyl esters, soaps and fatty acids. The oleochemical industry has become an important segment of the global chemical industry and particularly of the surfactants industry because of the promise of the products being natural, very biodegradable, toxicologically safe and renewable. Movements in the production and supply of fats and oils affects the economics of the surfactants industry. Traditionally petrochemical feedstocks have been cheaper alternatives but the environmental drive and causes which have led to price decline in the key oils for the surfactants industry—palm kernel oil and coconut oil—have made them preferred materials. Figure 4.3 gives surfactants that are derived from fats and oils.

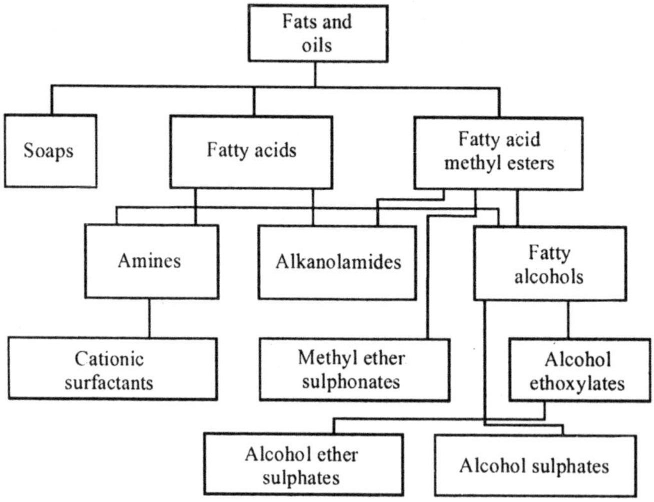


**Fig. 4.3.** Surfactants derived from fats and oils.

The paper industry provides two key surfactant raw materials, namely, tall oil and lignin sulphonates. Tall oil is a coproduct in the kraft pulping process for paper pulp using pine and similar woods. About 90 per cent of all tall oil (a mixture of unesterified oleic and linoleic acids with small amounts of rosin acids and unsaponifiables) produced is fractionally distilled to get tall oil fatty acids (TOFA) used as raw material for surfactants. Tall oil production is dependent on the growth of the paper industry.

Technically either of the two basic raw materials—hydrocarbons or fats—can be used to obtain majority of synthetic surfactants. The choice depends on the raw material availability and prices. Key surfactant intermediates worthy of attention are:

1. Linear alkyl benzene (LAB).
2. Detergent alcohols.
3. Alphaolefins.
4. Ethylene oxide/propylene oxide.
5. Fatty amines.

### Linear Alkyl Benzene (LAB)

Linear alkylbenzene is the most important industrial raw material for surfactants derived from petrochemical sources and is even otherwise the most dominant raw material of the detergent industry. Its level of consumption in any global region is an indicator of the scope for detergent products—laundry detergents, dishwashing detergents and materials used in the maintenance of buildings, offices, hospitals, schools, etc. LAB is the raw material for the production of alkylbenzene sulphonic acid and alkylbenzene sulphonate (LAS). Detergent formulators use both types of these intermediates. In many formulations, a salt of LAS such as Na-LAS (sodium salt) is added directly as a formulant. In other cases, the LAS salt is formed during the formulation process through the neutralisation of the acid with the desired base—caustic soda or triethanolamine.

The structure of LAB depends on two major factors :

1. The petroleum feedstocks used—normal paraffins and benzene in terms of range and purities.
2. The method of manufacture of LAB.

Because of the complexities involved in the selection of LAS for a particular formulation and end use, there have been immense improvements in the methods of manufacturing LAB, the parent material for LAS. The specifications for LAB in terms of the distribution of homologues and phenyl-position isomers makes it almost a 'designer molecule'.

### LAB Manufacturing Processes

Two basic routes employed for the manufacture of LAB are:

1. The alkylation of benzene with long-chain mono-olefins using HF and other catalysts.
2. The alkylation of monochlorinated benzene with *n*-paraffins using aluminium chloride catalysts.

The mono-olefins alkylation process is predominant and based on UOP technology. In the overall UOP process, internal mono-olefins obtained through the catalytic dehydrogenation of *n*-paraffins are used for the alkylation of benzene with HF catalyst to give LAB. UOP employs the PACOL process for producing the olefins from *n*-paraffins. In this process linear paraffins are dehydrogenated in the presence of selective hydrogenation catalysts, believed to be platinum, arsenic and lithium deposited on alumina. Operating temperatures in the PACOL reactor are in the range 400–600°C with dehydrogenation occurring in the presence of hydrogen. Some by-product light ends and hydrogen are separated from the reactor effluent and a part of this hydrogen-rich gas is recycled to the reactor. An equilibrium mixture of linear olefins together with unconverted *n*-paraffins and side reaction products like diolefins and aromatics is produced this way. If left unchanged, the diolefins will react with two moles of benzene to yield heavy diphenylalkane products or heavier polymer stocks, some of which are removed as by-product 'heavy alkylate' in the LAB refining column.

There have been several variations of the UOP process over the years. The most recent one is the DETAL process, a joint development of UOP and Petresa of Spain. The DETAL process employs a heterogeneous bed reactor with liquid feeds and utilises a solid acidic catalyst. This innovation reduces the investment of the overall plant and improves yields. Its greatest advantage is that it eliminates the use of HF and associated problems. In another of UOP's developments, the DEFINE process has the objective of increasing LAB yield and cutting down heavy alkylate production.

The equilibrium mixture from the PACOL process, termed as separator liquid, is charged to the DEFINE reactor alongwith make-up hydrogen for selective conversion of diolefins to mono-olefins. A stripping column separates dissolved light hydrocarbons, and the bottoms comprising a mixture of olefins and unconverted n-paraffins are charged together with benzene to the alkylation reactor, where LAB is produced. Benzene and unconverted paraffins are fractionated from the alkylation reactor effluent and recycled.

The final column fractionates LAB overhead from heavy alkylate bottoms. Reduction of dialkylates content improves the solubility of the ultimate sulphonated and neutralised product, although the presence of dialkylates contributes to improved detergency of the sodium salt.

High quality LAB with minimum isomers and side products has been obtained by varying temperature, concentration, reaction ratios, recycle points in the LAB manufacturing processes. Figure 4.4 shows schematic flow in the UOP process.

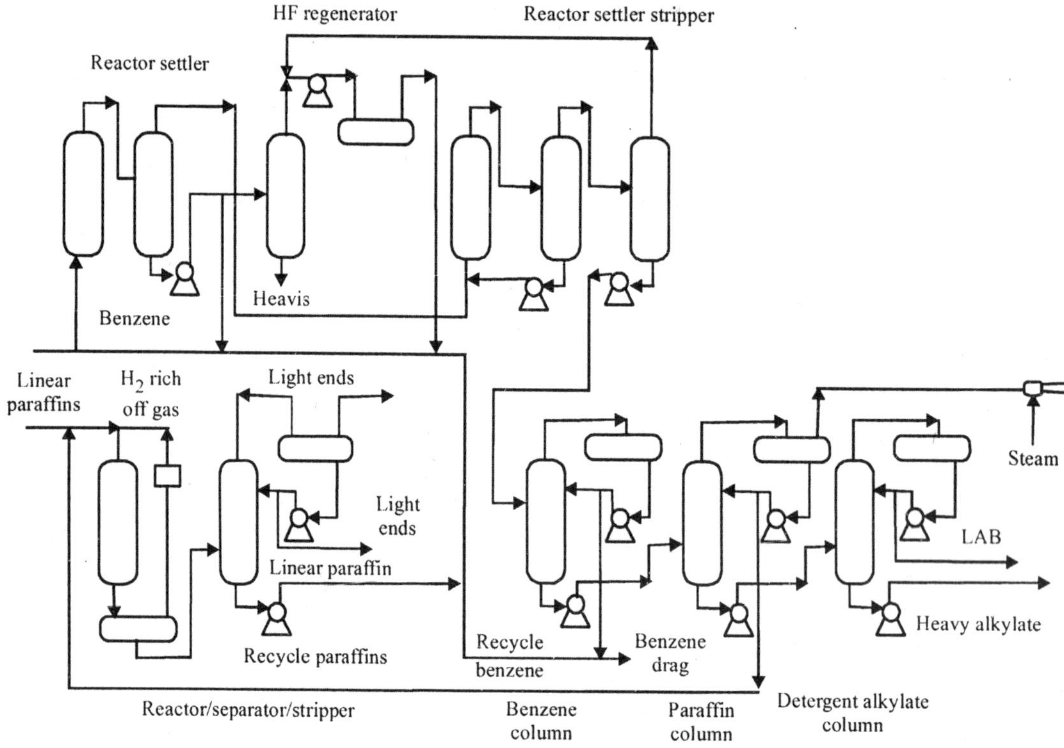


**Fig. 4.4.** Production of linear alkyl benzene schematic flow (UOP process).

Typical specifications for LAB from the UOP PACOL-DEFINE process are given in Table 4.2.

**Table 4.2.** Typical specifications for LAB from UOP PACOL 'DEFINE' process.

| | |
|---|---|
| Bromine No. | 0.02 |
| Saybolt colour | +30 |
| Doctor test | Negative |
| Water wt. % | 0.01 |
| Sulphonation wt. % | 98.5 |
| Biodegradability % | 95.0 |
| Paraffins wt. % | 0.3 |
| Indans and Tetralins | 2.0 |
| Normal LAB % | 93.0 |
| Appearance | Clear and bright |

## Detergent Alcohols

Detergent alcohols, as the name implies, are alcohols primarily produced for the manufacture of detergent intermediates, the most important of which are the surfactants such as alcohol ethoxylates, alcohol ether sulphates, alcohol sulphates and phosphate esters.

These alcohols are also of usage in the production of alkylpolyglucosides and have other markets in the form of free alcohols and derivatives. They are also commonly referred to as 'Fatty alcohols'. There are two basic categories of detergent alcohols, namely, the 'natural' and 'synthetic' alcohols. While both these types are interchangeably used for the manufacture of derivatives employed in several applications, the natural—oleochemical and renewable—resource alcohols are getting prominence on grounds of toxicity and environmental impact.

Derivatives of detergent alcohols are in the forefront of changes today and outpace LABS in value performance. This is due to evolutionary changes in the household laundry sector and revolutionary changes in the increased use of these products in the growing-type of detergent end-products. The evolutionary changes relate to:

1. Their ready application in detergent formulations without phosphates.
2. Suitability of the derivatives for detergent formulations which are effective at low washing temperatures—an aspect of importance as the washing loads are containing more and more sensitive, coloured synthetic fabrics.
3. Their superior hardness tolerance (calcium and magnesium ions present in water) in comparison with that of LABS.

The revolutionary changes are related to four growing types of detergent end-products, namely :

1. Low temperature compact powders.
2. High concentrated liquids.
3. 'One packet per wash load' products combining detergent, bleach and softener.
4. Growing concern for skin-safe low irritant products for dishwashing as well as personal care products.

These changes which vary from region to region as in United Sates, Western Europe and Japan are benefitting the derivatives manufacturers and also formulators.

Detergent alcohols are defined as alcohols containing 12 or more carbon atoms per molecule with a carbon backbone of high linearity mainly for the manufacture of surfactant intermediates. Anionics like alcohol sulphates, alcohol ether sulphates, phosphate esters, sulphosuccinates, nonionics like alcohol ethoxylates and alkylpolyglucosides and even amines for cationic surfactants can be made from these alcohols. Highly branched alcohols, of long chain lengths, are used mainly in the manufacture of plasticisers. Alcohol structures which are slightly branched are also used in surfactants manufacture. Detergent alcohols are commercially produced from natural raw materials, principally fatty acids and methyl esters and petrochemical sources, namely, ethylene and n-paraffins. The two routes divide the alcohols into the aforementioned 'natural' and 'synthetic' alcohols.

### *Natural alcohols*

Natural detergent alcohols, both saturated and unsaturated, are manufactured through the hydrogen reduction of fatty acids and their esters, typically by the Lurgi process, as well as the catalytic hydrolysis of methyl esters. The latter route is preferred nowadays for a variety of reasons including economy of investment. The methyl esters are, in turn, manufactured from triglyceride fats such as coconut oil, palm kernel oil, and tallow through transesterification processes. A continuous system for methyl ester production is given in Fig. 4.5.

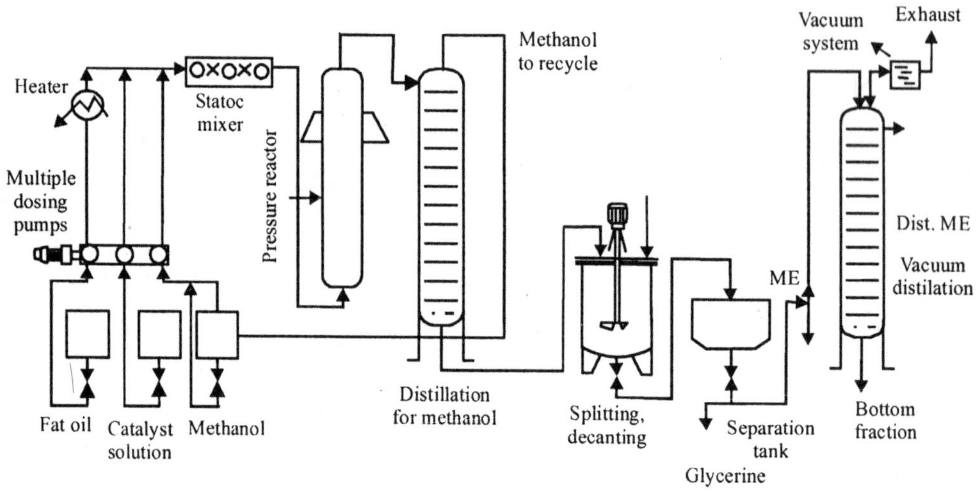


**Fig. 4.5.** Methyl ester production—continuous system.

### *Synthetic detergent alcohols*

Synthetic detergent alcohols are substantially obtained through modifications of Ziegler chemistry, often referred to as 'ethylene growth' processes. In these processes, ethylene is added to triethylaluminium to produce a mixture of alkyls which is then oxidised by air to the alkoxides and subsequently hydrolysed to the alcohols. The alcohols from Ziegler chemistry are always products with even number of carbon atoms, thus imitating the structure of the natural alcohols.

Internal olefins manufactured through normal paraffin dehydrogenation with special catalysts are also reported to yield slightly branched alcohols through oxo-chemistry. Secondary alcohols are made in Japan through an *n*-paraffin oxidation process.

### Alphaolefins (LAO)

Alpha olefins, often abbreviated as LAO, are linear mono-olefins with the double bond located at the terminal of a linear long-chain hydrocarbon. They differ from the 'internal mono-olefins' wherein the double bond is randomly located inside the hydrocarbon molecule. For detergent applications the carbon range of the raw materials used is $C_8$ to $C_{18}$. Apart from detergents, the alphaolefins figure in the production of oxo-alcohols for the plasticisers industry, in the manufacture of HDPE and LLDPE as comonomers, in polybutene production and as raw materials for polyalphaolefins (PAO) used ay synthetic lubricants.

### *Alphaolefins production*

The four main routes for the production of alphaolefins are:

1. Through wax-cracking.
2. Paraffin dehydrogenation.
3. Ethylene oligomerisation.
4. Shells SHOP process.

Shell is the sole operator in Western Europe using the wax-cracking process, that too partially, to produce alphaolefins. Chevron, which operated this process in USA, abandoned it in the mid-eighties. The cracking of wax occurs thermally and a wide band of alphaolefins are produced, substantial volumes of which have little application. The process requires a special type of wax with highly linear hydrocarbon waxes which are reported to originate only from Libyan and Indonesian crude oils. Shell has also operated a paraffin chlorination plant in USA to produce odd and even number $C_{11}$ to $C_{14}$ linear internal olefins of which a segment has been alphaolefins. They have been used for the manufacture of LAB and oxo-alcohols, but not for alpha-olefin based products.

By far, the most important of the processes to make alphaolefins are those based on Ziegler chemistry. In the classical process, an aluminium triethylene growth promoter (TEA)—catalyst—is used to build up long chains of low molecular weight ethylene polymers on the aluminium, the ethylene being inserted between the aluminium atom and the carbon. These are then displaced by more ethylene to yield a mix of alpha-olefins and TEA, the latter being recycled. The displacement involves dehydrometallisation followed by ethylene hydroalumination. Modifications of this process are the Gulf Process and the Ethyl Process. In the Gulf Process, the catalytic process is a one-step reaction producing a geometric distribution of alphaolefins. In the Ethyl Process, the one-step Gulf Process is changed to a two-step process carried out at lower and higher temperatures. By this method, using appropriate reaction conditions a stoichiometric distribution of alphaolefins of desired chain lengths can be produced.

Shell's SHOP process employed in plants in U.K. produces internal olefins for captive manufacture of oxo alcohols or LAB in Shell's plants. The alpha-olefins produced are for merchant sales. The SHOP Process is a variation of the Gulf Process with replacement of the aluminium catalyst by a proprietary nickel catalyst in a solvent like benzene or 1,4-butane-diol.

In India, Gujarat Godrej Innovative Chemicals whose plant is engineered by Lurgi, adopts a new route for alphaolefins. In this process, linear alcohols made from imported palm stearin are catalytically dehydrated and the alphaolefins produced are used as raw materials for alpha olefin sulphonates.

## Ethylene Oxide (EO)/Propylene Oxide (PO)

Ethylene oxide is an important building block of the petrochemical industry and is used to manufacture ethylene glycol (for polyester fibre and bottles, antifreeze compounds), polyethylene glycols (brake fluids, cosmetic intermediates, epoxides), glycol ethers (solvents) and surfactants. The EO based nonionic surfactants as well as other types of importance are alcohol ethoxylates (AE), alcohol ether sulphates (AES), alkylphenol ethoxylates (APE), ethoxylated fatty esters and acids and ethoxylated amines.

The glycols have been considered over the years as major outlets for ethylene oxide. However, currently nonglycol applications of EO are in the increase.

### *Manufacture*

EO manufacturing processes are: (i) chlorohydrin process; and (ii) direct oxidation of ethylene. The chlorohydrin process, in which ethylene is reacted with hypochlorous (chlorine and water) to give ethylene chlorohydrin which is treated with slaked lime or caustic soda to form EO, has been the traditional route to EO, but now this has been largely replaced by the direct oxidation process.

The direct oxidation of ethylene may be carried out over a silver catalyst in the vapour phase at about 245°C and 21 atmospheres of pressure. Either oxygen or air enriched in oxygen may be used for direct oxidation. The reaction may be represented as follows:

$$\underset{\text{Ethylene}}{CH_2{=}CH_2} + \underset{\text{Oxygen}}{\tfrac{1}{2}O_2} \longrightarrow \underset{\text{Ethylene oxide}}{\overset{\quad O}{CH_2{-}CH_2}}$$

Propylene Oxide (PO) is produced either by the chlorohydrin or hydroperoxide process. ARCO Chemical USA is the largest producer of PO world over. Various modifications of these processes are used for the production of PO.

## Fatty Amines

Nitrogen-based surfactants dominate the cationic types. Fatty amines comprising of a basic group of alkylamines—primary, secondary and tertiary, are important raw materials for these surfactants.

Primary, secondary and tertiary alkylamines—taking their names from the number of hydrogen atoms of the ammonia molecule substituted by one, two or three alkyl groups—are obtained by the reduction of nitriles which in turn are produced from fatty acids and ammonia in the presence of zinc oxide or manganese acetate catalysts. Alkyl group chain lengths may vary from $C_8/C_{10}$ to higher. Whereas the primary amines are only obtained from nitriles, the secondary amines are obtained both from nitriles and from primary amines through further alkylation. The tertiary amines, the most important of this group in terms of volume and diversity of down-stream products, are produced through several routes such as:

1. Reductive alkylation of a primary amine with formaldehyde and hydrogen.
2. From alkyl chlorides and dimethylamine.
3. From primary alkyl bromide and dimethylamine.

Tertiary amines are supplementally obtained from alpha-olefins.

Apart from the above types of oxygen-free amines, the fatty amines also comprise of oxygen-containing amines like ethoxylated amines.

The three groups of amines by themselves do not have any marked surfactant properties. About 20–25 per cent of the amines and their salts are used directly and the balance converted to derivatives. Primary amines are predominantly used for the production of tertiary amines. Other uses of these amines are in metal extraction, as froth flotation collectors, as anti-caking agents for fertilisers, as filming amines in steam evaporators etc.

Ethoxylated primary amines are used as cationic collectors in the flotation of zinc ore, as emulsification agents for pesticidal emulsions and in the petroleum industry. Secondary amines are the basic raw materials for the important range of dialkyl dimethyl quaternary ammonium salts used very largely as fabric softeners in home laundries and as textile conditioners in the textile industry. The major direct use of tertiary amines is in solvent extraction of metals. They are the basic raw materials for either classes of important nitrogen compounds, namely:

1. Alkanolamides
2. Quaternary ammonium compounds
3. Polyamides
4. Betaines
5. Imidazolinium compounds
6. Amphoteric surfactants
7. Amine oxides
8. Ethoxylated amines

All of these products are surface-active and are cationic in nature. The selection criteria for fatty amines and derivatives involve.

1. Solubility in water, non-aqueous media and two-phase systems.
2. Hydrophile—lipophile balance (HLB).

3. Surface tension.
4. Other surfactant properties like—Forming, Krafts point, cloud point.
5. Antimicrobial properties.
6. Various physical and chemical properties.

Amines with bridging amids groups termed 'Amidoamines' are also building block amines. They are derived from fatty acids, i.e. the fatty acid is the source of the alkyl radical rather than the nitrile. The important economic advantage of the amids bridging is that these amines can be prepared at far lower levels of investment than are the basic amines which require the use of high pressures and expensive equipment. Also the amids group improves the multifunctionality of these amines. Many of these materials and their derivatives perform as effectively as the basic amines. In direct use they find application as flotation agents, asphalt emulsifiers and corrosion inhibitors and compete with the basic amines and their derivatives.

## KEY FEATURES OF SURFACTANTS

### Major Surfactants

Amongst the various types of surfactants major ones can be identified either by their large-scale use in industry or by their special chemical structures and characteristics. The products highlighted in this section are:

1. Sulphonates—lignin sulphonates, petroleum sulphonates, alkylbenzene sulphonates, alkylnaphthalene sulphonates.
2. Sulphates—alcohol sulphates, alcohol and alkylphenol ethersulphates.
3. Polyhydric alcohol esters—sorbitan esters, glycerol esters, polyglycerol esters, glycols and glycol esters.
4. Ethoxylates—Ethylene oxide/propylene oxide block copolymers, alcohol ethoxylates, alkylphenol ethoxylates, polyhydric alcohol ethoxylates, lanolin ethoxylates, amine ethoxylates, polysiloxane ethoxylates.
5. Alkanolamides.
6. Fatty amine oxides.
7. Acetylenic alcohol derivatives.
8. Silicones.
9. Fluorosurfactants.
10. Alkyl polyglycosides.
11. Sulphosuccinates and sulphosuccinamates.
12. Phosphate esters.
13. Amine based surfactants—imidazolines, quaternary ammonium compounds, betaines.
14. Sarcosinates and taurates.

#### *Lignin sulphonates*

Although lignin sulphonates have low surface activity, they are used in many areas where a conventional surfactant can be used. They are obtained as by-products in the pulping of wood in the paper industry.

Lignin is one of the three major constituents of wood—the celluloses, pentagens and lignin. It represents 15–30 per cent of the wood depending on the species. There is no clear understanding of the structure of lignin, but in general, propane units are bound through ether linkages to phenols with the molecular weight of the order of tens of thousands.

When wood chips are treated with sulphite liquor the cellulose and hemicellulose are not affected, whereas the lignin is partially sulphonated on the aromatic ring, with some degradation to pentoses and hexoses (sugars). The sulphonates and the sugars are all water-soluble and thus are easily separated from cellulose pulp. The sugars in the sulphite liquor are destroyed by fermentation and the residue, together with surplus sodium or calcium sulphite (depending on the pulping method used) is spray dried to the commercial product. Unlike other products of recovery processes, the lignosulphonates do not consume sulphonation chemicals. Various refinements on the raw, dried product yield products of different molecular weights, cation and chemical composition.

### *Petroleum sulphonates*

The anionic petroleum sulphonates are obtained either as by-products of oil processing or by synthetic methods. Both materials are used in commerce competitively.

When high-grade petroleum oils (white oils) are prepared, reactive components and sulphonatable coloured bodies in the base oil have to be removed from the non-reactive portion, which is the desired material. One of the best ways of doing this is to sulphonate the oils with oleum or sulphur tri-oxide or both. The reactive components react with the acid to form sulphonates. The non-reactive components are naphthenic-paraffinic hydrocarbons. The sludge obtained in the sulphonation of base oils, termed 'green acids' contains the water-soluble sulphonates, excess acid and occluded oil. The products are largely disulphonated. After separation, the upper layer contains all unreacted oil and oil-soluble sulphonates termed 'mahogany' acids.

The oil is now given an alkaline wash and extracted with a low-boiling alcohol. The paraffinic-naphthenic fraction is not soluble in the alcohol whereas the sodium salts of mahogany acids are. When the alcohol is stripped from the lower layer, a high molecular weight sodium petroleum sulphonate results with some 30 per cent unreacted oil. The sulphonates by and large are monosulpho-nated products.

Synthetic petroleum sulphonates are byproducts obtained in the Friedel Craft reaction to get detergent alkylates from benzene such as:

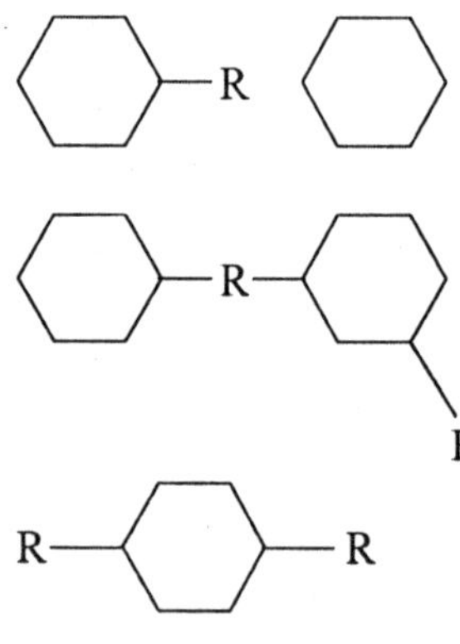

whose mixture is sulphonated with oleum. Both the natural and synthetic sulphonates are marketed as such or as overbased materials. The products marketed in largest volume are the alkaline earth salts produced by the double decomposition of the sodium sulphonate with an aqueous solution of barium, calcium or magnesium chlorides.

The petroleum sulphonates find application as emulsifiers for textile oils, dry cleaning solvents, degreases, soluble oils, metal working fluids, mold release agents and froth collectors in ore flotation. They are also used as dispersants (pigment dispersion in paints or sludge dispersion in fuel oil) and as rust inhibitors.

### *Alkylbenzene sulphonates*

The largest used surfactant across the world are alkylbenzene sulphonates. There are two broad types of these products, namely:

1. Linear alkylbenzene sulphonates (LABS).
2. Branched chain alkylbenzene sulphonates.

The branched chain product has poor biodegradability and as such does not figure in the household sector. The linear product is predominantly used in the household detergents and scourers, but the usage extends to Institutional and Industrial (I&I) cleaners and industrial sectors. The length of the alkyl chain has profound influence on the detergency and solubility of the product.

The most common method of sulphonating alkylbenzenes has been through oleum. The disadvantage of the process is the need for the disposal of the spent acid. Air-$SO_3$, used predominantly for the sulphonation of long-chain alcohols is increasingly being applied to alkylbenzenes. The detergency properties of LABS (dirt removal power) are superior to those of the alcohol ethoxylates and moreover it is a cheaper product. However, it's high foaming tendency and higher efficiency only at higher temperatures are disadvantages for its use in the household sector.

There are a number of LABS manufacturers. A substantial portion of the product is for captive use. Firms active in sulphonation also undertake toll manufacture. The main usage areas are commercial dishwashing, emulsion polymerisation, petroleum production, textile industries and in pesticidal sprays.

### *Alkylnaphthalene sulphonates*

Naphthalene sulphonic acid and salts are hydrotropes, i.e. materials which assist in solubilisation of associate actives without other impact on the system. In this respect they are similar to cumene, xylene and toluene sulphonic acids. They do not have any meaningful surfactant properties. With the introduction of alkyl side chains such as propyl, butyl and dipropyl surfactancy develops.

The sulphonates are prepared by condensing naphthalene sulphonic acid with formaldehyde. In this case the disulphonates are formed. If the homosulphonates are desired naphthalene, oleum and formaldehyde are reacted together.

Alkyl naphthalene sulphonates and the alkyl naphthalene sulphonates—formaldehyde condensates are used in applications where lignosulphonates can also be used. Most importantly they are used as dispersing agents and stabilisers wherein particularly the condensates are superior performers. The areas of application are:

1. As water reducers in the preparation of concrete (the formaldehyde and melamine formaldehyde condensates are the 'superplasticisers').
2. As solubilisers in the production of SBR lattices.
3. As pigment wetting and dispersion agents in the mixing of paints (sodium diisopropyl naphthalene sulphonic acid and condensates are used).
4. As low-foam surfactants and dye dispersion aids in the textile industry.
5. As pitch dispersants in paper industry.
6. As syntans—leather tanning agents.

7. As suspension and dispersion agents in the formulation of pesticide suspension concentrates and in the formulation of wettable powders.
8. As dispersing agents in cure pastes in the plastics and elastomer industries.
9. As compact dispersing agents in oil muds.
10. As free flow additives in fertiliser mixtures and other dry materials.
11. In carpet steam cleaners.
12. In tank soak metal cleaners.
13. In high foaming alkaline and acid cleaners.

### *Alcohol sulphates*

Alkyl sulphates or alcohol sulphates are the first of the world's synthetic detergents, pioneered in Germany during the Second World War. They are currently produced by the sulphation of linear primary alcohols with air-$SO_3$ or chlorosulphonic acids, followed by neutralisation with a variety of cations. Most of the alcohol sulphates are made from natural alcohols because of their large use in personal care products. The alcohol's used range from $C_8$ to $C_{16}$. In the United States, they are derived from animal tallow base alcohols and as such are the least expensive anionics. The most-used product is lauryl alcohol sulphate. Secondary alcohol sulphates are also used as surfactants in industrial applications.

In general the alcohol sulphates are more hydrophilic than the sulphonates. Sodium lauryl alcohol sulphate has strong wetting, detergent and emulsifying properties. They are good foamers but are sensitive to hard water, necessitating effective builder support in consumer products. The major outlets for alcohol sulphates are in laundry powders, personal shampoos, in emulsion polymerisation of rubber lattices for textile and paper coatings, in wool scouring, kier boiling of cotton, as dye levelling agent for anionic dyes, as wetters for fruit processing, as pigment dispersion aids in paints and inks, in foaming of gypsum board, in soaking and degreasing of leather and in binding compositions for non-wovens.

### *Alcohol and alkylphenol ether sulphates (AES/APES)*

The hydrophilicity of ethoxylate structures can be markedly enhanced by the sulphation/sulphonation of the—OH group. In the process the ethoxylates lose their nonionic nature and become anionic materials.

When the aromatic nucleus of a compound is sulphonated we have a 'sulphonated' product. If the reacting group in a hydroxyl group is an aliphatic compound, then the process is termed 'sulphation'.

In the sulphonation of an alkylphenol, while the primary reaction is sought at the active hydrogen site, some sulphonation of the aromatic nucleus also occurs. The ether sulphates are diversely known as ethoxysulphates or sulphated polyethers. AES and APES are used interchangeably in most technical applications.

The manufacture of the ether sulphates is based on sulphonation routes akin to those used for the manufacture of LABS. The heat of reaction is however higher and special types of reactors have to be employed. Common sulphonating agents used are oleum, chlorosulphonic acid and air-sulphur trioxide.

The alcohol ethoxylates, while soluble in water, produce turbidity at higher temperatures with likelihood of layer separation. AES overcome this problem. They are also high foaming products compared to the low foaming ethoxylates and are therefore of advantage where foaming action is required. In comparison with other anionic products they are excellently soluble, have good hard water tolerance and are more compatible with enzymes. They are extensively used in personal care products due to their mildness to the skin. In household detergents they are synergistic with LABS.

Virtually any sulphonator can make AES and APES. The basic markets for AES are in light-duty liquids, bath additives and personal shampoos. A sizeable quality is also used in household liquid surface cleaners. APES on the other hand is almost entirely consumed in the industrial sector. In most industrial operations the markets for AES and APES overlap. Major ones are in emulsion polymerisation of polyvinyl acetate and acrylic polymers, as foaming agents in the production of foam concrete and plasterboard and as foam booster in commercial dishwashing applications. The ether sulphates are also well known as dispersants for coating fillers in the paper industry, as foamers in drilling mud compositions, in the surfactant mixes for enhanced oil recovery and as wetting agents for primary backing compounds in the resins and elastomers industries.

### *Sorbitan esters*

The surfactant field comprises of several nonionics belonging to the ester class. A wide variety of esters derived from alkanoic acids by reaction with polyhydric alcohols—sorbitan, glycerol and polyethylene/polypropylene glycols have commercial significance. The sorbitan esters with a wide range of hydrophilic and lyophilic properties are commercially produced. Sorbitol is a hexahydric alcohol derived from high purity dextrose. Sorbitan which is a mixture of anhydrosorbitols mainly consists of.

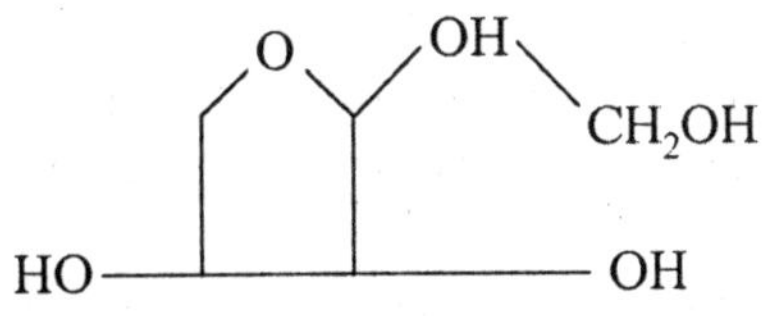


(1,4 Sorbitan)

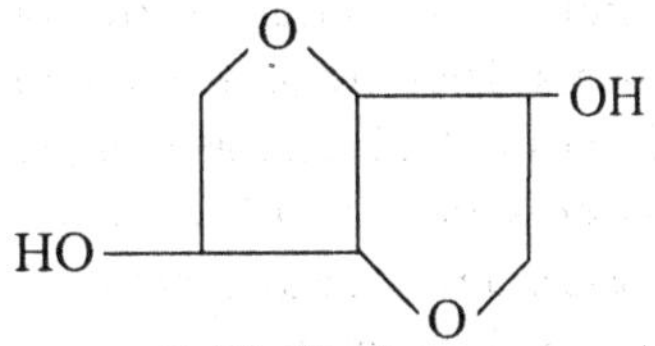


(1,4 Sorbitan)

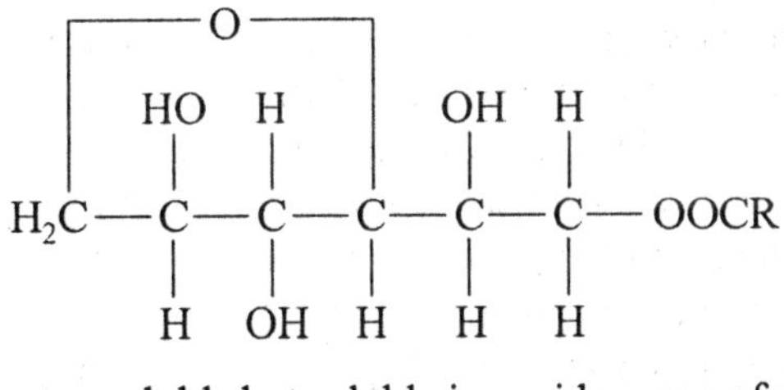


The sorbitan esters are not water-soluble but soluble in a wide range of mineral and vegetable oils, as sorbitan is not a strongly hydrophilic group. The lipophilic character of these esters aids tapping of water in oil to form water-in-oil (w/o) emulsions. Sorbitan mono, di and tri esters are derived by the reaction of sorbitan with fatty acids. The physical form and HLB are tailor-made by varying the type of fatty acid used and the degree of esterification. The products are approved for human ingestion and are used widely in foods, beverages and pharmaceuticals. Their lipophilic, emulsifying, solubilising, softening and lubricating properties are made use of in synthetic fibre manufacture, textile processing and cosmetics.

### *Glycerol esters*

Glycerol esters form another group of polyhydric alcohol based nonionic surfactants derived by the esterification of glycerol with saturated and unsaturated fatty acids. Mono- and di-glycerides are made both through the esterification of glycerol with the appropriate fatty acid and through 'glycerolysis', the latter gaining prominence. In its modern batch version, glycerolysis involves reacting a suitable animal or vegetable fat with excess glycerol catalysed by NaOH or KOH at around 250°C.

The reaction proceeds as follows:

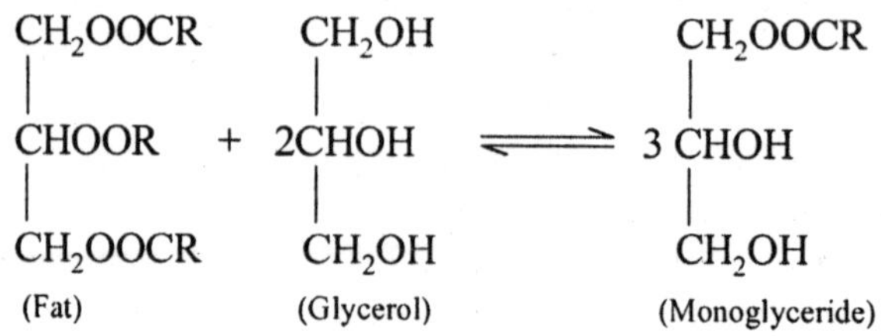

Excess glycerol is used to drive the reaction to the right and to slow down the reverse reaction. The position of the acyl radical—*alpha, beta, gamma*—depends mainly on the temperature of glycerolysis, but the *beta* product reverts to the *alpha* one on cooling. In place of fats, methyl esters are also used for glycerolysis. Fats used are soya bean oil, cottonseed oil, edible tallow and food grade triglycerides. Typically the commercial product analyses 40–48 per cent monoester, 30–40 per cent diester and 5–10 per cent triester with low percentages of free fatty acid and glycerol. Product concentrations can be changed through molecular distillation of the mixed ester product or a mixture can be enriched with a pure or enriched monoester.

Both the solid and liquid type glycerol esters are hydrophobic materials which can be used as w/o emulsifiers, co-emulsifiers, lubricants, rust preventive additives, mould release agents and as solvents for dyes and pigments. They are used as components in lubricants and softeners for the leather industry, as anti-icing additives in fuels and as rust preventive additives in compounded oils. Other application areas of glycerol esters are in textiles, metals and plastics.

Glycerol monostearate is used as an emulsifier and pacifier in cosmetics. The glycerol esters are effective in building up viscosity in cosmetic systems. The non-dispersible grades are used as bases for creams and lotions. By adding other non-ionics such as sorbitan esters or ethoxylated polyol esters, anionics such as soaps or cationics such as quaternary ammonium compounds, the glycerol esters can be readily dispersed. The self-emulsifying grades are used as primary emulsifiers and in anti-perspirant formulations.

An important use of monoglycerides and diglycerides is in foods. They are widely used in breads, cakes, other bakery products, candies, ice creams, yeast, butter, whipped toppings and icings. Table 4.3 gives some typical commercial glycerol esters and suppliers of monoglycerides. Other than the stearates, glycerol monocaprylate made from short chain coconut fatty acids is rather unique. Due to its structure, it is an efficient solubiliser dissolving water upto 20 per cent by weight. It has been used very effectively in the pharmaceutical industry as a solubiliser for vitamins, flavours and medicaments.

**Table 4.3.** Some typical commercial glycerol esters and their functions.

| *Ester type* | *Form* | *Mono %* | *Free glycerine* | *Acid value* | *Iodine value* | *HLB* |
|---|---|---|---|---|---|---|
| Monocaprylate | Liquid | 75 | 15 | 3 | 1 | 8.3 |
| Monooleate | Liquid | 40 | 4 | 4 | 95 | 3.4 |
| Monostearate | Flake | 40 | 6 | 6 | 1 | 3.8 |
| Distearate | Flake | – | – | 5 | – | 2.4 |
| Monostearate SE | Bead | 30 | 4 | 4 | 3 | 8.4 |
| Monolaurate | Solid | – | – | 4 | – | 4.9 |
| Dilaurate | Solid | – | – | 5 | – | 4.0 |
| Monococoate | Solid | 50 | 5 | 4 | 10 | 3.8 |

*(Contd ...)*

**Table 4.3.** Some typical commercial glycerol esters and their functions. *(Contd.)*

| *Functions* | *Application areas* |
|---|---|
| Lubricant | Textile staple fibres |
| Lubricant/Antistatic agent | In formulations for high speed processing of polyamide, polyester filaments. |
| Fabric softener | In textile industry |
| Emulsifiers/lubricants | Drilling and cutting oils, grinding and polishing pastes for metals. |
| Internal lubricant/thermal stabiliser/ wetting agent for additives | PVC processing |
| Antistatic agents | Polyethylene, polypropylene, foamed polystyrene. |

### *Polyglycerol esters*

Compounds derived from the condensation products of hydrophobic groups containing an active hydrogen and glycidol constitute an important family of nonionics. They have the general formula:

$$RX(C_3H_5O)_nOH$$

where R is a typical surfactant hydrophobic group and X is O, N or another functionality capable of linking the glycidol to the hydrophobe. The maximum value of '*n*' in the case of polyglycerol is ten as further reaction with the hydrophobic group becomes difficult. The existence of three possible isomers of diglycerol and increase in isomeric possibilities with each subsequent addition of glycidyl units leads to complexity of the products. Ranging from hydrophilic monoesters lipophilic diesters, the polyglycerol esters can be represented as:

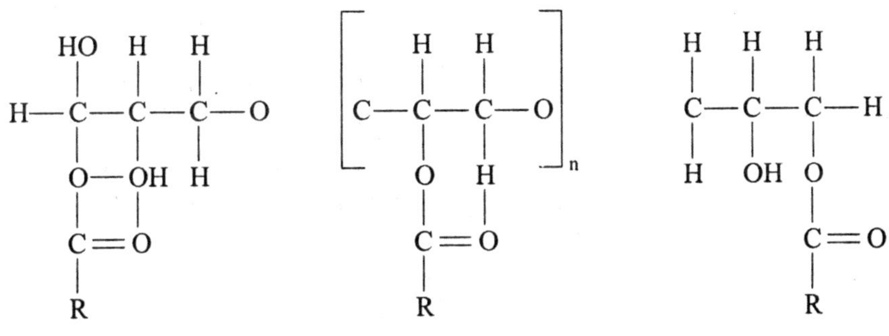

They are prepared by the esterification of polyglycerols with specific fatty acids. Polyglycerol can also be prepared by the alkaline dehydration of glycerol and then esterified with a fatty acid.

The polyglycerol esters are effective nonionic emulsifiers for both o/w and w/o emulsions. They are used in emollient creams and lotions, in pharmaceutical vehicles and due to their nontoxicity in food emulsions such as margarine and moisturising additives for bread and other baked goods. Typical polyglycerol esters are given in Table 4.4.

### *Glycols and glycol esters*

Polyethylene glycols are condensation polymers of ethylene oxide represented by the formula:

$$HO(CH_2CH_2O)_nH$$

where $n$ represents the average number of ethylene oxide units. Ranging from molecular weight 200 to 3000, represented as PEG 200, PEG 400 etc. they exhibit good water solubility. Water is miscible in all proportions with PEG 200 through PEG 600. The polyethylene glycols are chemical intermediates for other nonionics. Being themselves nonionics they find applications as solubilisers, in coatings, adhesives, lubricants, metal working, paper manufacture, petroleum production, ceramic making, printing, electronics, solvents, cleaning formulations, humectants and viscosity modifiers.

**Table 4.4.** Typical polyglycerol esters.

| *Ester* | *Form at 25°C* | *Hydroxyl value* | *Saponification value* | *Iodine value* | *HLB (+ 1.0)* |
|---|---|---|---|---|---|
| Decaglycerol tetraoleate (Drewmulse 10-4-O) | Liquid | 200–250 | 125–145 | 60 | 6.0 |
| Decaglycerol octaolete (Drewmulse 10-8-O) | Liquid | 25–50 | 150–180 | 80 | 4.0 |
| Decaglycerol decaoleate (Drewmulse 10-10-O) | Liquid | 20–60 | 160–190 | 90 | 3.0 |
| Decaglycerol decastearate (Drewmulse 10-10-S) | Solid | 30–60 | 150–190 | 3 | 3.0 |
| Hexaglycerol distearate (Drewmulse 6-2-S) | Solid | 270–310 | 100–130 | 5 | 8.0 |
| Triglycerol monostearate (Drewmulse 3-1-S) | Solid | 315–345 | 120–140 | 3 | 7.0 |
| Triglycerol monooleate (Drewmulse 3-1-O) | Liquid | 230–290 | 120–160 | 90 | 7.0 |

'Drewmulse' is the trade name of PVO International Inc. and corresponding Stepan Company products are designated 'Drewpol'.

Surfactants of the glycol ester group are monoesters of either ethylene or propylene glycol. The extreme hydrophobicity of the diesters render them unfit for use as surfactants. Since the simple glycol esters possess one unesterified hydroxyl group, they can be ethoxylated to increase their hydrophillic characteristics. Commercial polyethylene glycol esters contain varying proportions of monoesters, diesters and polyglycol. The physical properties of these esters vary with the type of fatty acid and the degree of ethoxylation.

The ethoxylates contain nearly 60 per cent by weight of ethylene oxide to enable the solubilisation of the fatty acid at room temperature. Their principle use is as emulsifiers in the textile industry for processing oils, antistatic agents and softeners. Other applications are as fibre lubricants, as detergents in scouring operations and as emulsifiers for pesticides and cosmetic preparations.

## Lanolin Ethoxylates

Ethoxylated lanolin derivatives are intended to provide the benefits of lanolin in finished cosmetic products without imparting the unctuous characteristics of lanolin. They are formed by ethoxylation of the hydroxyl groups present in lanolin or its hydrogenation product—giving rise to complex mixtures of ethoxylated sterols, lanoline alcohols, lanolin fatty acids and unaltered lanoline. This complexity arise due to the presence of water in the lanolin and the transesterification reactions taking place during the ethoxylation. Most members of this group are waxy solids, water soluble due to the high level of ethoxylation. Specific examples are POE 20 lanoline and PPG 12 POE 65 lanoline oil.

### Ethoxylated Polysiloxanes

The introduction of the first organosilican block copolymer made it possible to lower the surface tension of many organic liquids that are not influenced by hydrocarbon surfactants. Organosilicon block copolymers emerged as commercial products at a time when new polymer chemistry and its applications were taking form and the technology of polyurethane had just begun to unfold itself. Nonionic silicone surfactants are the dominant stabilisers of polyurethane foam. Foam ingredient solubility is imparted to the silicone by such hydrogen bonding groups as the ether oxygens and hydroxyls of polyoxyethylene and polyoxypropylene glycols.

Silicone—oxyalkylene block copolymers typified by Union Carbide 'Silwet' surfactants are of two distinct structural types. The major class is a polydimethylsiloxane to which polyethers have been grafted through a hydrosilylation reaction. This process results in an alkyl pendant type (AP type) copolymer in which the polyoxyalkylene groups are attached along the siloxane backbone through a series of hydrolytically stable Si–C bonds. These products have the general formula:

$$Me_3SiO\,(Me_2SiO)_x\,(Me\,\underset{\displaystyle PE}{\underset{|}{Si}}O)_y\,SiMe_3$$

wherein Me is $CH_3$ group and PE is $—CH_2CH_2O(EO)_m(PO)_nZ$. EO and PO are ethylene and propylene oxides and Z is either hydrogen or a lower alkyl group. The second class is a branched polydimethylsiloxane to which polyethers have been attached through condensation chemistry. This creates an alkoxy-end-blocked (AEB) type copolymer in which the polyoxyalkylene groups are attached to the ends of the silicone backbone through Si–O–C bonds. This linkage offers limited resistance to hydrolysis under neutral or slightly alkaline conditions, but breaks down quickly in acidic environments. These products have the general formula:

$$(MeSi)_{y-z}\,[(O\,Si\,Me_2)_{x/y}\,O—PE]_y$$

wherein Me is $CH_3$, PE is $—(EO)_m\,(PO)_n\,R$, R being a lower alkyl group.

A broad range of this class of surfactants has been developed by varying x, y, m and n values in the above formula. These products offer unique properties and performance that are not achievable with conventional organic surfactants. The ethoxylated polysiloxanes are especially used in the cosmetic industry as emulsifiers for silicones. The low surface tension, high wetting, superior dispersing, emulsifying, lubrication, sheen, gloss enhancing, static suppressing, moderate prefoaming, broad range of aqueous cloud points and solubility, good thermal stability and good compatibility with other surfactants and system components, are the various parameters taken advantage of in the applications of these products. Their low toxicity, environmental acceptability and their cost effectiveness due to their performance at low concentrations are added advantages. Other than cosmetics, end uses are in adhesives and sealants, agriculture, automotive specialties, coatings, fibres, hydrocarbon antifoams, mineral processing, paper, petroleum extraction and processing, pigments and dyes, plastics and rubber, pharmaceuticals, printing inks and textiles. A significant application is in their role as emulsifying agent in the formation of flexible urethane foams smoothening the reaction between polyol, isocyanate and water. In rigid urethane foams also they contribute to structural soundness.

Commercially available Union Carbide 'Silwet' brand products are indicated in Fig. 4.6.

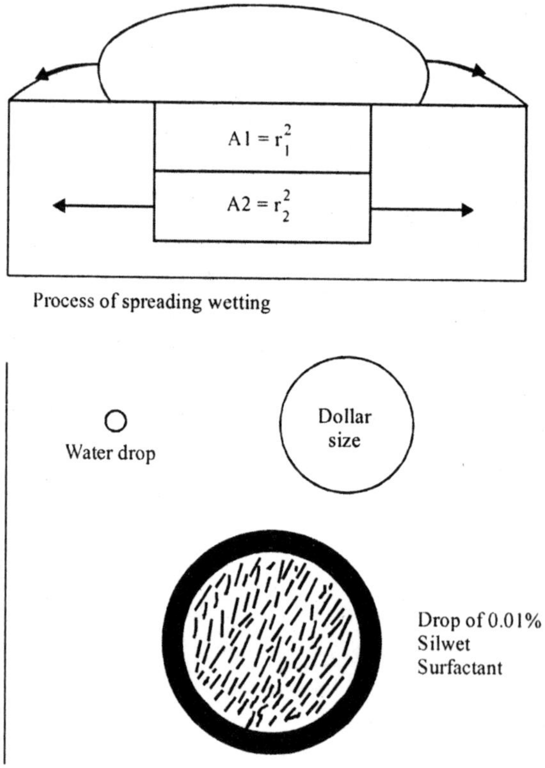


**Fig. 4.6.** Wetting and spreading of silwet surfactants.

## Alkanolamides

The alkanolamides are another class of major nonionic surfactants comprising of :

1. Monoalkanolamides.
2. 2:1 knitchevsky diethanolamides.
3. 1:1 superamides.
4. The mixed ratio amides.

Monoalkanolamides are made by condensing one mole of a monoalkanolamide with one mole of fatty acid or ester at 180°C according to the equation:

$$RCOOH + NH_2CH_2CH_2OH \text{ —R } R\,CO\,NHCH_2CH_2OH + H_2O$$

With monoisopropanolamine the corresponding monoisopropanolamide is obtained. This reaction also produces a mono-ethanolamide ester byproduct, thereby restricting the product purity to around 95 per cent. A 99 per cent purity product can be obtained by reacting two moles of fatty acid with one mole of monoethanolamine to give a fatty acid acrylaminoethyl ester and transesterifying this intermediate with a further mole of the amine. The superior hydrolytic stability and viscosity build up properties of monoisopropanolamide makes it the product of choice in shampoos in the United States. In Western

Europe the moderate foaming characteristics of monoethanolamides and their solid form boost their use in powder laundry detergents.

Diethanolamides of commercial interest are the 2:1 knitchevsky amides and the 1:1 superamides. The 2:1 products are made by reacting two moles of diethanolamine with one mole of fatty acid at 140–170°C. The use of fatty acid esters and sodium methylate catalyst also gives the 2:1 diethanolamides of mixed composition. The product purity is thus only 60–65 per cent. These alkanolamides are liquids and are used in water-based systems. The 1:1 superamides with purity around 90 per cent are obtained at 100–110°C and pressures atmospheric or lower by reacting equimolar quantities of methyl ester and diethanolamine using alkaline sodium methylate catalyst. In this process, methanol is continuously removed using explosion-proof equipment. Typical compositions of the 2:1 and 1:1 amides are as follows:

| *Composition* | *Percent by weight* | |
|---|---|---|
| | *2:1 knitchevskyn amide* | *1:1 superamide* |
| Diethanolamide | 65 | 90 |
| Free diethanolamine | 22 | 5 |
| Amine salt | 10 | Trace |
| Amide ester | 1 | 4 |
| Water | 2 | Trace |
| Methanol | – | 0.2 |
| Methyl ester | – | 0.8 |

The superamides are gels or soft, low melting solids with slight solubility in water. The viscosity building properties of these amides are directly related to their water solubilities and in this respect the lauric isopropanolamide is superior to the lauric diethanolamide.

The alkanolamides figure in a number of applications. In detergent and cosmetic formulations they basically perform as foam quality improvers, foam stabilisers, viscosity builders and detergency synergists. They are used in light duty washing liquids, heavy-duty laundry detergents, and home and institutional cleaners, including carpet cleaners. In the formulation of shampoos, bubble baths, synthetic toilet bars and institutional scrub soaps they are used directly. Ethoxylated alkanolamides and sulphosuccinates derived from their reaction with maleic anhydride are also used in cosmetics. Their asset in shampoos is due to their effectiveness in the presence of high levels of grease. The alkanolamides are compatible with other nonionic and anionic surfactants. Thus a 5 per cent addition of alkanolamide in a formulation containing 20 per cent anionics and 75 per cent builders enhances the detergency of the product. The superamides are excellent foam boosters and valuable constituents in handwash detergents and shampoos. The addition of fatty alkanolamides to surfactant formulations reduces the 'Zein number' of the formulation, rendering it thereby skin non-irritant. Other applications are as bases for deodorant sticks, as textile greasing agents, and as rust inhibitors in drilling and cutting oils. Lauric acid based alkanolamides dominate the market.

### Fatty Amine Oxides (FAO)

The substitution of the traditional alkanolamides as foam boosters in dishwashing liquids was the specific instance, which brought recognition to the nonionic fatty amine oxides. These are produced by reaction of tertiary amines and hydrogen peroxide or peroxyacids. Their general structure is given by:

$$R\,R'\,R''\,N^{+}\,O^{-}$$

wherein R, R' and R" may be alkyl, aryl or any or several other structures, with at least one being a fatty acid residue. Two typical reactions showing their formations are:

$$C_{12}H_{25}(CH_3)_2N + H_2O_2 \longrightarrow C_{12}H_{25}(CH_3)_2N \longrightarrow O + H_2O$$

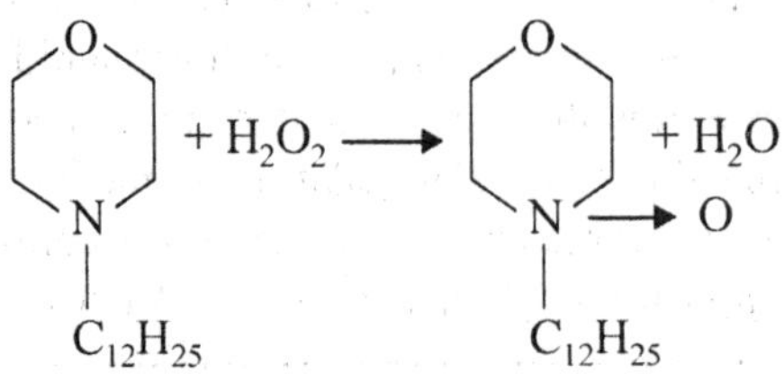

These materials while possessing a formal charge separation on the nitrogen and oxygen atoms behave as electrolytes. In aqueous solution FAO exhibit nonionic or cationic properties depending on the pH. At lower pH the weakly cationic form precipitates metathetically with anionic surfactants such as sodium dodecyl benzene sulphonate. Under neutral or alkaline conditions they exist in solution as non-ionised hydrates.

The most common amine used for the production of FAO is the alkyldimethylamine where the alkyl group contains 12 to 16 carbon atoms. In turn, the alkyldimethylamines are derived from either linear alpha-olefins or detergent alcohols. The synthesis of FAO involves the conversion of the alpha-olefin to the primary alkyl bromide which when reacted with dimethylamine gives the corresponding alkyldimethylamine.

FAO are good foam stabilisers, thickeners and emollients, emulsifiers and conditioning agents exhibiting mildness towards the skin. In dishwashing liquids they show superior foam stabilisation properties, when used along with alcohol ether sulphates in presence of grease, compared with the more commonly used fatty alkanolamides. Other applications are as rinse additives in laundering, shampoo formulations, waterless hand cleaners and hair conditioners.

### Acetylenic Alcohol Derivatives

Surfactants based on acetylenic diols are a series of products under the trade name 'Surfynols' marketed by Air Products Co. The core member of this group, Surfynol 104, is a nonionic molecule containing a hydrophilic portion in the middle of two symmetric hydrophobes. Chemically, it can be represented as:

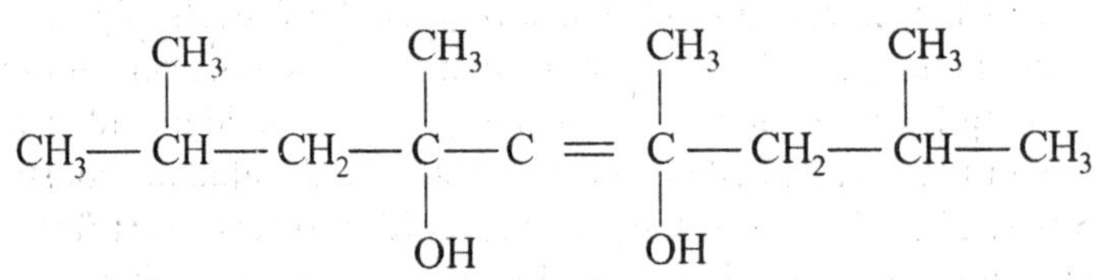

(2, 4,7,9-tetramethyul-5-decyn-4,7-diol)

This unique chemical structure allows Surfynol 104 to provide multifunctional properties such as surface tension and foam control. Other members of the Surfynol group are other acetylenic alcohols and diols and ethoxylated products. Surfynol 104 effectively reduces the surface tension of water to 33.1 dynes/cm under dynamic conditions at a concentration of 0.1 per cent. In this respect it is superior to the performance of Triton X. Surfynol 104 does not have a cloud point and as such functions as a defoamer over a very broad temperature range. Table 4.5 gives the application areas in which the wetting, defoaming and dispersant properties of some Surfynols are exploited.

**Table 4.5.** Application areas of 'Surfynol' surfactants.

| *Applications* | *Surfynol types* | | | | | |
|---|---|---|---|---|---|---|
| | *61* | *82S* | *104A* | *440* | *465* | *485* |
| *Coatings* | | | | | | |
| (Acrylic, alkyl, epoxy SBR, polyurethane) | WD | – | WD | W | W | W |
| Powder | – | WG | – | – | – | – |
| Radiation curable | – | – | W | W | W | W |
| Paper (SBR, PVA) | – | – | WD | WD | – | – |
| *Graphic arts* | | | | | | |
| Flexographic and gravure inks | WD | – | WD | W | W | W |
| Overprint varnish | W | – | WD | W | – | – |
| *Dyes/pigments* | | | | | | |
| Pigment dispersion | WD | – | WD | – | – | – |
| Dye milling | – | – | WD | – | – | – |
| Crystallisation and filtration | – | – | WDC | – | – | – |
| Adhesives (acrylic, PVA, phenolics, styrene butaeidne | – | – | WD | W | W | W |
| Foundry | – | – | WD | – | – | – |
| Emulsion polymerisation | – | – | WD | W | E | E |
| PVC production/plastisols | – | – | WD | – | – | – |
| Agricultural chemicals | W | W | WD | W | W | W |
| Oilfield chemicals | W | – | WD | WD | – | – |
| Hard surface cleaners | W | W | WD | W | W | W |
| Metalworking fluids | – | – | WD | W | W | – |
| Textiles | W | – | WD | W | W | – |
| Cement/Concrete | – | – | WD | WD | – | – |

W– low or nonfoaming wetting agents; D–defoaming, anti-foaming and/or deair entrainment agents; G–pigment grinding aid or dispersant; C–crystallisation enhancer; E–emulsifier; 61–3, 5 dimethyl-1-hexyn3-ol; 82-S – 3,6-dimethyl-4-octyne-3,6-diol; 104A–2, 4, 7, 9 tetramethyl-5-decyne-4,7 diol; 440–104 + 3.5 moles EO; 465–104 + 10 moles EO; 485–104 + 30 moles EO.

## Silicones

Like all true surfactants the silicone surfactants possess two moieties termed 'lipophilic' and 'lipophobic', rather than 'hydrophilic' and 'hydrophobic' as is the case with hydrocarbon surfactants. The variety of solubilising groups with which the silicones can be reacted permit tailoring a silicone surfactant for specific systems. The generalised structure of a silicone surfactant is

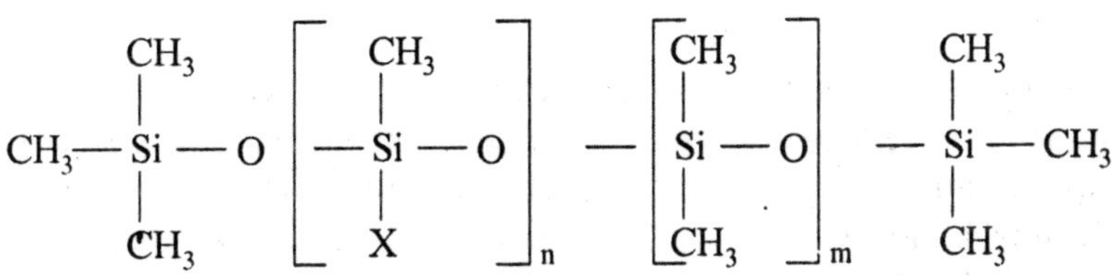

wherein X is any of the hydrophilic groups of anionic, cationic or nonionic character or organophilic groups such as esters, amides and polyethers. Unlike most common organic surfactants, silicone surfactants are multifunctional. The many possible combinations of siloxane molecular weight, functionality and polyether composition control the HLB and molecular geometry, the primary influences of surfactant behaviour. The silicone backbone can be considered, as a 'Superhydrocarbon' because of the density of methyl groups present, and is very stable and very flexible. Most silicone surfactants thus are liquids and remain so even at high molecular weights. Major applications of silicone surfactants are as wetters, emulsifiers for silicone fluids and concrete mould release agents. Other silicone based surfactants are fluorosurfactants and ethoxylated polysiloxanes, referred to in other sections.

### Fluorosurfactants

A fluorosurfactant consists of a hydrophobic and oleophobic perfluorinated tail and a hydrophilic or organophilic head group, structurally represented as:

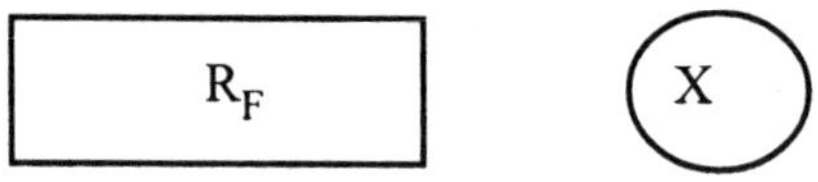


The fluorochemical tail, $R_F$, consists of the stable perfluorocarbon group. This group can be modified in length and structure to meet end use needs, and provides exceptional resistance to thermal and chemical attack. This fluorochemical portion is basically responsible for its capacity to dramatically reduce surface tension as well as being the major difference between fluorosurfactants, conventional hydrocarbon surfactants and silicone surfactants. The head group X is commonly a water- soluble group but can also be designed to be oil-soluble for use in non-aqueous systems. The nature of X can be varied to give anionic, cationic, nonionic or amphoteric surfactants such as:

$$R_FSO_3^-K^+;\ R_FNH_3^+Cl^-;\ R_F(CH_2CH_2O)_nH;\ R_FN^+H(CH_3)\ CH_2\ COO^-$$

The fluorosurfactants can be made commercially by the following processes:

1. *Simons electrochemical fluorination process (ECF):*

$$HF + C_8H_{17}SO_2F_{18} \longrightarrow C_8Fl_8SO_2F + H_2$$

followed by conversion of $C_8F_{17}SO_2F$ to various types of surfactants.

2. *Telomerisation process:*

ECF

$$C_2F_5I + n\ (CF_2\text{–}CF_2) \longrightarrow C_2F_5\ (CF_2CF_2)_n\ I$$

followed by conversion of the telomer iodides to surfactants

3. *Anionic oligomerisation of fluorine monomers:*

$$n\ (CF_2 = CF_2) \longrightarrow (C_2F_4)n$$

followed by reaction of the oligomers with alcohols, phenols, oleum or polyethylene glycols leading to various surfactants.

The advantages of fluorosurfactants are lower surface tension in aqueous systems, better effectiveness at very low concentrations, excellent stability in hostile environments and good surface activity in organic media. Areas of application include floor polish emulsions, photographic processes, production of fluoropolymers, electroplating bath additives, electronic coatings, as additives for oil/gas well stimulation, in battery electrolytes, in fire-fighting foams and as additives in paints, coatings, inks and adhesives.

## Alkyl Polyglycosides

The abundance of carbohydrates and the fact that they are both natural and renewable has evoked fresh interest in the last few years in carbohydrate derived surfactants. Nonionic alkyl glucosides, commonly known as alkyl polyglycosides (APG), have attracted attention as highly efficient and mild surfactants. They are obtained by the reaction of glucose with fatty alcohol, starch providing the source for glucose. APG has a generic structure:

$$C_nH_{2n+1}O\ (C_6H_{10}O_5)_x\ H$$

where x ranges from 1 to 4.

APG is reportedly more soluble than a typical alcohol ethoxylate or alkylphenol ethoxylate designed for household laundry applications. Detergent formulations based on APG are also more stable than the above-referred ethoxylates over a wider range of temperatures. Consequently its liquid detergent formulations require less solubilising agent and can be formulated with higher levels of soluble builders. Liquid dishwashing formulations incorporating it reportedly require no foam booster. Its mildness to the skin promotes its use in hand dishwashing personal care applications. APG can also be used effectively in combination with anionic surfactants. Henkel is the major producer of this type of nonionic surfactant.

## Sulphosuccinates, Sulphosuccinamates

The sulphosuccinates are made by forming a maleic ester and then sulphonating it. The starting raw materials for these anionics are maleic anhydride and a fatty alcohol, alcohol ethoxylate or an alkanolamide as the second reactant. If amines are used in the reaction replacing the alcohols, then sulphosuccinamate surfactants are produced. The products obtained in these reactions can be half esters or diesters. For the half ester an aqueous solution of sodium sulphite is used as the sulphonating agent and for the diester, sodium bisulphite. Typical half ester and diester sulphosuccinates are illustrated below:

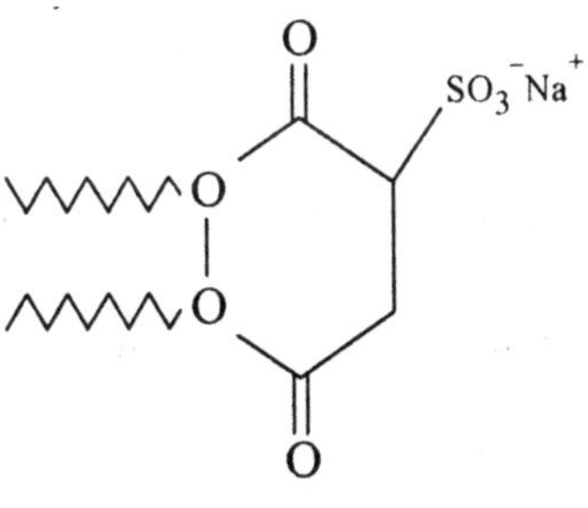

Diester sulphosuccinate

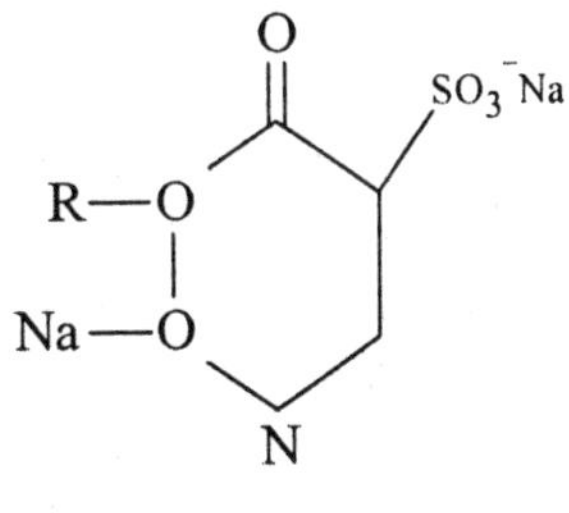

Half ester sulphosuccinate

The sulphosuccinamates are typically manufactured by the reaction:

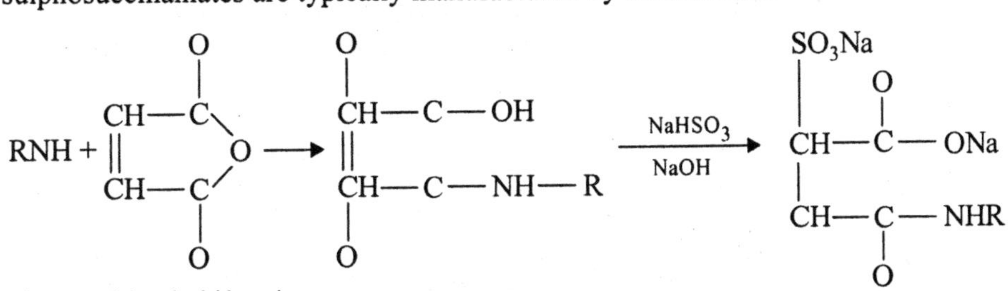

Primary maleic anhydride amine

involving a primary amine and maleic anhydride, and then adding sodium bisulphiate across the double bond.

The sulphosuccinates are good to excellent foamers and wetting agents. The half esters have application in household detergents, but the sodium dialkylesters are industrially used in the following areas:

1. Emulsion polymerisation of PVA emulsions and PVC dispersion resins.
2. Foam stabilisers in rug shampoos.
3. As components of solvent degreasers for metals.
4. As solubilisers in industrial dry cleaning formulations.
5. Dewatering of mineral slurries.
6. Mineral flotation.
7. Oil-spill chemicals.
8. Pigment dispersion and wetting agents in paints.
9. Foaming agents for secondary foam backings.
10. Coating of textiles, leather and leather-cloth.
11. Modification of plastisol viscosity.

## Phosphate Esters

Analogous to the sulphate esters, the phosphate esters are reaction products of phosphoric acid with alcohols or ethoxylates resulting in an anionic product. As it is difficult to phosphate an alcohol with orthophosphoric acid, pyrophosphoric acid is used. Likewise, oleum replaces sulphuric acid. It is possible this way to generate mono, di and polyphosphoric esters. Aromatic phosphate esters are also feasible and these can be further ethoxylated. The phosphated and ethoxylated esters form an important group of industrial surfactants.

The phosphate esters find application primarily in their exploitation as:

1. Detergents.
2. Dispersants.
3. Emulsifiers

As a detergents they are used in industrial and institutional cleaners. As dispersants they are used in paper coatings, paints and inks, textile dyeing and pesticide/herbicide suspension concentrates. As emulsifiers they are used in emulsion polymerisation of lattices, cutting oils, drilling aids, crop protection chemical concentrate. They also are used in dry cleaning, as corrosion inhibitors, and as antistats in textile cleaning formulations.

## Amine Based Surfactants

Amines are the building blocks for the entire train of cationic surfactants as illustrated below:

The amines find direct applications as surface-active agents in ore flotation, filming amines for corrosion inhibition and as anti-caking agents for granular and blended fertilisers.

### *Amine ethoxylates*

These are prepared by ethoxylation and the products contain 2–50 EO units. Their uses are as under:

1. Cationic collectors in ore flotation and frothing agents.
2. Part of filming amine formulations.
3. Additives for lubricating oils and gasoline.
4. Emulsifiers for agricultural sprays.
5. Emulsifiers for textile spinning oils.
6. Emulsion breakers.

7. Dye-levelers and textile finishing agents.
8. Corrosion inhibitors.
9. Antistats.

### Quaternary Ammonium Compounds (QAC)

The quaternary ammonium compounds are commercially the most important of the group of cationic and amphoteric surfactants accounting for 45–50 per cent of the total market for this class. There are over 200 QAC in commercial use. The QAC are made by 'quaternising' primary, secondary and tertiary amines with four different types of quaternising agents:

1. Methyl chloride or bromide.
2. Dimethyl or diethyl sulphate.
3. Benzyl chloride.
4. Ethylene oxide or diethanolamine.

Products obtained by the use of different quaternising agents have different properties and serve widely different markets. The alkyl, benzyl or ethoxy groups attach themselves to the nitrogen atom of the amine. The counterion is the halide or the methosulphate group. Alkyl chain lengths are determined by fatty acids used in preparing the amine and profoundly affect the properties of the product. If a symmetrical tertiary amine is the starting material and methyl chloride is used as the quaternising agent, a symmetrical QAC:

$$[R_3N(CH_3)]^+ Cl^-$$

is obtained. Distearyldimethyl ammonium chloride (DSDMAC) is the most important QAC and is used as fabric softener. It is also used in the preparation of organic-modified clays used in petroleum drilling, as thixotropic agents in paints and inks, and also for its bacteriostatic properties.

When monoalkyl type QAC are quaternised with benzyl chloride, compounds known as 'benzyl quats' are obtained. They have the structure:

$$[R\text{—}N\,(CH_3)_2.CH_2.C_6H_5]^+ Cl^-$$

These quaternaries form a special class of biocides, sanitisers etc. used in hospitals, restaurants, food processing centres and swimming pools.

The other groups of QAC, namely,

$$R\text{—}CH_2\overset{+}{N}(Me)[CH_2CH_2OH]_2\ Cl^-$$

$$\text{and } R\text{—}\overset{+}{N}(Me)[(CH_2CH_2O)_x]_2\ Cl^-$$

are obtained by quaternising with diethanolamine and ethylene oxide. They have a wide range of hydrophilicity. They are important in the formulation of asphalt/bitumens, in vehicle cleaning formulations and as emulsifiers in pesticide formulations.

### *Imidazolines and Polyamines*

The cationic properties of amines are greatly improved by the incorporation of additional nitrogen atoms into their structures, resulting in polyamines of the type:

$$H_2N\ CH_2\ CH_2\ NH_2\ CH_2\ CH_2\ NH_2$$

Diethylenetriamine

They are made by reducing the product obtained by the cyanoethylation of primary and other amines with acrylonitrile. Their difunctional character makes them very effective corrosion inhibitors.

Polyamines react with fatty acids and dimer/trimer acids to form the imidazolines such as:

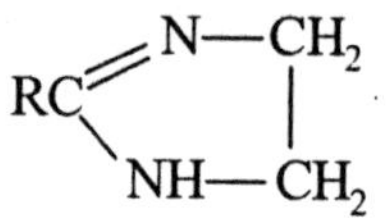


These, with various substituents, are used in fabric softeners, corrosion inhibitors in oil-field applications and other areas.

### *Betaines*

As already pointed out, the amphoteric betaines are special members of the ring opened imidazoline derived surfactants. They differ from other amphoterics with regard to their solubility and electrical nature in alkaline solution. They are compatible with anionics at all pH values. The carboxyl containing betaines are insensitive to water hardness. They are used as levelling and wetting agents in textile processing, detergents, scrubbing compounds, antistatic agents, as fabric softeners, as lime soap dispersants and in personal care products. Coconut fatty acid amidopropylbetaine is particularly used in hair shampoos, foam baths, shower foams and liquid soaps where a high, creamy foam is required. It has a good skin tolerance.

The most common route to carboxyl-betaines is through the quaternisation of long-chain alkyldimethyl tertiary amines with chloroacetic acid:

$$R(CH_3)_2\ N + ClCH_2COOH \longrightarrow R\ R(CH_3)_2N^+CH_2COO^-$$

There are several commercial producers of cocoamido-propylbetaine and other betaines. The betaines play an important role in the amphoterics market.

### Sarcosinates and Taurates

The sarcosinates and taurates are two fairly important groups of surfactant products with amino acid moieties in their structure. They were initially introduced by IG Farben.

The sarcosinates are condensation products of fatty acids with N-methyl glycine (sarcosine). They are manufactured from fatty acid chlorides and sodium sarcosine:

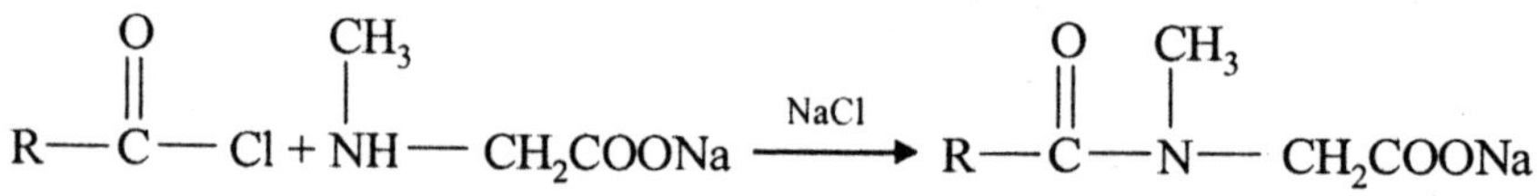


The alkyl groups in the sarcosinates are usually lauryl or oleyl, the former dominating. These products are anionic foam boosters and are used in hair conditioners and shampoos, rug shampoos, tooth paste, hand soaps and pharmaceuticals. They are also used in emulsion polymerisation and in window cleaning formulations. They are important as emulsifying agents for the removal of certain anti-corrosive agents on metal prior to their further processing.

The taurates, also referred to as taurides and N-acylmethyl taurates, are made in a manner similar to that of the sarcosinates with the reaction occurring between N-methyl taurine and fatty acid chlorides under alkaline conditions:

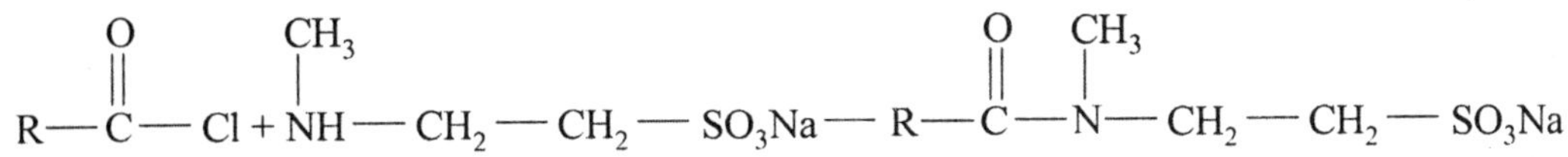

Coconut oil fatty acids and oleic acid, again, are the preferred fatty feedstocks but in this case oleic acid is the preferred fatty acid. Both sodium and potassium salts prevail. The taurates combine the properties of soap (foaming power and emulsifying property) with those of the alkylaryl sulphonates (greater insensitivity of hard water). The taurates are particularly water-soluble. They are used as additives in all areas of cleaning materials and in cosmetics and pharmaceuticals.

## TECHNOLOGIES FOR MANUFACTURE OF SURFACTANTS

### Important Physico-chemical Characteristics of Surfactants

When one considers the impact of surface science in general, and emulsions, dispersions, foaming agents, wetting agents etc. in particular, on our day-to-day existence, one begins to realise the overwhelming extent to which these areas of chemistry and chemical technology permeate our lives. Although surfactant science and technology is a wide field, some specific aspects of these materials related to their application are highlighted in this section. These are:

1. The HLB system.
2. Surfactant solubility and the Kraft temperature.
3. Emulsion stabilisation.
4. Foam stabilisation.
5. Wetting and related phenomena.
6. Dispersion.

#### *The HLB system*

Quantitative correlation of the chemical structure of surfactant molecules with surface activity to facilitate their choice in a given formulation, although has not been possible in respect of all their areas of applications, has to a large degree proved fertile in establishing such a criteria in the field of emulsions. The HLB of an emulsifier is an expression of its Hydrophile-Liphophile Balance (HLB) i.e. balance of the size and strength of the hydrophobic polar groups and the lipophilic nonpolar groups of the emulsifier. The HLB system developed by *Griffin* employed certain empirical formula to calculate the HLB number in a range of 0–20 on an arbitrary scale. Hydrophilic surfactants which possess high water solubility and generally act as good solubilising agents, detergents and stabilisers for oil in water (o/w) emulsions lie at the high end of this scale. Surfactants with low water solubility which act as solubilisers of water in oils and good water in oil (w/o) emulsion stabilisers lie at the low end of the scale. The effectiveness of a given surfactant in stabilising a particular emulsion system would depend on the balance between the HLB values of the surfactant and the oil phase involved.

#### *Surfactant solubility and the kraft temperature*

The nature of surfactant molecules, having both hydrophilic and hydrophobic groups, is responsible for their tendency to concentrate at interfaces and thereby reduce the energy of the system in which they interact. The primary mechanism for energy reduction in most cases will be adsorption at various interfaces. However, when all available interfaces are saturated the overall energy reduction may continue through other mechanisms. The physical manifestation of one such mechanism is the crystallisation or precipitation of the solute from solution. An alternative is the formation of molecular aggregates or 'micelles' that remain in 'solution' as thermodynamically stable, dispersed species with properties distinct from those of the monomeric solution.

For many ionic materials it is found that the overall solubility of the material, in water increases as the temperature increases. That effect is the result of the physical characteristics of the solid phase, i.e. the crystal lattice energy and heat of hydration of the material being dissolved. In the case of ionic surfactants, it is often observed that the solubility of the material will undergo a sharp, discontinuous increase at some characteristic temperature.

### *Emulsions stabilisation*

Emulsions are used in a large number of applications namely, food, cosmetics, medicinals, agriculture, paints and numerous industrial processes. In most of these cases, the emulsion form is used because it represents a convenient means of achieving delivery of some active material to this site, at some convenient concentration and frequently at some convenient area of coverage or controlled rate of release.

An emulsion can be defined as a heterogeneous system consisting of at least one immiscible liquid dispersed in another in the form of droplets whose diameters in general exceed 0.1 mm. Such systems possess minimal stability, which may be accentuated by additives such as surfactants. An emulsion may be either oil-in-water (o/w) or water-in-oil (w/o), where the first phase mentioned is the dispersed phase and the second, the continuous phase.

The major class of emulsifiers and stabilisers is that of monomeric surfactants that adsorb at interfaces and produce electrical, mechanical and steric barriers to drop coalescence, in addition to their role in lowering the interfacial free energy between the dispersed and continuous phases. Because of the role of the interfacial layer in emulsion stabilisation, it is often found that a mixture of surfactants with widely differing solubility parameters will produce emulsions with enhanced stability. In terms of HLB (0–20), at the high end of the scale lie hydrophilic surfactants which possess high water solubility and generally act as good solubilising agents, detergents and stabilisers for o/w emulsions. At the low end of the HLB scale are surfactants with low water solubility which act as solubilisers. The effectiveness of a good surfactant in stabilising a particular emulsion system will depend on the balance between the HLB values of the surfactant and the oil phase involved.

Recently, multiple emulsions of the type w/o or o/w have become of practical interest in controlled drug delivery, emergency drug overdose treatment, waste-water treatment and separation technology. Such systems are more unstable than conventional emulsion systems and the choice of surfactants to stabilise these systems is complex.

The sensitivity of emulsions to temperature led to the suggestion that 'Phase Inversion Temperature' (PIT) be used as a possible method for emulsion preparation. It is generally found that the same circumstances that affect the solution characteristics of nonionic surfactants (their micelle size, critical micelle concentration, cloud point etc.) will also affect the PIT of emulsions prepared with the same materials. The process of selecting the best surfactant for the preparation of an emulsion has been greatly simplified by the development of the more-or-less empirical approaches exemplified by the HLB and PIT methods. However, each method has its limitations and cannot eliminate the need for some amount of trial-and-error experimentation.

### *Foam stabilisation*

Foams and emulsions have a great deal in common with regard to the basic physical principles controlling their stability. Their major differences lie in the natures of the dispersed phase, liquid or gas, and in the fact that foams will generally involve a much higher volume fraction of the dispersed phase than that encountered in emulsions.

Foams have wide technical importance in the fields of fire fighting, polymeric foamed insulation, foamed rubbers, foamed concrete, mineral separation and electroplating operations. They also have certain aesthetic utility in many detergent and personal care products. Unwanted foams may be a nuisance in sewage treatment, coating applications and crude oil processing.

A foam is produced by the introduction of air or other gas into a liquid phase, during which time the bubbles become encapsulated in a film of the liquid. The thin liquid film separating two or more gas bubbles is referred to as a lamellar film indicting that the film structure has two essentially identical interfaces in the close proximity. The ability of a surfactant to perform as a foaming agent is dependent primarily on its effectiveness at reducing the surface tension of the solution, its diffusion characteristics, its properties with regard to disjoining pressures in thin films and the elastic properties it imparts to interfaces. The amount of foam that can be produced in a solution under given conditions will be related to the product of the surface tension and the new surface area generated during the foaming process.

Surfactants play a key role in foam formation, suppression as well as stabilisation. While all foams containing fluid phases are thermodynamically unstable, their degree of stability, commonly referred to as 'Resistance', can vary from seconds to weeks. More or less permanent foams can be produced in presence of extremely small amounts of surfactants. Maintenance of the foam is as important as its original formation. The static aspects of foam stability are related to the action of surfactant molecules and additives at the various interfaces in the system, while the dynamic characteristics are controlled primarily by the dilatational viscosity of the interfacial layers and the bulk rheological properties of the system. The choice of surfactant or surfactant formulations, therefore, must take these factors into account in the context of the system being prepared.

### *Wetting and related phenomena*

The adsorption of surfactants onto a solid surface from solution is an important process in many situations such as detergency, waterproofing and dispersion stabilisation. In these and many other related applications of surfactants, the ability of the surface active molecule to situate itself at the solid-liquid interface and produce the desired effect is controlled by the chemical natures of the components of the system–the solid, the surfactant and the solvent. A similar role can be played by surfactants on essentially infinite surfaces related to wetting, spreading, adhesion and lubrication. Classically, there are three classes of wetting phenomena—adhesional wetting, spreading wetting and immersional wetting. The distinctions amongst these, although may seem subtle, are significant from a thermodynamic and phenomenoligical point of view.

It has been found that surfactants with symmetrically located internal head group substitution and ortho-substituted alkylbenzene sulphonates are better wetting agents than their straight-chain and para-substituted analogs. The enhanced wetting activity of the nonlinear materials is generally associated with their more compact structure, allowing for further diffusion to the solid-liquid interface and greater relative adsorption efficiency. The presence of additional polar groups in the molecule like ester and amide linkages or POE chains usually results in loss of wetting power. Nonionic surfactants such as polyoxyethylenated alcohols and fatty acids show lesser and lesser wetting power as the POE chain is increased. For a given hydrophobic chain length, the maximum in wetting power will occur for the POE chain length for which the cloud point lies just above the test temperature. As a rule POE nonionics with an effective hydrophobic chain length of 11 methylene units and POE chain length of 6 to 8 will exhibit optimal wetting power. In addition, POE alcohols and thioethers are generaly found to be superior to equivalent POE fatty acids.

A number of external factors that can affect the wetting power of surfactants include temperaure, electrolyte content, pH and the addition of polar organics and cosurfactants.

Wetting plays an important role in many technologically important processes, and an understanding of the part played by surfactants can go a long way towards solving the problem of choosing the proper surfactant for the job. Some of these areas are detergency and soil removal, enhanced oil recovery (EOR) and suspension of solids in liquids.

### *Dispersion*

Dispersion is everywhere. Be it foods, pharmaceuticals, cosmetics, paints or printing inks and many such industries, no new products would have been developed without dispersion, i.e. a process of reducing particles into a uniformly distributed submicron size and distributing them also uniformly in a suitable liquid medium. The paint industry represents a vital area where dispersion processes play a crucial role. Irrespective of quality and quantity of surface treatment received by pigment/extender powders, and no matter how favourable the resin molecular architecture is for dispersion, the wetting agents or surfactants are indispensable to achieve complete dispersion.

The range of surfactants available are so large that it is almost impossible for a paint technologist to choose the right surfactant or surfactant mix and its dosage for achieving dispersion for the given formulation at optimum cost with optimum energy and efficiency. While the role of surfactant in dispersion process is well conceived, its requirements are rigid. The interesting classes of surfactants giving high performance in dispersion process are:

1. Chemical surfactants based on fluorine and silicon.
2. EO/PO polycondensates.
3. Surpynols (Acetylenic diol based surfactants).
4. Polymeric hyperdispersants.
5. Innovative class of hyperbranched and dendrite structured surfactants.

The evolution of various classes of surfactants became necessary as the dispersion of pigments in hi-tech resins posed problems. Epoxides, polyurethanes, polyesters and powder coatings do not offer ease of dispersion. Polymeric dispersants have been developed for sophisticated low molecular weight, low viscosity, high solid coatings. Unlike alkyds, for achieving dispersion in such difficult resins, dispersants must have anchoring groups with high functionality like carboxylic acids, sulphonic acids, amino acids, amine, pyridyl, acetyl etc. which will have strong bonding to the pigment surface and also soluble polymeric chains more or less identical to the main binder of the paint system. Dispersion has been a challenge to the paint industry from the beginning and this challenge persists with the emergence of new technologies.

## PROPERTIES OF SURFACE-ACTIVE AGENTS

The characteristic of surface active molecule is the change in surface properties as concentration of surfactant increases in aqueous solution. For example, as concentration rises the surface tension of an aqueous solution of, say, sodium dodecyl sulphate ($C_{12}H_{25}OSO_3Na$) falls rapidly, with corresponding changes in the physical properties such as interfacial tension, electrical conductivity, etc. At a certain concentration level a discontinuity occurs and surface tension and other properties no longer fall. The concentration at which this discontinuity occurs is called the *critical micelle concentration* (CMC).

It has been found that surface tension fell as the concentration of single anions increased (for instance $C_{12}H_{25}OSO_3^-$ in the example given) until at the CMC the single ions began to associate into groups which is called micelles. These micelles may be in the form of spheres of molecular size, in which the hydrophobic tails of the anions are oriented to the centre of the sphere, while the hydrophilic heads are at the outer surface. Thus a spherical micelle of sodium dodecyl sulphate would consist of a group of $C_{12}H_{25}$ tails pointing towards the centre of the sphere, with $OSO_3^-$ heads at the surface. This micelle would correspond very roughly to a droplet of dodecane of molecular size. In fact, micelles do have the property of dissolving water-insoluble organic matter. This phenomenon is called solubilisation and is one of the characteristics of surface-active agents important to the cosmetic chemist.

## Selection and Use of Surface-active Agents

### *Detergency*

Detergency is a complex process which involves the wetting of a substrate (hair or skin), the removal of greasy soiling matter, the emulsification of the removed grease and the stabilisation of the emulsion.

For skin cleansing, soap is still an excellent detergent. Custom dictates that a high level of foam is necessary, though it performs no function. Increased foaming can easily be achieved by superfatting with long-chain fatty acids (as in shaving soaps).

Hair washing is more complex and here foam volume does appear to play some part. Sodium lauryl ether sulphate (SLES) is a common component of shampoos, and foaming is frequently enhanced by the addition of alkanolamides. Ampholytic surface-active agents are used for specialised shampoos.

### *Wetting*

All surface-active agents have some wetting properties. Short chain ($C_{12}$) alkyl sulphates, alkyl ether sulphates and alkyl aryl sulphonates are all commonly used.

### *Foaming*

High foam volume and stable foams are generally achieved by the use of SLES reinforced by an alkanolamide.

### *Emulsification*

Usually a good emulsifying agent requires a slightly longer hydrophobic unit than does a wetting agent. Soap is still used as an emulsifying agent in cosmetic products, very often because of ease of preparation. If a fatty acid is incorporated in the oil phase and the alkali in the aqueous phase, then stable oil-in-water emulsions are easily formed *in situ* by simple mixing. Water-in-oil emulsions (such as in certain hair creams) are frequently stabilised by calcium soaps. Non-ionic surface-active agents are also of value in emulsions.

### *Solubilisation*

All surface-active agents above the CMC have solubilising properties. This is important when it is required to incorporate a perfume or an insoluble organic component into a clear product, for example, a shampoo. Soaps, alkyl ether sulphates and indeed most surface-active agents have been used for this purpose. It is of course necessary to use high concentrations to give good solubilisation.

All the above properties may be modified by the presence of electrolytes. In general electrolytes tend to lower the CMC and this should improve solubilisation. They may also tend to break emulsions and in general electrolytes should not be added to cosmetic products containing surface-active agents until their effects on surface-active properties have been fully checked.

### *Other properties*

All cationic products adsorb strongly at protein and other negatively charged substrates. They are thus used to modify the surface of a substrate, for example to improve the feel and appearance of hair. Cationics have some anti-microbial properties also and may be used as components of special shampoos and of mouthwashes (for example chlorhexidine).

Sodium N-lauryl sarcosinate is known to inhibit the enzyme *hexokinase* (which is involved in the glycolytic breakdown of sugars in the mouth) and has been used in toothpastes.

Anionics can have an effect on each other; for example, the foam produced by SLES can easily be destroyed by soap (both anionic). This property is made use of in the formulation of low foam detergents.

## Biological Properties of Surface-active Agents

As surface-active agents are adsorbed at the surface, hence they may have biological effects. All cosmetic products containing surface-active agents should be checked rigorously to ensure that they do not have harmful effects on users.

### *Dermatological effects*

Surface-active agents wet the skin and may remove grease from the surface of the skin. When wrongly used they may create chapping, cracking and dryness of the skin. The $C_{12}$ moiety seems particularly active in this respect and the $C_{12}$ sulphate, for example, is used to create chapping artificially. Fortunately the effects are easily reduced by mixture with sulphates of other chain length, by addition of ethylene oxide and by other means.

In general, all cosmetic products containing surface-active agents should be patch tested (and if appropriate eye-tested) to ensure that they produce no adverse reaction.

### *Biodegradation*

It is good practice to use only biodegradable surface-active agents. In some countries this is compulsory. Branched-chain alkyl aryl sulphonates are not biodegradable, but the corresponding straight-chain compounds are, as they are all soaps and alkyl sulphates.

### *Toxicological effects*

Surface-active agents are not, compounds of high toxicity. Nevertheless, since they may be ingested either accidentally or from a toothpaste or mouthwash, it is wise to check oral toxicity of cosmetic products containing them.

Cationics are the most toxic and have $LD_{50}$ values of the magnitude of 50–500 mg per kg body weight; anionics are roughly in the range 2–8 gram per kg and nonionics range upwards from about 5 g per kg[12]. Cosmetic products containing surface-active agents should therefore be reasonably safe from toxic hazards.

The various surface-active agents used in cosmetics are as under:

## Basic Surfactants

### *Alkyl ether sulphates*

Alkyl ether sulphates are the most widely used surfactants for cosmetics and toiletries due to their well-balanced properties. They are excellent foamers, although the foam structure is relatively coarse. Foaming is not affected by hard water and this, together with their low critical micelle concentration, makes alkyl ether sulphates ideal for use in foam-bath preparations, shampoos and shower baths.

Alkyl sulphates are excellent foamers, producing a rich and creamy foam, but are unstable in hard water. This is not a problem for shampoo or shower applications, but means that they cannot be used as the main component in foam baths. Instead, they must be combined with surfactants such as olefin sulphonates or sulphosuccinates, which are stable in hard water.

A very important application of alkyl sulphates is their use in toothpastes. Sodium lauryl sulphate is available in very pure form with low amounts of unsulphated matter. This results in a very neutral taste compared to the bitter taste of other surfactants.

$\alpha$-Olefin sulphonates are a mixture of alkene sulphonates and hydroxy alkane sulphonates prepared by sulphonation of $\alpha$-olefins followed by alkaline hydrolysis.

They are stable in hard water and show hydrotropic properties. This results in low cloud points on cooling, and high solubilising power for superfatting agents. The difficulty of thickening can be overcome by combination with other surfactants. For example, a mixture of 60 per cent olefin sulphonate and 40 per cent sulphosuccinate, which are individually very difficult to thicken, shows a thickening behaviour comparable to alkyl ether sulphates. Contrary to the alkyl sulphates and the alkyl ether sulphates, olefin sulphonates are stable at both acidic and alkaline pH values.

### *Other basic surfactants*

Linear alkyl benzene sulphonate and alkane sulphonate are very powerful surfactants with good detergency and foaming properties. However, due to their strong defatting action, they leave a harsh and dry feel on skin and hair. For very cheap formulations, small amounts can be used to reduce the costs of raw materials. Their main use is in dish-washing liquids and all-purpose cleaners.

## Mild Anionic Surfactants

Compared to the basic surfactants, a much wider group of mild anionic surfactants is used in cosmetics and toiletries. The purpose of mild or secondary surfactants is to improve skin and eye compatibility of the formulation. On the other hand, mild surfactants usually show reduced foaming and cleansing compared with basic surfactants.

### *Sulphosuccinates*

The sulphosuccinates have good skin compatibilities compared with the basic surfactants. In general, the amide-based sulphosuccinates are better than the alcohol-based types. Sulphosuccinates are good foamers and are relatively cheap. A disadvantage is their instability at both low and high pH values.

### *Cocoyl isethionates*

Cocoyl isethionate is mainly used as a surfactant in syndet bars. In normal soaps it improves skin compatibility and acts as a lime soap dispersant.

### *Acyl amides*

The two most important acyl amides used in cosmetics and toiletries are the sarcosinates and the protein fatty acid condensates. They are prepared by reaction of fatty acid chloride with *N*-methylglycine or with oligopeptides. The acyl amides can be considered as modified soaps with good water solubility. Due to their enlarged polar ends, they are stable in hard water.

### *Alkyl ether carboxylates*

Alkyl ether carboxylates are prepared by reaction of chloroacetic acid with fatty alcohol ethoxylates.

Unlike alkyl ether sulphates, alkyl ether carboxylates have no ester linkages and are stable at low pH-values. They are available as 100 per cent active material in the acidic form, or as neutralised solutions.

### *Magnesium surfactants*

The significant applicational property of magnesium surfactants is that they are practically unchanged compared with the sodium forms. From this point of view they belong to the group of basic surfactants showing high foaming and excellent cleansing properties.

### *Alkyl phosphates*

These are used for mild shampoos and shower-bath preparations. Both the alkyl phosphates based on fatty alcohols, and the alkyl phosphates based on low ethoxylated fatty alcohols.

## Amphoteric Surfactants

Amphoteric surfactants are surfactants where the charge changes as a function of the pH value of the formulation in which they are used. They are generally regarded as mild surfactants but this is not a simple matter and may not always be true. Amphoteric surfactants build complexes in combination with anionic surfactants and these complexes are milder than the individual surfactants.

### *Alkyl betaines*

Alkyl betaines are prepared by condensation of an alkyl dimethyl amine with sodium chloroacetate. Betaines can improve the foaming of a formulation, particularly the structure of the foam, which becomes finer and more creamy.

### *Alkylamido betaines*

Alkylamido betaines are prepared by condensation of fatty acids with dimethylamino propyl amino (DMAPA), followed by reaction with sodium chloroacetate. The application properties of alkylamido betaines are generally similar to those of the alkyl betaines. The coco-type is by far the most important.

### *Amine oxides*

Unlike betaines, amine oxides are never anionic. However, they do show cationic or non-ionic behaviour depending on the pH, and therefore behave quite similarly. As much, they are excellent foamers. In combination with anionics, small amounts can act as foam boosters and can improve the foam structure. Like betaines, they are good thickeners for anionic surfactants. In addition, amine oxides are good conditioning agents in hair rinses.

## Non-ionic Surfactants

Non-ionic surfactants are having poor foaming properties. These surfactants are rarely used as shampoo surfactants, although they are often added as solubilisers for perfumes.

### *Ethoxylated products*

Ethoxylation allows the production of a wide range of products. Variables are the hydrophobic part of the molecule, and the number of ethylene oxide units added.

### *Alkyl polyglycosides*

A relatively new range of commercially available surfactants is the alkyl polyglycosides (APG). Based on carbohydrates, APGs are mixtures of different isomers, (i) stereoisomers, (ii) binding isomers and (iii) ring isomers.

Surface active APGs are prepared by the glycosylation of starch or monomer glucose with fatty alcohols. This can be done directly or by transglycosylation using butylpolyglucosides as intermediates in a two-step process.

## Cationic Surfactants

Unlike the anionic or amphoteric surfactants, cationics are seldom used for cleansing applications. These surfactants show a high substantivity to surfaces such as skin and hair due to conditioning. They can act as emollients for the skin, or as conditioning agents on the hair.

### *Monoalkyl quaternaries*

The monoalkyl quaternaries are still of major importance in conditioning. The most important is cetyl trimethyl ammonium chloride (CTMAC) which shows a light to moderate conditioning intensity combined with excellent water solubility. The substantivity and conditioning properties are improved as the length of the carbon chain increases.

### *Dialkyl quaternaries*

The dialkyl quaternaries are well known as relatively strong conditioning agents. It exhibits strong conditioning properties, good detangling and combing, and adds manageability to unruly or damaged hair.

### *Trialkyl quaternaries*

An increase in the number of alkyl chains leads to considerable improvement in properties. A relatively new product is tricetyl methyl ammonium chloride (TCMAC), which shows superior detangling, static control, and excellent combing properties on both wet and dry hair.

### *Benzyl quaternaries*

Benzyl quaternaries are obtained by reaction of alkyl dimethyl amines with benzyl chloride. Based on short-chain amines, quaternaries with good anti-microbial properties are obtained.

### *Ethoxylated quaternaries*

They are more hydrophilic than normal quaternaries, but show relatively good conditioning properties regarding both wet-combing and, particularly, static control. As in the case of betaines and aminoxides, the ethoquats are compatible with anionic surfactants.

Chapter 5

# Fats and Oils in Paint, Varnishes and Related Products

## INTRODUCTION

Paint is a very broad word used to describe many different types of materials used for decorative and protective coating of various surfaces. The type of *binder*, such as alkyds, vinyls and epoxy, may be the basis for classifying paints. Since most binders are used in several different kinds of paint, however, paints also may be classified according to the properties of the products or end use. Alkyd enamels, for instances, are gloss paints with good abrasion resistance and good cleanability, while alkyd flat wall paints are characterised by a very low sheen and inferior abrasion resistance and cleanability.

Paints used as the final coat on a surface are referred to as *finish coats or topcoats*. Paint applied before the topcoat is called an *undercoat*. Undercoats often are classified according to use. *Fillers* are undercoats used to fill holes, pores, or irregularities to provide a uniform surface for the topcoat. Primers are used to aid the adhesion of the topcoat to a surface and to prevent absorption of the topcoat into a porous surface. Primers can also be used to prevent corrosion of metals that are to be painted. Surfacers are highly pigmented undercoaters used to make a surface more uniform and give adhesion to the final coat. Surfacers often are formulated so that they can be sanded smooth before the topcoat is applied. *Sealers* are clear or pigmented materials applied to a surface to prevent some materials in the surface to be painted, e.g. a dye, from migrating into the topcoat.

Pigment is any substance, usually in the form of a dry powder, that imparts colour to another substance or mixture. Most pigments are insoluble in organic solvents and water: exceptions are the natural organic pigments, such as chlorophyll, which are generally organosoluble. To qualify as a pigment, a material must have positive colourant value. Pigments may be classified as— Inorganic and Organic. *Inorganic pigments* includes metallic oxides (iron, titanium, zinc, cobalt, chromium); earth colours (ochers, umbers); lead chromates; carbon black etc. where as *Organic pigments* includes animal (rhodopsin, melanin); vegetable (chlorophyll, xanthophyll, indigo, flavone, carotene); synthetic (phthalocyanine, lithols, toluidine, para red, toners, lakes, etc.).

Varnish is an organic protective coating, similar to a paint, except that it does not contain a colourant. It may be comprised of a vegetable oil (linseed, tung, etc.) and solvent, or of a synthetic or natural resins and solvent. In the first case the formation of the film is due to polymerisation of the oil and in the second to evaporation of the solvent.

Varnish contains no pigment and the dry film is usually glossy, although sometimes ingredients are added to produce a film without gloss. Such a varnish is called a flat varnish. Varnishes are usually classed as spirit or oleoresinous. A spirit varnish is a solution of dispersion of a resin in a volatile liquid. It dries by evaporation of the volatile solvent. Spirit varnishes contain such solvents as methanol, toluene, ketones, etc. and often also thinners such as naphtha or other light hydrocarbon. Examples of spirit varnish are cellulose nitrate lacquer and shellac varnish. An oleoresinous varnish is a solution of a mixture (usually cooked) of a drying oil and resin, in a volatile liquid. It dries by evaporation of the volatile portion, followed by oxidation and polymerisation of the residue. Examples of oleoresinous varnish are spar varnish and floor varnish. A third type of composition, namely, thickened oil, is also called varnish, specifically, bodied oil varnish. The oleoresinous type is thus a hybrid. Examples of straight oil varnish are patent leather and lithographic varnishes.

*Lacquer* is a protective or decorative coating that dries primarily by evaporation of solvent, rather than by oxidation or polymerisation.

Paints also can be classified into two very broad groups, those which use water as the primary liquid of the paint and those which use other liquids, such as hydrocarbons or aliphatic or oxygenated compounds. Many of the same polymers and pigments can be used in both types of paint. The primary difference is that most non-water-dilutable paints are solution paints, where as the liquid is a solvent in which the polymer is dissolved. Water-based paints, however, are primarily latex paints, where the polymer particles are a discontinuous phase and water is in the continuous phase. *Organosols*, however, consist of a resin dispersed in a liquid other than water. There also are water-based paints which have a water-soluble binder instead of a latex dispersion.

In addition to the pigment, binder and liquid, a paint also may contain many additives, such as defoamers, thickeners, flow agents, catalysts, wetting agents and plasticisers to improve various properties of the paint. The selection of the pigments, binder and additives depends, of course, upon the properties desired in the paint to be made.

## MANUFACTURE

A primary concern in the manufacture of paint is the dispersion of pigments in the liquid portion of the paint. Pigments used in paint contain aggregates of fine particles, which are held together by strong forces of attraction. Work is required in the form of high shear to attrition applied to the pigment-liquid slurry in order to reduce the pigments to their ultimate particles size as determined by the manufacturer of the pigment.

Considerable work may also be required to wet the surface of the pigment and to displace any air or moisture that may be absorbed on the surface by the liquid. Wetting agents such as soaps or detergents in the water phase of latex paints are often used to help wet and disperse pigments.

After the pigments are dispersed, they must be prevented from coming together again and flocculating, or forming soft lumps or pigment and liquid. The nonpolar binder of a solvent-thinned paint will usually serve to envelope the pigment particles and prevent particles and prevent the electrostatic charges on the surface of the pigments from causing flocculation. In dispersions of water-based paints surfactants in the water phase will neutralise the charges on the pigments surfaces and protective colloids, such as water-soluble cellulosic polymers, are introduced into the water phase to prevent flocculation after dispersion. In practice, few pigments are dispersed to their ultimate particles size and paints usually contain many aggregates and flocculants (Fig. 5.1).

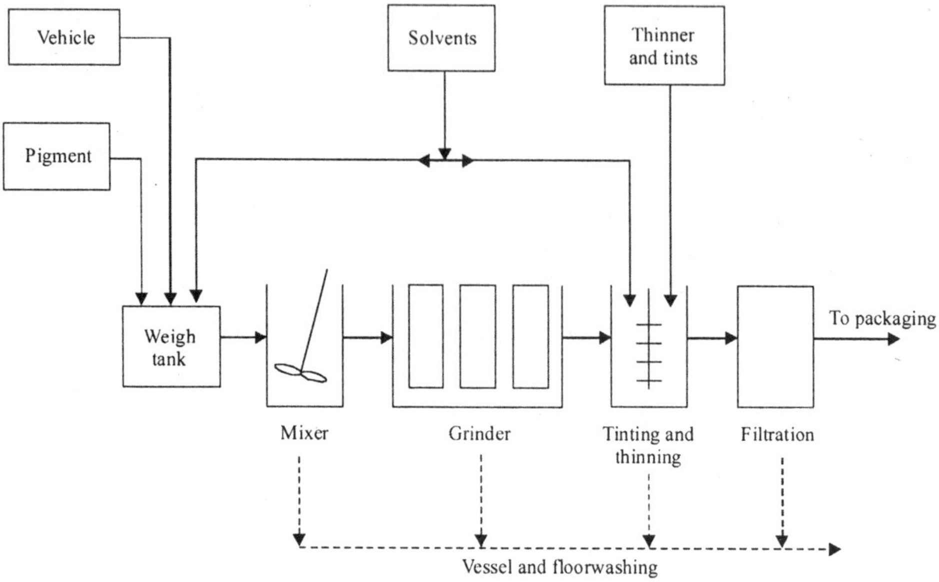


**Fig. 5.1.** Typical flow chart for paint production.

Since some pigments are harder to disperse than others, different types of dispersing equipment or mills are used. Dispersing pigments in a liquid is often called grinding although there is very little reduction in size of the original pigments particle during the dispersing operation of paint manufacture.

The high-speed stone mill consists of a stationary carborundum stone and a high-speed rotating stone. Pigment pastes are passed between these stones; the distance between the stones can be varied for more or less shearing action. These mills are suitable for high production rates of paints fairly easy to disperse, e.g. architectural paints, where very fine dispersion is not required.

Roller mills consist of steel rollers rotating in opposite directions at different speeds. The pigment-liquid paste is passed between the rolls, which can be adjusted to different clearances. Three-roll mills are the most widely used in the paint industry. Roller mills have relatively slow production rates and require skilled operators but are capable of producing fine dispersion.

Heavy-duty dough mixers, consisting of two roughly S-shapes blades which overlap and rotate in opposite directions, are sometimes used to disperse very heavy pastes.

Ball and pebble mills consist of larger cylindrical steel tanks which rotates around a horizontal axis. The mill is partly filled with steel or porcelain balls or pebbles and the material to be dispersed. Baffle bars are usually added to the sides of the tanks to help lift the balls for better dispersion. Steel balls are more efficient because of their greater density but cannot be used to produce white paints. Ball mills require little attention after they have been charged and are capable of producing good dispersion.

The sand mill consists of a cylinder containing coarse sand as a grinding medium. The pigment paste to be dispersed is fed into the mill and rotating impeller disks driven by a vertical shaft impart a circulation pattern to the sand-paste mixture. The difference in velocity between the particles near the surface of the impellers and the rest of the material develops a high shear action to disperse the pigments. A coarse screen allows the pigments slurry to pass through the mill while retaining the sand in the mill. Pigment slurrys can be passed through these mills for continuous operation. Production rates of sand mills can be fairly good and dispersion is quite good. These mills are often used for high-quality industrial finishes.

The high-speed disperser consists of a tank containing a circular impeller driven at high speed by a vertical shaft. Dispersion of the pigment-liquid mixture is achieved by high shear action developed near the surface of the impeller. High-speed dispersers are used where very fine dispersion is not required or whenever the pigments will disperse easily in the liquid. The production rate is very high and this type of equipment is used to manufacture most architectural paints.

After the pigment is dispersed and stabilised in the liquid, it is usually transferred to a thin-down tank, equipped with slower agitation, where the paint is adjusted to the desired viscosity and shaded. Materials can be added to the paint in the thin-down tank that are not required in the grinding operation and can be added with slow agitation. The binder of a solvent-thinned paint is usually added during the grinding operation; the binder of a latex paint is usually added in the thin-down tank. Thin-down tanks are usually on a floor below the grinding equipment to take advantage of gravity flow and the finished paint is usually dropped to a floor below the thin-down tanks for straining and filling.

## Application

In addition to the familiar brush and roller methods, paints can be applied with air or airless spray equipment; electrostatic, hot, or steam spraying; use of aerosol packaging; dip, flow and electrodeposition coating; roller coating machines; and powder coating.

Although most architectural paints are applied with a brush or roller, much paint is now being applied by profession painters with compressed air or airless spray equipment. With airless spray equipment the paint is atomised by forcing it through a very small orifice under very high pressure.

With electrostatic spraying the atomised paint is attracted to the conductive object to be painted by an electrostatic potential between the paint and the object. Very little paint is lost with this process and irregular objects can be coated uniformly. Hot spray application consists of heating the paint so that it is more fluid and higher-solids paints can be applied. With steam spraying, steam is used to atomise the paint. Two-component spray equipment consists of two material lines leading to the spray gun so that two materials, e.g. an epoxy and a catalyst, can be mixed in the gun just before application.

Aerosol is a method of packaging paint in a can containing a compressed gas so that the paint can be atomised through a small orifice opened by a push button. Many different methods are used for industrial application of paint, including most of the spray methods.

Dip application is a simple method where objects to be coated are suspended from a conveyor chain and dipped into a larger tank containing the paint. This method is often used for undercoating objects where paint uniformity and appearance are not important. In flow coating the paint is allowed to flow over the object to be painted, which is usually suspended from a conveyor. This process is similar to dip coating but is used where the object, e.g. a bed spring, is too larger for a tank.

Electrodeposition consists of depositing a paint on a conductive surface from a water batch containing the paint. The negatively charged paints particles are attracted to be object to be coated, which is the anode when an electric potential is applied. Paint can be applied to very irregular surfaces at very uniform thickness with little loss of paint. The system is limited to one coat of limited film thickness and equipment cost is high.

Roller coating machines are used to apply paint to one or both sides of flat surfaces, e.g. fibreboard or tin plate. The thickness of the coating can be controlled by the clearance between a doctor blade and the applicator rolls. Decorative effects, such as wood-grain patterns, can be applied with these machines. Flat sheets of wood, fibreboard, or metal and rolls of fabric, paper, or metal can be coated with these machines.

In powder coating paint in a dry powder form is applied on the surface of a heated or electrostatically grounded object to be coated. Following powder application, the object is heated to fuse and cure the coating.

### *Drying and curing*

There are two main mechanisms by which a wet paint applied to a surface becomes a dry solid coating. Some paints, e.g. lacquers and most latex paints, dry only by the evaporation of the liquids in the paint. The polymer binder is completely cured when the liquids evaporate and no chemical change is required to harden the polymer. In latex paints, the latex binder consists of very small particles of solid polymer separated by water, which is the continuous phase. When the water evaporates, the polymer particles touch each other and fuse together, or coalesce, into a continuous paint film. Pigment particles are also dispersed in the water phase and the dry paint film consists of a mixture of pigment and polymer particles fused together. If the latex particles are so hard that they will not fuse together when the water evaporates, coalescents must be added, e.g. carbitol acetate or dibutyl phthalate.

The other broad method of drying takes place by a chemical reaction to cross-link a soft polymer after the liquid have evaporated to produce a hard paint film. Some paint binders contain unsaturated compounds, such as linseed or soya drying oils, which react with oxygen in the air to form solid polymers. Other materials, such as isocyanates, react with water vapour in the air in order to polymerise.

Some materials, e.g. epoxy resins, must be cured by reaction with curing agents such as amines and polysulphide resins. The majority of these paints are supplied as two-package systems to separate the resins from the curing agent until just before the paint is applied.

Oxidative polymerisation is usually aided by polymerisation catalysts called driers. Lead octoate, cobalt naphthenate, or manganese isodecanoates are often used. Heat applied to the paint film is also used as an aid to polymerisation reactions.

## Solvent Paints

Many different types of vehicles, e.g. alkyds, chlorinated rubber, acrylics and nitrocellulose, used in solvent paints will be discussed under specific types of paints.

Because of its high refractive index, titanium dioxide is the primary pigment used in solvent paints to give opacity to white and light-coloured paints. Among other white pigments used is zinc oxide for hardness, drying and mildew resistance of exterior paints.

Extender pigments, such as talc, silica, mica, clay and calcium carbonate, are used to control many different properties of a paint (hardness, gloss, settling, rheological properties and rust prevention). Since the solvent evaporates and contributes nothing to the dried film, the least expensive combination of solvents can be used which will dissolve the polymer, give the desired viscosity and conform to air-pollution standards. Antiskinning agents are often used to prevent oxidation and surface hardening of the paints resins in the can. Oxime compounds, e.g. methyl ethyl ketoxine and substituted phenols e.g. guaiacol, are often used.

Scratch-and marproof additives, such as amino resins, wax, polyethylene and cellulose derivatives, are sometimes added to harden the paint film or develop a lubricated surface.

Flooding and floating is the separation of one or more pigments from the rest of the paint at the surface of the film, usually due to a difference in densities of the pigments. Antiflood and antifloating agents such as silicones can be used to correct this problem.

Additives which will attach to the surface of the pigments and form a pigment-vehicle bond for better grinding or dispersion are often used. Compounds which contain both hydrophilic and lipophilic groups, e.g. morpholine, methyl ethyl ketoxime and soya lecithins, are used so that the hydrophilic part will attach to the pigment and the lipophilic part will attach to the vehicle.

Flow or leveling agents are used to correct irregularities or defects in applied paint films. These defects can be caused by differences in concentration between the surface and the interior of the paint during evaporation of the solvent, over polymerised particles in the vehicle, an unclean surface, pigment flocculation, or solvent mixtures with a wide range of evaporation rates. Some of the materials used are silicone resins and unsaturated organic acids.

### *Drying-oil paints*

Drying oils, which have been used for years in paint binders or vehicles, are naturally occurring materials usually of vegetable origins, i.e. liquid triglycerides with three molecules of long-chain fatty acids to each glycerine molecule. The majority of these acids have 18 carbon atoms, many of which are unsaturated. The oil differ greatly in their drying properties, which depend upon the degree of unsaturation of the acids. Linseed, safflower, soya, tall oil cottonseed, tung and oiticica contains fairly large percentages of unsaturated acids, e.g. oleic, linoleic, linolenic, eleostearic and licanic acids.

Drying oils are often used in exterior architectural paints because of their durability, ease of application and moderate cost. Disadvantages are slow drying and poor chemical resistance.

### *Alkyd paints*

Alkyds are polyester resins made from polybasic acids and polyhydric alcohols. Glycerol and pentaerythritol are often used for the polyalcohol and phthalic anhydride and maleic acid are often used for the polycarboxylic acids. All the oils previously discussed are also used in alkyds by converting the fatty acid oils into monoglycerides and then reacting with a dibasic acids such as phthalic anhydride:

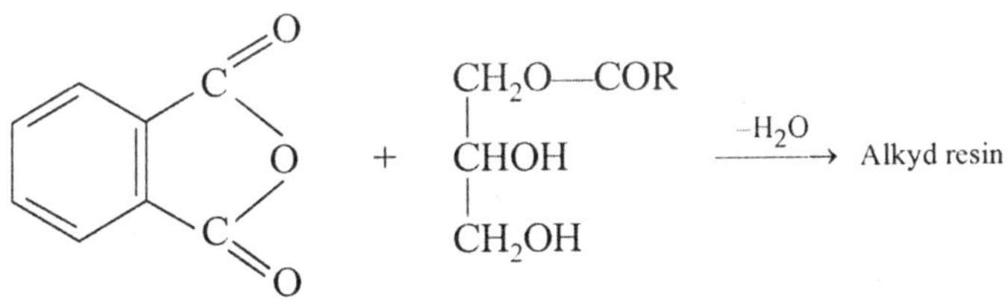


Phthalic anhydride    Fatty acid

Alkyd resins vary greatly in their properties because of the many different oils, alcohols and acids that can be used to make them.

Alkyds have faster drying, better gloss retention and better colour than oils. Most unmodified alkyds have low chemical and alkali resistance. Alkyds can be modified with rosin esterified in place of some oil acids. Phenolic resins, such as *o*- or *p*-phenylphenol, can also be used in order to produce greater hardness and better chemical resistance.

Styrene and vinyl toluene are also used to modify alkyds for faster dry, better hardness and toughness. Co-polymers of silicones and alkyds are often used for stove or heater finishes or for coating articles which may be subjected to heat up to about 450°F (232°C). They have good adhesion, hardness, flexibility, toughness, exterior durability and resistance to solvents, acids and alkalis.

Acrylic monomers can be co-polymerised with oils to modify alkyd resins for fast dry, good initial gloss, adhesion and exterior durability.

Aromatic acids, e.g. benzoic or butylbenzoic, may be used to replace part of the fatty acids for faster air dry, high gloss, hardness, chemical resistance and adhesion.

### *Vinyl paints*

Vinyl chloride–vinyl acetate co-polymers in solution form are often used in paints. These paints are resistant to alkali and organic acids, alcohols, oil and aliphatic hydrocarbons. They can be dissolved in ketones, esters and chlorinated hydrocarbons; have good water resistance, toughness and flexibility; are nonflammable; and have good clarity and good exterior durability. Vinyl co-polymers often are used in paints for railway car finishes and organosols.

#### *Epoxy paints*

Epoxy resins are prepared from epichlorohydrin $\underset{\diagdown O \diagup}{CH_2CH}CH_2Cl$ and a dihydroxy compound, usually a diphenol.

Two reactions are involved in the polymerisation: condensation to eliminate HCl and addition reactions to open epoxide rings along the chain to produce hydroxyl groups. The polymer has epoxide rings at each end and hydroxyl groups along the chain, which ensure good adhesion to polar surfaces such as metals. There are two types of expoxy resins, catalysed types and epoxy esters.

Catalysed epoxies must be converted to useful products by reaction with curing agents, e.g. amines, polyamide resins, polysulphide resins, anhydrides, metallic hydroxides, or Lewis acids.

Most of these materials are supplied as two-package systems to separate the material until just before application. Polymer curing takes place by reaction of a curing agent with epoxide rings to cross-link the polymer. Paints made from these polymers have excellent chemical resistance and hardness and are often used for maintenance coatings, trade sales specialities and industrial finishes. Chalking, with exterior exposure and the two-package system limit the use of expoxy finishes.

The epoxy resin can also be reacted with drying oils or fatty acids to produce expoxy esters, which cure by air drying or heat. Paints made with epoxy esters do not have as good chemical and solvent resistance as catalysed epoxies, but they are superior to oils and alkyds in this respect. They also exhibit chalking on exterior exposure.

### *Acrylic paints*

Acrylic resins can be divided into: thermoplastic and thermosetting types. Acrylic resins used in paints are mono-or co-polymers of acrylic acid or methyacrylate acid esters. Some of the common monomers are methyl methacrylate, butyl methacrylate, methyl acrylate, butyl acrylate, ethyl acrylate and 2-ethylhexyl acrylate. Thermoplastic resins become soft when heated and reharden when cooled. Paints made with these resins have excellent exterior durability and excellent gloss and colour retention. They are non-yellowing and can have good resistance to stain. Heating after application allows these paints to reflow, producing excellent appearance. Thermoplastic acrylic finishes are primarily used for automotive and product finishes. They have the disadvantages of poor adhesion and high price for both polymer and solvents. Thermosetting acrylic resins have at least one monomer belonging to the acrylic family which will react with itself or other resins at elevated temperatures to cross-link in order to cure. In addition to the acrylic monomers previously listed, acrylonitrile, acrylamide, styrene and vinyl toluene are often used in these polymers. Polymers which react to cross-link primarily because of hydroxyl groups are usually combined with an epoxy resins; those which react mainly with carboxyl groups usually are combined with an amine resins. Thermosetting acrylic paints, which are hard and stain-resistant and have high gloss, are often used for appliance finished. Tough flexible finishes can be formulated for coil coatings.

### *Cellulosic paints*

Cellulosic polymers, such as nitrocellulose, ethyl cellulose, ethyl hydroxyethyl cellulose, cellulose acetate and hydroxyethyl cellulose, can be dissolved in solvent mixtures which are primarily oxygenated liquids, such as esters, ketones, ethers and alcohols. Coatings made with these polymers dry only by loss of solvent, which is often referred to as lacquer drying. Cellulosic lacquers must often be plasticised with compounds such as dibutyl phthalate or butylbenzyl phthalate. Compatible resins, e.g. ester gum or acrylic or maleic resins, are often used to improve gloss, adhesion and elongation.

Cellulosic lacquers have very fast dry but low film build. They are used extensively for furniture finished and automobile refinishing.

### *Chlorinated-rubber paints*

Chlorinated rubber is made by chlorinating natural rubber, is soluble in aromatic hydrocarbons and is used to make paints with excellent chemical resistance, water resistance and fast-drying qualities. The polymer is rather brittle and must be plasticised with materials such as drying oils and chlorinated biphenyl. The disadvantage of these paints is poor resistance to organic solvents and oils; exterior exposure is fairly good. They are used primarily where good chemical or water resistance is required, as in industrial maintenance. Styrene-butadiene co-polymer resins are very much similar to chlorinated rubber in solubility and other physical properties.

### *Polyester paints*

Polyesters are unsaturated thermosetting polyester resins similar to those used for reinforced plastic. Although alkyds can be considered unsaturated polyesters, this term has been reserved for resins which have unsaturated compounds in the backbone of the polymer. These resins are made by reacting unsaturated dibasic acids, e.g. maleic anhydride, citraconic anhydride, fumaric acid, itaconic acid, phthalic anhydride and adipic acid, with polyhydric alcohols, e.g. propylene glycol. Styrene or some other aromatic vinyl monomer is added to the polyester resins, which is then solubilised and made into a paints. Inhibitors, e.g, hydroquinone, are added to prevent premature polymerisation in the can; organic peroxides or some other catalyst must be added to initiate polymerisation of the styrene monomer and the polyester resin for curing, which is often carried out at elevated temperatures.

Polyester finishes are very hard, tough, resistant to solvents and fairly heat-resistant. They are often used for furniture finishes. Adhesion of these paints in often poor.

### *Phenolic paints*

Phenolic resins as used in coatings are primarily made from phenol and *para*-substituted phenols reacted with formaldehyde to form methylol groups on the phenol ring.

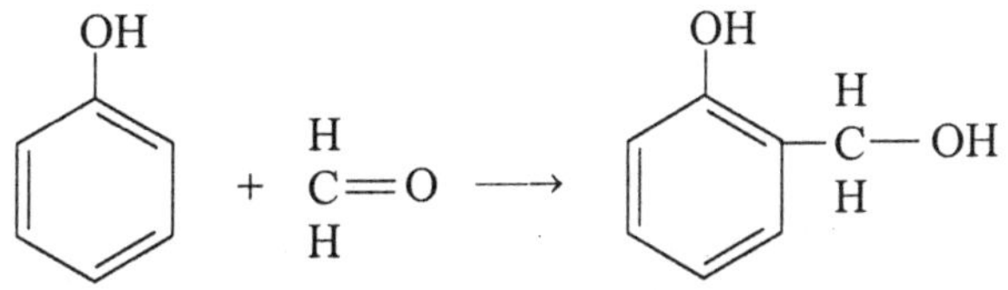


Phenol Formaldehyde Methylol phenol

Condensation polymers are then produced by reacting these groups with phenol. Phenolic coatings have fast dry, high build and good resistance to moisture and chemicals. Their poor initial colour and tendency to yellow after application limit their use. Phenolic coatings are often used for baked can coatings and oil-modified phenol-aldehyde finishes are often used for marine finishes and aluminium paints.

### *Polyurethane paints*

Polyurethanes are based upon reactions of isocyanates, RNCO. Urethane coatings have excellent solvent and chemical resistance abrasion resistance, hardness, flexibility, gloss and electrical properties. They are, however, rather expensive and the aromatic isocyanates yellow after applications.

Polyisocyanates such as toluene diisocyanate react with hydroxylated drying oils to produce resins analogous to alkyds and expoxy esters. Finishes made with these urethane oils air-dry by oxidation of the unsaturated oils. Polyhydroxy materials can be reacted with isocyanates with an excess of the isocyanate so that the polymer will contain NCO groups, which react with moisture in the air after the coating is applied to cross-link the polymer for curing. Moisture-cured urethanes are difficult to pigment since the pigments must be completely dry and nonalkaline. They are used primarily for clear coatings.

Phenol will react with an isocyanate to block the isocyanate or prevent it from reacting with hydroxylated materials in the system. When this type of coating is heated to 150°C after application, the phenol volatilises from the film, leaving the isocyanate free to react with the hydroxyl-bearing resin to cure the film. Two-package systems are also used where a catalyst, e.g. a tertiary amine, is added to cross-link the polymer.

### *Coumarone-Indene paints*

Coumarone-indene resins which are derived from coal tar are used widely to make aluminium paints since they aid leafing of the aluminium and minimise gas formation. They have a yellow colour, however and only fair durability except in aluminium paints.

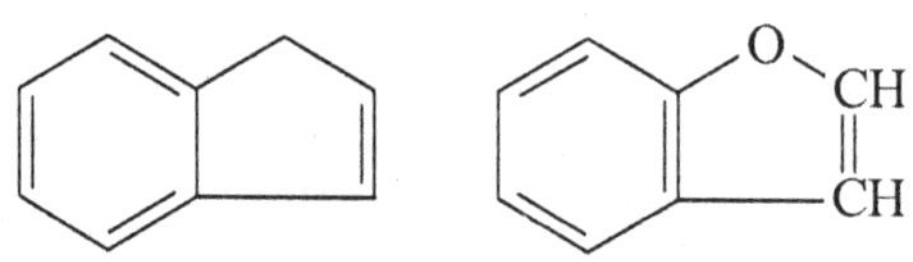


Indene Coumarone

### *Bitumen paints*

Bitumen resins are also used for aluminium paints as well as for maintenance and insulating coatings because of their low cost. They are dark in colour and have low solvent resistance and only fair durability except with aluminium.

### *Other materials for solvent paints*

Urea-formaldehyde, melamine-formaldehyde and other triazine-formaldehyde resins are all hard, glossy, colourless, brittle and chemically resistant thermosetting polymers used to modify basic coating vehicles, such as alkyds, acrylics and vinyls. Although rosin is sometimes used in making vehicles, such as maleic rosin alkyds, for use in consumer products or industrial finishes because of their moderate cost, it is being replaced gradually by other materials. Silicones sometimes are used to modify alkyds and polyesters for better exterior durability.

## Latex Paints

Rutile titanium dioxide is the primary pigment used in latex paints to obtain opacity, or hiding, in white or pastel paints. Semi chalking grades are used for interior paints, chalk-resistant grades for exterior paints and fine-particle-size grades for semigloss paints. Zinc oxide is sometimes used for exterior paints to help prevent mildew, but care must be taken in using this pigment because of its chemical reactivity.

A number of pigments, such as calcined clay and delaminated clay, can be used to advantage in latex paints as titanium dioxide extenders to increase opacity. These pigments have larger surface areas due to irregular surfaces or fine particle size and the latex vehicles will not cover all the pigment surface when the paint film is dry, leaving entrapped air in the film. The interfaces of air with pigment and vehicle increase the light refraction of the film and thus the opacity. Good hiding can be obtained with these pigments at a low cost, but the paint film often becomes porous and difficult to clean.

Most of the common extenders, e.g. mica, calcium carbonate, clay, talc, silica and wollastonite, can be used in latex paints. Since these pigments vary in particle size, shape, hardness, colour, surface treatment and water demand, they can affect viscosity, flow, gloss, colour, cleanability, scrubbability, enamel holdout, uniformity of appearance and even opacity to some extent. Extender pigments are selected to obtain the desired properties for each type of paint. Slightly soluble ammonium phosphate compounds are used as the primary pigment in intumescent fire-retardant paints.

*Surfactants* are used in latex paints to help wet and disperse pigments, emulsify liquids and function as defoamers. These materials have a balanced polar-nonpolar structure, which in water-base paints is usually referred to as a *hydrophile-lipophile balance*. The chemical composition of surfactants can vary greatly and they are usually only classified into anionic, cationic and nonionic types. Anionic surfactants (aryl alkyl sulphonates, sulphosuccinic acid esters, soaps, water-soluble amines and sulphonated oils) and nonionic surfactants (partial esters of polyhydric alcohols with long-chain carboxylic acids, long-chain alcohols with free hydroxyl groups and ethers of polyhydric alcohols with long-chain fatty alcohols) are used primarily in latex paints.

The more hydrophilic water-soluble surfactants are used to wet and disperse pigments. Surfactants with a lipophilic chain to dissolve in a polar liquid and hydrophilic group to dissolve in water are used to emulsify varnishes or oils in latex paints. Surfactants which are not water-soluble are used as defoamers.

Thickeners or protective colloids are used in latex paints to produce the desired viscosity and help stabilise emulsion and pigment dispersions. Water-soluble protein or casein dispersions and cellulosic polymers (carboxymethyl, hydroxyethyl and methyl cellulose) are the most commonly used. Soluble polyacrylates, starches, natural gums and inorganic colloidal materials have also been used.

Protective colloids can affect many properties of a paint, such as washability, brushability, rheological properties and colour acceptance. Since latex paints are susceptible to bacterial attack, they should contain preservatives. Several different types of preservatives can be used: phenolic, mercuric, arsenic, or copper compounds, formaldehyde and certain quaternary chlorinated compounds. Some of these compounds are chemically active and some are toxic.

### *Polymers used in latex paints*

Many of the polymers used in the plastics industry and in solution coatings previously described can also be obtained in latex form. An advantages of using polymers in this form is that high-molecular weight fully cured polymers can be made to flow well whereas in solution form they would have high solution viscosities.

Latex paints dry fast are easy to apply; equipment can be cleaned with water; and there is no fire hazard or air pollution during application. A latex paint does not lose its liquid gradually with a gradual increase in viscosity during drying as a solution paint does because of the inversion of the two-phase system when most of the water has evaporated. Sometimes this can produce application or flocculation problems. The tendency of latex polymers to coalesce when frozen and the fact that they seldom are made with more than 55 per cent solids also can be disadvantages.

The main types of latex polymers used in latex paints are styrene-butadiene, vinyl homo- or co-polymers and acrylic polymers or co-polymers.

### *Styrene-Butadiene latex*

This low-cost material can produce paints with good colour acceptance, low temperature coalescene, toughness, adhesion, cleanability and good sealing properties. The styrene-butadine co-polymer remains a little soft after the water has evaporated from the film, but the polymer contains some unsaturated groups which air-oxidise to cross-link the polymer and harden the film. This oxidation will cause the polymer to yellow slightly. Styrene-butadiene paint are used primarily for interior wall applications.

### *Vinyl latexes*

These materials are usually co-polymers of a vinyl monomer, e.g. vinyl acetate or vinyl chloride and some other monomer, e.g. ethylene or vinylidene chloride, acrylonitrile, methyl acrylate, methyl methacrylate, fumaric acid, maleic anhydride, or styrene.

The various vinyl latexes have different properties depending upon the monomers used, the particle size of the latex and the surfactant system. There are, however, a few general properties of vinyl latexes. They dry fast and are fully cured when the water evaporates from the paint; are nonyellowing; have good adhesion and good scrubbability; and may have good flow and leveling properties. Some vinyl latexes do not have very good freeze-thaw resistance, low-temperature coalescence, or colour acceptance. Vinyl co-polymers are usually plasticised with liquids such as dibutyl phthalate, butyl cellosolve acetate or tributyl phosphate. Vinyl latex paints are used primarily for interior and exterior architectural surfaces.

### *Acrylic latexes*

These materials are polymers or co-polymers which are composed primarily of monomers of the acrylic family, e.g. methyl methacrylate, butyl methacrylate, methyl acrylate and 2-ethyl hexylacrylate. Other monomers, e.g. styrene or acrylonitrile, can be polymerised with acrylic monomers.

Acrylic latexes vary greatly in their physical properties, according to the monomers used, the particle, size and the surfactant system of the latex. In general, acrylic latexes are cured by loss of water only, do not yellow, have good exterior durability, are tough and may have good abrasion resistance.

Acrylic polymers are fairly expensive and some latexes do not have very good colour compatibility. Acrylic latex paints can be used for concrete floors, interior flat and semigloss finishes and exterior surfaces.

### *Other latex materials*

Hydrocarbon resins or drying oils are also available in emulsion or latex form, but these materials are used primarily to modify the properties of the three major types of latex just described. Thermosetting or catalysed polymers can be made in latex form for industrial paints. Alkyd varnishes or drying oils sometimes are emulsified in water-base paints to modify the properties of the latex, e.g. plasticising it, or to add desired properties to the paint, e.g. adhesion to exterior chalky surfaces.

## Water-Soluble Paints

Water-soluble paint binders, such as egg albumin, gum arabic and casein, have been used for many years. Most of these materials have serious limitations, including water sensitivity and poor durability.

Many of the synthetic polymers now used in solvent or latex paints can be solubilised in water. Carboxylic, hydroxyl, epoxy, or amine groups on a polymer in conjuction with coupling solvents, such as alcohols, alcohol ethers, or glycol ethers, are the primary mechanisms by which resins are solubilised.

Maleic or fumaric acids can be reacted with drying oils to produce resins with some carboxy groups which can be solubilised with ammonia or amines. Alkyds can be solubilised by leaving a reactive carboxylic group on the resin instead of terminating the reaction with a monobasic acid or drying oil acid. Water soluble resins have been used for air-dry and low-bake industrial primers and finishes, coating applied by electro deposition and architectural semi-gloss paints. Contamination and mechanical and chemical stability can be a problem with these resins.

## RELATIONSHIP OF FATS AND OIL TO PAINT—COATING INDUSTRY

An organic liquid which, when applied as a thin film, readily absorbs oxygen from the air and polymerises to form a relatively tough, elastic film. The oxidation is catalysed by such metals as cobalt and manganese. Drying oils are usually natural products such as linseed, tung, perilla, soyabean, fish and dehydrated castor oils, but are also prepared by combination of natural oils or their fatty acids with various synthetic resins. The drying ability is due to the presence of unsaturated fatty acids, especially linoleic and linolenic, usually in the form of glycerides. The degree of unsaturation of an oil and hence its drying ability is expressed by its iodine number. The drying oils have the greatest capacity for iodine and the nondrying oils the least.

The drying oils owe their value as raw materials for decorative and protective coatings to their ability to polymerise and cross-link, or 'dry', after they have been applied to a surface, to form tough, adherent, impervious and abrasion-resistant films. Their film-forming properties are closely related to their degree of unsaturation, since it is through the unsaturated centres or double bonds that polymerisation and cross linking take place. With one exception (to be noted later), the oils used in paints, varnishes and similar products are relatively high in iodine value. In any given product, there is an optimum degree of reactivity in the oil; the speed with which the oil dries must be balanced against such factors as elasticity and durability in the paint film. In general, however, unsaturation is at a premium in paint and varnish and the oils in greatest demand are those in which drying takes place most readily.

The form in which drying oils are used in coating applications has gone through an evolutionary change over time. The simplest and most primitive way is to use them directly as the film former of a coating. In the case of fast-drying oils with conjugated double bonds, such as tung, oiticica and dehydrated castor oil, heat treatment is necessary to 'gas-proof' them, so that the oils do not dry into undesirable wrinkled and/or frosted films. The oils are not necessarily used in their original form of triglycerides for coating applications. It has become a common practice to hydrolyse them first and the free fatty acids are then used to synthesise coating resins with certain advantages. Therefore, the term fats and oils includes fatty acids for the purpose of the discussions in this chapter, unless otherwise specified. Through progress, people found that sometimes mixtures of different oils could be used to greater advantages and that natural gums could be added; thus oleoresinous varnishes were born. Thereafter, human creativity started to make rapid and diversed progress in the development of new coating materials many of which have departed completely away from the drying oil base.

When the drying characteristics of oils were relied on as the sole (or major) cause for a varnish-based coating film to dry, those oils belonging to the linolenic or conjugated acid groups, such as linseed, perilla, tung, oiticica and highly unsaturated winterised fish oils were of the prime interest to coating formulators. With the advent of synthetic resins, particularly, alkyd resins, it has become possible to make considerable use of oils with poorer drying characteristics. Semidrying oils such as soyabean oil, safflower oil and sunflower seed oil have become viable as raw materials for making 'drying' paints. Nondrying oils, such as coconut oil, are also used in coating materials. However, their function is primarily that of a plasticiser rather than of an active component for air drying. In addition, castor oil a nondrying oil, has been converted chemically by dehydration to give excellent drying property.

## FILM DRYING PROCESS OF OIL-BASED COATING MATERIALS

### Drying of a Nonconjugated System

As mentioned earlier, it is through the unsaturated centres or double bonds of the fatty acids in drying oils that polymerisation and cross-linking, i.e. drying, takes place. Hence, oils are conventionally classified, based on their iodine values into three groups: drying, semidrying and non-drying. The generally accepted demarcations are, respectively, > 140, 140–125 and < 125. These numbers are, more or less, arbitrarily assigned. When the oil contains conjugated double bonds, the iodine values determined are usually low due to incomplete halogen absorption. Such a classification can only be used for a rough guidance.

The chemical mechanism of drying has been established as an oxidative radical chain reaction process, which has been summarised as follows:

1. A period of induction at the beginning of the reaction during which no visible change in physical or chemical properties in the oil is noticed; natural antioxidant compounds are consumed during this period.
2. The reaction becomes perceptible and oxygen uptake is considerable; discrete interaction of oxygen and olefins takes place followed by the formation of hydroperoxides.
3. Conjugation of double bonds occurs accompanied by isomerisation of *cis* to *trans* unsaturation.
4. The hydroperoxides start to decompose to form a high free-radical concentration; the reaction becomes autocatalytic.
5. Polymerisation and scission reactions begin and yield high molecular weight cross-linked products and low molecular weight carbonyl and hydroxy compounds; carbon dioxide and water are also formed and are present in the volatile products of film formation.

It is now generally believed that the induction is slow at first but is autocatalytic and the rate increases steadily. The rate depends on the reaction conditions such as temperature, light and traces of heavy metals or inhibitors in the oil or coating.

Chemically, the air-drying of a nonconjugated oil such as linseed is characterised by the absorption of 12–16 per cent by weight of oxygen. The reactivity of drying oils is based on the mesomeric stabilisation of the radical intermediate: the unpaired electron is delocalised over several carbon atoms and less energy is required to eliminate the proton as illustrated below:

| *Triglycerides* | *Mesomers* | *Activation energy (kJ/mol)* | *Relative rate of oxidation* |
|---|---|---|---|
| Stearate | [a] | 415 | 0 |
| Oleate | 2 | 335 | 1 |
| Linoleate | 5 | 289 | 120 |
| Linolenate | 11 | 168 | 330 |

[a] Saturated molecules.

## LINSEED OIL

Linseed oil, the best-known member of the vegetable drying oil family, is obtained from the seed of the flax plant (*Linum usitatissimum*). Flax is also the source of linen. The same plant, however, cannot be used to produce both linen and oil, as that grown for fibre must be cut before the seed is ripe. The flax plant flourishes in a wide range of climates, the principal countries being the United States, Canada, Russia, India and Argentina.

In the United States, the legal weight of a bushel of flaxseed is 56 pounds. The percentage of oil in the seed ranges from 32 to 43, the average being equivalent to somewhat less than 2½ gallons of oil per bushel of seed.

### Separating the Oil

Two methods account for the production, these are hydraulic pressure and Anderson Expeller.

#### *Composition and properties*

The composition of the acids of linseed oil is:

| | | |
|---|---|---|
| Palmitic | $C_{15}H_{31}$, COOH | 8 to 11 |
| Stearic | $C_{17}H_{35}$, COOH | – |
| Oleic | $C_{17}H_{33}$, COOH | 5 to 19 |
| Linolic | $C_{17}H_{31}$, COOH | 23 to 63 |
| Linolenic | $C_{17}H_{29}$, COOH | 24 to 50 |

The amount of unsaponifiable matter in linseed oil is of the order of 1 to 2 per cent and consists of compounds of phosphorus, calcium and magnesium with non-oleaginous matter, probably carbohydrates (mucilage). If allowed to age, the mucilaginous matter will separate and along with foreign matter will settle as 'foots' to the bottom of the container. Heating the oil to 450° F or higher accelerates the separation and the oil is said to 'break'.

The characteristic property of the 'drying' oils is due to the high degree of unsaturation of their fatty acids. Oxygen from the air is absorbed at the double bonds giving rise to a sequence or concurrence of oxidation, polymerisation and association reactions that finally results in the conversion to a solid or gel. The role of each of the reactions is influenced by such factors as temperature, free fatty acids, light and the presence of certain metallic and other compounds. In fact, polymerisations may be brought about by heating the oil in a vacuum. The solid product is called linoxyn. Among other products are carbon dioxide, water and low molecular weight fatty acids and aldehydes. In this films, the conversion, called drying, requires a period of 48 hours or more, depending upon temperature, relative humidity and the amount of light. By the addition of small percentages of compounds of lead manganese or cobalt, the drying may require as short a time as 8 hours or less.

To a certain extent, the drying property of linseed oil is indicated by its iodine value. According to the specifications of the American Society for Testing Materials for Raw Linseed Oil, this should be not less than 177. The value depends upon the type of flax and the place where it is grown. At one time the average iodine value of North American seed was much higher than that of today's wiltproof and heavier bearing flax.

The amber colour of raw linseed oil varies according to its origin, degree of maturity of the seed and to the care used in separating it. Sunlight bleaches the oil and the yellow colour imparted by the oil to exterior house paints fades quickly after application of the paint. On the other hand, the colour is often intensified in diffused light and in the dark. Much of the yellowing is due to changes in the highly unsaturated acids such as linolenic. For example, soyabean oil, containing less of such acids, yellow sless, where as perilla oil containing more, yellows more. Suitable refining reduces the tendency to yellow.

Tremendous amounts of linseed oil are used in the raw unrefined state in paint and linoleum. Lesser amount are used in such products as brake linings, gaskets, fibre board and foundry cores. For many products, refined, boiled or thickened types are better. Among these products are certain types of paint, varnish, ink and patent leather.

### Refined Linseed Oil

Refining methods range from simple refrigeration and filtration to effect removal of the solid fats, foods and a small portion of the break, to more rigorous methods involving treatment with acids, alkalis and bleaching earths.

#### *Acid refined oil*

Lots of 500 to 1,000 gallons of oil contained in lead lined tanks are treated with strong sulphuric acid, 60° to 65° Be. About 1 per cent of the acid based on the weight of the oil is added in a fine stream to the oil which is vigorously agitated during the addition and for an hour or so afterward. The mucilaginous 'break' and much of the colouring matter are coagulated in the form of a muddy-greenish sludge which is allowed to settle over-night. The oil is drawn off and washed with water and steam until free from sulphuric acid. The water is allowed to settle and is then drawn off. In some cases, additional bleaching and clarification may be effected by means of fuller's earth.

Acid refining causes some hydrolysis of the oil, whereby the amount of free fatty acids is increased. The extent of the increase is controlled by the strength of the acid and the temperature during treatment. High acid oil wets pigments readily and is much used as a grinding oil in making paint.

#### *Alkali refined oil*

In preparing this oil, the acid value is reduced to 0.5 or less by neutralisation with aqueous 10 per cent caustic soda solution. The amount of caustic solution is estimated from the original acid value of the oil. The oil is first warmed slightly and the caustic solution is then added in fine streams to the vigorously agitated oil. After the precipitate has coagulated, it is allowed to settle. The oil is then drawn off into a tank and after being heated to drive off residual water, it is treated with bleaching clay. Alkali refined oil is also known as varnish maker's oil.

### Drying of a Conjugated System

Tung oil, whose dominant feature is the conjugated *cis, trans, trans*-9, 11, 13-octadecatrienoic acid, α-eleostearic acid, dries to a coherent film with absoption of only 5 per cent by weight of oxygen.

## VARNISHES

Varnishes are of two types–spirit and oleoresinous. A spirit varnish is a solution or dispersion of a resin or other film-forming substance in a volatile solvent. It dries by evaporation of the solvent. Certain synthetic types may require baking. Examples of spirit varnish are shellac and cellulose nitrate lacquer. An oleoresinous varnish is a solution of a mixture (usually cooked) of drying oils and resin in a volatile solvent. This varnish dries by evaporation of the volatile portion, followed by oxidation and polymerisation of the residue. Examples are spar varnish and floor varnish. Typical flow diagram for varnish manufacture is shown in Fig. 5.2.

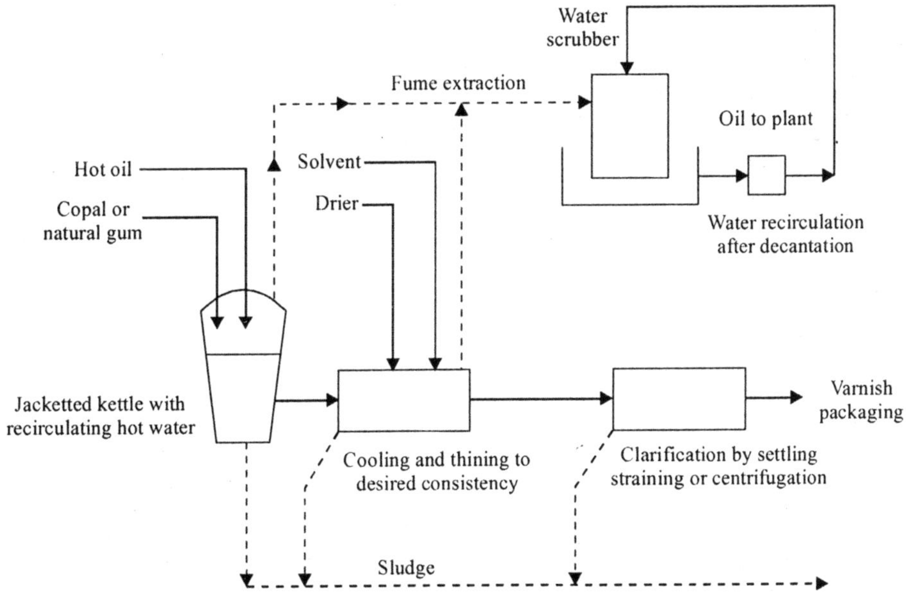


**Fig. 5.2.** Typical flow chart for varnish manufacture.

## Raw Materials

The basic raw materials used in the manufacture of varnishes include the many drying oils, dozens of resin and scores of solvents and driers.

### *Drying oils*

Practically all of the commercially available drying oils and some of the semidrying oils are used in the manufacture of varnishes, the chief ones being tung, linseed, perilla, oiticica, dehydrated castor, menhaden, sardine and soyabean oils.

### *Resins*

Until the third decade of the twentieth century, the natural products dominated the resin field, but since that time synthetic resins of many types have been developed and used in large amounts.

The principal natural resins entering into varnish manufacture are Kauri, Congo, Pontianak, East India, Batu, Manila and Damar. The first two are used almost exclusively in oleoresinous varnishes; the last two almost exclusively in spirit varnishes. In addition, there is rosin, widely used as such and also in the form of its glyceride, ester gum. This resin was the first important synthetic resin to be used in varnishes, for the coumarone resins. The next important synthetic resin was a blend of ester gum and phenol-aldehyde resin. Today synthetic resin for the varnish industry include straight oil-soluble phenol-aldehyde resins, alkyds, maleic acid compositions, chlorinated rubber, polymerised rubber, vinyl derivatives, penta-erythritol esters, terpene polymers, urea-aldehyde, melamine and others. The cellulosic derivatives—cellulose nitrate, cellulose acetate, cellulose acetobutyrate and ethyl cellulose—may also be included. Strictly speaking, some of the materials should not be classed as resins and for want of a better name they are classed as film-forming compounds or resinoids. The word 'filmogen' used in medicine is hereby suggested as a short word to replace 'film-forming compounds'.

### *Solvents and thinners*

The requirements of spirit varnishes for solvents and thinners are more varied than those of oleoresinous varnishes. Most of the needs of the latter are met by turpentine, petroleum spirits and coal tar naphta. Most natural spirit varnish resins are likewise dissolved by these liquids. However, other solvents are needed for lac and for the cellulosic and other spirit type filmogens. Ethyl alcohol is the solvent for shellac. Butyl acetate, amyl acetate, cellosolve, butyl alcohol and acetone are among those used in cellulosic lacquers. Numerous other solvents of lesser importance are used in small amounts.

### *Driers*

These are used only in oleoresinous varnishes. They include inorganic compounds of lead, manganese and cobalt, such as the oxides, acetates, carbonates and borates, which are converted to oil-soluble forms when cooked into the varnish. They include also compounds of lead, manganese and cobalt soluble in oil without further treatment such as linoleates, resinates and napthenates.

## Manufacture of Spirit Varnishes

This is a simple operation, although considerable knowledge is necessary to select the most suitable and economical ingredients and their proportions in certain types. This refers especially to the cellulosic lacquers and other synthetic filmogens. In the case of a simple product consisting of resin and solvent, the two ingredients are placed in a barrel or other container and are agitated by stirring or by other means until the resin is completely dispersed or 'cut'. Best results are obtained by adding the resin to the solvent so that the individual pieces of resin are wet completely.

In other cases, it is desirable to heat the batch to improve dispersion. For example, what is known as gloss oil is made by reacting rosin at approximately 400°F or higher with slaked lime, $Ca(OH)_2$. When the reaction is complete, the hot calcium resinate is thinned with petroleum spirits.

Natural resins, especially the lower grades, contain more or less foreign matter which must be removed by settling, filtration or centrifuging. The practice of designating the amount of resin in a spirit varnish as the number of pounds of resin dispersed in a gallon of solvent still prevails. A four-pound 'cut' of shellac is made by 'cutting' four pounds of dry shellac in one gallon of alcohol. However, since the presence of dirt and other insoluble matter makes it necessary to determine the actual amount of non-volatile matter, there is a growing tendency to express the amount of resin on a percentage basis.

Spirit varnishes constitute a large proportion of all varnishes made. They include shellac varnish for floors, natural resin coatings for coating paper, cellulosic automotive lacquers, a large amount of sheet metal coatings, based on alkyd, vinyl, urea and other resins.

Spirit varnishes made with synthetic resin are usually more complex than those made with such natural resins as shellac and Manila. Cellulose nitrate lacquer, which is discussed in a separate section, is an example. Take the case of the recently developed urea-formaldehyde and melamine filmogens. Baking is necessary to develop their excellent property of hardness, but this hardness must be tempered with plasticisers in order to impart satisfactory adhesion. Alkyds are widely used for this purpose.

Vinyl resin, chlorinated rubber and rubber hydrocarbons are other filmogens that must be modified by the addition of plasticisers or other resins before satisfactory coatings can be made with them.

## Manufacture of Oleoresinous Varnish

This is a process which still retains some of the aspects of an art. It consists in cooking resins and drying oils together until they are permanently miscible or until the viscosity has been raised to the desired value. The 'bodying' (thickening) of drying oils by heat or by a combination of the heat and air belongs to this field. The processing is done in both set and portable kettles, most of the oleoresinous combinations being processed in the latter.

Portable kettles are of various sizes and shapes, but the usual type has straight sides, is from three to four feet in diameter and from three to four feet deep and has a capacity of some 200 gallons. Early kettles were made of copper, but today aluminum, stainless steel and Monel metal are extensively used. The kettle is mounted on a low truck by which it can be moved as needed. A thermometer for indicating the temperature and a long metal paddle for stirring are standard accessories.

The source of heat is usually a coke, oil or gas fire, located in a pit below the floor level where the kettle truck can straddle it. Each fire is provided with a flue or stack for carrying off the products of combustion and the fumes from the kettle. In some localities where civic regulations forbid the discharge of fumes into the air, disposal systems must be provided.

### *Cooking the varnish*

The desired amounts of resin and oil are weighed or measured into the kettle, which is then rolled over the fire pit. The heat is applied, usually gently at first, until the resin is melted and then more intensely to bring the batch to the desired temperature. It is usually necessary to pull the kettle off the fire before the top heat is reached, as the exothermic reaction in the varnish releases heat for the last 10 to 25 degrees rise. At the top temperature or some lower point, the batch is held until permanent dispersion of the resin in the oil is assured or until the batch has been thickened (bodied) to the desired degree. The dispersion is tested by noting the clarity of a cooled drop. The body is gaged by the nature of the drip from the paddle, or by the length of the 'string' drawn away by a finger applied to a cooled drop. A more scientific method is to withdraw samples, thin them as required and then determine the viscosity with an apparatus such as the Gardner Holdt Bubble Tubes. Another method is to note the time required for a drop of the batch to form a gel when placed on a hot plate. However, there often is not enough time available for these tests and the skill and art of the practical experienced varnish cooker is still needed.

When the varnish has been cooked sufficiently, it is pulled to a safe distance from the fire, where it is allowed to cool naturally, or it may be cooled by water directed on the outside of the kettle or even on the hot varnish itself. When the temperature has been reduced to 475°F or less, the thinner is added cautiously and is thoroughly mixed with the varnish base. The thinned varnish is then clarified by being passed through a filter press or a centrifuge. Finally it is pumped to storage tanks to await packaging.

Driers may be added at some stage of the cooking or may be reserved until the varnish is thinned or cooled. If the varnish has a tendency to form a skin over the surface, the driers may not be added until just before packaging. The commonly used compounds are litharge, basic carbonate of lead, lead acetate, cobalt acetate, manganese acetate, the metallic soaps and the napththenates. In order to inhibit the skinning of some types of varnishes in partially emptied containers in the user's hands, antiskining agents are often added. These are often phenolic compounds such as guaiacol, catechol, or their derivatives.

The harder types of fossil resins are not miscible with drying oils and must be 'run' before they can be used in oleoresinous vanish. This 'running' is a severe heat treatment, resulting in a degradation of highly polymerised molecules into simpler ones. A moderate amount of fossil resins is run by the resin manufacturer but probably most of it is run by the varnish manufacturer as he uses it.

The resin is put into a varnish kettle and heated quickly to a temperature of 600°F or higher. Voluminous foaming takes place and copious fumes (amounting to about 20 to 30 per cent of the resin) are given off. The running is complete when the foam has subsided and the melt is free from lumps. In some cases, a small percentage of low-melting-point resin, such as rosin, or some of the oil added to the batch promotes the running. Some effort has been made to use the fumes as a source of chemicals or solvents, but in the great majority of cases they are discharged into the air, unless civic regulations forbid it. While the resin is being run, the oil portion of the varnish is being heated in another kettle and at the proper time is added to the resin. From this point, the procedure is the same as for synthetic resin or recent resin varnishes.

### *Indirect heating*

The intense heat on bottom-fired kettles superheates the layer of varnish next to the bottom and causes considerable decompostion of the varnish. One way of avoiding this condition is to heat with electricity. The heaters may be the immersion type or may be elements placed against the outside of the kettle. In both cases the kettle is thoroughly lagged. The temperature is controlled by a thermostat. By taking advantage of non-peak load rates, the cost of electric heating becomes fairly attractive.

Another method is to circulate hot vapours of compounds such as diphenyl (Dowtherm) around jacketed kettles. Kettles holding as much as 1500 gallons have been installed. Coils for circulating cold diphenyl are provided in case cooling is necessary. Electrically and vapour-heated kettles are usually equipped for vacuum or inert atmospheres above the liquid and fume disposal systems. However, fume losses are low becasue decomposition of the varnish is negligible. Very pale products of low acid value are obtained. Moreover, the kettles are equipped with condensers so that losses accompanying thining at high temperatures are negligible. On account of the initial expense, manufacturers have been slow to adopt these methods and by far most oleoresinous varnishes are still cooked over direct fires.

### *Storing the varnish*

Varnish storage tanks are located inside buildings where they can be kept warm in the winter. In some installations, an atmosphere of inert gas is kept over the varnish to prevent skinning or thickening due to oxidation. This gas may be carbon dioxide or, as is more often the case, it is flue gas produced from special burners.

### *Reactions in the varnish kettle*

The chemical reactions occurring in drying oils during varnish cooking are still somewhat obscure, but belong mainly to the polymerisation type, the reactions taking place at the double bonds of the fatty acid radicals. The course of the reactions is altered by the presence of resins and driers and by the oxidation which takes place at the double bonds of the fatty acid radicals. The course of the reaction is altered by the presence of resin and driers and by the oxidation which takes place at the surface of the hot varnish. There is some evidence of interchange of acid radicals between the oil and resin molecules of the glyceride or acid types. This interesterification occurs more readily before the acid groups become locked up in polymerised molecules. At any rate, in most cases better miscibility usually results if the resin and oil are cooked and bodied together than if the resin is added to the separately bodied oil.

Tung oil reacts in an interesting manner with ceratin phenolic resins of the heat-advancing type. One of the characteristics of raw tung oil is its tendency to form a wrinked film when it dries. This tendency is reduced by heat treatment at temperatures of the order of 550° to 600°F, but during this heat treatment the oil rapidly thickens and the treatment must be stopped before the oil is converted to a gel. However, mixtures of tung oil and certain heat advancing phenolic resins may be cooked at much lower temperatures (450°F) to yield varnishes that do not wrinkle. Oiticica oil resembles tung oil in its varnish-making properties, but does not equal it in resistance to water and alkali. Dehydrated castor oil is obtained by heating castor oil with a catalyst, where by the hydroxyl group attached to the twelfth carbon atom and hydrogen from an adjacent carbon atom are evolved in the form of water. A doubly conjugated group is thereby formed. The resulting product has some of the properties of tung oil.

Perilla oil resembles linseed oil, but bodies more rapidly when heated or blown with air. It is dried faster and harder than linseed oil. Soyabean oil bodies vary slowly in the varnish kettle and if cooking is continued until the usual body is reached, losses by volatilisation are likely to be high. Soyabean oil varnishes tend to dry more slowly than most other types, but if sufficient time is available before adverse conditions are encountered, excellent service may be expected.

Sardine oil is another oil that tends to give relatively soft, slow-drying varnishes, but, as in the case of soyabean oil, the varnishes are very serviceable.

## Types of Varnish

The proportions of oil and resin in an oleoresinous varnish are usually expressed as the number of gallons of oil per 100 pounds of resin. Thus a 10 gallon varnish is made with 10 gallons of oil and 100 pounds of resin. If the proportion of oil to resin is, say, 12 gallons or under, the varnish is classed as a 'short' oil type. A 'long' oil varnish is one containing 25 or more gallons of oil to 100 pounds of resin.

### *Spar varnish*

On account of its suitability for coating the spars, masts and rails of ships, long oil water-resisting varnishes are called spar varnishes. Early types were made with fossil resins and linseed oil, but today straight or modified phenolic resins and tung oil are widely used.

### *Interior varnish*

These are usually of medium oil length and do not have the elasticity requirements of long oil varnishes when exposed out doors.

### *Furniture varnish*

These are short in oil and are made with a wide variety of resins.

### *Baking varnishes*

These are not readily classified as to oil length or type of resin. They are hardened by baking and the drier content is greatly reduced or eliminated entirely. Baking temperatures range from 150°F upwards to over 300°F. If artificial heat below 150°F is used, the operation is called "forced drying" rather than baking.

## Oleoresinous Varnishes

Coating systems were advanced from oily vehicles to oleoresinous varnishes for improved performances. These are basically oils that have been 'hardened' or modified by treatment with one or more suitable resins, natural or synthetic.

The oils and resins are combined, usually by heating together at temperatures of 250°C or above, until a homogenous mixture is formed.

In most cases, it is simply a case of dissolving the resins or resin in the oil. In some cases, chemical reaction may have taken place between the resins and the oil such as that between the methylol groups of a 'heat-reactive' phenolic resins and the double bonds of a drying oil in forming a chroman ring structure as shown below:

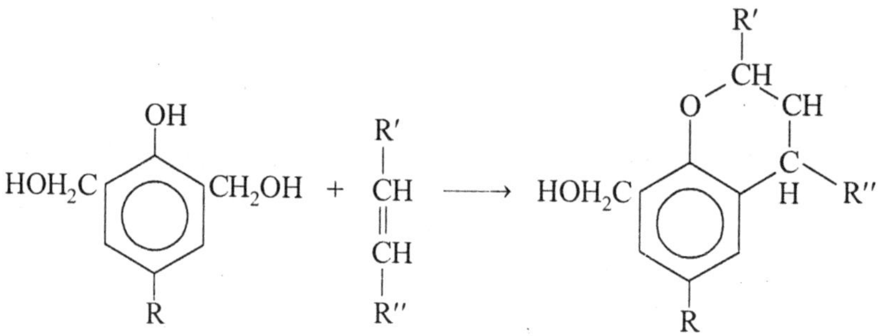

Chroman ring structure

The major improvements obtained by incorporating resins into drying oils are faster drying, greater film hardness, higher gloss, better water and chemical resistance and greater durability. The degree of properly change depends on the type and the amount of the resins incorporated in the oil.

From oleoresinous varnishes, the coatings industry progressed into alkyd resins. While one might say that this was only an evolutionary change, it nevertheless did open a new horizon for coating technologists and has been responsible for the longevity of oil-based coating materials. It behooves us to take a more comprehensive look at the various aspects of alkyd resins.

## ALKYD RESINS

Alkyd resins are thermosetting hydroxycarboxylic resins, produced by the esterification of a polybasic acid with a polyhydric alcohol. A widely used member of this series is prepared from phthalic anhydride and glycerol. Phthalic acid imparts hardness and stability. Alkyd resins made with maleic acid have a higher melting point. A less brittle and softer resin is obtained with azelaic acid. Adipic acid and other long-chain dibasic acids yield resins of considerable toughness and stability. Pentaerythritol or glycerol may be used instead of glycerol. Advantages of alkyd resins include high adhesion to metals, ease of colouration, transparency, toughness, heat and chemical resistance and good dielectric strength. The alkyds are easily blended and consequently are used widely in paints and finishes. The alkyds may be reacted with fatty acids, oils and other resins, such as melamine and urea, to provide them with the compatibility required by drying oil. High impact alkyd compounds, which have a high glass content, find applications in switchgear, electrical terminals, relay housings and electronic encapsulations. Mineral-filled grades are applicable to automotive ignition parts, small switch housings and electronic insulation. The properties of alkyd resins are summarised in Table 5.1.

Alkyd resins can be considered as polyester resins that have been plasticised from within by fatty acids. Common plastification rates vary from 25 to 75 per cent.

Depending on the application required, fatty acids can be saturated or unsaturated and the most common types available range from $C_{14}$ to $C_{18}$.

**Table 5.1.** Properties of alkyd resins.

Specific gravity: 1.20–2.30, depending upon type and amount of filler used.

Bonding: After degreasing or sandblasting moulded alkyd parts can be bonded with epoxy bonding resins. Common cements and glues are not effective.

Moisture absorption: Low at room temperature, an advantage for electronic parts. Resins with electrical-grade fillers may absorb as low as 0.05 per cent moisture.

Colourability: Usual colour produced are white, blue, green, medium tan, red and gray. A strong pigment, such as titanium dioxide, makes opaque light shades possible.

Types of moulding compounds:

Fibrous, compounded with long glass fibres (~0.5 inch) to impart strength

Rope (extruded)

Granular, compounds with short (1/16 inch) glass, cellulose, or absestors fibres.

| | *Impact strength ft-lb/inch of notch* | *Tensile strength psi* |
|---|---|---|
| Fibrous | 8–10 | 6000–7000 |
| Rope | 1–4 | 4000–9000 |
| Granular | 0.3–0.5 | 3000–5000 |
| | *Dielectric constant* | *Continuous max. service temp.* |
| Mineral filler | 5.5–6.0 | 135°C, (275°F) |
| Mineral and cellulose filler | 5.8–6.5 | 121°C, (250°F) |
| Glass filler | 5.2–6.0 | 149°C, (300°F) |

Design: Moulded parts can be held to dimensional tolerances of ± 0.001 inch/inch. Postmould shrinkage usually is small. Sectional thicknesses less than 0.040 inch are not recommended.

## Production Principles

High number hydroxylated polyester resins are esterified by fatty acids. This process is commonly known as the fatty acids process. It is the simplest. It generally consists of only one stage. The basic polymer is produced *in situ*. Esterification is then carried out by heating polyalcohol, polyacids and fatty acids to temperatures at which they react. With oils an operation in two distinct stages is necessary as triglycerides is chemical compound that do not have reactive site for forming polyester chains.

The first stage consists of producing a functional diol like a monoglyceride. The second stage consists of polycondensing a polyacid of minimum functionality 2, to form the macromolecular chain.

An example applied to glycerine:

Triglyceride + 2 glycerines = 3 monoglycerides

This reaction takes place at 205–210°C with polyols such as glycerine, trimethylolpropane, diethylglycol, diethanolamine and at 248–250°C with polyols such as pentaerythritol.

The catalyst used can be soda, potassium, lime or calcium salts, lithia or lithium salts, etc.

The reaction takes place very fast and produces an unaltered mixture of monoglycerides, diglycerides and triglycerides, more and more complex as polyols used can be diols, triols or tetrols. The proportions in which they combined cannot be analysed by stoichiometry as this is governed by the relative proportion of oil and polyol in the final alkyd resin.

The reaction is verified by observing the solubility in ethanol or methanol in the proportion 1/1, 2/1 or 3/1 depending on the base polyol. In the second stage this polyol of general functionality is polycondensed with a polyacid to form the base chain. Polycondensation takes place from 180°C onwards and depending on the phthalic anhydride concentrations can go up to 220–240°C.

Isophatic acid reacts at higher temperatures: 230–270°C. Tetraphatalic acid will not be used according to this process and it is only through means acidolysis or methyl esters that this acid will be combined.

During polycondensation water is released which can be removed either naturally with the help of a condenser or through a vacuum to accelerate the reaction or by the formation of an azeotropic solution/ solvent with the help of an aliphatic or aromatic hydrocarbon.

The advantage of this procedure is that the presence of solvent ensures a regular temperature during reaction as well as a permanent removal of all polyacid sublimations in the reactor; finally since it creates an oxygen-free atmosphere saturated with vapour, polymers obtained have an extremely light colour.

The reaction is analytically monitored by regular measurement of the acid value which decreases with progress in condensation and the viscosity which increases gradually with change in molecular weight of the formed polymer. When the conditions in which the reaction takes place are good.

Log viscosity = F (time) and Log (Acid value) = F (time) curves obtained are straight lines, which enable a good characterisation of polymers to be obtained.

### *Composition and nature of components*

Various types of oils and fatty acids are used. Based on their main characteristics, however, they can be divided into two broad categories, with driers and semi-driers in the first one and non-driers in the second one. The first group gives system for paints drying in air, for forced drying at 60–80°C or for oven reticulation. The second group gives plasticisers or paints for oven reticulation only.

The most commonly used oils or fatty acids are given in Table 5.2.

**Table 5.2.** Most commonly used oils and fatty acids.

| *Oil* | *Fatty acid* |
|---|---|
| Linseed oil | – |
| Soyabean | Soyabean fatty acid |
| Sunflower oil | – |
| Corn oil | – |
| Grapeseed oil | – |
| Safflower oil | – |
| Rapeseed oil | – |
| Peanut oil | – |
| Cottonseed oil | – |
| Coconut oil | Coconut fatty acid |
| Castor oil | – |
| Dehydrated castor oil | Dehydrated castor fatty acid |
| Hydrogenated castor oil | 12-Hydroxystearic acid |
| | Tall oil fatty acid |

Using fatty acids makes it possible to use the desired polyol, a simple production process and a composition which corresponds to the objective. Use of oils leads to larger quantities of glycerine but results in more economical products.

Polyacids must have a minimum functionality of 2 to create the macro-molecular chain, in the chain: 1 diacid, 1 diol, 1 diacid. The most commonly used polyacid is orthophatalic anhydrides; isophtalic and terephtalic isomers are also used in the production of inks or varnishes that require high thermal resistance.

### *Properties of alkyd resins*

Alkyd resins attain a molecular weight that is much higher than the same oils more or less polymerised or co-polymerised with vinylic monomers; this gives perfectly adapted drying periods which can be adjusted at will (Table 5.3).

**Table 5.3.** Properties of alkyd resins.

| *Description* | *Oil* | *Stand oil* | *Modified oil* | *Alkyd (65% oil)* | *Alkyd (25% oil)* |
|---|---|---|---|---|---|
| ASTM drying, hr. | 36 | 24 | 12 | 6 | 2 |
| Persoz hardness at 100, 24 hr. in seconds | – | – | 5 | 50 | 100 |
| Characteristics molecular weight | 980 | 10,000 | 15,000 | 30,000 | 50,000 |

Stock oils offer complementary properties which are given in Table 5.4.

**Table 5.4.** Properties of stock oils.

| | |
|---|---|
| Linseed oil | Air drying and hardness<br>Shining<br>Yellowing |
| Sunflower oil | Air drying<br>Does not yellow |
| Rapeseed oil | Stable when oxidised<br>Heavy plasticiser |
| Castor oil | Polyfunctional alkyd (OH)<br>Cross-linking with etherified melamine |
| Hydrogenated castor oil | Polyfunctional alkyd for cross-linking with polyisocyanate |

### *Analytical characteristics*

Like all polymers, condensation polymers are characterised by:

1. Terminal groups.
2. Molecular characteristics (number and weight).

Acid and hydroxyl values are two very important features of terminal groups. Molecular mass or its distribution can be characterised by GLC but also by viscosity at different concentrations in a solvent.

Finally, as we are dealing with a group of ester bonds, the saponification value enables definition of the formed polymer. A more detailed gas chromatography analysis reveals the composition of methyl esters, which will give carboxylic acids and acetates, in turn will give polyols.

## BINDERS

The protective properties of a coating are determined primarily by the binder. In the early days of paint technology, binders were limited to materials of natural origin, such as drying oils, congo resins and asphalts. These still find some usage in the protective-coating industry.

Coatings in liquid form can be divided into two types, *solutions* and *dispersions.* In solution systems the resin binder is dissolved in a solvent. In dispersion systems the resin is in the form of tiny spheres (usually 10 microns or less in size) suspended in a volatile liquid carrier. If the liquid in a dispersion system is water, the system is called an emulsion; if it is an organic material, it is called an organosol. When the liquid evaporates, a mixture of soft resin and pigment is left behind and fuses into a continuous film. The general types of resins used are listed below. Mixtures of two or more types may sometimes be used to improve certain properties.

1. *Oils:* easy application; soluble in aliphatic solvents.
2. *Alkyds:* all purpose; combined with other resins; most are soluble in aliphatic solvents.
3. *Cellulosics:* (nitrate and acetate)–used in lacquers; fast dry.
4. *Acrylics*: good colour and durability.
5. *Vinyls:* good durability; abrasion resistant.
6. *Phenolics:* good chemical resistance; yellow colour.
7. *Epoxies:* good chemical resistance.
8. *Polyurethanes:* good flexibility; abrasion resistance.
9. *Silicones:* good heat resistance.
10. *Amino resins:* (ureas and melamines)–blended with alkyds for baking finishes; tough, good colour.
11. *Styrene-butadiene*: low cost; alkali resistant.
12. *Polyvinyl acetates*: low cost; good colour retention.
13. *Acrylics*: good colour and durability.

## Binding Materials for Surface Coatings

### *Pure oils*

For the last thirty years, use of vegetable oils as binding agents has been declining. There is an increasing demand, however, for oil paints and brayed and distempered white paints. There is also a positive upward trend in the sales of coloured paints, fine paints for artists and fat-based inks for letter press and offset printing.

The mechanical performance of these products cannot, however, compare favourably with that of alkyd polymers. They also take much longer to dry. The single most commonly used oil as binding agent in oil paints is linseed oil, due to its quick drying properties. Semi-drying oils such as soyabean, safflower, sunflower and grapeseed oils never form the base of oil paints. They are used as additives in paints based on alkyd resins to facilitate application ('brushability'). They also improve the flexibility of the dry film and increase the dry-matter content of the film.

Linseed oil is also an ideal vehicle for insecticides and fungicides used in treating wood. It penetrates deep into wood, the chemicals it contains are therefore properly absorbed and protect the substrates against insects and funguses. Apart from hydrophobic properties that arise from its fat content triglyceride is an excellent solvent of active material which spreads evenly throughput wood.

Elimination of solvents in building paints has been a way to protect, man and environment. Vegetable drying oils (linseed) or semi drying oils (sunflower, safflower) are used as a kind of substitutes for part of the solvent. Thanks to their own solvent properties and their low viscosity, they form part of the binding materials as they contribute to the formation of the film.

Siccative and semi drying vegetable oils are a kind of reactive solvents for building paints. Building paints have evolved to 'high solid content' systems where the solvent ratio must be as low as possible. This is easily ensured by using semi-siccative vegetable oils as solvents and as binding material in combination with alkyd polymers.

### *Polymerised or modified cooked oils*

Polymerisation of oils with or without metallic salts or co-polymerisation of the same with vinyl monomers are means of 'fattening' triglyceride that do not attain a molecular weight of 1000.

Depending on their polymerisation rates, molecular masses thus obtained can weigh up to 10–12,000. This represents a considerable change in their properties, but they are substantially weaker in comparison with alkyd resins which attain 30–40,000 or with vinyl polymers which go well beyond 1,000,000 when dispersed in water.

Like raw vegetable oils these products are not often used as binding material in building paints. On the contrary, research has shown the need to more towards products that have a low solvent content. It promotes the use of these products designed above which are not very polymerised and have 'high solid content'. The building paint industry is perhaps a good outlet for this range of products.

## MODIFICATION OF ALKYD RESINS BY BLENDING WITH OTHER POLYMERS

### Modified Alkyds

Alkyd or polyester resins modified by fatty acids are polymers with several reaction sites that favour the formation of new polymers. The lateral hydroxyl group, with the help of an isocyanate, forms polyurethanes which feature chemical resistance to abrasion, hardness and fat drying.

The same hydroxyl group can be esterified by carboxylic acrylic polymer and thus form a modified alkyd: acrylic, vinyltoluene, styrene, etc.

Transesterified by methylols of silicone polymers, a modified alkyd-silicone chain is formed with their combined hardness characteristics and the drying properties of the alkyd.

When esterified with a tri- or tetrafunctional acid, the polymer obtained consists of a number of free carboxyl groups and forms a polyelectrolyte soluble in water by salification. The presence of ethylene bonds on fatty acid allows fixing of a group of chemical compounds with different properties: water-soluble acid groups with maleic anhydride, acrylic styrene with modified vinyl monomer modifications.

As mentioned earlier, one of the important attributes of alkyds is their good compatibility with a wide variety of other coating polymers. This good compatibility comes from the relatively low molecular weight of the alkyds and the fact that the resin structure contains, on the one hand, relatively polar and aromatic backbone and on the other hand, many aliphatic side chains with low polarity. The alkyd resins involved in a blend with another coating polymer may serve as a modifier for the other film former, or it may be the major film former and the other polymer may serve as the modifier for the alkyd to enhance certain properties. The following describes some of these compatible blends.

Nitrocellulose-based lacquers often contain a fair amount of short- or medium-oil akyds to improve flexibility and adhesion. The most commonly used are short-oil non-drying alkyds. Amino resins or urethane resins with residual isocyanate functional groups may be added to cross-link the coating film for improved solvent and chemical resistance. The major applications are furniture coating, top lacquers for printed paper and automotive refinishing primers.

Amino resins are probably the most important modifiers for alkyd resins. Butylated urea– or melamine–formaldehyde resins are compatible with alkyds. They react with the free hydroxyl groups of the alkyd to effect cross-linking and to impart hardness, mar resistance, chemical resistance and durability to the coating. Short- or medium-oil alkyds of both drying and non-drying types are frequently used. Colour and colour retention requirements often dominate the choice of the alkyd. Many industrial baking enamels, such as those for appliances, coil coatings and automotive finishes (especially refinishing enamels), are based on alkyd–amino resins blends. Some of the so-called catalysed lacquers for finishing wood substrate requires low bake or no bake at all.

Chlorinated rubber is often used in combination with medium-oil drying type alkyds. The alkyd gives better toughness, flexibility, adhesion and durability and the chlorinated rubber contributes to faster dry and better resistance to water and chemicals. The major applications are highway traffic paint, concrete floor and swimming pool paints.

Vinyl resins of the type that are the co-polymers of vinyl chloride and vinyl acetate and that contain a fair amount of hydroxyl groups (from the partial hydrolysis of vinyl acetate) and/or carboxyl groups (e.g. from co-polymerised maleic anhydride) may be formulated with alkyd resins to improve application properties and adhesion. The blends are primarily used in making marine topcoat paints.

Synthetic latex house paints sometimes contain emulsified long-oil or very long oil drying alkyds to improve adhesion to chalky painted surfaces.

Silicone resins with high phenyl contents may be used with medium- or short-oil alkyds as blends in air-dried or baked coatings to improve heat and weather resistance, whereas the alkyd component contributes to adhesion and flexibility. Major applications include insulation varnishes, heat-resistant paints and marine coatings.

## Chemically Modified Alkyd Resins

While blending with other coating resins provides a variety of ways to improve the performance of alkyds, or vice verse, chemically combining the desired modifier into the alkyd structure would eliminate the compatibility problem and give a more uniform product. Several such chemical modification of the alkyd resins have gained commercial importance and are discussed below.

Vinylated alkyds are alkyd resins that have been incorporated with a significant amount (20–60 per cent by weight) of vinyl monomers (such as styrene, vinyl toluene and methyl methacrylate) by grafting the monomers through a free-radical mechanism onto unsaturated reaction sites in the resin molecules. The modified resins embodies the good attributes of ease of application, good wetting and adhesion from the alkyd as well as fast solvent release, hardness and weather resistance from the vinyl modification. The common objective of such a modification is for achieving a drying rate comparable to that of lacquers materials. The reaction sites on alkyd resin molecules are primarily the allylic carbons on unsaturated fatty acids chains and the double bond of α, β-unsaturated dibasic acids. Free radicals, generated from the thermolysis of such free-radical initiators as benzoyl peroxide, dicumyl peroxide and *di-t*-butyl peroxide are usually required to kick off the reaction. Ideally, the initiating species would attack the active sites on the resin molecules and all of the added monomers would be evenly distributed in grafted side chains. In reality, it is inevitable that part of the monomers would engage in homopolymerisation and some of the resins molecules would remain unmodified. The presence of a large amount and high molecular weight homopolymer of the vinyl monomer would lead to incompatibility and result in a hazy product. Methods that have been used to minimise homopolymerisation include using fatty acids having conjugated diene structure, using maleic anhydride as part of the dibasic acids for the alkyd, choosing initiators such as peroxides or hydroperoxides that tend more to extract allylic protons than to add to double bonds, avoiding initiators that would decompose at very low temperatures, adding the monomer gradually along with an appropriate amount of the initiator, choosing monomers that would have a more favourable tendency to co-polymerise with the active species on the alkyd resins and properly maintaining the reaction temperature. Chain transferring is the preferred mechanism for terminating chain growth from the addition polymerisation of the monomer. Usually, the solvent, the fatty acid chain and the monomer are effective chain-transfer agents. If an additional transferring agent is used, care must be exercised, or too much of it could cause the formation of a larger amount of very low molecular weight homopolymers and would result in poor film properties. Occasionally, vinylation is first performed on the fatty acids or the oil before the alkyd reactions.

It should be emphasised that the presence of a large amount of either conjugated fatty acids or maleic anhydride in the alkyd formulation gives rise to a high degree of probability of premature gelation during the alkyd reaction. An allowance must be made in the alkyd formulation and the polyesterification is frequently terminated at a relatively high acid number (about 15), to avoid gelation. It has been reported that the optimum amount of maleic anhydride in the alkyd is an amount having a maleic group in one-third of the resin molecules.

A common procedure for the preparation of vinylated alkyds is as follows:

First, a base alkyd resin is brought to the desired end point. The resin is then cooled to about 160°C and often diluted with aromatic thinner. Next, the desired monomers is added, usually at about 20–60 per cent based on the final product, followed by an appropriate amount of a free-radical initiator. Alternatively, a premix of the monomer and the initiator is added at a controlled rate over most of the reaction. Then the reaction is brought to monomer reflux, until the residual monomer content has dropped below a specified level. The residual monomer, if any, is stripped away before the product is diluted in a solvent, filtered and packaged.

Silicone alkyds are etherification products of alkoxy-polysiloxane oligomers and the free hydroxyl groups of alkyd resins. The property improvements and applications are similar to those of the alkyd–silicone blends, with the added advantage of incorporating the stable –Si–O–C– structure into the alkyd molecules. The preferred silicone oligomers are those with high phenyl contents and the alkyds at long-or medium-oil length based on polyols with primary hydroxyls. To improve the thermal stability effect imparted by the silicone modification further, one can use isophathalic alkyds rather than the phthalic type. The etherification reaction may be carried out on an alkyd resins designed for the purpose or on the polyol before it is used in alkyd preparation. The silicone content of the modified alkyd lies usually between 20 and 60 per cent of the total product.

Urethane alkyds, or uralkyds, are alkyds with a part or even all of the dibasic acids replaced by diisocyanates. The isocyanate group, –N=C=O, reacts with the hydroxyl group of a polyol, at low temperature, to form a urethane linkage, –NHC(O)O–, without spitting out water as a by-product. Toluene diisocyanate (TDI) is commonly used for such modification. It is commercially supplied as an 80:20 mixture of 2,4- and 2,6-isomers. The –NCO group *para* to the methyl group has about 8 times greater reactivity than the one on the *ortho* position, which aids greatly the control of the reaction. Since, the NCO groups is reactive with labile protons, water must be excluded from the reaction system.

The esterification reaction of the base alkyd must be brought to the desired end point with the by-product water removed and the temperature lowered to about 100°C to prevent any continuation of the esterification before the introduction of the NCO reactant. TDI is highly toxic and is usually handled in a closed system under a dry inert gas blanket. Metallic soaps, such as dibutyltin dilaturate, stannous octoate and calcium naphthanate, are used as reaction catalysts. The reaction is vigorous and exothermic. Therefore, the reaction temperature is maintained under 135°C and great care must be exercised to bring the reaction under proper control.

Uralkyds have superior adhesion, hardness, abrasion resistance, durability and chemical resistance to the unmodified alkyds. They find major applications in wood floor finishes, marine coatings, metal primers and maintenance paints.

Phenolic resins are well known for their contribution in improving hardness, gloss and water and chemical resistance in oleoresinous varnishes. Those based on *p*-alkyl-substituted phenols and with heat-reactive methylol groups have also been incorporated into alkyd resins. The reaction has not been well studied. Presumably, the methylol group would react with the unsaturation functionality on the fatty acid chain to form the chroman structure, similar to what is believed to have occurred in the varnish. Etherification between the methylol group and free hydroxyl of the alkyd resins, catalysed by the residual acidity in the resin, would be another possible reaction.

Polyamide modified alkyds show a special rheological behaviour—they are thixotropic. Typically, the polyamide resin would be of the type based on dimer acids, i.e. dimerised unsaturated fatty acids and aliphatic diamines, such as ethylene diamine. These would react to form polyamide resins with low acid and amine values. The alkyd resins would be a medium-or long-oil drying alkyd. The reaction products from the polyamide and the alkyd are gel like materials that undergo a time-dependent shear thinning and recover to the gel like stage after the shearing action is stopped. This allows the preparation of 'no-drip' paints, which are easy to brush and can be applied at high film thickness from a single coat with little or no danger of sagging. Pigment settling during storage of the paint is also minimised.

The major applications are flat oil-based architectural paints and maintenance paints. Generally, up to 10 per cent of the polyamide based on the weight of the alkyd is added to the alkyd and heated at normal alkyd reaction temperature under agitation. Ester interchange reaction takes place and fragments of the polyamide resin become chemically bonded to the alkyd. The modified alkyd serves as a compatibiliser for the mutually insoluble unreacted polyamide and alkyd to form a gellike structure. The stiffness of the structure decreases with the increasing amount of the compatibiliser. If the reaction is allowed to continue and there is no unreacted polyamide left in the system, little or no thixotropy will be exhibited. Therefore, the reaction must be precisely controlled to give the desired degree of thixotropy. Aromatic solvents such as xylene tend to destroy the thixotropic structure. Therefore, they must be reduced to less than 0.5 per cent in the product. Polyamide modified alkyd resins are available commercially to be used as additives for making thixotropic alkyd paints.

## High Solids Alkyds

There has been a strong trend in recent years to increase the solids level of all coating materials, including alkyds, to reduce solvent vapour emission. To raise the solids level and still maintain a manageable viscosity, the molecular weight of the resin must be reduced. Consequently, film integrity must be developed through further chain extension and/or cross-linking of the resin molecules during the 'drying' step. A high cross-linking density necessitated by the lower molecular weight of the resin would build a high level of stress in the film and cause it to be prone to cracking.

Therefore, adequate flexibility should be designed into the resin structure. This means that the distance between the hydroxyl groups of the polyol and the carboxyl groups of the dibasic acid would need to be lengthened by linear linkages. In addition to the manipulation of resin molecular structure for increasing coating solids, the use of more active, though more expensive, oxygenated solvents also serves to reduce the viscosity of resins solutions.

Chain extension and cross-linking of this solids alkyds resins are typically achieved by the use of polyisocyanato oligomers or amino resins. An adequate amount of excess hydroxyl groups must be designed into the alkyd structure to provide reaction sites for these modifiers. To limit the molecular weight of the alkyd resins, the molar ratio between polyols and dibasic acids should be greater than 1. The hydroxyl functionality of the formulation should be controlled by a careful selection of polyols to avoid an over presence of free hydroxyl groups in the product, which would adversely affect water resistance and other properties of the coating film. Most of the high solids alkyd systems are used in industrial baking finishes. For air drying applications higher doses of driers are usually needed to achieve acceptable drying rate.

### Water-Reducible Alkyds

Replacing solvent-borne coatings with water-borne coatings would not only reduce solvent vapour emission but also improve safety against the fire and health hazards of organic solvents. Alkyd resins may be rendered water-reducible by either converting the resins into an emulsion form or by incorporating 'water-soluble' groups in the molecules.

The most common approach for imparting water solubility in an alkyd resins is to leave enough pendent carboxyl groups in the resin and to neutralise them with fugitive base, such as ammonia or low molecular weight amines, to build ionic characteristics into the resins.

Nonfugitive base materials, such as caustic soda, would leave the salt in the coating film and damage its water-and corrosion-resistance. Trimellitic anhydride (TMA) is the most frequent choice of ingredient to provide the pendent carboxyl groups. It is reported that glycerol gives resins with poor hydrolytic stability. Therefore, polyols with primary hydroxyl groups are preferred for the preparation of water-soluble alkyds.

The recommended procedure for the preparation of water-soluble alkyds is to hold off the TMA in the initial stage of the alkyd reaction so that the high functionality of TMA would not be a cause of gelation. When the reaction has progressed to a desired low acid number, i.e. the building of the polymer chain is completed, the temperature is lowered to 180°C; the TMA is then added and maintained at that temperature until a desired acid number, usually about 50–60, is reached. At such a temperature, only the anhydride group of the TMA would react to form half esters and the remaining two carboxyl groups would essentially remain unreacted. If one desires to have the TMA participating in the backbone structure of the resin, a part of the TMA is charged in the beginning of the alkyd reaction, often with the presence of an appropriate amount of a monohydric alcohols, such as benzyl alcohol, to balance the functionality of the system.

The remaining TMA is then added in the same manner as describe to 'end cap' the resin and to provide pendent carboxyl groups for water solubilisation.

The finished resins are usually dissolved in oxygenated coupling solvents, such as glycol ethers, to improve the solubilisation of the resin in aqueous media and the handling of the resin. Water and the base are premixed and added to the resin solution when needed. The coupling solvents usually have higher boiling points than water.

During the drying process, the solvent would be enriched in the coating film as water evaporates preferentially. The resin molecules would become better solubilised, i.e. molecular chains would be extended and result in better formation and integrity of the film.

## SAFETY AND ENVIRONMENTAL PRECAUTIONS

The manufacturing of alkyd involves a wide variety of organic ingredients. While most of them are relatively mild with low toxicity, some of them such as phthalic anhydride, maleic anhydride, solvents and many of the vinyl (especially acrylic) monomers are known irritants, or skin sensitisers and are poisonous to human.

Person involved should be thoroughly familiar with the hazard potential of each and every one of the chemicals by consulting the material safety data sheets provided by the suppliers and practicing the recommended safety precautions in handling the materials. The use of personal safety equipment such as protective goggles, gloves, clothing and respiratory devices should be diligently observed. Since large quantities of highly flammables solvents are routinely handled in an alkyd plant, fire safety should be the utmost in everyone's mind. Electrical equipment and power circuity in the plant should conform to all applicable codes and all equipment should be properly grounded. The areas for the reactors and storage tanks should be separated by fire walls and must be adequately ventilated. The storage tanks should be blanketed by inert gas. A slight positive pressure of inert gas should be maintained in the reactor or storage tanks during discharge of the resin or resin solution to prevent air from being sucked into the vessel to form an explosive mixture with the solvent vapour.

With the ever increasing awareness of the need for environmental protection, the emission of solvent vapours and organic fumes into the atmosphere should be prevented by passing the exhaust through a proper scrubber. The solvent used for cleaning the reactor is usually consumed as part of the thinning solvent. Aqueous effluent should be properly treated before discharge.

Chapter 6

# Fats and their Derivatives in the Leather Industry

## INTRODUCTION

Since the time when man used animal hides, one has tried to make them imputrescible through tanning and then to give them flexibility adapted to their utilisation, by fat liquoring. Fats have been used, right from the outset of dressing leather, to give back mobility to fibres which constitute the skin, mobility which is lost during tanning.

Today, fats of all kinds and their derivatives are used for different purposes and under various forms. Quantities used vary depending upon the leather: 0.5–1 per cent for 'flower oil'—fats emulsified or not—applied on the skin, on the flower side, in order to prevent breaking of vegetable tannins by air, but 30 per cent also be used as in the case of 'tallowing' or 'oiling' of vegetable, mineral or mixed tanning leather.

Fats, extracted from fatty tissues of animals, seeds and vegetable pulps, are used in practically all the stages involved in dressing leather—and in this case either in the stage of oil which is said to be virgin—or after transformation by simple chemical reaction: sulphating, sulphonation, esterification, amidification, sulphorisation, oxidation, chlorination, etc. or through reaction on their fatty acids or alcohols obtained by hydrolysis and transformed into wetting agents, emulsifying agents, plasticisers, lubricants, waterproofing compounds, that is to say, transformed into auxiliary products (Table 6.1). It in their capacity as lubricants that fats are most widely used and this is known as 'fat liquoring'.

**Table 6.1.** Main fats used in tannery (virgin oils).

| *Origin* | |
|---|---|
| *Animal* | *Vegetable* |
| Mammal oils | Non-drying oils |
| Flare | Castor |
| Horse | Olive |
| Neat's foot | Coconut |
| | Palm kernel |
| Tallows | Peanut |

*(Contd ...)*

**Table 6.1.** Main fats used in tannery (virgin oils). *(Contd.)*

| *Origin* | |
|---|---|
| *Animal* | *Vegetable* |
| Woolgrease lanoline | Jojoba |
| Fish oils | Semi-drying oils |
| Cod liver | Cottoneseed |
| Herring | Soyabean and lecithin |
| Shark oil (squalene) | Rapeseed |
| Birds | Vegetable waxes |
| Egg yolk (phospholipids and lecithins) | Carnauba |
| | Candelila |
| Insects | |
| Bees wax | |

## FAT LIQUORING OF LEATHER

The nature of fats or derivatives, used during the fat liquoring of leather, depends on the properties that we want to give to them and also on the method retained for blending. In fact, there are various ways of blending fat into leather: we can distinguish four groups each of which requires appropriate equipments.

### Blending Method

#### *Blending of raw fat in dry or humid leather*

The fats appear in liquid or melted form and the operation is carried out either by cold or hot treatment: this is the case of dressing leather—on tripe skin—tallowing—on dry leather—it is sometimes done under vacuum to accelerate fat penetration.

#### *Humid leather blending in the form of water emulsion in oil (w/o)*

This is done by oiling or fat liquoring in a bath for a very short time, for which a mixture of fats or their derivatives are used having a certain hydrophilic character—oxidised fatty acids, fatty amides of amino alcohol, etc.—generally accompanied by liquid or solid fatty matters, fish oil, tallow, etc.

#### *General oiling formula*

1. Moellon (oxidised fish oil with 12 per cent 'degragene' oxidised fatty acid): 1/3.
2. Pressed or raw tallow: 1/3.
3. Fish oil (herring, cod, menhaden): 1/3.

These quantities depend on the weight of leather to be treated, which generally is of vegetable or mixed tanning and varies from 15 to 30 per cent.

Certain furs—e.g. of wild fowl—are also treated in this manner. Compositions which are quite complex are based on lanoline, phospholipids and other fats are liable to reveal the suppleness, lightness, un-tying and brilliance of highly appreciated fur. Main derivatives of fats used in leather industry are given in Table 6.2. Application of fat derivatives in tannery are shown in Table 6.3.

**Table 6.2.** Main derivatives of fats used in the leather industry.

*Derivatives of soluble or emulsionable fats*

*Anionics*

Soaps, sulphated oils and fats sulphonated, hydroxysulphonated, α-sulphonated and sulphochlorinated:

$$R{-}OSO_3Na,\quad \underset{\displaystyle OH}{\underset{|}{R}}{-}SO_3Na,\quad R{-}SO_3Na,\quad R{-}SO_2Cl$$

Sulphated derivatives of amino alcohol amides (ethanolamine, propanolamine) amino sulphonic acid amides, sulphated fatty alcohols, sulphated non-ionic sulphosuccinic ester derivatives, phosphorus derivatives (alcohols, diglycerides).

*Cationics*

Substituted polyamino and amino salts, amino alcohol esters, amino alcohol amides and Kritchewski derivatives, substituted diamine amides, diamine polyethylene amides (imidazolines), quaternary ammonium salts and hydrates:

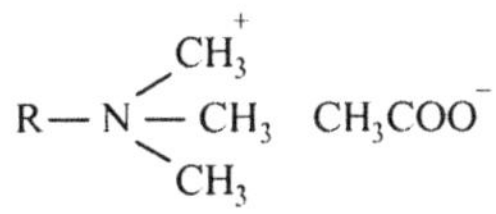

Most of these compounds are obtained from saturated fatty acids (by softening)

*Non-ionics*

Polyalcohols, polyglycols, polyalcohol and polyglycol esters, oxyethylated and oxypropylated derivatives, polyethoxyethers of oxyalcoyl amides.

*Amphoteres*

Phospholipids (naturals and synthetics) derivatives of amino acids, sulphated, sulphonated and sulphited cationics.

*Others*

Anionic and cationic compounds, peptised re-dispersed by an excess of one among them.

*Derivatives of unsoluble fats*

*Obtained by scission of natural fats:*

*Hydrolysis*

- Fatty acids: stearic, oleic, palmitic, lauric, etc.
- Fatty alcohols
- Glycerol

*Cracking*

- Castor oil derivatives

*Obtained by reaction with carboxyl group of fatty acids*

- Esters, fatty amides, fatty oxyethylamides, amines

*Obtained by transformation of fatty chain:*

- Hydrogenated, halogenated, oxidised, epoxidised.

**Table 6.3.** Application of fat derivatives in tannery.

| *Operation* | *Property of derivative* | *Ionic character* | *Main effects sought after* |
|---|---|---|---|
| Conservation | Fungicide | Anionic | Stopping micro-organism development |
| | Bacteriostatic | Cationic | Preservation of collagen fibre |

*(Contd ...)*

**Table 6.3.** Application of fat derivatives in tannery. *(Contd .)*

| *Operation* | *Property of derivative* | *Ionic character* | *Main effects sought after* |
|---|---|---|---|
| Soaking | Wet<br>Cationic | Non-ionic | Increase in the rate of restoration of skin moisture |
| Deliming | Wet | Anionic | Regularisation of the operation |
| Pickling | Emulsifying<br>Grease-removing | Non-ionic | Removing grease from the non-solvent pickle |
| Tanning | Wet | Anionic | |
| Covering | Dispersing | Cationic | |
| Fat liquoring in tanning bath | Fat emulsifying | Amphoteric | Dispersion of tanning metallic salts, resins for filling and fats. Accelerated penetration of baths in the skin. Deacidification. |
| Re-tanning and neutralisation | Wet | Anionic | |
| Pre-liquoring | Fat emulsifying | Amphoteric | Neutralisation of fibre loads |
| Dyeing | Wet<br>Dispersable | Anionic<br>Anionic<br>Cationic<br>Non-ionic | Deep lubrification. Preparing penetration for re-tanning Improvement in yield, uniformity and heat of the colour as auxiliary dyeing agents |
| Fat liquoring | Fat emulsifying and fat surfactant | Anionic<br>Cationic<br>Amphoteric<br>Non-ionic | Lubrification of leather fibres.<br>Contribution of particular properties (impermeability, anti-dirt, fire proofing) |
| Finishing | Wet emulsifying plastifying | Antionic<br>Non-ionic<br>Cationic | Plasticisation of pigment layers, regularisation of finishing agents, penetration. Improvement in dry and humid resistance to friction. |

### *Blending of oily emulsion in water (o/w), i.e. 'fat liquouring in bath'*

This is done for various articles such as shoes, clothes, glove making, fine leather crafts, furniture etc. and consists of getting penetrated, by more or less marked chemical affinity, an emulsion that is obtained with the help of anionic, cationic, amphoteric fats or non-ionic (emulsifiers) and non-soluble bodies and derivatives of (emulsified) aliphatic hydrocarbons. Usable quantities vary according to the articles: 2–3 per cent 'shoe quality' calf leather: up to 15–20 per cent leather used for furnishings. The concentration of emulsion is from 1 to 30 per cent and more if need be. Blending is done in a drum, vat or screening machine but the emulsion can also be applied with the help of pulverising ramps or mechanical brushes, etc.

### *Blending of fats or fat derivatives solubilised in an organic solvent*

This method known as 'impregnation' has not been developed in spite of certain industrial attempts. Advantages obtained such as a high suppleness by penetration of dissolved fats in the centre of the skins—even the thickest ones (several millimeters)—have not compensated for the inconveniences generated by the use of inflammable and volatile solvents.

## Objectives of Fat Liquoring

Regardless of more or less intense and deep lubrication of leather fibres: suppleness, resilience, solid, 'falling', nature and importance of the penetration of fats or their derivatives bring about other predominant qualities for the presentation and usage of a given article. Specifically, this concerns the improvement of physical and mechanical properties: resistance to traction, tearing along stitch lines, or specific characteristics such as water-proofing, inoxidisability, resistance to friction, fire-proofing, or smooth finish which will increase the commercial value of the leather products. This is the main reason for a multitude of marketed compounds which are derived from lipogenesis and lipochemistry.

### Fats Used

Fats, animal or vegetable (Table 6.4) are divided into two groups:

1. Glycerides: whose typical oil is that of neat's foot, is replaced these days by reconstituted oleic glycerides or by a mixture of animal or vegetable fats.
2. Cerides: whose typical oil is the sperm oil type, is replaced currently by long chain monoalcohols, monoacid esters (C14, C18) or by monoacid glycol diesters (C14, C18) which may now give place to jojoba oil. The nature of fats used evolves equally with the place of production of leathers and the industries using them are themselves influenced by local oleicole productions. For example olive oil is used more frequently in Spain and Italy than in France where rapeseed oil is preferred. This is also the case with oils and fats of animal origin.

**Table 6.4.** Application of crude fats.

| *Operation* | *Type of article treated* | *Fats* | *Obtained effects* |
|---|---|---|---|
| Fat liquoring | Furniture, equipment, shoes, fur, common fur (fox, etc.) | Tallow, wax, suintine | Fullness, suppleness, impermeability |
| Packaging | White and pastel tawed leather, gloves | Egg yolk, vegetable lecithin | Suppleness, fullness |
| Shining | Gloves, clothes | Animal oils, lard neat's foot | Shining, silky touch |
| Impregnation | Top leather and others | Unoxidisable fats, wax | Suppleness, shiny finish |
| Flower oil | Leather from vegetable tanning, soles, | Mammal, fish oil equipment | Suppleness of flower, mechanical resistance |
| Chamoising | Chammy | Oxidisable fish oil | Tanning of fats by oxidative degradation products |
| Finishing | All articles | Animal and vegetable oils | Plasticisation of finished products and pigmentary films |
| Varnishing | Top leather, leather articles, varnish | Siccative oils (linseed) | Polymerisation of films |

## CHEMICAL AND PHYSICAL PHENOMENA OBSERVED DURING LEATHER FAT LIQUORING

As an amphoteric body, the skin is capable of fixing anions or cations following the pH of the medium in which it is found. Every time that an ionised compound reacts with leather by electrovalent bonding, its ionic balance point is modified.

The variation in the ionic balance point of the skin has formed the subject of studies since ancient times but today we know its approximate value according to treatments done during improvement of leather and which can be easily defined by colorimetric methods. This parameter is determinant as much as the pH on which operations have been carried out and electrovalent reactions occur between the skin and the chemical component which are present for re-tanning, dyeing, fat liquoring, etc. The pH at the ionic balance point, the skin still conserves a potential of reactivity which is due to a great variety of amino acids of which it is composed and each of which possesses different isoelectric points.

This 'potential of reactivity' whose lowest value is found in pH of ionic balance point will increase, more or less, as soon as the pH of medium will be modified, towards acidic medium for auxiliary anionic compounds and basic compounds for cationics.

Variation of ionic equivalent point which greatly conditions the stratigraphic and topographic distribution of fats and their derivatives in leather is given in Table 6.5.

**Table 6.5.** Ionic balance point of the skin during its transformation.

| *Treatment* | *Ionic balance point* | *Variation in ionic balance point* | |
|---|---|---|---|
| | | *Decrease* | *Increase* |
| | | Functions of collagen entering into combination | |
| | | Basic | Acids |
| Native collagen | 7 | | |
| River | 4.7 | | |
| Painting | 4.5–4.9–5.2 | | ⟶ |
| Bathing | 5.5–5.8 | | |
| Pickling | 6 | | |
| Chromium tanning | 7–8 | | ⟶ |
| Masked chromium tanning (masking anions) | 4.5 | ⟵ | |
| Vegetable tanning | 5 or < 5 | ⟵ | |
| Sulphite vegetable tanning (Chestnut and vegetable tan solubilised by sulphitation) | 3 | ⟵ | |
| Chromium vegetable tanning | 5 or < 5 | ⟵ | ⟶ |
| Formol tanning | 4.5 | ⟵ | |
| Zirconium tanning | 4.3 | ⟵ | |
| Silicon tanning | 4.7 | ⟵ | |
| Dressing | 4.1 | ⟵ | |
| | | Fixing of anions | Fixing of cations |

Note: Anionic fat liquor decreases the ionic balance point of the skin while cationic fat liquor increases it.

Fixation, by salification reaction of the emulsifying part of fat liquoring composition applied in bath mainly conditions the deposit of fatty molecules but, other type of reactions can also take place: semi-polar, hydrogen bonds, etc. At this stage of operation, the surfactant properties of the deposited fatty phase will make it possible to shift the acqueous phase which bathes the leather fibre. An adequate rapport of surface and interfacial tensions of the two phases is necessary (Table 6.6).

Penetration in interfibrillar spaces is thus controlled by chemical, mass and superficial properties of fat liquoring and fat auxiliary agents which compose them: ionic charge, melting point, viscosity, interface tension, etc. and also by different ways in which they are applied: duration of the bath, temperature, pressure. We have compared not without reason the way fat liquoring penetrates leather to form a chromatographic system in which composite elements are divided according to their polarity, as against the polarity of the different layers of skin, the latter are considered superposition of adsorbants in which molecules are displaced according to their own polarity and the polarity of other elements present: re-tanning agents, colouring agents, emulsion resins, which are applied earlier or later.

**Table 6.6.** Deposit and penetration phases of fats and auxiliary products during fat liquoring of leather.

| *Blending mode* | *Remarks* | *Orientation of fatty molecules-perpendicular view of the fibre surface* |
|---|---|---|
| I[st] stage:<br>Fat liquoring in long bath | O/W emulsion, weak concentration fatty molecules on the fibre surface | 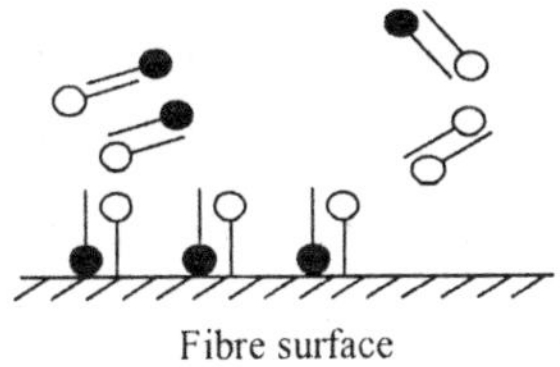 |
| 2[nd] stage:<br>Fat liquoring in short bath | O/W emulsion increase in concentration of fatty molecules on fibre surface | 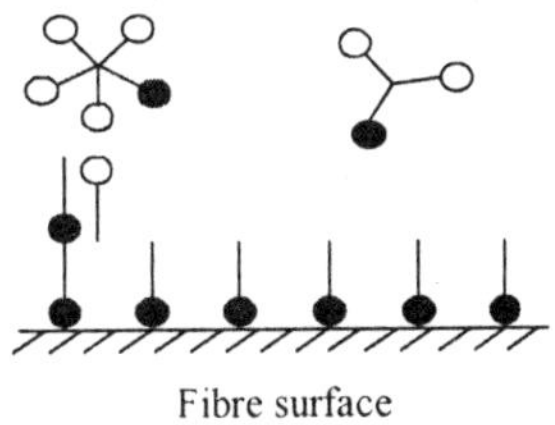 |
| 3[rd] stage:<br>Fat liquoring in bath<br>Oiling | Emulsion 'water in oil' (w/o)<br>High concentration of fatty molecules on fibre surface. | 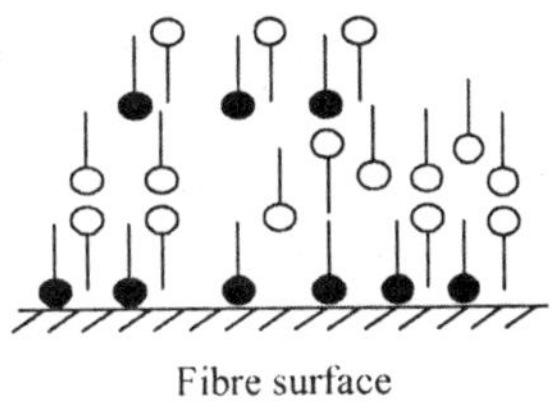 |
| 4[th] stage:<br>Fat liquoring in bath<br>Oiling<br>Tallowing<br>Impregnation | Separated oily and water phases<br>Spreading of the organic phase on the surface of the fibre which drives the water phase by an appropriate connection between the FCH-FCA and FAH interface tension | 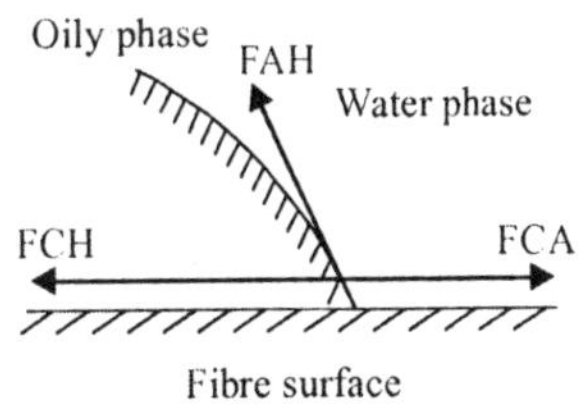 |

○— Emulsionable ionised molecules
●— Emulsioned molecules

FCA = leather/water interface tension
FCH = leather/oil interface tension
FAH = water/oil interface tension

Aqueous emulsions do not always have sufficient stability in the presence of leather medium: acidity, salification. Due to this a lot of operational difficulties are caused, which bring in defects in treated leather and at worst, ends up in non-repeatability of results.

## NEW CONCEPTS: EMULSIONS AND LEATHER

Fat emulsification has greatly evolved over the last few years. Studies carried out in the field of aqueous dispersions have led to new concepts not only in the case of removing fat from skins but also for feeding leather, much more than depollution, in the case of industrial wastes is directed towards gradual suppression of organic solvents which were earlier almost partially used for removing fat and from time to time for food 'solvant oil'. Toxicity of industrial effluents, generated by an excess of harmful salts, have similarly evolved production methods of leather towards elaboration 'of intermediary technologies' for minimum pollution.

The latter which can be achieved with the help of more or less toxic salts substitutes of chromium salts used during pre-tanning or tanning allow classification in choice and in thickness of the animal skin before tanning. They are called 'stabilised dry white' and 'stabilised wet white'. This development leads to work modification and a change in work orientation.

## Degreasing and Fat liquoring in Bath

These two operations are linked by an inversion of results sought after: In the first case, we try to make fat penetrate leather and then leave it, in the second case, we try to eliminate excess fat resulting from the presence of adipose cells in precise places: dorsal parting, collar for example on sheep, caprine, reptile or bird skins.

Methods and means employed are thus contrary, degreasing being carried out at relatively low temperatures to avoid denaturation of the helical structure of the untanned collagen and warm fat liquoring so as to re-activate leather more with emulsifiers provided and bring down surface and interfacial tensions of the fat phase.

Non-ionic compounds with minimum skin reactivity are used in normal operating conditions when it comes to degreasing. For fat liquoring, anionic or cationic emulgators capable of reacting with ionised leather sites with the aim of causing emulsion rupture by progressive fixation of the emulsifying agents is used. Be it for degreasing or fat liquoring, emulsion finess is essential because it favours either natural fat extraction from skin by making it cross thick beds of dermic fibres, or fat emulsion penetration in tanned fibres.

## Micro-Emulsions

Most fats or their derivatives which can be used for fat liquoring leathers give oil emulsions in water and water in oil. In order to increase stability of the latter, it is necessary to make some adjustments by increasing its alkalinity or by adding in the middle certain compounds capable of changing the balance of phases towards a water micro-emulsion in oil, the form in which the commercial product comes.

Since, many years, it is known that sulphated oils present homegenous zones in their phase diagrams, whose surface evolves according to pH, salt and water content, free acidity temperature and the presence of co-surfactant subsances, etc. Most commercial products are limpid at ordinary temperatures in spite of their high water content. In this case we have a W/O micro-emulsion.

When W/O fat liquoring emulsions are diluted (pure fat liquoring oils), we observe right from the first addition of water to the mixture, the appearance of a limpid, clear, more viscous product which retains its initial aspect to gradually become an opalescent product, which then more or less quickly turns into a milky cloud. Its stability will vary, it will even be precarious, depending on temperature and concentration. In the meantime, we pass from W/O micro-emulsion to an O/W emulsion.

Separation of phases takes place with the passage of time, according to 'Winsor' classification. It is very often an oily phase (reverse emulsion) which floats on the surface and an aqueous phase. Their composition is different and it is known that the emulsifying agent used for obtaining the emulsion shares its molecules in the presence of both the systems and this depends on their hydrophilic and hydrophobic affinity.

In this system, the W/O micro-emulsion zone is larger than that of O/W micro-emulsion. This goes on to say that the product presented in a concentrated form will be stable and in good shape, whereas in a diluted form, we can see more or less fast, phase separations accelerated by increase in temperature.

These two observations are caused due to solubility variations and hydrophilic and lipophilic affinity level of the emulsions:

'Winsor I' (two phases): micro-emulsion floating oil phase.

'Winsor II' (two phases): balanced aqueous phase with an O/W micro-emulsion.

'Winsor III' (Three phases): W/O micro-emulsion between an oily and an aqueous phase.

'Winsor IV' (one phase): W/O or O/W; micro-emulsion.

It has been said that in the case of sulphated or sulphited oils that molecules with hydrophilic arrangements of:

$$—O—SO_3—M \text{ ou } —SO_3—M$$

concentrate in the aqueous phase and disappear almost completely in the oily phase.

It has been shown that in normal operating conditions of fat liquoring preparation, there was no loss of combined $SO_3$ for sulphonated or sulphated oil and expansion of the oily part was simply caused by demixion rather than by more or less rapid hydrolysis of the emulsifier.

## Bicontinous Micro-Emulsions

Micro-emulsions are defined as emulsions whose particles remain invisible to the naked eye and whose average diameter is 500 Å or 50 nm. In practice, they are found in a more or less bluish transparent surrounding. The bicontinuity of a micro-emulsion depends on the flow of W/O emulsion to O/W emulsion without any break and a gradual decrease of the O/W ratio while the oil-emulsifier ratio remains constant.

A graphic representation of these structures has been imagined by various studies and a simplified model has been proposed (Fig. 6.1).

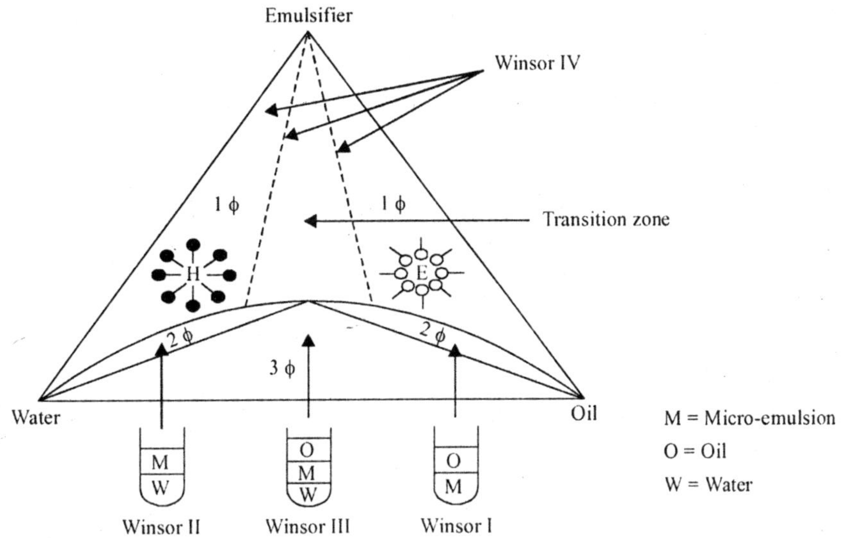


**Fig. 6.1.** Simplified graphic representation of the phase diagram.

## Role of Co-surfactant Substances in Emulsion Bicontinuity

As their name denotes, co-surfactant substances are agents which increase the properties of surfactant substances when they are added to the latter. They are generally ionic with a strong polarity and are made up of a hydrophobic chain which is more or less long, more or less branched and aligned to a hydrophilic site such as:

$$—CH_2OH,\ —CHOH,\ —CO_2H,\ —NH_2$$

Surfactant substances modify the emulsion structure by lowering interfacial water-oil tension within required limits. Their molecules are directed to interface in the same way as surfactant substances and settle between them, causing a bend and flexibility of the interface which favours the coming together of the oily and aqueous phase.

Generally, surfactant substances help in the formation of only one type of micro-emulsion; depending on whether they are more hydrophilic than hydrophobic, we obtain 'oil in water' micro-emulsion and vice-versa.

A strong hydrophilic site, joined to a very small hydrophobic structure leads to concavity of the interfacial film towards the aqueous phase and surrounds it in the oily phase. In practice, the more soluble the surfactant substance, the stronger are the chances of forming an oil micro-emulsion in water. Branching of the hydrophobic chain of the co-surfactant agent leads to a reverse bend favouring formation of water micro-emulsion in oil.

### *Bicontinuity of fat liquor micro-emulsion*

If we examine the phase diagram of traditional fat liquoring, we notice that very often, at the preparation temperature of the pre-emulsion, we are already in a quite fragile stability zone. This is the case for ordinary sulphated oils, sometimes for sulphited oil, where the dilution line—for a commercial oil containing 2.5–3 per cent combined $SO_3$, i.e. 15–20 per cent of the emulgator present in the oil—runs parallel to the demixion line marking the passage of the 'Winsor IV' micro-emulsion phase to the 'Winsor I' phase, the oily phase observed floating on a micro-emulsion.

The more intense the dilution, from the time the pre-emulsion introduced into the fuller containing skins and bath, the more acute are the demixion risks (Fig. 6.2).

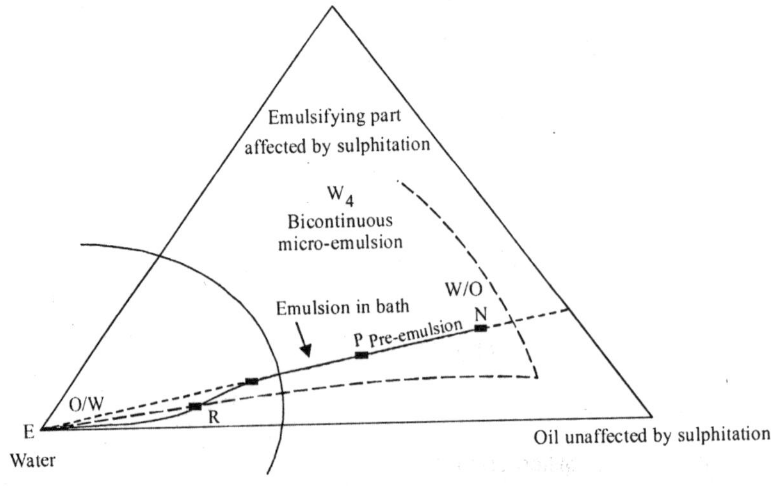


**Fig. 6.2.** Sulphited oil adsorption.

N = Point of auilibrium of fat liquor before dilution.
P = End point of pre-emulsion preparation.
C = Dilution point of the mixture at operation concentration according to the quantity ratio of fat liquor length of bath.
R = Break point of pre-emulsion in the presence of leather, beginning of the formation of demixion.

### *Appearance of Winsor I phase, micro-emulsion in equilibrium with the organic phase*

Lines:

NP = Fat liquor pre-emulsion zone.
PR = Oil micro-emulsion zone in the presence of leather in the fat liquor fuller's earth.
PE = Fat liquor emulsion zone (O/W micro-emulsion in the absence of leather).

Dilution and also adsorption of the emulgating phase of the sulphated or sulphonated molecules by salification reaction with the amino sites of the skin all account for it. These risks increase when the pH value is farther towards an acid medium from the ionic balance of leather in this state and with the fast disappearance of the emulgator (Fig. 6.3).

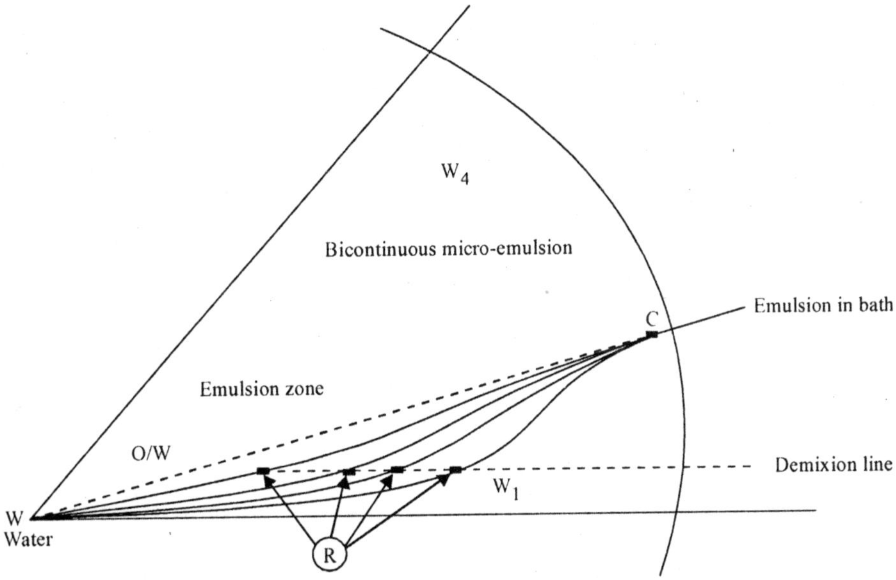


**Fig. 6.3.** Substantive variation of the fat liquoring emulsion (sulphited oil) according to the pH of the bath.

Thus, in order to master the fat liquoring operation, it is indispensable to bother about only one parameter at a time: 'reactivity potential of leather' which was considered only rarely and to use a composition which in all dilution phases appears in the form of a bicontinuous micro-emulsion, fat deposit on the leather fibre is controlled only by fixation of the emulsifying part.

The craftsman has to be assured that at fixed dilutions, he is in front of micro-emulsion.

If it is contrary to this, he will simply add a minute quantity of surfactant substance appropriate to its preparation. This practice which is very simple can avoid difficulties that crop up at times.

It is natural that fat liquoring emulsions possess good stability during the entire operation when fat is transferred to leather. Otherwise, we will witness a rapid deposit of fat phase on leather. This in itself does not always destroy the quality of articles obtained but causes the concentration of sulphated and sulphonated molecules in the residuary bath, thereby considerably increasing the COD of waste products.

One can no longer ignore the pollution caused by industrial operations. Use of fat and its derivatives in the leather industry does not avoid pollution; progress achieved in the use of emulsions contribute to its reduction.

Chapter 7

# Fats and Oils in Textile Industry

## INTRODUCTION

Very varied in its activities the textile industry is a large consumer of technological auxiliaries and various auxiliaries. Some of these auxiliaries are used for giving textiles their final properties whereas the others play a role only during different transformation operations without finally appearing in the finished article.

In France, the textile industry is divided into four main fibre categories:

1. Fibres of animal origin—mainly wool.
2. Fibres of vegetable origin.
   (a) Mainly cotton
   (b) Linen
3. Artificial fibres (production dropping constantly).
   (a) Obtained by chemical transformation of wood: viscose, acetate, etc.
4. Synthetic fibres.
   (a) Polyester.
   (b) Polyamide.
   (c) Polyacrylic.

Whatever the category, the production process right from the origin of the fibre to the finished article, passes through three essential stages in the textile industry:

1. Fibre production.
2. Fibre transformation.
3. Processing of finished articles.

## FIBRE PRODUCTION

From the basic material, natural fibres, wood cellulose and petrochemical-derived raw materials, fibre producers manufacture fibres or yarns. Production requires successive operations with different technological auxiliaries, adapted to the types of fibres and operations.

### Wool Processing

#### *Sheep-shearing*

Sheep-shearing: a purely mechanical operation.

### *Wool-washing*

Wool-washing: The aim is to remove stains (straw, thistle, droppings, earth) and natural fats from wool.

It is done in a washer (Leviathan) with cold or warm water in the presence of non-ionic surfactants.

### *Carding and combing*

The purpose is to completely eliminate plant residues, to straighten out and equalise the fibres and eliminate the shortest ones.

Carding is done on a 'card' with a non-ionic lubricant (wetting agent, oiling, antistatic) with a high fat content. Combing, performed with a comber, requires a rehumidification product and an oiling agent (non-ionic agent in an aqueous medium).

### *Spinning*

This is in fact the production of yarns, which involves the following successive operations: (i) homogenisation and parallelisation of fibres; (ii) rolling of fibres; and (iii) drawing and twisting the yarn.

It is necessary to add a lubricant which will give cohesion to the fibres, while promoting lubrication (normally a non-ionic grease with a high fat content is used).

Technological auxiliaries such as (i) detergents; (ii) lubricants; (iii) antistatic agents; and (iv) spinning auxiliaries, are responsible for consumption of about 3–5 per cent of the weight of fibres used. They do not remain on the finished objects and are subsequently eliminated.

## Cotton Branch

This process does not require any technological auxilliary in the initial stages, the natural wax of cotton acts as lubricant.

## Artificial and Synthetic Fibre Branch

Whether artificial or synthetic, the manufacturing process is based on a common principle:

1. To obtain, through a solution (artificial) or fusion (synthetic), a product likely to pass through a drawing plate in order to be drawn into filaments.
   At the outlet the filaments are collected as they are (continuous yarn), or cut in such a way as to obtain fibres (flock) which would subsequently form yarns. These successive operations use numerous technological auxiliaries.
   (a) Coagulation bath auxiliaries (for artificial fibres).
   They regularise the orientation of macromolecules. (These are non-ionic surface agents like ethoxyl fatty amine.)
   (b) Yarn lubricants (synthetic fibres).
   It helps in lubricating the yarns and facilitates the drawing of filaments of desired dimensions. These lubricants are non-ionic surface agents like ethoxyl fatty acid esters (for polyester and polyamide) and cationic, modified fatty amides (for polyacrylics).
   (c) Finishing oiling (artificial and synthetic fibres).
2. Process to facilitate subsequent spinning operations.
   Lubrification is carried out inside a bath (aqueous solution) or by pulverisation. These agents are generally non-ionic surfactants.

### Winding

Whatever be the type of yarn produced, it is wound so that it can be stored and used later. Other auxiliaries (coining oils) are used during these operations which are carried out at high speeds and require good lubrication.

## FIBRE TRANSFORMATION

Fibre producers present their products to the customers in two forms: (i) continuous yarn on a bobbin; and (ii) stock fibres (cut fibres) in balls. These materials can be used as they are for subsequent operations or can be dyed before use (package dyeing, stock dyeing).

### Spinning

In the case of stock fibre yarn, the fibre has to undergo successive spinning operations identical to the procedures for wool.

We find spinning oiling which help in the cohesion of fibres and lubrication when the fibres pass over the spinning equipment. The products should give high fibre/metal friction co-efficient (cohesion) and low fibre/metal friction coefficient (lubrication), this avoids breaking and production of static electricity.

### Weaving, Knitting

Whether the yarns are dyed or not, the main work of the transformers consists of producing cloth which will be subsequently used for making dresses.

In order to be able to resist the constraints created by the high speeds of knitting or weaving, the yarns (fibres, yarns) should undergo certain preparatory operations.

#### *Sizing*

Sizing increases the resistance to breaking (thread with polymers which blocks the different fibres among themselves). These auxiliaries are starch, carboxymethylcellulose, acrylates or polyvinyl alcohols.

#### *Overwaxing*

Treatment with wax or oils in emulsion which makes them more slippery over the looms. These compounds should be soluble or emulsifiable in water. They are esters of fats.

#### *Weaving*

This is carried out at high speeds on looms where the weft is placed across the warp yarn, either with the help of a shuttle or through hydraulic impulsion. The lubricants required for the proper manufacture of cloth are also products derived from the chemical processing of fats. Such as ethoxyl esters, sulphated oils.

#### *Knitting*

The knitting machine is different from that of the loom. Knitting is carried out from a single yarn (no weft/warp). A very slippery yarn should be used in order to obtain a regular knit. The yarn to be used should be paraffined.

### Processing the Cloth

The cloth or knitted fabric previously produced undergo a series of essential operations which condition the quality of the articles subsequently made. Most of the defects appearing during making up are due to insufficient preparation.

### *Singeing*

A purely mechanical procedure which removes the fibres present on the surface of the fabric.

### *Desizing, degreasing, deoiling*

The purpose of these operations is to eliminate all the technological auxiliaries used earlier. They are carried out in an alkaline or enzymatic medium with the help of wetting surface agents, detergents generally of non-ionic type, which do not give much lather, such as weakly ethoxylated fatty alcohols.

### *Kirr boiling*

Alkaline processing which helps in eliminating natural substances found along with cotton (semi-cellulose, wax, etc). The processing is more effective if hydrophile wetting agents, which are low foaming and stable in an alkaline medium, such as ethyoxlated fatty alcohols, phosphoric esters.

### *Bleaching*

The purpose of this operation is to eliminate the different coloured dust matter found on the cloth. Bleaching is carried out with oxidants: oxygenated water, sodium hypochlorite, etc. in the presence of specific surfactants, stabilising wetting sequestrants, whether non-ionic or anionic.

### *Mercerizing*

Mercerizing makes the fibres shiny, increases their resistance as well as their affinity to colours. This operation is carried out in a medium with a strong sodium hydroxide concentration, in the presence of anionic surfactants such as alkylsulphates.

### Dyeing or Printing

Printing deposits a coloured paste on the surface of the cloth, according to a precise motif. The paste is deposited by rollers, one roller per colour and motif or by printing screen.

Dyeing gives the fabric a uniform colour over its full length and thickness. Dyeing is carried out through a series of operations, each of which requires a specific surfactant.

1. Dispersing coloured pigments or dye.
2. Levelling agents, which avoid the presence of spots or unevenly coloured areas.
3. Wetting agents which make fibres more hydrophile (polyamide).
4. Carriers enabling a better penetration of pigments and dye in synthetic fibres.
5. Dye fixers for fabric which is a mixture of synthetic and natural fibres, thus improving the fixing power of dye during washing.
6. Soaping agents which eliminate the dye and non-fixed pigments from the fibres.
7. Defoaming agents which prevent foam formation.

### Softening

This operation has many purposes:

1. Modifying the sensation of touch, the appearance, the softness of the fabric.
2. Improving its mechanical properties and making it antistatic.

Softening is done by treating the fabric in an aqueous medium with softeners, cationic or non-ionic surfactants, which remain fixed on the fabric.

Average quantity: 0.2 per cent of the weight of the fabric.

### Finishing

The operation consists of treating the fabric with products which will give it the final desired characteristics: (i) crease resistance; (ii) water-proofing; (iii) rigidity; (iv) fire-proofing; and (v) antistaining.

Products such as latex polymers are used in the presence of surfactants which help in fixing them better.

The quantity of dressing is only a few per cent but may attain 20 per cent for cotton fabric. For their industrial use, transformers use about 1 per cent, the weight of fibres treated with technological auxiliaries, the majority of which are fat derivatives. Most of the operations carried out in an aqueous medium require an adhoc surfactant.

## PROCESSING THE FINISHED PRODUCTS

Finished articles, hosiery, knitwear, clothing, etc. also use technological auxiliaries during the course of different operations which they undergo: (i) dyeing auxiliaries; (ii) bleaching auxiliaries; and (iii) stabilisers.

The textile industry consumes bulk quantaties of technological auxiliaries most of which are fat chemistry derivatives.

Chapter 8

# Fats and Oils in Pharmaceuticals

## INTRODUCTION

In topical pharmaceuticals, the skin permeability and drug release from the vehicle determine the essential diffusion processes. These are a product of the physicochemical properties of drugs and of the vehicles. This naturally excludes the possibility of a universal vehicle. Therefore, a theoretical and experimentally corroborated design for the composition or choice of vehicles is necessary for efficient local therapeutic agents. Fats and oils play a key role in this formulation design. An appropriate selection of the fats or oils will not only provide the vehicle with cosmetic elegance, better moisturing or better protective properties, but also enhance the therapeutic benefits of the drug.

For example, it is well recognised clinically that an occlusive vehicle with a high concentration of oils or fats will enhance the therapeutic efficacy of topical corticosteroids. It is probably due to the fact that the fats or oils will increase skin hydration, which then enhances the bioavailability of the drugs through the skin. Since the target site of topical pharmaceuticals is local, i.e. the skin, an optimised topical product is one that delivers the drug to the target skin site and achieves therapeutic local concentration for as long as it takes to achieve cure or amelioration of disease symptoms. On the other hand, for sun care and cosmetic products, an optimised topical product is one that is formulated to leave the active ingredient on skin for as long as is necessary with minimal absorption into skin. For super moisturisers and/or antiaging products (skin wrinkle removers), ideal delivery of the therapeutic ingredient (s) may require delivery and localisation into the stratum corneum and/or into the epidermis, thereby putting in question the regulatory status (drug or cosmetic) of these products.

All definitions so far mentioned concentrate on optimising the topical dosage form from a 'availability' (and subsequent site clearance) or a 'lack of availability' point of view. In reality, however, true optimisation of a topical drug product using fats, oils, or even solvents must also include the following considerations in the optimisation process:

1. Are the physicochemical properties of the drug in question sufficiently well understood?
2. Are the pharmacological models used in assessing and/or predicting the pharmacological activity of the drug relevant and well conducted?
3. What research vehicles were used to screen the drug pharmacological activity?
4. Were these relevant and appropriate?

5. Is the target tissue for the drug known? Is this the epidermis, dermis, or some specific cellular group within these strata?
6. Has drug delivery and specifically drug uptake and retention within the skin layers been adequately evaluated?

Human skin, as a heterogeneous membrane, is an essential organ of the human body. The main function of the skin is to separate and protect the internal organs and tissues from physical or environmental attacks such as from microbes, chemicals and various forms of light radiation. It also prevents the loss of the internal fluids and ingredients from within the body to the environment. In general it is difficult for most chemicals to permeate through the skin. However, since literally thousands of different substances produce delayed hypersensitivity reactions in apparently normal skin, it is evident that the protective permeability barrier of the skin is not perfect.

Briefly, the skin consists of three layers, the epidermis, the dermis and the subcutaneous tissues. The epidermis can be further divided to stratum corneum and viable epidermis. The stratum corneum, or horny layer, is the principal location of skin resistance to permeation. Before any topically applied drugs can act either locally or systemically, they must penetrate the stratum corneum–the 'barrier layer' of the skin. This is the necessary first step, both for the therapeutic action of applied drugs and also in producing any injury by exogenous substances (Chemical agents or antigenic substances). The stratum corneum consists of corneocytes, the keratinised end product of epidermal cell differentiation and associated lipids and their by-products. Keratin, deposited within the corneocytes, provides strength and chemical resistance. Resistance to transport through the stratum corneum depends on the properties and arrangement of its alternating hydrophilic and hydrophobic layers, as well as their thickness, all of which vary from species to species and even from place to place on the same individual.

Topical formulations usually contain excipients that partition into the skin according to their physicochemical properties. Certain excipients alter the physical barrier function of the stratum corneum. When this occurs, the solubility and permeability of compounds in the horny layer are affected. Vehicle design plays a crucial role in the development of topical pharmaceuticals. In fact, an inappropriate vehicle will diminish a drug's therapeutic potential. Conversely, a suitable base can significantly increase the bioavailability of actives and thereby improve the drug's efficacy.

In skin care products fats and oils are commonly used excipients, as skin protectants providing a hydrophobic layer to protect skin from environmental insults. In cosmetics fats and oils are used as emollients to keep skin at an appropriate level of hydration. Petrolatum, mineral oil, and silicone oil are examples of fats and oils recognised to be the most effective skin protectants and emollients used in cosmetic products. In topical pharmaceuticals, fats and oils are used as the key excipients to provide the occlusivity of the vehicle, which is critical to the bioavailability of the active drugs. Clinically an occlusive vehicle enhances the therapeutic efficacy of active drugs possibly by increasing skin hydration, which enhances the penetrability and bioavailability of the drug through skin.

Some components, including drugs, such as nitroglycerine (in ointment base) occasionally evaporate, and their disappearance systemically alters drug activity in the remaining vehicle. For chemically unstable drugs, exposure to air and light can be a problem. In view of such issues, drug delivery into and retention in the skin are often not well predicted by steady-state flux data. Therefore, drug localisation within the skin depends on the unique properties of the formulations and may not relate well to *in vitro* flux data as measured in diffusion cells.

For topical products, an optimal drug buildup in the skin with little or no drug flux through the skin is ideal. For example, sun-screening agents increase the power of the horny layer to protect the viable tissues from ultraviolet (UV) irradiation; topical antibiotics and antibacterials protect a damaged barrier from infection; and emollient ointments, creams, and lotions restore skin pliability and elasticity after low humidity environments and/or detergents have produced a desiccated horny layer.

This Chapter is a brief review of the basic and applied science of topical pharmaceuticals, limited to dermatological products only, a review of the current status of regulatory and research activities, and lastly speculation on the future trend of topical pharmaceuticals. Another objective is review of the importance of fats and oils in topical pharmaceuticals.

## FACTORS AFFECTING THE ABSORPTION OF TOPICAL PHARMACEUTICALS

Design of topical dermatologic agents raises some difficult questions: What is the nature of the barrier layer? How do molecules get through it? Is the stratum corneum a uniform barrier or is one part more permeable than another ? How fast does penetration occur and how is this affected by the chemical nature of the penetrating molecules? Can the permeation rates of substances be modified by using different vehicles or by hydration? What concentration levels are produced in the viable layers of the skin by the topical application of a drug? Is the blood perfusion rate a controlling factor in percutaneous absorption? Answers to these and related questions are of obvious importance for dermatologic practice, in the formulation of topical preparations and with regard to skin physiology in general.

When a topical preparation is applied to diseased skin, the clinical response arises from a sequence of three processes:

1. Release of the drug from the vehicle.
2. Followed by its penetration through the skin barriers.
3. Its activation of the desired pharmaceutical response.

Effective topical therapy optimises these processes as they are affected by the drug, the vehicle, and the skin. Factors that influence these three processes are discussed below under separate headings of rate limiting steps, biological factors and physicochemical factors.

### Rate Limiting Step

The purpose of a topical formulation is to deliver an effective amount of therapeutic agent to the target site. The target site may be on the skin surface (soaps, insect repellents, sun screens) or on the epidermal basal cells (antiviral agents, antifungals). Products designed to function on the skin surface work best if absorption into the skin is minimal. In this case, the limiting factor is the release rate of the active drug from vehicle. However, many drugs can function only if they are able to reach the viable tissues. Examples are corticosteroids such as hydrocortisone, used to treat a variety of skin ailments; local anesthetics such as lidocaine, for relief of itch or local pain; and retinoic acid used in the treatment of acne. The clinical effectiveness of topical pharmaceuticals depends on both the actives' inherent potency and more importantly, amount of applied dose reaching the target tissue. Consequently, the limiting factor of the whole process resides in the stratum corneum. In some incidences, drugs that are active when given orally or by injection have failed to work when applied to the skin, probably due to failure to penetrate the stratum corneum.

### Biological Factors

#### *Condition of skin and skin diseases*

The process of penetration in both normal and diseased skin depends on the state of the barrier function. If the barrier function of the skin is in abnormal disease state or is compromised, the limitations placed on the penetration do not apply. Diseases characterised by defective horny layer, such as dermatoses exhibiting parakeratosis (i.e. psoriasis and eczema) are accompanied by reduced barrier function of skin and increased percutaneous absorption. A partial destruction of the barrier appears in parakeratosis usually resulting in increased permeation, which approximately corresponds to incomplete stripped skin in barrier function. Exogenous agents such as mustard gas, acids, alkalies, etc. which also injure the skin, increase its permeability.

### *Age, sex and ethnicity*

Absorption through skin of infants is considered to be more rapid than through adult skin. Also the permeability of normal skin varies between sexes and among different age and ethnic groups.

### *Regional skin sites*

Large permeation differences have been demonstrated among differing skin sites. For steady-state diffusion of simple, small molecules through skin the extant data suggested the following ordering:

Plantar > Palmar > Dorsum of hand > Scrotal and Postauricular > Axillary and Scalp > Arm, Legs, and Trunk.

### *Species variation*

Human beings and animals display wide differences in physical skin characteristics such as the number of appendageal openings per unit area and thickness of the stratum corneum, which affect the penetration pathways and penetration resistance of skin. The skin of common laboratory rodents is more permeable than human skin. The average permeability order is:

Rabbit > rat > guinea pig > human, though the relationship between species is not consistent for different substances.

### *Skin biotransformation*

Topically applied drugs can be degraded on the skin or during diffusion through the stratum corneum by the influence of light, oxygen, or the chemical and microbial environment on the skin. They also can be enzymatically metabolised in the skin during the penetration process. The skin is a site of extrahepatic metabolism of foreign compounds. The importance of drug metabolism in the skin for epicutaneous therapy depends on whether a local or a systemic effect is intended.

## Physicochemical Factors

### *Hydration*

Skin hydration is known to increase the permeability of both polar and nonpolar substances. The water content of the horny layer can be influenced by the choice of vehicle. Fats and oils are most effective excipients providing vehicle occlusivity. The markedly increased permeation obtained with occlusion can also be explained by increased hydration of the horny layer.

### *Temperature*

Higher temperatures can increase penetration in two ways:

1. By increasing the rate of diffusion of any substance in contact with skin surface by increasing molecular motion.
2. By inducing vasodilation of skin vessels.

### Drug concentration

This can be examined by Fick's law of diffusion, which states that the driving force causing the transfer of a substance from a region of high concentration to one of low concentration is proportional to the concentration gradient.

### Effect of pH

Skin may be regarded as an amphoteric protein gel of varying degrees of hydration with a pH value around 5. Deviation of the pH above a definite value causes irreversible damage to the stratum corneum, thereby increasing the permeability to extreme levels. At higher pH values, a significant increase in the water diffusion rates and decreases in water-binding capacity were observed. The rates of absorption of acidic and basic drugs are strongly influenced by the pH of the vehicle, since it affects the degree of ionisation of the electrolytes. The nonionised moiety usually shows higher permeability. It is also known that the permeation rates are not necessarily proportional to the concentration of the nonionised drug in the vehicle.

### Effect of vehicle

Vehicles are normally chosen from among mild or innocuous bases. The influence of the vehicle on the release and penetration of the active drugs has been a focus of pharmaceutical scientists for the last two decades. It is now clear that vehicles can affect the permeation of skin even if the stratum corneum *per se* is not affected. Also, the rate of release is governed by the partition coefficient of actives between the vehicle and stratum corneum. It has also been suggested that the permeation of a drug molecule from a saturated or near saturated vehicle should be maximal and independent of the vehicle used. Apparently, most vehicles are absorbed to some extent by the stratum corneum and produce direct effects on it. These lead to changes in the solubility and diffusivity of drug actives in vehicle and/or skin and, accordingly, in their permeation rate. Other important drug–vehicle factors are the particle size of suspended drugs, viscosity of vehicle, and drug concentration in the vehicle.

### Solubility characteristics of permeants

The solubility characteristics of a substance greatly influence its ability to permeate biological membranes. The concentration presented at the absorption site is determined by the aqueous solubility of a drug. The partition coefficient of the drug between the aqueous and lipid phases of the vehicle strongly influences its rate of transport across the absorption site.

### Molecular characteristics of permeant

According to the Stokes–Einstein equation ($D = CM^{1/3}$), an inverse relationship exists between the absorption rate and molecular weight. Small molecules generally permeate more rapidly than large molecules; but within a narrow range of molecular size, there is little correlation between size and permeation rate. The specific effect on the permeation rate of the size and shape of the permeating molecules can be determined only if the effect of size and shape can be separated from the effect of solubility characteristics.

### Permeation enhancers

One method frequently used to increase the rate of absorption is adding materials that can combine with or dissolve in the structure of substances making up the barrier. These agents are known as absorption enhancers or sorption promoters. For example, aprotic solvents such as dimethylsulphoxide are capable of increasing permeation rates of most active drugs markedly. The mechanism lies in the swelling of keratin and disturbing the structure of the stratum corneum, thus reducing the diffusional resistance and increasing the permeability.

### *Surfactants*

The surface tension of a vehicle may affect the absorption of incorporated substances by its ability to spread over the surface of the skin. It is also known that, in addition to reducing the surface tension of the vehicles, a surfactant also enhances the solubility and dispersibility of the drug actives. Surfactants can penetrate through the skin and alter the permeation of drug actives. Among the surfactants, anionic surface-active agents such as sodium lauryl sulphate show the best effect, followed by nonionics and cationics. The penetration enhancement caused by the anionic surfactants was explained in terms of altered barrier function of the stratum corneum. In some cases, anionic surfactants could also increase water loss through the human skin. Unfortunately, these surfactants in them selves may constitute a considerable hazard in the topical application. The most common adverse influence of surfactants or detergents on the skin is well known as eczema, a clinical symptom of primary irritation of the skin of the hands.

## VEHICLE SYSTEMS FOR TOPICAL PHARMACEUTICALS

Generally, in the development of any new drug preparations, guidelines exist with particular respect to the behavior at the site of absorption. The following points should be considered:

1. The chemical form of the drug employed (acid, base, ester, salt, etc.).
2. Physical form (dispersion, hydration, solubility, etc.).
3. Choice of vehicles.
4. Drug formulation, dependent on the route of application.
5. Formulation procedure.
6. Interaction with other substances administered simultaneously (food, drugs, etc.).
7. The skin disease to be treated (dermatitis, psoriasis, eczema, dry skin, etc.).

With these guidelines, vehicle selection is considered to be a critical step for topical pharmaceuticals to be effective. The following describes the most common types of vehicle being used in topical pharmaceuticals. The knowledge for the use of fats and oils in these vehicles such as ointments, oil-in-water or water-in-oil emulsions, sticks and others is a very important element for the formulation design.

### Types of Topical Vehicle Systems

A drug may be applied as such (single-component system) or, as is usually the case, incorporated into a suitable vehicle before application. A multicomponent vehicle system exists when more than one chemically defined substance is included in a homogeneous or heterogeneous mixture. A disperse system consists of substances irrespective of form and size distributed evenly in an other substance-external phase, which is termed the embedding, continuous, or dispersion medium. Disperse systems may vary in consistency from fluid to elastic to semisolid and even to solid. The suitable vehicles for topical pharmaceuticals include solutions, suspensions, gels, ointments, emulsions, powders and solids.

### *Solutions*

Solution systems are the simplest to prepare and offer many flexible advantages for the formulator. They allow incorporation of a wide range of drugs and accommodate problem materials such as low solubility drugs or ionic drugs to be dissolved and the desired drug release and penetration parameters desired. Also, mixed solvent systems may allow for fine tuning of the drug delivery parameters. Disadvantages of these systems include messiness of application and the associated uncertainty of dosing. These problems may, if drug compatibility is not an issue, be improved by the use of thickeners to make the solution less fluid.

### *Suspensions*

In cases where the drug substance cannot be readily dissolved in a pharmaceutically appropriate solvent, it may be topically delivered by suspending it uniformly in a suitable vehicle. Suspensions may be formulated in a variety of materials, but present manufacturing and physical stability concerns the formulators. Drug uniformity, crystal growth, settling, low bioavailability and consumer acceptability are all particular problems for preparation of suspensions. Manufacturing uniform batches without floccculation or agglomeration requires careful attention to processing and requires almost care and appropriate controls. To prevent settling the systems employed are usually semisolid gels or emulsions. Elimination of crystal growth in suspensions is a highly complex issue. Where suspensions of solid materials such as physical sun screens are involved, consumer aesthetic attributes are a concern. Grittiness and tackiness are issues that may require formulation ingenuity to control them to levels acceptable to the uses.

### *Gels*

These materials consist essentially of liquid phases that either dissolve or stably suspend the drug substance in a matrix formed by use of a gellant that enclose domains of liquid in a three-dimensional framework. The liquid may be water or alcohol, oil, glycol, or mixtures of these. The gellant may be polymeric such as hydroxypropyl cellulose or polyvinyl alcohol or inorganic such as bentones or silica aerogels. Gels may also be prepared with surfactant systems. Gel systems offer many options in drug delivery and represent an increasingly employed vehicle choice. Gels prepared without emulsifiers are of particular interest in the case of topical liposomes. Care must be taken to ensure that the gellant does not interfere with drug release from the system as many of these materials have very high surface areas and adsorb materials tightly.

### *Ointments*

These vehicles have long occupied a prominent position for delivery of topical drugs. In premodern times natural fats and oils were exclusively used and more recently hydrocarbons and petrolatum have been extensively used. However, we now know that many drugs are not well released from these systems, and for this reason plus their greasiness they are now used less frequently. They do, however, represent a desirable approach to drug delivery especially in cases where the normal skin barrier has been compromised and emolliency and moisture retention are desirable additions to the drug effect itself. Also the occlusiveness of these materials tends to increase the hydration of the stratum corneum, which has a significant effect on penetration of many drugs.

A variety of ointment systems have been identified:

1. Hydrocarbons and petrolatum.
2. Natural fats and oils.
3. Polyethylene glycols.
4. Silicone ointments.

The last two in the list are of particular interest, respectively, because they allow handling of more hydrophilic drugs (polyethylene glycols) and because they are very inert with chemically unstable drugs and offer high occlusivity and unique solubility environments for cases where release and penetration control are necessary.

## *Emulsions*

By far the most common topical drug vehicles are emulsions. These consist of two phases: water and materials soluble in it and oils and fats plus materials soluble in them. Fluid emulsions are generally referred to as lotions and nonflowing ones as creams. Whether fluid or semisolid, emulsions are generally of two types: those in which the lipid is dispersed as fine (diameter = 0.5 to about 15 μm) droplets in a continuous aqueous phase are descried as oil-in water (o/w) emulsion, and those in which the aqueous phase is similarly dispersed in a continuous lipid phase are referred to as water-in-oil (w/o) emulsion. Of the two, o/w is preferred mainly due to its cosmetic attributes.

There are cases of mixed emulsions where the phase relationships are ambiguous as well as multiple emulsions of various types such as water-in-oil-in-water. These mixed emulsions are uncommon and do not at present represent commercially significant options. Technically they are also of problematic interest for a variety of practical reasons, particularly consumer aesthetics and physical stability.

Since emulsions represent enormously flexible drug delivery systems, they can be prepared with a wide range of properties. They can be prepared with either water thin or very thick, across a wide range of pH conditions, with a diverse set of aqueous and lipid component and over a considerable range of phase volume fractions. The continuous phase may be either aqueous or oil in nature. All of these factors play roles in various drug delivery scenarios. Hence, emulsions can successfully solve many of the problems that may present themselves to the formulators. In addition the emulsion is often a much more aesthetically acceptable option for the patient than a thick, greasy ointment.

### *O/w emulsions*

This category can vary widely in formulation and also generally has the most pleasant attributes. Depending on the emulsifier used, the pH of an o/w can range from as low as 3 to as high as 10. The o/w phase volume ratio can easily range from 5 parts or less of oil in 95 parts water at one extreme all the way to 70 parts or more of oil in 30 parts water. Preparation of these emulsions is usually relatively easy and their physical stability tends to be easily achieved. Drug solubilisation is aided by the range of phase volumes possible as well as by the considerable variety of lipids ranging from very low polarity to quite polar, which can be successfully emulsified. The o/w emulsion is relatively nonocclusive and, with care in selection of the emulsifier and excipients, is also very mild for use on inflamed, compromised, or injured skin.

### *W/o emulsions*

Water-in-oil emulsions are less flexible in formulation than their o/w counterparts, but offer the advantage of moderate occlusiveness as in petrolatum or hydrocarbon ointments but having aesthetic properties that, while inferior to o/w systems, are superior to pure petrolatum. Water-in-oil emulsions are reasonably easily prepared but are more demanding and harder to stabilise than o/w ones. These emulsions are also prone to phase separation during freeze–thaw cycling and therefore necessitate protection during shipping and storage to avoid freezing temperatures.

### *Other Emulsions*

Multiple emulsions are not yet employed in approved preparations mainly due to their impracticality and physical instability. The preparation of these emulsions is very complex and has not been demonstrated to result in pharmaceutically meaningful delivery systems. Additionally, multiple emulsions are inherently very prone to physical instability.

### Powders

Delivery of drug substances as powders is an ancient practice that persists in a few cases today. This method allows an intractable drug, usually having poor solubility, to be presented to the delivery site in a system that also features the possible benefit of reduction of friction and/or management of excess moisture as from perspiration at the site of application, for instance, on the feet or in intertriginous locations. Manufacture of these preparations must carefully address batch uniformity, and the drug substance must be carefully prepared and handled to avoid agglomerates that both perturb uniformity and may result in unacceptable grittiness during application by the patient. The use of cornstarch also requires careful attention to manufacturing conditions during handling to avoid the hazard of dust explosions.

### Solids

Topical application of drug active via solid, e.g. stick preparations, is limited to special cases such as oral areas where emollient preparations alleviate dryness and maceration to spot area application of ultraviolet absorbing materials and to the axilla for odour control. These vehicles are essentially ointments that have been solidified to a near waxy state. Stick preparations are usually lipid based except for a few that are based on ethanol solidified with sodium stearate or similar soap and are used to deliver antimicrobials to the axilla for control of odour. Those based on volatile cyclic silicone preparations solidified primarily with fatty alcohols are used for control of underarm perspiration and odour with aluminum–zirconium complex salts. Manufacturing of stick preparations requires careful control to avoid settling and resultant drug nonuniformity during formation of the final putup.

## Emulsifying Agents for Topical Pharmaceuticals

In all emulsions the successful combination of the aqueous and liquid phases in physically stable systems depends on use of emulsifying agents. Emulsifying agents are the most commonly used ingredients in topical pharmaceuticals. They are polar substances containing both hydrophilic and lipophilic groups. They belong to the group of substances of surface-active agents (surfactants) and include detergents, dissolving intermediaries and suspending or antifoaming agents. The technology of emulsification is extremely complex and entirely too large a subject to be fully addressed in the scope of this Chapter. Attention is drawn to the emulsifiers which are of relevance to topical pharmaceuticals.

Emulsifiers form a surface adsorption film around the liquid globule of the internal phase. The hydrophilic and lipophilic groups are so oriented that the original interfacial tension between the external and disperse phase is eliminated and two new interfaces are formed, whereby the original, high interfacial tension is reduced. Hydrophilic groups are oriented toward the aqueous phase by the interaction with secondary valencies, leading to binding of water molecules and solvation of the hydrophilic portion of the emulsifier in this phase. The lipophilic groups orient toward the oily phase and become similarly dissolved. The emulsifier film already mentioned is thus formed at the interface of the two phases. The resultant reduction of surface-free energy allows the formation of high surface area systems having small particle diameters of disperse phase.

The emulsion type is determined by relative affinity of the emulsifier to the aqueous and oily phase. If an emulsifying agent is more soluble in water than oil, the binding of water molecules by its hydrophilic groups exceeds the binding of oil molecules by its lipophilic groups. Therefore, the interfacial tension of the oil layer is dominant, i.e. the force of the interface molecule inward is stronger, and the lipid globules formed are pushed toward the centre to become enveloped by water molecules. An o/w emulsion ensues. More simply, the phase of an emulsion is the outer phase in which the emulsifying agent most readily dissolves. If an o/w emulsion is displaced with increasing quantities of a w/o emulsifying agent, then an emulsion inversion occurs from o/w to w/o emulsion.

There are three primary types of emulsifying agents being used in topical pharmaceuticals, ionogenic (anionic or cationic), nonionogenic (lauryl, cetyl, stearyl alcohols and their oxy ethers and ethoxylates), and complex emulsifiers (nonionogenic and ionogenic combinations). Also of potential usefulness are amphoteric materials having both anionic and cationic moieties attached to the same molecule. Lastly there are stearic techniques for stabilising emulsions that employ submicron-sized particulates or polymeric materials that are drawn to the oil–water interface and prevent droplet merging.

Emulsion selection involves many factors. Major ones include appropriateness for the type and thickness emulsion desired, compatibility with the drug to be delivered, and whether or not penetration enhancement is desired. In general, mixtures of emulsifiers are more efficient in producing physically stable emulsions than are single emulsifiers. It is desirable to use as little emulsifier as possible to avoid unwanted side effects such as irritation at the delivery site. Probably the most useful system for selection of emulsifiers is known as the hydrophile/lipophile balance (HLB) system. This method is based on the proportion of hydrophilic and lipophilic parts of the emulsifier. A low HLB value indicates that the emulsifier is more lipophilic in overall nature and a high value indicates that the emulsifier is more hydrophilic. The following provides a rough guide for selection of emulsifier systems having HLB values appropriate for various applications: w/o emulsions, 4–6; wetting agents, 7–9; o/w emulsions, 8–18; detergents, 13–15; and solubilisers, 15–18.

Experience has shown that in actual practice various materials require specific HLB emulsifier systems for optimal results in making stable emulsions. This has been addressed by the development of 'required HLB values' for various emollients. These are usually empirically determined by preparing test emulsions with different emulsifier blends having increasing HLB values. The blend that produces the most uniform and stable emulsion is identified as the required HLB for the material being tested. The following are required HLB values for a variety of fats and oils of pharmaceutical interest: cottonseed oil, 5.5; lanolin, 9.0; mineral oil (light), 10.5; peanut oil, 6.0; petrolatum, 7.0; and silicone oil (dimethicone), 5.0.

The preparations actually of interest may contain mixtures of fats and oils. Calculation of the appropriate emulsifier blend for mixtures is done by taking the individual required HLBs and adjusting to obtain the proportion based on the percentage of each material in the final blend. The adjusted values are then summed and the resulting number is the estimated required HLB for the mixture. Much experience has shown that, while this system is very useful and time saving, the best results are obtained by preparing several different mixtures having the estimated required HLB and determining which of these produces the best final result. Fine tuning of the blends and quantity of emulsifier used in the final formula may be needed to reach an optimum condition that provides both good physical properties and stability of the emulsion as well as optimum pharmaceutical properties such as drug delivery.

## OILS AND FATS AS INGREDIENTS IN TOPICAL PHARMACEUTICALS

As already noted, the formulation of topical pharmaceuticals is inextricably entangled with the use of oleaginous/fatty mixtures, which serve as components of their delivery systems. These vary widely in composition and form in response to particular pharmaceutical needs. Worldwide these material represent a dermatologic preparations business. Table 8.1 provides a sampling of formulations that employ fats and oils in topical preparations. The formulator must prepare a vehicle that, presents the active principle to the patient in a quantity and rate suitable for the intended pharmaceutical activity. The composition must be uniform and consistent in properties at all times so that the final composition is reliable when used whether in an ethical or over-the-counter situation.

**Table 8.1.** Examples of formulations employing fats and oils.

| *Body cream* | *Body oil* | *Lip balm* | *Suntan butter* |
|---|---|---|---|
| Canola oil | Hazelnut oil | Methyl gluceth 20 distearate | Ceraphyl 140A |
| Butyl oleate | Borage oil | Candelila wax | Mineral oil |
| Cetyl alcohol | Microcrystalline wax | Castor oil | Cocoa butter |
| Emulsifying wax NF | Coconut oil | | Octyl methoxycinnamate |
| Propylene glycol | Isopropyl palmitate | | Ozokerite |
| Water | Dimethicone | | Coconut oil |
| Preservative | | | |
| Fragrance | | | |

In every region of the world these concerns are addressed by the establishment of standards for both the drug active substances and for many of the excipients used in preparation of drug vehicle formulations. Compendia of these standards are established and maintained to ensure that the materials used are suitable in quality, strength, purity and identity for their intended use and that all concerned can depend on the materials meeting the established requirements.

In addition to the usual tests for industrial characterisation of such materials, the monographs usually include tests that specifically identify the material and that ensure freedom from adulterants that might affect its usefulness in drug preparations. The user is cautioned that the monographs are subject to continuous revision and updating. It is therefore necessary before using these materials to inspect the latest edition and/or supplement for the most current specifications.

In practice there are two rather distinct usages of fats and oils in topical pharmaceuticals, ethical pharmaceuticals, and over-the-counter (OTC) preparations. Distribution of ethical preparations are controlled and limited to physician prescription. Formulation of ethical preparations utilises a much more limited range of oleaginous excipients. Those which have been very well characterised and described in the official compendia are mush favoured. Novel materials having more limited recognition are sometimes utilised. However, due to the intense scrutiny given them in the review process, they tend to be less favoured unless their inclusion satisfies a compelling pharmacologic necessity. Ethical drugs must be approved by the Food and Drug Administration (FDA) prior to distribution. This approval involves submission of extensive data including manufacturing information and specifications for each component in the finished composition. FDA requirements and scrutiny of excipients is widely known to be increasingly close and demanding in recent years. This scrutiny does not preclude use of non-compendial materials, but it is widely believed that use of such materials results in less difficulty in clearing the so-called regulatory hurdle during review of the new drug application.

Over-the-counter drugs, available to the consumer on the basis of self-diagnosis, are a large, more varied business employing a greater range of excipients in formulation.

It must be recognised that there may be a considerable difference from country to country in terms of which materials are included in the local compendium as well as the properties and specifications requirements. In the present age of global business focus, these variations place considerable strain on the task of meeting the differing local requirements. There are efforts underway to harmonise these differences but the problem is by no means resolved nor is it likely to be eliminated in the near future. The user of these materials is cautioned to consult the regulations of those countries in which a particular finished drug is to be marketed to ensure that any fats and oils utilised are in conformity with appropriate requirements.

In general the use of fats and oils in pharmaceuticals makes use of the same properties that make them useful in nondrug topical preparations. Use in drug product formulas, however, calls for very strict control of quality and uniformity beyond that which may be required in other uses. Additional controls may be employed or emphasised for pharmaceutical excipients such as low arsenic and heavy metals limits. When use in a pharmaceutical is contemplated, the formulator must consult the appropriate compendia to determine any requirements that may apply to the specific situation being faced.

Over-the-counter drugs are an active growing segment in the skin care market. The rising cost of health care is causing pressure on the skin care pharmaceutical sector as well as elsewhere to identify conditions that can be reliably self-diagnosed by the patient and for which treatment drugs are available whose safety and efficacy profiles permit making them available OTC. Many companies are responding to this opportunity and intense competition for this business is a result. Since in most cases the drugs are limited in scope of activity and effectiveness, claims marketers are resorting to packaging and aesthetic differences as means of competing for consumers. However, in the case of ethical preparations, the selection of drug is almost entirely the provence of the attending physician whose selection is not usually a consequence of patient acceptablity of the vehicle but more likely on effectiveness in treating the patient's disease.

When elegant preparations are desired, the techniques of the cosmetic formulator are needed, but the range of materials available to the drug formulator are needed, but the range of materials available to the drug formulator is less varied.

A few materials are recognised in the United States as having specific drug properties when used at stated concentrations in suitable formulas. Lanolin, petrolatum, and mineral oil are recognised as skin protectants; when they are used within the established concentration ranges, the products containing them are permitted specifically worded claims. This practice is not, however, universally accepted. Outside the United States other requirements will be in effect. Table 8.2 presents the use of these materials in various types of preparations and the purpose served in each.

**Table 8.2.** Use of lanolin, petrolatum, and mineral oil in various preparations.

| *Material* | *Preparation* | *Function* |
|---|---|---|
| Lanolin | Baby creams and ointments | Hydrophobic skin protectant |
| | Dry skin preparations | Emollient, protectant, and retain moisture |
| Petrolatum | Diaper protectants | Hydrophobic skin protectant |
| | Night creams | Emollient retain moisture |
| | Sun care products | Wash off resistance |
| Mineral oil | Baby oil | Mild cleansing and retain moisture |
| | Cleansing creams | Oily soil removal |
| | Dry skin preparations | Emollient |

Oils and fats have a variety of skin care benefits. This is particularly true of the more general publications of interest of cosmetics and general skin care preparations producers. A growing trend in the marketplace over the last few years has been increasing emphasis on treatment preparations. These are often presented with effectiveness claims that are sometimes difficult to differentiate from those of prescription drugs. The term *cosmeceutical* is often applied to such products. Given the intense competition among manufacturers for consumer attention and the general need for self-medication, it is most likely that this overlapping of drug and cosmetic will continue.

## TECHNICAL PROBLEMS OF TOPICAL PHARMACEUTICALS

First, it must be understood that drug penetration into the skin is a relatively slow process with a low uptake of substance, in comparison to oral uptake. Difficulties are encountered in attempts to apply the generally accepted principles of pharmacology and pharmacokinetics to the field of external therapy, since these are defined primarily for the systemic application and subsequent distribution of drugs. Thus, modern terms, such as bioavailability of drugs, which describes the presence of active ingredient in central compartments, cannot be applied in the field of external therapy such as topical pharmaceuticals, where the sites of application and action are identical.

It is well known that alterations of the inert contents of drug tablets cause considerable variations in blood levels. It is also true that vehicle variation can influence the efficacy of topically applied drugs. Thus, in contrast to drugs administered orally, which in general exhibit uniformity in preparations by individual manufacturing companies, topically applied drugs have a range of preparations for different therapeutic requirements. Ointments, creams, lotions, powders, pastes, and suspensions containing the same drug, lead to considerable variations in dosage requirements. Since it is comparatively simple to determine the clinical effectiveness attained in the skin by a definite drug preparation in comparison to other preparations, any disadvantages of a particular carrier may be compensated for by a corresponding increase in dosage. However, in most cases, the optimal vehicle to drug reactions are not realised, meaning that different preparations with identical concentrations of a drug are encountered. Only in the search for the minimal dosage necessary for topical application of powerfully active drugs is importance attached to the penetration promoting or inhibiting properties of carriers. These latter studies illustrated the fundamental problem encountered in the external application of drugs; the carrier in this case is not purely auxiliary but has therapeutic activity of its own.

Skin permeability and drug release from the vehicle determine the essential diffusion processes. These are a product of the physicochemical properties of drugs and of the vehicles. This naturally excludes the possibility of a universal vehicle. Therefore, a theoretical and experimentally corroborated design for the composition or choice of vehicles is necessary for efficient local therapeutic agents. For example, tailored vehicles have been sought for some steroids, notably in the extensive experimental and clinical studies of fluocinolone acetonide and its ester. The solubility, release of substance from the vehicle to isopropyl myristate and the partition coefficient were investigated using model systems. The vehicle compositions under consideration were compared using the vasoconstrictor test. Thus, the optimal base was determined from which the most powerful and long lasting vasoconstrictive activity of fluocinolone acetonide acetate could be obtained.

As a result of such investigations the following general aspects were observed in the choice of such vehicle for substances to be applied locally:

1. The substance should be dissolved at near saturation levels in a phase of an emulsion or in a minimal quantity of a solvent (e.g. propylene glycol for fluocinolone acetonide acetate) added to a single-phase ointment.

2. At higher drug concentrations, the substance need only be adequately soluble in one phase of the vehicle or in the single phase vehicle, since further substance can become dissolved during the penetration process.

These considerations were substantiated by the comparatively high penetration rates of various 1 per cent hydrocortisone ointments, whereby the penetration is not as strictly vehicle-dependant as in the case with fluorinated steroids, which are usually applied at one-tenth of the hydrocortisone dose.

## REGULATORY REQUIREMENTS FOR TOPICAL PHARMACEUTICALS

*In vitro* investigations of drug availability or liberation from ointment base are considered today to be prerequisites for adequate diffusion to the ointment–skin interface. If a drug is poorly soluble, the higher the ratio of drug dissolved in the vehicle to the amount suspended, the greater the liberation attained. The drug concentration present should, however, not be lower than the theoretical saturation concentration, i.e. the amount that is soluble in the vehicle, because otherwise an unsuitable partition coefficient between the ointment and membrane or horny layer might be attained. This coefficient is ideally unity. Therefore, if the drug concentration equals or is slightly higher than the saturation concentration in the ointment, diffusion alone is rate limiting for its liberation.

Investigation of dermal and transdermal absorption of topical dermatology pharmaceuticals is essential for the evaluation of drug safety. Dermal absorption studies in animals should occur before human clinical trials. Absorption in humans is studied before marketing. Generally speaking, it is not possible to assume that absorption studies will define clinical efficacy. The efficacy evaluation is mostly based on *in vivo* clinical trials.

Preclinical studies should normally include toxicity testing in two mammalian species (rodent and nonrodent) using the formulation(s) intended for clinical trials. The duration and frequency of dosing and other details of toxicity testing are determined by the nature of the product and the proposed clinical trials. In all cases, short-term skin irritation studies usually in the rabbit, and skin sensitisation tests, normally in the guinea pig, will be required. Preparations intended for use on the face or scalp should also be tested for eye irritation in animals. In appropriate case photosensitivity testing may be required.

Pharmacokinetic studies should be conducted in the two animal species chosen for toxicity testing. Absorption, distribution, metabolism, and excretion (ADME) of the drug and its metabolites, where practicable, should be determined using chemical assays or radiolabeling techniques. If absorption of the drug is negligible (<1 per cent), then further studies by the topical route may not be required, i.e. repeat dose toxicity testing and reproductive testing. However, for drug with a potential teratogenic side effect, the reproductive testing of the topical preparations should be conducted. The fate of the active ingredient following absorption into the systemic circulation may of course be studied following parenteral administration.

During the clinical studies a formal pharmacokinetic analysis in humans will be required. It is desirable to conduct such an analysis as early as possible during clinical development in order to determine the significance of the animal studies. However, the pharmacokinetic studies do not affect the number of clinical trials or number of patients needed to establish clinical safety and efficacy. The formulation studied should be that intended for marketing. Plasma samples and fecal and urinary collection should also be taken. In human and animal studies monitoring of local and systemic effects should be undertaken in all toxicity and pharmacokinetic studies. The degree of absorption in human will determine, as in animals, the number and type of studies needed to establish safety.

Bioavailability of a systemic preparation is defined as the amount of substance available in the inner compartment within a definite time after application. It depends on the release of active ingredient from the vehicle and absorption into target sites. The influence of the pharmaceutical carrier on both *in vitro* release and *in vivo* availability of active ingredient is thereby investigated. Such investigations have become mandatory, since marked pharmacological and pharmacokinetic differences have been noted after administration of identical amounts of active ingredient in different preparations. If new formulations of established ingredients are developed, then, in most cases, a crossover bioavailability study comparing the new and old formulations should be undertaken.

Bioequivalence of topical pharmaceuticals is important to practicing clinicians, pharmacists, the pharmaceutical industry, and regulators. For regulators in the FDA, a simple, reliable and relatively inexpensive method of determining bioavailability and bioequivalence is critically needed. For clinicians, potency and safety are key considerations in selecting appropriate topical pharmaceuticals from among the many different agents, strengths and vehicles available. For the pharmaceutical industry, simple, reliable and inexpensive nonclinical tests, *in vitro* or *in vivo* animal models, represent not only cost savings in R & D expenditures but also improvement in the screening process for formulation optimisation.

At present, generic topical products must be pharmaceutically equivalent to the corresponding pioneer product before an Abbreviated New Drug Application (ANDA) can be accepted in the Office of Generic Drugs of the FDA. To be pharmaceutically equivalent to its innovator counterpart, any generic drug must contain the same active ingredient, in the same concentration, be intended for the same route of administration, and have comparable labeling. For a topical generic product, the vehicle in which the drug is suspended or dissolved should be the same as that of the innovator product. When administration of a drug does not produce concentrations in the body sufficiently high for measurement, other methods to measure bioequivalence, in descending order of preference, include: (i) an *in vivo* pharmacodynamic study, (ii) a comparative clinical trial, (iii) a comparative blood level study in animals, (iv) an *in vitro* study, and (v) any other method the agency believes to be acceptable to document bioequivalence.

Since no accepted nonclinical models or approaches to predict or determine the bioavailability and bioequivalence of topical pharmaceuticals are now available, the bioequivalence assessment of test and reference products is based on studies with clinical end points or pharmacodynamic measurement.

Difficult questions have also been raised about how to ensure bioavailability and bioequivalence when (i) minor formulation changes and (ii) process changes are involved during scale-up. Can *in vitro* and/or *in vivo* studies of drug uptake in skin answer these questions? Reliable data are not available to support any such approach. With improvements on the analytical assay and modern pharmacodynamic methods, it may be possible to develop a bioequivalent model to predict drug retention based on experimental information for a given drug.

It should be emphasised that these requirements are to be interpreted flexibly. For example, much more care and consideration should be utilised when evaluating the potential absorption of a potent new drug such as corticosteroid, antibiotic, antifungal, or a potential teratogenic retinoid compared with a cosmetic, moisturising, soothing, or protective agent.

## DRUG DELIVERY

### Microencapsulation

Microencapsulation is the envelopment of small solid particles, liquid droplets or gas bubbles with a coating. Contents of microcapsules are called active agent or actives or internal phase (IP). The coating material is known as shell or wall or membrane.

The unique feature of a microcapsule is its size; i.e. between 1 and 1000 μm. Capsules smaller than 1 μm are called nanocapsules. The use of microemulsions in microencapsulation offers several advantages:

1. Microemulsions protect peptides from degra-dation and enhance peptide absorption.
2. Higher rate of drug encapsulation (86 per cent) into nanocapsules than when prepared with coarse emulsions (55 per cent).
3. Uniform aqueous droplet size (200 nm) eliminates need for high energy ultrasonification or homogenisation.

In Fig. 8.1a, reactants dissolve in the dispersed phase diffuse to the interface formed by an immiscible suspending medium, where they react with coreactants dissolved in the suspending medium. The dispersed phase to be encapsulated can be water or water-immiscible solvent.

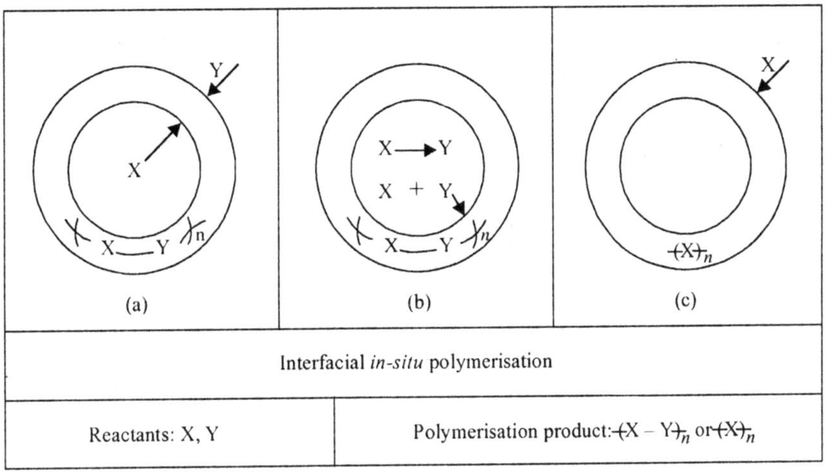


**Fig. 8.1.** Reactants dissolve in different phases.

In Fig. 8.1b, isocyanate polymerisation in a dispersed solvent phase is used to form the capsule wall. Isocyanate groups partially react with water to give amines and $CO_2$. These amines and the remaining isocyanate groups then react to form a polyurea capsule wall.

In Fig. 8.1c, an aqueous phase is emulsified in a water-immiscible solvent and *n*-alkyl cyanoacrylate added to the organic phase. It diffuses to solvent water interface where it polymerises spontaneously to form a poly (*n*-alkyl cyanoacrylate) capsule wall.

### *Microemulsion drug delivery systems*

*Characteristics*

1. Long storage stability.
2. Stabilised by surfactant (in conjunction with cosurfactant).
3. Good penetrating power.
4. Highly adhesive to vertical surfaces.
5. Easy to prepare with low cost.

### *Applications*

With oil in water microemulsions it is possible to incorporate hydrophobic drugs, thereby enhancing their solubility as well as bioavailability.

A futuristic biphasic dosage for insulin combines the advantages of microemulsion drug delivery systems with those of nanocapsules. Here insulin is nanoencapsulated into a medium (base), which itself is a microemulsion of insulin. The micro-emulsified insulin will cause rapid release of insulin into the circulation while nanoencapsulated insulin will cause a more extended release. This is because the microemulsion will act immediately when the patient takes the dosage, whereas the insulin will diffuse out of the nanocapsules because of concentration gradient. Thus, the gap between two dosages is increased (Fig. 8.2).

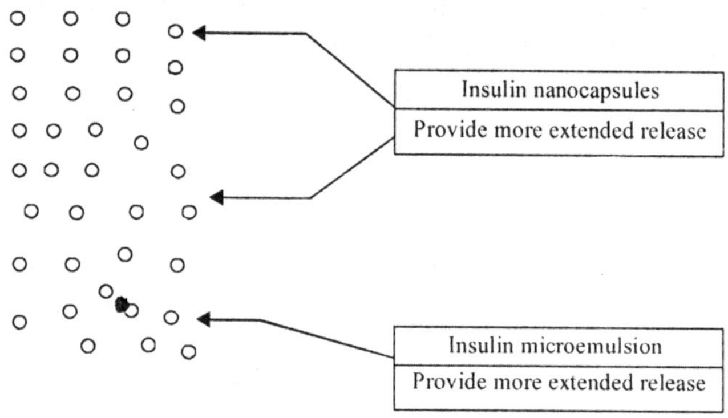


**Fig. 8.2.** Biphasic dosage to improve bioavailability of insulin and tamoxifen using microemulsions.

Chapter 9

# Fats and Oils in Natural Waxes

## INTRODUCTION

Similar to vegetable or animal fats, natural waxes such as, for example, beeswax also have been used for a long time. Even now they have numerous industrial and cosmetological applications. However, some of them, on account of their rarity or cost, are increasingly being replaced by synthetic waxes with equivalent characteristics. Traditionally, waxes are substances whose properties are close to those of beeswax, which is the 'reference wax'.

Physically, with very few exceptions as spermaceti oil, or jojoba oil, waxes are amorphous, opaque, compact, greasy solids. They have a high melting point (between 40 and 90°C), and are shiny, penetrable, slippery and of a consistency varying with each source (ranging from a soft and greasy consistency in the case of beeswax to a dry and brittle one for Carnauba wax). They are generally soluble in aromatic and chlorinated solvents, ethers, esters and ketones and are partially soluble in aliphatic hydrocarbons and alcohols. Of course, like any other fat from natural sources, they have widely varying properties (and compositions) depending on the soil, climate, their age, degree of oxidation, impurities and possible adulterations as well as the treatments (extractions, refinings, winterisations, etc.).

Functionally, the term 'wax' refers to numerous synthetic substances used as substitutes of natural waxes in some preparations as well as formulated products: furniture waxes, shoe polish, etc.

Chemically, waxes are substances containing higher mono-and difunctional alcohols and fatty acids with high molecular weight, sometimes free but mostly combined through esterification.

In order to remain within the subject matter of this work dealing with fats, this chapter will, thus eliminating especially the hydrocarbon-based waxes:

1. Paraffin and microcrystalline waxes, extracted from petroleum crudes during refining operations.
2. Ozocerite (or ozokerite), ceresin and Utah wax, extracted from bituminous soil.

Waxes are widely represented in nature and have origins in each of the following three classifications: mineral, animal and vegetable. The aim of this chapter is to indicate the source, extraction mode, chemical composition, physical and chemical characteristics and main uses of each wax.

## NATURAL WAXES OF MINERAL ORIGIN

Mineral waxes, mainly made up of fatty esters, are montan wax, lignite waxes and peat waxes. These fossil waxes represent an evolutionary stage in the slow transformation of decomposing vegetable matter into hydrocarbons.

## Montan Wax

This wax is obtained through solvent extraction of certain lignite varieties after grinding and drying. While term montan is traditionally reserved for certain deposits in East Germany, equivalent products have been obtained from lignites in California, West Germany and more rarely in Czechoslovakia and Russia.

The chemical composition (Table 9.1) and the characteristics (Table 9.2) of montan wax are very much affected not only by its origin but also by the nature of extraction solvent (usually a mixture of ethyl alcohol and aromatic solvent). The composition is mostly a mixture of free fatty acids and fatty esters based on fatty alcohols of C22 to C30 and fatty acids of C22 to C32. It also contains resinic derivatives (di- and triterpenes, more or less aromatic polycyclic hydrocarbons, etc.), asphaltenes and ketones (unbranched aliphatic methyl ketones and symmetric ketones of C25 to C35, especially odd numbered ones).

Montan wax is a solid, its colour ranging from light brown to dark brown (depending on its resin content), and is very tough (cannot be easily scratched with the finger nails). It can be refined with partial elimination of resins and asphaltenes by means of a vacuum distillation, a steaming or a solvent extraction (ether, acetone, methanol, aliphatic hydrocarbons or esters, etc.) and then, if required, treated with bleaching earth. Its colour then becomes lighter (amber to light brown) and it hardens further.

**Table 9.1.** Typical chemical compositions of crude montan wax.

| *Composition (% of wax)* | *Ex-East Germany* | | *California* | |
|---|---|---|---|---|
| Free fatty acids | 17 | | 17 | |
| Free fatty alcohols | 2 | | 1 | |
| Fatty esters | 53 | | 57 | |
| Ketones | 3 | | 3 | |
| Resins | 22 | | 14 | |
| Asphaltenes | 3 | | 8 | |
| *Composition of carbon chain (% of each component)* | *Acid* | *Alcohol* | *Acid* | *Alcohol* |
| C20 | – | – | – | – |
| C21 | 1 | – | – | – |
| C22 | 2 | 1 | 7 | 19 |
| C23 | 2 | – | – | – |
| C24 | 6 | 5 | 13 | 12 |
| C25 | 3 | – | – | – |
| C26 | 14 | 13 | 11 | 35 |
| C27 | 5 | 1 | – | – |
| C28 | 21 | 23 | 43 | 34 |
| C29 | 6 | 2 | – | – |
| C30 | 24 | 29 | 26 | – |
| C31 | 3 | 2 | – | – |
| C32 | 9 | 18 | – | – |
| C33 | 4 | 6 | – | – |

**Table 9.2.** Average characteristics of mineral waxes.

| *Type origin* | *Montan (Ex-GDR)* | *Montan (California)* | *Lignite (Devon)* | *Peat (England)* | *Peat (Scottland)* |
|---|---|---|---|---|---|
| Content in the mineral,% | 4–14 | 4–14 | 4–6 | 5–10 | 3–12 |
| Density at 20°C | 0.99–1.03 | 1.03 | 1.04 | 1.03 | – |
| Melting point, °C | 80–86 | 84–87 | 72–84 | 63–68 | 64–73 |
| Acid value | 26–33 | 43–50 | 29–34 | 50–55 | 43–51 |
| Saponification value | 66–92 | 105–114 | 72–77 | 116–119 | 107–116 |
| Ashes, % | 0.2–0.4 | 0.3–0.5 | 0.6–0.8 | – | – |
| Resins, % | 14–23 | 11–20 | 40–45 | 20–23 | 20–25 |
| Asphaltenes, % | 3–5 | 8–12 | – | 10–15 | – |

It is employed in considerable quantities as a low and more regulated cost substitute of Carnuba wax, when the colour is immaterial, especially in 'polishes'. It is also used for manufacture of carbon paper, in hydraulic fluids, electrical insulation materials, as a covering agent for phytosanitar products.

### Lignite Wax

Both lignite and montan waxes have the same origin. However the former comes from less noble deposits. It usually has a higher resin content (at times may be as high as 50 to 95 per cent), depending on the quality of lignite, extraction solvent and the temperature reached during extraction (Table 9.2.) It is of dark brown colour, very tough and brittle, odourless and can be refined as in the case of montan wax. Its applications are identical to those of montan wax.

### Peat Wax

English, Scottish, Irish or New Zealand peats are solvent-extracted (e.g. benzene) after drying and grinding. They yield 3–12 per cent of a wax of dark brown to black colour, which is tough, brittle, more easily breakable than montan and lignite waxes. Its acid and saponification values are higher than those of above waxes (Table 9.2), but its applications are the same.

## NATURAL WAXES OF ANIMAL ORIGIN

Animals, ranging from one of the tiniest insects to one of the biggest marine mammals, produce numerous waxes. Foremost among these waxes are: Beeswax, Chinese wax, gum-lac wax, spermaceti wax (or white of whale) and lanolin (or refined wool grease or wool wax).

### Beeswax

Beeswax is secreted by eight wax-producing glands located on the ventral surface of worker bees, *Apis mellifica* at a proportion of one part beeswax for eight parts honey. Its use is to construct after mastication honeycombs of bees. These are fused and mixed with boiling water for eliminating honey. The floating fatty matter consist of raw beeswax of colour range from light yellow to brown, depending on the nature of the flowers frequented by bees, age of honeycombs and proportion of some natural contaminants.

This crude wax is then subjected to a refining process again consisting of fusion and washing and if necessary, earth followed by a decolouration on diatom earth and/or on activated carbon and a filtration.

It can also be bleached by oxidising agents (potassium permanganate, bichromate, peroxides, etc.) or by exposure to sun.

Beeswax can be easily saponified and emulsified on account of the sizeable presence of free fatty acids on one hand and by diols and hydroxy acids and their esters on the other. Its main constituents are: palmitate, palmitoleate, hydroxypalmitate and oleate of higher alcohols (mostly myricil C30 and lacceryl C32), cerotic acid ($C_{25}H_{51}$ COOH) and odd-numbered hydrocarbons (heptacosane and nonacosane mainly) composition of a refined beeswax per cent of the wax is shown in Table 9.3 and composition of carbon chain per cent of each componet is shown in Table 9.4. Its minor constituents are: sterols (cholesterol as well as lanosterol, beta sitosterol ) in ester form, terpenoides and pheromones of honeybees (geraniol, farnesol, etc.), volatile compounds responsible for the characteristic odour of beeswax (aldehydes in C8, C9 and C10, furfural, benzaldehyde, etc.) and lactones (myristo and palmitolactones). Its composition and characteristics fluctuate in a noticeable manner among the 16 varieties of *Apis mellifica* and also among the other types of honeybees *(Apis florea and apis cerana).*

Natural beeswax has a characteristic honey smell, is tough and brittle when cold. However it becomes plastic when held in hand. It is fully compatible with the entire group of fats (glycerides, waxes, hydrocarbons, alcohols and fatty acids, etc.) and is actually often adulterated by them (tallow, stearic acid, paraffin, microcrystalline wax, ceresine, Japan wax). Its physical and chemical characteristics are summarised in Table 9.10.

Among all natural waxes, beeswax is undoubtedly the most famous and the oldest. It was used as a binder by the Lascaux painters and as a preserving agent by the embalmers of ancient Egypt. It is still used in making candles for religious customs and services.

It is used in cosmetic and pharmaceutical applications (creams, lotions, ointments, etc.), noble ingredient of industrial and service products (shoe polish, furniture polish, etc.) and for traditional uses (mainly candles).

**Table 9.3.** Composition of a refined beeswax (percentage of the wax).

| *Constituent (in crude form)* | *Characteristic value* | *Range* | *Constituent (after saponification)* | *Characteristic value* | *Range* |
|---|---|---|---|---|---|
| Free acids | 13 | 8–16 | Fatty acids | 33 | 23–37 |
| Free alcohols | 1 | 0–2 | Hydroxyacids | 13 | 5–17 |
| Total esters | 71 | 68–80 | Fatty alcohols | 31 | 30–35 |
| Comprising: | | | Diols | 3 | 1–4 |
| Monoesters | 44 | 23–57 | Sterols | 1 | 0–2 |
| Hydroxyesters | 12 | 8–24 | | | |
| Di and triesters | 14 | 10–16 | | | |
| Esters of sterols | 1 | 1–2 | | | |
| Hydrocarbons | 13 | 10–17 | Hydrocarbons | 13 | 10–17 |
| Lactones, flavons terpenes, pheromones | 1 | 0–2 | Lactones, flavons terpenes, pheromones | 1 | 0–2 |
| Water and impurities (pollen, bee glue, pigments) | 1 | 1–4 | Water and impurities (pollen, bee glue, pigments) | 1 | 1–4 |

**Table 9.4.** Composition of carbon chain (per cent of each component).

| *Chain* | *Hydrocarbons* | *Free fatty acids* | *Fatty acids in esters* | *Total hydroxy acids (*)* | *Free alcohols* | *Alcohols in esters* | *Total diols* | *Fatty monoesters* | *Hydroxy esters* |
|---|---|---|---|---|---|---|---|---|---|
| C12:0 | – | – | 0–1 | 0–1 | – | – | – | | |
| C14:0 | – | 0–1 | 1–2 | 1–2 | – | – | – | | |
| C16:0 | – | 0–5 | 63–74 | 60–72 | – | – | – | | |
| C16:1 | – | | 17–21 | | – | – | – | | |
| C17:0 | – | – | t | 0–1 | – | – | – | | |
| C18:0 | – | 0–3 | 1–2 | 4–10 | – | – | – | | |
| C18:1 | – | | 7–12 | | – | – | – | | |
| C19:0 | 0–1 | – | – | – | – | – | – | | |
| C20:0 | – | | 0–1 | | – | 0–1 | t | | |
| C20:1 | – | 0–1 | 0–1 | 0–4 | – | – | – | | |
| C21:0 | 0–1 | t | – | – | – | – | – | | |
| C22:0 | t | 2–4 | t | 2–10 | t | 0–2 | t | | |
| C23:0 | 2–4 | 0–1 | – | t | – | – | – | | |
| C24:0 | t | 28–45 | 0–2 | 3–8 | 0–25 | 4–15 | 15 | | |
| C25:0 | 6–8 | 0–1 | – | t | – | – | – | | |
| C26:0 | 0–1 | 10–15 | 0–1 | 4–9 | 0–5 | 6–10 | 20 | | |
| C27:0 | 24–32 | – | – | t | – | – | t | | |
| C28:0 | 0–1 | 10–14 | 0–1 | 1–2 | 0–5 | 13–15 | 39 | | |
| C29:0 | 16–19 | – | – | t | – | – | 2 | | |
| C29:1 | t | – | – | t | – | – | | | |
| C30:0 | 0–1 | 8–14 | 0–1 | 0–1 | 40–90 | 33–41 | 15 | | |
| C31:0 | 13–19 | – | – | – | – | – | 2 | | |
| C31:1 | 4–6 | – | – | – | – | – | | | |
| C32:0 | t | 7–13 | 0–1 | – | 10–30 | 21–28 | 6 | | |
| C33:0 | 4–8 | – | – | – | – | – | – | | |
| C33:1 | 10–15 | – | – | – | – | – | – | | |
| C34:0 | t | 8–16 | 0–1 | – | 0–5 | 4–5 | – | | |
| C35:0 | 1–2 | – | – | – | – | – | – | | |
| C35:1 | t | – | – | – | – | – | – | | |
| C36:0 | – | 1–6 | t | – | – | 0–1 | – | | |
| C38:0 | – | t | – | – | – | – | – | | |
| Total C40 | | | | | | | | 12–19 | 12 |
| Total C42 | | | | | | | | 9–15 | 14 |
| Total C44 | | | | | | | | 9–17 | 13 |
| Total C46 | | | | | | | | 20–32 | 26 |
| Total C48 | | | | | | | | 20–36 | 26 |
| Total C50 | | | | | | | | 5–15 | 8 |
| Total C52 | | | | | | | | 0–2 | 1 |
| Total C54 | | | | | | | | – | – |

* Mainly ω–1–hydroxy (e.g: 15–hydroxy palmitic acid) with traces of ω-and ω–2–hydroxy., t: traces or less than 0.2 per cent.p6

## Chinese Wax

Chinese wax (or insect wax) is secreted by the insects of the *Coccus cerriferus* family, which deposit it over the tree branches which form their habitat. They are found in China (1,500 insects are required to produce 1 to 2 gram wax). This is scraped and refined by fusion in hot water, skimming, filtration and if required, bleaching.

Chemically it is mainly made up of ceryl cerotate ($C_{25}H_{51}$— COO — $C_{26}H_{53}$). Its average composition, far simpler than that of beeswax, is shown in Table 9.5 and its specifications are summarised in Table 9.10.

**Table 9.5.** Composition of a refined Chinese wax (% of the wax).

| *Constituent (on crude)* | *Content* | *Constituent (after saponification)* | *Content* |
|---|---|---|---|
| Free fatty acids | 0–1 | Fatty acids | 46–52 |
| Free fatty alcohols | 0–1 | Fatty alcohols | 44–50 |
| Fatty esters | 93–95 | | – |
| Hydrocarbons | 2–3 | Hydrocarbons | 2–3 |
| Resins | 0–1 | Resins | 0–1 |

*Nature of hydrocarbon chain (% of each component)*

| *Chain* | *Hydrocarbons* | *Total alcohols* | *Total acids* | *Esters* |
|---|---|---|---|---|
| C14 | – | – | 7 | |
| C16 | – | – | 2 | |
| C18 | – | – | 3 | |
| C20 | – | – | t | |
| C22 | – | – | 1 | |
| C24 | – | 2–7 | 14 | |
| C25 | 4 | – | – | |
| C26 | t | 63–75 | 49 | |
| C27 | 5 | – | – | |
| C28 | 1 | 21–28 | 16 | |
| C29 | 8 | – | – | |
| C30 | 3 | 2–4 | 6 | |
| C31 | 42 | – | – | |
| C32 | 1 | – | 2 | |
| C33 | 29 | – | – | |
| C34 | t | – | – | |
| C35 | 6 | – | – | |
| C46 | | | | 1 |
| C48 | | | | 2 |
| C50 | | | | 11 |
| C52 | | | | 37 |
| C54 | | | | 25 |
| C56 | | | | 14 |
| C58 | | | | 8 |
| C60 | | | | 2 |

t: Traces or less than 0.2%.

Chinese wax is yellowish white, glossy and crystalline. It is mainly used for making candles and in maintenance products such as furniture polish.

## Shellac Wax

Shellac is a resinous substance produced by a cochineal insect found in India, namely *Tachardia Lacca.* It is used in making varnish. During its production there is a phase of solubilisation in alcohol or in an alkaline solution. This causes precipitation of a glossy by-product known as shellac wax.

Shellac consists of an interesterified hydroxyacid-based resin and of 5–10 per cent wax. This wax comprises of linear fatty esters, which are longer than those of Chinese wax (Table 9.6) and, depending on the quality of separation contains more or less sizeable proportions of resins and hydrocarbons.

**Table 9.6.** Composition of a shellac wax (% wax).

| *Constituent (on crude)* | *Content* | *Constituent (after saponification)* | *Content* |
|---|---|---|---|
| Free fatty acids | 1–4 | Fatty acids | 21–36 |
| Free fatty alcohols | 8–14 | Fatty alcohols | 61–77 |
| Fatty esters | 70–82 | | – |
| Hydrocarbons | 1–6 | Hydrocarbons | 1–6 |
| Resins | 0–4 | Resins | 0–4 |

*Nature of hydrocarbon chain (% of each component)*

| *Chain* | *Hydrocarbons* | *Alcohols* | *Acids* |
|---|---|---|---|
| C14:0 | – | – | 1–5 |
| C14:1 | – | – | 0–2 |
| C16:0 | – | – | 3–17 |
| C16:1 | – | – | 0–2 |
| C18:0 | – | – | 0–4 |
| C18:1 | – | – | t |
| C19:0 | – | – | 0–1 |
| C20:0 | – | – | 0–2 |
| C22:0 | – | – | 0–1 |
| C24:0 | – | 0–1 | 0–1 |
| C25:0 | 0–3 | – | t |
| C26:0 | 0–1 | 0–1 | t |
| C26:1 | – | t | t |
| C27:0 | 38–42 | t | 0–1 |
| C28:0 | 2–4 | 62–67 | 6–19 |
| C29:0 | 30–36 | 0–1 | 1–2 |
| C30:0 | 1–3 | 21–22 | 7–26 |
| C31:0 | 11–13 | t | 1–2 |
| C32:0 | t | 9–10 | 18–28 |
| C34:0 | 2–3 | 1–3 | 18–25 |
| C36:0 | t | t | 0–2 |

t: Traces or less than 0.2%.

Shellac wax is hard, its colour ranging from light brown to dark brown depending on the method of separation from resin. Thanks to its toughness and gloss it is similar to Carnauba wax.

It is mainly used as a component of electrical insulation materials and as a substitute for Carnauba wax in its different applications (grease, paint stick, carbon paper, shoe and furniture polish, etc.).

## Spermaceti or White of Whale

Spermaceti (or white of whale or cetaceum or spermaceti wax) is a wax extracted either from a large cavity located in the head of sperm-whale *Physeter macrocephalus,* or from its adipose tissues.

1. This frontal organ probably serves as an ultrasound echo receiver as well as a float regulator. It consists almost exclusively of a spongy mass (2.5 to 3 t for a 15 m long, 39 t sperm-whale), from which head oil (or spermaceti oil or spermaceti) is extracted, on the one hand, and a solid portion, the white of whale (or spermaceti wax or again just spermaceti), on the other hand.
2. Melted flare of sperm-whale (9 to 10 t in the preceding example) also yields white of whale (about 10–12 per cent) after cold filtration and spermaceti oil, with a higher triglyceride content than head oil, (Spermaceti oil derived from the head or from the flare can also be hydrogenised to obtain synthetic spermaceti).
3. The raw white of whale is then refined by boiling with diluted soda lye, separated and washed with hot water. Finally it is pressed into cakes.

The chemical compositions of the different fats derived from sperm-whale are summarised in Tables 9.7 and 9.8. Spermaceti wax is almost fully made up of fatty esters, which are mainly cetyl palmitate base ($C_{15}H_{31}$— COO — $C_{16}H_{33}$). Some authors mention the presence of triglycerides, their concentration depending on the quality of separation from spermaceti oil (their main constituents being myristic and palmitic acid), and presence of a small percentage of cholesterol esters.

Processed white of whale comes in the shape of glossy and odourless large, white and brittle crystals. It can be reduced to powder, and it is compatible with beeswax and fats (stearic acid, oils).

**Table 9.7.** Composition of spermaceti wax, 'white of whale' (percentage of wax).

| *Constituent (on crude)* | *Content* | *Constituent (after saponification)* | *Content* |
|---|---|---|---|
| Fatty esters | 90–95 | | |
| Free fatty acids | 0–2 | Total acids | 49–55 |
| Free alcohols | 1–3 | Total alcohols | 45–50 |
| Triglycerides | 0–5 | Hydrocarbons | traces |
| Hydrocarbons | traces | | |

*Nature of carbon chain of spermaceti wax (percentage of each component)*

| *Chain* | *Total alcohols* | *Total acids* | *Chain* | *Esters* |
|---|---|---|---|---|
| C10:0 | – | 0–1 | Total C26 | 0–2 |
| C11:0 | – | t | Total C27 | 0–1 |
| C12:0 | – | 10–15 | Total C28 | 3–19 |
| C13:0 | – | 0–2 | Total C29 | 1–2 |

*(Contd...)*

**Table 9.7.** Composition of spermaceti wax, 'white of whale' (percentage of wax). *(Contd .)*

| *Chain* | *Total alcohols* | *Total acids* | *Chain* | *Esters* |
|---|---|---|---|---|
| C14:0 | 3–6 | 26–37 | Total C30 | 25–36 |
| C14:1 | – | t | Total C31 | 2–3 |
| C15:0 | 1–2 | 1–2 | Total C32 | 23–42 |
| C16:0 | 58–80 | 30–40 | Total C33 | 1–3 |
| C16:1 | 1–2 | 1–4 | Total C34 | 8–18 |
| C17:0 | 2–3 | 1–4 | Total C35 | 0–1 |
| C17:1 | – | t | Total C36 | 2–7 |
| C18:0 | 12–29 | 5–16 | Total C37 | t |
| C18:1 | 0–8 | 0–6 | Total C38 | 0–3 |
| C19:0 | – | 0–1 | | |
| C20:0 | 0–2 | 2–5 | | |

**Table 9.8.** Composition of spermaceti oil (percentage of oil).

| *Constituent (on crude)* | *Content* | *Constituent (after saponification)* | *Content* |
|---|---|---|---|
| Fatty esters | 66–74 | | |
| Free fatty acids | 0–3 | Total acids | 52–60 |
| Free alcohols | 3–5 | Total alcohols | 38–48 |
| Triglycerides | 21–34 | Hydrocarbons | 0–1 |
| Hydrocarbons | 0–1 | | |

*Nature of the carbon chain of spermaceti oil (percentage of each component)*

| *Chain* | *Total alcohols* | *Total acids* | *Chain* | *Esters* |
|---|---|---|---|---|
| C10:0 | – | 0–1 | Total C26 | 1 |
| C12:0 | – | 2–4 | Total C27 | t |
| C12:1 | – | 0–1 | Total C28 | 4 |
| C13:0 | – | t | Total C29 | 1 |
| C14:0 | 1–3 | 4–7 | Total C30 | 9 |
| C14:1 | 0–1 | 3–5 | Total C31 | 1 |
| C15:0 | 0–1 | 0–1 | Total C32 | 19 |
| C15:1 | – | t | Total C33 | 3 |
| C16:0 | 28–36 | 8–14 | Total C34 | 28 |
| C16:1 | 3–9 | 16–21 | Total C35 | 3 |
| C16:2 | t | 0–1 | Total C36 | 19 |
| C17:0 | 1–2 | 0–1 | Total C37 | t |
| C17:1 | 1–2 | 1–2 | Total C38 | 8 |
| C18:0 | 4–5 | 1–2 | Total C39 | t |

*(Contd...)*

**Table 9.8.** Composition of spermaceti oil (percentage of oil). *(Contd .)*

| *Chain* | *Total alcohols* | *Total acids* | *Chain* | *Esters* |
|---|---|---|---|---|
| C18:1 | 40–48 | 27–37 | Total C40 | 2 |
| C18:2 | 0–1 | 1–2 | | |
| C19:0 | t | – | | |
| C19:1 | 0–1 | t | | |
| C20:0 | 0–1 | 0–1 | | |
| C20:1 | 3–8 | 10–14 | | |
| C20:2 | 0–1 | t | | |
| C22:0 | t | t | | |
| C22:1 | – | 2–4 | | |

t: Traces or less than 0.2 per cent.

It is used mainly in cosmetic and pharmaceutical formulations. Today it has become virtually unavailable due to the ban on whale hunting. It has hence been replaced with synthetic whites of whale (based on cetyl palmitate).

## Wool Grease and Lanolin

Wool wax or wool grease is secreted by the sebaceous gland of sheep. It is extracted by centrifuging the detergent wash-solutions of raw wool. It is found in the form of a sticky wax having a colour ranging from light to dark brown and having a more or less foul smell. It is then refined by means of an aqueous alkaline solution for eliminating foreign materials and free fatty acids. It is then bleached, deodourised and dried to get lanolin.

Lanolin, given its very complex fatty ester based compsition, is technically a wax. But, on account of its properties and uses, it is often considered as a traditional animal fat. It has the speciality of having as its main constituents esters formed by aliphatic alcohols, which are more or less branched, triterpenic alcohols and sterol with fatty acids and hydroxylated acid, which are more or less branched (Table 9.9).

**Table 9.9.** Lanolin composition (percentage of wax).

| *Constituent (on crude)* | *Content* | *Constituent (after saponification)* | *Content* |
|---|---|---|---|
| Free acids | 0–1 | Fatty acids | 29–38 |
| Esters of triterpenic alcohols and sterols | 29–44 | α-Hydroxyacids<br>ω-Hydroxyacids | 11–15<br>1–3 |
| Esters of aliphatic alcohols | 14–24 | Polyhydroxy acids | 3–6 |
| Monohydroxyesters of sterols, aliphatic and triterpenic alcohols | 15–20 | Aliphatic alcohols<br>α-Diols<br>Polyols | 5–15<br>4–6<br>3–8 |
| Free aliphatic alcohols | 6–20 | Cholesterol | 11–20 |
| Free triterpenic alcohols and sterols | 4–5 | Lanosterol<br>Dihydrolanosterol | 1–8<br>1–5 |

*(Contd...)*

**Table 9.9.** Lanolin composition (percentage of wax). *(Contd .)*

| *Constituent (on crude)* | *Content* | *Constituent (after saponification)* | *Content* |
|---|---|---|---|
| Di-and polyhydroxy esters and free diols | 7–9 | 7–ketocholesta–3,5–diene<br>3–β–hydroxy–7–keto-lanosta–8 ene | 2–4 |
| Oxidation derivatives | 2–5 | | 1–4 |
| Hydrocarbons | 0–1 | Oxidation derivatives | 2–5 |
| | | Hydrocarbons | 0–1 |

*Nature of the lanolin carbon chain (% of each component)*

| *Chain* | *Fatty acids* | *α-Hydroxyacids* | *Fatty alcohols* | *1–2–Alcane diol* |
|---|---|---|---|---|
| C10 n | t | t | – | – |
| i | t | – | – | – |
| C11 n | t | t | – | – |
| ai | 0–1 | – | – | – |
| C12 n | t | t | t | t |
| i | 0–1 | t | t | – |
| C13 n | t | t | – | – |
| ai | 0–1 | t | – | – |
| C14 n | 2–4 | 2–3 | t | t |
| i | 2–5 | t | t | t |
| C15 n | t | 0–1 | – | – |
| ai | 3–10 | 0–1 | t | t |
| C16 n | 4–5 | 62–71 | t | 2–19 |
| i | 3–11 | 1–2 | 0–1 | 1–2 |
| C16 : 1 | 0–1 | – | – | – |
| C17 n | t | 0–1 | – | t |
| ai | 3–6 | 0–1 | 1–4 | t |
| C18 n | 2–4 | 1–3 | 0–1 | 1–4 |
| i | 4–7 | 8–13 | 0–1 | 20–28 |
| C18 : 1 | 0–1 | – | – | – |
| C19 n | t | 0–1 | – | t |
| ai | 4–7 | t | 1–3 | 2–5 |
| C20 n | 0–1 | t | 1–3 | 1–5 |
| i | 4–7 | 0–1 | 7–18 | 9–12 |
| C21 n | t | t | t | t |
| ai | 4–6 | t | 9–25 | 5–10 |
| C22 n | 0–1 | t | 1–3 | 1–2 |
| i | 2–3 | 0–1 | 3–6 | 12–22 |

*(Contd...)*

**Table 9.9.** Lanolin composition (percentage of wax). *(Contd .)*

| *Chain* | *Fatty acids* | *α-Hydroxyacids* | *Fatty alcohols* | *1–2–Alcane diol* |
|---|---|---|---|---|
| C23 n | t | t | t | t |
| ai | 2–3 | 1–2 | 2–4 | 4–14 |
| C24 n | 2–6 | t | 2–9 | 0–1 |
| i | 2–3 | 1–2 | 3–5 | 3–12 |
| C25 n | t | t | 0–1 | t |
| ai | 5–6 | 0–1 | 6–8 | 1–2 |
| C26 n | 2–5 | t | 1–7 | – |
| i | 3–6 | t | 6–13 | 0–1 |
| C27 n | t | t | t | – |
| ai | 3–6 | t | 9–16 | t |
| C28 n | 1–2 | t | 0–2 | – |
| i | 2–3 | t | 1–7 | t |
| C29 n | t | t | – | – |
| ai | 1–4 | t | 2–4 | t |
| C30 n | 0–1 | t | t | – |
| i | 1–3 | t | 1–3 | t |
| C31 n | t | t | – | – |
| ai | 1–4 | – | 0–1 | – |
| C32 n | t | t | t | – |
| i | 0–2 | – | 0–1 | – |
| C33 n | – | – | – | – |
| ai | 0–2 | – | t | – |
| C34 n | – | – | – | – |
| i | 0–1 | – | t | – |
| C35 n | – | – | – | – |
| ai | t | – | t | – |

t: Traces or less than 0.2 per cent; n = normal; i = iso; ai = anteiso.

This composition as well as the physical and chemical properties vary depending on the sheep's race, sheared body part, season and processing.

Lanolin is a greasy, amber-coloured paste and has a slight, characteristic smell. It melts at 36–42°C, is insoluble in water moderately soluble in cold ethanol, and soluble in hot ethanol, ether and chloroform (Table 9.10). Like beeswax, lanolin too has been used from very ancient times for its softening and lubricating properties in cosmetic and dermatological fields. It has also got some industrial applications (treatment of tissues, printing ink, lubrication, rust prevention, etc.).

However it also finds application in the form of its fractions (lanolin alcohols, lanolin acids, lanolin oil and lanolin wax obtained by crystallisation) and its derivatives: acetylated, ethoxylated, propoxylated, hydrogenised, hydroxylated, isobutylated or even quaternarised lanolin (by condensation with amines of dimethylaminopropylamine type.

**Table 9.10.** Average characteristics of animal waxes.

| *Characteristics* | *Beeswax* | | *Chinese wax* | *Shellac wax* | *Spermaceti wax* | *Spermaceti oil* | *Lanolin* |
|---|---|---|---|---|---|---|---|
| | *Europe America* | *Far East* | | | | | |
| Density at 25°C | 0.95–0.98 | 0.97–0.99 | 0.93–0.97 | 0.97–0.98 | 0.91–0.96 | 0.88–0.89 | 0.88–0.96 |
| Melting point (°C) | 61–65 | 61–65 | 65–80 | 74–82 | 42–50 | 17–20 | 36–40 |
| Refractive index | 1.439–1.445[1] | – | 1.467[2] | – | 1.440[3] | 1.470[4] | 1.480[2] |
| Acid value | 15–23 | 5–10 | 3–13 | 8–24 | 0–1 | 0–2 | 1–4 |
| Ester value | 70–82 | 70–102 | 77–80 | 60–110 | 116–134 | 126–144 | 90–114 |
| Saponification value | 85–104 | 76–111 | 80–93 | 69–126 | 116–135 | 126–146 | 91–118 |
| Iodine value | 6–12 | 8–11 | 1–2 | 2–9 | 1–4 | 77–91 | 20–38 |
| Hydroxyl value | 10–15 | – | – | 47–52 | 2–5 | 4–6 | 100–144 |
| Unsaponifiable matter (%) | 50–56 | – | 49–50 | 68–76 | 45–50 | 38–48 | 44–53 |
| Hydrocarbons (%) | 10–17 | – | 2–3 | 1–3 | Traces | 0–1 | 0–1 |
| Free acids (%) | 8–15 | – | 0–2 | 1–4 | 0–2 | 0–3 | 0–1 |
| Total acids (%) | 33–52 | – | 46–52 | 21–36 | 49–55 | 52–60 | 48–55 |

Refractive index: 1 at 75°C. 2 at 40°C. 3 at 60°C. 4 at 20°C.

## VEGETABLE WAXES

A large number of plants, especially in tropical countries and desert regions, secrete wax on the surface of their stems, leaves or berries in order to reduce evaporation.

The best known vegetable waxes are Carnauba and Candelilla, and jojoba oil. Other lesser known sources of wax are cotton, the 'Esparto' reed, the 'Douglas' fir, linseed, 'Ouricury', rice-husk, sugar cane, etc.

Finally, even though several fats possess the physical properties of real fatty ester based wax, they have a glyceridic composition. Bayberry wax and Japan wax belong to this category but these are dealt very briefly in this section.

### Carnauba Wax

Carnauba wax is secreted during the dry season (from June to December) by the leaves of the *Copernicia prunifera cerifera* palm called the Tree of 'Life by the Brazilians'. This palm tree produces wax only in the semi-arid regions of north and north-eastern brazil at the rate of about 100 grams per tree in a year. The leaves are cut, dried in the sun and beaten to extract the wax in a powder form. This powder is melted, filtered and moulded after cooling.

It is mainly composed of a mixture of free monohydroxylated and dihydroxylated alcohols (C28 to C34) and of monoesters and polyesters of these alchohols with acids and hydroxyacids (C20 to C30) (Table 9.11).

The peculiarity of Carnauba wax is its hardness and its high melting-point (82–88°C). Its properties change according to the region of origin (Parnahyba, Piaui, Ceara and Bahia) and the age of the leaf at the time of cutting (Table 9.14). Four grades are usually defined according to the colour (pale yellow to brownish green) but they also depend on various chemical properties (darker waxes having a higher acid number and a lower iodine value).

**Table 9.11.** Composition of refined carnauba wax (percentage of wax).

| *Constituent (in raw wax)* | *Content* | *Constituent (after saponifaction)* | *Approximate content* | *Nature* |
|---|---|---|---|---|
| Free fatty acids | 3–6 | Fatty acids | 24 | C24–C28 |
| Free fatty alcohols | 10–15 | ω-Hydroxyacids | 6 | C22—C28 |
| Total esters | 80–85 | Fatty alcohols | 39 | C24–C34 |
| | | α ω-Diols | 21 | C24–34 |
| Comprising | | | | |
| Monoesters | 36–40 | Hydroxycinnamic acid | 7 | |
| Cinnamic acid diesters | 26–34 | P. methoxycinnamic acid | 2 | |
| ω-Hydroxyesters | 10–14 | | | |
| Hydrocarbons | 1–3 | Hydrocarbons | 1 | C27–C31 |
| Moisture and impurities | 0–1 | Moisture and impurities | 0 | |

Carnauba wax is traditionally used in floor wax and various types of polish (for furniture and shoes) where the resistance and gloss of the wax film are improved by its inclusion. It is also used as a protective coating for fruit, in the manufacture of printing ink, as a hydrophobic agent and, very often, also as an additive in other waxes in order to raise their melting-point.

## Candelilla Wax

Candelilla wax is found in the form of scales on the stems of the *Euphorbia Cerifera, Euphorbia Antisiphilitica* and *Pedilatus Pavonis* plants which grow wild in North Mexico and South Texas. The wax is extracted by immersing the plant in boiling acidulous water. It is then skimmed off the surface of this solution, drained and dried. It is usually remelted, filtered and bleached to obtain refined candelilla wax.

Half of Candelilla wax is composed of hydrocarbons (mainly hentricontane ($C_{31}H_{64}$) and the other half is a mixture of fatty acids, aliphatic and triterpenic alcohols and their esters (Table 9.12).

**Table 9.12.** Composition of candelilla wax (percentage of wax).

| *Constituent (in raw wax)* | *Content* | *Nature* | *Major element* |
|---|---|---|---|
| Free fatty acids | 7–10 | C30–C34 | |
| Free fatty alcohols | 10–14 | C20–C32 | |
| Fatty esters | 22–32 | C42–C64 | |
| Hydrocarbons | 40–65 | C29–C33 | $C_{31}H_{64}$ |
| Resins (triterpenes) | 3–6 | | (80 to 90%) |

Refined Candelilla wax is hard and glossy and has a spicy smell. Its colour varies from pale yellow to grey-brown depending on the extraction and refining processes used, the nature of the plants and the year. It is compatible with fats and all animal and vegetable waxes.

Candelilla wax is widely used in household cleaning materials such as furniture and shoe polish and candles, beside its industrial applications (electrical insulators, rubber, varnish, etc.). It is used less frequently for cosmetic purposes except where its characteristics are sought: although it is not as hard as Carnauba wax, it is used to increase the hardness and melting point of paraffins. It is also a long-standing ingredient of chewing-gum.

## Jojoba Oil

Jojoba oil is a liquid wax present in a very high proportion (45 to 60 per cent) in the seeds of the *Simmondsia chinensis* bush which grows naturally in the semi-desert areas of North Mexico, south Arizona and California. This bush is now also grown in plantations. Jojoba oil is extracted by the same technique as glyceridic vegetable oils.

It consists almost entirely of fatty acids-based aliphatic esters and mono-unsaturated fatty alcohols, the major constituents being eicosenol docosenol eicosenoate and ($C_{19}H_{37}COO$ — $C_{20}H_{39}$ and $C_{19}H_{37}COO$ — $C_{22}H_{43}$) (Table 9.13).

**Table 9.13.** Composition of jojoba oil (percentage of wax).

| *Constituent (in raw wax)* | *Content* | *Constituent (after saponification)* | *Content* |
|---|---|---|---|
| Esters | >98 | Total acids | 47–49 |
| Triglycerides | 0–1 | Total alcohols | 50–52 |

*Nature of carbon chain in jojoba oil (% of each component)*

| *Chain* | *Total acids* | *Total alcohols* | *Chain* | *Esters* |
|---|---|---|---|---|
| C14:0 | t | t | Total C34 | t |
| C16:0 | 0–2 | t | Total C36 | 1–2 |
| C16:1 | t | t | Total C38 | 6–7 |
| C18:0 | t | t | Total C40 | 26–30 |
| C18:1 | 10–13 | 1–2 | Total C42 | 49–52 |
| C20:0 | t | t | Total C44 | 10–11 |
| C20:1 | 66–71 | 39–46 | Total C46 | 1–2 |
| C20:2 | t | 1–2 | Total C48 | 0–1 |
| C22:0 | 0–1 | 1–2 | Total C50 | t |
| C22:1 | 14–20 | 45–52 | | |
| C24:0 | t | 0–1 | | |
| C24:1 | 1–3 | 6–11 | | |
| C26:0 | t | – | | |
| C26:1 | 0–1 | t | | |

t: Traces or less than 0.2 %.

Jojoba oil is clear and does not become rancid easily, even under harsh conditions. Because of its softening qualities, it is very popular as a raw material for a wide range of cosmetics, from tanning lotions to shampoos. Its derivatives also have cosmetic and industrial applications.

In its sulphonated form it is used as an extreme-pressure additive and emulsifier in cutting oils. Its hydrogenated form is a white wax with a melting point between 65 and 75°C. Its properties are practically identical to those of spermaceti wax (which is on the verge of disappearing) and it has the same applications. Hydrolysation and ethoxylation produces non-ionic tensioactive elements whose characteristics are much in demand due to their long mono-unsaturated aliphatic chain. Characteristics of vegetable waxes is shown in Table 9.14.

**Table 9.14.** Average characteristics of vegetable waxes.

| *Properties* | *Carnauba wax* | *Candelilla wax* | *Jojoba oil* |
|---|---|---|---|
| Density at 20°C | 0.99–1.00 | 0.97–0.99 | 0.86–0.88 |
| Melting point (°C) | 80–91 | 65–73 | 6–7 |
| Refraction index at 40°C | 1.469–1.472 | 1.456–1.458 | 1.459–1.461 |
| Acid value | 2–10 | 12–22 | 0–3 |
| Ester value | 75–85 | 30–47 | 90–92 |
| Saponifaction value | 78–95 | 43–65 | 92–95 |
| Iodine value | 7–20 | 15–44 | 82–88 |
| Unsaponifiable fraction (%) | 50–55 | 65–75 | 38–52 |
| Hydrocarbons (%) | 1–3 | 40–60 | t |
| Resins (%) | 4–6 | 10–20 | t |

t: traces or less than 0.2%

## Waxes Obtained from Various Other Sources

Table 9.15 summarises the sources and main physical and chemical properties of various vegetable waxes that are less important than the three studied above.

All these waxes are used in the manufacture of shoe and furniture polish, printing inks and carbon-paper. They are however gradually being replaced by synthetic waxes which are much cheaper.

There are in addition about 50 other lesser known waxes which have been the subject of research and some, such as apple-skin wax, may even be treated as laboratory curiosities.

Some examples are given below (M = melting point):

1. Caranday wax (Brazilian palm) (*Copernicia australis*) M = 74–85°C.
2. Colombian palm wax (*Ceroxylon andicola*) M = 93°C.
3. Raphia wax (*Raphia pedunculata*) M = 82°C.
4. Canacu wax (*Calathea lutea*: Amazon) M = 80–86°C.
5. Madagascar wax (*Euphorbia stenoclada*) M = 88°C.
6. Pisang wax (*Musa simiarum: Guinea*) M = 78–81°C.
7. Retamo wax (*Ceri nimbi*) M = 76–78°C
8. Sorghum wax, hemp wax, Atoca wax (cranberry or moorberry wax), Cauassu wax and floral waxes (rose, mimosa, jasmine, etc.).

## Triglyceridic Waxes

Obviously this name does not correspond to the chemical definition of the word 'wax' as it has been described in the introduction. Nonetheless, several substances having a waxy or greasy appearance are traditionally classified as waxes, e.g. Japan and Bayberry wax, while others with very similar characteristics and composition, e.g. cocoa and Ucuhuba butter, are clearly classified as fats (glyceridic vegetable fats).

**Table 9.15.** Waxes of various origins.

| *Common name wax* | *Esparto wax* | *Fir wax* | *Linseed wax* | *Cotton wax* | *Ouricury wax* | *Cuban palm wax* | *Rice husk wax* | *Sugar cane* | *Retamo* |
|---|---|---|---|---|---|---|---|---|---|
| Synonym | Reed wax | Bark wax | | | Palm wax | | | Cachaza wax | Cerinimbi |
| Origin | Esparto | Douglas reed fir | | Green cotton | Syagrus coronata | | | | |
| Production area | Southern Spain North Africa | North America | | | Brazil | Cuba, brazil | | Cuba, Java, Louisiana | Argentina |
| Production method | Beating of fibres and extraction of powder | Extraction from bark | Extraction from stems | Extraction | Scraping of leaves, filtering | beating of leaves extraction of powder | Cold fractionation from rice-husk husk oil | Solvent extraction from pressed sugar cane | Immersion of stems in boiling water |
| Colour | Greenish brown | Honey to dark brown | | Yellow | Light to dark brown | | Honey to light brown | Straw yellow to very dark brown | Light brown |
| Melting point (°C) | 67–81 | 65–67 | 67–70 | 76–80 | 81–85 | 82–86 | 76–82 | 75–82 | 76–78 |
| Density | 0.98–0.99 | 1.02 | 0.94–0.99 | 0.98–1.00 | 0.97–1.05 | | | 0.98–0.99 | 0.98–0.99 |
| Acid value | 22–33 | 25–30 | 18–23 | 21–28 | 8–20 | 6–16 | 1–10 | 6–32 | 45–50 |
| Ester value | 33–55 | | 59–65 | 23–50 | 70–90 | | 73–110 | | |
| Saponification value | 62–79 | 145 | 78–84 | 50–76 | 80–105 | 70–101 | 75–120 | 75–95 | 85–90 |
| Iodine value | 8–23 | | 21–29 | 20–27 | 6–10 | | 5–10 | 5–10 | 5 |
| Hydroxyl value | | | | 48–83 | | 55–59 | | | |
| Unsaponifiable matter (%) | 60–65 | | 70–80 | 57–68 | 50–55 | | 39–57 | | |
| Hydrocarbons (%) (major) | ($C_{31}$ $H_{64}$) | | | 7–9 | 1–3 ($C_{31}$ $H_{64}$) | 0–1 | | 3–9 ($C_{27}$ $H_{56}$) | 15–20 |
| Free fatty acids (%) (major) | | | | 25–26 | 4–9 | | | 7–15 ($C_{27}$ $H_{55}$–COOH) | |
| Aliphatic esters | | | | 46–48 | 23 | | | 78–82 | |

Information about Japan wax and myrtle wax is summarised in Table 9.16.

**Table 9.16.** Japan wax and Bayberry wax.

| *Common name* | *Japan wax* | *Myrtle wax* |
|---|---|---|
| Synonym | Sumac wax | Bayberry wax<br>Candleberry wax |
| Origin | Rhus succe danea | Myrica cerifera and<br>Myrica carolinensis |
| Place of production | China, Japan | Colombia, S. Africa |
| Method of production | Steaming or pressing of fruit,<br>filtering and bleaching | Boiling in water |
| Colour | Cream | Greenish white |
| Melting point (°C) | 50–56 | 46–50 |
| Density at 25°C | 0.97–0.99 | 0.97–0.98 |
| Refraction index at 80°C | 1.455 | 1.436 |
| Acid value | 6–20 | 2–4 |
| Ester value | 210–225 | 205–215 |
| Saponification value | 217–237 | 206–217 |
| Iodine value | 4–15 | 1–10 |
| Hydroxyl value | 25–31 | – |
| Unsaponifiable fraction (%) | 2–4 | 0–1 |
| Hydrocarbons (%) | 0–2 | t |
| Free fatty alcohols (%) | 1–2 | t |
| Free fatty acids (%) | 3–5 | 0–1 |
| Triglycerides (%) | 94–96 | 98–99 |
| Nature of fatty acids | C14 2% | |
| | C16 88% | C14 59% |
| | C18 4% | C16 35% |
| | 5–7% diacids | C18 6% |
| | in C18–C20–C22 | |

Thus all natural waxes–animal, vegetable or mineral–have been used for a long time in a variety of applications, depending on their physico-chemical characteristics: brightness, gloss, softness, hardness, smell, compatibility with fats, solubility of pigments in the wax, emulsifying properties, etc.

The best known and most commonly used are beeswax and lanolin (and its derivative) which are moreover chemically the most complex. Nevertheless, other waxes like Carnauba wax and jojoba oil have firmly established themselves and are difficult to replace. For various reasons (such as the near extinction of the sperm whale, the source of spermaceti, supply problems for Cuban palm wax, high cost of ouricury wax) more and more waxes have given way to substitutes based on synthetic derivatives (e.g. cetyl palmitate instead of spermaceti) or mixtures of other waxes and/or hydrogenated compounds.

Chapter 10

# Fats and Oils in Cosmetics

## INTRODUCTION

A wide range of chemical and natural materials are used in the manufacture of cosmetics. Some of the most commonly used materials include—water, surfactants, shampoos and bath additives, oil components, silicone oils, cream bases, oil-in-water (o/w) emulsifiers, water-in-oil (w/o) emulsifiers, humectants, aerosol propellants, antioxidants, colours and perfumes.

Cosmetics owe their origin to the use of fats. Even our neolithic ancestors used animal fat to protect their skin and keep it smooth. In ancient times, sources of fat became more diverse with the appearance of vegetable oils like sesame seed, olive and almond oil. In addition, there was tallow, lanolin and even crocodile fat. They were scented and rubbed on the body.

Apple pulp gave its name to pomade, from the French *pomme*, for apple. Later, technical advances brought refined fats (which are less vulnerable to oxidation) and synthetic surface active derivatives with variable intrinsic properties (melting power, foaming agents, detergents, solubilisers, emulsifiers). The appearance of these surfactants enabled the cosmetics industry to progress rapidly.

Surfactants are surface active agents which reduce the surface tension of the liquids. Anionic, cationic and non-ionic surfactants are used in cosmetics and toiletries due to their balanced properties. They are excellent foamers. The basic or primary surfactants are the backbone of the cleansing products. They include alkyl ether sulphates, linear alkyl benzene and alkane sulphonates. Mild anionic surfactants are sulphosuccinates, cocoyl isethionates, acyl amides, alkyl ether carboxylates and magnesium surfactants. Amphoteric surfactants include, alkyl betaines, alkyl amide betaines, amine oxides and ethoxylated products. Alkyl polyglycosides are non-ionic surfactants. The list of cationic surfactants include monoalkyl, dialkyl, trialkyl, benzyl and ethoxylated quaternaries. The large variety of surfactants obtained from fats fully meets the cosmetics industry's rquirements. Fats constitute the base of the cosmetics industry. They cannot be dissociated from it as they act on two levels:

1. As lipidic phase in many cosmetics.
2. As a source of transformation producing a large number of multi-functional derivatives.

## LIPIDS THAT ACT IN A COSMETIC PACK

Oils and fats are diverse group of essential constituents found in all living plants and animals. Although more than two and half lakh plants exist, only a minuscule proportion produces seeds that bear oil in sufficient quantities for commercial exploitation. Of all the vegetable oils commercially developed, a dozen of them actually contribute more than 90 per cent of vegetable oil production for use. Animal and plant lipids have been traditionally used for more than a thousand years for human consumption as food and in the manufacture of non-food items for personal grooming and adornment. Now-a-days, plant fats and oils are also used in manufacturing chemical derivatives and oleochemicals.

The most important class of plant lipids used in cosmetic formulations are the triacylglycerol or triglycerides (Fig. 10.1). Triacylglycerols have been used in personal care formulations, colour cosmetics and hair-dressing products, for their cosmetic and pharmaceutical properties.

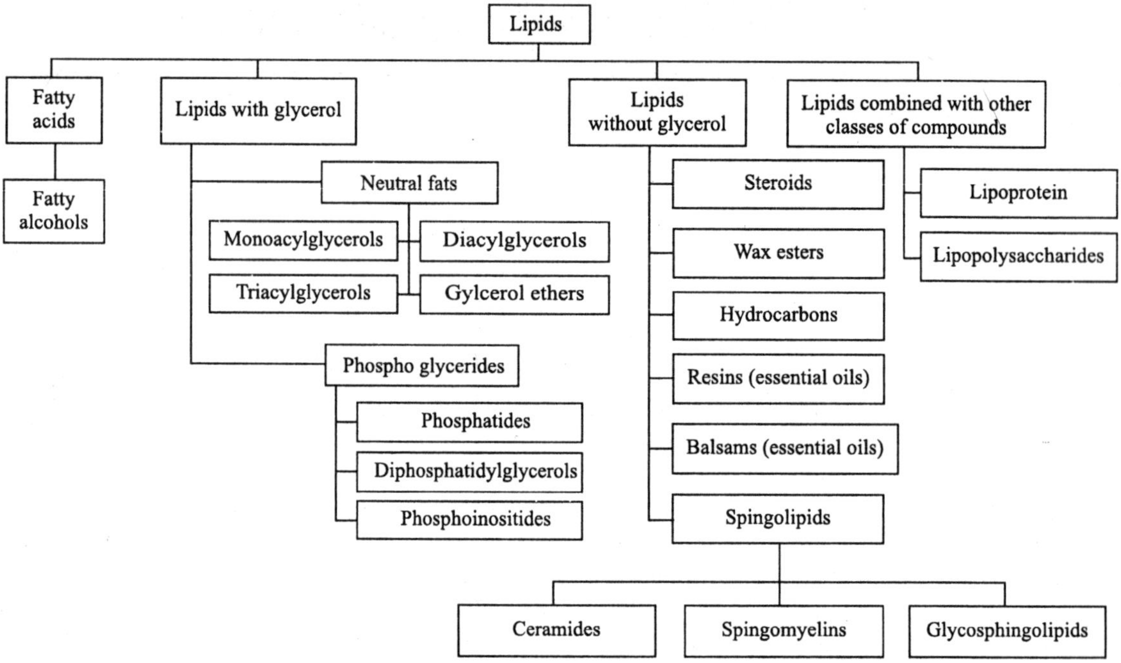


**Fig. 10.1.** Some important lipids used in cosmetic applications.

Triglycerides, structurally, can be split to give one molecule of glycerol and three molecules of fatty acids. Fatty acids used in cosmetic products have varying chain lengths, which are all straight chain, containing an even number of carbon atoms. There are about 40 different fatty acids occurring in nature, the largest is the group containing 18 carbon atoms.

Natural oils and fats vary widely in their physical properties even though they consist of similar or same fatty acids. This is largely because individual fats and oils differ over a large range in the component fatty acid proportion. The structure of the individual triglyceride components may also vary. Moreover, the two factors are interdependent and are largely responsible to the overall effects observed in individual fats and oils.

A cosmetic formulation consists of a number of ingredients. Ingredients are generally selected to take advantage of their physical, rheological and functional properties, as diluents or even simply for marketing claims. Vegetable fats and oils, wax esters from lanolin and jojoba oil, phospholipids, ether lipids, sphingolipids are some lipid classes that are becoming popular in many personal care products largely due to their skin conditioning, emolliency and emulsifying properties. Market dynamics indicates that the use of naturals and its derivatives in personal care products will only increase in times to come, making these actives easily available for regular use.

## Fatty Acids

Fatty acids normally used in cosmetic preparations are naturally occurring straight chain monocarboxylic acid with an even number of carbon atoms. The fatty acid molecules are either saturated or unsaturated. Of the various fatty acids used, the most important ones are given below. Natural lipids used in cosmetic applications are generally in the form of fatty acid derivatives where the fatty acid is condensed to fatty alcohol or long chain derivatives:

| | |
|---|---|
| Lauric acid | 12 Carbon atoms |
| Myristic acid | 14 Carbon atoms |
| Palmitic acid | 16 Carbon atoms |
| Stearic acid | 18:0 Carbon atoms |
| Oleic acid | 18:1 Carbon atoms |
| Linoleic acid | 18:2 Carbon atoms |
| Linolenic acid | 18:3 Carbon atoms |
| Ricinoleic acid | 18:1 Carbon atoms |

## Lipids with Glycerol

Neutral fats and natural oils, fats and waxes were some of the earliest cosmetic ingredients to be used in cosmetic preparations. Fatty acid esters of glycerol that are distributed in plant and animal oils can be either liquid or solid in nature, depending on the chain length and their unsaturation. Triacylglycerols can have two or more fatty acid groups in each of the position or can also have the same fatty acid group.

| | |
|---|---|
| $CH_2—OH$ | $CH_2—OCO(CH_2)_{14}CH_3$ |
| \| | \| |
| $CH—OH$ | $CH—CO(CH_2)_{14}CH_3$ |
| \| | \| |
| $CH_2—OH$ | $CH_2—OCH(CH_2)_{14}CH_3$ |
| Glycerol | Triacylglycerols |

Triacylglycerols are insoluble in water but are soluble in most organic solvents. On hydrolysis, they produce glycerol and the combined fatty acids. Triacylglycerols lipid rich natural oils have a beneficial effect on skin with a good water barrier, reducing *trans*-epidermal water loss (TEWL). TEWL from our skin leads to dry skin that can be corrected by use of cosmetic application containing oil rich in essential fatty acids.

## Natural oils

Natural oils of plant origin provide very good emmolliency in creams and lotions. Oils are also used as carrier of other ingredients to be delivered to skin in a cosmetic formulation. Some natural fats are prone to oxidation due to the presence of unsaturated bonds. This is more if the neutral fat breaks down during processing and manufacture. Oxidation products like aldehydes, peroxides, etc. interact with the fragrance and other active material giving rise to colouration and rancidity. Super refining of these oils removes these oxidation products for use in cosmetic applications without any problems.

### *Essential fatty acids*

Essential fatty acids metabolised in our body are required for an unimpaired condition of the epidermis. Essential fatty acid deficiency can be due to poor diet, diseases like psoriasis, atrophic eczema, skin dermatitis, dry skin caused due to ageing, modern lifestyles and use of harsh soaps and detergents. Human body cannot synthesise essential fatty acids repeatedly at short intervals and they have to be supplemented, either by diet or by topical application. Naturally occurring oils that have triacylglycerol lipids include cocoa butter, shea butter, babassu, almond, sesame, shorea butter, mango seed oil, and these are often used in cosmetic formulation. However, all these oils are often expensive and difficult to use in mass market cosmetic formulations. The best way to get around the situation is to use alternative fat derivatives that are less expensive, but equally functional. The simplest method to do this would be the use of hydrogenated and fractionated fat additives.

### *Hydrogenated and fractionated fat additives*

Hydrogenation is a simple process by which the properties of vegetable oils can be altered. During hydrogenation, the unsaturated form of the fatty acid in the vegetable oil is converted to their saturated form. Hydrogenation is an inexpensive way for changing the property of a specific oil, as the degree of hydrogenation can be adjusted and controlled according to our requirements. Hydrogenated fat can also be fractionated further into different fractions (stearin fraction, olien fraction, etc.). The degree of fractionation and the technique used generally determines the property of the obtained fat fraction.

Although any vegetable oil or fat is suitable for hydrogenation and subsequent fractionation, generally inexpensive oils that are abundantly available are only used as starting material for this purpose. Some typical vegetable oils used for this purpose include, coconut oil, soyabean oil, rapeseed oil, palm kernel oil, etc. An important factor that cannot be ignored is that all these fats are used in a cosmetic formulation with additives like hydrocarbon, actives, emulsifiers, etc. that may significantly influence the performance properties and behaviours of the finished products.

### *Glycerol ethers*

Glycerol ethers are also known as alkyl glycerol or ethyl lipids. Shark liver oils contain large amount of glycerol ethers when compared to their availability in human tissues. They are generally found as esterified fatty acids. Butyl alcohol that is widely found in animal tissue is a typical glycerol ether.

Natural glycerol ethers, due to its scarcity, are used in a minor proportion in cosmetic applications. Semi-synthetic materials that have been prepared have become available for use in the recent past. Anti fungal and anti bacterial activity, UV protection and skin conditioning are some therapeutic benefits credited to its use in cosmetic applications.

$$\begin{array}{l} CH_2{-}O{-}CH_2(CH_2)_{16}CH_3 \\ | \\ CH{-}O{-}CO(CH_2)_m\,CH_3 \\ | \\ CH_2{-}O{-}CO(CH_2)_n\,CH_3 \end{array}$$

Ether lipid

$$\begin{array}{l} CH_2{-}O{-}CH_2(CH_2)_{16}CH_3 \\ | \\ CH{-}OH \\ | \\ CH_2{-}OH \end{array}$$

Butyl alcohol

### *Phosphoglycerides*

Phosphoglycerides are an important group of lipids found in almost all-living cells. The basic difference between phosphoglycerides and other glycerol compounds is the presence of one hydroxyl group esterified to phosphoric acid. Phosphoglycerides on hydrolysis give fatty acids, glycerol and phosphate. Phospholipids are both hydrophilic and hydrophobic in nature.

This amphipathic property enables phosphoglycerides help in linking both the polar and non-polar molecules—an essential requirement in cell membrane functioning.

Phosphoglycerides include phosphatidylcholin (lecithin), phosphatidyl ethanolamine, phosphatidyl serine, etc.—all derivatives of phosphatidic acid.

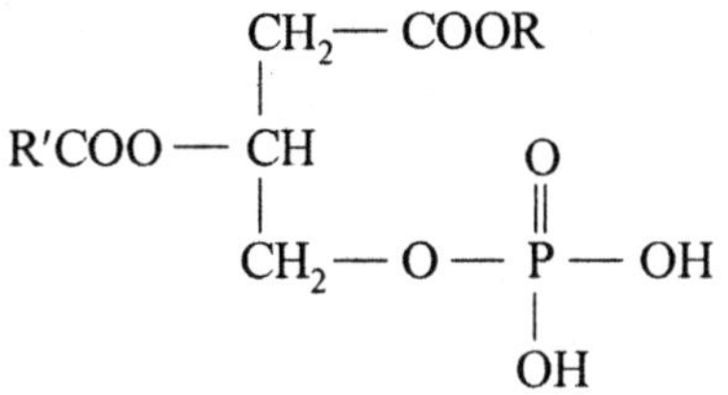

Phosphatidic acid

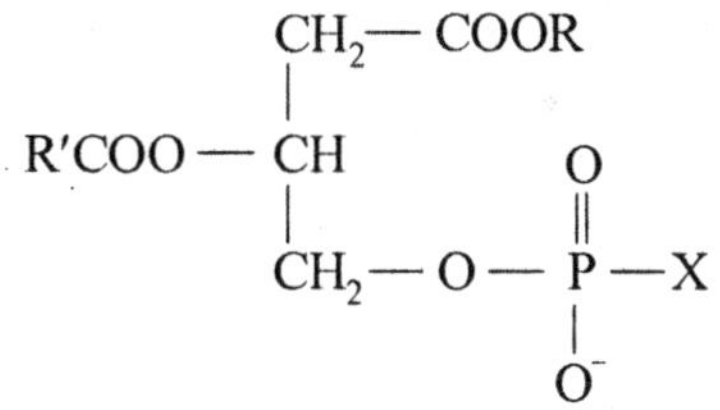

Phosphatide

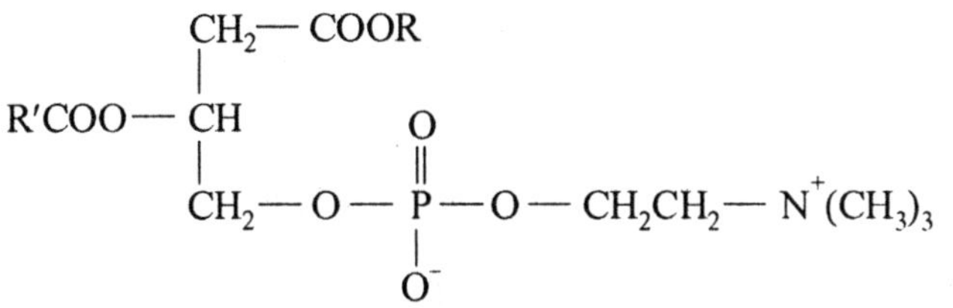

Lecithin

## Lipids without Glycerol

### *Sphingolipids*

It consists of amino alcohol like sphingosine, N-acylated by fatty acid. The parent compound is ceramide. Ceramides are found in plant and animal tissues.

Water loss in human skin is controlled by a five micron thick lipid structure present at the base of the stratum corneum. The lipid structure, which acts as a water barrier, is made up of ceramides that are a mixture of complex sphingolipids. Two important ceramide structures are spingosine, N-acylated with long chain saturated fatty acid ($C_{16} - C_{30}$) and a linoleate bearing esters linked to spingosine with long chain hydroxy acid. When there is a fatty acid deficiency in human body, due to poor diet or disease, the linoleate is replaced by oleate. This replacement causes the lipid bilayer to change in its physical structure, resulting in the skin becoming permeable to water. Thus to maintain proper water balance in the skin the stability of the linoleate-rich ceramides bilayer is very important. Although ceramides are found in plant and animal tissues, it is not available in large amounts.

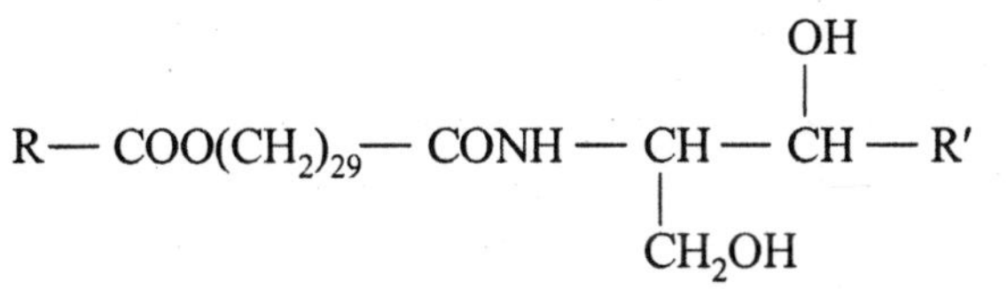

RCO = Linoleate

$R' = CH_3CH_2CH = CH_2—$

Linoleate based ceramide

However, synthetic ceramides, derived from natural intermediates is now available as functional alternatives. Complex sphingolipids are compounds formed by the esterification of the primary hydroxyl group of simple ceramides. Some common examples are sphingomyelin, glucosphingolipid, cerebroside, and ganglioside.

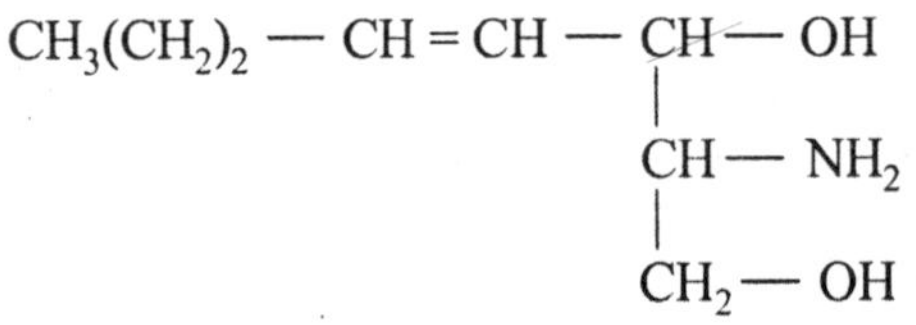

Sphingosine

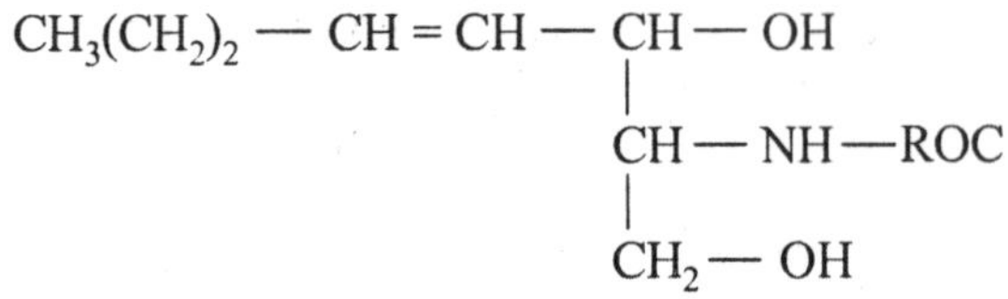

Ceramide

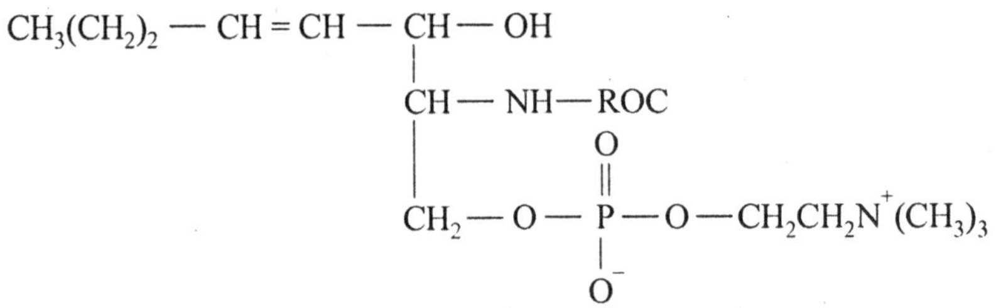

Sphingomyelin

### Wax Esters

Wax esters have been traditionally used in cosmetic formulations. The important ones are esters obtained from living animals like spermaceti (whales), lanolin (sheep), orange roughy oil (deep-sea fish, *Hoplostethus atlanticus*) and from vegetable sources like Jojoba oil. Wax esters obtained in nature are mixtures of long chain fatty esters with the carbon chain length between 20 and 40. They are resistant to hydrolysis and are often found in nature as water repellent coating on plant and animal surfaces. When fatty acid is esterified to obtain an aliphatic primary alcohol, it is termed as a wax ester. Wax esters can be either solid or liquid in nature. Now-a-days wax esters from whale spermaceti has been completely replaced by hydrogenated Jojoba oil that possesses very similar composition as spermaceti. Jojoba oil is a vegetable oil and renewable. Orange roughy oil is a by-product obtained from marine food industry. However, the largest source of wax esters is woolgrease refined to get lanolin that contains important lipids and cholesterol esters occurring in human skin.

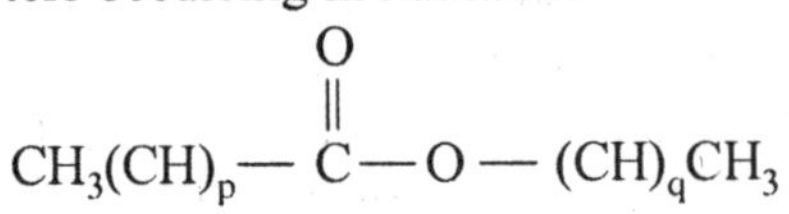

Wax ester

Natural lipids that are complementary to human skin lipids are very important in formulating cosmetics. Traditionally we know that lipids help in correcting imperfect skin condition and maintain it in order. With skin care product market on an increase, use of lipids in these products will only show a corresponding rise. Fats derived from plants, along with lanolin and its derivatives, are good ingredients on cosmetic applications, imparting emolliency with good skin compatibility. Lipids without doubt could provide unique active materials for formulating innovative products tailored to meet specific consumer requirements in the near future and beyond.

## FATS USED AS LIPID PHASE IN COSMETICS

### Role of Fats at Physio-Cutaneous Level

Fats are naturally present in the skin. They are found on the surface of the epidermis and in it (Fig. 10.2) in the form of (i) sebum; and (ii) epidermal lipids.

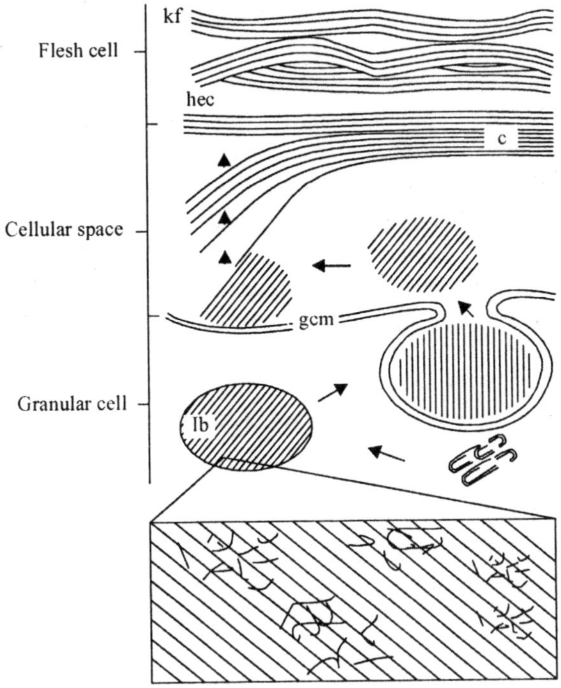


**Fig. 10.2.** Lipid distribution in the epidermis—in lamellar form in granular cells and in—an unorganised form in the corneal layer.

### *Sebum*

Sebum is secreted by sebaceous glands. In an adult human being, it is composed of—60 per cent triglycerides, 25 per cent esters, 12 per cent squalene and less than 1 per cent cholesterol.

Lipids are present in sebum in an unorganised form.

#### *Epidermal lipids*

Epidermal lipids are produced by keratocytes. But these lipids are present in a highly organised form. They are arranged in multiple lamellar layers which have an excellent barrier effect.

Fats in the skin have an essential role. If they are eliminated, the skin becomes incapable of retaining water. Transepidermal water loss increases considerably and the skin becomes rough and cracks.

External application of fats on the skin is important because it can overcome the skin's natural deficiencies and restore its role as a barrier. Further, some fatty acids like linoleic and linolenic acids and their higher derivatives called essential fatty acids are vital for the cells. Any deficiency causes excessive scaling, an increase in the loss of water and adulteration of lipids. Since they are not synthesised by mammary glands, they have to be introduced from an external source.

## MAIN FATS USED

Among the fats used as softening agents, following categories can be distinguished (i) mineral fats; (ii) animal fats; (iii) vegetable fats; and (iv) synthetic fats.

### Mineral Fats (or Hydrocarbons)

Mineral fats for cosmetic use are high-boiling fractions obtained from crude oil distribution that are purified and refined by treatment with sulphuric acid. Two types are of importance (i) liquid paraffin (viscosity 110 to 230 mPas); and (ii) light liquid paraffin (viscosity 25 to 80 mPas). Mineral oils are colourless, clear and odourless liquids, insoluble in alcohol or water. They are excellent cosmetic emollients because they are inert and do not penetrate into the skin. They therefore have excellent skin compatibility and little or no comedogenic potential. Since they are not considered 'natural', mineral oils have been attacked repeatedly but are still the most widely used oil component in skin-care formulations.

The fatty character of mineral oils help to form a film on the skin which increases hydration by blocking the normal evaporation of water. Combinations with synthetic esters such as isopropyl stearate are recommended in order to open the film, and to guarantee the right balance for water evaporation.

#### *Vaseline*

Vaseline or petroleum jelly is defined in the French Pharmacopoeia as 'a mixture of hydrocarbons obtained by treating the heavier fractions of certain types of petroleum'. The number of carbon atoms in hydrocarbon constituting vaseline vary from 22 to 35. They are saturated hydrocarbons, either acyclic having a straight or branched chain or naphthenic.

Vaseline is essentially used in dermatological preparations as an active ingredient because of its inertia and its occlusive nature permitting percutaneous penetration.

However, it is less commonly used in the cosmetics industry because it is not very pleasant to touch.

#### *Paraffin oil*

Paraffin oil is classified under two different grades in the French Pharmacopoeia and is defined as 'a purified mixture of saturated liquid hydrocarbons obtained from petroleum' (Table 10.1).

**Table 10.1.** Different grades of paraffin oil.

| *Description* | *Fluid paraffin oil* | *Thick paraffin oil* |
|---|---|---|
| Density (20°) | 0.835–0.855 | 0.860–0.885 |
| Viscosity (mPa.s) | 20–37 | > 37 |

The two grades differ from one another due to their density and viscosity. Paraffin oil is mainly used in cosmetics industry in manufacture of beauty-care products (emulsions and oils) and in certain make-up preparations (mascara, eyeliner, compact powder).

It has the advantage of being (i) inert; (ii) nonoxidisable; (iii) easily available; (iv) easily emulsifiable; and (v) cheap. However, since it does not have a pleasant feel, it is normally used with another fat.

#### *Ozocérite and ceresin*

Ozocerite is a fossil wax found in the vicinity of oil-fields. Ceresin is refined ozocerite. They are characterised by their high melting point, > 60 and 78°C, respectively. For these reasons they are used as hardening agents in lipsticks and other cosmetic preparations in the form of sticks and also as a fatty phase thickening agents in some E/H emulsions to reduce exudation.

### Animal Fats

They differ from vegetable oils on account of:

1. The presence of a substantial amount of fatty acids with an odd carbon number.
2. A low unsaponifiable rate containing mainly cholesterol and very few natural antioxidants.

Due to the rise and spread of movements for the protection of animals and environment, animal fats are used less and less as softening agents in cosmetic preparations. It is, however, possible to mention a wax and hydrocarbon derived from animal fats.

#### *Lanolin*

Lanolin is obtained from refined lambswool-grease. It is one of the fats listed in the French Pharmacopoeia where it is defined as 'a very complex mixture of esterified fatty acids and free alcohol, including cetylic alcohol, cholesterol, lanosterol and dihydroxylanosterol. The ratio of total fatty acids is about 60 per cent'.

Lanolin is hardly ever used in its natural form because it is very sticky and also because the label is required to mention that the 'product contains lanolin' as it was found that some people are sensitive to it. It is thus preferable to use its derivatives.

### Squalane or Perhydrosqualene

Squalane is produced by the hydrogenation of squalene obtained from shark liver oil. It is characterised by a saturated hydrocarbonic chain in C30.

It is stable, it does not become rancid, it is an excellent softening agent and has a good tolerance. But for ethical reasons preference is given to perhydrosqualene, a synthetic product which is largely used in cosmetic preparations being soft to touch, creamy and not sticky.

Triglycerides derived from animal fats like mink oil and tortoise oil may also be used.

### Vegetable Fats

A large number of different vegetable oils are used in the manufacture of cosmetics. There has been considerable progress in the improvement of their quality in terms of (colour, odour, reproductivity, stability) etc.

Vegetable oils are chosen for their feel, their specific properties (related to the presence of unsaponifiable matters, essential fatty acids, etc.) and the exotic nature of some oils.

Several studies have demonstrated the beneficial effects of unsaponifiable matters for (i) elasticity of the skin; (ii) cell metabolism; and (iii) restoring the cutaneous structure.

Vegetable oils are used in a large variety of cosmetic preparations such as day-creams, night-creams, body lotions, lipsticks, shampoos, shower gels, etc.

Doses for use vary from 0.1 to 20 per cent. In case of general beauty-care preparations, vegetable oils are associated with other lipid phases for reducing their greasiness and making them more emulsifiable.

#### *Sweet almond oil*

It is extracted from the kernel of the *Prunus amygdalus communis*. It is a well-known traditional oil due to its dermatological properties related to its high oleic acid and vitamin E contents.

It is used in small proportions in formulations, especially in baby-care preparations. But it is often substituted by oils extracted from kernels of other fruits because they are cheaper.

### *Kernel oil*

It is obtained from kernels of rosaceous fruits like apricot, almond, peach and plum. Its composition and properties are very similar to those of the oil of sweet almonds.

### *Wheat germ oil*

As its name indicates, it is extracted from wheat germ.

Major components of wheat germ oil are glycerides of linoleic, palmitic and oleic acids together with a significant amount of vitamin E. It is used in a proportion of 0.5–1 per cent in some foamy preparations.

### *Copra oil*

It is composed of 99 per cent triglycerides, mainly lauric and myristic acids.

Copra oil is one of the major sources of fat for surfactants.

On the other hand, it is not widely used as a softener in cosmetic preparations, except as Monoi oil (name given to copra oil in which Foamflowers have been macerated) in suntan lotion formulations with an exotic appeal.

### *Castor oil*

It is extracted from the seeds of *Ricinus communis.*

Castor oil is composed of triglycerides whose fatty acids contain 87–90 per cent ricinoleic acid which gives its special properties. It is the only vegetable oil to have such a high density and viscosity and is therefore seldom used; in its pure form. It is used in lipstick to solubilise pigments.

### *Safflower oil*

It is obtained from the seeds of *Carthamus tinctoris* and is mainly composed of linoleic acid triglycerides.

Because it is so rich in essential fatty acids, it is widely used in skin-care products.

### *Avocado oil*

Avocado oil is extracted from the fruit of *Persea americana*. Its major components are linoleic and palmitic acids and unsaponifiable substances whose proportion may be as high as 10 per cent. The unsaponifiable matters in avocado oil are mainly sterols and branched chain hydrocarbons.

Various studies have shown that unsaponifiable matters in avocado oil applied percutaneously to rats over a period of 15 days are able to modify some constituents of dermal conjunctive tissues, giving more elasticity to the skin. That is why avocado oil is used in bath oils, lipstick, creams, capillary conditioners, etc.

### *Shea (karite) butter*

Karite nut or shea butter is obtained from the fruit of *Butyrosporium parkii.*

It is different from other triglycerides because it is solid and very rich in unsaponifiable matter (5–7 per cent) consisting mainly of triterpenic alcohols.

Several studies have revealed UV anti-erythemic activity which helps tissue regeneration and softening of the skin. This is why shea butter is a well-known and appreciated ingredient in cosmetology, especially for suntan and post-suntan lotions.

### *Jojoba oil*

It is a liquid, unsaturated wax from the esters of long carbon-chain fatty acids and long carbon-chain unsaturated alcohols. Jojoba oil has a very fatty character and shows excellent cosmetic properties.

Jojoba oil is extracted from the seeds of *Simmondsia chinensis*. It is quite unique in the vegetable kingdom because it is a liquid wax. The composition of jojoba oil is very similar to that of spermaceti for which it is an excellent substitute. Jojoba oil is used as a superfatting agent in shampoos to add gloss to hair. Since, its composition resembles that of sebum, it can reinforce hydrolipid film after several applications. It is thus widely used in nourishing creams, body lotions, lip salves, etc.

### *Others*

#### *Vegetable oils*

A large number of other vegetable oils are also used in cosmetics. They are chosen not only for their efficacy but also because their novelty and exotic appeal provide strong marketing arguments. Some examples are evening primrose, borage and grapeseed oils.

Vegetable fats are also found in the form of waxes such as carnauba and candillila. They are not commonly used because they have a high melting point. They are mostly found in lipstick and depilatory waxes.

#### *Synthetic oils*

Synthetic oils are esters, usually obtained by direct reaction of fatty acids with alcohols. A wide range of combinations to produce different synthetic esters is possible, and the list of synthetic oils available to the cosmetic industry is long.

#### Isopropyl esters

The most important synthetic ester for cosmetics is isopropyl myristate. It has a very dry character and shows good spreading on the skin. It can open films of mineral oil or other fatty oils in order to let water evaporate, and it is an excellent solvent for perfumes, ultraviolet filters, etc. The isopropyl esters can cause difficulties when used in formulations that are filled into polystyrene or polyethylene containers since it is good solvent.

#### Ethylhexyl esters

Ethylhexyl esters of palmitic acid or stearic acid have become important products for the cosmetic industry. Compared with isopropyl esters, the molecular weight is higher, but spreading properties are similar.

#### Oleic acid esters

The two most important oleates are decyl oleate and oleyl oleate. The character of these esters is relatively fatty and, as with unsaturated esters, they can become rancid. The advantage of the oleates is that, even at high molecular weight, the esters are liquid.

#### Caprylic/capric acid esters

The most important caprylic/capric esters are the triglyceride and the propylene glycol diester. The glycerol ester is mainly used as a replacement for natural triglycerides, with the advantage that it is stable against oxidation.

#### N-Butyl stearate

N-butyl stearate is an excellent skin-compatible ester with very good emollience. The character is not very fatty compared with the isopropyl esters.

### Isocetyl stearate

Isocetyl stearate is a saturated liquid ester of high molecular weight and good emollience. Water can evaporate easily through the ester and even small amounts can perforate a hydrophobic film of mineral oil.

### Octyldodecanol

Octyldodecanol is not an ester but may be used as a cosmetic oil component. It combines a fatty character with good spreading properties.

### Di-isopropyl adipate

Di-isopropyl adipate is an ester of a dicarbonic acid. Contrary to most other oils, it is soluble in alcohol and is therefore used, for example, as a refatting agent in aftershave formulations.

### Pentaerythritol tetraisostearate

Pentaerythritol tetraisostearate is a high molecular weight, highly branched ester of pentaerythritol and isostearic acid. It has a very fatty character and shows no comedogenic effects. Consequently, it is an ideal ester for skin-protection formulations such as baby products and, due to its water-repellent properties, water-resistant suntan preparations.

## Synthetic Fats

Natural oils from vegetable sources are mainly glycerol esters based on mixtures of fatty acids. The most important are the saturated lauric, myristic, palmitic and stearic acids and the unsaturated oleic and linoleic acids. These fatty acids are derived from natural sources and typically have even numbers of carbon atoms in their chains. Important characteristic values for triglycerides are the saponification number and the iodine number. The latter indicates the amount of unsaturated fatty acids present.

Triglycerides are relatively fatty and spread very little on the skin. They generally show moderate comedogenicity, exceptions being sunflower oil and safflower oil where no comedogenic effects have been found. Castor oil is used in lipsticks and also in alcoholic preparations.

### *Triglycerides*

It is possible to obtain by synthesis triglycerides on saturated chains which do not present oxidation or rancidity problems and have a pleasant feel. Some of the commonly used monoglycerides are along with their applications shown in Table 10.2.

**Table 10.2.** Commonly used monoglycerides.

| *Identity* | *Uses* |
|---|---|
| Hydrogenated palm glyceride | Emulsifying agent<br>Suspending agent |
| Glyceryl cocoate | Skin conditioning agent |
| Glyceryl stearate | Emulsifier |
| Glyceryl oleate | Skin care agent<br>Emulsifier |

List of triglycerides of varying melting points and textural properties used in cosmetic are given below:

| | |
|---|---|
| Caprylic/capric/lauric triglyceride | Trihydroxystearin |
| Caprylic/capric/linoleic triglyceride | Trilaurin |

Caprylic/capric/stearic triglyceride
Caprylic/capric triglyceride
Lauric/palmitic/oleic triglyceride
Oleic/linoleic triglyceride
Oleic/palmitic/lauric/myristic/linoleic triglyceride
Triarachidin
Tribehenin
Tricaprin
Tricaprylin
Trilinolein
Trilinolenin
Trimyristin
Trioctanoin
Triolein
Tripalmitin
Tripalmitolein
Tristearin

### *Fatty acid and fatty alcohol esters*

A wide range of these esters can be obtained by esterification of a fatty alcohol with a fatty acid because the length of the hydrocarbon chain can vary from C3 to C22 (sometimes even longer chains can be found). All esters from this range have one common characteristic, namely a softening effect.

They spread very easily and penetrate the skin, making it soft, velvety and supple. These properties can sometimes be modulated depending on the chemical structure of the ester and the physico-chemical characteristics resulting from this structure.

For example, esters having low surface tension are the easiest to spread. Further, these esters have the advantage of not being susceptible to oxidation. They do not become rancid and are easy to emulsify.

Because of all these reasons, they are highly appreciated as constituents of lipidic phase of a large number of cosmetic preparations. They are used either by themselves or in association with other fats (mineral or vegetable) in order to improve their texture.

Among the most commonly used esters, mention may be made of: isopropyl myristate, isopropyl isostearate, 2–ethylhexyl adipate, cetostearyl octanoate, etc. Due to the great diversity of esters it is possible to satisfy different requirements. As a result, they have been introduced in a large number of formulations e.g. beauty creams and lotions, suntan lotions, powder compacts, lipstick, mascara, etc.

### *Silicone oil*

The high molecular weight organo polysiloxanes (dimethicone) are hydrophobic oils with good skin protection and non-sticky skin-feel. They show very high spreading and are used in small amounts in stearate creams to avoid the soap-up effect. To prevent eye-sting, non-volatile oils should not be used in products that are applied to the eye area. Due to their water-repellent properties, silicone oils are important for waterproof sun products. They are usually classified by viscosity.

It is possible to distinguish two families of products in this range:

#### *Volatile silicone oils*

Volatile silicone oils used in cosmetics are cyclomethicones.

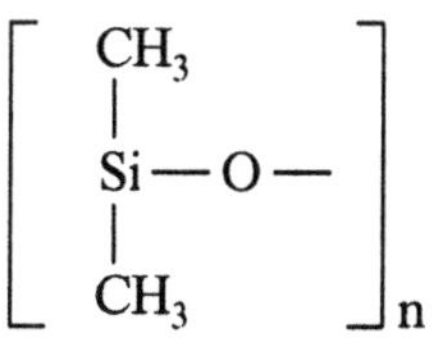


n = 4, 6, 8,

Their solubility in many solvents or excipients, which is allied to their volatility, makes it possible to use them in many preparations, especially antiperspirant sticks.

### *Non-volatile silicone oils (dimethicones)*

Non-volatile silicone oils belong to the dimethicone family.

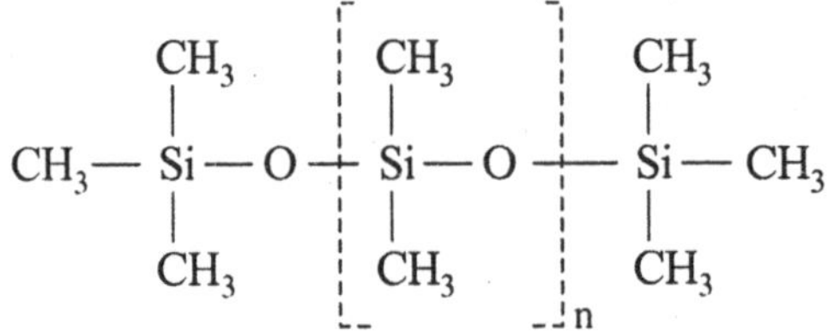

These are oils with straight chains offering a vast range of viscosity.

They are characterised by their water-repellant quality. Because of their chemical inertia, their resistance to heat and ageing and their low surface tension, easily spreadable, they are commonly used in a wide variety of cosmetic preparations e.g. mascara, lipstick, beauty creams and lotions, waterproof suntan lotions, etc. They are used in association with other fatty phases because they are difficult to emulsify.

As a rule, the neutral petroleum-derived lipids (paraffin oils) are easier to emulsify and less costly than the natural lipids. Although the use of mineral oil as a cosmetic emollient has persisted for many years, the cosmetic industry today utilises natural lipids as ingredients for their cachet. The popularity of these substances rests in part on the mystique of using a natural lipid on the body, a mystique that is enhanced by the efforts of segments of the cosmetic industry to convince users that the application of a particular nature-derived lipid is good for the skin or hair and provides special benefits. Although there are some specific lipids that play a proven role in skin physiology, the generalisation that natural lipids are good for you because they are natural is not based on sound scientific evidence.

## BENEFITS ASSOCIATED WITH THE USE OF NATURAL FATS AND OILS IN COSMETICS

The incorporation of natural fats and oils in cosmetic products designed to beautify and cleanse the body is widespread. Table 10.3 identifies a variety of finished cosmetic products in which animal or plant fats and oils are used. The desired beneficial effects resulting from the use of natural lipids on the skin include physical and biochemical effects.

**Table 10.3.** Major cosmetic products[a] containing natural fats and oils.

| *Cosmetic product* | *Approximate fat/oil content (%)* | *Reason for use of fat/oil* |
|---|---|---|
| Hand and body lotion/cream | 2–15 | Emolliency, moisturisation |
| Face cream (night) and facial oils | 5–100 | Occlusivity lipid replacement |
| Cleansing lotion/cream | 5–15 | Solvency, emolliency |
| Hair conditioners | 2–5 | Lubricity |
| Hair dressing (pomade, brilliantine) | 5–100 | Gloss, holding |
| Powder | 1–3 | Lubricity, compression aid |
| Make-up (stick, pan, liquid) | 15–40 | Gloss, adhesion |
| Bath products (dispersible and oil) | 2–100 | Emolliency, moisturisation |
| Suncreen | 3–10 | Hydrophobicity, emolliency |

[a] All products may in addition contain lipids that are not generally recognised as natural.

Some of these are uniquely associated with animal or plant-derived lipids, while others may merely be the result of the use of a well-formulated cosmetic product.

The cosmetic use of natural lipids—which are often oxidatively unstable and difficult to emulsify would not have achieved continuing popularity unless they could be claimed to perform some beneficial physiological function. Aside from the impact of the 'green movement' on the selection of cosmetic ingredients, natural oils provide at least two major clinically important benefits—occlusivity and the presence of essential fatty acids.

## Physical benefits

Natural fats and oils are used without significant chemical modification for differing effects in a variety of cosmetic products. They remain on the skin, and their presence results in several important physical effects.

### *Emolliency*

Emolliency is the effect produced by an emollient on human skin. Emollients are defined as substances that impart softness, smoothness, and flexibility to skin and that have the ability to maintain these conditions for some time. Water alone can produce these effects, but the perceived benefits are rapidly lost as the water evaporates. The term emolliency in practice is rarely differentiated from the term moisturisation, with the latter implying that the water content of the substrate (skin) is raised. It is well established that the application of a water-oil mixture in a simple emulsion causes a temporary increase in the skin's water content. Most of this water is lost within a few minutes, but the residual effect of the non-volatile lipid produces emollient benefits. The lipid film on the skin surface may help to retard the evaporative loss of water when the emulsion dries on the skin.

### *Occlusivity*

Occlusivity in cosmetic and pharmaceutical practice refers to the ability of a substance to create a film on the skin surface that interferes with the evaporation of water from the skin surface to the environment. Any reduction of transepidermal water loss (TEWL) raises the level of water retained in various strata of the epidermis. The occlusive character of fats and oils is therefore utilised to increase the water content of skin. This increase alters the viscoelasticity of stratum corneum and makes the skin more supple. Occlusive substances impede the escape of water vapour; these materials, as a rule, can also act as water repellants, i.e. interfere with access of water.

When applied in emulsion form, the emulsifiers may help the lipid to spread by increasing skin wetting. Co-emulsifiers (fatty alcohols, polymers, etc.) may modify the lipid and may decrease greasiness. It is apparent that most natural triglycerides cannot be used in large amounts in modern cosmetics.

### *Gloss*

Gloss is the ability of a surface to reflect incident light. Many fats and lipids increase specular reflection of skin or hair by providing a smooth surface through the process of coating.

### *Adhesion*

Adhesion to skin or between hair fibres can result from the application of many lipids. In the case of hair, lipids help to align fibres parallel to each other by lateral adhesion. Adhesive qualities are also required for the retention of pigmented substances and some drugs on the skin.

### *Lubricity*

Lubricity is related to emolliency and is due to the ability of all types of fats and oils to reduce frictional effects between surfaces.

### *Solvent*

Solvent properties of fats and oils can be used in cosmetics for the removal of lipid soil in cleansing preparations. The ability of fats and oils to mix with other lipid soil or to dissolve other organic substances, such as organic dyestuffs, contributes to the popularity of natural fats in cosmetic cleansing preparations.

Finally, natural fats and oils have been claimed to be useful replacements for skin lipids that may have been inadvertently removed from the skin by typical lipid solvents (acetone or hydrocarbons) or by excessive exposure to detergents.

## Biochemical Benefits

The content in vegetable oils of polyunsaturated fatty acids, especially of the so-called essential fatty acids (EFA), is an important driving force behind the growing use of these lipids in skin products. One of the extensively documented benefits of the use of natural lipids is the presence of EFA.

Essential fatty acid deficiency in human causes two primary skin defects: scaling and epidermal hyperplasia. Teleologically, epidermal lipids are constructed to provide much of the barrier function of the skin and to create a membrane that resists the passage of water and other substances. The ceramides, together with the remaining lipid components, have been shown to form water-retaining lamellar structures within the stratum corneum. Table 10.4 provides a simple classification of cosmetically useful lipids by chemical type, while Table 10.5 is a more comprehensive listing of natural lipids used in cosmetics.

**Table 10.4.** Major classes of natural lipids found of use in cosmetics.

| *Type* | *Comment* |
|---|---|
| Fatty acids | Generally derived from natural triglycerides |
| Triacylglycerides | Includes natural and synthetic products |
| Mono and diacylglycerides | Derived from natural lipids |
| Phosphoglycerides | Isolated from natural sources; synthetic phosphatides have role in vesicle formation |
| Sphingolipids | Rarely isolated pure from animal sources; major components of brain or spinal column lipids |
| Wax esters | Generally of plant or fish origin |
| Sterols | Isolated from plant or animal sources; major constituent of unsaponifiables |

**Table 10.5.** Natural fats and oils used in cosmetics.

| *Common name*[a] | *Source* | *Obtained*[b] |
|---|---|---|
| Almond oil | *Prunus amygdalus* | E |
| Animal skin lipids | Various species | |
| Apricot kernel oils | *Prunus armeniaca* | E |
| Avocado oil | *Persea americana* | E |
| Babassu oil | *Orbignya oleifera* | E |

*(Contd...)*

**Table 10.5.** Natural fats and oils used in cosmetics. *(Contd .)*

| *Common name*[a] | *Source* | *Obtained*[b] |
|---|---|---|
| Bayberry wax | *Myrica cerifera* | M, Extr. |
| Beeswax wax | *Apis mellifera* | M |
| Borage seed oil | *Borago officinalis* | E |
| Butter | | Cow's milk, M |
| Candelilla wax | *Euphorbiaceae spp.* | M, Extr. |
| Camellia oil | *Thea sinensis (leaf)* | E |
| Canola oil | *Brassica spp.* | E |
| Carnauba | *Copernicia cerefera* | M, Extr. |
| Carrot oil | *Daucus carota sativa (seed)* | E |
| Cashew nut oil | *Anacardium occidentale* | E |
| Castor oil | *Ricinus communis (seed)* | E |
| Cherry pit oil | *Prunus (various species)* | E |
| Chia oil | *Salvia hispanica (seed)* | E |
| Cocoa butter | *Theobroma cacao* | E |
| Coconut oil | *Cocos nucifera* | E |
| Cod liver oil | *Gadusmorrhua* | E |
| Corn germ oil | *Zea mays* | E |
| Corn oil | *Zea mays* | W.M. |
| Cottonseed oil | *Gossypium spp.* | E |
| Colostrum | Animal species | M |
| Cucumber oil | *Cucumis sativus* | E |
| Egg oil | Chicken | Extr. |
| Evening primrose oil | *Enethera biennis* | E |
| Fish liver oil | Various species | M, Extr. |
| Grape seed oil | *Vitis vinifera* | E |
| Hazelnut oil | *Corylus spp.* | E |
| Hybrid sunflower seed oil | *Helianthus annuus (variety)* | E |
| Illipe butter | *Madhuca latifolia* | E |
| Japan wax | *Rhus succidanea* | E |
| Jojoba oil | *Simmondsia chinensis (seed)* | E |
| Kukuinut oil | *Aleurites moluccana* | E |
| Lanolin | Wool | M, Chem. |
| Lanolin oil | Fraction of lanolin | Ph |
| Lanolin wax | Fraction of lanolin | Ph |
| Lappa seed oil | *Arctinium lappa* | E |
| Lard | Hog | R |
| Lecithin | Animal or plant | W.M., Extr. |
| Lesquerella oil | *Lesquerella fendleri (seed)* | E |

*(Contd...)*

**Table 10.5.** Natural fats and oils used in cosmetics. *(Contd .)*

| *Common name*[a] | *Source* | *Obtained*[b] |
|---|---|---|
| Linseed oil | *Linum usitatissinum* | E |
| Macadamia nut oil | *Macadamia ternifolia* | E |
| Mango seed oil | *Mangifera indica* | E |
| Meadowfoam seed oil | *Limnanthea alba* | E |
| Menhaden oil | *Brevoortia tyrannus* | M, Extr. |
| Mink oil | *Mustela (skin)* | M, R, Extr. |
| Mink wax | Fraction of mink oil (q.v.) | Ph |
| Moringa oil | *Moringa oleafera* | M, Extr. |
| Mortierella oil | *Mortierella isabellina* (seed) | E |
| Montan wax | Lignite | Chem. |
| Neatsfoot oil | Cattle (hooves) | Extr. |
| Oleostearine | Fraction of beef tallow | M |
| Olive husk oil | *Olea europaea* (husk) | E |
| Olive oil | *Olea europaea* | E |
| Omental lipids | Bovine omentum | M, R |
| Orange roughy oil | *Hoplostethus atlanticus* | M, Extr. |
| Ouricury wax | *Syagrus coronata* (leaves) | M, Extr. |
| Palm kernel oil | *Elaeis guineensis* (seed) | E |
| Palm kernel wax | Fraction of palm kernel oil (q.v.) | M |
| Palm oil | *Elaeis guineensis* | E |
| Peach kernel oil | *Prunus persica* | E |
| Peanut oil | *Arachis hypogaea* (seed) | E |
| Pengawar djambi oil | Cyatheaceae | E |
| Pentadesma butter | *Pentadesma butyracea* (nut) | E |
| Phosphatidyl choline | Purified lecithin | Extr. |
| Pistachio nut oil | *Pistacia major* | E |
| Placental lipids | Animal placenta | M, Extr. |
| Rapeseed oil | *Brassica campestris (seed)* | E |
| Rice bran oil | *Oriza sativa* | E |
| Rice bran wax | Fraction of rice bran oil | M |
| Safflower oil | *Carthamus tinctorius* | E |
| Salmon oil | *Oncorhyncus* | M, Extr. |
| Sesame oil | *Sesamum indicum* | E |
| Shark liver oil | *Galeorhinus and Hypoprion spp.* | M, Extr. |
| Shea butter | *Butyrospermum parkii* (seed) | E |
| Shorea butter | *Shorea stenoptera* | E |
| Soyabean oil | *Glycina max* | E |
| Soyabean lipid | Fractionated soyabean oil | M, Extr. |

*(Contd...)*

**Table 10.5.** Natural fats and oils used in cosmetics. *(Contd .)*

| *Common name*[a] | *Source* | *Obtained*[b] |
|---|---|---|
| Spent brain wax | Barley fermentation residue | M, Extr. |
| Sunflower seed oil | *Helianthus annuus* | E |
| Sweet almond oil | *Prunus amygdalus dulsis* (seed) | E |
| Tall oil | Wood pulp | Chem. |
| Tallow | Animal tissue | R |
| Vegetable oil | Plant seeds | E |
| Walnut oil | *Juglans spp.* (nuts) | E |
| Wheat bran lipids | *Triticum aestivum* | M, Extr. |
| Wheat germ oil | *Triticum aestivum* | E |

[a] C.T.F.A., nomenclature.
[b] Common commercial isolation method: Chem. = chemical separation; E = expeller pressed; Extr. = solvent extraction; M = mechanical; Ph = physical separation; R = rendering; and W.M. = wet milling.

## PRODUCTION AND PURIFICATION

The natural lipids used in the food industry are identical to those used in cosmetics. Depending on the source, lipids may be directly expressed from nuts or seeds. Drying and crushing may precede the expression to improve the yield.

Fleshy plant sources (e.g. avocado) are sliced and dried before expression. Solvent extraction is generally restricted to the recovery of lipids from sources low in oil.

The raw lipid (obtained by expression or by extraction and solvent stripping) requires further refining for cosmetic or food use. The complete removal of non-glycerides (sterols, pigments, glucosides, proteins, fatty acids, etc.) may not be mandatory for all food and cosmetic applications. In fact, the presence of plant sterols, vitamins (A, D, and E), and other impurities may be desirable.

The removal of pigments (potential sensitisers) and of polar materials (resulting from peroxide formation) is mandatory for most cosmetic applications. In some instances, the refining steps required for food lipids may be insufficient to make them acceptable for cosmetic use. In light of this, the demands of the cosmetic industry resulted in special refining methods to yield colourless oils that are not only odourless but also contain almost no polar substances. This type of refining is achieved by proprietary adsorption chromatography processes during which polar constituents in natural oils are retained, while the non-polar triglycerides or wax esters are eluted. An alternative process of providing highly purified cosmetic oils employs supercritical fluid extractions.

This latter procedure can yield specific fractions of lipids from a natural mixture, depending on the pressures and temperatures employed. It has been found that avocado oil, carnauba wax, cocoa butter, lanolin fractions, shark liver oil, and lecithin can be processed by supercritical fluid ($CO_2$) extraction for ultimate cosmetic use.

The adsorption processes have no effect on the lipids or the lipid fraction's chemical characteristics, although impurities are clearly removed (Table 10.6). The adsorption chromatographic procedure is reported to materially improve the oxidative stability of natural oils as determined via the Rancimat system. Some typical results from induction period studies of cosmetic oils and the impact of antioxidants is shown in Table 10.7.

**Table 10.6.** Characteristics of lipids before and after adsorptive refning.[a,b]

| Description | Sesame oil | | Almond oil | | Orange roughy oil | | Shark liver oil | | Olive oil | | Borage oil | | Babassu oil | |
|---|---|---|---|---|---|---|---|---|---|---|---|---|---|---|
| | *b* | *a* | *b* | *a* | *b* | *a* | *b* | *a* | *b* | *a* | *b* | *a* | *b* | *a* |
| Colour (APHA) | 225 | 25 | 160 | 15 | 150 | 10 | >500 | 80 | >500 | 35 | >50 | 70 | 70 | 5 |
| Acid value | 0.11 | 0.06 | 0.17 | 0.05 | 0.17 | 0.03 | 0.56 | 0.06 | – | – | – | – | – | – |
| Sap. value | 187 | 189 | 190 | 190 | 100 | 99 | 173 | 173 | – | – | – | – | – | – |
| Hydrox. value | 7.5 | 6.3 | 11 | 8.4 | 13 | 11 | 6.5 | 6.3 | – | – | – | – | – | – |
| Iodine value | 120 | 114 | 102 | 99 | 32 | 30 | 111 | 107 | – | – | – | – | – | – |
| POV[c] | 2 | 1 | – | – | 12 | 1 | – | – | 15 | 2 | 9 | 3 | 3 | 1 |

[b] = before; a = after.
[c] POV = peroxide value.

**Table 10.7.** Effect of antioxidants on stability of cosmetic oils.

| | Induction period (h) | | | | | | |
|---|---|---|---|---|---|---|---|
| | | Antioxidant (1000 ppm)[a] | | | | | |
| *Oil* | *No inhibitor* | *A* | *B* | *C* | *D* | *E* | *F* |
| Apricot kernel | 4.3 | 20.8 | 9.1 | 8.7 | 9.6 | 24 | 24 |
| Sunflower | 1.3 | – | 9.7 | 5.6 | 8.6 | 8.5 | 24 |
| Wheatgerm | 9.1 | 13.5 | 10.6 | 10.3 | 11.0 | 32.1 | 21.2 |
| Borage | 4.1 | 11.9 | 7.1 | 4.2 | 6.7 | 15.2 | 15.5 |
| Evening primrose | 3.1 | 8.2 | 7.2 | 3.5 | 5.9 | 16.2 | 10.2 |

[a] Tests run at 100°C; A = ascorbylpalmitate; B = butylated hydroxytoluene; C = δ-tocopherol; D = butylated hydroxyanisole; E = *t*-butylhydroquinone; and F = propylgallate.

Improvement of long-term storage stability of lipids generally requires inclusions of antioxidants. Rigid exclusion of heavy-metal contaminants and use of sequestrants, storage in the dark, and elimination of air during filling are required for long-term stability of finished cosmetic products containing natural lipids. In addition, cosmetic formulators must be concerned with oxidative stability during use. Application of a stable or stabilised natural lipid in a finished cosmetic to the skin exposes the lipid to light and oxidation. This *in situ* exposure can result in free radical formation via H-abstraction or may induce a type II reaction due to formation of singlet oxygen. The latter is catalysed by ultraviolet (UV) light absorbing photosensitisers, e.g. chlorophyll. Type II reactions are not controlled by the usual (free radical inhibiting) antioxidants. Singlet oxygen scavengers and UV light absorbers are the most promising candidates for stabilising natural lipids in the skin. For these reasons, the use of carefully puified natural lipids has become customary in the cosmetic industry.

## FORMULATION REQUIREMENTS

The utility of a lipid in a cosmetic composition depends on a number of factors. Some of these arise from sensory (tactile and odour) concerns; others relate to the presence of specific fatty acid components that are believed to provide skin benefits; still others relate to the stability of the incorporated lipid and of the finished cosmetic; finally, perceived benefits from the inclusion of a specific lipid or lipid fraction play an important role in consumer acceptance. The results of balancing these factors are a number of ground rules for the use of lipids in cosmetic formulations:

1. The use of a specific lipid depends on its tendency to leave the skin (or other body tissue) without an oily greasy feeling. The formulator, therefore, restricts use of oils to more penetrating oils, which leave a satinlike finish on the skin. If heavier (greasier) lipids are used, their concentration must be limited, or they must be blended with other (drying) components.
2. Only lipids of high purity can be used. Off-odours and the presence of peroxides (or their decomposition products) may make commercially available lipids unacceptable.
3. The lipid must be stable or easily stabilised against peroxidation.
4. The nature of lipids used in cosmetics—which require label declaration—must have broad consumer acceptance. Sometimes, the label declaration may include a lipid used by the supplier to achieve a particular characteristic (use of illipe butter in cocoa butter). The current tendency of consumers to reject animal products or unhealthy appearing ingredients plays a critical role in lipid selection. For example, the use of cholesterol—a required natural constituent of human skin—may elicit some adverse feeling. The use of tallow or even hydrogenated tallow in cosmetics is steadily decreasing. Instead, formulators turn to various hydrogenated vegetable oils or tristearine as replacements.
5. The lipids used in cosmetics should contain components that have demonstrable (health) benefits. The presence of specific (poly) unsaturated fatty acids plays a role in consumer acceptance, especially if this fatty acid has been shown to contribute to human health.

## FATS USED AS SYNTHESISING AGENTS

Fats are widely used in cosmetics industry due to their ability to act as synthesising agents and produce a large number of multifunctional derivatives including surfactants.

### Obtaining Basic Derivatives

Basic derivatives used in synthesis of surfactants are divided into four categories: (i) fatty alcohols; (ii) fatty acids; (iii) fatty amines; and (iv) methyl esters.

They are obtained from triglycerides contained mainly in tallow, copra oil, palm oil or palm kernel oil. These derivatives are commonly obtained by cross-esterification.

### Fatty Acids and their Derivatives

#### *Fatty acids*

Fatty acids and especially stearic acids are used in the form of soap by reacting with a base like caustic soda, potash or triethanolamine.

Fatty acid from tallow, neutralised by triethanolamine and sold under the name of triple pressure stearic acid (Table 10.8), is used as an emulsifying agent in beauty creams and shaving foams.

**Table 10.8.** Composition of triple pressure stearic acid.

| | |
|---|---|
| Stearic acid | 45% |
| Palmitic acid | 55% |
| Unsaturated fats | Eliminated |

Though it was very widely used in the past, it is now substituted by its derivatives because its high pH is incompatible with the acid pH of human skin.

### *Fatty acid derivatives*

*Ethoxylated fatty acids*

Ethoxylated fatty acids produce a series of hydrophilic molecules labelled according to the number of grafted ethylene oxide molecules. The fatty acid most commonly used for this purpose is stearic acid. They are used as emulsifying agent along with other more lipophile of fatty acids.

*Fatty acid esters*

In this series it is possible to find a large number of compounds which vary because of:

1. Length of the fatty acid (lauric, myristic, palmitic, stearic and oleic acids).
2. Nature of the alcohol (polypropylene glycol, glycerol, sorbitanne, polyglycerol, sucrose, etc.)

Depending on their structure, they act either as emulsifying or solubilising agents or as a consistency factor. The most important category in this series is composed of sorbitanne esters and their ethoxylated derivatives. Sorbitanne stearates or oleates when used with their ethoxylated counterparts (20 moles) become remarkable emulsifying pairs. They are very effective, have an excellent tolerance and pleasant feel. They are found in many cosmetic preparations like creams, lotions, shaving foams, etc. The only obstacle to their use is, as in other esters, their very acid or alkaline pH. Their use is therefore avoided in depilatory and oxidising creams.

*Fatty acid chloride derivatives*

By reacting fatty acid chloride with an amino derivative (protein or amino acid) it is possible to obtain acylamino acids.

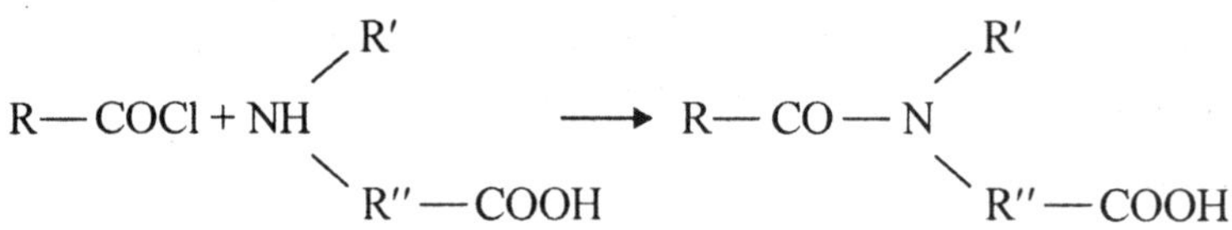


(R = C12 in general because this is optimum for foam level)

N-acylamino acids are anionic foaming surfactants very commonly used in shampoos, shower gels and liquid soaps because of their intrinsic properties. On account of their softness, they cleanse the skin without excessive delipidation. Their skin tolerance is excellent. They also help to lessen the irritation caused by some preparations. Further, presence of an aminic sequence in acyl amino acids makes the hair light and lustrous and gives it body.

## Fatty Alcohols and their Derivatives

*Fatty alcohols*

Cetylic and stearylic alcohols are used as a thickening agent in fatty phases. They have to be used in small amounts to avoid excessively slow penetration.

### *Fatty alcohol derivatives*

*Ethoxylated fatty alcohols*

They are obtained through the condensation of ethylene oxide on fatty alcohols.

Depending on the length of the fatty chain (lauric, palmitic, stearic and oleic acids) and the degree of ethoxylation, it is possible to obtain emulsifiers or solubilisers.

The presence of an ether bond makes them highly resistant to hydrolysis. They are used without any difficulty as emulsifying agent in depilatory, bleaching and capillary creams and to solubilise many ingredients.

### *Alkylpolyglucosides*

Alkylpolyglucosides are obtained by reacting a fatty alcohol with glucose.

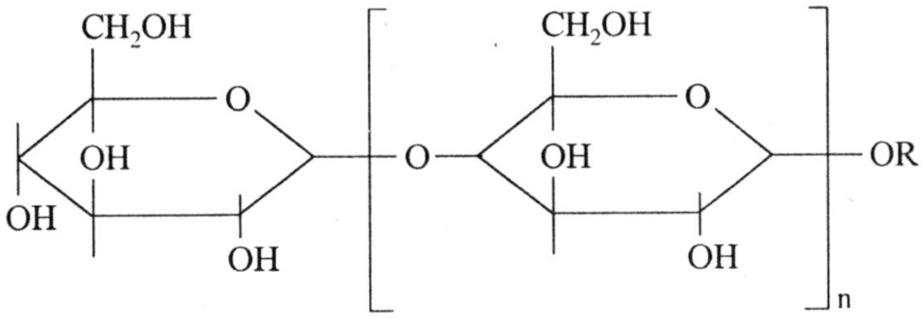


R: $C_{10}$— $C_{12}$— $C_{14}$ n: 1 to 3

It is thus possible to obtain non-ionic foaming agents which are very stable under hydrolysis.

This range is quite unusual because it produces non-ionic foaming agents from natural products which are biodegradable and very soft on the skin.

The foam obtained in this manner is very stable. The optimum at the foam level is obtained on chains in $C_{10}$. Decylpolyglucoside satisfies the requirements of many cosmetic preparations like shower gels, foaming gels for the face, shampoos, etc.

### *Alkylsulphates and alkylethersulphates*

These anionic surfactants are obtained by sulphating fatty alcohol. A distinction can be made between alkylsulphate and alkylethersulphate families. In this case, the presence of ethylene oxide improves the solubility of the product and its resistance to cold.

$$R—O—SO_3H \qquad R—O—(CH_2—CH_2—O)_n\,SO_3H$$

$$n = 1 \text{ to } 3$$

Alkylsulphate Alkylethersulphate

They are used as a base in foaming preparations on account of their cleansing and foaming properties. The optimum foaming effect is obtained with chains in $C_{12}$–$C_{14}$.

## Fatty Amines and their Derivatives

### *Diethanolamides*

This product is a result of the reaction of diethanolamine with a methyl ester of coconut oil fatty acid.

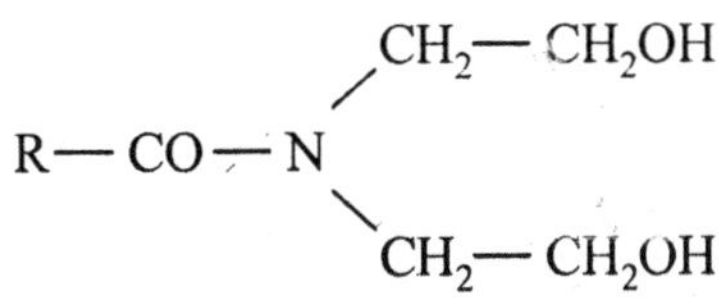


This non-ionic agent is used in small amounts in foaming preparations because of its thickening, superfatting and stabilising effect on foam.

### *Methyl ester derivatives*

Derivatives of methyl esters act as synthesising agents in the manufacture of some amphoteric surfactants like betaine and sulphobetaine. Methyl esters of copra oil fatty acids are the most commonly used because they produce the maximum foaming action surfactants.

These amphoteric derivatives are very often used in association with anionic derivatives because of their thickening action in acid pH. In these conditions, they considerably improve foam when confronted with hard water and stains. Further, in acid pH, they act as a conditioner and can be profitably used in shampoos.

This list is far from exhaustive. It is only intended to give an idea of the different surfactants that can be obtained from fats for use in the cosmetics industry, keeping in mind that the properties that are most desired are: (i) foaming; (ii) cleansing; (iii) emulsifying; (iv) solubilising; and (v) softening.

Thus fats form the very basis of the cosmetics industry. They are used at two levels. Firstly, they are used in their natural form, because of their softening action and even their 'nutritive' value, in the majority of cosmetic preparations. Among the large number of fats of diverse origins (animal, vegetable and synthetic) available to us, the ones that are most likely to satisfy the physical and sensorial criteria of the end products are selected by cosmetics manufacturers. Secondly, fats are also used in the form of synthetic derivatives resulting in multifunctional compounds.

The inclusion of a hydrophilic part produces a large number of surfactants whose diverse intrinsic actions such as foaming, moisturising, cleansing, emulsifying and solubilising, satisfy the requirements of several cosmetic preparations like shampoos, shower gels, creams for disentangling hair, beauty creams, suntan gels, lipsticks, mascara, microemulsions, etc. The use of fats is indispensable and cannot be dissociated from cosmetics.

Chapter 11

# Recovery and Refining of Glycerine

## INTRODUCTION

Glycerine is clear, colourless, odourless, syrupy liquid having sweet taste. It is hydroscopic, specific gravity (anhydrous) 1.2653, melting point 18°C, boiling point 290°C. It is soluble in water and alcohol but insoluble in ether, benzene, chloroform and in fixed and volatile oils. It is combustible, flash point 320°F (160°C).

Glycerine is the name of the commercial product consisting of glycerol and small amount of water. Glycerol is actually trihydric alcohol $C_3H_5(OH)_3$, which is more accurately named 1,2,3- propanetriol.

The glycerol in fats is usually recovered in either soap lyes or sweet waters. Normally, these materials are purified and most of the water they contain is evaporated under conditions which do not vapourise any significant quantity of glycerol to give respectively soap lye crude glycerine on saponification, or hydrolyser, crude glycerine. Crude glycerine is normally purified by distillation is sometimes followed by bleaching and other treatments. Alternative processes which use ion exclusion and/or ion exchange instead of distillations are also discussed in this chapter.

Glycerine can also be recovered from the by-products of methyl esters (from fats and oils) and fatty alkanolamides; the amount produced from these sources has been relatively small but has increased significantly due to worldwide construction of fatty alcohol plants.

## PROCESSING PRINCIPLES AND DETAILS

### Sources of Glycerine

Natural glycerine is essentially a by-product of certain processes producing products out of animal or vegetable fats:

1. Fat splitting (hydrolysis) under high pressure and temperature to obtain fatty acids and sweetwater glycerine.
2. Saponification of fats with caustic soda, yielding spent soap lyes containing glycerol, water, salt (NaCl) and other impurities; depending on the process for saponification and soap washing, the concentration of the glycerol varies greatly.
3. Transesterification of fats, typically with methanol, using sodium methoxide as a catalyst, to produce methyl esters and glycerol.

The chemical reaction for this process is as follows:

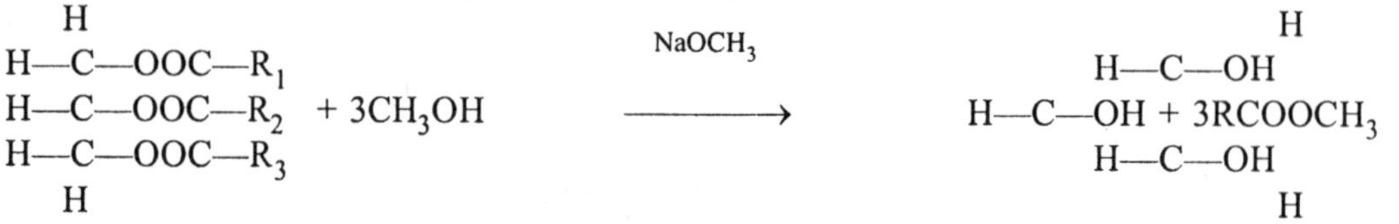

Triglyceride Methanol Catalyst Glycerol ME

While the initial chemical reactions for the above processes yield high theoretical concentrations of glycerol, subsequent washing steps dilute the glycerol to typical values which are given below:

| | |
|---|---|
| Fat splitting | 12–18% |
| Saponification, kettle | 10–15% |
| Saponification, continuous | 20–25% |
| Transesterification | 25–30% |

Synthetic glycerine is produced (Dow process) by chlorinating propylene to allyl chloride, converting this to dichlorohydrin, which is then converted to glycerine through the addition of small amounts of dilute NaOH and $Na_2CO_3$.

## TREATMENT OF SOAP LYES AND SWEETWATERS

Soap lyes are commonly neutralised with fatty matter before they leave the soapmaking department. They still contain some alkali, as well as soapy matter and other organic impurities and a substantial percentage of salt (around 10 per cent). If the older system of cooling and skimming, but not neutralisation, is used, the lyes contain a higher percentage of free alkali than in neutralised lyes. Sweet waters from high pressure splitters operated without catalyst only contain organic impurities of a fatty nature, plus some other compounds from the fats used. If the fats are split with the aid of a zinc oxide, or other alkaline, catalyst, the zinc, or other metal, leaves in the fatty acids, as a soap and the sweet waters are little affected. Twitchell sweet waters contain some sulphuric acid, or sulphates, as well as organic impurities. The purification, or treatment, of the lyes, or sweet waters, is intended to remove organic compounds other than glycerol as completely as possible; and the success with which this is achieved is reflected in the matter organic non-glycerol (MONG), or organic residue, of the crude glycerine produced.

The steps in the usual treatment of soap lyes are:

1. *Acidification* to a pH of about 4.6 – 4.8, although this is often omitted if the lyes have been well neutralised with fatty acids. Acidification is preferably with hydrochloric, rather than sulphuric, acid to avoid introduction of sulphates into the system. It liberates fatty acids and those which are insoluble in water, although of relatively low molecular mass, separate on standing. They are removed from the settler, often periodically and are sometimes returned to a low-grade soap, perhaps after filtration. These skimmed fatty acids are poor soapmaking material; they are low in molecular mass and contain hydroxy acids produced by oxidation and they tend to give sodium salts which have little value and/or again dissolve in the next lyes.
2. *Purification* by addition of a soluble iron or aluminium salt, a stage normally called *first treatment*. The pH is kept at, or adjusted to, about 4.5–5.0 (often 4.6 – 4.8); and the mix is well agitated by compressed air blown in through perforations in a pipe near the base of the tank. This air also ensures that if iron is used, it is oxidised to the ferric state and blowing should, if necessary, be continued until the mix is brown rather than greenish black. Oxidation is particularly necessary,

if sodium hydrosulphite is used in the soapmaking process. The precipitate may be ferric hydroxide, or aluminium hydroxide, on which fatty acids, proteinaceous matter and other impurities are adsorbed; or basic salts of the low molecular mass fatty acids may be present, the precise chemistry is obscure. The two main precipitants are solutions of ferric chloride, $FeCl_3.6H_2O$, or of aluminium sulphate, sometimes called alum, $Al_2(SO_4)_3 \cdot 18H_2O$. Ferrous chloride with excess acid, which is the residue left after hydrochloric acid has been used to pickle steel, can be used, preferably after the iron has been oxidised to the ferric state by bubbling in chlorine gas carefully through a porous diffuser. Excess acid can be reduced in concentration by allowing the spent pickling acid to stand in contact with scrap iron wire when additional iron goes into solution. It is preferable to use chlorides, rather than sulphates, in all additions to soap lyes, because otherwise sodium sulphate will build up in the recovered salt which is re-used in soapmaking and because sulphates are liable to increase problems with scale in the evaporators. Aluminium chloride is, however, not a convenient material to handle; and when aluminium treatment is used the aluminium is usually added as a sulphate. Sodium aluminate can also be used as a source of aluminium, sometimes in conjunction with ferric chloride. Opinions differ as to the relative merits of iron and aluminium for the purification of soap lyes, but iron has the advantage that it can usually be added as chloride. The quantity of precipitant needed varies with the composition of the lyes, but about 0.1–0.2 per cent of the hydrated ferric salt (added as a concentrated solution) is common, although more, up to about 0.4 per cent may be required. The completion of the first treatment process is often checked by the filtration of a small sample and addition of a few drops of diluted precipitant to the clear filtrate. If the treatment process is complete, no precipitate should be produced, but a good end-point is not always obtained; the pH at which the test is carried out is important. The adequacy of the treatment is proved ultimately by the MONG of the crude glycerine produced. Lyes from soaps with a high proportion of good tallow are more easily treated than those from high lauric oil blends, or those from low-grade fats. A very high glycerol concentration in the liquor can also make effective treatment more difficult.

3. The first treatment mud is removed by filtration, normally with filter presses or leaf filters. It is often advantageous to allow the tank contents to settle, to filter first the lower layer and to put the nearly clear upper layers through a clean press, or presses.
4. The filtrate is made slightly alkaline with caustic soda, in what is known as the *second treatment*. A pH of about 9 is common when the precipitant is iron, but a slightly lower alkalinity may be preferred with aluminium. Mixing and oxidation are again with air. A relatively small quantity of precipitate separates in the second treatment. It contains the excess iron or aluminium, silica and a little organic matter.
5. Finally, the treatment process is completed by a second filtration.

Variations in the details of the treatment process are possible, including estimation of the quantity of precipitant required by small scale tests. Single treatment processes have been devised in which, after precipitation of impurities with iron, or aluminium, as in the double treatment process, the mix is made slightly alkaline with sodium carbonate, or sodium bicarbonate, before being filtered once only.

Calcium chloride, followed by a second stage using sodium carbonate, has been used to treat soap lyes, but the purification is less good than with iron, or aluminium.

The purpose of treatment for spent soap lyes is to prepare the feedstock for glycerine recovery by removing impurities that would cause operating problems in subsequent processing steps, resulting in a poor-quality finished product. In addition, in handling soap lyes, care must be taken to prevent fermentation, which causes serious losses of glycerine due to the formation of trimethylene glycol, gases and acids during decomposition.

Finally, excess impurities in the lye can act as nucleation sites for crystal growth in the evaporator, spawning the growth to a large number of small crystals and discouraging the growth of larger, more easily separated salt crystals.

Spent soap lyes, as they are draw from the continuous saponification area or soap kettles, consist primarily of glycerine, sodium chloride and water as well as small quantities of caustic soda, sodium sulphate, sodium carbonate, soap and fatty acids and also some albuminous and oleaginous matter. To remove these impurities the soap lye is treated in batches as follows:

1. A cooling and settling step is used to separate the soap and fatty acids. These materials are less soluble at lower temperatures and rise to the surface, where they can be removed by skimming. Other heavy materials will settle to the bottom and remain in the tank until it is cleaned periodically. This is best done in the soap plant before the lyes are transferred to the treatment plant (Fig. 11.1).
2. The skimmed and settled lye is transferred to the first treatment tank. It is good practice to heat the lye to 50–60°C before this tank or in it; the treatment process is most efficient at this temperature. In this first tank, the lye is acidulated to pH 4.6–5.0 with sulphuric or hydrochloric acid. Hydrochloric acid is recommended to avoid the formation of sulphates. This step liberates fatty acids, which may be skimmed as most are insoluble in water.
3. A coagulant is added to cause the residual fatty acids to form insoluble soaps, which coagulate. The pH is adjusted to about 4.5. Also, compressed air is bubbled into the tank and is used for mixing and to ensure oxidation. The coagulant is usually either aluminium sulphate (alum) or ferric chloride. If ferric chloride is used, the air agitation is particularly needed to oxidise the iron to its ferric state. While ferric chloride is more difficult to handle than aluminium sulphate, is strongly corrosive to iron and is more expensive, it is usually preferred because the salts formed using alum also contains sulphates. The precipitates of either aluminium hydroxide or ferric hydroxide will adsorb other impurities such as fatty acids and proteinaceous matter.

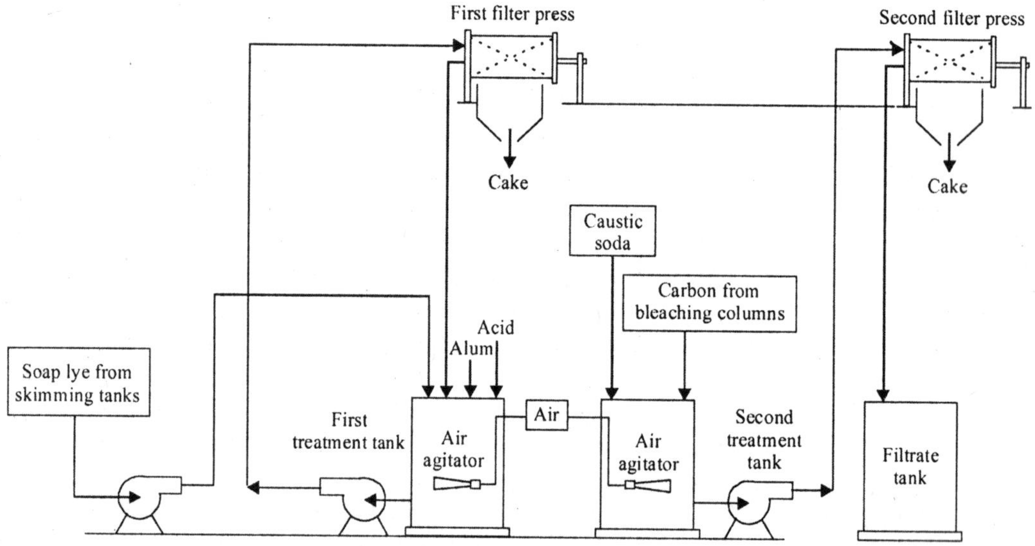


**Fig. 11.1.** Batch soap lye treatment flowsheet.

### Treatment of Sweet Water

Treatment can also be carried out using a continuous or semicontinuous operation. In each case, accurate, in-line pH measurement is needed to control the addition of the treatment chemicals. Filtration can be accomplished on a semicontinuous basis using plate and frame filters by sizing the treatment tanks to allow surge capacity for filter clean out intervals, or a continuous vacuum drum can be utilised. These processes lend themselves to the larger capacity plants.

Treatment of sweetwaters is less involved than the treatment of soap lyes as described above. The purpose for treating sweetwaters is to remove any unreacted fats and to prevent bacteriological growth in the glycerine. Typically, the sweetwater is settled for several hours to allow any free fatty acids to rise to the surface, where they are skimmed off the top. Then, using air agitation an excess of lime is added to neutralise the fatty acids still present. This excess is calculated as 0.23 per cent of CaO and the batch must be adjusted appropriately to obtain this excess. The sweetwater is then filtered. Sodium carbonate is sometimes used to remove the calcium excess as carbonate. Some processors treat using barium chloride and sulphuric acid, which is a more effective and expensive process. Also, it has been shown that addition of a small amount of polyelectrolyte, such as used in water treatment, will significantly improve the quality of the crude glycerine.

## EVAPORATION OF PURIFIED LIQUORS TO CRUDE GLYCERINE

Many variations are possible in the design of plants for evaporation to crude glycerine. Points to be considered in the design of an evaporation plant to produce crude glycerine include :

1. The capacity required in terms of water to be evaporated, crude glycerine produced and for soap lyes, salt to be separated.
2. Thermal efficiency and maximum consumption of steam.
3. Type of heating unit.
4. Arrangements for separation of salt, when relevant.
5. Avoidance of undue losses of glycerol.
6. Vacuum equipment and other ancillaries.
7. Method of control.
8. Materials of construction.

### Evaporation

Following treatment, the soap lyes or sweetwaters are concentrated to 80–88 per cent glycerol through an evaporation process. This is an energy-intensive process and depending on the plant's capacity and energy costs, can have either a single-, double-, or even triple-effect evaporator. The primary design consideration is the amount of water to be evaporated. With soap lyes, the secondary consideration is the method for handling the salt that is 'dropped' during the concentration process.

The amount of water to be evaporated is calculated from the amount of liquor and its glycerol concentration. Table 11.1 shows the concentrations that occur, starting with a dilute soap lye and concentrating to 80 per cent crude glycerine; losses are ignored.

Two general kinds of evaporators are used and their names refer to the type of circulation used to transfer heat to the liquor for evaporating the water. Natural circulation evaporators rely on a thermosyphon to circulate liquors while forced circulation units use a pump to achieve the required circulation. The heating tubes may be inside or outside the evaporator body, but most designs, especially the older calandria style evaporators, utilise internal tubes for heating (Fig. 11.2).

**Table 11.1.** Example of concentrations in soap lye and crude glycerine.

| *Substance* | *Percent* | *Weight (kg)* | *Percent* | *Weight (kg)* | *Change* |
|---|---|---|---|---|---|
| Glycerol | 10 | 100 | 80 | 100 | None |
| Salt | 11 | 110 | 9 | 14 | 96 kg salt |
| Water | 79 | 790 | 11 | 16 | 774 kg water |

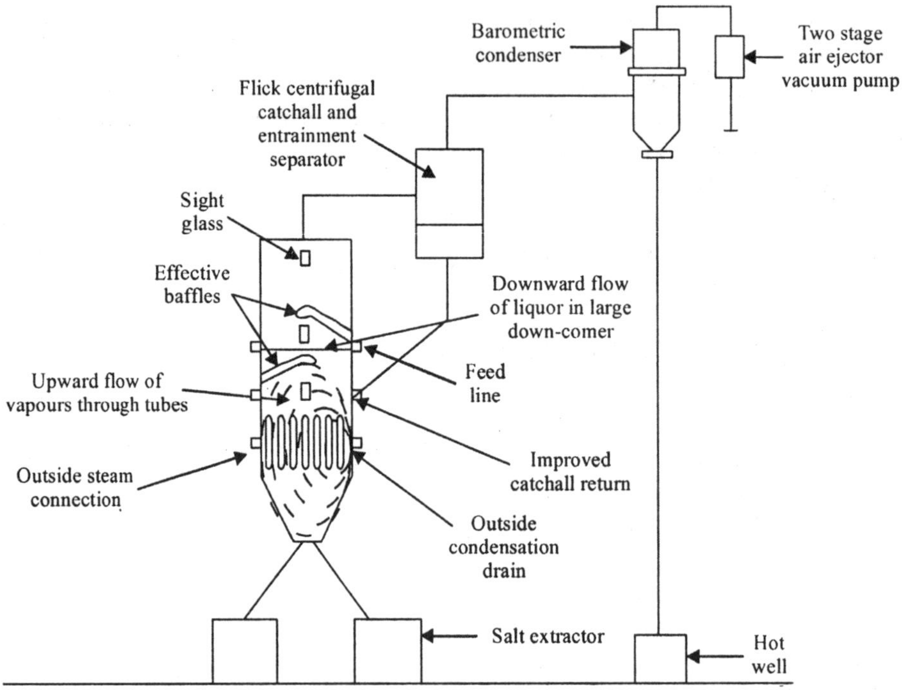


**Fig. 11.2.** Calandria-style single-effect evaporator with flick separator.

Forced circulation heating is used for most new evaporation plants for several reasons. Primarily, pumping the liquor allows the selection of a higher flow rate and velocity then can be achieved by natural convection, which improves heat transfer, thus reducing the surface area required for heating. Higher velocities also reduce fouling and scaling on the tubes. One drawback is the extra energy needed for the pump and motor. Also, if the velocity is too high there can be problems with salt crystal growth.

Evaporation can be either single effect or multiple effect, depending on the number of evaporation stages used. Single-effect evaporation offers the simplest process and minimum equipment investment. In a single-effect evaporator a batch of dilute glycerine can be evaporated to a half crude state with concentration of 40–50 per cent and stored. When sufficient quantities of half crude are available, it is processed to full crude (80 per cent concentration). The steam economy for a single effect evaporator is 0.8–1.0 kg of water evaporated for each kilogram of steam used. Single-effect evaporators, operating as a semicontinuous unit, have been used with some frequency in the past, but are now primarily used for small plants because of the economy and convenience of operation inherent in continuous, multiple effect plants.

Multiple-effect evaporation is accomplished by joining two or more evaporators in series and using the heat in the waste vapour from one effect to heat the subsequent effect. Live steam is added in the first effect only. The difference in the working pressure in each effect (each effect working at a lower pressure than the previous one) allows the use of lower pressure heating steam in the following effects. This pressure difference also allows transfer of liquor from one effect to the other.

As a general rule, it can be assumed that approximately 0.8 kg of water can be evaporated per kilogram of steam for each evaporator effect. Thus, one effect will evaporate 0.8 kg of water per kilogram of steam, two effects will evaporate 1.6 kg and three effects will evaporate 2.4 kg.

Multiple-effect evaporators are most often used in glycerine recovery plants of medium to large size, with the typical plant having a two-effect evaporator. The user must calculate and compare the savings in steam use for operating additional effects against the additional equipment and maintenance costs as well as space and operational complexity for additional effects.

Additional economy can be obtained through the use of thermal vapour recompression (TVR). Water vapour from the evaporator is captured and recompressed in a thermocompressor, using high pressure motive steam, so the vapour can be condensed in the effect's heat exchanger. The pressure after the thermocompressor is between that of the water vapour and motive steam. Only a portion of the vapour is recompressed in the thermocompressor, with the remaining vapour used to heat the subsequent effect. The thermocompressor can be used on a single-effect evaporator or on the first-effect of multiple-effect units. This equipment favours low $\Delta T$ in the heat exchanger to minimise the compression ratio. The thermocompression effect will add an additional economy equivalent to at least half of an additional evaporation effect at a modest extra cost, typically just a fraction of the cost of an additional effect. Recovered condensate can be economised by using it to preheat incoming liquor.

The operation of a two-effect, continuous evaporator is shown in Fig. 11.3. Treated soap lye or sweetwater is fed continuously and regeneratively heated with condensates coming from the heaters. Water is evaporated in the first evaporator chamber, which is kept at a pressure slightly above atmospheric pressure (4–6 psig), using heat provided by the thermocompressor. The thermocompressor takes part of the vapour from the first effect heater and recompresses it to increases its temperature and pressure. The remaining vapours go to heat the second effect. In the first evaporator, about 60 per cent of the water is evaporated. (In a three-effect evaporator the first effect evaporates about 45–50 per cent of the water, with the remaining two effects roughly splitting the amount of water evaporated.

The liquor from the first effect is then fed by gravity and by pressure difference to the second effect, which is under a slight vacuum (115 mm Hg absolute). There it is concentrated to 80 per cent for soap lyes or up to 88 per cent for sweetwater crude (containing no salt) and then pumped to crude glycerine storage. As soap lyes are concentrated, the concentration increases until the dissolved salts begins to come out of solution and form crystals. These crystals grow and settle to the bottom of the evaporator, where they are removed as described below. Some salt crystals will drop out in the first effect, but most of the salt will drop in the second effect. The proportions are determined by several factors, including the operation of the evaporators, the characteristics of the soap lye and the concentrations reached in each effect. The solubility of sodium chloride in glycerol is relatively unaffected by temperature, however.

### Thermal Efficiency

In addition to steps to reduce the quantity of water to be evaporated, methods are available to minimise the usage of steam required to carry out the necessary evaporation.

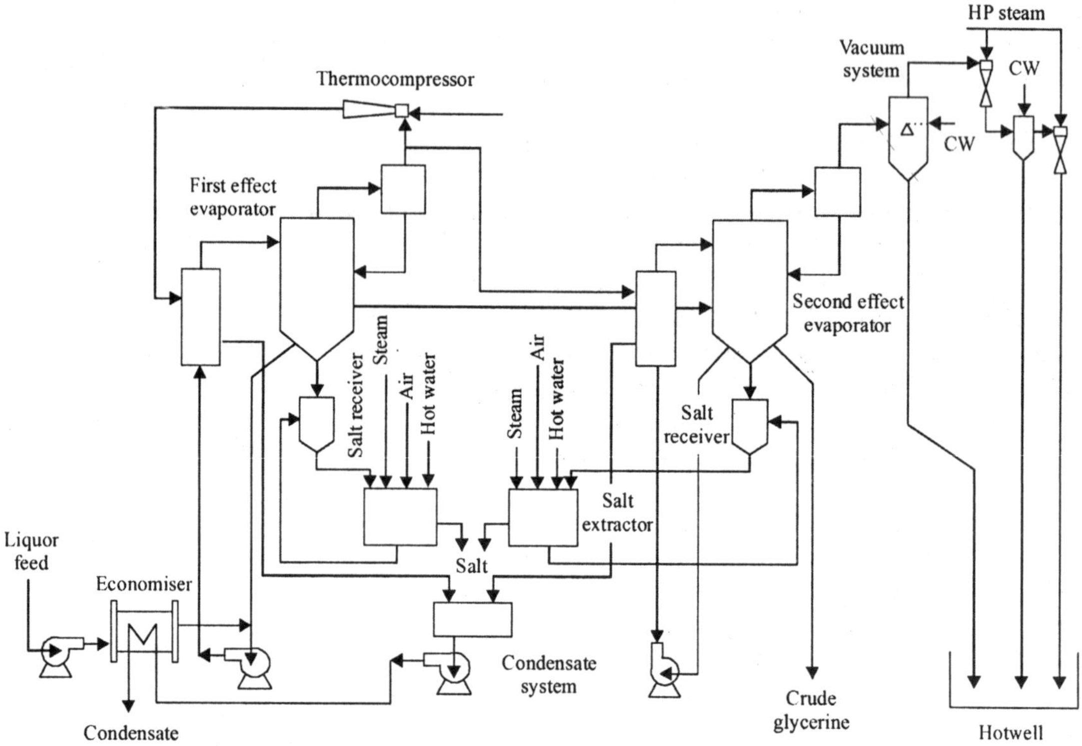


**Fig. 11.3.** Double-effect evaporator with thermocompressor flowsheet.

The boiling point of water, or any other liquid, falls as the pressure under which it is held is reduced. These simple facts provide the key to several ways in which the thermal efficiency of an evaporation system can be improved. Thermal efficiency is often and conveniently expressed in terms of kg of water evaporated per kg of steam supplied.

### *Improving thermal efficiency*

The main ways in which thermal efficiency can be improved are :

1. *Multiple effect operation* in which the vapour from one evaporating vessel, or 'effect', becomes the heating steam in another effect operating under a lower pressure. In this way, the latent heat is used more than once.
2. *Vapour recompression* in which the vapour from a vessel is compressed, by a steam jet or by mechanical compressor and returned as heating steam to the same evaporating unit. It will be appreciated that for the latent heat to be reusable the pressure must be increased; and that a simple addition of superheat to the vapour will not do.
3. Selection of the optimum *method of feeding* the liquor to the evaporator; and/or the use of the heat exchangers. As discussed later, heat exchangers can be used internally between effects, or to recover heat from outgoing streams of vapour, condensate, or product.
4. Integration with other plants in the factory.

Before these procedures are considered, it is necessary to discuss boiling point elevation and temperature differences. Boiling point elevation is the difference between the boiling point of the solution in an effect and the boiling point of water under the pressure in the vapour space of the effect. It is due to two causes:

1. *Boiling point elevation due to dissolved substances* in relation to fatty acid distillation, which increases as the concentration of dissolved substances increases. For semi-crude soap lye glycerine at 40–45 per cent glycerol it is about 15°C; and for soap lye crude at 83 per cent glycerol about 37°C.
2. *Boiling point elevation due to hydrostatic head.* When a liquid is heated at a distance below its surface, the pressure is that in the vapour space plus that equivalent to the head of liquid above the point at which the heat is supplied.

When conditions permit satisfactory recompression, a useful improvement in thermal efficiency is achieved without increase in other costs and for only a very small additional capital cost for the jet (Fig. 11.4).

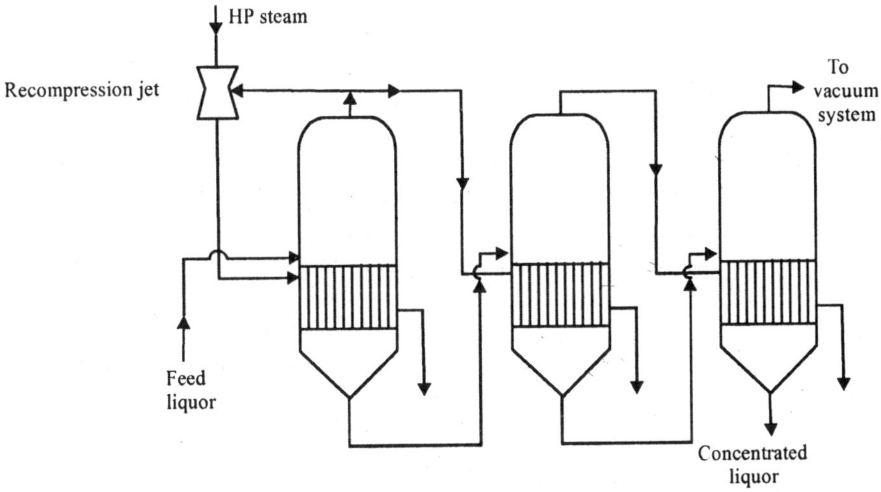


**Fig. 11.4.** Principle of triple effect evaporation with vapour recompression.

Parallel feed, that is separate feed to each effect, is suitable for the evaporation of brine to produce salt, but not for glycerine liquors in which the concentration of glycerol increases very substantially as the evaporation proceeds. Feed can be forward, that is 1 – 2 – 3 etc. the effects being numbered in the direction of heat flow; or backward which is the reverse direction, for example 3 – 2 – 1. The direction of feed often has an important influence upon steam consumption, depending upon the temperature of the feed liquor. If the feed is cold, it has to be heated to its boiling point under the pressure in the effect to which it goes. With forward feed, the temperature of the whole of the feed must be raised to the relatively high boiling point in the first effect by new steam and this sensible heat does not produce any vapour in that effect, although the heat in the partly concentrated liquor does produce some vapour by flashing when it is transferred to the second and subsequent effects which operate under lower pressures than the first effect. With backward feed, a cold liquor is heated in stages as the unevaporated portion is transferred from effect to effect. Only the final portion which reaches the first effect is heated with new steam and the sensible heating in the second and subsequent effects is done by vapours from an earlier effect; that is by latent heat which has already been used one, or more, times in the evaporator.

If the feed liquor is very hot, for example above the boiling point in the first effect, no sensible heat is required even with forward feed; and flash vapour is produced in each effect and forms part of the steam used in the next effect. With backward feed, the hot liquor flashes to the pressure in the last effect and the latent heat is not used again, at least in the evaporator. The residual liquor then has to be re-heated to the temperature in its next effect, as for cold feed. These considerations show that it is beneficial thermally to use forward feed with hot liquor and backward feed with cold liquor; at some intermediate temperature the factors balance and there is no difference between backward and forward feed. The thermal advantage of backward feeding with relatively cold feed liquor can nearly be achieved by the use of forward feed with interstage preheaters as indicated in Fig. 11.5. The feed is preheated stage by stage with vapour at as low a temperature as possible to make the maximum re-use of latent heat. A preheater cannot operate without a positive temperature difference in the required direction; and with modern vacuum equipment it is often not possible to use a preheater in the vapours from the last effect (number 1 in the diagram), but any heat which can be recovered from the final vapours is clear gain. Heat transferred to the feed in interstage preheaters is not available for use in the next effect, but the overall result is beneficial with cool feed for the reasons discussed in relation to forward and backward feed. Heat can also often be recovered from other streams leaving the effects, particularly the condensates. These condensates can be handled in several ways, bearing in mind the fact that condensate leaves the first effect heater at a higher temperature than that from the second effect and so on :

1. Condensates can be allowed to flash on reduction in pressure on transfer from one effect heater to the next, or in separate vessels. The flashed vapour is used as steam at a suitable point in the system and the residual hot, or warm, water remains available. This condensate can be re-used as process water in the soapmaking or fat-splitting process, as well as a source of heat.
2. Unflashed, or flashed, condensate can be used in a heat-exchanger.

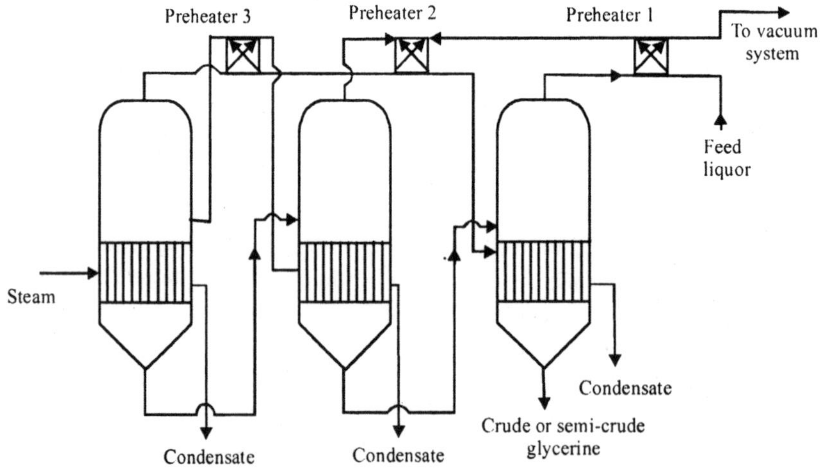


**Fig. 11.5.** Evaporation with forward feed and interstage preheaters.

### *Factors in the selection of a system of feeding*

Other factors to be considered in the selection of a system of feeding are:

1. With forward feed the most concentrated liquor is evaporated at the lowest temperature, so that its viscosity is higher than that of the fully concentrated liquor using backward feed. This adversely affects heat transfer, particularly with natural circulation heaters.

2. More sensible heat leaves in the crude, or semi-crude, with backward than with forward feed; but some of this may be recoverable in heat exchangers.
3. Liquor can normally be transferred from effect to effect by pressure difference with forward feed, although a pump may be needed to introduce the feed liquor into the first effect. With backward feed it is normally possible to suck the feed liquor into the last effect, but pumps are needed for transfer from effect to effect, unless unusual levels are adopted which provide sufficient gravity head to more than offset the adverse pressure differences between the vapour spaces.
4. The salt crystals which separate during the evaporation of soap lyes are rather fine; and they tend to become finer with rise in the temperature at which they are separated and with increase in the glycerol content of the liquor. Fine crystals are more difficult to wash effectively than coarser ones. The same factors also tend to induce 'salting' of the heater tubes, a phenomenon discussed later. Salting can be a serious problem with natural circulation soap lye evaporators under some conditions and forward feed may then be advisable.

When the feed liquor is relatively cool it is usual to use forward feed with interstage preheaters. For special reasons, it is also possible to use other arrangements.

As mentioned earlier and for the reasons just discussed in relation to methods of feeding, it is commonly a serious error to assume 1 kg of live steam will evaporate *n* kg of water in *n*-effect evaporator, unless the feed is exceptionally hot or a recompression jet is used. A good estimate of the performance to be expected from an evaporator, including its evaporative capacity, can be made with the aid of detailed heat balances over each stage of the process. Because of the dynamic equilibria maintained within a unit, the temperatures and pressures adjust themselves automatically to meet both heat balance and heat transfer criteria in each section of the plant. For a plant which does not exist, it may be necessary to make provisional assumptions as to intermediate temperatures, or likely heat transfer coefficients; and to make adjustments, if necessary, to reconcile the two sets of criteria. Such calculations are not difficult, but manual calculation can be tedious, particularly when several effects are involved. Plant manufacturers will provide estimates in response to serious enquiries, doubtless by the use of computer programmes, but it is sometimes useful for an operator to estimate what is likely to result from changes such as the inclusion of an additional heat exchanger, or a different feed arrangement. In making such estimates, it must be remembered that any change is likely to induce other changes to maintain the dynamic equilibria.

As a first approximation it is sometimes assumed that an *n*-effect evaporator costs *n* times as much as a single effect plant for the same duty, because the overall temperature difference is split between the effects with a consequent need for additional heating surface. This is also subject to considerable error. Estimates obtained some years ago from a well-known plant manufacturer for single and double effect units for the same, rather small, duty showed that the cost of the double effect was substantially less than twice that for the single effect. At least in part this was because of the larger final entrainment separator, condenser and vacuum equipment needed for the single effect unit in which the whole of the water evaporated passes to the condenser.

The discussion of thermal efficiency has so far considered the evaporator alone, but it is sometimes worthwhile to involve other plant. Steam from another unit which is not at high enough pressure to be used as first effect steam may be able to be introduced between effects. Different plant units commonly do not synchronise at all times, so that it is usually necessary to provide an alternative steam supply, probably through a reducing valve. Surplus heat from a high pressure fat splitting plant can often be used in the evaporation of the sweet water.

## Types of Heating Unit

Long tube climbing film heaters, in which much vapourisation takes place in the tubes, are used for sweet waters, but are not suitable for soap lyes, because the tubes would block with salt. Traditionally, short tube natural circulation evaporators were useful for soap lyes. These have a bundle of vertical tubes (say 1–2 m long) arranged in a steam chest to give what is called a calandria. Commonly calandrias built into the main evaporator bodies and provided with downcomers of various designs, such as a sector, a large central hole, or a series of tubes larger than the boiling tubes, are used. The liquor circulates up to the small tubes and returns through the downcomers. It is also possible to have natural circulation evaporators in which the liquor rises through the vertical tubes in an external heater and returns via the liquor reservoir in the main body. In order to obtain the maximum rate of heat transfer and hence of evaporation, it is necessary to work with a static level of liquid below the top of the tubes, so that some vapourisation takes place in the tubes giving an increased rate of circulation due to the vapour lift effect. With soap lyes care must be taken to avoid too low a static level, as otherwise salt will deposit in the tubes in effects in which the liquor is saturated with salt. Salting of the tubes can sometimes occur when an evaporator is being operated in a way which would normally be satisfactory; this is presumably due to some small changes in the composition of the purified lye. It is also sometimes possible to operate natural circulation evaporators with a static level well above the upper tube plate of the calandria, particularly when finishing batchwise to crude strength. In this case, the liquid is heated in the tubes to above its boiling point at the pressure in the vapour space and flashes as it rises to a region of lower hydrostatic pressure. It is usually necessary to avoid operation at some intermediate levels at which 'priming', or boiling over, is liable to occur.

Evaporator effects in which salt separates usually need to be washed out at regular intervals with water, or with other liquor which is not saturated with salt. When a plant is operated semi-batch, rather than continuously, the charge of concentrated crude, or semi-crude, containing some solid salt, is discharged from the last effect; and it is common practice to start the next batch by refilling with unsaturated lye, or even water, to dissolve deposited salt. If the tubes become blocked with salt, it is necessary to soak with water and to boil as long as is needed to clear them. Mechanical methods are sometimes used, but care is necessary to avoid damage and this procedure is best avoided if possible. Various insoluble scales can build up slowly in the tubes of a glycerine evaporator and these normally have to be removed mechanically. Scales can contain silica, but they are particularly common when lye, sweet water, or other liquor containing sulphate and calcium is introduced, for example, if the plant is washed out with hard water. Calcium carbonate scales are also encountered; these can sometimes be removed by careful use of acid. The modern tendency is to use external calandrias with forced circulation by pump. This system improves heat transfer and so permits the use of less heating surface. External heaters with forced circulation minimise the risk of salting and scaling; and an external heater can be arranged to permit mechanical descaling more readily than an internal calandria. As mentioned earlier, forced circulation is particularly useful for the final effect of an evaporator in which crude glycerine is produced.

Whatever type of heater is used, it is necessary for satisfactory heat transfer to remove effectively both condensate and non-condensible gases, mainly air. When the pressure in the steam chest is above atmospheric, ordinary steam traps can be used for the condensate; and air can be discharged, with a little steam, through a slightly opened vent valve, or cock. For steam chests at below atmospheric pressure, the air must be vented through a small connection to a vapour space at a sufficiently low pressure; and a pumping trap, or pump, used for the condensate. A convenient system for the condensate is shown in Fig. 11.6. The centrifugal pump runs continuously, but discharge of condensate is intermittent. The condensate is virtually at its boiling point at the pressure in the steam chest and can only be pumped when there is a sufficient head of liquid in the receiver to prevent vapourisation (cavitation) in the pump. The level in the receiver rises until the pump starts to discharge and then falls until flow virtually ceases; the cycle being repeated indefinitely.

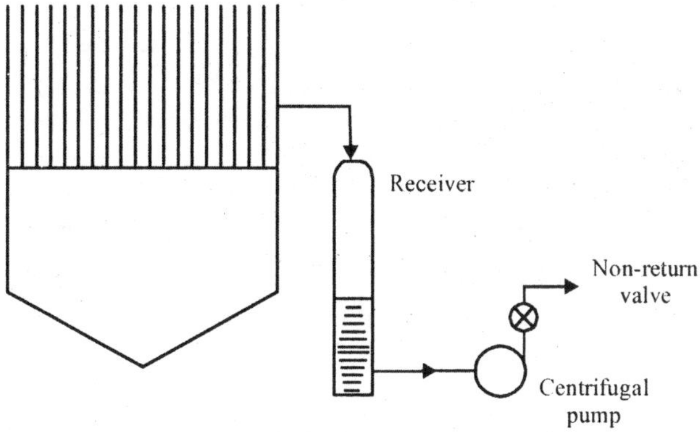


**Fig. 11.6.** Arrangement for discharge of condensate from a vessel at sub-atmospheric pressure.

### *Entrainment separation*

Some kind of entrainment separation is necessary to reduce losses of glycerine entrained in the vapour stream from each evaporator. Control of the vapour velocity by proper evaporator and duct sizing will minimise the amount of entrainment. An entrainment separator should be installed at the outlet of each evaporator body and the glycerol recovered should be returned to its evaporator. The separator is usually of the cyclonic type, which can be equipped with internal impingement baffles (Flick type). Mesh- or baffle-type separators are possible, but care must be taken to prevent any buildup on the internal components, which would cause excessive pressure drop.

As the glycerol concentration grows during evaporation, salt crystals begin to form as the liquor saturates. These salt crystals, as they grow, will tend to sink to the cone in the bottom of the evaporator. Proper treatment and filtration of the soap lye is important to ensure that there are not excessive amounts of impurities in the lye that can hamper the growth of the crystals. There are several methods to recover the salt from the evaporators and all of the methods address then need to wash as much glycerol from the salt as practical. Washing is done with water and sometimes lyes, with the purpose of recovering glycerol that is removed with the salt. The wash liquor is subsequently returned to the treatment plant. Obviously, any water that is added must be reevaporated, so a balance must be achieved between glycerol loss and water to be evaporated.

## Removal and Recovery of Salt in the Evaporation of Soap Lyes

All effects in which the lyes have become saturated and solid salt separates, need a facility for the removal of the solid as soon as possible after it is formed and the crystals have grown to a reasonable size, otherwise the liquor may become unduly thick. Effects which do not normally produce salt may be fitted with simple vessels to collect sludge and any abnormal salt. Three systems are, or have been, used:

1. Salt filters, or dry salt boxes.
2. Wet salt boxes.
3. Devices through which the effect contents are circulated, a thickened slurry is discharged and substantially clear liquor is returned to the effect.

Dry salt boxes are messy, labour intensive and do not permit adequate removal of glycerol from the salt to meet the requirements of efficient modern soapmaking. In the larger and the more efficient, factories they were replaced many years ago, but they are still in use in some places. The other two systems remove the solid salt from the evaporator in the form of a slurry in a more, or less, concentrated glycerine liquor; and the separation process is completed in some form of centrifuge.

The operation of dry and wet salt boxes can be explained by reference to Fig. 11.7. In both types salt settles into the quiet zone of the effect body below the calandria and down into the salt box. A sufficiently deep cone and settling zone is important; and, if the plant is to be operated semi-batch, sufficient liquor volume is necessary in the last effect to hold an adequate batch of crude, or semi-crude, glycerine. Consider now a dry salt box as shown in Fig. 11.7a. The salt settles onto the grid and filter gauze. When sufficient salt has collected, valve A is closed, valve B is opened and the liquor is drawn back into the effect, helped by the pressure of the steam introduced through the valve at the top of the box.

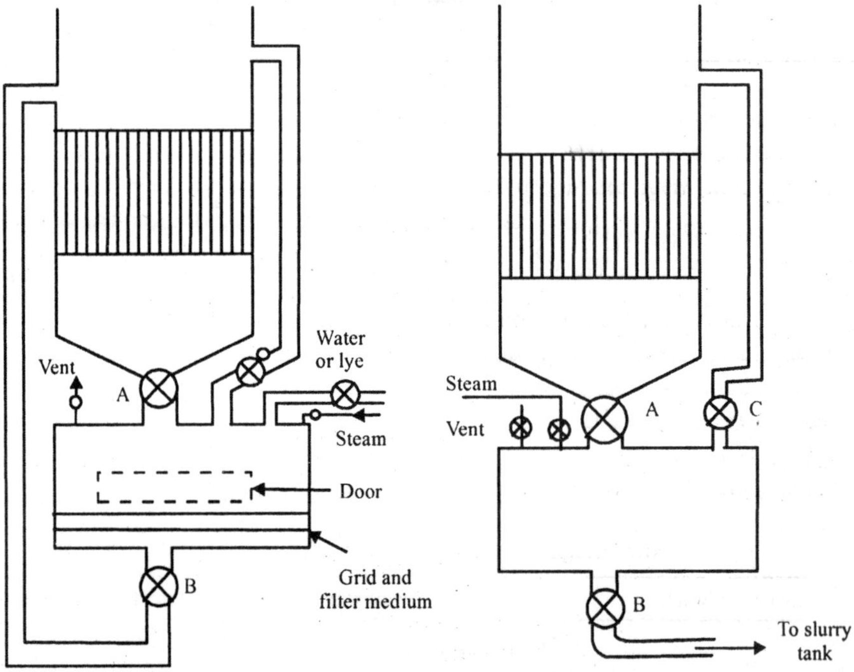


**Fig. 11.7.** Operation of salt boxes. (a) Dry salt box, or salt filter (b) Wet salt box.

The bed of salt is washed with lye and/or water and the liquor is again displaced by steam. The steam valve and valve B are closed, the vent to atmosphere opened and the door on the box opened. The salt is shovelled out for return to the soapmaking department as solid salt, or as brine. The door of the box is closed, the box is evacuated through the balance valve C and valve A opened to commence another cycle. It is particularly important to ensure that the box is fully evacuated before valve A is opened; otherwise expansion of the air from the box will cause the liquor in the effect to prime. It is also important to ensure that air cannot leak in through the valves; lubricated cocks have been found useful for these duties.

Wet salt boxes can be cylindrical drums, or vessels of any other convenient shape. They are operated in a manner similar to, but somewhat more simply than, the dry boxes. When sufficient salt has accumulated, as judged by inspection through sight glasses, or in accordance with a time schedule, valve A is closed, valve B and the vent are opened and the slurry is run out and/or blown out with steam after the vent has been reclosed. The box is evacuated via valve C before it is reconnected to the effect.

Pairs of boxes, dry or wet and used alternately, are common on effects in which large quantities of salt are produced. If only one box is fitted, salt accumulates in the cone of the effect whilst the box is disconnected. Discharge by continuous pumping from the base of the effect is arranged as shown in Fig. 11.8 later for a evaporator which uses a baffled settling vessel. Thickened slurry is discharged from the base of the vessel and the rest of the liquor is returned to the effect and/or transferred to the next effect. Sufficient pressure is maintained in the circulation loop by the pump and the valves in the return and transfer lines to enable the discharge of thickened slurry to be controlled by a valve without another pump. Hydroclones (liquid cyclones) instead of the baffled vessels have been found very satisfactory separating devices for systems of this kind. Care must be taken not to interfere with the circulation through the heater tubes if a continuous pumping system is introduced into a natural circulation evaporator. An attempt to do this many years ago produced a well thickened slurry, but led to unacceptable salting of the calandria tubes. Return of the circulated liquor to the base of the effect, with sufficient volume below the calandria, may prevent this difficulty. The system works well with forced circulation effects and is probably best in modern practice.

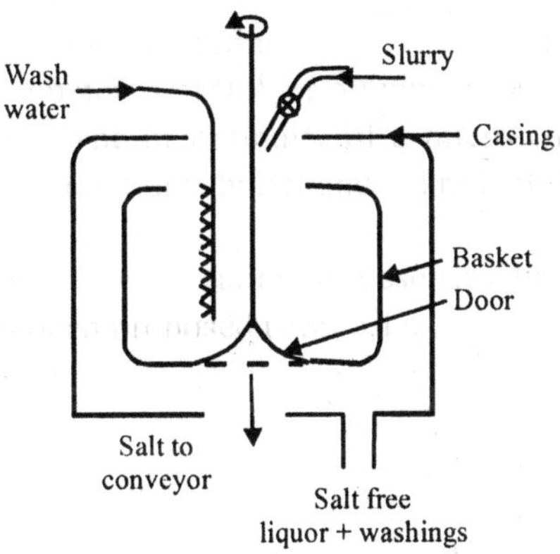


(a) Centrifuge arranged for the discharge of solid salt

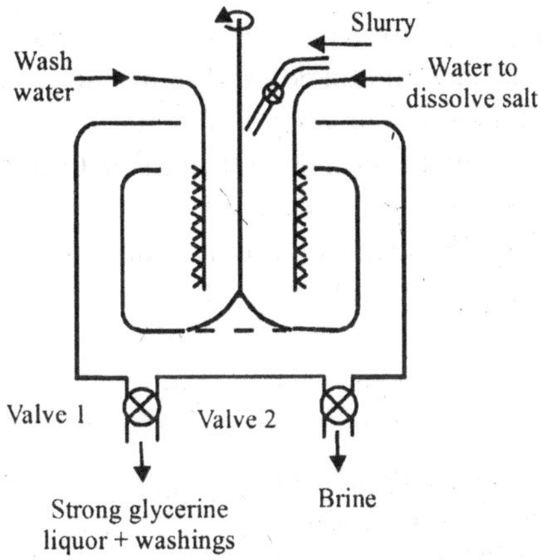


(b) Centrifuge arranged for dissolution of salt *in situ*.

**Fig. 11.8.** Batch salt centrifuge, (a) arranged for the discharge of solid salt, (b) arranged for dissolution of salt *in situ*.

It is necessary to separate the solid salt in the slurry removed from the evaporator from the liquor and to wash the salt as free from glycerol as is economically feasible. Some washing with lye may be possible, but mainly fresh water is used; and this water has to be evaporated, which sets a limit to the amount of washing which is worthwhile. Traditionally, basket centrifuges operated batchwise are used and the salt is discharged as a crystalline powder. It is returned to the soapmaking department as such, or after solution in water, preferably condensate, to form a near saturated brine. Modern soapmaking plants use only brine; and systems have been developed in which the salt is dissolved in the centrifuge itself. Alternatively, the salt is dissolved in a separate unit. When sulphate is introduced into the system, by the use of aluminium sulphate or sulphuric acid, it is necessary to discard some recovered salt from time to time to prevent undue build up of sodium sulphate in the salt. The batch centrifuges used are normally of the suspended basket type; and some points regarding their use are discussed in relation to Fig. 11.8.

1. Perforate basket machines are used; and the filter medium is a fine weave metal gauze backed by a coarse gauze to facilitate drainage. A high speed to give a high centrifugal force is desirable to remove the liquors from the solid salt as completely as possible.
2. The slurry should be fed with confidence in intermittent bursts having a high rate of flow. A 4 inch pipe with a control cock operated with a long handle is convenient. This rate of feed makes the slurry run up the basket wall to the top and gives a salt cake of fairly uniform thickness. Too high a feed rate, or unduly long bursts, cause slurry to overflow the rim of the basket and to carry solid salt over into the centrifuged liquor. Actually some salt always gets into the clear liquor and arrangements must be made to deal with it; as also from any vessels in which soap lye crude, or semi-crude, is held. Too low a feed rate gives a very uneven salt cake, with a large thickness at the base and a small one at the top; this is because most of the liquor has time to penetrate the cake before the slurry can rise far up the basket wall.
3. When most of the liquor has been removed, the salt cake is washed with a controlled quantity of water and spinning is continued to remove the wash water as completely as possible. Sometimes two short washes, separated by a period of spinning, are given. If a fixed perforated pipe is used for wash water, care must be taken to ensure that it does not interfere with the discharge of the salt and a spray on a flexible hose is an alternative. Timers to help the operatives to adhere to a cycle laid down by the management are useful. In order to reduce the glycerol content of the salt to the required level with a minimum use of wash water, careful control is essential, particularly of the time during which the water is sprayed.
4. With the system shown in Fig. 11.8a, the machine is stopped, the door over the bottom outlet raised and held up on a hook and the salt dug out to fall into a conveyor, or receptable below. A narrow spade with the cutting edge cut at an angle is useful for this work. When lower speed and less efficient, centrifuges were used, a rope placed round the bottom of the basket before the slurry was fed provided a simple means to dislodge the salt cake. The end of the rope had a hook which held it to the upper rim of the basket; and this end was pulled to bring down the salt at the end of the cycle. However, well washed and dried salt in modern machines is normally too hard for this technique to be used.
5. With the system shown in Fig. 11.8b, valve 1 is open and valve 2 closed during most of the cycle for the glycerine liquor and the washings to flow away. When the salt has been washed and dried, valve 1 is closed, valve 2 opened and a controlled flow of water introduced to dissolve the salt. With proper adjustment of the conditions the salt can substantially be dissolved to produce nearly saturated brine. Care is necessary to ensure that the casing is arranged to allow the two streams of liquid to drain with minimum contamination one with the other. Also that the valves close tightly and that the operators control the valves and sprays with care. With proper precautions, this system can give very satisfactory results without the heavy manual labour required with system (a). In large factories, the modern trend is to automatic centrifuges. Several types have been used, but some have not been very successful because either they could not handle fine glycerine salt satisfactorily, or because they need too much wash water to reduce the glycerol in salt to the required level. The best type is probably one in which the operation is very short cycle, automatic batch. The salt is discharged with a knife; and it is a

useful refinement for the knife mechanism to be arranged so that the cut leaves a slightly different thickness of salt in consecutive cycles in order to minimise the formation of a 'glaze' of very fine broken crystals which may impair drainage.

The glycerol left in salt varies considerably with the system employed, the quantity of wash water used and the care with which the operation is carried out. With centrifuges 0.4–0.6 per cent glycerol on salt should be possible, but figures from 0.2–0.1 per cent are quoted. With dry salt boxes, 1 per cent is sometimes claimed, but 2–4 per cent is probably more usual (corresponding figures calculated on saturated brine are about one quarter of those for salt).

## Vacuum Maintenance

Glycerine evaporators now normally operate at 2–4 inches Hg absolute pressure (28–26 inches Hg vacuum with 30 inches Hg barometric pressure); and a high-level direct contact condenser with barometric leg is commonly used. The non-condensible gas can be removed with a two-stage ejector system, with an intermediate condenser, or by a mechanical vacuum pump. Because of the level of pressure needed in a glycerine evaporator, no booster jets are normally installed. The direct contact main condenser can be replaced by a surface condenser if the condensate from the last effect is required, for example, for use with other condensates in the splitting section of a fatty acid plant.

Despite a somewhat higher equipment cost, the surface condenser system offers several significant advantages over a barometric condenser:

1. Any glycerine in the vapour that is condensed is recovered and is not lost into the hot well water.
2. The surface condenser is cooled with clean water, typically circulated from a cooling tower.

In a barometric system, the spray water becomes contaminated, requiring a separate cooling tower system from the 'clean' cooling water used elsewhere in the plant. The 'dirty' cooling tower water must be periodically replaced to prevent excess buildup and fouling in the cooling system.

Following the condensing section and depending on the utility situation at the plant site, either a liquid ring vacuum pump or a two-stage steam jet with intercondenser is used. A booster ejector is not required to maintain the vacuum level required for evaporation.

## Minimisation of Losses of Glycerol

During evaporation, liquor containing glycerol and salt can be lost by being carried out in liquid/solid form with the vapours and discharged in a condensate. The carry-over can be as droplets, foam, or, if a serious prime occurs, as bulk liquid.

To minimise this loss:

1. Adequate vapour space should be provided above the level of the boiling liquid.
2. Suitable separators and external 'catchalls', usually cyclonic, should be included in the system. It is useful to provide sight-glasses in external separators and/or the return lines, so that the plant operator can see immediately when any serious carry-over occurs and take remedial action.
3. The liquor levels in the effects should be controlled with care, automatically or manually.
4. The treatment of the lyes, or sweet water, should ensure that the liquors have no undue tendency to foam.

External catchalls are sometimes fitted only to the effect in which the concentration of glycerol is highest. Devices which measure roughly the electrical conductivity of the condensate from the evaporated vapours are very useful indicators of carry-over. They actually indicate the presence of salt, or other electrolyte, but this is relevant as salts and glycerol are carried over together. The main problem is to obtain a continuous flow of condensed vapour from the last effect when, as is common, a direct contact condenser is used in the vacuum system; and the condensate is mixed with cooling water. It is not essential to obtain a representative sample of the vapours plus any entrained liquor; all that is necessary is to get a sample which will indicate when any abnormal amount of salt is present in the vapours and this can often be achieved by the use of a small partial condenser in the vapour line. For example, a ring carrying cold water placed above a collecting tray, in a vertical vapour main, as shown in Fig. 11.9, can be used.

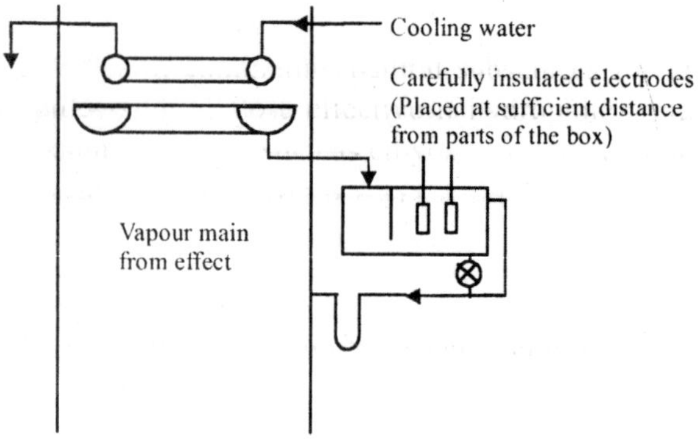


**Fig. 11.9.** Partial condenser for entrainment indicator.

There is a continuous flow of condensate through the box containing the electrodes which are connected to a suitable low A/c voltage and a measuring instrument or, better, recorder. An audible alarm can be included to draw immediate attention to any substantial increase in conductivity.

Losses of glycerol during evaporation can occur in other ways, for example, if a leak develops in a calandria. The effects are normally left containing water during shut-downs (to dissolve deposited salt) and, if a leak is present, some salt solution can be drained from the steam chest of the calandria after the plant has stood for a period of time.

Fig. 11.10 shows the flow of glycerol through a soap factory with glycerine recovery; and indicates the sources of loss. The diagram is drawn for a double-effect evaporator with direct contact condenser; and refers to closed coils in soap pans. Such coils are now obsolete, but they can be a serious source of loss if present. The diagram can readily be modified to suit other plant arrangements and similar ones prepared for the splitting route. It is useful to determine the recovery in each section of the factory, particularly if the overall loss is high, but, this is usually not done easily; and, for routine purposes, those with limited resources may find it best to concentrate on checks on the controllable losses and the important glycerol in re-cycled salt. Care is necessary to ensure that truly average samples of materials such as muds are tested. Misleading samples of treatment mud can be taken from near the point at which water is introduced into the filter presses.

## REFINING OF GLYCERINE

### Treatment of Crude Glycerine

The quality of soap lye crude glycerine can be improved by steaming it in an acid condition, so as to remove volatile fatty acids. This can be done by acidification before the later stages of evaporation, but, with normal equipment, corrosion problems are likely. The crude glycerine is acidified to about pH 3.5

with sulphuric, or hydrochloric acid. Sulphuric acid is cheaper than hydrochloric, but has the usual disadvantage that it adds sulphates to the system, although this is less important than when salt to be re-cycled can be affected. Both mineral acids have similar effects on corrosion, because substantial concentrations of chloride ions are already present.

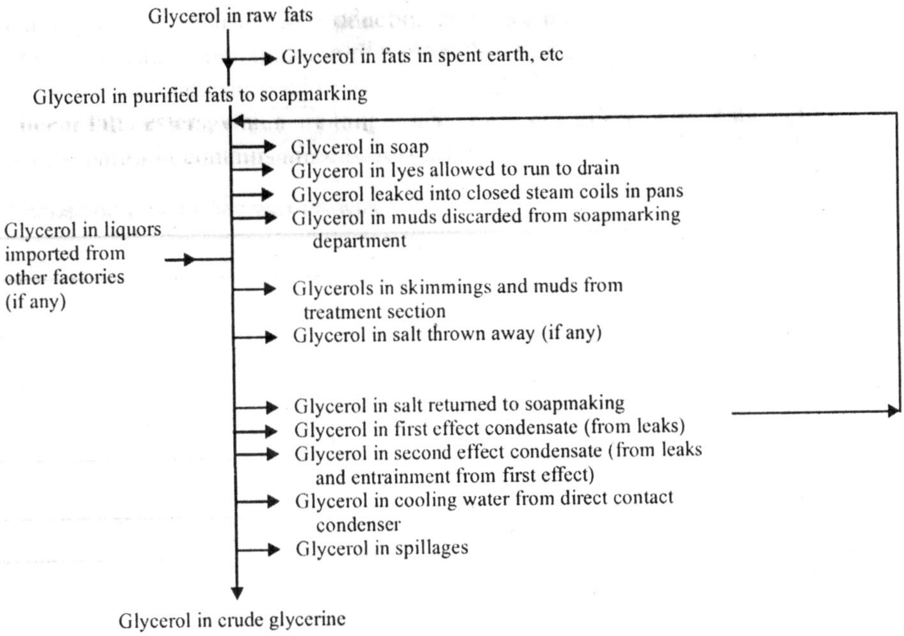


**Fig. 11.10.** Flow of glycerol through a soap and glycerine factory.

Type 316 stainless steel (16–18 per cent Cu, 10–14 per cent Ni plus 2–3 per cent Mo) has been found to give a reasonable plant life, although some pitting occurs. The main design and operating problem is to adjust conditions so that a substantial part of the low molecular mass fatty acid is removed without the loss of too much glycerol. A continuous unit which has been used satisfactorily is shown diagrammatically in Fig. 11.11. The process can also be carried out batchwise.

## Distillation of Glycerine

Under atmospheric pressure pure glycerol boils with decomposition at about 290°C and, as for fatty acids, it is necessary to use reduced pressure and/or open, or carrier steam to reduce the boiling temperature in glycerine stills. Principles of distillation discussed for fatty acids apply to glycerine distillation, but there are also important differences between the processes for fatty acids and for glycerine. In particular:

1. The nature of the non-volatile residue, which form soap lye crude glycerine includes large proportions of salt and soapy matter.
2. The fatty acid vapours contain a considerable range of compounds of varying volatility, whereas in glycerine distillation the vapours are predominantly glycerol and water, although the relatively small concentrations of organic impurities are important.
3. Glycerol and water are completely miscible in the liquid phase, whereas fatty acids and water are not. Also glycerines do not solidify at temperatures likely to be encountered in still condensates.

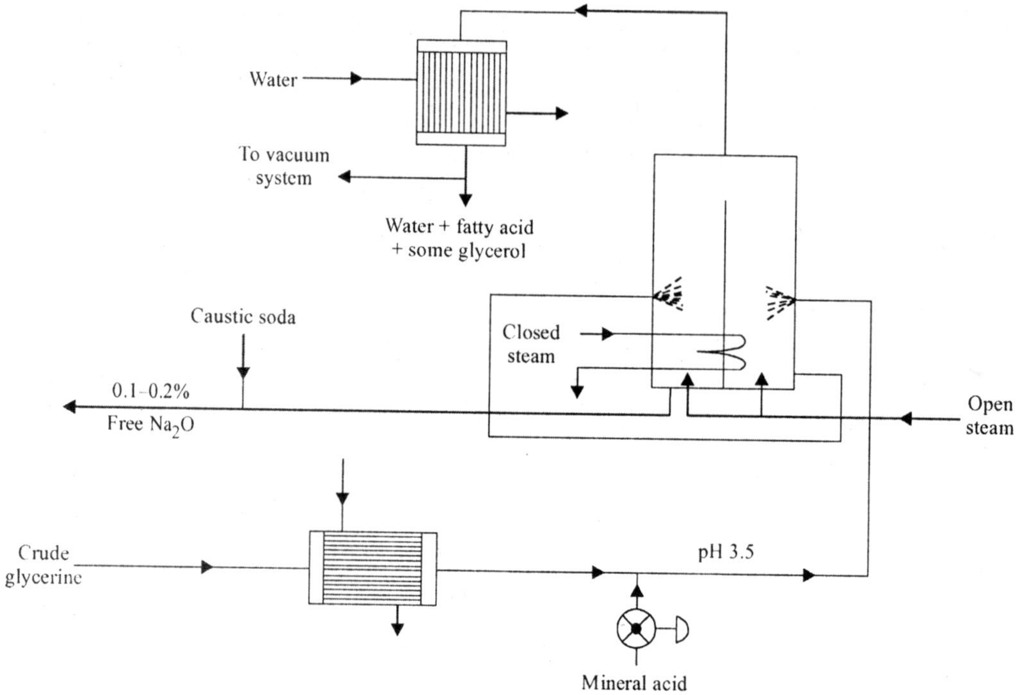


**Fig. 11.11.** Removal of volatile fatty acids from soap lye crude glycerine.

### *Chemical aspects of glycerine distillation*

Several important, but undesirable, reactions tend to occur in glycerine liquors held at still temperatures:

1. Formation of volatile glyceryl esters by reaction between glycerol and the soaps of low molecular mass fatty acids:

$$\underset{\text{Glycerol}}{C_3H_5(OH)_3} + \underset{\text{Soap}}{R.COONa} \longrightarrow \underset{\text{Mono-glyceride}}{C_3H_5(OH)_2\text{—}O.CO.R} + \underset{\text{Caustic soda}}{NaOH}$$

2. Formation of polyglycerols.
3. Formation of objectionable organic nitrogen compounds from proteinaceous residues in the crude glycerine.
4. Decomposition of glycerol to highly volatile compounds normally lost in the vacuum system.

Esters, fatty acids (if any) and the volatile nitrogenous bodies effect adversely the quality of distillates in which they are condensed. Polyglycerols are substantially non-volatile under distillation conditions; but they are undesirable because they represent a loss of glycerol and because they modify the residue in ways which can be adverse.

The condensation reactions by which polyglycerols are formed are not reversible and the ether links they contain are not broken under still conditions. Careful materials balances determined for glycerine distillations normally show a small unaccounted for loss, which may be associated with radical decomposition of glycerol to gaseous products.

Ester formation is a balanced reaction which is suppressed by the presence of caustic soda in the still vapouriser; and polyglycerol formation is catalysed by alkalis, with caustic soda more potent than the milder alkalis sodium carbonate and soaps. It is, therefore, important to carry out the distillation with as little caustic soda in the still as will permit the production of distillates satisfactory as regards content of esters and fatty acids (measured by the saponification equivalent, but also likely to affect taste and odour). Physical conditions of distillation are very important. Increases in the time for which and the temperature at which, glycerine is held in the still vapouriser both increase the formation of polyglycerols. There is an interaction between the two conditions, so that a very short time at a relatively high temperature may be less damaging than a longer time at a lower temperature. More severe physical conditions also increase the influence of alkaline catalysts, so that mild alkalis may not be very significant if the physical conditions are mild, but become important when they are more severe. It has also been shown that with a given crude glycerine the distillate is much lower in saponification equivalent when distillation is very rapid.

Similar factors probably affect the formation of volatile nitrogeneous bodies. The introduction of a small quantity of water into the still in, or with, the crude glycerine may also help to suppress polyglycerol formation, even though the water is rapidly vapourised. Under some conditions, the use of open steam may also help to reduce ester formation. When distillates need to be re-distilled, it is important to saponify any esters in them before the second distillation. This can be done by agitating the hot distillate with air after addition of caustic soda; and a suitable quantity of caustic soda is needed in the still during re-distillation to suppress re-formation of esters.

Formation of polyglycerols is commonly a major source of loss of yield in glycerine distillation; and it may be useful to estimate the polyglycerol in the still residues, which are often called foots. Elaborate separation procedures have been developed for special investigations, but the polyglycerol content is usually deduced from a periodate and an acetin, or dichromate, test on the foots. As explained the periodate test does not respond to polyglycerols and, unless sugar or other unusual substances are present, records only glycerol. The acetin test includes polyglycerols, so that the difference between glycerol estimated by acetin and glycerol estimated by periodate is a measure of the polyglycerol content of the sample. To convert this difference to the glycerol from which the polyglycerols were formed, it must be multiplied by a factor which depends on the composition of the polyglycerols. If di-glycerol is assumed, the factor is 6/4 = 1.5 and if *tri*-glycerol, it is 9/5 = 1.8. To obtain the general relationship, when $n$ moles of glycerol condense, $(n-1)$ ether links are formed with a loss of $(n-1) \times 2$ hydroxyl groups. Hence, 3 $n$ hydroxyl groups form a polyglycerol containing 3 $n$–2 $(n-1)$ hydroxyls. In practice, a factor around 1.8 is probably reasonable.

Similarly, the difference between glycerol percent by dichromate and glycerol percent by periodate is a measure of the polyglycerols present and in this case no factor is required. Some stills give a very poor distillation yield when found to be operating at 60–70 mm Hg absolute pressure in the still instead of the 10–14 mm Hg for which it was designed. This raises the boiling liquor temperature to 170–180°C, substantial quantities of caustic soda had to be used and the yield is only about 75 per cent. A sample of liquid residue run out of the still, not a true average sample of the foots, tested 7 per cent glycerol by periodate and 37 per cent by bichromate. It, therefore, contained 7 per cent glycerol and 30 per cent polyglycerol calculated as glycerol; and showed qualitatively that major conversion of glycerol to polyglycerol had occurred and was a major cause of the very bad yield. In order to relate the polyglycerol found in foots to distillation loss due to this cause, it is necessary to test a true average sample of the residue and to find the quantity of this residue formed from a known quantity of crude glycerine by measurement, or by calculation from the salt contents of the two materials.

### *Reduce recovery and disposal*

The residue that accumulates at the bottom of the still contains some glycerol, polymerised glycerol, aldehyde resins, organic products of decomposition and salt, which must not be allowed to become too concentrated as buildup of residue will cease quality and capacity problems in the plant. If the residue is allowed to concentrate in the still, it displaces volume for incoming crude glycerol, thus reducing the still's capacity. At least two methods of removing the residue are used:

1. A residue receiver vessel placed below the still accumulates residue, which is periodically discharged to the treated lye tank for reprocessing.
2. Residue is removed continuously from the still and redistilled to recover the remaining glycerine; the concentrated residue from the bottom of the foots still is discharged to a drum for disposal, typically to land fill as solid waste.

Flow sheet of Wurster and Sanger glycerine refining plant which is of the continuous distillation process used for sweetwater or soap lye crude glycerine is shown in Fig. 11.12.

The crude glycerine is preheated regeneratively by hot distilled glycerine. The liqour then enters the still heater where it is further heated to about 165°C and circulated by means of a circulation pump. The circulated liquor is partially vapouried through the aid of vacuum (6 mm Hg) and sparging steam in the flashing chamber. The vapour rises through an entrainment separating section and then enters the condensing section. Here the glycerine vapours are condensed in a layer of packing, wetted by recirculated, cooled, distilled glycerine. The remaining vapours enter the scavenging condenser where remaining traces of glycerine are condensed and recovered as 80–90 per cent substandard glycerine, which is sent to intermediate storage. The substandard glycerine is re-refined after sufficient quantities have been collected for a 2- to 3- dry run every month. Typically, the substandard glycerine is processed into a lower grade of glycerine, such as high gravity or dynamite grade.

The residue in the bottom of the still is continuously discharged from the crude still to the foots still. The residue should be kept rich in glycerol (>25 per cent) to improve handling and distillation characteristics. A small amout (0.5–1 per cent) of phosphoric acid is added to keep the residue soft by lowering the pH to retard the formation of polyglycerol. Here it is reheated by recirculation through an external heater to about 175°C and partially vaporised, under vacuum and with about 25 per cent stripping steam. Most of the vapours are condensed in the foot still's condenser and the condensate is returned to the crude still. Any remaining glycerine is recovered where the vapour pass through the scavenging condenser. The residue at the bottom of the foots still is discharged to a drum for disposal.

The distilled glycerine from the crude still is re-evaporated in the deodouriser at about 130–140°C, again in the presence of high vacuum and stripping steam and with external heating, to ensure optimal removal of odouriferous materials and residual moisture. The vapour passes through a packed section where the incoming feed condenses the vapoursied glycerine. Proper reflux rates are needed to ensure removal of close boiling impurities. Residual volatile are condensed in the scavenging condenser.

The deodourised glycerine is cooled before flowing through columns of activated carbon, where colour impurities and trace odouriferous materials are removed. The bleached glycerine is filtered to remove trace carbon particles, cooled further and sent to storage.

The Mazzoni glycerine refining process is similar to the process described above, but certain details are treated in a different fashion (Fig. 11.13). Most notably is the use of a scraped surface heat exchanger for recovering glycerine from the foots. Some processors use air-cooled condensing units that consist of large tubes through which air, rather than cooling water, is allowed to circulate. This design is seldom used in new plants.

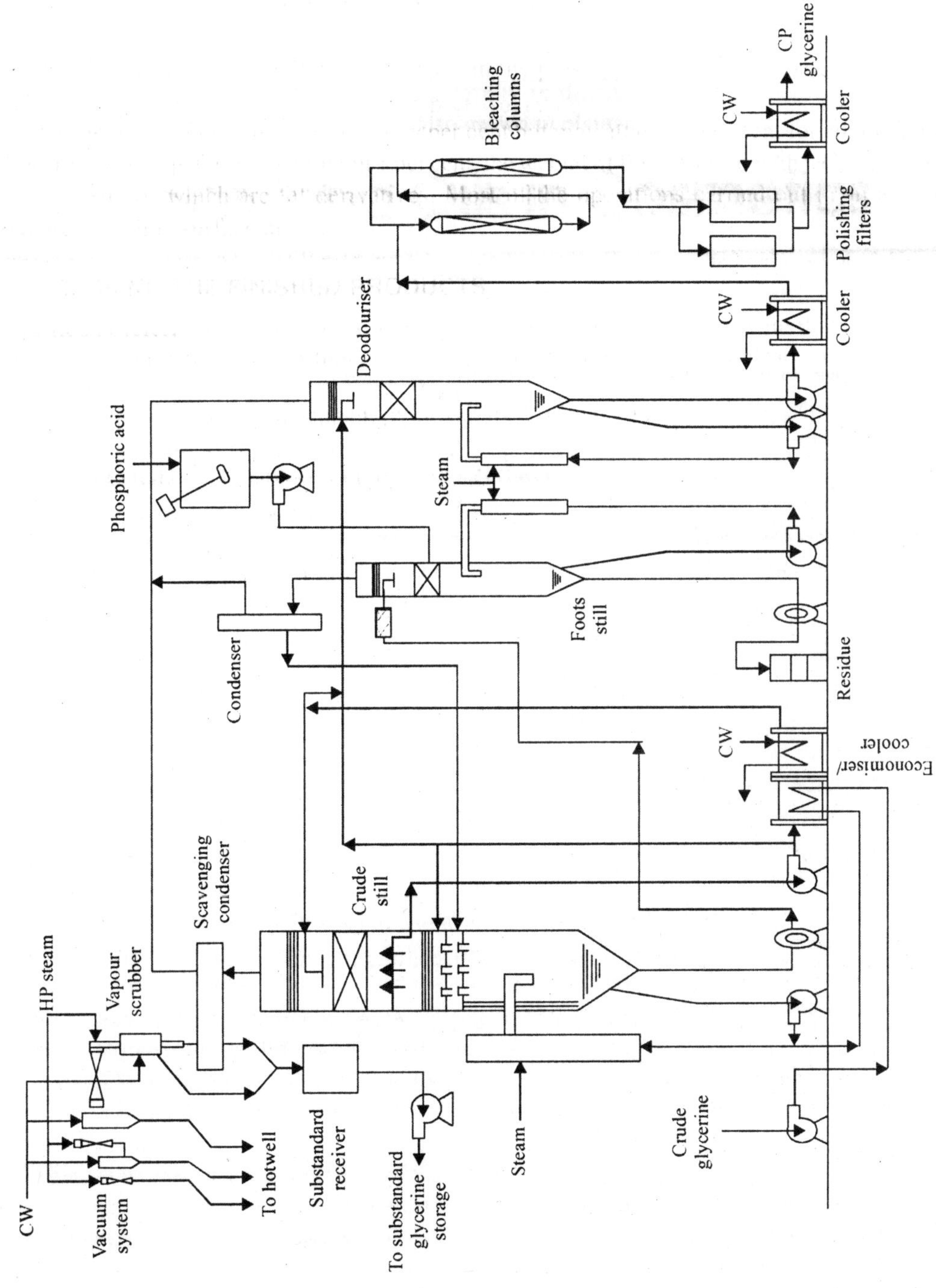


**Fig. 11.12.** Flow sheet of Wurster and Sanger glycerine refining plant.

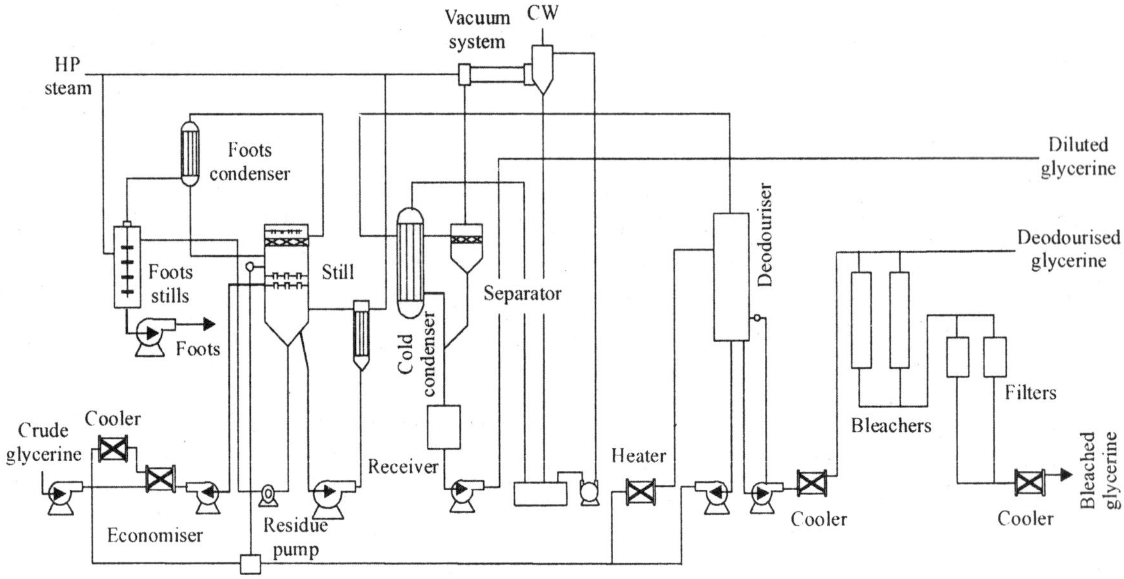


**Fig. 11.13.** Flowsheet of Mazzoni glycerine recycling plant.

## *Scraped surface heat exchanger*

An alternate method of distilling glycerine is through the use of an agitated thin-film drying system (Figs. 11.14 and 11.15). The process uses the countercurrent method to separate vapour flowing up through the thin-film dryer from the liquid feed, which is flowing downward. The thin-film dryer consists of two major assemblies: a heated body and a rotor. Preheated crude glycerine enters the dryer tangentially at the top, above the heated zone and is distibuted evenly over the inner circumference of the body wall by the rotor. Product spiral down the wall while bow waves developed by the rotor blades generate highly turbulent flow and optimum heat flux.

## *Storage and stability*

Dilute and crude glycerines contain certain amounts of suspended material (precipitates, salt, etc.) that tend to settle out during storage. Therefore, to avoid introducing these materials into the process when liquor is drawn, it is recommended that discharge nozzles be located above the lowest level of the tank and that the tank is emptied and cleaned periodically.

Dilute solutions of glycerol (< 50 per cent) are subject to fermentation, which reduce yield and introduce breakdown products that degrade the glycerol. Holding the glycerol above 70°C and/or at higher concentrations will alleviate this problem. Concentrated glycerine is difficult to pump at lower temperatures because of its high viscosity. It is recommended that glycerine be pumped at 40–50°C; lower temperatures make pumping difficult and higher temperatures can affect the colour. If heating coils or steam tracing are used, it is important to use low pressure steam so as not to overheat the glycerol and cause breakdown products.

Stainless-steel vessels or stainless-steel lined vessels are recommended to prevent the formation of colour complexes, especially if moisture or residual fatty acids are present in a carbon-steel tank. Since glycerine is hydroscopic, care should be taken to exclude moisture from the refined glycerine storage.

tank. Glycerine subjected to heat should not be stored in vessels containing copper or iron, as iron and copper salts will catalyse oxidation of glycerine under those conditions.

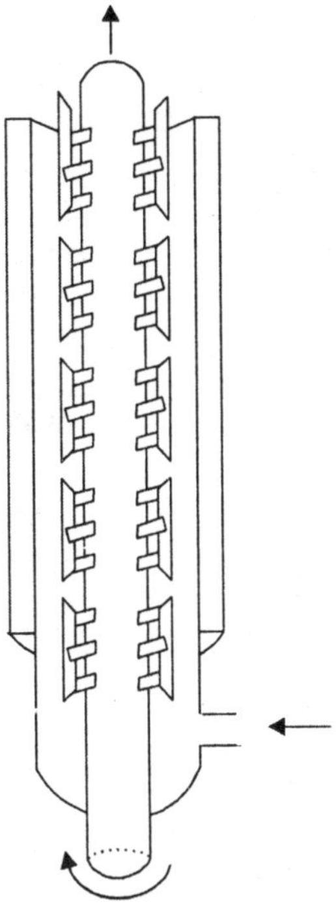

**Fig. 11.14.** Conceptual representation of a vertical agitated thin-film processor.

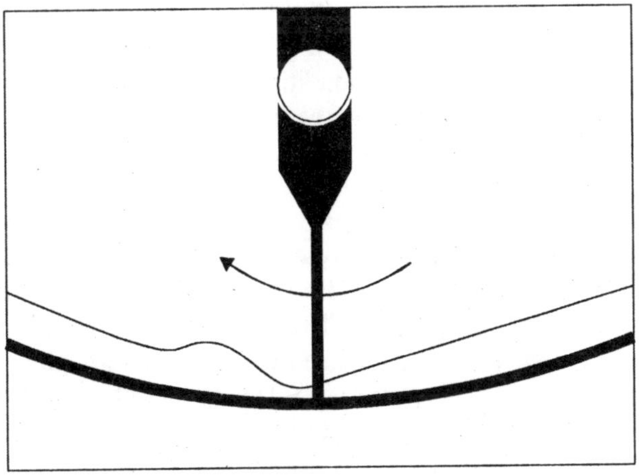

**Fig. 11.15.** Detail of agitated thin-film processor.

## Vapourisation Equipment

Soap lye crude glycerine contains about 10 per cent of salt and soapy matter; and the disposal of this non-volatile material influences the choice of vapourisation equipment which can be used and makes it usual with many types to operate semi-batch, rather than continuously. This problem is much reduced with saponification crudes, particularly those made from high pressure split sweet waters which contain little inorganic matter. For this reason, saponification crudes are often considered superior to soap lye crude, but, especially when produced from low grade fats, they are liable to contain nitrogenous and other impurities which can lead to problems with the quality of the distillates made from them.

As a batch of soap lye crude glycerine is distilled semi-batchwise, with additional crude fed to replace the glycerol vapourised, the concentrations in the residue of sodium chloride, other inorganic materials and matter organic non-glycerol (MONG) increase; and the viscosity of the residue rises. The recovery of as much glycerol from this residue as is economically worthwhile is a major problem; and several procedures are possible:

1. Traditionally the main distillation is stopped whilst the residue is still liquid enough to flow out of the still. The residue is then diluted and purified as much as possible using treatments similar to those adopted for lyes or sweet waters. The purified liquor is re-evaporated to give 'foots crude', which is distilled separately to produce a low grade distillate. A variant is to return the still foots to the soap lyes before treatment. This is simple, but is not recommended because of the build-up of impurities in the system, despite some purification in the re-treatment.
2. Some plants remove residues from the main still in liquid form for the rest of the glycerol to be stripped in a separate foots still. This enables the main still to be operated more, or less, continuously, although it needs to be washed out from time to time. Recently the stills offered are of the type in which solid salt is separated in a centrifuge through which the liquor is passed on its way from the main still to the foots still. Apart from the cost of the additional equipment (offset by the additional capacity of the main still), the use of a foots still tends to increase the time during which the material is exposed to distillation temperatures. Separation of salt and other solids in a centrifuge helps the final stages of vapourisation, but exposure of hot liquor to air may entail some risk of oxidation of glycerol.
3. For many years it has often been found best to continue the main distillation until little more glycerol can be vapourised. With many still designs this involves a loss of glycerol in foots discarded to drain of at least about 2 per cent of that fed. If a reasonably good quality soap lye crude glycerine has been distilled and conditions are such that only a little polyglycerol is formed, the residue is so stiff that it cannot be run out of the still; and it has to be dissolved by being boiled with water. If the quality of the crude glycerine is poor, polyglycerol formation is high, or the final steaming has not been complete, some of the residue can be run out as liquid foots. When liquid foots are formed, investigation is desirable to ascertain the cause and, if possible, to arrange remedial action for future distillations.

In most stills the residue after as much glycerol as possible has been vapourised from a soap lye crude is a toffee-like material, but, as described later, it has been found recently that under some circumstances a dry, powdery, residue can be obtained from a once-through vapouriser.

When re-distillation of once distilled glycerine is used to produce chemically pure grade, a procedure which is no longer common, a liquid residue containing glycerol, polyglycerols, caustic soda and soapy matter is left. From time to time, this needs to be transferred to a first distillation of crude glycerine.

Three major types of vapouriser are used in the distillation of crude glycerine:

1. Pot stills and variants, in which open and closed steam are used to provide natural circulation vapourisation.
2. Stills with external heaters and forced circulation.
3. Scraped wall, once through, vapourisers.

Types 1 and 2 can be used for saponification as well as soap lye crudes, but type 3 cannot, because the necessary dry, powdery, residue is not produced, unless a liquid residue can be discharged satisfactorily.

*Natural circulation pot vapourisers* are used in Van Ruymbeke stills, a type which was very widely employed for many years; and in various other designs. In essence, they comprise pots fitted with closed heating coils (usually steam) and a perforated ring, or cross, near the bottom for the introduction of open steam. The Van Ruymbeke still operates at around 60 mm Hg; and hence required about 1–1.5 lb open steam/lb glycerol vapourised, with a boiling temperature around 170–175°C. Steam in the closed coils at 175–200 psig (condensing at 192–198°C) gives a reasonable temperature difference to induce heat flow and a good rate of distillation. One major installation with several Van Ruymbeke stills for the distillation of soap lye crude glycerine, plus two smaller stills for re-distillation to produce chemically pure grade, operated satisfactorily for many years with steam from Lancashire boilers at 150–160 psig (185–188°C) which gave a rather small temperature difference. Special arrangements to make the best use of the available steam are shown diagrammatically in Fig. 11.16. Features are as under:

1. A good flow of steam through the coils, with efficient removal of condensate and air, was induced by use of surplus steam which was recovered from the 90 psig collector.
2. Superheated steam was used in the crude still coils. This is contrary to the usual view that superheat in the steam used in heating coils reduces rather than increases heat transfer, but this use of superheat was found to increase distillation rate, presumably because it increased the temperature of the upper parts of the coils which were in contact with froth, as well as more compact liquor. Saturated steam was found advisable for re-distillation, when the boiling point elevation was lower.

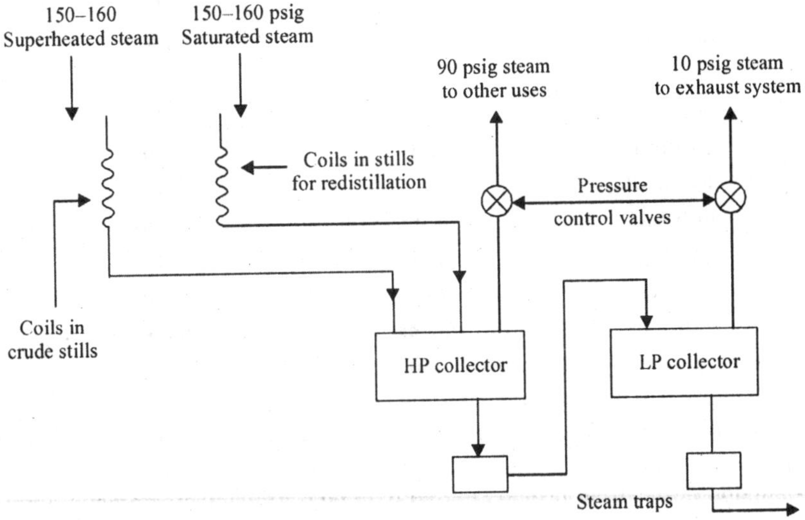


**Fig. 11.16.** Steam supply arrangements for a group of Van Ruymbeke stills.

Classical Van Ruymbeke stills includes a 'superheater' for the open steam. This comprised a tube and shell heat exchanger in which the flow of open steam towards the perforated distributor in the still is exposed to heat from full pressure steam. The superheater is presumably intended to avoid the introduction of liquid water into the hot glycerine liquor where it might cause primes. It can, however, be shown by theoretical analysis, using the Fanno line technique, that expansion of steam through a control valve and pipe is very different from that through a nozzle, because of the 're-heating' effect of pipe wall friction as the linear velocity becomes very high; and that liquid water is unlikely to be formed. Later on it was found that these superheaters were unnecessary and they were removed without any ill effect.

In addition to its role as a carrier fluid, an important function of the open steam in pot-still vapourisers is to provide agitation, with proper circulation of the liquor over the closed coils without local overheating. Some plant manufacturers have introduced the open steam through jet-type vapour-lift devices to improve the circulation; for example upto the centre of the pot and down over the annular heating coils. This is particularly useful when the still is designed to work at a relatively low absolute pressure and with a low open steam ratio.

It is important to ensure that flow passages through condensers and ducts are adequate to take the maximum requirement of open steam without undue pressure drop. Stills have been constructed in which this was not done, so that an adequate flow of open steam could not be used without a substantial increase in still pressure.

Several important modern stills use *external heaters and forced circulation* which permit operation at low pressures of about 2–4 mm Hg. Often no open steam is used during the main part of the distillation, but it is also possible to inject some open steam into the vessel to and from which the liquor is circulated; in which case a rather higher still pressure is usual. One early type was a Scott still for which the soap lye crude is dried before being fed into the still in which a pressure of about 2 mm Hg is maintained. This still is operated with a liquor temperature of 177°C, which is surprisingly high for a pressure of 2 mm Hg, even with no water in the crude glycerine as fed to the vapouriser. Still of this type requires a large addition of caustic soda to produce distillate with an acceptable saponification equivalent and that the yield was not good. The large amount of caustic soda used and the polyglycerol formed, would tend to account for the high boiling temperature, plus the excessive caustic soda, for the poor yield.

Recently very successful stills with vapourisers of this type have been developed. Additional stills of this type and similar vapourisers have been used. These stills operate at 3 – 4 mm Hg without open steam, but with a little water left in the crude glycerine. The boiling liquor temperature is around 155°C and the yield is superior to that with Van Ruymbeke stills. To maintain an absolute pressure in the vapouriser of 3–4 mm Hg, the vapour system, spray arresters, ducts and condensers, must be designed to have very low vapour velocities and hence low resistance to flow. The booster jets are used to compress the water vapour and non-condensible gas leaving the last glycerine condenser; and the capacity of the booster jet, or jets, is often the factor which limits the capacity of the still. This means that the water content of the crude glycerine should normally be kept low. The significance of the presence of a little water in the crude glycerine as fed to the vapouriser on polyglycerol and ester formation is not clear, but it may be important. To enable the distillation to be completed in the main vapouriser, without a foots still, closed and open steam coils are fitted in the vessel, together with a damper in the circulation pipe, as shown in Fig. 11.17. During the main part of the distillation the butterfly damper is open and the liquor is circulated at a high rate by the centrifugal, or axial flow pump through the external heater. It is returned to the vessel through a tangential inlet onto a ledge round which it swirls; vapour flashes off and the residual liquor runs back into the reservoir to be recirculated with the new feed. The closed and open steam coils in the still vessel are not used during the main part of the distillation. When sufficient crude has been fed and the liquor worked down until its viscosity becomes too high for continued operation in this way, the damper is closed and the liquor in the circulating system is transferred by the small pump to the main reservoir. The residue is then worked to dryness using the closed and open steam coils, as for a pot still. During this final stage the pressure in the still and the liquor temperature are higher than for the main run.

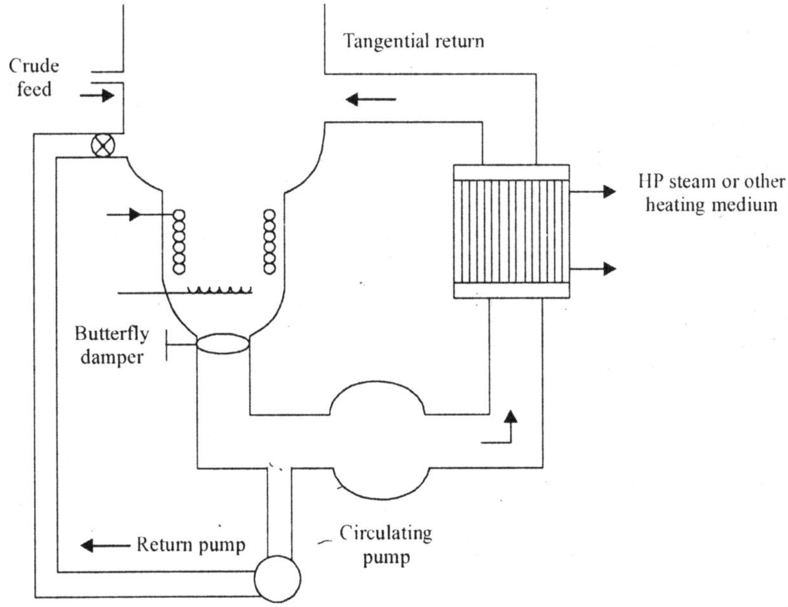


**Fig. 11.17.** Forced circulation vapouriser.

A more recent development is the use of a *'once-through' scraped wall vapouriser* for the distillation of good quality soap lye crude glycerine. When conditions are right the glycerol is vapourised very rapidly and almost completely, without formation of polyglycerols and a dry, powdery, residue is discharged from the base of the unit. In the Gianazza still the residue is collected in a large vessel from which it is said to be discharged once, or twice, a week. It is also possible to remove the residue continuously through an automatic air-lock system and screw conveyor. The vapouriser is heated with steam at say, 250 psig (208°C); or with hot oil, or other high temperature heat transfer medium. A low pressure of a few mm Hg is maintained as for other stills not using open steam.

Because of the very short residence time it seems probable that the working pressure may be less critical than with other types of vapouriser. Because of its almost instantaneous vapourisation, this type of unit has great advantages when it works well. The yield can be exceptionally high, because virtually no polyglycerol is formed and vapourisation of glycerol is nearly complete. The formation of esters and other undesirable compounds is also substantially lower than with other vapourisers using the same crude glycerine. On the other hand, it can only work if a dry, powdery residue is formed; patches of toffee-like material can cause major problems.

This requires the unit to be kept in good mechanical condition and capable of uniform heat transfer. It also requires a good soap lye crude glycerine with a sufficiently high ratio of salt to MONG; at least 3:5 is said to be a useful criterion. It has also been found that particularly good crude glycerines, which produce a very dusty residue at the base of the vapouriser, can cause some difficulties due to carry over of dust, because there is little organic matter to bind the particles of residue. The dry residue produced by this type of vapouriser may be more readily disposed of than the rather foul smelling and polluting liquors obtained when other types of vapouriser are washed out.

## Spray Arresters

Traces of liquor, or solid residue, carried over in the vapours from the still can spoil a distillate, particularly the first (and best) when more than one fraction is collected. The strict limits for chlorides make this particularly critical with soap lye crudes. The main problem is to remove spray without excessive flow resistance, especially in stills which operate at very low pressures. Thin beds of rings having a large cross-section, so as to minimise linear velocity and hence resistance, are now usual. Such beds are often situated in the upper part of the still vessel; and they should be provided with water sprays to enable them to be washed when the still is cleaned. Older stills, such as the Van Ruymbeke, which operated without booster jets, commonly used catch-all vessels. Particularly when foots are worked down to dryness and considerable boiling out with water is needed to remove the residue, carry over is liable to occur during the wash out. Some stills include a large valve in the vapour line which can be closed to isolate the vapouriser from the condensing system during the washout. Such valves can be valuable; but care must be taken to avoid constriction of the vapour passages and the possibility of in leakage of air.

## Odour and Colour

Apart from any carry-over of droplets, the vapours leaving a glycerine still contain glycerol and water, plus small concentrations of esters, TMG and other organic impurities. The proportion of water to glycerol depends upon the quantity of open steam used, including water in the crude glycerine as fed to the vapouriser. Because of the large difference in volatility of water and glycerol, separation of these two components by fractional condensation is relatively easy; and is the main way in which refined glycerines are produced. Fractionating columns with reflux and reboil are not usual; and are generally undesirable because of the pressure drop they introduce and the extra time for which the alkaline crude glycerine is kept at distillation temperature.

Colour and odour problems can be mostly avoided by using a high quality raw material, treating and storing crude glycerol properly and avoiding high temperatures for an extended time period. Impurities in crude glycerine, especially MONG, affect the quality and quantity of distilled glycerine. If the MONG content is high (3–5 per cent), odour, taste and colour reversion problems may exist in the final product. Trimethylene glycerol, which is present in MONG, can affect the colour of glycerine and also lead to problems in storage. A properly designed deödouriser or 'stripper' vessel, located after the distillation column and operating at high vacuum and with open or 'sparging' steam will remove close boiling impurities and most odour bodies. Remaining colour bodies are removed by filtration with activated carbon.

Acrolein may be formed in the presence of iron, iron salts, acid or neutral salts:

$$C_3H_5(OH) = C_3H_4O + H_2O$$

Acrolein odour is distinctive and is detected in small quantities.

It has been found that glycerine left inside processing vessels, especially carbon columns, can degrade and develop hard-to-remove odours. Therefore, process equipment should be cleaned completely after shutdown and carbon columns blown down with air and washed with water. Also, fresh carbon (and ion exchanger resins) can contain odour-causing impurities. They can be washed with distilled glycerine that can be subsequently redistilled or sold as a lower grade.

## BLEACHING AND THE TREATMENT OF DISTILLATES WITH ION-EXCHANGE RESINS TO UPGRADE QUALITY

Traditionally, distilled glycerines are bleached batchwise by an addition of 0.2–2.0 per cent of activated carbon. The mixture at about 80°C is stirred, held for 1–2 hours and filtered through a press. The carbon retains a high proportion of glycerol even after it has been steamed; and it may be worthwhile to slurry the used carbon with condensed water and to re-filter, wash, steam and/or air blow in the press to recover weak glycerol. It is also possible to re-activate the washed carbon by stoving it in lidded crucibles in an oven (although to a lower level of activity than that of new activated carbon). This recovery process is not necessarily economic. Some bleaching can also be achieved by pumping distillates through the carbon cake in a filter press. It is possible to bleach glycerines in continuous columns which contain beds of granular carbon.

A relatively recent process is to improve the quality of distillates by the use of ion-exchange resins instead of, or after, activated carbon. At one time it was thought necessary to dilute the distillate to about 80 per cent glycerol and to reconcentrate by evaporation of water after treatment with the resins, but it has since been found that useful purification can be achieved by the use of appropriate resins on undiluted distillates. Macrorecticular resins (which have large discrete pores capable of removing relatively large molecular mass compounds from the liquid) are normally best for this duty which demands the removal of organic, rather than ionic, impurities.

Deodourisation, using heat and open steam, under vacuum, is often necessary after a distillate has been passed through beds of resins to remove odours imparted by the resins.

Selection of the best combination of resin beds, possibly following an activated carbon treatment, requires experiments and/or consultation with a resin manufacturer.

## PRODUCTION OF REFINED GRADES OF GLYCERINE WITHOUT DISTILLATION

### Ion Exclusion

The ion exclusion process for the partial purification of glycerine liquors is based on the fact that in a bed of granular resin (50–100 mesh) of a suitable type, such as 8 per cent cross-linked medium porosity resin consisting of sulphonated styrene-divinyl benzene co-polymer fed with an impure glycerine solution. There are three phases; the resin, the liquor within the resin granules (called the resin liquor) and the surrounding liquor (called the interstitial liquor). Most low molecular mass substances of nonionic, or slightly ionic, nature (such as glycerol) diffuse freely in and out of the resin liquor and the interstitial liquor. Ionic substances, such as sodium chloride and sodium sulphate, are repelled by the ionic charges of the resin and exist at a much lower concentration in the resin liquor than in the interstitial liquor. Thus, when a mixture of ionic and non-ionic solutes is passed through a resin column and is followed by a water rinse, ionised solutes are excluded from the resin granules and leave the column first. The nonionic material is held back and appears in the rinse water. By the use of properly balanced flows of impure glycerine and of rinse water a partial purification can be achieved, as shown in Fig. 11.18. Purification can be completed by ion-exchange. It is said that Indian Pharmacopoeia grade can be produced from saponification crude glycerine without any distillation.

### Ion Exchange

Glycerine liquors can be purified by the use of ion exchange resins in a process similar to that used for the de-mineralisation of water. The cations, $Na^+$, $Ca^{2+}$ and $Mg^{2+}$ are exchanged for $H^+$ in a bed of cation resin (regenerated with hydrochloric or sulphuric acid); and the anions $Cl^-$ and $SO_4^{2-}$ (or their acids) are removed by an anion resin (regenerated with caustic soda). Organic impurities are also removed. Because of the quantities of resins and of regenerating chemicals needed, the process is impracticable for soap lyes, but it was at one time considered at least marginally economical for soap lye crude glycerine diluted to 25 – 30 per cent glycerol (much of the salt having been removed in the evaporation of purified lyes to crude glycerine). The weak base anion resins are most economical for the removal of the strong acids, hydrochloric and sulphuric, but they need to be 'backed' by a strong base anion resin to remove weak organic acids.

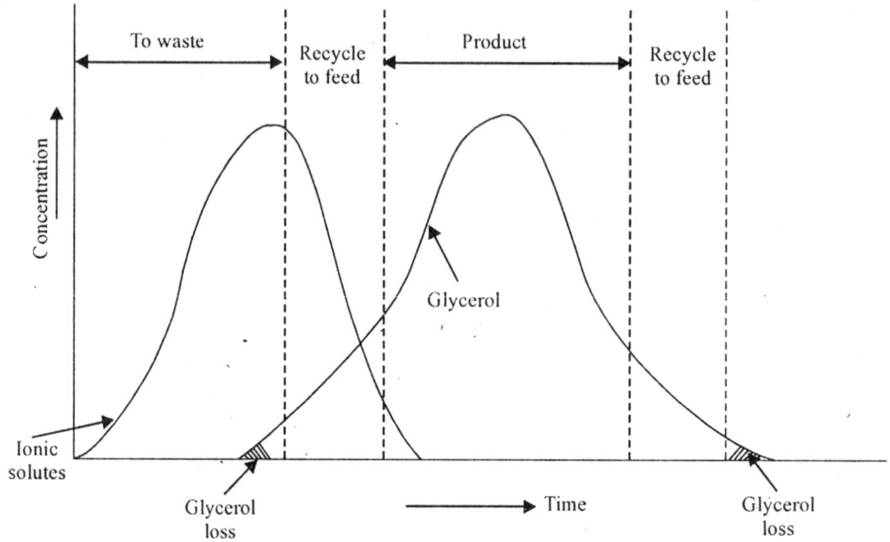


**Fig. 11.18.** Partial purification of glycerine liquors by ion-exclusion.

### Reverse Osmosis Using Semi-permeable Membranes

Experiments have shown that weak glycerol liquors can be concentrated some ten-fold by forcing water through a suitable cellulose acetate membrane under a pressure of 1500 psig (103 bar), with a recovery of 96 per cent in the concentrate. This is an application of reverse osmosis in which preferential sorbtion of water by the membrane, as well as flow through capillary pores, is involved. This use of reverse osmosis might, perhaps, provide a basis for the recovery of glycerol from liquors which would otherwise be too weak to be processed.

## PROCESSING PLANTS

### Layout Considerations

Plant layout considerations for a glycerine recovery plant are similar to those for other processing plants: adequate space must be allowed for equipment installation, operation and maintenance as well as operator safety. Since some of the operation requires some material handling, consideration must be given to these specific requirements. These operations include batch feeding of treatment chemicals (most of which are dangerous to handle), disposal of filter cake and charging of the bleaching columns with carbon. It generally makes sense to arrange the equipment in the order of process flow, as piping and other connections can be made much more conveniently. Also, if process tanks are located to allow gravity feeds, some pumps can be eliminated. Locating equipment on several levels also helps reduce the overall footprint, or floor space requirements.

All areas should be equipped with wash-down equipment and all levels must have adequate floor drains with the floor surfaces properly pitched. Care must also be exercised when selecting flooring materials, as glycerine is slippery; this is a real concern, because spills and leaks inevitably occur. Provisions must be made in the plant for efficient cleaning procedures for the equipment, piping and building.

## Materials and Construction

Careful selection of the materials of construction is important to ensure adequate life of equipment in the glycerine processing plants and to ensure a high quality finished product. This is especially true when recovering glycerine from soap lyes, where the presence of sodium chloride, in crystal and solution forms, greatly increases the corrosion potential.

### *Treatment plant*

When treating sweetwater glycerine, ductile iron (DI) can generally be used for pumps, although lined DI and stainless-steel pumps can be used with an expectation of longer life. More care must be used when selecting material for treating spent lye. Treatment vessels can be either lined with carbon steel or stainless steel. Fibreglass reinforced polyester (FRP) is a good choice for both tanks and piping in the treatment plant, but care must be taken to ensure that the temperature range of the FRP material is observed and that the pipelines are correctly supported. Also, steam blowing of pipes should not be allowed as the high temperature can damage the FRP piping.

The open plate and frame filter presses used typically have cast iron frames with polypropylene frames that give excellent life. Filter clothes are usually constructed of synthetic materials such as polyporopylene felt, rather than natural fibres such as cotton. Plastic tools must be used to clean cake from the cloths to avoid damage.

A plate and frame heat exchanger is sometimes used to preheat incoming soap lye, as treatment is more effective with warm lye. The plates can be of 316L stainless steel, or even titanium for effective corrosion resistance.

### *Complete stills*

The various possible elements considered in previous sections can be selected and combined in many ways to form complete glycerine stills. Some plant manufacturers have used widely differing designs at various times, so that mention of a plant manufacturer does not necessarily indicate the nature of a still. In this section, only two stills of particular historical importance, plus some designs from recent manufacturers are outlined.

Figure 11.19 shows one form of *Van Ruymbeke still*. The vapouriser is a pot with closed and open steam coils. Traditional stills did not include a scrubbing tower; and the weaks from the cold water condensers had to be evaporated to recover glycerol. The inclusion of a packed scrubbing tower, fed with a small quantity of cold water sufficient to reduce the temperature of the vapours leaving the tower to an appropriate level, recovers practically all the glycerol leaving the air cylinders, the weaks are then discarded to sewer.

Figure 11.20 shows one version of the *Garrigue still* which was widely used. The vapouriser is similar to that of the Van Ruymbeke still. Surface condensers are used to produce product by fractional condensation. Weaks, sometimes called sweet waters, are concentrated in a built-in single effect evaporator which provides the open steam for the main vapouriser. This open steam is superheated in a heat exchanger which also serves as the first product condenser. A steam heated concentrator is included to bring the third product fraction up to the strength required for dynamite ('high gravity') grade.

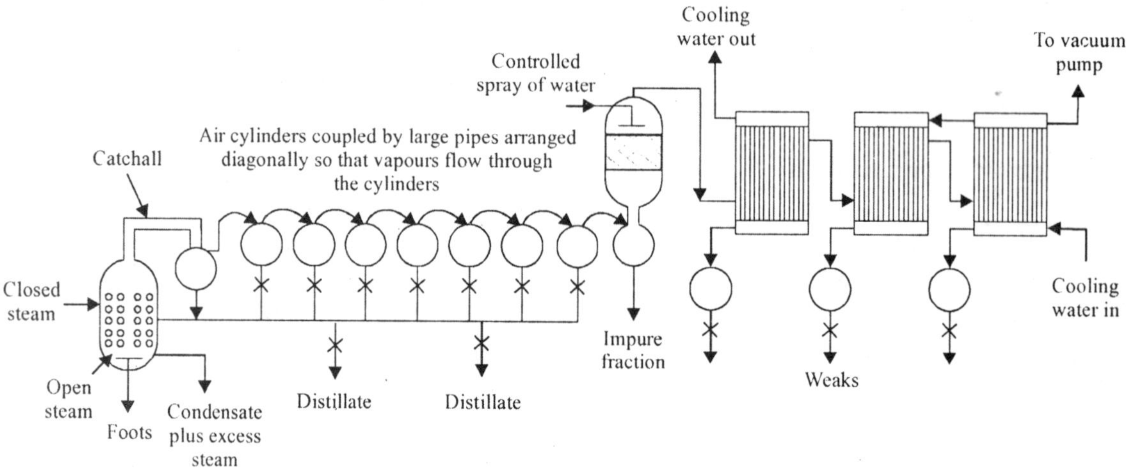


**Fig. 11.19.** Van Ruymbeke still.

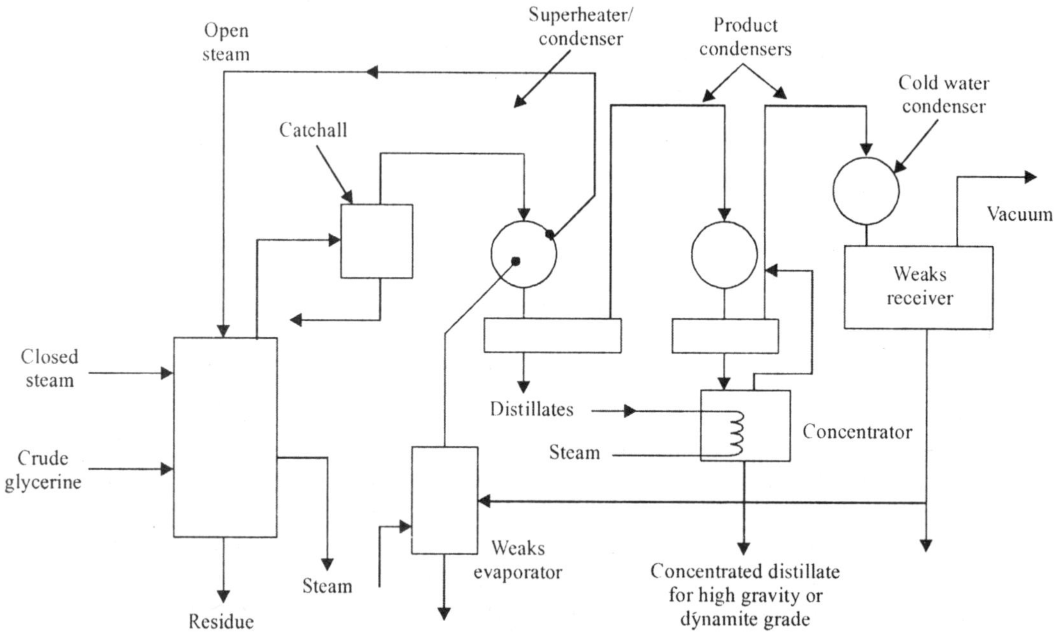


**Fig. 11.20.** Garrigue still.

The *Gianazza* HPS still which is very similar to the *Frazer Scott*, glycerine is dehydrated before it is fed to the forced circulation vapouriser. Salt slurry, which separates as the crude glycerine loses water in the dehydrator, is returned to the crude glycerine department. Condensation is in surface condensers which use a circulated fluid cooled in auxiliary coolers. The *Gianazza GDTF 'Turbotherm'* still includes a scraped-wall, once-through, vapouriser, otherwise it is similar to the Gianzza type HPS.

In recent process developed by *Lurgi,* the *Lurgi* still preheated crude glycerine is vapourised in a vessel with one, or more, legs, each equipped with closed heating elements over which the liquor rises by the vapour lift principle induced by open steam injected at the base of the leg. With this steam also acting as carrier fluid, vapourisation is at about 10 mm Hg total pressure. This vapouriser is similar in principle to that used in Lurgi fatty acid stills. In some large units the multi-compartment vapouriser distils soap lye crude continuously, a liquid residue being withdrawn to be worked down in a separate foots still. Distillate from the foots still is redistilled. Other Lurgi stills of this type are operated on the semi-batch principle. The main condenser is of the direct contact type using glycerine cooled slightly in a heat exchanger in which it boils water. In addition to the stripping which occurs in the condenser, the product leaves through a separate stripping section of the tower in which it contacts the vapours from the vapouriser. Practically all the glycerol left in the vapours leaving the main condenser is separated as a 90–94 per cent glycerol fraction in a cold glycerine direct contact condenser. This fraction, which needs to be reprocessed, amounts to 6–10 per cent of the total glycerol fed.

### Materials of construction for glycerine treatment and evaporation plants

Traditionally, mild steel, wrought iron, or cast iron tanks and pipes have been used for soap lye treatment plants, with cast iron filter presses clothed with cotton cloths. Particularly for the early stages of treatment carried out under acidic conditions, the presence of chlorides and blowing with air give highly corrosive conditions; and the life of tanks is often short. The lining of such tanks with rubber, cement, or some acid resistant plastic coating, is highly desirable. Cast iron filter presses have a reasonably long life, although it is often useful to scrape the grooves in the plates from time to time. Expensive polypropylene filter plates are now sometimes used; and it may be desirable to assess the economics of using polyester cloths. For many years, cast-iron was used for the bodies of soap lye evaporators, with arsenical copper tubes in the calandrias and heat exchangers. Such evaporators lasted a very long time; examples are known where the cast-iron bodies were very difficult to break-up after a life of about forty years, although they had been re-tubed 3 – 4 times during this period. These evaporators used exhaust steam, plus make-up, at about 4 psig in the main and less in the actual calandrias. The adverse effect of higher temperatures, resulting from the use of higher pressure steam, is shown dramatically by experience with the quadruple-effect evaporator mentioned earlier. The arsenical copper tubes in the first effect calandria lasted only 1–2 years; and the rate of corrosion fell with reduction in temperature from first to last effect. Experiments showed that monel metal was a particularly good material for tubes, but that the less expensive cupro-nickels (including the cheapest 90:10 type) were adequate for the duty. Special cast irons, such as Ni-resist, have also been used for the bodies of soap lye evaporators. Copper has been used successfully for the bodies of sweet water evaporators.

These traditional materials are still recommended by some plant manufacturers, but other alternatives, include mild steel. Sacrificial anodes of aluminium have sometimes been found beneficial in reducing the corrosion of mild steel, or cast-iron, evaporators.

The baskets of centrifuges used for salt separation have to withstand high stresses as well as corrosive conditions. Mild steel was normally used, but, with modern more efficient machines, failures, particularly at welds, are known. High nickel alloys (or, perhaps certain stainless steels), with the necessary mechanical properties as well as corrosion resistance, should normally be selected. Design to avoid high stresses is very important, whatever metal is selected; and rivetted, rather than welded, side steams may be advantageous.

### Evaporation plant

Materials selection for the evaporation plant is the greatest challenge in the glycerine recovery plant, again, especially if there is salt present. For evaporating sweetwater, carbon-steel vessels might be used, but better results will be obtained using stainless steel. The corrosive nature of the brine–glycerol solution makes carbon-steel plate and most grades of stainless steel a poor choice for evaporators handling soap lye. The heavy walled, cast-iron calandria that have been used extensively for soap lye evaporations in the past have long lives; however, economic considerations discourage new evaporators of that style from being built.

The material of choice during the past 20 years or so for welded plate evaporators (and related components handling salt and/or brine) is copper–nickel alloy. The two principal alloys to consider are 9010 (90 per cent copper–10 per cent nickel), which has the best corrosion resistance to the glycol–brine solution and 7030, which is more expensive.

An interesting class of materials that should be considered for evaporator service are the duplex stainless-steel alloys. These are high performance stainless-steel alloys that have a 'duplex' structure, in that the alloy contains both the austenitic and ferritic phases in the grain structure at the same time. These materials have shown high resistance to stress corrosion cracking in chloride-bearing environments and also possess generally high corrosion and erosion resistance. Other more expensive alloys such as Ferralium 225, Hastelloy and Monel have been used in various applications.

Pumps for the evaporation plant must stand up under the hot glycerol–brine solution and are also generally of the centrifugal design. Cast stainless steels (CF8M alloy) are often used for this application, but a superior material for pumps is a chromium–nickel–copper–molybdenum stainless steel known as CD-4M, which should be considered when choosing the pump material. The extra cost is generally not great. Seals for pumps are generally mechanical. If the fluid contains salt crystals, which are abrasive, it is recommended that water be used to flush the seal. Both seal faces on the pump side should be of a hard material; using carbon for one of the seal halves has not proven to be successful.

### *Refining plant*

Materials for the refining plant can be selected on the basis of the impurity content of the glycerol being handled. The initial distillation vessel or residue still, if it is handling crude glycerine with salt in it, should be a high grade of stainless steel. The L grades generally exhibit slightly better corrosion resistance than their base versions. The 316 and 316L stainless steels are recommended for stills handling soap lye crude glycerine, while 304 stainless steel is adequate for sweetwater stills, vapour scrubbers, condensers, deodourisers and bleaching equipment where salt is not present. The 316L is especially recommended for foot stills where salt is present.

In some cases, more corrosion-resistant materials have been used, the life of such equipment is exceptionally long.

Pumps for handling glycerine can be centrifugal as long as the temperature and viscosity of the product is satisfactory for the pump being used. For lower temperatures and for pumping residue, a positive displacement pump with a pressure relief loop (internal) should be used.

## INSTRUMENTS AND CONTROLS

Control techniques for glycerine plants have not changed much over the years, as it is necessary to control pHs, levels, temperatures and vacuums in the various portions of the plant. Most operations can be accomplished manually; and traditionally, manual operation of these plants have been the norm. However, the development and refinement of digital single-loop controllers and associated sensing equipment have increased the ease and accuracy of control. Reliable in line pH monitoring has made treatment plants easier to operate.

A caution must be given, however, to select instrumentation that can handle the sometimes severe conditions in the glycerine recovery plant. The crude glycerine in the evaporation and crude refining sections contains both dissolved and suspended salt, while the residue in the refining plant contains salt as well as other impurities that make the viscosity of the residue high. Vane-type flow meters have been used successfully for measurement for difficult flows such as residue from the crude still. Rotometers often will be plugged by suspended solids and are recommended only for flows that are 'clean' and will maintain a constant viscosity over time.

Materials of construction for wetted portions of the instrumentation must be selected using similar considerations to those used for the equipment itself. In addition, plastic diaphragm seals are available to cover and protect the pressure-sensing faces of immersed level transmitters. This allows the use of stainless steel in applications where, if unprotected, the corrosion rate would be unacceptably high on the thin face of the instrument. Where pressure gauges are installed in fluid lines that contain any suspended or entrained solids, it is recommended that a diaphragm seal be installed to protect the instrument.

## Quality and Testing

### *Glycerine grades*

Several general grades of glycerine have been established or are in common use in US and other European countries. US Pharmacopoeia (USP) or chemically pure (CP) glycerine standards have been established for the highest grades of glycerine, water-white in colour and with glycerol contents of not less than 95 per cent. It conforms to standards given in US Pharmacopeia. It is used by the food and pharmaceutical industries because of this high purity. Both natural and synthetic glycerines meet these specifications (Table 11.2).

**Table 11.2.** Refined glycerine specifications.

| *Characteristic* | *99.7% USP* | *99.5% USP* | *99.0% USP* |
|---|---|---|---|
| Glycerol, % minimum | 99.7 | 99.5 | 99.0 |
| Specific gravity, 25/25°C | 1.26092 | 1.26073 | 1.25945 |
| Colour | 25 hazen maximum, APHA 10 maximum | 25 hazen maximum, APHA 10 maximum | 25 hazen maximum, APHA 10 maximum |
| Ash | 0.01% maximum | 0.01% maximum | 0.01% maximum |
| Chlorides | 10 ppm maximum | 10 ppm maximum | 10 ppm maximum |
| Sulphate | 20 ppm maximum | 20 ppm maximum | 20 ppm maximum |
| Arsenic | 1.5 ppm maximum | 1.5 ppm maximum | 1.5 ppm maximum |
| Heavy metals | 5 ppm maximum | 5 ppm maximum | 5 ppm maximum |
| Readily carbonisable | – | Not darker than matching fluid H | Not darker than matching fluid H |
| Chlorinated compounds | 30 ppm maximum | 30 ppm maximum | 30 ppm maximum |
| Acrolein, glucose and ammonia compounds | Not yellow, no ammonia odour | Not yellow, no ammonia odour | Not yellow, no ammonia odour |
| Fatty acids and esters | 1.0 ml 0.5 N NaOH maximum | 1.0 ml 0.5 N NaOH maximum | 1.0 ml 0.5 N NaOH maximum |
| Acidity or alkalinity | Neutral | Neutral | Neutral |

High gravity glyceride is a commercial grade of glycerine, near white in colour and with high glycerol content (>98.7 per cent with specific gravity minimum of 1.2595 at 25/25°C). It conforms to Federal specification O-G-491C and also to ASTM D1257. Dynamite glycerine meets all high gravity specifications, except that a darker colour is allowed. Saponification (88 per cent) and soap lye (80 per cent) crude are crude, unrefined grades of glycerine offered for sale to glycerine refiners. Saponification crude is concentrated sweetwater from fat splitting, while soap lye crude comes from soap making and contains some salt.

## ANALYTICAL TESTS AND SPECIFICATIONS

Some of the main analytical tests used for glycerines and implications of the results which they give are discussed below.

### Glycerol Content

Glycerol content can be estimated by several methods. The standard method is now oxidation with periodic acid (*sodium periodate* in acidic solution). Glycerol is oxidised to formic acid according to the equation:

$$\begin{array}{ccccccc} \underset{\text{OH}}{\overset{|}{CH_2}}-\underset{\text{OH}}{\overset{|}{CH}}-\underset{\text{OH}}{\overset{|}{CH_2}} & + & 2H10_4 & \longrightarrow 2H10_3 & + \; 2H.CHO & + \; H.COOH & + \; H_2O \\ \text{Glycerol} & & \text{Periodic acid} & \text{Iodic acid} & \text{Formaldehyde} & \text{Formic acid} & \text{Water} \end{array}$$

The quantity of glycerol originally present in the sample is determined by titration with standard caustic soda of the formic acid formed under standard conditions. This test estimates as glycerol any compound which contains groups $-\underset{\text{OH}}{\overset{|}{CH}}-\underset{\text{OH}}{\overset{|}{CH}}-\underset{\text{OH}}{\overset{|}{CH}}-$ with three adjacent carbons each with a hydroxyl group attached. Such compounds include sugars and sugar alcohols (sorbitol, etc.) which are sometimes found in glycerines which have been contaminated with glycerine substitutes used in some food industries, or added in an attempt to defraud. On the other hand, TMG (trimethylene glycol, 1,3 dihydroxypropane, or propane 1,3 diol) and polyglycerols are not estimated as glycerol by this test. As considered later, TMG appears in some crude glycerines made from low-grade fats which have been stored badly, or, more particularly, from lyes which have been stored under conditions which permit destruction of glycerol by micro-organisms. As well as indicating a destruction of glycerol, TMG in a crude glycerine can cause problems in the refining operations. Polyglycerols are formed by heating glycerol under certain conditions, particularly in the presence of alkaline and other, catalysts.

$$\begin{array}{ccccc} \underset{\text{OH}}{\overset{|}{CH_2}}-\underset{\text{OH}}{\overset{|}{CH}}-\underset{\text{OH}}{\overset{|}{CH_2}} & + & \underset{\text{OH}}{\overset{|}{CH_2}}-\underset{\text{OH}}{\overset{|}{CH}}-\underset{\text{OH}}{\overset{|}{CH_2}} & \longrightarrow & \underset{\text{OH}}{\overset{|}{CH_2}}-\underset{\text{OH}}{\overset{|}{CH}}-CH_2-O-CH_2-\underset{\text{OH}}{\overset{|}{CH}}-\underset{\text{OH}}{\overset{|}{CH_2}} + H_2O \\ \text{Glycerol} & & \text{Glycerol} & & \text{di-glycerol} \end{array}$$

The reactions are of the type Tri-glycerols and higher members of the polyglycerol family, are formed by similar condensation reactions in which a molecule of water is eliminated between two hydroxyl groups to form an ether linkage. The reaction can occur between two primary alcohol, —$CH_2OH$, groups as shown, or a secondary alcohol group >CH–OH can be involved, so that many isomeric polyglycerols are possible.

Formation of polyglycerols, or, more strictly, avoidance of their formation, is important in glycerine distillation and estimation of polyglycerols in still residues (foots) is often useful.

For many years before the development of the periodate method, the *acetin method* was standard for the estimation of glycerol. This involves acetylation of the glycerol with acetic anhydride and determination by hydrolysis with standard caustic soda, of the quantity of the estertri-acetin formed. The acetin method includes as glycerol other hydroxyl compounds likely to be present, such as sugar, TMG and polyglycerols. Because of the elimination of hydroxyl groups when polyglycerols are formed, the acetin method does not return polyglycerols as the full amount of glycerols from which they were produced. Thus:

1. Two molecules of glycerol contain 6 hydroxyl groups, but the diglycerol formed from them has only 4 hydroxyl groups.
2. Similarly tri-glycerol produced from three molecules of glycerol with 9 hydroxyl groups, contains only 5 hydroxyl groups.

The third important procedure for the estimation of glycerol is the *dichromate method*. This method, used particularly for soaps and lyes, involves removal of soapy and other, impurities, followed by estimation of oxidisable matter calculated as glycerol. It includes all oxidisable compounds not removed in the preliminary purification of the sample. Because the condensation reactions in which polyglycerols are formed by elimination of water do not change the oxygen requirement for complete oxidation to carbon dioxide and water, the figure obtained for glycerol percent includes any polyglycerols expressed in terms of the mass of glycerol from which they were formed.

For many purposes, the glycerol content of refined glycerines is estimated by determination of *specific gravity* at 15°C, or *relative density* measured at 20°C and reference to tables giving the properties of pure glycerol/water mixtures. The presence of other substances can interfere with the accuracy of results obtained.

### Non-volatile Inorganic and Organic Residues, Particularly in Crude Glycerines

The content of inorganic matter is determined as *percentage ash*. In the case of refined grades a little sulphuric acid is added during the determination to give *sulphated ash*.

For many years, important organic impurities in crude glycerines were estimated by determination of *total residue* per cent, the residue left after vapourisation of the glycerol under carefully controlled conditions in an oven at 160°C. The *organic residue* (OR) was then total residue per cent – ash per cent.

More recently, organic residue has largely been replaced by *MONG* per cent (matter organic non-glycerol) which is defined as 100 – (ash + glycerol + water) per cent. Glycerol is determined by periodate and water by the Karl Fischer method. Determination of total residue is tedious and rather subject to error, so that MONG is more convenient. The figures found for OR and MONG are very similar for many crude glycerines, but there are differences which are significant under some circumstances:

1. TMG is included in MONG, but not in OR, because it is more volatile than glycerol. Hence, MONG should equal OR + TMG.
2. Any volatile alcohols, occasionally present in samples of glycerine including material from the alcoholysis of fats to form methyl, or ethyl, esters of the fatty acids, are included in MONG, but not OR.
3. Sugars and other polyhydroxy compounds which increase the apparent glycerol content of a sample, artificially reduce MONG.
4. Polyglycerols and other relatively non-volatile organic compounds should be included in both MONG and OR, but substantial proportions of diglycerol, which is slowly volatile under total

residue conditions, may make difficult the determination of the total residue end point (that is the stage at which the residue is deemed to have reached constant weight).

### Nitrogen

Organic nitrogen which comes from proteinaceous materials is sometimes found in quantities up to 0.5 per cent as nitrogen in poor quality saponification crude glycerines made from low grade animal fats, although 0.025–0.15 per cent is more usual. Crude glycerines with a high organic nitrogen content tend to give distillates with objectionable odours and colours. Many nitrogenous compounds have been found at low concentrations in such distillates, with 2-piperidone, a toxic substance, the most important. Washing the fats with sulphuric acid before they are split does not help much, but earth bleaching is often useful.

### Arsenic

A limit for arsenic content is found in many glycerine specifications. At one time it was troublesome because of use in the factories which produced the crude glycerines of sulphuric acid contaminated with arsenic; but there is now seldom any problem.

### Sodium Chlorate

Sodium chlorate is an unusual and unexpected, constituent of crude glycerines, but it has sometimes been found in significant quantities which have caused explosions during distillation in some types of still. It arises from the use of caustic soda made in electrolytic cells which are incorrectly designed or operated. If crude glycerine containing sodium chlorate has to be processed, the chlorate can be reduced under strongly acidic conditions, for example with spent pickling acid derived from steel works which contains ferrous chloride and hydrochloric acid.

### Silver Tests for Reducing Substances in Refined Glycerines

Most specifications for refined glycerines include a test for reducing substances, such as aldehydes, which is based on the darkening which occurs with ammonical silver nitrate. In some cases comparison is made with the darkening produced by a standard quantity of dextrose (glucose). Different silver tests vary in sensitivity, so that conditions laid down in a particular specification must be followed precisely. Substances such as soluble phosphates derived from some activated carbons, can interfere with certain tests.

### Fatty Acids and Esters in Refined Glycerines

The fatty acid and ester content of a refined glycerine is usually important; and most specifications include a test to measure the quantity of caustic soda reacted when the sample is boiled with an excess under standard conditions. For chemically pure grade this was known as soda absorption expressed in ml of N/10 NaOH absorbed in a standard test. Now it is commonly called saponification equivalent for all grades of refined glycerine and expressed as percentage $Na_2O$ absorbed, or as milli-equivalents per 100 gram. Fatty acids and esters in a distillate usually affect its odour and taste adversely.

### Chlorides in Refined Glycerines

Most specifications include a test for sodium chloride based on reaction with silver nitrate. This may be expressed as maximum parts per million permitted, or in terms that the turbidity produced in the test is nil, or less than in a control. The sensitivity of this test makes important the almost complete exclusion of entrainment of still liquor when soap lye crude containing a high percentage of salt is being distilled.

For specifications of glycerine Ref. BIS specification No. 12590–1988 and Glyceryl monostearate Ref. BIS specification No. 4236–1985.

### Other Tests

Apart from tests for colour, odour, sugars, lead and other heavy metals included in general specifications, some purchasers of glycerine impose tests specially relevant to the purpose for which they use the material.

## LOSSES

Losses are a natural part of any processing plant and control and reduction of the losses will, in part, determine the economic effectiveness of the plant's operation. All areas of the glycerine recovery plant can produce losses. In storage, fermentation can cause serious losses of glycerine due to the formation of trimethylene glycol, gases and acids during decomposition.

In the evaporation plant especially, glycerine can be carried out with the vapour during evaporation and lost in the vacuum system's condensing water. The amount of carryover can be minimised in several ways:

1. Entrainment separators, or 'catchalls' and mist eliminator pads can be installed where the vapour is removed form the vessel by the vacuum system; the recovered product can be returned to the process at an appropriate point.
2. Control the vapour velocity and maintain an adequate vapour space at the top of the vessel; it becomes important to control the liquid level to maintain the required vapour space.
3. Use a surface condenser to recover any additional condensable that escapes into the vacuum system.
4. The treatment process should be done carefully to control any tendency to foam.

### Other Sources of Glycerine Losses

Other sources of glycerine losses are as follows.

1. Glycerine is carried out with salt coming from soap lyes; proper washing of the salt is required to minimise this loss.
2. The filter cake and skimmings from the treatment plant.
3. Spent bleaching carbon contains glycerine of the highest concentration produced in the plant. It should be allowed to drain and then the columns should be cleared with compressed air or nitrogen to extract as much free glycerine as feasible. Then, the spent carbon can be added to the first treatment tank, where additional glycerine will be washed from the carbon. The carbon is separated in the filter press and disposed of with the rest of the filter cake.

Residue removed from the distillation process contains a certain percentage of glycerine; much of the glycerine can be recovered from this residue by a separate distillation step.

Glycerine leaks into heat exchangers.

There is spillage from all sources; careful plant design and operation are needed to minimise spills from overfilling tanks, equipment malfunctions and other operational errors.

Opening of equipment for maintenance can result in loss of product. Where possible, the liquid should be blown into the next process vessel or carefully drained into a bucket for recovery. Any product that is recovered that is not badly contaminated can be returned to a storage tank upstream in the process. If a 'substandard' glycerine tank is provided in the refining section, it can be equipped with an easily removable cover.

### Piping Considerations

In addition to the few suggestions on material selection given earlier, there are other issues that need to be considered when designing or evaluating the piping for a glycerine recovery plant. As in any piping system, the piping must be properly supported, be given correct slopes for proper flow and self-drainage, accommodate thermal expansion and be designed with adequate by-passes for equipment maintenance and proper operation control. In addition, consideration should be given for sampling points, drains and for blowout–wash-down connections. Fluid velocities should be maintained within standard design limits. Another area of concern may be the possible problem of electrolysis due to dissimilar materials used together in the piping system.

### Waste Management

The three primary waste streams from a glycerine recovery plant are skimmings and filter cake from the treatment plant, contaminants in the vacuum system condensing water and residue (foots) from the glycerine refining plant. Filter cake discharge is typically sent to a solid land fill. The concentrated residue from a foots still when allowed to cool will typically solidify and must be disposed of as required by the local environmental authority.

Containments that get into the vacuum system's condensing water supply are a double problem. This is glycerol and other condensables that are carried over from the vapour stream from the evaporation plant. First, the glycerol represents a loss of product. Second, these contaminants will foul the condensing water supply, which is typically recirculated through a cooling tower. A certain amount of the condensing water is generally bled away and replaced with makeup water to maintain an acceptable water quality level. An important alternative here is the inclusion of a closed-loop system that incorporates a surface condenser cooled with clean water recirculated from a cooling tower.

The surface condenser, usually a horizontal shell and tube heat exchanger, replaces the barometric condenser in the vacuum system and while using clean cooling water, allows any condensables to be recovered.

Ion exchange plants theoretically do not produce any waste streams, but in practice large amounts of acid and alkali waste-water are produced when the resin beds are regenerated. In addition, the resins themselves must be periodically replaced, thus causing a disposal problem of the exhausted resins.

## USES OF GLYCERINE

### Applications Related to its Key Properties

Its wide range of applications is in part related to a few of its key properties:

1. It is natural product, nontoxic and generally recognised as safe (GRAS) for human consumption.
2. It is an excellent humectant, emulsifier and plasticiser.
3. It is compatible with a wide variety of materials and mixes well.
4. It possesses antioxidant properties.

### Prominent or Interesting Uses of Glycerine

1. Adhesives, used for plasticising and penetrating properties.
2. Agriculture, used in sprays, dips and washes.
3. Antifreeze: properties include low freezing point and excellent compatibility.
4. Cleaners and polishes, used in a wide range of cleaners and polishes for the home and automotive markets.

5. Corrosion prevention, used in gums and resins for surface coatings for metals.
6. Cosmetics, widely used for cosmetic applications, acting as a bodying agent, emollient, humectant, lubricant and solvent; present in skin creams and lotions, shampoos and hair conditioners, soaps and detergent.
7. Dental creams, up to 50 per cent of typical dental creams, used as humectant, to ensure good dispersion and to serve as a vehicle for dyes and flavourings.
8. Explosives, a large amount is consumed in the manufacture of nitroglycerine-based explosives.
9. Food and beverages, used in a wide variety of applications; serves as a solvent, carrier, emulsifier, conditioner, freeze preventer and coating; used in wine, liqueurs, chewing gum base, confectioners and chewy bars; kosher producers exist.
10. Leather, used in tanning and finishing processes.
11. Metal processing, widely used for pickling, quenching, stripping, electroplating, galvanising and soldering.
12. Paper, acts as a humectant, plasticiser, softening agent and barrier against grease and solvents; used for greaseproof and glassine papers.
13. Pharmaceuticals, used in salves and dressings, antibiotic preparations, capsules and suppositories; used as a vasodilator for angina pectoris (nitroglycerine).
14. Photography, properties include wetting and plasticising.
15. Resins, includes ester gums, phthalic acid and maleic acid resins, polyurethanes and epoxies.
16. Textiles, facilitates printing and dyeing, lubricating and snag proofing; use for treatments for antistatic, antishrink, anticrease, waterproofing and flameproofing.
17. Tobacco, used as a humectant, softening agent and flavour retainer.

Chapter 12

# Oil Processing and Quality Assurance

## INTRODUCTION

Oil processing includes all aspects of production from unit operations (refining, bleaching, etc.), quality assurance and process control. Oil processing has changed considerably with the adoption of programmable controls in the early 1980s, evolving to almost fully computer-controlled processing.

This chapter discusses both processing and quality assurance from the traditional to the most current practice. The commonly used analytical methods are reviewed and the quality management organisation is discussed.

## OIL PROCESSING

Edible fats and oils may be derived from a number of plant sources (soyabean, cottonseed, sunflower, etc.) and from various animal sources (tallow, lard and fish). They may be differentiated by their source and by their degree of saturation, chain length of fatty acids and physical form. The relatively long-chain unsaturated seed oils in their natural form (e.g. soyabean oil) are generally grown in temperature climates and are liquid at room temperature. At the other extreme, relatively short-chain saturated oils in their natural form (e.g. palm oil) are grown in tropical climates and may be either solid or liquid at room temperature. These basic chemical and physical differences between the unsaturated and saturated oils bring about unique processing, product performance and quality assurance opportunities. Because of their widespread use, it is believed that soyabean oil and palm oil best represent both ends of the fats and oils spectrum.

Whatever the source, crude oil must be separated from its surroundings and refined further to remove free fatty acids (FFA) and other impurities that impact flavour, odour and appearance of the finished product. The most common steps of oil processing include (i) recovery of the crude oil; (ii) crude oil filtration and degumming; (iii) alkali or physical refining; (iv) bleaching; (v) hydrogenation; (vi) winterisation or fractionation; (vii) blending; and (viii) deodourisation. The oil must be properly handled throughout these processing steps. Numerous analyses are performed to ensure finished oil quality.

### Preparation of the Oilseed

Seed preparation is required before extracting the oil. The preparation steps include (i) sorting and grading the seed; (ii) cleaning to remove foreign materials; (iii) tempering or conditioning to obtain optimum moisture content; (iv) cracking and dehulling; (v) heating to plasticise the seed meats; and (vi) flaking to ensure rapid, through contact with the extraction solvent.

A recently developed alternative to flaking before oil extraction is the introduction of extrusion cookers, or expanders. These processes are applicable to the preparation of soyabean, cottonseed and rice bran oils. During extrusion or expansion, the oilseed is ground and heated with steam addition. The cooked meats are extruded into strands or ropes, then cut into collets or pellets. The pellets have a higher density than flakes and result in improved solvent contact that allows for higher yields of extracted oil and phospholipids and less solvent retention following solvent extraction.

### Expelled Oil

The earliest used oil extraction method was pressing. Hydraulic pressing was developed nearly two centuries ago. Since the hydraulic press was labour intensive and slow for processing large volumes of oilseed, continuous physical extraction methods were more recently developed, including the French screw press and Anderson expeller. These devices function effectively for oil sources with relatively high (30–50 per cent) oil content, such as corn germ and sunflower seed, but are not as effective for lower (20 per cent) oil content oilseed, such as soyabeans. Pressing or expelling may be used before and in conjunction with solvent extraction techniques.

### Solvent Extraction

Solvent extraction is routinely used to remove the oil from soyabeans, canola sunflower and other oilseeds. The original batch methods of solvent extraction has been replaced, for the most part, by continuous systems with much higher extraction efficiencies and capacities. Today, *n*-hexane is the most common solvent used. Alternative solvents that are less flammable than hexane have been studied with some degree of success and include methylene chloride, trichloroethane, trichlorotrifluoroethane, ethanol, isopropanol and mixed solvent systems.

Perhaps the most exciting alternative extraction solvent is supercritical carbon dioxide. The feasibility of supercritical extraction has been demonstrated commercially only with relatively small volume systems. The challenge, technically, will be to design continuous systems and extractors large enough to accommodate the large volumes of soyabeans and other oilseeds commonly handled.

### Crude Oil Filtration and Degumming

The fats and oils obtained from the expelling or extraction of oilseed are called 'crude' fats and oils. These crude oils contain relatively small amounts of non-triglyceride materials (such as partial glycerides, meal fines, etc.), that must be removed in a series of processing steps to make the oils edible and salable. For example, crude soyabean oil contains small amounts of FFA, colour bodies, phosphatides, flavours, odours, pesticides and trace metals, which are removed through a series of refining steps before the oil is in its finished form. Some of the trace materials, such as tocopherols, are considered desirable in the oil. The oil is processed in such a manner to maintain these components.

The purpose of filtration and degumming is to remove solids consisting mostly of residual meal, metal fragments, dirt and gums (hydrated phospholipids) from the crude oil. The filtered oil is later either alkali or physically refined. The gums may be recovered, dried and further processed into lecithin, a co-product that is widely used in food, in feed and as an industrial ingredients. Before filtration, the desolventised crude oil is transferred from an agitated bulk storage tank and slurried with diatomaceous earth (DE), commonly known as filter aid. The heated oil slurry (c. 80°C) is passed through either manual or automated self-cleaning filters that use the DE to remove the particulate matter.

The filtered oil is next transferred, batchwise, to a hydration tank where the oil is mixed with small amounts of softened water, depending on the phosphatide content of the crude oil. In the case of soyabean oil, the phosphatide content is approximately 2.5 per cent. The filtered oil is allowed sufficient time for the phosphatides to hydrate and then is passed to a degumming centrifuge, which separates the denser wet gums from the less dense oil. The degummed oil may be either vacuum dried and cooled to about 38°C for storage or passed directly on to the alkali refining step. At this point, phosphoric acid also may be added to increase the refining yield. While in storage, the degummed oil must be agitated to prevent separation of residual phosphatides. The degumming step may be entirely eliminated in some refineries. The phosphatides are totally removed along with soap in the alkali refining step. However, the degumming step is mandatory for successful physical refining.

An important co-product from the degumming of certain oils, particularly soyabean, corn, sunflower, rapeseed and peanut oils, is the lecithin. This material is produced by drying the gums, usually by a thin-film evaporator, to moisture levels less than 1 per cent. The lecithin may be sold as is, as a dark plastic material, or further processed or bleached to a lighter coloured fluid material. Bleaching is performed by treating the gums or dried lecithin with bleach, normally hydrogen peroxide, to lighten the colour. This is followed by addition of free fatty acid, vegetable oil and/or divalent metal salts to lower the viscosity. The resulting product has numerous food and industrial applications, including emulsification and release functions.

## Alkali Refining

Alkali or caustic refining of either crude or degummed oil refers to the reaction of dilute sodium or potassium hydroxide with the FFA, phosphatides and some proteinaceous and mucilagenous substances. The alkali solution reacts preferentially with FFA to form water soluble soaps. Residual phosphatides present in degummed oil or all the phosphatides of crude oil react with the caustic solution to form oil insoluble hydrates. The level of caustic solution added is optimised to prevent hydrolysis of triglycerides or neutral oil loss yet maintain acceptable oil quality. It is not possible to remove all the impurities in the alkali refining step. For example, with crude palm oil, which is relatively low in phosphatides, the main purpose of alkali refining is to remove FFA.

The caustic is prepared by dilution of sodium hydroxide with softened water to a specific gravity of about 17–18° Bé. The amount of alkali solution added to the oil is known as the 'treat', which can be measured based on the percent FFA in the degummed oil. The correct excess treat will enable the highest efficiency and allow for minimum neutral oil loss. The excess treat depends on the quality of the feedstock being alkali refined and must be constantly adjusted. A rule of thumb is to target the excess treat at 0.13 per cent above the quantity of NaOH or KOH required to neutralise the feedstock FFA and keep the neutral oil loss to a minimum. This is variable due to the quality of the crude oil feedstock.

The alkali solution is added slowly to the oil and allowed to mix thoroughly. The fatty acid soaps and phosphatides form a stable thick emulsion with the oil. This emulsion may be partially broken by heating the oil–soapstock –water mixture and then passing it through a refining centrifuge where the lighter density oil separated from the higher density soapstock–water portion. Factors that have an influence on separation efficiency include (i) differences in density between the phases; (ii) feedstock temperature with higher temperatures improving efficiency; (iii) viscosity differences; (iv) centrifugal force; and (v) centrifuge residence time.

A water-washing step is normally used to remove the remaining soap from the once-refined oil. The soap content of once-refined oil may be in the range of 200–600 ppm soap, decreasing to less than 50 ppm after water washing. The alkali refined oil, at a temperature of approximately 82°C, is water washed by mixing with 93°C softened water at a rate of about 15 per cent of the oil flow. Most of the water associates with the soap. The higher temperature of the softened water provides an enhanced separation due to the 'thermal shock'. In addition, phosphoric acid may be included in the wash-water to improve separation and help balance the pH. The water-washed alkali refined oil may be either vacuum dried prior to bleaching or sent directly to the bleaching process. Newer methods for soap removal involve adsorption onto hydrated silica, modified bleaching earths, or rice hull ash. Oil losses are reportedly reduced by the adsorption process.

The co-product of alkali refining is the soapstock, which may be used in poultry and livestock feeds or the fatty acids distillation industries. It may be sold to the end users as it is or acidulated with sulphuric acid to a neutral or slightly acidic pH. This process converts the soapstock into a two-phase FFA–water system, which requires further separation.

### Physical Refining

Physical or steam refining is especially efficient in removing FFA from low phosphatide content oils, including palm and coconut oils. It is critical that the phosphatide content be low before physical refining to prevent colour and flavour degradation. Oils, such as soyabean that have been degummed in conjunction with phosphoric acid to phosphatide levels of less than 20 ppm phosphorus (sometimes known as 'super-degummed' oil) may be effectively physically refined.

To remove the volatile FFA and carotenoid components, the low phosphorus oil is steam sparged under a 4–6 mm Hg vacuum. Due to the high acidity of the FFA, the reaction vessel must be constructed of corrosion-resistant stainless steel. The level of FFA is reduced in degummed soyabean oil from about 1.5 per cent to less than 0.05 per cent (typically 0.01–0.03 per cent) in the finished oil. The process is similar to deodourisation, except a longer residence time is used to allow maximum FFA removal.

A new process known as modified physical refining has been developed using special amorphous silica hydrogel. It has been reported that the hydrogel eliminates the need for the alkali-refining and water-washing steps. In this process, heated degummed oil is slurried with the silica powder and passed through a filter. The filtered oil possesses decreased levels of FFA, phosphatides and colour bodies. The oil is then vacuum deodourised. The method has been shown to be effective for both soyabean and palm oils. As with regular steam refining, the phosphorus content of the oil must be low. Acid treatment is required to remove non-hydratable phosphatides from the feedstock.

### Bleaching

The bleaching process removes colour bodies, residual soaps and phosphatides from alkali refined oil. The colour bodies are mainly chlorophyll, carotenoid and xanthophyll. The hot (105°C) vegetable oil is slurried, usually with acid activated bleaching earth or clay. The bleaching clay is generally calcium montmorillonite or natural hydrated aluminium silicate, commonly called bentonite. Non-acid activated bleaching clay (attapulgite), silicon dioxide and activated carbon may also be used as absorbents. The slurry is placed into a vacuum bleacher, which removes excess moisture and allows the absorption of the pigments, soap and phosphatides onto the clay surface. After a residence time of about 20 minutes the dried slurry is pumped from the bleacher to plate-and-frame or pressure leaf filters that have been precoated with DE to help separate the oil from the clay. Residual oil in the clay (up to about 35 per cent) may be partially removed and recovered by treatment with steam or nitrogen pressure. The bleached oil may be sent either to bleached oil storage, to hydrogenation, to a blending tank, or to deodourisation.

## Hydrogenation

Hydrogenation is the process by which a fat or oil is reacted chemically with a hydrogen gas in the presence of a catalyst. The hydrogen reacts at the unsaturated double bonds of the fatty acids. The hydrogen addition reduces the degree of unsaturation, which dramatically changes the chemical and physical attributes of the fat or oil. Chemically, hydrogenation increases the oxidative and thermal stability of fats and oils. Physically, hydrogenation converts liquid oils to semisolid or solid forms for a greater range of uses in such food applications as shortenings and margarines.

Both batch and continuous hydrogenation have been explored. Almost all edible oil hydrogenation is now batch processed. Batch hydrogenation is conducted in a pressurised vessel or convertor in the presence of nickel catalyst. Since the chemical reaction is highly exothermic, it is necessary to cool the reaction vessel during hydrogenation. Typical reaction temperatures ranging from 121 to 128°C and reaction pressures of about 0.7–4.2 kg/cm$^2$ are used. When the desired degree of hydrogenation is achieved, the oil is slurried with DE and passed through a filter that has been precoated with DE to separate the catalyst from the oil. The oil may also be treated with citric acid to help chelate and remove residual catalyst. Levels of residual nickel in edible oils are of particular concern in European countries.

The extent of hydrogenation is readily controlled. If soyabean oil is only slightly or 'brush' hydrogenated, the oil remains fluid at room temperature. Continued hydrogenation brings about further hardening of the oil. Complete hydrogenation causes the oil to become a hard and brittle stearine at room temperature. Preparation of a hydrogenated oil to exact chemical and physical characteristics can theoretically be achieved by selective hydrogenation using the proper conditions of temperature, time, pressure, type of catalyst and starting base stock. Selective hydrogenation can have a major effect on the content of *cis* and *trans* isomers produced in the final product.

## Blending

It is not always possible or practical to obtain specific physical and chemical properties of fats and oils by selective hydrogenation alone. It is usually more efficient to prepare standard hydrogenated base stocks that are blended proportionally to achieve the desired properties including solid fat index (SFI), fatty acid composition (FAC), melting point and flavour. The base stocks may be composed of different oils such as soyabean and palm, or combinations of the same oil type hydrogenated to varying degrees. For example, the addition of cottonseed oil to partially hydrogenated soyabean oil helps impart a finer crystalline structure when the blend is processed into shortenings. When hydrogenated cottonseed oil is used, it increases the melting point. In the blending process, which may be computer assisted, the base stocks are heated to above their melting points to be flowable, pumpable and well mixed before being deodourised.

## Winterisation–Fractionation

Winterisation refers to removing the high melting triglycerides or high melting solids that either are found naturally in the oil or are formed as a results of partial hydrogenation. In some products, such as corn oil, the solids may be high melting waxes.

Winterisation is achieved by slowly cooling and holding the oil for a specified period of time in a cooled vessel, usually at about 4°C. The temperature used depends on the desired use of the finished product. After the solids crystallise, the oil is winterised by passing it through a filter that may be coated with DE to separate the solids from the oil. Vacuum filters are also used. A winterised salad oil should remain clear when stored at refrigerator temperature, while a winterised liquid frying shortening should remain fluid at room temperature. Winterisation may be done in batch units or semicontinuously, depending on the type of oil. The semicontinuous process is normally preferred because it has greater throughput potential and less time restraints than does batch processing. An oil that will be latter hydrogenated is normally not winterised.

Similar to winterisation principles, palm oil may be fractionated to separate the lower melting components (olein) from the higher melting stearin. This may be accomplished by dry fractionation or, to a lesser extent, by detergent fractionation. In dry fractionation, palm oil is heated to 70°C for about 30 minutes under agitation and cooled gradually to about 20°C. The oil is then filtered, using vacuum filters, separating the solids from the liquid oil. Detergent fractionation uses heating and cooling steps as in dry fractionation but includes the addition of detergent to help improve separation efficiency of the solids from the liquid oil. Centrifuges are used in place of filters. Modern dry fractionation plants can recover approximately 75 per cent olein from unhydrogenated or lightly 'brush' hydrogenated oils. With detergent fraction of olein, the recovery is a comparable 80 per cent. Because of lower operating costs, dry fractionation is more commonly used. Palm oil fractionation can be done either batchwise or continuously, with the latter being more efficient.

## Deodourisation

Deodourisation is usually the last step in edible fats and oils processing. The purpose of deodourisation is to remove the remaining traces of volatile components, including flavour and odour compounds (mostly aldehydes and ketones), colour bodies, FFA, pesticides and herbicides, moisture and dissolved gases. With few exceptions, virtually all edible oils, whether non-hydrogenated or hydrogenated, pure or blended, are deodourised in a similar manner before reaching the consumer. The goal of deodourisation should be to obtain a nearly flavourless and odourless product with low FFA (normally less than 0.05 per cent) and light colour.

In deodourisation, steam is sparged into the deaerated oil. External heating plus the steam heats the oil to more than 200°C under high vacuum (< 10 mm Hg). As the oil passes through several trays or compartments with a specified residence time, the volatile components are stripped away and separately condensed. The deodourised oil is then cooled as its leaves the deodouriser. Salad oils are cooled to about 38°C and hydrogenated oils are cooled to about 5°C above their melting points to remain fluid. Deodourisation may be done batchwise, semicontinuously, depending on volume demands and equipment. Deodourisers are made more energy efficient by reducing the amount of steam used by utilising thin-film design to provide better steam–oil contact.

Chelating agents, including phosphoric acid citric acid and lecithin, may be added to deodourised oil to deactivate trace metals. Depending on the end product, stability may be enhanced not only by using proper deodourisation conditions but also by adding antioxidants, including TBHQ, BHA, BHT, ascorbyl palmitate, natural extracts, or tocopherols. Nitrogen blanketing may also be used to protect the heated oil from oxidation during storage.

An important co-product of deodourisation is the deodouriser distillate. The distillate is the condensate of volatile components removed while deodourising oil. The distillate may be further separated into tocopherols for vitamin E and sterols for pharmaceutical and cosmetic applications.

## Tank Storage

Vegetable oils are kept in storage tanks between each step of processing as well as when finished. A storage tank facility, commonly known as a 'tank farm', may be quite larger and complex. Ideally, a tank farm should be enclosed if located in a temperate or cold climate. If this is not possible, oil and utility lines and tanks must be heat traced. Proper drainage allows for adequate cleaning. A tank farm should also be surrounded by retaining walls that will keep the contents of a spilled storage tank confined to the premises. The tanks must be connected so as to minimise potential cross-contamination between products.

Regardless of the particular oil being stored, there are common guidelines to follow. Storage tanks, preferably vertical, should be equipped with agitators and/or circulators to allow mixing and blending. This eliminates the problem of stratification, especially during storage of crude, degummed and hydrogenated oils. Agitation is essential for blending operations. To maximise oxidative stability, properly stored oil should have minimal exposure to heat, air, light, moisture and heavy metals. Ideally, vegetable oils, either hydrogenated or non-hydrogenated and co-products should be kept warm enough only to ensure they are flowable and pumpable. The lower storage temperatures minimise heat degradation and save energy. Exposure of oil to air may be avoided by storing oils under high purity (> 99 per cent) nitrogen in airtight tanks. Exposure to light can be minimised by keeping tank lids covered and by using opaque tanks. Moisture buildup, sometimes a problem in cold weather due to condensation, is minimised by putting low moisture materials into storage under climate-controlled surroundings and by having a rapid inventory turnover rate. Heavy metals, including iron and especially copper, in the presence of heat, air and light can result in rapid deterioration of oil quality and must be avoided. If possible, only stainless steel and plastic materials should be used around oil products.

### Process Control

Modern oil processing plants run efficiently and operate cost effectively. Good process control is essential. It is important to monitor inventory and operating conditions in each refining step, preferably with automated equipment and instrumentation. The use of electronic equipment such as properly calibrated flow meters and tank load cell ensures accurate quantity measurement of oils and auxiliary ingredients. Exact measurements are needed to determine realistic and precise yield and loss calculations.

Computers are increasingly being used to control oil processing plants. Conditions such as pressure, temperature, agitation and flow rates may be continuously monitored and controlled by the computer connected to such devices as pneumatic valves and temperature sensors. The computer can be programmed to respond immediately to critical control points as well as use feed-forward and feedback loops in conjunction with strategically placed electronic sensors throughout the process. Thus, the guess work is taken out of oil processing, allowing the modern refinery to turn more consistently and with better quality.

## QUALITY ASSURANCE

Quality encompasses all aspects of production from finished product specifications including purchasing, sales and service. To the user, product quality means the product supplied meets or exceeds specifications. Today, process control reduces the amount of process inspection. The term partnering between the supplier (oil process) and the customer (user) is often used to describe today's quality environment.

The newer approaches to quality have moved away from inspection to a step-by-step process control. The processes are audited to ensure they remain within established guidelines. Continual improvement of the system through training and implementation is essential to total quality management (TQM). Newer aspects of quality management include select supplier programmes, ISO 9000, statistical process control (SPC), good manufacturing practices (GMP) and total quality management (TQM).

Most quality management programmes are somewhat obvious and have been in place. The ISO 9000 programmes, from the European Economic Community, are being implemented through the International Standards Organisation (ISO). Five international standards (ISO 9000, ISO 9001, ISO 9002, ISO 9003 and ISO 9004) on quality management and quality assurance have been published. The ISO 9002 is most adaptable to oils refining and address preventing, determining and correcting problems during production.

## Oil Refinery Quality Team Management

The TQM initiatives have not been robust. With oils, the cost of poor quality is hidden sometimes by reblending of out-of-spec products, commodity competitiveness and focus on operations.

The oil refining industry has expertise in most of the tools of TQM, including flow charts, Pareto diagrams, histograms, run charts, control charts and process capability charts. All are part of statistical process control (SPC) and enhance problem-solving capabilities.

Quality teams are integral to TQM. These teams may represent sales, marketing, quality assurance, technical service, maintenance, or sanitation. The nature of teams and their composition are continually evolving.

## Quality Assurance in the Oil Refinery

One of the accepted facts of edible oil refining is that crude oil based stock quality fluctuates. This fluctuation can cause process upsets and product quality deviations. The role of quality control is to monitor regularly the material being produced at each processing step by performing routine control tests for each oil type. The values are compared with target values for each process step and from these data, operations are able to adjust the process appropriately to meet the product specifications. The ultimate goal is to achieve an end product within the correct chemical and physical parameters desired by the end user and to be of consistent quality. Other concerns of the quality assurance department of the oil refinery include making blend calculations, conducting oil flavour panels, monitoring waste-water, tracking statistical process control and record keeping, implementing plant good manufacturing practices (GMP) and safety rules and addressing customer complaints.

## Quality Analysis for Oil Products

Numerous analytical tests are routinely performed in an oil refinery laboratory (Table 12.1). Besides requiring trained chemists and laboratory technicians, special methods and equipment are required to measure the oil quality properly. Over the years, numerous methods have been developed by professional organisations, including the American Oil Chemists' Society (AOCS), The Association of Official Analytical Chemists (AOAC), and the American Society for Testing Materials (ASTM) as well as by much collaboration by private companies, universities and laboratories. The following summarises the significance of each analytical test and the general procedures and equipment required. Each is typically found in an edible oil quality control laboratory.

Colour of vegetable oil and co-products is contributed by colour bodies or pigments, including carotenoids and chlorophylls. Colour measurement is generally done using a colour comparison techniques. The Lovibond colorimeter, either manually or automatically operated, is the most commonly accepted analytical instrument for colour determination. Lovibond colour is an indication of the red and yellow colour components (carotenoids and chlorophylls, respectively) of oil. A Gardner colorimeter is more suitable for darker substances, such as fluidised lecithin and measures colour intensity on a scale of 1 (light) to 18 (dark).

**Table 12.1.** Routine quality control tests for soyabean oil.

| *Determination* | *AOCS method* |
|---|---|
| Lovibond colour | Cc 13e–92 |
| Karl Fischer moisture | Ca 2e–84 |
| Oven moisture (moisture and volatile matter air oven method) | Ca 2c–25 |
| Fatty fatty acid | Ca 5a–40 |
| Soap in oil (titrimetric method) | Cc 17–79 |
| Peroxide value (acetic acid–chloroform method) | Cd 8–53 |
| Solids (sediment; insoluble impurities) | Ca 3a–46 |
| Phosphorus content in oil | Ca 12–55 |
| Phospholipids in oil (nephelos) | Ca 19–86 |
| Iodine value | Cd 1–25 |
| Refractive index | Cc 7–25 |
| Capillary melting point | Cc 1–25 |
| Dropping point | Cc 18–80 |
| Solid fat index | Cd 10–57 |
| Solid fat content (by NMR) | Cd 16b–93 |
| Active oxygen method | Cd 12–57 |
| Oxidative stability (by OSI) | Cd 12b–92 |
| Fatty acid composition (by GLC) | Ce 1–62 |
| Metals by atomic absorption | Ca 15–75, Ca 18–79 |
| Flavour–odour | Cg 1–83 |
| Metals (by ICP) | AOAC 985.01[a] |

[a] Association of Official Analytical Chemists.

Moisture measurement may be done by Karl Fischer moisture analysis for low moisture (< 5 per cent) oils and lecithins. For higher moisture materials such as wet gums and soapstock, a vacuum oven or moisture balance is more applicable.

The FFA analysis indicates the efficiency of removing free fatty acids in the alkali refining and deodourisation steps. In use, the higher the FFA in a frying oil, the lower the smoke point of the oil. The FFA is determined via titration with an acid-base indicator.

Soap analysis indicates the efficiency of alkali refining and water washing. Soap is measured by titrating to a specified end point.

Peroxide value measure the extent of primary oil oxidation (hydroperoxides). It may also indicate the amount of residual bleaching material (normally hydrogen peroxide) left in lecithin. The oil is reacted with iodine for a set period of time and then titrated with sodium thiosulphate, in the presence of starch, to an end point colour.

Phosphorus (phosphatide) measurement indicates the efficiency of the degumming and alkali refining steps in removing gums. Phosphorus content is determined by ashing the sample and colorimetrically measuring the phosphorus by UV spectrophotometry against known standards. Direct injection using plasma emission spectrophotometry (ICP or DCP, inductively or direct coupled plasma) can also be used, giving faster and more accurate feedback, normally in about 15 minutes. Alternately, phosphorus can be estimated by measuring turbidity of an oil–acetone mixture using a ratio turbidimeter. The turbidity is correlated to standards of known phosphorus levels, providing rapid in-process results. This nephelometric technique is most suitable for crude, degummed and once refined oils and has been shown to be effective for soya and corn oils. It is not effective for phosphorus below 2.0 ppm.

Iodine value is a measure of oil saturation, defined as the amount of iodine (in grams) required to react with 100 g of oil. Oils with higher iodine values are less saturated than oils with lower iodine values. The analysis is done by titrating with WIJS solution to a specified end point. Iodine value can be quickly estimated by measuring an oil's refractive index, using a refractometer and correlating the refractive index to standards of known iodine value. Refractive index (RI) is a variable that depends on temperature. Care must be taken only to compare oils at the same temperature.

Melting point depends on the chain length and degree of saturation of the fatty acids in the oil. Melting points may be performed using open-ended or close-ended capillary tubes placed in a water bath. Historically, the Wiley melting point was used. Because these methods are not adaptable to automation, melting point determinations have been largely replaced by the so-called dropping point determination. The dropping point apparatus uses an automated instrument that measures the minimum temperature at which the oil flows through a defined orifice (dropping point). This value correlates with the Wiley melting point.

Solid fat index (SFI) or solid fat content (SFC) is used in applications where fat plasticity is required, such as with margarines and shortenings. Both are apparent indicators of the solid fat present over a defined temperature scale. Solid fat index is measured by dilatometry and solid fat content is measured by pulsed nuclear magnetic resonance (NMR). The SFI requires a series of temperatures-controlled baths at 10, 21.1, 26.7, 33.3 and 40°C and glass dilatometers for determining volume of the sample at each temperature. Alternatively, the SFC solids may be measured directly by a fully automated NMR.

Oxidative stability can be predicted by the oil stability instrument (OSI), which is a computerised version of the active oxygen method (AOM). Oil samples are subjected to high temperature and air exposure and the resulting volatiles are continuously monitored over a period of time. The older AOM and Rancimat methods have, for the most part, been replaced by OSI.

Fatty acid composition is used for oil identification and preparation of specialised oil blends. It is measured by analysing the methyl esters of the fatty acids by gas chromatography.

Trace metals in oils can acts as catalysts to promote oxidation under certain conditions. Metal analyses may be accomplished by ashing the oil sample and measuring metals content by atomic absorption or measuring the metals in the oil directly by ICP.

Oil flavour and odour are due mainly to volatile aldehydes caused by partial oxidation. Flavour analyses can be accomplished in several ways. In practice, the trained laboratory analyst may informally taste oils on a pass–fail basis for routine process samples. Taste panels made up of trained panelists in a round table format or at separate flavouring booths also may be used on a formal basis for finished oils. The flavour analyst grades the sample against descriptors of known standards and assigns an numerical score on a scale of 1 (least acceptable) to 10 (most acceptable). An acceptable oil is given a score of 7 or higher and may have such satisfactory descriptors as nutty, bland and buttery. Oils with poor flavour with a score less than 7 have less desirable descriptors such as rancid, burnt, painty and fishy. New methods have been studied as alternatives to evaluate flavour including head space analysis by gas chromatography and electronic sniffers.

### Crude Oil Quality

It is important that only good quality crude oil is put into the process initially. Major analyses include neutral oil loss, Lovibond colour, moisture, FFA, solids, phosphorus and iodine value. The oxidative stability can be predicted by measuring peroxide value. Certain analytical results are used to defined the treatment for the next processing step. The phosphorus content of crude oil determines the level of hydration necessary for proper degumming.

### Degummed Oil Quality

Degummed oil is normally checked for moisture, phosphorus, FFA and colour. The FFA level of the degummed oil will determine the amount of caustic treat used in the alkali refining step.

### Alkali Refining –Water Washing Oil Quality

Once-refined oil that has been treated with alkali is usually analysed for FFA, phosphorus, moisture, soap and colour. Compared with degummed oil, once refined oil has reduced FFA and phosphorus and increased moisture and soap levels. The soap content decreases from 200–600 ppm before water washing to less than 50 ppm following water washing. Reduced soap content is necessary for proper bleaching to occur.

### Bleached Oil Quality

Bleached oil analyses may include colour, peroxide value, soap, phosphorus, moisture, FFA and clarity. A significant reduction in colour is expected following bleaching. A visual clarity check or filtration check ensures the absence of bleaching clay in the oil. Peroxide value may increase, since bleached oil is more vulnerable to oxidation.

### Hydrogenated Oil Quality

Analytical tests that may be conducted on hydrogenated oils include solid fat index, iodine value or refractive index (before, during and after the hydrogenation reaction), peroxide value, solids content, melting or dropping point and fatty acid composition. These tests establish the characteristics of the particular base stock and the information on solids is used when calculating blend formulations.

### Deodourised Oil Quality

Finished oils have the following analyses performed: flavour, colour, moisture, peroxide value FFA and oxidative stability. It is assumed that such necessary information as iodine value, solid fat content, fatty acid composition and drop point is obtained from the base stock before deodourisation.

### Co-products Quality

If lecithin is manufactured in the process, the following analyses are normally conducted: acetone insoluble (AI), hexane insolubles (HI), acid value (AV), Brookfield viscosity and phospholipid composition by high performance liquid chromatography (Table 12.2). Deodouriser distillate and soapstock are typically analysed for total fatty acid (TFA). The analytical tests for lecithin, deodouriser distillate and soapstock are all either official methods or recommended practices of the American Oil Chemists' Society.

**Table 12.2.** Routine quality control tests for co-products and waste-water.

| *Determination* | *AOCS method* |
|---|---|
| Acetone insolubles | Ja 4–46 |
| Hexane insolubles | Ja 3–87 |
| Gardner colour | Ja 9–87 |
| Karl Fisher moisture | Ja 2b–87 |
| Acid value | Ja 6–55 |
| Brookfield viscosity | Ja 10–87 |
| Peroxide value | Ja 8–87 |
| Phospholipid composition (by HPLC) | Ja 7b–91 |
| Total fatty acid | G 3–52 |
| pH | G 7–56 |
| Biological oxygen demand | AOAC 973.44 |
| Chemical oxygen demand | AOAC 973.46 |

### Waste-water Analyses

A growing environment concern for vegetable oil refineries is the quality of waste-water effluent. Oil processing plants use and manufacture materials that may be potentially detrimental if released into the environment. Part of the quality assurance function is to regularly monitor waste-water samples for compliance to local, state and federal regulations. Typical analytical tests include biological oxygen demand (BOD), chemical oxygen demand (COD) and pH.

## STATISTICAL PROCESS CONTROL (SPC)

With the widespread development of personal computers and availability of software packages, it has become relatively simple for quality control and operations to monitor quality trends for each oil processing step (Table 12.3). Past and present data may be charted and used to predict future product quality.

**Table 12.3.** Quality control tests for each soyabean oil processing step.

| *Processing step* | *Major tests commonly done* |
|---|---|
| Crude oil | Colour, moisture, FFA, Solids IV, phosphorus, neutral oil loss |
| Degummed oil | Moisture, phosphorus, FFA, colour |
| Alkali refined oil | FFA, phosphorus, moisture, soap, colour |
| Bleached oil | Colour, PV, soap, phosphorus, FFA, moisture, clarity |
| Hydrogenated oil | IV (or RI), PV, solids, FAC, melting (or dropping) point |
| Blended oils | IV (or RI), PV, solids, FAC, melting (or dropping) point |
| Deodourised oil | Flavour, colour, moisture, PV, FFA, OSI |
| Lecithin | AI, colour, moisture, viscosity, HI, viscosity, phospholipid composition |
| Soapstock | TFA |
| Deodouriser distillate | TFA, tocopherols |
| Waste-water | BOD, COD, pH |

With the ability to tie both quality control equipment and electronic process sensors into one central computer, a processing and quality record may be kept and updated continuously. The data can be used by the computer to control processing parameters. SPC also helps the laboratory to ensure complete product compliance with the end user. Customer complaints, rejected shipments and rework material are minimised. The oil refinery functions more efficiently with reduced operating expenses.

## QUALITY CONTROL RECORD KEEPING

The life blood of a good quality control programme is adequate record keeping. Whether the analytical results are kept physically in a notebook or electronically sorted by computer, proper records must be kept and reviewed on a regular basis. Quality control needs to communicate information promptly to operations and this can be expedited using the computer. Important information includes date, time of day when the sample was taken, batch or lot number, analytical results and any special comments or instructions such as whether a resample was obtained or if a parameter was out of specification. Proper quality control data are invaluable for timely resolution in a recall situation. As mentioned, electronically stored data can be used in conjunction with statistical process control for trending purposes.

## SAMPLING

It is necessary to obtain representative samples of base stocks, auxiliary ingredients, in-process material and finished products. This may be achieved in a number of ways, including (i) samples taken directly from the process steam in continuous systems; and (ii) dip, bomb, or tryer samples taken from materials made batchwise or from finished products. For better representation, it is safest to take several dip samples from different parts of the storage vessel and composite them. Bomb and tryer sampling devices are designed to take a more representative cross-section so compositing of samples may not be necessary. In all cases, it is most critical that the material be homogeneous before sampling.

### Safety and GMP

In a vegetable oil processing plant, good manufacturing practices (GMP) and safety go hand in hand. Part of the responsibility of quality assurance may be to make certain that GMP and safety policies are adhered to by all employees in the plant. Areas of concern may include such topics as oil spill cleanup procedures, entry procedures, lockout of equipment during maintenance, etc. Experience has proven that a clean and safe plant will operate more effectively than one that has poor GMP and is unsafe.

## FUTURE OF PROCESSING AND QUALITY ASSURANCE

The basic unit operations in processing of food oils (refining, bleaching, etc.), have not changed significantly. Improvement in equipment design has occurred. Most noteworthy has been the adoption of in-plant computerised process controls. More sophisticated computerised product analysis is being utilised in the laboratory. In an oil refinery, almost all aspects of oil production may now be visualised in a series of process control computer screens. Product in-process, inventories, shipping and laboratory analysis are in the spread sheets. Laboratory quality assurance has remained with the process control and inspection function. The future for quality assurance will evolve with total quality management.

Chapter 13

# Storage, Stability, Packaging and Transport of Fats and Oils

## INTRODUCTION

Damage to oils during storage and the resulting loss in quality and yield, can occur due to deterioration—adverse chemical changes due to the intrinsic nature of oils; contamination—from natural substances associated with the source of the oil, traces of pesticides and that introduced by oil processing or in transit; and adulteration—deliberate admixture of one oil with another.

Hydrolysis, the reaction of fats and oils with water and oxidation, the chemical reaction in which oxygen combines with another substance with the liberation of heat, are the two basic processes that result in the deterioration of oils and fats. Oxidation is responsible for much more of the deterioration of fats and oils than hydrolysis.

Moisture promotes the splitting of triglycerides to form free fatty acids that can increase refining losses directly related to the free fatty acid of oils and fats. Essentially, hydrolysis is the reverse of making a fat molecule. The deterioration of a fat or oil in the presence of oxygen is termed oxidative rancidity. The initial step in the oxidation of an oil or fat is the addition of oxygen at or near the double bond of a fatty acid chain to form unstable compounds generally designated as peroxides.

When the oxidation of an oil is followed experimentally, either by measuring the amount of oxygen absorbed or determining the peroxide value, the course of oxidation is defined by the oxygen absorbed in the oil. During the initial or induction phase, oxidation proceeds at a relatively slow and more or less uniform rate. Peroxides are formed during this period at a much faster rate than they are destroyed so that their content increases more or less parallel with the oxygen absorption. Thus, after a certain critical amount of oxidation has occurred, the reaction enters a second phase, characterised by a rapidly accelerating rate of oxidation. The point at which the sample begins to smell and taste rancid coincides with the beginning or early part of the second phase. As oil oxidation continues, with time, the peroxides that are formed decompose to generate volatile and nonvolatile compounds that contribute to flavour and odour deterioration of oils and fats. The extreme stages of oxidation, polymerisation and degradation are accompanied by rapid increase in the viscosity of the oil. There is considerable variation among different fats in the manner in which their oxidation and accompanying flavour deterioration proceeds. The amount of oxygen that must be absorbed to produce rancidity will vary according to: (i) composition of the oil; (ii) presence or absence of antioxidants and prooxidants; and (iii) conditions of oxidation.

Generally, oxygen absorbed will amount to about 15–150 per cent of the oil by volume or 0.02–0.20 per cent by weight. Fats high in oleic acid and low in linoleic acid will become rancid after the absorption of less oxygen than fats in which the ratio is reversed.

The relative oxidation rates of pure fatty acid esters are shown in Table 13.1. This table is based on a common factor of 1 that has been arbitrarily assigned for the oxidation rate of oleic acid.

**Table 13.1.** Relative reaction rate of fatty acids with oxygen.

| *Fatty acid* | *Approximate relative oxidation rate* |
|---|---|
| Oleic acid (C18:1) | 1 |
| Linoleic acid (C18:2) | 10 |
| Linolenic acid (C18:3) | 25 |

Oxidation has an important effect on the colour of fats and oils. While oxidation bleaches the carotenoid pigments, it tends to develop the colour of other types of colouring material and in certain cases, even develops coloured compounds of quinoid nature from the fatty acids of glycerides of the oils. Cottonseed and, in certain cases, soyabean oil are subject to marked darkening upon oxidation.

## CONTAMINATION OF FATS AND OILS

Contamination begins in the fields and is possible at all stages right to the consumer's plate. The sources of contamination are several and some are to be found before the primary products are processed. All agricultural products are likely to contain evidence of the environment from which they come; dirt and dust, fragments of vegetable matter, insect parts and so on. Other insoluble impurities appearing in crude oils will come as precipitates from the oils themselves. Most of the farm rubbish and the materials from the original plant and from the oil itself will have been removed before the oil is shipped, but some may carry over and some extra precipitation from the oil is possible.

Pesticide residues are the most common agricultural chemicals that contaminate oils and are almost inevitable. Although edible oils are secondary products derived from the processing of crops, there will be some carryover of pesticide residues into the oil.

Some oilseeds are dried in the field with hot gases from oil burners. Copra is often dried in ovens or kilns fired by coconut shells. In each case contamination with diesel fuel fumes and/or polyaromatic hydrocarbons is inevitable.

So long as uncoated mild steel tanks are accepted, there is a danger of mechanical contamination by rust flakes or powder or of the formation of iron salts by interaction with fatty acids in acid oils. Accordingly, such tanks are often coated with inert materials (plastics for food contact). The dangers of metal contamination can be eliminated altogether by using stainless steel.

### Flavour, Odour and Stability

Many fats and oils develop off-flavours during storage, which has been called reversion, that is, oil flavour reverting to the original flavour of the crude oil. This is more accurately called retrogression since the flavour retrogrades from a better to a worse condition. Also, aged oils do not develop off-flavours identical with that of the crude oils, but for any given type of oil, it is always specific and characteristic. While all oils develop off-flavours, some do so at a faster rate than others. Soyabean and marine oils develop 'fishy' flavours, lard and tallow develop 'animal' flavours, rapeseed and linseed oils develop 'grassy/painty' flavours and palm oil develops 'painty/rancid' flavours. All of the efforts to prevent the development of off-flavours in oils are aimed at slowing down the rate at which they develop. Multivalent metals, such as iron or copper, act as prooxidants when present with triglycerides. Recent research has indicated that citric acid addition to canola oil is important for flavour stability in the oil.

### Additives and Stability

Perhaps the most important additive to edible oils to enhance stability during storage and transport is citric acid or a similar metal chelating agent. Traces of metals in a reactive form can act as catalysts that promote the autoxidation of fats and oils. Concentrations of iron as low as 0.1 ppm or copper as low as 0.01 ppm can accelerate the deterioration of soyabean oil. If an oil has higher concentration of metals, it is necessary to first lower the content by a vigorous acid wash of the oil, for example, with very dilute phosphoric acid or by use of activated bleaching earth to absorb metals and metal complexes from the oil. A small amount of citric acid (i.e. 0.05 per cent w/w citric acid/oil) can be added to enhance the effectiveness of a bleaching earth. Some other acids that have been used to chelate metals are ascorbic, phosphoric, tartaric and ethylenediamine tetraacetic acid. It should be noted that citric acid does not improve the stability of olive oil.

Antioxidants effectively improve the stability of fats and oils with lower levels of natural antioxidants, such as tocopherols and with lower levels of polyunsaturation. With oils that have a high content of polyunsaturated fatty acids with significant levels of tocopherols, such as soyabean oil, antioxidants minimise the accumulation of peroxides but do not improve the flavour stability. The desired qualities of an antioxidant used to stabilise edible fats and oils are: (i) use in a product is safe; (ii) contributes no odour, flavour or colour to the product in which it is used; (iii) effective in the product at low concentrations; (iv) easily incorporated into the product; (v) retained after cooking processes such as baking and frying; and (vi) available at low cost for the application. Lauric, olive and palm oils contain only small amounts of tocopherols compared to soyabean, corn and cottonseed oils. Synthetic antioxidants were shown to be effective in the stabilisation of virgin olive oil. Butylated hydroxyanisole (BHA), butylated hydroxytoluene (BHT) and tertiary butyl-hydroquinone (TBHQ) all improved the oxidative stability of the olive oil during storage at 50°C (120°F) in the dark but were not effective in preventing photooxidation of olive oil. It should be noted that addition of antioxidants to oils, which are already deteriorated, will not improve the quality or stability of the oil.

## STORAGE AND HANDLING

Bulk storage and handling facilities are needed by (i) extraction plants and refineries; (ii) industrial users of fats and oils; and (iii) tank farm operations.

### Crude Fats and Oils

Crude fats and oils may contain significant amounts of non-triglyceride materials that should be removed prior to bulk storage, handling and transport. These include gums, phospholipids, fines, moisture, seed fragments, meal and proteinaceous material. Seed fragments and cell tissue contain lipases that, in the presence of moisture, hydrolyse triglycerides into free fatty acids. Also, the combination of moisture, seed fragments and mucilaginous material provides an ideal system for microbial growth that can lead to putrefaction and development of objectionable odours. These factors, coupled with the fact that crude fats and oils are usually stored at temperatures conducive for enzymatic and microbial growth, emphasise the importance of removing non-triglyceride materials that may be subject to microbial attack before storage.

### *Removal of fat-soluble materials*

Phospholipids (1–3 per cent) represent the bulk of the fat-soluble materials in most crude vegetable oils and are particularly troublesome with soya oil and beans. Soyabean oil is the major commercial source of lecithin. Removal of phospholipids is desirable because crude oil stored for extended periods is apt to deposit sludge in the bottoms of storage tanks, barges or rail tank cars. Even when an oil is degummed it should be filtered prior to storage.

### *Storage in tank farms*

Storage of crude fats and oils is necessary to maintain sufficient quantities of oil to supply the refining operation. The objectives of such a bulk storage system are to maintain the quality of the incoming crude oil and to supply a uniform product to the refining operation. Key features of bulk storage system include: (i) storage tanks and piping for each type of crude oil to be stored and handled; (ii) an inventory control system that allows an accurate measure of crude oil coming into the refinery and to the refining operation; and (iii) a spill prevention, containment and control system.

## Quality Deterioration during Storage

The most serious types of quality deterioration occurring in bulk storage of crude fats and oils include comingling, contamination, colour increases and chemical changes.

### *Comingling*

Comingling occurs through the unwanted mixing of different types of crude oils and can occur either through human error, faulty equipment design, or both. Comingling can be prevented by separate tanks and piping systems for each major types of crude oil handled. Also, the use of large storage tanks rather than small ones will minimise the chance for comingling.

### *Contamination*

Fats and oils with appreciable amounts of phosphatides and other easily hydratable materials, such as from soyabean, cottonseed, corn, rapeseed, etc. can be contaminated with moisture. Conditions of high humidity, warm oil and cool temperatures can produce sweating within crude oil storage tanks. In effect, water dripping into the oil causes an *in situ* degumming of the oil with an accumulation of sludge deposits on the bottom of the tank. The effect can be lessened by employing vent systems that restrict the flow of air in and out of the tank.

In the course of long-term storage, crude oils may be contaminated by oxidised or hardened oils that eventually coat the tank walls. This can be prevented by periodic cleaning of the tanks.

### *Chemical changes*

Primary consideration during crude oil storage and handling include preventing refinery losses and oxidation of the crude oil.

Storage of crude oil in the presence of moisture promotes hydrolysis, which in turn leads to increase in free fatty acids and lower yields of refined oil. Hydrolysis can be best prevented by ensuring that tanks are completely dry prior to pumping crude oil into storage and by keeping the oil as cool as possible throughout all handling operations.

Oxidation of crude oil may be prevented by excluding oxygen and air from the oil. One way is to fill storage tanks from the bottom, thereby minimising exposure to the air as it enters the tank. A second but more expensive way to preserve crude oil quality is to blanket storage tanks with an inert gas such as nitrogen. However, inert gas blanketing is generally used more in conjunction with finished oil storage and handling.

Trace metals, in particular iron and copper, accelerate the oxidative deterioration of fats and oils. Crude oil should not be in contact with copper or brass valves. One brass valve in a soyabean oil refinery can lower the flavour stability of deodourised products.

### *Colour fixation—Increases*

Oxidation of crude fats and oils can also lead to *colour fixation*. The most notable examples of colour fixation are in cottonseed and palm oils, where oxidation of gossypol and carotenoid pigments are involved.

Although not particularly sensitive to colour fixation, the lecithin in crude soyabean oil may darken excessively if the oil needs to be melted prior to unloading from rail tank cars or trucks. The use of high-pressure steam on heating coils should be avoided. Instead, low-pressure steam or, better yet, hot water should be used for this purpose.

## Equipment for Storage and Handling of Crude Fats and Oils

### *Tanks*

Crude oil storage tanks are usually constructed of welded carbon steel. Their sizes vary throughout the industry depending on the refining capacity of the particular plant and the availability of incoming crude oil.

Crude oil storage tanks may be located in the open or enclosed. If located outside, they should be set on concrete pads to protect them from surface water that eventually can cause rust and corrosion. Most vegetable oils will not usually freeze in cold weather when held in large storage tanks. However, because storage tanks may be used for harder oils with higher melting points, the tanks should be equipped with heating coils to facilitate melting of bulk stored fats and oils when required.

Successful refining is, in part, dependent on the supply of uniform crude fats and oils to the operation. Thus, storage tanks for crude fats and oils should be equipped with agitators to allow blending of crude oils and to assure uniformity.

## Unit Processing Operations

The purpose of this section is to outline the basic unit processing operations required to convert crude fats and oils into finished products such as cooking-salad oils, margarine and shortening base stocks. Edible fats and oils are handled within the unit processing operation by pumping them through pipes. All edible fat and oil processing steps consist of independent operations that are usually not connected with each other by continuous flow. Between each processing step, there may be one or more storage tanks. Generally, in each processing step the oil is heated to process temperature, held for reaction and cooled to preserve oil quality. Between processing steps, the oil is allowed to cool to a minimum or ambient temperature while being maintained in a fluid state to facilitate material handling by pumping.

The basic processing steps consist of (i) crude oil storage; (ii) degumming and/or caustic refining; (iii) bleaching; (iv) hydrogenation; (v) fractionation and/or winterisation; (vi) deodourisation; and (vii) finished oil storage. Each step removes specific components from the oil. These materials may be classified as:

1. *Naturally occurring*: gums, phospholipids, prooxidant metals, pigments, colour bodies, tocopherols and free fatty acids.

2. *New compounds formed during processing:* soaps, oxidation products, hydroperoxides, polymers and their decomposition products; colour bodies, isomers and high-melting triglycerides.
3. *Processing adjuncts:* hydrogenation catalysts, bleaching clay, metal scavengers that precipitate from finished oils, i.e. citric acid.
4. *Contaminants introduced by processing*: moisture, trace metals, carbonaceous materials and oil-insoluble materials.

To produce an edible fat (an oil having the desired colour, flavour and oxidative stability and functional properties), naturally occurring compounds must be removed as efficiently as possible consistent with preventing the least damage to the naturally occurring antioxidants and neutral oil fraction. Formation of new compounds, precipitated processing adjuncts and contaminants are all undesirable. The goal of handling and processing within each operation is to prevent and minimise their formation or to remove them from the oil.

A summary of typical handling operations and storage conditions for edible fats and oil products within the unit operations is shown in Table 13.2.

**Table 13.2.** Storage and handling of soyabean oil within the unit processing operations.

| *Unit processing operation* | *Substance(s) removed* | *Handling operation(s)* | *Process reaction temperature °C (°F)* | *Oil storage* | |
|---|---|---|---|---|---|
| | | | | *Temperature °C(F)* | *Atmosphere* |
| Crude oil storage | Oil insolubles | Pumping: tank cars, trucks or barges to storage tanks | 54 (130) ambient | Ambient | Air |
| Degumming | Phosphatides<br>Trace metals | Pumping: Storage tank to mix tank to centrifuge to storage or refining | 70 (158) | Ambient | Air |
| Caustic refining | Free fatty acids<br>Phospholipids<br>Trace metals<br>Pigments—colour bodies | Pumping: Storage tank to inline mixer to centrifuge | 74 (165) | 50 (122) ambient | Nitrogen or air |
| Water washing | Soaps | Pumping: From primary centrifuge to water wash centrifuge | 88 (190) | 50 (122) ambient | Nitrogen or air |
| Drying | Moisture | Pumping: From water wash centrifuge to vacuum dryer to storage tank | 82 (180) | 50 (122) ambient | Nitrogen or air |
| Bleaching | Pigments, soaps | Pumping: Storage tank to bleaching unit to filter to storage tank | 105 (224) | 60 (140) ambient | Nitrogen or air |
| Hydrogenation | Olefinic bonds | Pumping from storage tank to hydrogenation converter | 140–225 (284–437) | Approx. 10°C (180°F) above melting point | Nitrogen or air |
| Winterisation | High melting triglyceride | Pumping from converter to chill tank through filter | 4–5 (30–35) | 60–66 (140–151) | Nitrogen or air |

*(Contd...)*

**Table 13.2.** Storage and handling of soyabean oil within the unit processing operations. *(Contd .)*

| *Unit processing operation* | *Substance(s) removed* | *Handling operation(s)* | *Process reaction temperature °C (°F)* | *Oil storage* | |
|---|---|---|---|---|---|
| | | | | *Temperature °C(F)* | *Atmosphere* |
| Filtration | Spent bleaching clay | Pumping from converter to filter to storage tank | | 60 (140) ambient | Nitrogen or air |
| Deodourisation | Pigments<br>Unsaponifiables<br>Free fatty acids<br>Odour and flavour compounds | Pumping from storage tank to deodouriser | 204–274 (400–525) | 60–66 (140–151) | Nitrogen or air |
| Polishing filtration | Bleaching earth<br>Citric acid<br>Carbonaceous material | Pumping from deodouriser to polish filter to oil cooler to storage tank | 50–70 (122–158) | 60–66 (140–151) | Nitrogen or air |
| Finished oil storage | Oil insolubles | Pumping from storage tank to polish filter to packaging line or to tank cars or tank trucks | | 60–66 (140–151) | Nitrogen or air |

Virtually all edible fats and oils exhibit the best oxidative stability in their crude state. With increasing processing, oxidative stability decreases and reaches a minimum after bleaching. However, deodourisation and treatment of the oil with metal inactivators and/or autoxidants cannot be relied upon to correct faulty upstream handling and processing techniques since peroxide buildup during crude oil storage, degumming, refining or bleaching will ultimately have a detrimental effect on the storage properties of the finished oil. Storage and handling practices should be designed to compensate for the decreased oxidative stability. Refined-bleached oil is the least stable and extended storage of this oil is not recommended.

### Finished Oil

A finished oil is defined as that fat or oil product ready for use alone or as an ingredient without further processing. Assuming that raw materials are of high quality and previous operations are satisfactory, the result is a light-coloured product, free from odours, flavours, hydroperoxides and contamination. The goal of finished oil storage and handling is to preserve and maintain these quality factors. A general scheme for the storage and handling of finished oils is shown in Fig. 13.1. Oil coming from the deodouriser is passed through a heat exchanger and a polishing filter to remove any solid materials. After the polishing filtration, the oil is pumped through a cooler to a storage tank. Before packaging, the oil is generally filtered again to remove any solids picked up while in storage. Paper is generally used for this filtration.

Autoxidation increases markedly with temperature. It has been estimated that for salad oils in general the rate of oxidation doubles for each 15°C (59°F) increase in temperature within the range 20–60°C (68–140°F). From this standpoint, it is desirable to handle the oil at temperatures as low as possible during discharge from the deodouriser into the storage tank. In practice, however, it is not desirable to cool the oil below 65.6°C (150°F) prior to the polish filtration for the following reasons: (i) moisture is apt to condense in the oil; and (ii) at lower temperatures the filtration rate decreases.

Further cooling before storage is generally considered optimal. Where storage tanks are not blanketed with nitrogen, cooling is desirable. On the other hand, the use of nitrogen-gas blanketing eliminates the need for cooling.

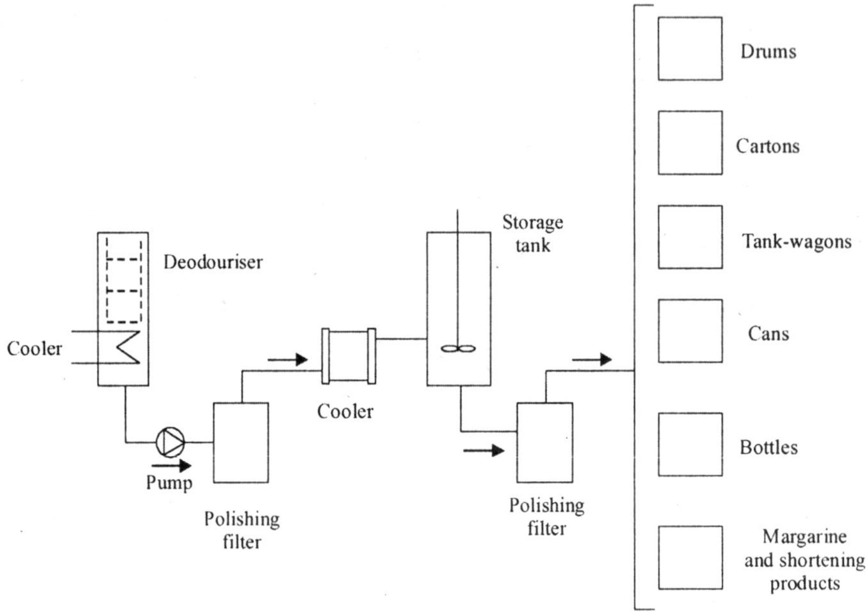


**Fig. 13.1.** General scheme for the storage and handling of finished oils.

## Storage Tanks

Storage tanks for finished fats and oils are generally constructed of iron, although more expensive stainless steel is sometimes used as is aluminium. Finished oil storage tanks vary considerably in size. Typically, they are designed to hold a given number of standard rail tank carloads of oil.

Finished product storage tanks are usually closed and fitted with tank tops, internal heating or cooling coils and an agitator to promote heat transfer and to prevent localised overheating.

Factors most likely to affect the quality of finished products to such an extent that they would require reprocessing or use as a subquality product include: (i) contamination from atmospheric adulterants; (ii) internal contamination from water and soaps; (iii) overheating; and (iv) exposure to air and oxygen.

Atmospheric adulteration is avoided by storing the product in completely closed tanks. Finished products held in storage tanks can be contaminated by water leaking from steam or cooling coils. Moisture in the oil can promote hydrolysis, particularly at elevated storage temperatures. Properly maintained coils are necessary. Overheating of bulk stored oil is undesirable because soaps and prooxidant metals are much more active at elevated temperatures. Finished oil storage tanks are equipped with automatic temperature controllers to prevent overheating.

## Nitrogen Blanketing

Oxidation has the most detrimental effect on the quality of finished products held in bulk storage. Thus, exclusion of oxygen during storage represents a highly desirable and practical method for preventing quality deteriorations.

The usual procedure involves replacing oxygen with nitrogen. Finished product is delivered from the deodouriser to the storage tank under a complete nitrogen blanket. Nitrogen can be supplied from a tank of liquid nitrogen or from commercially available nitrogen generators. Liquid nitrogen is distributed at a purity of 99.998 per cent. The tanks can be equipped for either gaseous or liquid withdrawal. For gaseous applications, a line of vapourisers—atmospheric, electric, steam or hybrids—can be used to convert the liquid to gas, which is then discharged at the required pressure to use points.

## Industrial User Sector

Industrial users of fats and oils, such as those involved in the manufacture of fried foods and baked goods, often purchase their supplies in bulk because of cost advantages. Among these are elimination of cost of shipping containers, reduced shipping rates and reduced labour costs due to more efficient unloading and handling of products within the user's plant. Other advantages include a reduction in product waste and sanitary problems associated with small containers. Bulk handling also may reduce the amount of storage space needed and this factor can be of economic importance in some plants.

On the other hand, bulk handling offers some disadvantages, including sizable capital investment for the purchase and installation of pumps, tanks and other essential equipment. Operation of a bulk handling system requires the attention of more skilled personnel than the simple unloading and handling of smaller containers. A further disadvantage is that the product may deteriorate before being used. This is particularly true of shortenings that must be held in a molten state. Melted products are generally more susceptible to deterioration than packaged products.

Many types of fats and oils are amenable to bulk handling, including all salad and cooking oils, all types of frying fats, bread shortenings and products normally fluid and pumpable at ordinary temperatures, such as 'fluid' bread and cake shortenings.

Several types of products are not amenable to bulk handling because their physical or functional properties are dependent on proper plasticising, tempering and resultant crystal structure, all of which are altered by heat or melting. These include: (i) margarines and water-shortening emulsions that will break down under heat, with a separation of the fat and water phases; (ii) baking shortenings used in the manufacture of cakes and icings; (iii) shortenings containing significant amounts of lecithin; and (iv) shortenings used in Danish pastry.

## Equipment for Bulk Storage and Handling

Equipment required for bulk handling include: (i) tank cars or tank trucks for transporting bulk shipments; (ii) equipment for unloading the fats; (iii) storage tanks; and (iv) equipment necessary to deliver the product to the point of use within the user's plant.

### *Sparging techniques*

Sparging represents a practical method for protecting finished oils from oxidative deterioration during shipment from refinery to destination. The technique is particularly useful when finished oils are loaded into tank cars and trucks. The principle involved is saturation of the oil with nitrogen while it is completely free of air and oxygen, i.e. after deodourisation. A sparger introduces tiny bubbles of nitrogen into the oil stream; as the saturated oil falls into the tank car or truck, the effusing gas sweeps the headspace and thus removes most of the air and oxygen from the vessel. Another benefit of nitrogen sparging is the reduction of oxygen picked up from natural leakage of air into the tank car or truck. The gradual flushing action of the sparge gas reduces the reabsorption rate of oxygen because of the pressure differential between the liquid and the headspace above it. Nitrogen blanketing and sparging techniques have many other applications in protecting oil quality (Table 13.3).

**Table 13.3.** Summary of treatment areas with nitrogen to protect edible oils.

| *Area of treatment* | *Method of introducing nitrogen* | *Classification* | *Approximate usage ($ft^3$)* |
|---|---|---|---|
| Manufacturing pumping from deodouriser | In-line between cooler and storage tank | Sparging | Approximately 0.125 $ft^3$/gal |
| Bulk oil storage | From sparger directly into | Blanketing headspace | Enough to maintain positive pressure |
| Filling to tank cars | In-line between storage and tank car | Sparging | 1000 $ft^3$/8000-gal car |
| Tank car or truck | Into headspace of car after | Blanketing | Undetermined filling |
| Customer's plant pumping from tank car to oil storage | Sparging in-line during pumping | Sparging | Approx. 0.125 $ft^3$/gal |
| Storage in tanks | Directly into headspace plus | Blanketing nitrogen from sparger | Enough to maintain positive pressure |
| Pumping from storage to filler or header | Sparging in-line during pumping | Sparging | Approx. 0.125 $ft^3$/gal |
| Filler bowl or header | Entrance into closed filler bowl or header pressure | Blanketing | Maintain slight positive |
| Closing or capping machine | Shroud or purge techniques | Blanketing or purging | Undetermined |

## Precautionary Measures in Bulk Handling

In bulk storage systems where positive-displacement pumps are used, special precautions are necessary on both sides of the pump. On the suction side, a double-side, double-check system should be used to avoid pumping the fat from a closed vessel, thereby preventing collapse of the tank and piping. Similarly, on the discharge side, checking for closed valves is recommended to prevent danger to operating personnel, damage to equipment and loss of product from blown fittings.

When it is necessary to heat fats to keep them fluid, heat should be applied to line, pumps and tanks. Although insulation may be used to retain the necessary amount of heat, the ability to apply heat in emergency situations is most desirable.

Occasionally, even when shortenings are held above their apparent melting point, hard fractions will separate or 'seed' out. In effect, the composition of the fat changes in different levels in the tank and possibly can block the lines. Application of heat and agitation may be necessary to correct the problem. It is recommended that experimentation on the product be carried out at the projected temperatures and times to ensure that this problem will not occur under operating conditions in the plant.

Close control of temperature within handling systems is necessary because pumping rates are dependent on viscosity of the fat, which in turn is temperature dependent. Similarly, where volumetric means are used to control weights, it is absolutely essential to control temperature because the specific gravity is related to temperature.

## Handling and Packaging of Institutional Shortenings and Margarines

Industrial and commercial shortenings are packaged in 22.3-kg (50-lb) cans and cubes, 49.1-kg (110-lb) cans and 169.6-kg (380-lb) open-end steel drums. Commercial shortenings are also packaged in corrugated paper boxes measuring 30.5 cm (1 ft) on a side. Polyethylene film is used to line the box and prevent leakage. This type of packaging is popular because it is cheap, the contents are easily removed and the package can be readily disposed of by folding flat. Storage problems are minimal because the package takes up little space. Plastic cube liners are usually coloured because they melt in hot fat if they are accidentally dropped into a fryer and the colour makes them easy to detect and retrieve. A potential disadvantage of cube packaging is that melted cube liners cause foaming, darkening and smoking of the frying fat.

Drum-packaged pourable shortenings may be emptied upright with a hand-operated pump or emptied by gravity through a spigot with the drum placed horizontally on a rack.

Fully hardened fats are sold as flakes or powdered materials and are packaged in multiwalled paper bags holding 22.3 kg (50 lb) or in fibre drums holding 44.6–59.3 kg (100–200 lb).

### Margarines

Margarines intended for institutional and industrial trade are often packaged in friction-top cans holding 13.4–22.3 kg (30–50 lb) and in cubes. Open-end steel drums [208.7 L (65 gal)] hold 181.5 kg (420 lb) of margarine.

### Storage of Shortenings—Effect on Functional Properties

Shortenings may be classified as solid, fluid, liquid and powdered. Five factors govern the quality of shortenings; (i) triglyceride and emulsifier composition; (ii) processing conditions; (iii) tempering conditions; (iv) usage temperature; and (v) storage conditions.

#### *Solid shortenings*

Solid shortenings are the most generally useful products of their class since they may be used for both frying and baking. The crystal structure is an important consideration in solid shortening compositions. For example, for optimum creaming ability (or the ability to incorporate air into a cake batter), the shortening must be stable and in the β form and characterised as being smooth and creamy. These products can be adversely affected by improper and nonuniform storage conditions. As reflected by pound cake volume, general-purpose solid shortenings will show no loss in creaming ability over storage temperatures of 4.4–21.1°C (40–70°F). Beyond that, a marked loss of creaming ability is observed.

Prolonged storage of solid shortenings at high temperatures causes the lower melting fat fractions to liquefy; upon cooling, they solidify into crystal forms that are less functional.

Solid shortenings require no refrigeration during storage. However, they will absorb odour. Solid shortenings should be stored in a cool, dry place away from odour-producing materials.

#### *Fluid shortenings*

Fluid shortening are pourable, opaque products containing suspended solids that are hard fats or emulsifiers, depending on whether the product is to be used for frying or baking. Like solid shortenings, proper storage temperatures are important in preserving the functional properties of the product. Storage of fluid shortenings at temperatures below 18.3°C (65°F) will cause the product to set up and lose fluidity. Loss of fluidly can be reversed by warming. Storage at temperatures beyond 35°C (95°F) will cause either partial or complete melting of the suspended solids. This, however, is not a reversible phenomenon because cooling will cause formation of large crystals that will not remain in suspension. If the product is formulated with hard fats as the suspended solids, improper storage might be of little consequence because they will merely settle to the bottom of the container upon cooling. On the other hand, if the solids are emulsifiers settling out will upset the balance of the product, with the upper portion underemulsified and the bottom part overemulsified. Thus, the product would not perform as intended.

### *Liquid shortenings*

The liquid shortenings include regular cooking oils, salad oils and products resulting from fractionation of semihard fat. These products are not ordinarily emulsified. Ordinary cooking and salad oils pose no unusual storage problems because they possess no crystal structure to protect and have no suspended solids. If they are stored at temperatures low enough to cause solidification, they can easily be returned to clear liquids by returning the product to ambient storage conditions.

## Factors that Contribute to Oxidative Deterioration of Finished Products

Because of the importance of preventing oxidation in finished products, a more thorough discussion follows. It should be pointed out that the effects of atmospheric oxidation apply to all fats and oils regardless of their stage of processing.

Five factors contribute to the oxidative deterioration of fats and oils: (i) oxygen or air; (ii) heat; (iii) light; (iv) prooxidant metals; and (v) time.

Although the elimination of air prevents oxidative deterioration, complete elimination of air is impractical. The solubility of oxygen in soyabean oil is quite high (3.2/100 mL) and is sufficiently soluble to yield a peroxide value of 18, assuming complete reaction. Flavour deterioration in soyabean oil occurs at peroxide values considerably lower than 18.

Even when using nitrogen blanketing, sufficient oxygen already may be present in the oil to promote oxidative deterioration. A more practical approach is to minimise the incorporation of air into the oil by proper handling procedures.

Faulty handling procedures to be avoided are:

1. Allowing a liquid fat to cascade or fall through air into storage or holding tanks. As mentioned previously for crude fat and oil handling, a more desirable procedure is to fill the tank from the bottom with subsurface entry.
2. Sucking of air into the suction side of pumps or lines caused by faulty pump seals or fittings.
3. Whipping air into a fat through improper agitation within storage tanks. Creation of whirlpools or vortices should be avoided.
4. Minimise blowing of lines with air, which in turn may bubble air through the fat being held in the storage tank.

Heat accelerates the reaction of atmospheric oxygen with edible oils. For deodourised products, it is estimated that the speed of oxidation is doubled for each 15°C (59°F) increase in temperature over the range 20–60°C (68–140°F). A rule of thumb for handling is to keep the fat no warmer than necessary to facilitate handling by pumping. For hydrogenated fats, a holding temperature 5.6°C (10°F) above the melting point is considered sufficient. Design of bulk storage systems should take into account the use of low temperature whenever possible. For example, short insulated lines, as opposed to long unheated ones, will permit lower storage temperatures and thus prolong the quality of bulk-stored fats.

Localised overheating is detrimental to fat quality and should be avoided. All storage tanks with heating devices should be equipped with a mechanical agitator. Power agitation will not only minimise fat damage from localised overheating but will save time and heating costs as well. If agitation is temporarily out of service, the temperature differential between the fat and the heating medium must be kept minimal.

The importance of avoiding metallic contamination cannot be overstressed. Copper and iron are strong prooxidants capable of lowering the flavour and oxidative stability for fats and oils at levels of 0.01 and 0.1 ppm, respectively. Copper or copper-containing alloys should never be used in equipment used for handling and storing fats and oils. The prevention of iron contamination is somewhat more difficult, because most of the industry uses black iron for construction of tanks, pumps and lines. However, through proper treatment and cleaning of black iron equipment, iron contamination can be kept to a minimum.

All edible fats and oils deteriorate under the effects of light. Normally, light-induced deterioration is not of concern in the oil processing operation or at the user level because processing, handling and storage are carried out within closed systems.

Any fat or oil will deteriorate even if stored and handled under ideal conditions. Unhydrogenated or lightly hydrogenated oils, which do not require heat for keeping them liquid, have greater resistance to deterioration than do shortenings. The latter, however, will keep for 2–3 weeks in a melted condition. Shortenings should not be allowed to solidify and then be reheated during use. Bulk storage systems should be designed with a maximum turnover time of 2–3 weeks or within the storage life of the product. The mixing of fresh shipments with those already in storage should be avoided. Small quantities of old products in a tank may hasten the deterioration of new product mixed with it.

## Transport of Fats and Oils

During transport the quality of oils may be impacted by oxidation, adulteration from other oils, moisture condensation, contamination with foreign matter (dirt, rust, mould, etc.) and overheating—induced or spontaneous. Transport of fats and oils is designed to preserve quality in the delivery to the product formulator in the most cost-effective manner. The advantages and disadvantages of bulk handling of fats and oils are given below.

### *Advantages*

1. Container packs unnecessary.
2. Lower freight charge per tonne.
3. Lower labour handling costs in transit and at destination.
4. Less loss due to smaller overall holdup in handling facilities (average loss estimated at 0–5%).
5. No cost for package disposal.
6. Cost savings in smaller storage area.
7. Easier hygiene situation (hence cost saving).

### *Disadvantages*

1. Greater capital investment in handling facilities.
2. More highly skilled (paid) individual operators
3. A larger quantity purchased at one time
4. Holding time of the commodity is longer and, possibly as a melted solid, places demands on planning.

## Temperatures during Shipment and Unloading of Fats and Oils

Extremes of temperature of an oil during shipment can have a deleterious effect on quality since oxidative deterioration and hydrolysis are accelerated by temperature. The International Association of

Seed Crushers has established recommended safe limits of temperature during the voyage and on discharge for the principal oils in transport (Table 13.4).

**Table 13.4.** Recommended schedule of temperatures (°C): principal oils.

| *Oil* | *During voyage* | | *On discharge* | |
|---|---|---|---|---|
| | *Minimum* | *Maximum* | *Minimum* | *Maximum* |
| Sunflower | Ambient | Ambient | Ambient | 20 |
| Soyabean | Ambient | Ambient | 20 | 25 |
| Safflower | Ambient | Ambient | Ambient | 25 |
| Peanut | Ambient | Ambient | 20 | 25 |
| Rapeseed | Ambient | Ambient | Ambient | 20 |
| Maize | Ambient | Ambient | Ambient | 20 |
| Palm | 32 | 40 | 50 | 55 |
| Palm stearine (RBD) | 40 | 45 | 60 | 65 |
| Palm olein (RBD) | 25 | 30 | 50 | 55 |
| Coconut | 27 | 32 | 40 | 45 |
| Fish | 20 | 25 | 30 | 35 |
| Palm fatty acid distillate | 42 | 50 | 67 | 72 |

## Deterioration during Extended Storage

Prime concerns in storage of crude and once-refined soyabean oils include increases in moisture and volatile matter, colour after refining and/or bleaching, peroxide value, free fatty acids and refining losses. Aside from possible effects on flavour and oxidative stability, many of these quality factors also affect the commercial value of oil in trading channels.

### *Moisture and volatile matter*

For all oil types, moisture and volatile matter ranged from 0.02 to 0.2 per cent during the test period with an average of $0.0686 \pm 0.0347$. Because the results were similar for all storage conditions, higher temperatures and limited access to air did not affect moisture and volatile matter. Accordingly, because variation of temperature of oils held in field tanks is less than that of oils stored in drums, it was concluded that generally moisture and volatile matter would not change during commercial storage.

### *Colour after refining*

In control drums, the colours of once-refined oils and oils refined from crude and crude-degummed samples decreased in relation to time and temperature. The magnitude of the effect is dependent on initial colour, extent of processing and storage conditions of time and temperature. Once-refined oils show the largerst colour decreases, followed by crude-degummed; the least effect is noted in crude oils.

Since, the colour decrease is accompanied by autoxidation as measured by peroxide value, it was suggested that the two phenomena are interrelated. Evidence for this relationship stems from the decrease in refined colour, which is less when access to air is limited and greater when temperatures are high. Limited access to air decreases the rate of oxidation, whereas high temperatures promote it. Furthermore, the magnitude of the loss in refined colour coincides with the relative oxidative stability of the oil with respect to processing, i.e. crude most stable—least decrease; once-refined least stable—most decrease. The study indicated that a small decrease in refined colour is a good indication of a small decrease in quality of oils held in storage.

### *Peroxide value*

Peroxide values taken at the end of 4 years of storage showed that once-refined oils were the least stable to oxidation, followed by crude-degummed, with crude oil being the most stable. The conditions of storage had a marked effect on peroxide development. Oils stored in no-breather drums had the lowest peroxide values, followed by control drums, with oils stored in half-filled drums having the highest peroxide values. The implications of the latter results are obvious: oxygen or air should be excluded during storage.

### *Free fatty acids*

Storage of once-refined oil in control drums had only a slight effect on free fatty acid content after the 4-year storage period. From initial values of 0.058 per cent, they increased to 0.135 per cent. As observed with peroxide formation, the increase in free fatty acids is dependent on access to air and to temperature within the storage drum. For example, oils stored in no-breather drums increased from 0.058 to 0.07 per cent free fatty acids after 4 years of storage, whereas oils stored in half-filled drums increased to 0.21 per cent. It would be expected that these small increases in free fatty acids would be much slower in large field tanks because the temperature range is much lower.

It was concluded that the effects of storage on the free fatty acid content of crude and crude-degummed oil were variable. In some samples, no increases were evident after 4 years of storage, but in others small increases were observed.

As would be predicted, significant relationships were found between the cup refining loss and the free fatty acid content of crude and crude-degummed oils. It was suggested that the refining loss after storage can be estimated based on the free fatty acid content.

## Colour Reversion

Under certain conditions, soyabean oil can be refined, bleached and deodourised to yield a pale light-coloured product that becomes darker upon storage. This phenomenon, known as colour reversion, has been studied in detail by Japanese scientists. The degree of colour change is dependent on the nature of the raw soyabeans, refining and deodourisation conditions, storage conditions, temperature, exposure to light and contact with air during storage. The tendency to undergo colour reversion is variable. Some oils require only a few hours to revert, whereas others require several months. In either event, the colour becomes paler after it reaches a peak.

The moisture content of the soyabeans, going to the extractor, appears to be a key factor in promoting colour reversion in fully refined oils. The tocopherol content of extracted oil is markedly affected by the beans' moisture content. Over the normal moisture range (7–12 per cent), the tocopherol content of crude soyabean oil remains normal, but in wet beans the tocopherol content is markedly reduced. Subsequent work showed that tocored [2, 7, 8-tri-methyl-2 (4′, 8′, 12′ -tridecyl-2)-chromane, 5, 6-quinone] is the precursor of the reversion colour and that this material reaches a maximum in soyabeans with moisture contents of 15–18 per cent.

Other work indicated that no change in the α-tocopherol content of the crude oil was observed by elevating the moisture content of soyabeans, but a decrease in γ- and δ-tocopherols occurred. The decrease in γ- and δ-tocopherols was accompanied by an increase in dimeric oxidation products. The latter effect could be reversed by decreasing the moisture content of the beans. Thus, it appears that colour reversion involves oxidation of γ-tocopherol to a dimeric product that, in turn, is converted to tocored. The following mechanism was suggested for colour reversion:

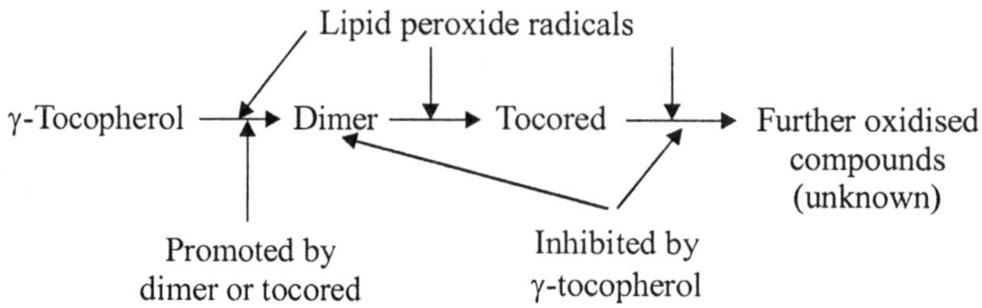


## PACKAGING

Packaging systems for oil and fat products have become more complex as the demands for containing, protecting and identifying products have increased for manufacturers. The packaging system must be designed to produce an economical product that satisfies regulatory guidelines, consumer demands and production requirements.

It is necessary to have defined goals and objectives to accomplish the successful integration of all packaging issues. Early communication with various parties (production, vendors, marketing, engineering firm, construction firm, etc.) becomes critical in defining project schedule, design criteria, process flow diagrams, general arrangement and desired equipment.

### Product Characteristics

To determine the type of equipment and packaging, the designer must determine and define characteristics of the product or products in the conceptual phase. For example, oils, oil bases and fats contain a large percentage of molecules sensitive to oxidation. These products (liquid oils, margarine, shortening, etc.) often have points of unsaturation or double bonds within their molecules due to the absence of hydrogen atoms. These double bonds are susceptible to oxidation. Exposure to reactive environmental compounds, such as oxygen, liberates off-flavours or undesirable compounds. This type of product deterioration can often be prevented or minimised by limiting the exposure of the product to oxygen or metal ions through the selection of the packaging materials. Thus, the requirements of the product dictate the package construction and consequently the equipment.

A design criterion provides a basis and rationale for sizing and selecting the packaging equipment. Since, the requirements of any two plants are not exactly alike, there is no one ideal plant layout. However, some guidelines exist for edible oil packaging areas.

### Packages, Containers, Closures

No packaging system can be functional without a compatible material that will deliver both good machine throughput and provide a sufficient packaging barrier. Any packaging material that is used for an edible oil packaging project must have a low initial microbial load (must be clean) and provide a barrier compatible with the oil product. Product shelf life and stability are directly related to the inherent ability of a packaging material to affect the transmission of oxygen and moisture. It is important to determine the desired shelf life of the product and then select the material that will provide that shelf life. For example, it may be preferable to use a low cost material with a low oxygen barrier property when packaging edible oils for institutional trade. In this case, the turnover is rapid and the desired shelf life is only a few weeks. The same oil product packaged in a high oxygen barrier material would allow a retail shelf life of several months. Therefore, although the same products are processed in a like manner and the packaging systems are nearly the same, the final products will have completely different shelf life characteristics based on the packaging material used.

Another consideration is the product–container compatibility. For an oil product intended for an extended shelf life, an appropriate container choice would be a multiple layer container, as the layers are composed of high or low density polyethylene on the outside, ethyl vinyl alcohol in the middle and high density polyethylene on the inside. High density polyethylene on the inside is a suitable plastic for contact with edible oil and provides an adequate moisture barrier. Low density polyethylene is not a suitable inner layer material due to the high rate of migration allowed. The middle layer is the oxygen barrier layer and the outer layer should be resistant to water. Another multiple layer container is the combination of paper and plastic materials. There are a variety of combinations possible for these containers, and their use depends on the properties of the product, container and packaging equipment. This combination is commonly used in composite cans for shortening. Normally, the equipment vendor has experience in selecting the appropriate packaging material and the packaging barrier is developed between supplier and manufacturer. The primary packaging materials that are successfully being used include styrene, polyethylene, polypropylene, polyvinylchloride, Saran, paper, aluminium foil and combinations of these materials.

Two processes widely used in forming barriers are plastic coextrusion and lamination. Plastic coextrusion is the combining of two or more polymer layers during a one-step process of film or sheet extrusion. Lamination involves an adhesive or extrusion-bonding process that brings together two or more materials that have been preformed.

There are three major forms of plastic packages used in oil packaging projects: blow-moulded containers, such as plastic oil bottles; thermoformed packages, such as trays, tubs, cups and inserts and injection-moulded fitments, closures and lids.

## Edible Oil Operations

Edible oil operations use glass or plastic containers in a variety of sizes and shapes. Glass containers are rarely used in this market category. Plastic containers dominate this packaging operation.

### *Low viscosity products*

Plastic bottles are either received in bulk or in reshippers, whereas glass bottles are usually received in reshippers. A reshipper is a shipping case that is used for both returning the empty containers and shipping the filled containers. Plastic and glass reshippers are decased by inverting the reshipper unscrambler to orient and convey the containers. Containers for plastic bottles that are preformed outside the plant are received in bulk handling containers. The bulk handling containers either are pallet loaded, which requires a depelletiser, or are in bulk boxes, which require an unscrambler. Another option would be to have an in-house blow-moulding operation. The containers pass through an unscrambler or depalletiser for loading onto a single-lane conveyor. The container can be oriented so that a label can be placed on a side of the container, or the label may be placed after filling. These containers are then processed through a bottle washer or bottle inverter, which usually is 'of' an airjet type. The containers are filled with or without nitrogen sparging, capped, labelled and conveyed to secondary packaging.

### *Thermoset fat products*

Margarine, mayonnaise, shortening and other high fat products are filled as high viscosity products or at elevated temperatures. Margarine packages consist of wrapper sticks, squirt bottles and tubs. Margarine packed as four individually wrapped 0.25-lb (0.133-kg) sticks is still a common product. The line components for soft margarine are dispenser, former, wrapper, cartoner, labeler, case packer and palletiser. Wrapping and cartoning are usually done by a single machine. This unit must be adjusted based on the characteristics of the margarine, such as product temperature and emulsion firmness. Direct injection systems for forming the margarine have product recirculation loops to return the oil through a heat exchanger to control product temperature. The correct product filling pressure is controlled by the margarine oil formulations and the filling moulding temperature. The speed of a filler varies between 90 and 300 packages/minute. Increasing the product pressure during filling–moulding results in a more dense fill–mould. The margarine mould is wrapped by either paper or foil. The four sticks are cartoned by pushing them against the carton blank. A dense mould can be wrapped and cartoned easily; however, too much density will result in a crumbly mould or stick. Too soft a mould will alter under this pressure, and the carton will be irregular.

Mayonnaise is usually packaged in glass containers, whereas peanut butter and salad dressings are usually packaged in plastic containers, in the range of 0.2–1.0 litre (8, 16 and 32 fluid oz) and sometimes in 3.79 litre (1 gal) for institutional trade. Mayonnaise is still packaged in glass to allow the product to be retorted (heated to preserve) after container sealing and because of the sensitivity to oxidation. However, some of the small sizes for mayonnaise have been converted to plastic. Other oil and oil-based products are also sensitive to oxidation; thus after being sealed in the container, some products will oxidise until the headspace oxygen is used up. Therefore, headspace is often flushed with an inert gas such as nitrogen. The extent of oxidation after flushing is not sufficient to result in off-flavour formation. However, a loose lid or seal that has been warped or overheated will leak and admit oxygen from the atmosphere, causing rancidity. Oxidation or moisture loss from the product through such leakage will also cause the product to be rejected by the consumer. Filling equipment is normally volumetric volume. A product that has been flushed with a nitrogen–carbon dioxide mixture may create a headspace vacuum when the carbon dioxide is absorbed by the product. Overfilling the container will cause the lid to force the product onto the closure threads or the heat seal to fail. Obviously, overfill is to be avoided.

Metal, composite or plastic containers received on unitised pallets are usually packed with a shrink or stretch wrap shroud to maintain integrity of the load. Manual or automated depelletisers are available that orient the cans onto single-lane conveyors. The cans are inverted for cleaning and enter the filler. Filling is accomplished by timed flow or piston-type, volumetric action fillers. The end or side of the container may be ink jet printed with information such as code date, product code and/or inspection establishment number. If a seaming operation is used, then the roller must be kept precisely adjusted to maintain package integrity. If a heat sealing method is used, then precise temperature pressure controls are required to monitor integrity. Upon leaving this operation, the containers pass onto a single-lane conveyor, the intent of which is to provide sufficient control for labelling. The containers are packed manually or automatically into shipping cases, then pelletised.

## Bulk Packaging Operations

These systems have primarily packaged products such as shortening. The concept of bulk packaging of raw ingredients significantly reduces packaging costs and provides other benefits, such as increased warehouse and transportation efficiencies, reduced storage costs and reduced disposal problems of the spent container. There are a number of factors to be considered in bulk handling installations. Bulk deliveries must be in a liquid or semiliquid state to be easily removed from the carrier. Tank trucks or rails tankers usually have no facilities for heating the shortening and must be insulated to prevent the product from solidifying. An insulated tank filled with hot shortening cools at the rate of 0.2–1.0°C (0.5–2.0°F)/hr. and will remain fluid for 15–20 hr. The length of time depends on filling temperature, environmental conditions, oil crystallisation point and discharge methods.

On receipt, the melted shortening or liquid oil is pumped into storage tanks and held until needed. The bulk product should be maintained at a temperature of about 6°C (9°F) above its melting point. The product held in the tank should be flushed with nitrogen gas to prevent the slow deterioration and colour darkening of the product while in storage. The low storage temperature and nitrogen blanketing help retard deterioration, but the storage life of a heated oil product is limited. The bulk storage tank of product should be used in about 2 weeks for best results. The capacity of the storage tank should be based on this use rate. Storage tanks are fabricated in many shapes, round or square, conic or flat-bottomed, tall or short. The design selected depends on space requirements, which are often critical in a general equipment layout. A stainless-steel tank for food products is recommended. Other oil base products can be stored in tanks or fibreglass fabrication.

Thus, processing and packaging systems should be examined in regard to consumer requirements, new technologies and energy use.

## Future Developments

Energy issues will become increasingly important in plant design, processing and packaging. Plant site selection and material of construction will incorporate the advantages of energy conservation such as passive solar design. Other energy studies will optimise material flow through the plant. On-line testing of oil product for foreign substances with greater sensitivity will also be a future trend. For oil production operation, flexible processing and bulk packaging will become increasingly important, since this technology offers lower processing cost, less space and lower container cost. Containers and other packaging materials will be designed for a secondary usage, e.g. incineration for energy or recycling into other usable plastic products.

Chapter 14

# By-products Utilisation

## INTRODUCTION

This chapter discusses the by-products resulting from various oil seed meals, oils and their uses in various industries and in human foods. Some of the important by-products of seed processing discussed are—soyabean, rapeseed, canola, sunflower, safflower, cottonseed, palm, coconut, groundnut, sesame, olive, linseed. Lecithin, an edible by-product of oil processing possessing a variety of useful functionalities, is primarily a mixture of phospholipids such as phosphatidyl-choline, phosphatidylethanolamine, phosphatidylserine, phosphatidylinositol and phosphatidic acid and contains minor quantities of other water-soluble or hydratable components such as glycolipids and oligosaccharides. Refining by-products lipid, commonly referred to as RBL, results from the refining of crude oil, where continuous mixing of crude oil with a dilute sodium hydroxide solution produces a by-product consisting of free fatty acids, hydrolysed phosphatides and unsaponifiable materials. Free fatty acids are the valued component in RBL, the composition of which varies with the oil source, oilseed conditions at crushing, the oil removal method used, the solvent employed, the extent of extraction and the refining conditions. Larger quantities of RBL are produced as the oil becomes more refined.Refining by-product lipid has been considered as a growth medium for micro-organisms. It appears to provide a satisfactory supply of nutrients for growth, due in part to its high sodium content and its suitable trace element composition. Despite its residual oil content and high pH, many species of micro-organisms will grow on this substrate.

## BY-PRODUCTS OF SEED PROCESSING

### Soyabean

The by-products resulting from the processing, i.e. solvent extraction, of soyabean in the major producing countries—Argentina, Brazil and the United States are defatted protein meal and to the extent that the beans are dehulled prior to extraction, hulls. A portion of the separated hulls may also be added back to the meal. 85–90 per cent of the protein meal obtained from processing in the United States is used directly as livestock feed, after appropriate heat treatment (toasting) to inactivate trypsin inhibitors and other antinutritive factors. Relatively minor quantities are milled into flour or grits, primarily for edible applications, or used in the preparation of protein concentrates and isolates having food, feed and industrial applications.

### *Utilisation of soyabean meal in animal feeds*

Soyabean meal is the most extensively used of the oilseed meals and serves as a protein supplement for all classes of animals. It has become the standard to which all other protein sources are compared and its quality, acceptance and reputation are widely known. The meal contains from 45 to 50 per cent crude protein and from 2500 to 2750 kcal of metabolisable energy per kilogram, depending on the amount of hull present and the species of animal being fed. Dehulling increases the metabolisable energy values by about 5 per cent for cattle and 10 per cent or more for pigs and poultry.

Soyabean meal is an excellent protein supplement for lactating dairy cattle and for calves following weaning, as it is highly palatable and well digested. It also serves well as the supplemental protein in rations for growing and fattening cattle. However, in many ruminant feeding applications, soyabean meal may not be cost competitive with other proteinaceous and nonprotein-nitrogenous ingredients. Extensive research related to the development of 'protected' or rumen bypass protein from soyabean meal has resulted in the development of commercial products, which should increase the utilisation of soyabean meal in ruminant feeds.

Soyabean meal is the most economic high-quality protein available to feed manufacturers; hence it assumes a dominant role. Cereal-based ratios for pigs and poultry may contain soyabean meal as the only protein supplement, as a previous requirement for the inclusion of some animal or marine origin protein has been supplanted by the addition of lysine, methionine and vitamin $B_{12}$, where economics dictate.

Soyabean meal is used effectively in the formulation of pet foods, particularly for dogs, where simple corn–soyabean meal mixtures perform as well as complex diets containing high levels of animal protein and at substantially lower costs. Rapid growth in aquaculture over the last 10–15 years has resulted in significant new opportunities for utilisation of soyabean meal in finfish and shrimp diets. Whether additional heat processing of soyabean meal to further reduce levels of antinutritional factors or supplementation with lysine or methionine or both is beneficial to performance appears to be species dependent.

### *Utilisation of soyabean hulls*

Soyabean hulls are high in cellulose but low in lignin and therefore, highly digestible by ruminants. In fact, the digestible energy content of soyabean hulls, for ruminants, approaches that of grain. Consequently, soyabean hulls may be used, to economic advantage, in high-forage diets in lieu of grain, with various additional functional advantages. For growing cattle and sheep, replacing grain with soyabean hulls eliminates the risk of acidosis and reduces the negative effect of starch on fibre digestion. In the case of lactating cows or ewes, soyabean hulls can replace a significant portion of the grain in a grain–forage diet with no reduction in fat content or milk yield. Soyabean hulls are seeing increasing use in human food as a source of dietary fibre. Fibre-enriched pasta and white break appear to be the most popular vehicles for inclusion of soyabean hulls at this time.

### *Edible products derived from soyabean meal*

The primary edible products derived from soyabean meal/flakes are flour/grits (at 50 per cent), protein concentrates (containing 65–70 per cent or more of protein) and protein isolates (90 per cent protein). The nutritional quality, availability, price and functionality of these products has resulted in substantial usage in a wide variety of food and feed products.

Soyabean flour and grits are essentially ground soyabean meal/flake products differing primarily in particle size, both having received heat treatment appropriate to their required functionality and expected end use. Flours and grits are primarily employed as fat and water binders in baked good and pet foods and as raw material in the manufacture of soya sauce. Residual beany flavour and poor mouthfeel limit the use of soya flour and grits in many applications. Soya flours containing lecithin or various levels of fat (soyabean oil) are also commercially available.

Protein concentrates with improved flavour and functionality relative to soya flour may be prepared from undenatured soya flour/flakes by a variety of processes including acid or alcohol leaching, alkaline extraction, heat denaturation followed by aqueous leaching and membrane filtration. Products are ultimately spray or dispersion dried. Currently, acid-and alcohol-washed products dominate the marketplace and are available in a variety of functionalities for use wherever the nutritional, fat and water binding, emulsification, foaming, or viscosity modifying characteristics of casein or nonfat dry milk would otherwise be exploited. Typical applications of soya protein concentrates include comminuted meat products, baked goods, baby foods, cereal, milk replacers, pet foods and snacks. The expanding market for meat extenders and substitutes employs extrusion-texturised soya concentrates that, when rehydrated posses to a remarkable extent the chewiness and mouthfeel of meat.

Protein isolates are produced from undenatured soya flour/flakes by dilute alkaline extraction and subsequent acidification of the protein extract to the isoelectric point, approximately pH 4.5. The precipitated protein curd is recovered by centrifugation, washed, slurried in water and usually neutralised prior to being spray dried. Ultrafiltration or other protein recovery techniques may also be employed in isolate production. Soya protein isolates are relatively expensive but highly functional. Their binding, emulsification, foaming and nutritional properties are exploited in a variety of products including comminuted meats, dairy-type products and infant formulas. Due to their inherent insolubility at acidic pH, pectin-aggregated soya isolate particles show potential as clouding agents in citrus beverages. Soya isolates may be modified by hydrolysis, addition of substituent groups, or texturisation (spinning), thereby expanding their range of functionalities and applications.

### *Industrial uses of soyabean protein products*

Soya flour has seen significant use in glues for plywood and other laminated wood products. These have been displaced by petroleum-based glues that exhibit superior microbial and water resistance. Textile fibres prepared from soya isolate or flour never saw commercial production due to poor wet strength and an unpleasant odour when wet. Soya protein isolate is currently used in paper-coating applications.

Potential new uses for soya protein are driven primarily be environmental concerns and the search for new value-added uses for agricultural commodities. In addition, prices for petroleum-based polymers have increased relative to prices for agricultural products. Examples of products currently under development include biodegradable plastics, edible soya protein films and soya protein–carbohydrate films intended to reduce packing waste and soya flour-recycled paper composite material with the appearance of granite and the handling characteristics of hardwood. In addition, soya protein-based glues and adhesives and textile fibres, with functional characteristics superior to products from earlier efforts, are being developed. Still another use is as a feedstock for fermentations.

## Rapeseed/Canola

Rapeseed has become an important crop in the temperate zones of the world. The early nutritional experiences with the meal were probably not encouraging as the meal was unpalatable. Glucosinolates present in the meal release goitrogenic factors, such as oxazolidinethione, isothiocyanates and thiocyanates, when hydrolysed by the enzyme, myrosinase, also present in the seed/meal. These compounds interfere with iodine uptake and thyroxine synthesis by the thyroid gland. Under certain conditions, highly toxic nitriles may also be produced. Processing techniques that inactivate myrosinase and therefore, prevent glycosinolate hydrolysis have become standard operating practice around the world. Typically, seed containing 6–10 per cent moisture is rapidly heated to 80–90°C. A second deterrent to the use of rapeseed meal as a nutritional supplement is its high fibre content. Most rapeseed varieties have a dark, hard seed coat containing a condensed polyphenol-based complex that contributes a substantial amount of fibre to commercial rapeseed meal.

The reduction of glucosinolate levels in rapeseed through extensive plant breeding programmes has provided a major breakthrough in the utilisation of rapeseed meal. The meal is becoming increasingly available as a protein supplement for animal nutrition as more of the genetically improved varieties, known as canola, are grown. Canola is a trademark name and a generic term to distinguish specific seed varieties containing less than 2 per cent of erucic acid in the oil fraction and having a solid component containing less than 30 μmol/g of glucosinolates. Successful feeding of canola meal still requires knowledge of its glycosinolate status as well as the age and class of animal involved.

### *Composition of canola meal*

Canola meal contains 35–38 per cent crude protein and a favourable assortment of essential amino acid in its protein. It is higher in crude fibre, at 12 per cent, than is soyabean meal and consequently has lower metabolisable energy values (1900–2300 kgcal/kg, depending on the species being fed). The crude fibre is largely in the hull fraction, which comprises 16–25 per cent of the meal. Hulls are poorly digested, especially by nonruminants and are largely responsible for the relatively low metabolisable energy values. Dehulling would improve metabolisable energy values, but hull utilisation and loss of oil and cotyledon material in the hulls remains problematic.

### *Uses in animal feeds*

#### *Pigs*

Canola meal is a proven supplemental protein source in pig diets. Although poor palatability and reduced digestibility with young (6–20 kg) pigs prevent its use at levels higher than 25 per cent of supplement protein, research has confirmed that canola meal is a desirable supplemental dietary protein for growing, finishing and reproducing swine. Canola meal can be used at 50–75 per cent of the supplemental protein source in diets for growing (20–60 kg) pigs and as the sole supplementary dietary source for finishing pigs (60–100 kg) and dry and nursing sows.

#### *Poultry*

Canola meal is widely accepted for use in poultry diets. It is usually limited to 10 per cent of the diet in layer feeds because of increased bird mortality at higher levels. Long-term studies of egg production show no effect of canola meal on the number of eggs produced per bird. However, there is a small decrease in egg size due to slightly lower feed intake. Canola meal and rapeseed meal have an interesting effect on brown-shelled layers. These birds are deficient in an enzyme responsible for the breakdown of trimethylamine. As canola and rapeseed have high levels of choline and sinapine (precursors of trimethylamine), the eggs of these layers have a fishy taint. For this reason, a maximum of 3 per cent of canola meal may be used in the diets of brown-shelled egg layers. There is no concern about increased mortality at high canola meal inclusion rates for growing poultry such as broilers, turkeys, pullets, or waterfowl. When grower diets are appropriately balanced for energy and levels of digestible amino acids, canola meal can be effectively used as the major supplemental protein source.

### *Beef and dairy cattle*

Canola meal has gained widespread acceptance as a protein supplement in beef and dairy rations. Research has shown its effectiveness in a variety of production and management situations. Lactation trials have demonstrated that canola meal will maintain or slightly improve milk production relative to soyabean meal-based rations. Improved milk production may in part reflect the amino acid content of the bypass protein fraction of canola meal. In beef cattle rations, animal performance has been shown to meet or exceed industry standards when canola meal is incorporated. Canola meal can be used as the sole protein supplement in rations for growing and finishing cattle.

### *Use as a human food*

Rapeseed, as a protein source for humans, has many obstacles to overcome. The glucosinolate and fibre contents require applications of new processing technology to eliminate antinutritional qualities. Rapeseed flours, protein concentrates and isolates are lower in protein but higher in crude fibre and ash content than corresponding soyabean products. Rapeseed flours are comparable to soyabean flour in water adsorption and give much higher fat adsorption, oil emulsification and whippability values. Rapeseed flour viscoamylograph curves exhibit high viscosities but poor gelation properties. The utilisation of rapeseed products may also be limited by green or brown colours in aqueous systems.

### *Other uses*

Rapeseed or canola meal has been used as fermentation substrate. It has been included as an additive in compost for commercial button mushroom (*Agaricus bisporus*) production with good success. Canola meal has also been tested as a substrate for xylanase production by *Trichoderma reesei*. Results from this work indicate that the hydrolysis of canola meal by this enzyme system might be useful in converting this material to fermentable sugars that could be further processed to expensive end products such as solvents and chemicals.

## Sunflower

### *Sunflower meal*

The vegetable oil extraction industry produces three types of sunflower meal: undehulled meal containing 28 per cent protein and 25–28 per cent fibre, partially dehulled meal containing 35–37 per cent protein and 18 per cent fibre and double-dehulled sunflower meal containing 40–42 per cent protein and 12–14 per cent fibre. Thus, the composition of sunflower meal is dependent on the efficiency of the dehulling process.

The protein concentration and amino acid composition of sunflower meal also vary with the source of seed and high-temperature processing may have deleterious effect on its lysine content. Generally, however, sunflower meal exhibits a well-balanced amino acid composition with an essential amino acid index of 68, compared to 79 for soyabean meal and 100 for whole egg.

The energy content of sunflower meal compares favourably with that of other oilseed meals and increases as the residual oil content increases and as the fibre content decreases. Sunflower meal also compares favourably with other oilseed meal as a source of calcium and phosphorus and is an excellent source of water-soluble B-complex vitamins, namely nicotinic acid, thiamine, pantothenic acid, riboflavin and biotin.

Sunflower meal contains the polyphenolic compound, chlorogenic acid, which results in a yellow-green colouration following oxidation in the presence of alkali. The production of protein isolates and concentrates from sunflower meal/flour would require the removal or inactivation of chlorogenic acid.

### *Use as animal feed*

Sunflower meal can be fed to all classes of livestock. Most sunflower meal is fed to ruminants and is comparable, nutritionally, to cottonseed meal. High levels of sunflower meal are used in dairy, beef and sheep rations.

For swine, low-fibre sunflower meal is inferior to soyabean meal as the sole source of supplemental protein. This inferior performance is the result of lower palatability and nutrient content. In swine rations, with 20–30 per cent of the protein from sunflower meal, rates of gain are similar to soyabean meal, but larger quantities of meal are required. Lysine supplementation reduces this requirement. Studies have indicated that sunflower meal can effectively replace 50 per cent of the soyabean meal in growing–finishing swine rations. Higher rates of utilisation are possible as animals increase in weight because of the decreased requirement for essential amino acids.

Rations for laying chickens could incorporate low-fibre sunflower meal, to a level of 50 per cent of the protein concentrate portion, without significantly reducing egg production. Higher levels of addition are possible with lysine supplementation but may cause egg staining due to the presence of chlorogenic acid in the meal. Metabolisable energy trials with laying hens yielded a value of 2205 kcal/kg (dry matter basis) for dehulled, solvent-extracted meal.

The use of sunflower meal is often limited by its availability. Adequate volumes, available on a sustained and consistent basis, would ensure a high utilisation of sunflower meal in animal nutrition. Despite its dark appearance, lower energy and higher fibre content, as compared to soyabean meal, sunflower meal is a competitive product with potential for continued improvement through the use of tail-end screening to further reduce its fibre content.

### *Use in human foods*

Confectionery sunflowers have a history of use in the snack trade. The roasted seed has a pleasant nutty flavour. Dehulled and roasted sunflower kernels can be used as a nut substitute in many confectionery and bakery formulas. Physical and organoleptic analysis of colour, flavour, texture and acceptance indicate that a 10–15 minute roast at 177°C is the most desirable processing technique.

Chemical and physical analysis of hexane-extracted sunflower meal indicates that discolouration due to the oxidation of chlorogenic acid is a problem. An attractive cream colour, relatively bland flavour and excellent stability are possible if the processing conditions do not induce chlorogenic acid oxidation. Unfortunately, the conventional methods of making protein isolates promote the oxidation of chlorogenic acid. Studies have shown that organic solvents produce good extraction of polyphenols from sunflower seed and meal. The physicochemical properties of proteins from such extracted meals indicate no significant differences in amino acid content and only slight changes in nitrogen solubility due to protein denaturation. Sunflower flour and protein isolates have excellent emulsion and whipping properties and thus have great potential as functional agents and protein supplements in human food products, provided the polyphenols are removed or care is taken to prevent their oxidation.

### *Sunflower hulls*

#### *Chemical composition of sunflower hulls*

The hulls, a by-product of oil extraction, comprises 22–28 per cent of the total weight of oilseed sunflower and may be removed before or immediately following oil extraction or may remain in the meal. Sunflower hulls contain: 4 per cent crude protein; 5 per cent lipid material, including wax, hydrocarbon, fatty acids, sterols and triterpenic alcohols; 50 per cent carbohydrates, principally cellulose and lignin; 26

per cent reducing sugars, of which the majority is xylose; and 2 per cent ash. The high fibre content and low protein and energy content of sunflower hulls reduce their nutritional value.

### *Use in animal feeds*

Hulls can be used in ruminant feeds when finely ground and mixed with other ingredients. When adequate energy is provided, sunflower hull pellets may be used as portion of the roughage component in ruminant rations due to their high content of cellulose and lignin. They are used to add bulk to concentrated rations and to absorb liquids such as molasses. Sunflower hulls are sold to feed manufacturers and livestock feeders at prices comparable to those of other ingredients.

### *Hulls as a source of fuel*

The utilisation of sunflower hulls as a source of fuel has been studied. The heat value of hulls alone is 19.2 MJ/kg, whereas the heat value of hulls and meal combined is 23.6 MJ/kg. The higher heat value suggests that the combination of hulls and meal makes a better fuel. In many countries, the burning of sunflower hulls offers an alternative to higher priced fuels. The resulting ash has a high percentage of potassium and can be used as a fertiliser. Hulls have been pressed into cylinders with wood waste and sold as fire logs.

It has been reported that because of their high content of reducing sugars, it is possible to produce furfural and ethyl alcohol from sunflower hulls. Sunflower hulls also represent a source of lignocellulosic material for acid hydrolysis and fermentation. As a lignocellulosic waste material, sunflower hulls can be hydrolysed with acid to yield material suitable for the production of single-cell protein. Purple-hulled sunflowers contain anthocyanin, which may be useful as a natural red food colourant. Economic analysis suggests that the processing of these unique hulls may be economically justifiable.

### ***Sunflower stalks***

Finely chopped and dried stalks could be used as deburring and polishing abrasives in the metal manufacturing industry and are a replacement for peat moss in plant starter mix. It has been reported that stalks are easily processed and decolourised by existing pulping and fibre processing techniques. The processed material can be made into acoustical tile. This material weighs less than 60 per cent of standard acoustical tile and has better sound absorbency and strength. Sunflower heads and stalks also represent a potential source of low-methoxyl pectin for use in low-sugar jams and jellies.

## Safflower

Safflower is a minor oilseed crop limited in production by environmental constraints and by the plants's spiny nature. Unless the seed is well dehulled, the oilcake resulting from oil extraction will have a high fibre content. Undecorticated oil cake has a protein content of 20–22 per cent and an end use as manure. In contrast, removal of the hull improves the protein content to 40 per cent, making it acceptable as cattle feed despite low lysin levels. Leftover hulls and husks are added to cattle feed or are used to manufacture cellulose, insulation and abrasives.

## Cottonseed

Gossypol is a yellow-green polyphenolic pigment contained in discrete bodies in cotton leaves, stems, roots and seeds. This form of gossypol is readily extractable with 70 per cent aqueous acetone. The glands are ruptured during processing and the released pigment structures are highly reactive with other cottonseed components such as protein. The gossypol binds to the biologically available lysine, effectively reducing the concentration available to an animal. Gossypol also causes toxic effects in monogastric animals, including humans. An additional complication related to the presence of gossypol is the production of dark-coloured pigments in the oil and meal that cannot be removed by conventional refining and bleaching operations.

The adverse physiological effects of free gossypol on monogastric animals may be counteracted by making free gossypol a bound form during processing, either by precooking, followed by pressure and shearing in a screw-expeller, or by binding to ferrous salts. A polar solvent, such as aqueous acetone, acetone–hexane–water, or hexane–acetic acid, may then be used to extract residual free gossypol from the meal. The presence of bound gossypol reduces the protein efficiency ration, presumably by reducing the availability of lysine. Care is required to prevent thermal damage to protein, which would further decrease the nutritive value of the meal.

Utilisation of glandless cottonseed stains is an alternative to the extensive seed treatment necessary to lower the gossypol content. The goal of plant selection programmes has been to minimise the total gossypol concentration in the raw material. Presently , the glandless strains that have been developed are not widely produced due to concerns related to the influence of these gossypol-free characteristics on the yield and the quality of the commercially more valuable cotton fibre. Studies indicate that this concern is unfounded. An additional concern that has been addressed is that gossypol and related terpenoids are natural insecticides, such that the use of glandless cotton may encourage insect preference for the glandless cotton. It has been shown that insects do not prefer either strains.

Despite the presence of gossypol, interest in the cottonseed cake has developed as a result of its high content of protein, the valuable component in cottonseed by-products. A number of commercial products from defatted cottonseed have been extensively used in the past. Proflo flour contains 55–60 per cent protein and 4.5 per cent fats. This product was a nonallegenic dietary protein source contributing functionlities such as emulsification, antioxidation and water absorption to bakery-type products. Commercial production was suspended due to the limited market, but it is still produced for nonfood industrial purposes.

### *Cottonseed meal*

Cottonseed meal is second only to soyabean meal with respect to the quantity produced worldwide. This by-product of oil extraction is used in rations for cattle, sheep, goats, horses and mules. Neither glandless nor normal cottonseed meal is palatable to young pigs.

Broiler poultry feeds often contain cottonseed meal, with the potential to cause depressed weight gain and reduced feed efficiency. Cottonseed meal is not used in layer feeds since the gossypol produces a yellow discolouration of the yolks and whites of the eggs. Ferrous iron is added to most poultry diets containing cottonseed. A recent study of low-gossypol cottonseed meal found that it successfully replaced soyabean meal in the diet of broiler chickens with no harmful effects. A number of alternative uses for cottonseed meal exist. Adhesive and fibre production have used cottonseed meal as protein source. Plastics that contain cottonseed meal in equal parts with cottonseed hulls and phenolic resin have excellent flow properties, a short curing cycle, water resistance and strength.

### *Cottonseed hulls*

Hulls are used as roughage in animal feed and as mulch and soil conditioner. Additional uses for cottonseed hulls include fuel, insulation and a xylose and furfural source. Raffinose derived from cottonseed hulls is used in culture media.

## Palm

The major producers of oil palm products are located in the equatorial tropics and include Malaysia, Nigeria, Indonesia, China, Zaire and Cameroon. Palm fruit, when pressed, yields, approximately 43 per cent of crude palm oil and 57 per cent of press cake, which consists of 35 per cent pericarp (fibre) and 65 per cent nuts. Palm nuts consist of 83 per cent shells and 17 per cent kernels, which, when pressed yield approximately 50 per cent of each of palm kernel oil and palm kernel cake.

### *Palm kernel cake*

Palm kernel cake (PKC) protein is of average quality, which, at a level of 19 per cent, is the lowest of the commercial oil cakes. The positive characteristics of PKC are a valuable calcium to phosphorous ratio, a 48 per cent carbohydrate level, a 5 per cent oil content and a 13 per cent fibre content.

The gritty texture of PKC limits its use in feed for monogastric animals. It is used in poultry diets at an optimum level of 15 per cent in broiler and 20 per cent in layer diets and provides 1500 kcal/kg of metabolisable energy. Pig diets may contain PKC if blood meal is added as a supplement. Pigs will consume 20–30 per cent of PKC in their rations if it is introduced gradually to young animals, which otherwise find it distasteful.

The high fibre of PKC is necessary for dairy cattle to prevent metabolic and digestive problems. Each adult animal requires 2–3 kg each day. The level of fibre found in PKC prevents deficiency problems in lactating cows and may increase the fat content of the milk.

### *Palm fibre*

Palm press fibre or pericarp fibre includes not only palm press fibre from the oil extraction process but also empty fruit bunches and kernel shells. Its high fibre and lignin content, comparable to wood and low palatability limit its use in animal feed. It exhibits very slow digestion in the bovine rumen such that processing to increase its nutrient content is required prior to use. Supplementation with molasses, urea and vitamins allows palm press fibre to be used as fibre source. Urea produces an alkaline effect on the fibre and adds nitrogen to the feed.

The predominant use of palm press fibre is as biomass fuel for oil mill plants. The palm press fibre and kernel shells are burned to produce steam for generation of electricity. The potential heat energy of palm fibre is 4420 MJ/kg and of the shell 4848 MJ/kg. One tonne of shells and fibre used for fuel produces 578 kg of steam. The ash from the burning of this solid waste does not contain sufficient nutrients to be used as a fertiliser and dumping creates an air borne hazard and pollutant. It has been incorporated in concrete as a replacement for cement with a slight increase in the setting time but within American and British standards. Palm shells are composed of 20 per cent free carbon and are suitable for production of charcoal or activated charcoal. Empty bunch waste can also be used as a field mulch.

## Coconut

Coconut palms have the greatest economic value and distribution of all the palms and are considered the most useful of all plants, after grasses.

### *Copra cake*

Copra cake is a by-product of oil extraction from coconut. The dehusked coconut is split and the meat is scraped from the kernel cup and dried. The oil is extracted via expeller or solvent processes from the dried coconut meat copra. This product is available throughout the year, making it a cheap local source of animal feed. The cake is ground to meal for use in feed for poultry, cattle, sheep and swine. Copra cake can be used as a substitute for fish meal in swine feed but may cause constipation.

There are problems associated with the use of copra cake in feed. As the amount of cake in the feed increases, it palatability decreases. Copra meal tends to be less digestible than fresh coconut meal. Despite a protein content of greater than 20 per cent, the addition of methionine and lysine improves growth and feed utilisation. The method of oil extraction does not appear to influence the quality of meal produced. Neither expeller-extracted nor solvent-extracted copra meal, at levels of 10–14 per cent in poultry diets, caused any difference in egg production, mortality or efficiency of feed conversion.

### *Coconut flour*

Coconut flour is produced from the shredded kernel, dried in a continuous countercurrent drier and subsequently extracted with solvent to remove the residual oil. The white meal produced contains 25 per cent protein and 65 per cent carbohydrate, as well as various minerals and vitamins. This coconut product is used as high-protein additive to enrich other flours such as wheat, rice and corn flour. Coconut flour has properties comparable with those of wheat flour in the preparation of bread, biscuits and other food products. The protein quality of coconut flour may be very high (lysine 19 per cent; cystine, 8–9 per cent; histidine, 5–6 per cent; methionine, 3–9 per cent), making it desirable for use in baby food and convalescent food drinks. Its content of essential amino acids may be reduced, however, if excessive heat is generated during mechanical oil extraction. The flour also possesses a relatively high crude fibre content.

### *Coconut shell*

The coconut shell comprises 27 per cent of a dehusked nut by weight. This by-product of oil production has many local uses. Coconut shell has a composition similar to that of hardwoods but has a slightly higher lignin content and a lower cellulose content. In southern India and Sri Lanka, the shells are used directly as fuel in villages and small holdings and by local industries such as laundries, bakeries and iron foundries.

Charcoal is manufactured using a simple process. The air surrounding burning shells is limited, encouraging slow carbonisation rather than ash production. This process takes place in a kiln over a 3-day period with careful consideration given to the balance of conditions and time. The final product represents 30 per cent of the original weight of the shells. Charcoal is a preferred fuel as it produces no waste material when burned. Hot embers emit infrared wavelengths that are valuable in the cooking of foods such as fish, meat, or tubers.

Coconut shell produces heat energy at a level of 23 MJ/kg whereas shell charcoal produces 30 MJ/kg. Shell charcoal is also used to manufacture calcium carbide and the carbon electrodes of electric batteries. Both shell and charcoal generate producer gas. Reactors utilising this product are sold for refrigeration, water pumping and ground and marine vehicle operation.

Destructive distillation of the shell produces some interesting substances. Upon exposure to very high heat in the absence of air, the shell forms products from all three phases (gaseous, liquid and solid) as non-codensible gases, pyroligneous liquor, settled tar and retort charcoal will be generated. The pyroligneous liquor, a dark red and odourous liquid, yields acetic acid, methanol (locally called 'wood naptha'), and a variety of other products. The liquor may be used as boiler fuel and noncondensible gases may be compressed in cylinders for use as domestic cooking gas.

Coconut shell is the source of two other products, coconut shell flour and activated charcoal. Powdered coconut shell is used in the plastics industry as a compound filler for synthetic resin glues. It is also used as a filler and extender of phenolic moulding powders that give a smooth and lustrous finish to moulded articles, thereby improving their resistance to moisture and heat.

Activated charcoal is an adsorbent for toxic agents. It has been used in gas masks, but can also be used to remove odours and industrial stench. As well, this by-product is a contact catalyst used to facilitate some industrial chemical reactions.

### *Coconut husk*

The husk represents 35 per cent of the intact, mature coconut by weight. A number of products are derived from this by-product of oil production. Particles of husks may be consolidated with little or no adhesive. Pith is chipped with the husk material to produce the best bonding characteristics. The resulting self-bonding chips can be formed into boards of varying densities that are strong, durable, water repellent and fire retardant. These boards are used as low-cost construction products such as roofing panels. Boards with a density of less than 400 kg/m$^3$ are used for thermal insulation whereas medium-density boards of 500–900 kg/m$^3$ are used in construction and furniture. A combination of 25 per cent finely ground coconut husk and 75 per cent andesitic sand by volume has been used as a potting medium in nurseries.

### *Coir*

Coir is a valuable and versatile fibre derived from the coconut husk. The best quality coir is produced from green coconut, which is more difficult to harvest and has a lower copra yield than more mature coconut. The amount of copra and the quantity of coir produced are inversely related. Husks must be retted to manufacture coir. This process involves holding the husks under water, away from air, with mud and leaves for a period of few months to a year. Fermentation is accomplished by short rod bacteria such as *Pseudomonas, Rerobacter* and *Bacillus*. The microbial process is a polyphenolic degradation in which the pectic substances are decomposed. Slow moving and slightly saline water in a nature source speeds the process and produces a better quality fibre.

### Groundnut

Groundnut is comparable in nutritional value to more expensive animal-derived foods. Protein from this oilseed has the highest quality of the vegetable proteins, equivalent to casein. The oilseed cake from commercial-grade groundnuts is used for animal feed, whereas solvent-extracted edible-grade groundnut cake is milled into flour. The digestibility of groundnut meal is high when it is well dehulled. Groundnut meal is lower in available lysine than soyabean meal but contains a greater quantity of sulphur amino acids. Peanut flour has an amino acid content comparable to that of raw and roasted peanuts, indicating that moderate heat treatment does not alter the amino acid compositions.

A new process for peanut flour production has been developed that requires treatment of the seed with both heat and moisture prior to oil extraction. The flour end product is white and bland, contains 65 per cent protein and is devoid of peanut flavour. This allows the addition of this product to a variety of food products without disturbing colour, flavour, or texture. A high-quality flour has been used in the treatment of hemophiliac patients.

Protein produced from solvent-extracted meal has been used as a thickening agent in soups, baby foods, high-protein foods, institutional meals and meat products. Groundnut protein have also been used to manufacture a soft, wool-like, cream-coloured fibre, adhesive products such as plywood glue and wettable glue, and for paper coating.

Fermentation of peanut protein with micro-organisms increases the level of some essential amino acids. *Rhizopus oligosporus* is widely used to produce a tempehlike product. A similar product, oncom or ontjom, is produced in Indonesia using the fermenting agent *Neurospora sitophila*. The difference between the two products is the red colour of oncom, derived from the red hyphae of *N. sitophila*.

The seed coat of peanut is a commercial source of tannins and thiamin and sees limited use in feeds, primarily as a bulking agent in reduced-calorie pet foods. The presence of the seedcoat in meal may lead to a decrease in the availability of lysine. This apparent waste product is also used as mulch, fuel, litter for poultry houses and in the production of a high-grade activated charcoal.

The hull of the peanut is low in crude protein and exhibits low digestibility. This has limited the utilisation of this by-product. Physical and chemical treatment of the hulls have been applied to promote digestibility, with little success. Hulls are generally considerable waste products.

### Olive

Olive oil is produced in warm-temperature and subtropical regions where the olive tree grows well. The fruit develops maximum oil content during the mid-November to February or March period in the Mediterranean basin and from May to June in the Southern Hemisphere. Ninety percent of the olives grown is used to produce oil.

Initial extraction of oil from oil fruit produces a cake of fruit skin, pulp and kernel known as olive pomace or orujos. The value of this primary by-product of oil extraction depends on its oil and water contents, which are in turn, determined by the method of oil extraction employed and the operating conditions. Pressure extraction yields a residue containing 4–5 per cent oil, whereas classical presses leave 8–12 per cent oil in the pomace. Pomace flours are used as animal feed due to their high content of protein, which is also of high quality.

Secondary oil extraction from the pomace produces olive pomace oil containing large amounts of free fatty acids and is considered an inferior oil compared to virgin olive oil. Pomace must be dried in long revolving horizontal cylinders through which hot air is passed before the oil can be extracted.

The exhausted olive pomace is called kernel wood or orujillo. This product has few uses due to its low protein content and its high content of woody and cellulosic materials. Kernel wood is predominantly used as fuel for operation of the processing plant. Ash from kernel wood is used as fertiliser because of its potassium, phosphorus and calcium content. The low value of this by-product negatively influences the overall value of olive products, contributing to the high price for olive oil.

### Sesame

Sesame has declined in international trade due to a market preference for other oilseeds that are cheaper and easier to produce, such as groundnut. There is great value in dehulling if the product is to be used as a foodstuff, as removal of the hull lowers the oxalic and phytic acid levels in the meal. The presence of these components may decrease the bioavailability of calcium, magnesium, zinc and perhaps iron. The protein content, acceptability and enzymatic digestibility of the meal will also increase. There are problems associated with dehulling, however. It is difficult to remove the hull from sesame and there will be an associated loss of minerals. Whole press cake has a bitter taste due to the presence of the testa and is best used for manure and soil conditioning.

Cake or meal from oil extraction contains 40–50 per cent protein when processed in a screw press and 56–60 per cent protein after solvent extraction. Sesame products have a pleasant flavour and contain high levels of methionine and cysteine. The flour produced from sesame meal has a high nutritive value compared to other oilseed flours.

Sesame has specific uses in confectionery products such as halva, sesame seed cake, candies and as a garnish on bread and rolls. Microatomised protein food for feeding unweaned babies represents another use for sesame and other oilseed flours when enriched in vitamins and minerals.

### Linseed/Flax

Linseed is grown predominantely for its quick drying oil and for the fibre from its stalks. Whole seed is often shipped without processing at the location of harvest. Oil cake contains approximately 30 per cent protein and is used in feed for sheep, horses and dairy and beef cattle. A high mucilage content endows many positive properties to the meal. Linseed meal is comparable to soyabean meal in composition but energy and protein digestibility are lower than for most other oilseeds. Linseed meal endows 'bloom' and mellowness to the hide, which is most valued in the preparation of animals for shows. These qualities can be attributed to residual oil in the cake, slight laxative effects, and appetite stimulation. Care must be taken when linseed meal is used as feed for poultry. It contains $B_6$ antagonist, N-(γ-L-glutamyl)-amino-D-proline, such that supplemental vitamin $B_6$ must be used to prevent any detrimental effects from occurring. Low levels of lysine and methionine must be balanced using supplements before linseed meal can be used for swine feed. In addition, the possible presence of hydrocyanic acid should be monitored since it varies in concentration according to growing conditions. The cake has seen limited use as a soil conditioner.

## BY-PRODUCTS OF OIL REFINING

### Lecithin

Lecithin, an edible by-product of oil processing possessing a variety of useful functionalities, is primarily a mixture of phospholipids such as phosphatidyl-choline, phosphatidylethanolamine, phosphatidylserine, phosphatidylinositol and phosphatidic acid and contains minor quantities of other water-soluble or hydratable components such as glycolipids and oligosaccharides. The degumming of oil with water or a degumming agent such as citric acid, phosphoric acid, oxalic acid, acetic anhydride, or maleic anhydride yields a lecithin sludge and a degummed oil. Continuous agitated thin-film evaporation removes the approximately 50 per cent water content of the lecithin sludge, resulting in a highly viscous semiliquid or powdered product.

Soyabean is the predominant vegetable source of lecithin due to its availability and outstanding functional characteristics. Lecithin products from rapeseed and sunflower are increasing in market. Glandless cottonseed and corn are also potential commercial sources. The composition of rapeseed lecithin is very similar to that of soyabean but is considered to be of lower quality with respect to colour, flavour, taste and general appearance. This product has been predominantly used as a dust control agent in rapeseed meal and is added to livestock and poultry feeds. The development of canola or lowerucic-rapeseed varieties and specialised refining processes have increased the use of rapeseed lecithin. Glanded cottonseed contains more phospholipids than any other oilseed with the exception of soyabean. Solvent extraction methods, however, cause toxic gossypol to bind to the phospholipids. The resulting lecithin is dark brown in colour, limiting its use in food products.

#### *Purification and fractionation of crude lecithin*

Purification consists of the removal of nonlecithin components such as carbohydrates, proteins and other containments. Vegetable lecithin is more difficult to purify compared to other sources of lecithin because of its higher viscosity. The crude product obtained during degumming is contaminated with fines derived from seed material, hulls and other seed impurities. Regulatory standard require that these containments be removed because of the high levels of iron and heavy metals in the impurities, which interfere with the oxidative stability of the products to which lecithin is added. Methods used for purification include filtration, which can be performed on the crude oil, lecithin, or miscella, partitioning between organic solvent and water or salt solution, dialysis and adsorption on cellulose or Sephadex columns.

Fractionation takes advantages of the varying solubilities of different phospholipids in different organic solvents. Many options are available for large-scale production of marketable lecithin. The de-oiling process is one fraction system in which the separation of neutral and polar lipids is based on the insolubility of polar lipids in acetone; more then 60 per cent of lecithin consists of acetone-insoluble substances. Oily lecithin is combined with excess acetone and mixed vigorously, thereby solubilising triglycerides in the acetone. This process is repeated, after which the polar lipids material is dried and sold as a light yellow powder or in the form of granules. De-oiling on a smaller scale can be accomplished using adsorption of lecithin, in a hexane solution, on a silica column. Also, a new technology has been developed for treating the lipid mixture with a supercritical fluid. For example, carbon dioxide at a temperature of 40°C and a pressure of 300 bars has a solubility similar to that of liquid acetone. The gas may be recovered and reused. Oxidation of the lecithin does not occur since oxygen is displaced by carbon dioxide. The absence of solvent residues in either the oil or the lecithin eliminates flammability and environmental concerns. This process, however, has a number of drawbacks and its commercial feasibility has not been established.

Crude fractions can be obtained by solvent fraction treatment utilising lower alcohols, such as ethanol, or alcohol–water mixtures. The product is a soluble fraction rich in phosphatidylcholine, whereas phosphatidic acid and phophatidylinositol predominate in the insoluble fraction. The shift in the ratio of phosphatidylcholine to phosphatidylethanolamine improves the emulsification and antispattering capabilities of the soluble fraction. The products of this process can be used as they are or can be further purified with adsorbents. The soluble fraction is an excellent oil-in-water emulsifier and is predominantly used in margarine. The acidic phospholipids of the insoluble fraction are used in water-in-oil systems. The chocolate manufacturing industry uses this fraction to increase the viscosity of chocolate masses, thereby reducing the requirement for cocoa butter.

It is not possible to obtain pure phospholipids from lecithin by solvent fractionation. Chromatographic adsorption processes are capable of separating this complex mixture but cannot be used, in practice, to generate large quantities of the pure phospholipids. Chromatographic adsorption methods include aluminium oxide with ethanol or chloroform/methanol, silica gel with a variety of solvent system and a diethylaminoethyl–cellulose system. A high price is demanded for purified phospholipids.

### *Functionality and utilisation of lecithins*

There exist a number of functional groups in lecithin that can be modified. Phospholipid products derived by hydrolysis, hydroxylation and acetylation are the most likely to be used commercially. Hydrolysis can be performed with phospholipase A, acid or alkali. The resulting lecithin contains 56 per cent or more of acetone-insoluble material. Products of hydrolysis have improved hydrophilic and emulsifying properties. Hydrolysed lecithins are highly viscous or pasty fluids and tend to be light brown to brown in colour. The products resulting from hydroxylation of lecithin have improved oil-in-water emulsifying properties and improved water dispersibility. Acetylation of phosphatidylethanolamine creates improved fluid and emulsifying properties, as well as improved water dispersion. Modification of the polar phosphoric acid ester or the glyceride moiety of phospholipids is legally restricted in the food industry.

The wide range of functionalities of lecithin has been applied in a variety of industries, including pharmaceuticals, cosmetics and food. Lecithin components have both lipophilic and hydrophilic groups that respond to changes in pH and differences in ionic strength. These charged surfactants stabilise emulsions of water and oil. The selection of the appropriate raw material and the use of techniques such as fractionation, modification and compounding can be combined to generate the phospholipid characteristics best suited to a desired application. Different phospholipids and fatty acid combinations influence solubility, emulsifying quality, the type of emulsion (whether oil–water or water–oil), instantising properties, dietetic value and sensitivity to oxidation.

Lecithin acts as a lubricant and release agent between solids by coating the surface of the solids. A reduction of the surface tension and particle attraction occurs when lecithin coats solids in a solid-liquid mixture, which allows for stable dispersion and suspension of the solid in the liquid. Lecithin also reduces surface tension and enhances emulsification in immiscible liquid mixtures.

Lecithins makes up 0.5–1.0 per cent of the composition of many cosmetic products. The surfactant properties are most valued for the 'skin feel' that is produced. Longer wearing cosmetics evolved from pigments and particulates coated in lecithin since they have smoother surfaces, increased adhesion to the skin, and improved colour stability. Due to film adhesion, the presence of lecithin reduces the oily or greasy feeling of many products and reduces the transfer of cosmetics of clothing. The emulsification properties, ease of spreading and wetting ability of lecithin are also utilised in cosmetic products.

Lecithin may become a useful component of magnetic recording equipment due to its ability to act as surfactant, thereby facilitating dispersion of magnetic particles on pigment surfaces. The magnetic and physical properties of the recording tape are enhanced as a result. For lecithin to be active in this capacity, it must function in a variety of solvents including methylethylketone, tetrahydrofuran, cyclohexane and toluene. Another valuable property of lecithin is its ability to absorb to the surface of a variety of pigments such as iron oxide, chromium dioxide, iron metal and barium hexafurite.

Industrial coatings, paints and inks utilise lecithin because of its pigment dispersal characteristics. Lecithin binds to pigments surfaces allowing wetting of the pigment by the vehicle in which it is being dispersed. Lecithins with different functionalities can function in both oil-based and water-based formulations.

The food industry relies on lecithin in bakery, beverage and confectionery product development. The lecithin functionalities of emulsification, release, mixing and blending and instantising many of which were discussed above, are put to use in many aspects of food production. Lecithin is used in baking as a dough conditioner for cookie, cake and doughnut mixes with many positive effects including improved handling, a drier and more elastic dough, improved pan release, more uniform colour, texture and grain and decreased mixing time. Lecithin acts as a dispersor, aiding the mixing of unlike ingredients such as fats and flour or sugar. The activity of lecithin as a surface-active agent has been shown to retard the rate of staling in yeast-leavened products. Lecithin plays a valuable role in the instantising of beverage and food mixes by promoting the incorporation of powders into aqueous environments. Dry edible powders such as cake mixes, nutritional supplements and milk powders can be quickly integrated into liquids such as milk or water with the aid of lecithin.

The confectionery industry utilises the emulsification, antistick and viscosity properties of lecithin and benefits from the concurrent effects of shelf-life extension, texture improvement and decreased production costs. A product such as caramel will not blend correctly in the absence of lecithin. Uniform dispersion of fat, aided by lecithin, will decrease stickiness and provide tenderness for ease of cutting. The natural antioxidant properties of lecithin slow the decay of any product in which it is incorporated. Viscosity is very important in the chocolate industry where shape is often a requirement for consumer acceptability. High concentrations of butter, such as cocoa butter, impart high viscosity, which in turn makes production more efficient. Alternatively, lecithin may be used to provide a portion of the viscosity requirement and to eliminate greasiness in the finished product.

### Refining By-products Lipid (Soapstock)

Refining by-products lipid, commonly referred to as RBL, results from the refining of crude oil, where continuous mixing of crude oil with a dilute sodium hydroxide solution produces a by-product consisting of free fatty acids, hydrolysed phosphatides and unsaponifiable materials. Free fatty acids are the valued component in RBL, the composition of which varies with the oil source, oilseed conditions at crushing, the oil removal method used, the solvent employed, the extent of extraction and the refining conditions. Larger quantities of RBL are produced as the oil becomes more refined. The concentration of free fatty acids, gums and impurities and the efficiency of refining, influence the amount of RBL formed. Refining by-product lipid tends to be the item of lowest value produced in oil refining. Acidulation of RBL using sulphuric acid stabilises it and reduces its weight for shipment. Debris such as phosphatides, proteins and mucilaginous substances are present in varying quantities depending on the quality of degumming and refining and may cause emulsification to occur and prevent effective acidulation. The acidulation of RBL is the greatest waste-water producers in the refining system. Disposal of the effluent or acid water requires expensive treatment measures to comply with environment regulations. The use of nutrient chemicals in the refining, acidulation and neutralisation steps produces an acid water suitable for use as a liquid fertiliser.

Vegetable oils are refined with caustic potash (KOH) instead of caustic soda (NaOH) and acidulation with sulphuric acid is followed by neutralisation with ammonia rather than NaOH. This is an innovative method that closes the loop of agriculture processing. Animal feed is the dominant sink for RBL. Refining by-product lipid can be returned to meal to increase its weight and its fat content. This has been done to a level of 0.9 per cent for cottonseed meal and 0.4 per cent for soyabean meal. Soyabean RBL provides 6694 kcal/kg of digestible energy and 6599 kcal/kg of metabolisable energy to pigs. Refining by-products lipid not only increases caloric content but also provides essential fatty acids and increases food utilisation. Refining by-product lipid may also be added to feed for purposes of dust control, appearance, ease of handling and improved pelletability.

Raw and acidulated RBL are combined in different ratios with animal tallow to produce soaps of varying characteristics. Palm oil and coconut oil are the dominant fatty acids sources for soap manufacturing. Coconut oil and tallow are complementary in fatty acid composition such that in combination they provide the ingredients of toilet soap. There has been speculation regarding the use of safflower and sunflower RBL in this capacity if alterations were made to processing methods. Cottonseed and soyabean RBLs are available in larger quantities but the cost of upgrading these to the quality necessary for use in toilet soap inhibits their use.

Refining by-product lipid has been considered as a growth medium for micro-organisms. It appears to provide a satisfactory supply of nutrients for growth, due in part to its high sodium content and its suitable trace element composition. Despite its residual oil content and high pH, many species of micro-organisms will grow on this substrate.

## Spent Bleaching Earth

Spent bleaching clay from oil refining contains substantial absorbed oil (20–40 per cent w/w). This product is both a problem and a potential source of recoverable oil. The problems of spent clay are well known. Fat-containing clays are prone to spontaneous combustion when in contact with air, particularly if the bleaching earth contains highly unsaturated oils. Spent clay also represents an environmental concern, both as a fire hazard and as a threat to ground waters through fat-containing runoff when discarded in landfills. In addition, there is an economic loss of oil in the clay.

Technical solutions to the recovery oil from spent clay can be categoried into four areas:

### *Steam treatment*

This method, whereby steam is blown through the cake, is commonly practised in refineries to reduce the oil content of the cake to approximately 20 per cent.

### *Aqueous extraction*

In its simplest form, this procedure involves pumping 95°C water through the cake for approximately 30 minute to reduce its oil content. The oil must then be separated from the water. A procedure employing extraction with sodium carbonated, where spent clay was mixed with an aqueous 5 per cent sodium carbonate solution. The slurry was heated to 95°C and stirred slowly for 30 minute. Although the procedure is simple, hot carbonate water sometimes failed to displace the fat. The quality of the resultant oil was low and disposal of residual slurry [containing clay, water and salt (NaCl)] was difficult. Suggested methods of slurry disposal include inclusion in cement manufacture or blending into sandy soils to improve soil structure. Boiling the spent clay in a water suspension containing 1.5–2.5 per cent concentrated sodium hydroxide as a surface-active agent has also been described. This procedure yielded a dark-coloured oil that could only be used for technical purposes. After extraction, the residual slurry was centrifuged and the liquid effluent appropriately treated. The solid material was used as a landfill material or as a replacement for earth or sand for the covering of refuse dumps.

### *Solvent extraction*

Spent clay oil recovery by solvent extraction with hexane may be accomplished after filtration, either directly from the cake in the filter or after the cake has been removed from the filter. Any exposure of the cake to air prior to extraction will cause rapid deterioration of the fatty material. Depending on the intended use of the oil, perchloro-ethylene or methyl chloride may be effectively employed as solvents.

### *Pressure extraction*

The use of pressure (5–30 bars) in combination with water and sodium hydroxide has been shown to produce acid oils from spent clay that can be easily separated by decanting.

Other options for recovery of oil from spent clay include mixing the spent clay filter cake with milled oilseeds *en route* to solvent extraction. This procedure is used in some refineries having associated crushing and refining plants and is convenient if the fire hazard of the spent clay can be overcome and the level of addition is small enough to not significantly alter the mineral content of the meal.

Evaluations of spent bleaching clay as a feed supplement indicate that, for poultry diets, inclusion rates of up 7.5 per cent spent clay in diets produced no deleterious effects on feed intake, growth rate, or feed efficiency. These results suggest that spent clay could be added to poultry feed at 0.5–2.0 per cent which is similar to the amount of bentonite clay currently used as pellet binder in poultry diets. The metabolisable energy (ME) of spent clay was determined to be 2870 kcal/kg (dry matter basis) but would vary with oil content. Other studies have also demonstrated the feeding value of spent clay.

Spent clay and its associated disposal problems represent a concern for all refineries. Additional research may yield new and more valuable uses for this by-products material.

## Deodouriser Distillates

Deodouriser distillate is the material collected from the steam distillation of vegetable oils, as it occurs in the process of deodourisation or physical refining. The material has a sludge like appearance and consistency so is often referred to as scum oil.

The use and value of deodouriser distillate is dependent on its composition. Deodouriser distillate is a complex mixture of tocopherols, sterols, esters of sterols, mixed fatty glycerides, hydrocarbons and other materials contained in a substantial amount of fatty acids. If the material is high in unsaponifiable components, the tocopherols can be used in a manufacture of natural source vitamin E and sterol for drug manufacture. The quality of the distillate is dependent on the feedstock oil composition, processing equipment and operating conditions.

Comparisons of the tocopherols and sterol contents from various oils indicate that some oils have appreciably higher contents of specific tocopherols and sterols. For example, sunflower is high in α-tocopherols, whereas soyabean is higher in γ-tocopherol. As deodourisation strips tocopherols and sterols from the oil, different feedstock oils yield different concentrations and types of tocopherols.

Direct contact cooling of the deodouriser discharge vapour with a stream of circulating distillate is the most common method of condensing distillate vapour. Although designs of distillate recovery towers and their position in the deodourisation system vary, the purpose remains the cooling of the vapour sufficiently to condense most of the distillates. The operation of the deodouriser has a direct impact on distillate composition and quality. In general, higher deodourisation temperatures, longer exposure times and lower vapour pressures will increase the yield of distillate and decrease the tocopherol and sterol content remaining in the oil. Distillate is frequently collected and sold, representing value to the refinery. The demand and value is based on the total tocopherol content, as it relates directly to the economics of vitamin E production. End users of the distillate use a series of chemical and physical treatments such as saponification, esterification and molecular distillation to separate tocopherols and sterols. Research has been conducted on the use of supercritical fluid extraction to separate and concentrate tocopherols and sterols from the sludge.

With the increasing interest in natural source antioxidants, such as tocopherols and the use of both stigmasterol and β-sitosterol as raw materials for the production of progesterone and esterone by the pharmaceutical industry, it is likely that deodouriser distillates will continue to be an important by-product for some vegetable oil processors.

## Spent Catalyst

Catalysts are required in the hydrogenation of fats and oils. The most commonly used catalyst is finely divided nickel having a large surface area and used in a flaked or granular form. The nickel component and clay support each make up 50 per cent of the catalyst in a surrounding hydrogenated fat medium. Spent catalyst is removed from the hydrogenated oil using filtration methods such as plate and frame or pressure leaf-type filter presses. The recovered catalyst product contains nickels, clay materials and residual fat or oil, as well as any filtration media used to aid in filtration of the physically fine spent catalyst. The heat content of spent catalyst ranges from 4500 to 8700 kcal/kg.

The use of spent nickel catalyst in its posthydrogenation form is limited. Landfills have been the most popular depositories of spent nickel catalyst. Environmental concerns regarding the impact of nickel and conservation efforts to preserve nickel supplied have stimulated recycling and reclamation of the nickel component. Solvent extraction of organic material from the nickel is most effective with polar solvents such as isopropanol and methyl ethyl ketone.

Incineration is another reclamation technique in use. The multiple-hearth furnace can operate under designed conditions for years with low maintenance and low energy consumption. The fluid-bed incinerator is also very effective. Tests have shown that all organic compounds are decomposed and most carbon residue oxidised. The ash is of high nickel value being predominately composed of nickel oxide and silica. The energy input of these reclamation processing methods is fractional compared to the total heat content of spent catalyst. Generation of thermal energy through these processes has the potential for electricity generation as well. Spent nickel catalyst is a product with limited potential for further utilisation. In the future, however, the entire composition of the spent catalyst, including the organic fraction, may be found to have value beyond heat generation.

Chapter 15

# Antioxidants and their Importance in Fats and Oils

## INTRODUCTION

Antioxidants are compounds used in fat and foods to prevent deterioration, rancidity, or discolouration caused by oxidation. They are substances that can prevent or delay oxidation of a molecule. However, the simple term—antioxidant—belies the complex action these compounds have in preserving food against adverse changes initiated in the presence of oxygen. Adverse changes caused by oxidation lead to deterioration of flavour and odour of foods. Since these are important factors in consumer acceptance, many methods are used to prevent oxidative deterioration. Improvements in food preparation, refrigeration and packageing, all help to retard oxidation. These types of improvements, however, may not be sufficient to prevent oxidation, so processors address oxidation by adding antioxidants directly into a food product.

The most commonly used antioxidants include the phenolic compounds such as butylated hydroxyanisole (BHA), butylated hydroxytoluene (BHT), or tertiary butylhydroquinone (TBHQ). Compounds such as ascorbic and citric acids are also used in countering oxidative discolouration in meats and fruits and other foods susceptible to enzymatic oxidation. Other additives serve dual roles; sulphur dioxide, for example, can function as both a preservative and an antioxidant. Besides the more traditional antioxidants, a variety of compounds with oxidation inhibition activity have been studied and proposed for use in foods. These compounds include the natural flavourings such as extracts of rosemary and sage and compounds found in wood smoke. More recent research in this area has investigated items such as rice bran oil and the water-soluble and fat-insoluble polyphenolic catechins present in green tea for use as food antioxidants.

Currently, there is increased interest in antioxidants for their nutritional and health benefits. Similar to the reactions that occur in foods, oxidation can also cause deteriorative changes in the human body. In the body, oxidation is a metabolic process that provides energy for vital cell activities. This normal metabolism of oxygen in living cells leads to the unavoidable production of oxygen-derived free radicals. Scientists have compared these free radicals as roughly equivalent to the automobile exhaust emissions that results from incomplete combustion of fuel. While it is normal to have a balance of pro-oxidant free radicals and antioxidants, disease or poor nutrition can upset this balance. It is postulated that an upset oxidative balance can be a contributing factor in a broad spectrum of disease. These include such conditions as rheumatoid arthritis, bacterial or viral infections, heart attack and stroke, Parkinson's disease, Alzheimer's disease, cancers and others. Some studies seem to suggest that antioxidant vitamins (such as vitamins E and C and beta carotene) may lower the risks of heart disease and cancer. Another reason for the widespread interest in antioxidants indicates that few Americans (less than 10 per cent) meet the dietary recommendations for consuming fruits and vegetables. Fruits and vegetables are the primary source of two of the major antioxidant nutrients—vitamin C and carotenoids. Some individuals believe that antioxidant supplements could be of benefit in meeting nutritional recommendations. The nutritional antioxidant area and the effect of antioxidants on health is of active interest and research.

This section will cover the use, properties and regulation of antioxidants as additives to inhibit oxidation in foods. It will also look at the hunt for alternatives to the more traditional antioxidants. Finally, a sidebar looks at the health and nutritional claims for antioxidants.

## ACTION OF OXIDATION

To understand how antioxidants work, it may help to review the mechanism of oxidation. Oxidation is a key reaction in foods; such reactions can be either chemical or enzymatic. Lipid oxidation, which is a usual cause of deteriorative changes in fat-containing foods and the one we will be most concerned with, is a chemical oxidation reaction. Proteins also undergo chemical oxidation resulting in cross-linking and subsequent changes in the physical and functional properties. Enzymatic oxidation includes the oxygenation of unsaturated fatty acids by lipoxy-genase, the browning reaction catalysed by polyphenol oxidase and the conversion of glucose to lactose by glucose oxidase.

Of primary interest in this section is the oxidation of unsaturated lipids. Oxidation in lipid-containing foods may occur either by autoxidation or by photosensitised oxidation. Autoxidation is a free-radical chain reaction involving initiation, propagation and termination steps (Fig. 15.1). In the first of these, the initiation step, fat free radicals are formed when loosely held hydrogen atoms are lost from the fatty acid group. The resultant fat free radicals react with oxygen to form peroxyl free radicals. These peroxyl free radicals act as strong initiators or catalysts of further oxidation by extracting a hydrogen from another molecule triggering propagation.

In propagation, the peroxyl radicals remove a hydrogen atom from a lipid to form a relatively stable hydroperoxide and a new unstable fatty radical. The unstable fatty radical will then react with oxygen to form another new reactive peroxyl radical. As a final step in autoxidation, the hydroperoxides split into smaller short chain organic compounds such as aldehydes, ketones, alcohols, and acids which cause the off-odours and off-flavours characteristic of rancid fats and oils.

The auto-oxidative process is ended when two unstable radicals react or when a fatty radical reacts with a stable antioxidant radical. Antioxidants can function in one of two ways. They can donate a hydrogen atom to the fat free radical to reform the fat molecule, and thereby terminating oxidation in its first step. This delays the autoxidation process until the antioxidants are used up. Antioxidants also terminate oxidation by donating a hydrogen to a peroxide free radical to form a hydroperoxide and an antioxidant free radical. This antioxidant free radical is much more stable than the reactive fatty radical because of the electron resonance structure in the aromatic ring of the antioxidant. This breaks the chain reaction.

### Types of Antioxidants

Oxidative inhibition by antioxidants is thought to occur through their donation of electrons or hydrogen ions to a fat containing a free radical and the formation of a complex between the antioxidant and the fat chain. To be effective, they must be added as early as possible in the manufacturing process or to the finished fat. Antioxidants cannot reverse the oxidation of rancid oils. No single antioxidant will work to prevent oxidative deterioration for all food products. The selection of the proper antioxidant is determined by compatibility with and resulting effectiveness in certain fats; its application as approved by US federal regulations; its solubility in the fat or water phase of the product; good dispersibility throughout the food; and its stability of 'carry through' to post processing. 'Carry through' is an antioxidant's ability to survive a processing step such as frying or baking and still stabilise the finished food. Food and Drug Administration (FDA) regulations require the antioxidants and their carriers be declared on the ingredient labels of products and that this be followed by an explanation of their purpose. For example, 'BHA added to preserve freshness'.

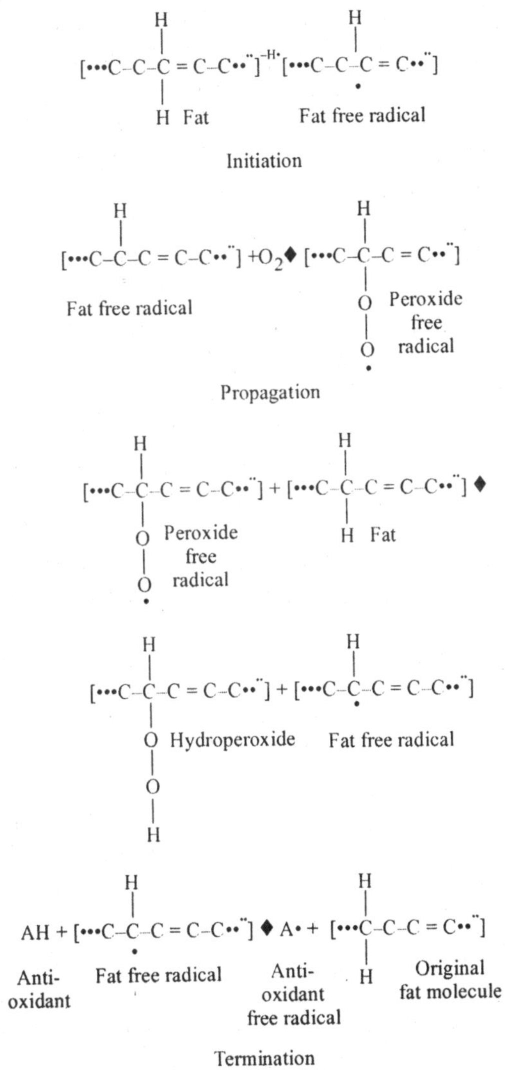


**Fig. 15.1.** Initiation, propagation and termination reactions of oxidation.

## Free-radical Terminators

Free-radical terminators are believed to interrupt the free-radical chain of oxidative reactions by contributing hydrogen from the phenolic hydroxyl groups. These include BHA, BHT, TBHQ, PG and the tocopherols. BHA or butylated hydroxyanisole is a mixture of two isomers. The 3-isomer is the better antioxidant; commercial BHA generally has at least 90 per cent 3-isomer content. BHA is a white waxy solid that is sold in either flake or tablet form. It is effective in animal fats and relatively ineffective in vegetable oils. BHA provides good 'carry through', especially in animal fats in baked goods. BHA is highly fat soluble and is insoluble in water. It is volatile and steam distillable. This makes it useful for addition to packageing. BHA is used in granola bars, dry breakfast cereals, animal fats, vegetable oils, potato flakes, active dry yeast, chewing gum, sausage snack foods and other products. These are discussed further in the end of this chapter. BHT or butylated hydroxytoluene is a white crystalline solid with properties similar to BHA. BHT provides good synergism when used in conjunction with BHA. BHT is widely used and relatively inexpensive. It is used in many of the same products as BHA. TBHQ or tertiary butylhydroquinone is a beige powder. It is the most effective antioxidant for most fats and oils, especially vegetable oils. TBHQ has poor 'carry through' in baking applications, but good 'carry through' in frying applications. It has adequate fat and oil solubility and does not form coloured complexes with iron and copper. TBHQ is used in stabilising highly unsaturated vegetable oils. Propyl gallate is a white, crystalline powder and is very slightly water soluble. It functions particularly well in stabilising animal fats and vegetable oils, but to a lesser extent than TBHQ in vegetable oils. With a melting point of 148°C, propyl gallate loses its effectiveness under heat conditions and is therefore not used in frying applications which involve temperatures higher than 190°C. Because it chelates iron ions to form an unappealing blue-black complex, propyl gallate is always used with the chelator citric acid to eliminate the pro-oxidative iron and copper ion catalyst. It is used in animal fats, vegetable oils and meat products.

Figure 15.2 illustrates that hindered amine antioxidants undergo the same two basic reactions as their fellow radical trapping phenolic counterparts. In addition, the amines can enter into a cyclic process to continuously trap radicals without being consumed. This can yield a longer lifetime at elevated temperature, provided that the amine antioxidant is not lost by volatilisation, extraction or other physical means. Comparison of various amine and phenolic antioxidants are shown in Tables 15.1. and 15.2.

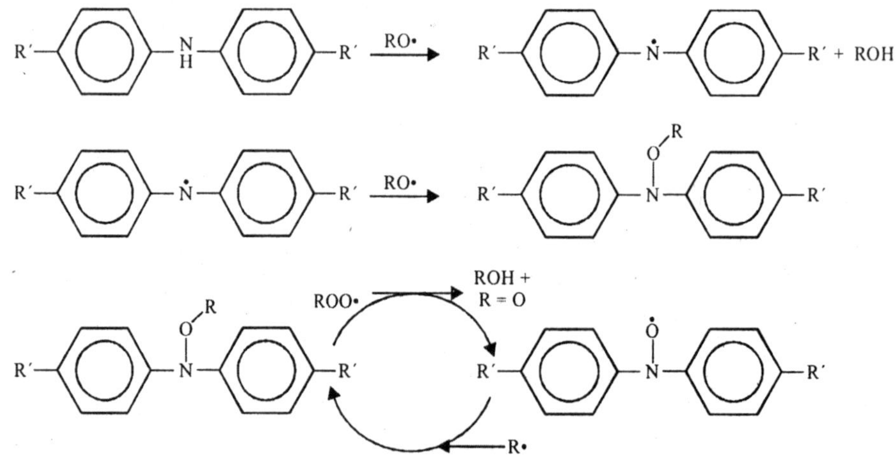


**Fig. 15.2.** Special feature of amine antioxidants.

**Table 15.1.** Comparison of amine antioxidants.

| *Class* | *Heat resistance* | *Flex resistance* | *Metal inhibition* | *Ozone activity* | *Staining and discolouration* | *Volatility* | *Solubility* | *Chemical stability* |
|---|---|---|---|---|---|---|---|---|
| Aryl naphthylamines (PBN) | Good | Good | Good | None | Considerable | Low | No bloom | Oxidises |
| Ketone amine condensates | | | | | | | | |
| 1. Acetone-diphenylamine condensates | V. good | Good<br>V. good | None | None | Considerable | Moderate | No bloom | Oxidises |
| 2. Polymerised 1,2-dihydro 2,2,4-trimethylquinoline | V. good | Moderate | None | None | Slight | Low | No bloom | Slight oxidation |
| Substituted diphenylamines e.g, 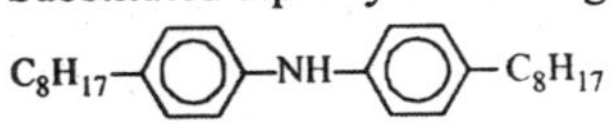 | Good | Moderate | None | None | Slight | V. low-low | No bloom | Slight oxidation |
| substituted *p*-phenylene diamines 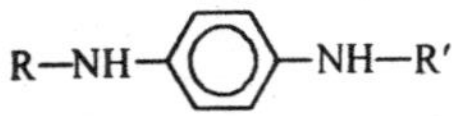 | | | | | | | | |
| 1. R = R′ = aryl, e.g. β-naphthyl | V. good | Weak | V. good | Some<br>V. good in CR | Considerable | V. low-low | Possible | Oxidises |
| 2. R = R′ = alkyl | Good | Moderate | V. good | Good<br>V. good | Severe | V. low-low | No bloom | Oxidises |
| 3. R = alkyl, R = aryl | V. good | V. good | V. good | V. good | Severe | V. low moderate | No bloom | oxidises |

**Table 15.2.** Comparison of phenolic antioxidants.

| *Class* | *Heat resistance* | *Flex resistance* | *Ozone activity* | *Staining and discoloration* | *Volatility* | *Solubility* | *Chemical stability* |
|---|---|---|---|---|---|---|---|
| Simple hindered monohydric phenols e.g. 2,6-di-*t*-butyl 1-4-methylphenol | Moderate | Poor moderate | None | Slight | Low moderate | No bloom | Stable |
| Phenol alkane *bis*-penols e.g. 2,2-methylene-*bis*- (4-methyl-6-*t*-butylphenol) | Moderate-V. good | Poor-moderate | Negligible | Moderate | Very low-low | No bloom | Slight oxidation |
| Phenol sulphide *bis*-phenols e.g. 4,4′-thio-*bis*- (6-*t*-butyl-3-methylphenol) | Moderate-good | Poor-moderate | Negligible | Moderate | Very low-low | No bloom | Slight oxidation |
| Polyhydric phenols | Moderate-V. good | Poor-moderate | Negligible | Slight | Very low | No bloom | Slight oxidation |

## MAJOR FOOD ANTIOXIDANTS

### Butylated Hydroxyanisole (BHA) and Butylated Hydroxytoluene (BHT)

Properties of major antioxidants are shown in Table 15.3. BHA is an effective antioxidant in fats from animal sources and in food containing animal products. As with most phenolic antioxidants, it is of limited effectiveness in unsaturated vegetable or seed oils. BHA maintains its activity during processing and is effective in baked goods. The latter is due to its stability to mild alkaline conditions. It is volatile, which makes it desirable to use in food packageing materials but undersirable in deep-fat frying fats.

**Table 15.3.** Physical properties of major chain-breaking antioxidants.

| *Characteristics* | *BHA* | *BHT* | *TBHQ* | *Propyl gallate* | *Ethoxyquin* |
|---|---|---|---|---|---|
| Chemical name | 3-Tertiary-butyl 4-hydroxy-anisole 2-tertiary-butyl-4-hydroxyanisole | 2,6-Ditertiary-butyl-4-methyl phenol | Tertiary butyl-hydroquinone | Esters of 3,4,5-tri hydroxybenzoic acid (octyl and dodecyl esters also used) | 6-Ethoxy-1,2-dihydro-2,2,4-trimethyl-quinoline |
| Molecular weight | 180.25 | 220 | 166.22 | 212.20 | 212.20 |
| Physical appearance | White waxy tablets or flakes | White granular crystal | White to light tan crystals | White crystalline powder | Brownish red liquid |
| Boiling range, 733 mm, °C | 264–270 | 265 | 300 | Decomposes above 148 | 123–125 |
| Melting range, °C | 48–63 | 69.7 | 126.5–128.5 | 146–150 | Liquid |
| Odour | Slight | Slight | Slight | Slight | Medium strong |
| Solubility | | | | | |
| Lipids | Soluble | Soluble | Soluble | Poorly soluble | Soluble |
| Water | Insoluble | Insoluble | Slightly soluble | Slightly soluble | Insoluble |
| Ethanol | Soluble | Soluble | Soluble | Soluble | Soluble |

BHT is the only common phenolic antioxidant that is not substituted *ortho*- or *para*-polyphenol. BHT does not have an optimum concentrations. It has similar properties to BHA. Combination of BHA and BHT show marked syncigiso in plain fats, but this effect has not been demonstrated to carry through to baked products. While BHA is a synergist for propyl gallate, BHT is not. BHA and BHT may impart odour in foods when used at high temperatures (frying temperatures) for extended time periods.

BHA is more effective in suppressing oxidation occurring in animal fats than that occurring in vegetable oils. Among its multiple applications, BHA is particularly useful in protecting the flavour and colour of essential oils; in fact, it is the most effective of all food-approved antioxidants for this applications. BHA is particularly effective in controlling the oxidation of short-chain fatty acids, such as those contained in coconut and palm kernel oils, which are typically used in cereal and confectionery products.

TBHQ is regarded as the best antioxidant for protecting frying oils against oxidation and like BHA and BHT, it provides carry through protection to the finished fried product. Increased oxidative stability and colour improvement can also be achieved via hydrogenation, a process that saturates the double bonds of liquid oils, thereby converting them to semi-solid plastic fats. The hydrogenated end products are suitable for use in shortenings and in other applications requiring higher melting fats. This process is, however, expensive and is limited in its hydrogenation capacity.

### Gallic Acid and the Gallates

Gallic acid (3,4,5-trihydroxybenzoic acid) is widely distributed in the vegetable kingdom, mostly as a component of the vegetable tannins, although free gallic acid has been reported to occur in tea. The lower gallates such as ethyl, propyl and butyl gallate remain slightly soluble in both water and fats, but the higher octyl, decyl and dodecyl esters are practically insoluble in water but easily dissolved in fats and oils. The antioxidant properties of gallic acid were described by *Golumbie* who found this compound peculiar in that it behaved as a primary phenolic antioxidant and as a synergist. Others have found propyl gallate effective in retarding lipoxidase oxidation of linoleate, but it has little effect in preventing hematin-catalysed oxidative fat rancidity. Like the tocopherols, the esters of gallic acid show optimum concentrations for antioxidant activity and may act as pro-oxidants when used at high levels.

Propyl gallate is the only ester of gallic acid permitted in foods in the United States, but several European countries allow the use of higher alkyl gallates. The various gallates have approximately equal antioxidant activities when used in equimolar concentrations, but the increased fat solubility of the higher esters is a definite advantage. The gallates are not carry-through antioxidants and their power to stabilise fried foods, baked pastries and crackers prepared from fats containing these compounds is low. The gallates form colour complexes with non-chelated metal ions (iron and copper).

### Chelating Agents

Chelating agents are not usually classified as antioxidants, yet they play a valuable role in stabilising foods. Often classified as synergists, chelating agents complex with pro-oxidative metal ions such as iron and copper. An unshared pair of electrons in their molecular structures promotes the complexing action. Among the most commonly used chelators are citric acid and its salts, phosphates and salts of ethylenediaminetetra-acetic acid (EDTA).

### Secondary Antioxidants

Thiodipropionic acid and dilauryl thiodipropionate are unique in their activities relative to the other preservatives. They decompose the hydrogen peroxide elicited during lipid oxidation into stable end products. Although both components are approved by FDA for use as chemical preservatives. *Lindsay* reported that these preservatives are used more in stabilising polyolefin resins than foods.

### Synergist

Synergists enhance or prolong the antioxygenic activity of primary antioxidants. Often these compounds by themselves have little effect, if any, on retarding fat oxidation. A significant synergistic effect often occurs between primary and secondary, or preventive, antioxidants. Many low molecular weight hydroxy and amino acids posses marked synergistic activity. However, synergism may also occur between primary antioxidants. Synergism of the primary antioxidants BHA and BHT is important in practical use; it results from increased hydrogen donation greater than that expected from the sum of two antioxidants. It has been proposed that, in the presence of BHA, BHT oxidation is enhanced while BHA is regenerated. Propyl and other gallates are effective synergists with both BHA and BHT.

The mechanism of synergism is certainly not the same for compounds. Synergists may function as regenerating agents of primary antioxidants, as inactivates of metallic pro-oxidants, as hydroperoxide decomposers and as other means of sparing primary antioxidants. The synergistic effect of tocopherols with ascorbic acid and its esters is of particular interest. This phenomenon was studied using α-tocopherol and ascorbic acid or ascorbyl palmitate. α-Tocopherol acts as the chain breaking antioxidant and the resulting α-tocopheryl radical reacts with ascorbic acid to regenerate α-tocopherol.

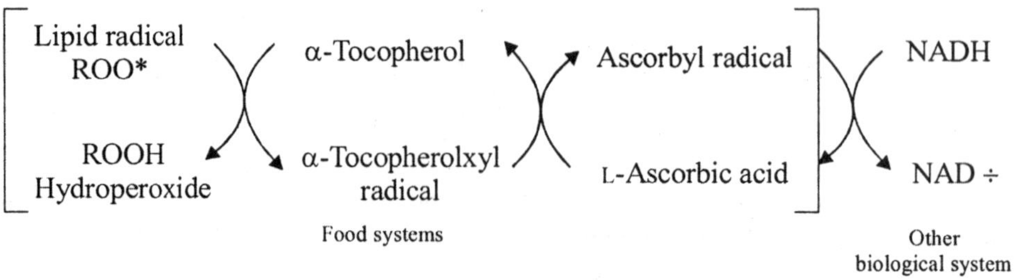


Support for this mechanism was reported by *Niki*, who demonstrated that in an *in vitro* oxidation system using methyl linoleate ascorbic acid was consumed first while α-tocopherol remained unchanged until most of the ascorbic acid was depleted. It was suggested that α-tocopherol scavenges peroxyl radicals much more rapidly than ascorbic acid but that the α-tocopherol radical reacts with ascorbic acid to regenerate α-tocopherol.

Synergism of the primary antioxidant BHA and BHT is important in practical use; it results from increased hydrogen greater donation than the expected sum from the two antioxidants. It has been proposed that in the presence of BHA-BHT oxidation is enhanced while BHA is regenerated.

Many of the flavonoids and related compounds have strong antioxidant characteristics in lipid—aqueous and lipid food systems (Fig. 15.3).

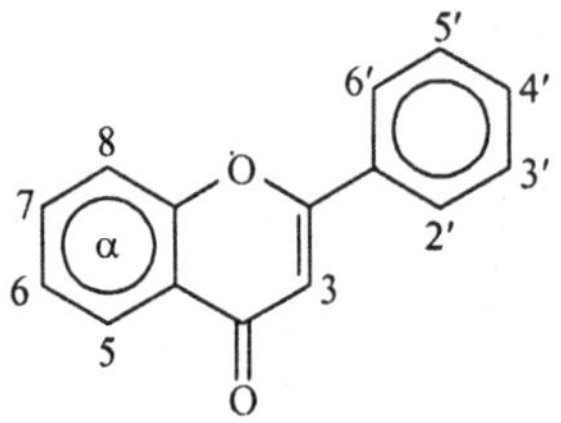


Flavones

| | |
|---|---|
| Quercetin | 3,5,7,3',4'-penta OH |
| Fisetin | 3,7,3',4'-tetra OH |
| Luteolin | 5,7,3,4-tetra OH |
| Quercetin | 5,7,3,4-tetra OH |

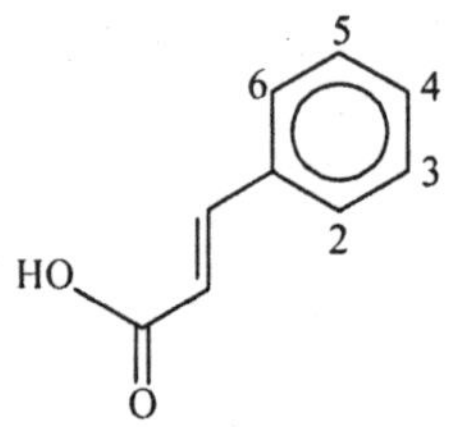


Cinnamic acids

Caffeic acid 3,4- di OH
Ferulic acid 4-OH, 3-OMe

**Fig. 15.3.** Structures of flavones and cinnamic acid derivatives.

Certain flavones, flavonols (Table 15.4), flavonones (Table 15.5), flavononals and cinnamic acid derivatives (Table 15.6) have considerable antioxidant activity. The low solubility of these compounds in lipids is often considered a disadvantage and has been considered a serious disadvantage if an aqueous phase is also present. However, flavonoids suspended in the aqueous phase of lipid-aqueous system offer appreciable protection to lipid oxidation.

It has been found that flavonols were effective antioxidants when suspended in lipid systems. Sources of natural antioxidants are as follows: algae; alma; browning products; citrus pulp and peel; cocoa powder or shell; heated products; herbs and spices; microbial products; oat flour; olives; osage orange; plant (extracts); protein hydrolysate; resins and soya products.

Perhaps the major reasons for interest in natural antioxidants is that there is a question as to the safety of the presently used synthetic antioxidants. However, natural compounds usually are not subjected to the scrutiny and scientific evaluation that the synthetic compounds (BHA, BHT, TBHQ) are. Therefore, caution must be taken in the use of natural compounds. Their potential as mutagens, carcinogens, teratogens, and/or other pathogens must be investigated. Certain problems other than toxicity may be

**Table 15.4.** Antioxidant activity of flavonols.

| *Compound*[a] | *Hours to reach a peroxide value of 50* |
|---|---|
| Control (stripped corn oil) | 105 |
| Aglycones | |
| Quercetin (3,5,7,3′,4′-pentahydroxy) | 475 |
| Fisetin (3,7,3′,4′,-tetrahydroxy) | 450 |
| Myricetin (3,5,7,3′,4′,5′-hexahydroxy) | 552 |
| Robinetin (3,7,3′,4′,5′-pentahydroxy) | 750 |
| Rhammnetin (3,5,3′,4′-tetrahydroxy 7 methoxy) | 375 |
| Glycosides | |
| Quercitrin (quercetin 3-rhamnoside) | 475 |
| Rutin (quercetin 3-rhamnoglucoside) | 195 |

[a] $5 \times 10^{-4}$ M in stripped corn oil.

**Table 15.5.** Antioxidant activity of flavones.

| *Compound*[a] | *Hours to reach a peroxide value of 50* |
|---|---|
| Control (stripped corn oil) | 105 |
| Aglycones | |
| Naringenin (5,7,3′-trihydroxy) | 198 |
| Dihydroquercetin (3,5,7,3′,4′-pentahydroxy) | 470 |
| Hesperitin (5,7,3′-trihydroxy-4′-methoxy) | 125 |
| Glycosides | |
| Hesperitin (hesperitin 7-rhamnoglucoside) | 125 |
| Neohesperidin (hesperitin 7-glucoside) | 135 |

[a] $5 \times 10^{-4}$ M in stripped corn oil.

**Table 15.6.** Antioxidant activity of some polyphenolics.

| *Compound*[a] | *Hours to reach a peroxide value of 50* |
|---|---|
| Control (stripped corn oil) | 110 |
| Hesperidin methyl chalcone | 135 |
| D-Catechin | 410 |
| Chlorogenic acid | 505 |
| Caffeic acid | 495 |
| Quinic acid | 105 |
| Propyl gallate | 435 |
| *p*-Comaric acid | 120 |
| Ferrulic acid | 145 |

[a] $5 \times 10^{-4}$ M in stripped corn oil.

associated with the use of natural antioxidants: (i) the amount of active ingredient may vary with the source and method of extraction; (ii) some may impact off colours and off odours; and (iii) the cost of extraction may often be nearly inhibitive and undesirable reactions may occur with nutrients in the product.

Antioxidants have been isolated from or detected in most edible plants, but they are not always in the edible portions. The same compound, or group of compounds, are not always present throughout the plant. Natural antioxidants occur in all higher plants and in all parts of the plant (wood, bark, stems, pods, leaves, fruit, roots, flowers, pollen and seeds). The antioxidants are usually phenolic or polyphenolic compounds. Typical compounds that possess antioxidant activity include tocopherols, flavonoids, cinnamic acid derivatives, phosphatides and polyfunctional organic acids. The flavonoids include flavones, flavonols, isoflavones, catechins, flavonones and chalcones. The cinnamic acid derivatives include caffeic acid, terralic acid and chlorgenic acid.

Methods of incorporating antioxidants into food systems is of utmost importance. They must be thoroughly blended in the separation or in the lipid portion. Type of food, processing conditions and equipment available all influence the method of addition. Antioxidants may be added directly, as a concentrate in oil or fat, as a food-grade solvent, or in emulsified form. Antioxidants and concentrate solutions may be proportioned in a continuous-flow system, sprayed, or added by various other methods such as through the packageing system or by pasting on the surface of fresh meat and fish products. Meat and fish products present different problems, because antioxidants can only be applied uniformly to the minced or ground product.

## ANTIOXIDANT FORMULATIONS

Antioxidants used in fats and oils are usually in the form of one of many solution (liquid) formulations that have been devised to satisfy specific needs and take advantage of the different properties of the antioxidant compounds. For example, combinations of BHA and BHT may be used to achieve maximum carry-through effect in baked goods. Formulations with TBHQ or propyl gallate combined with BHA and/or BHT and including citric acid may provide improved shelf life to a fat or oil (due to the TBHQ or propyl gallate) and increase the oxidative stability of baked goods prepared with the fat or oil (due to the BHA or BHT), while at the same time taking advantage of the metal chelating and synergistic properties of the citric acid. Combinations of antioxidants, such as BHA and TBHQ, may be used to obtain maximum allowable treatment levels in meat and poultry fats within certain restraints imposed by regulations. These are just a few of many reasons to use formulations, with major considerations being the following:

1. They combine antioxidants to take advantage of their different types of potency.
2. They enable better control and accuracy in applying antioxidants.
3. They provide the synergistic effects of antioxidant and acid combinations.
4. They enable more complete distribution or solution of antioxidants in fats and oils.
5. They minimise discolouration problems sometimes associated with specific antioxidants.
6. They are more convenient to handle than the pure antioxidant.

TBHQ has no marked effects on the liver at high doses and has little or no effect on rodent stomach hyperplasia. It is unlikely to cause tumors in these organs.

Tocopherol toxicity is low. At extremely high levels, $\alpha$-tocopherol impairs blood clotting by interacting with vitamin K and may cause toxic effects in the liver. Intake of tocopherols from food antioxidant use is low compared either with total intake from the diet or with levels that may cause adverse effects over long periods.

Toxicity of L-ascorbic acid and its derivatives is low. High levels of 1 gram/day may cause adverse effects, including oxalate formation and excretion. Because intake from its use as an antioxidant is low compared with daily dietary intake, no maximum acceptable intake has been established. It should be noted, however, that stereoisomers of ascorbic acid such as D-isoascorbic may have little or no vitamin C activity and their antioxidant and safety characteristics are different. In many countries, approval is limited to L-ascorbic acid and its derivatives.

## EVALUATION OF ANTIOXIDANT POTENCY

Many techniques have been developed and used to detect and measure the extent of oxidation in fats and fat-containing foods and to ascertain the resistance of fats to oxidative deterioration. Misleading results may be obtained because the mechanism of oxidation changes under the test conditions used to accelerate oxidation, e.g. under high temperature. Some of the more important methods include the following.

### Shelf Storage Tests

The most realistic estimate of the effectiveness of antioxidants in prolonging the life may be obtained by storing the test materials under conditions identical or similar to those in commercial practice. Periodic evaluation, by sensory and chemical tests, is used to determine the onset of oxidation. Because this method, known as the shelf storage test, is time-consuming, several other methods have been developed to enable rapid evaluation of antioxidants.

### Oxygen Absorption Tests

Oxidations involves measurable uptake of oxygen. Oxygen absorption tests have been developed based on the quantitation of oxygen, using manometric and bomb calorimetric techniques. Some tests simply measure the change in the weight of the sample.

### Oven Storage Test

Heat is used to accelerate formation of peroxides. Testing is usually conducted in electrically heated convection ovens at 60–63°C and is often referred to as the Schaal oven test. While sensory criteria such as odour and flavour are commonly used, chemical determination of peroxides and other factors may also be used to determine rancidity.

### Active Oxygen Method

The active oxygen method (AOM) is widely used to measure oxidative stability of fats and to estimate the potency of antioxidants in them. It has been referred to as the swift stability test and involves bubling air through the heated test sample to accelerate oxidation. Periodic analyses of peroxide content are conducted to determine the time required for the fat to oxidise under the AOM conditions.

### Other Methods

Other methods include measuring persistence and intensity of carotene in an agar plate diffusion system, oxygen uptake by electrode or gas chromatographic reactor and detecting and measuring antioxidant components in foods by thin-layer chromatography. Several methods have been developed that measure the coupled oxidative potency of β-carotene and linoleic acid (or other PUFA) when emulsified with the test antioxidant.

## THERMAL DEGRADATION OF PHENOLIC ANTIOXIDANTS

Antioxidants escape from the frying medium by volatilisation and steam distillation due to the high temperatures and the large amount of water that is boiled out. Loss of antioxidant during frying has often been reported as a factor contributing to the loss of antioxidant potency. It has been reported that BHT was not effective in frying operations. The ineffectiveness of BHA and BHT has been observed under frying conditions; TBHQ was seen to be relatively more effective under the same conditions. It was postulated that some phenolic antioxidants are ineffective during frying due to thermal destruction and loss through steam distillation.

*Peled* and co-workers found that the low effectiveness of BHT in retarding thermal deterioration of the oil in frying could be attributed to BHT losses due to volatilisation and/or destruction by oxidation. They also found that when the oil was heated in the presence of air, the antioxidant destruction was lessened by the presence of water in the system, but a large portion of BHT was lost through steam distillation.

### Controlling Oxidation

The deteriorative process of oxidation occurs naturally in all foods, not just those with high fat content. This makes the addition of antioxidants useful in most fat-containing foods. BHA, BHT, TBHQ and propyl gallate may be used in a variety of applications to prevent lipid oxidation. Spice extracts, wood smoke and other compounds may also be valuable in preventing oxidative degradation. Antioxidants are recognised as a means to ensure the quality and nutritional quality of the food supply.

## ANTIOXIDANTS: BODY'S GUARD AGAINST DISEASE AND AGEING

Some fruits, e.g. apple and vegetables like brinjal, on cutting and exposure to air turn brown. This browning has been attributed to oxidation of phenolic substances, present in plant material by air. Polyphenoloxidase enzymes, which are also present in most of the plants, catalyse the oxidation.

Similarly many freshly extracted fruit juices turn brown on exposure to air. Edible lipids, particularly those rich in polyunsaturated fatty acids, become rancid on exposure to air. Rancidity is due to oxidative decomposition of fat resulting in foul smell, bad taste and making the fat unfit for human consummation.

Some synthetic organic compounds, e.g. propyl gallate, butylated hydroxytoluene (BHT) and butylated hydroxyanisole (BHA), when added at a very low concentration to fruit juice or edible oil can slow down their air oxidation, preventing the development of brown colour and rancidity respectively. Such substances are referred as 'synthetic antioxidants'. There has been some controversy regarding the use of synthetic antioxidants in food, because some food scientists and medical personals do not regard them as completely safe and non-toxic food additives.

However, long-term experiments have now amply proved that at the permitted concentration of ~0.1 per cent, these antioxidants are completely harmless to human beings and protect against harmful air oxidation of food. Natural occurrence of such antioxidants in plants like rosemary has shown that these act as a natural defense of the plant against air oxidation under adverse conditions.

## Naturally Occurring Antioxidants

Vitamin C (ascorbic acid), vitamin A (retinal) and its precursor β-carotene and vitamin E, occurring in plant or animal tissue are known to act as antioxidant for the plants in which these occur and for the animals or human being using such plants for food. Antioxidant nature of vitamin C can be realised from the fact that whereas apple juice turns brown in air, lemon and orange juice, retains the original colour due to natural presence of antioxidant vitamin C.

Similarly vegetable oils rich in vitamin E, do not easily became rancid. Sulphur compounds (thiol and mercaptans), present in garlic and onions are also antioxidants. Thick and dark coloured bark of tropical trees, e.g. kher, babool and neem have bioflavonoids and their polymerised products, tannins in their bark, which acting as antioxidants, protect the plant against air oxidation, by oxygen, when it gets activated by UV-radiation present in bright and harsh sunlight. Seed coat of most plants, e.g. wheat and corn bran and legume husk are rich in antioxidants, which protect them from oxidation, when they need to be stored in contact with air, for a long period, till used as food or germinated to grow into a new plant.

Human body has its own antibodies and antioxidant enzymes, which act as defense against disease. Antioxidant enzymes have metal ions (copper, zinc, manganese, selenium and iron) in their prosthetic groups. These enzymes, acting as redox systems, mop up reactive form of oxygen species, which cause oxidation of vital constituents in body. Additionally, the dietary antioxidants boost their action.

## Free Radicals as Reactive Intermediates in Oxidative Degradation

Certain highly reactive type of atoms or organic molecules, which are called free radicals, are the most common intermediates in harmful air oxidation of plants or human tissues. Most of the normal molecules in our surroundings have electrons, constituting the chemical bonds, in paired form and in lower energy level. Unlike these free radicals have at least one unpaired electron.

A free radical is formed when an electron pair bond, in a molecule breaks in such a way that each of the fragments of the parent molecule carries an unpaired electron of the original bond. Chemists call it 'homolytic-fission of a bond'. Free radical has a tendency to react with other molecules and abstract an electron of an electron pair of a normal chemical bond in the molecule, thus creating a new free radical. In other words the generation of new free radicals does not stop and a chain reaction is set-up. Termination of free radical chain reaction can take place by a disproportion reaction or by a mutual reaction of two free radicals.

## Reactive Oxygen Species

Our body is always exposed to molecular oxygen (at ground energy level or the triplet state), which is present in the air at a concentration of about 22 per cent by weight. Molecular oxygen is essential for all living organisms and this is the form of oxygen, which we breathe regularly and to which most living things (animals and plants) are exposed. In case of plants, to a large extent, air oxidation of plant phenolics prevents harmful oxidation of more vital constituents of living plants.

Under certain conditions, e.g. on exposure to UV-radiations, smoking and in polluted atmosphere, molecular oxygen can get converted into reactive oxygen species (ROS) e.g. singlet state, which can cause harmful oxidation in living organisms and of biomaterials.

Formation of certain amount of free radicals, via ROS, in human body is a natural, metabolic process. Even the air we breathe has oxygen in the form of paramagnetic molecule, which is weakly reactive in free radical form.

Molecular oxygen, which is a weak *bis*-radical, on activation by light or heat energy become highly reactive and harmful. Thus, molecular oxygen on reacting with a fat molecule, at the site of unsaturation produce peroxide or hydroperoxide, which on decomposition sets up a free radical chain reaction. An antioxidant can quickly terminate such chain reaction.

Antibodies, present in human body, are thought to be defense guard against infection, but exact mechanism of their action was not known. Only recently it has been shown, by a team of scientists at the California Institute of Technology that antibodies catalyse formation of ROS in human body, to cleanse it of poisonous invaders.

Limited amount of free radicals produced via ROS in human body, control bacterial infection by destroying them. However, if produced in large excess, they can damage cells and tissues of the body and damage the cell wall, making it permeable to toxic substances and pathogens.

Damage of DNA in cell nucleus and its replication can result in cell mutation and disease such as cancer. Dietary antioxidants maintain a balance between the free radical required for the safety of the body from that of toxic invader and those produced in excess of such requirement.

## Mechanism of ROS Production by Antibodies

Antibodies were earlier thought to act as a defense mechanism in body by temporary binding of the pathogenic invader (antigen), till immune cells moved in to kill them. In the year 1990 it was shown by *Detty* and *Gibson* that antibodies could produce $H_2O_2$, a precursor of free radicals, which can kill an antigen bound to it.

The California team headed by *Wentworth* later showed that all the antibodies are capable of producing $H_2O_2$ by catalytic oxidation of water. It has been shown that as much as 500 mole equivalents of $H_2O_2$ can be produced per antibody molecule, without decrease in its catalytic activity. ROS, $H_2O_2$ produced by reaction of water and $^1O_2$ (singlet oxygen), has been shown to be an intermediate in production of hydrogen peroxide.

## Dietary Antioxidants

Dietary antioxidants are normally present in human food and these include vitamin A, C and E, plant pigments e.g. bioflavonoids and carotenoids, e.g. beta-carotene present in carrot, yellow fruits (pumpkin, mangoes and papaya) and green vegetables. Carotenoid, precursor of vitamin A, is also present as the **red pigment** (lycopene) in tomatoes and red peppers.

Vitamin C is present in citrus fruits and green vegetables. Bioflavonoids (plant pigments, other than green chlorophyll and yellow carotene) antioxidants are widely distributed in plants (tea, coffee, cocoa, gambier leaves, marigold flowers and kutch) and the seed coat of most of the food grains.

Herbal tea and herbal infusion have been used in the traditional systems of medicine (Ayurvedic and Unani). Healing power of a decoction, has been attributed to the presence of antioxidants, thus providing a scientific basis for their action. Vegetable oils, e.g. sunflower oil, peanut oil and soyabean oil, are rich in antioxidant vitamin E.

Trace metals, e.g. selenium, copper, iron, manganese and zinc, found in shellfish and nuts, also act as antioxidants, to neutralise harmful effect of free radicals. Just how important is to supplement dietary antioxidants, is a subject of debate, because often the claims are exaggerated. Common sources of dietary antioxidant are shown in Table 15.7.

**Table 15.7.** Dietary antioxidants and their source.

| *Antioxidant type* | *Source* |
|---|---|
| Vitamin C (ascorbic acid) | Citrus fruits, green vegetables, strawberries, blackcurrants, guavas |
| Vitamins A (retinal) | Milk products, eggs, meat, fish, liver |
| Vitamin E | Vegetable oils (soyabean, peanut, sunflower) nuts, peanuts, whole grains, legumes, almonds |
| Carotenoids | Yellow fruits (mangoes, papaya, pumpkin), carrots |
| β-Carotene lycopene | Red peppers, tomatoes |
| Bioflavonoid | Fruits, lemon, cherries, blackberries, blackcurrants, tea, coffee, cocoa |
| Minerals | |
| Selenium | Shell fish, fish, herring, cod fillet, dairy products, avocado |
| Manganese | Plant food |
| Iron | Pigmented fruits, meat, egg, sea food |
| Copper | Liver, shell fish, cocoa, nuts, mushrooms, vegetables |
| Zinc | Grains, meat, fish, oyster, eggs, dairy products, vegetables |

## Food Preservatives vs. Antioxidants

Packaged food contains added preservatives (bisulphite, benzoic acid, sorbates etc.) to increase their shelf life. The preservatives are only effective against microbial spoilage of food, but when in contact with air food is also spoiled due to air oxidation. Toxicity of packaged food can result in a type of food poisoning called botulism. To guard against such infection antioxidants are added to packaged food. Some food processor prefer the use of natural antioxidants, which can be herbal extracts.

According to them, herbal extracts can be economical, yet safer, than synthetic antioxidants. Thus, the solvent extracted pigments from tamarind seed coat contains antioxidant suitable for preventing rancidity of oils and improving shelf life of fried foods. Bioflavonoids from catechu (catechin) also contains antioxidants suitable for use in food. Rice and corn bran oils are rich source of antioxidant vitamin E.

It has been mentioned that UV-radiation can activate oxygen (singlet state), which can damage skin tissues and cause skin disease (cancer) on long exposure to sun. To prevent this, cosmetic companies use vitamins E and C is some skin-care products. Dietary food fibres, which are soluble, non-starch polysaccharides, are depolymerised by free radicals into harmless products (sugars and lower polysaccharides). Thus, they act as free radical scavengers or antioxidants. Water-soluble as well as fat-soluble antioxidants are known and appropriate choice of an antioxidant should be made, based on the nature of food.

### Infection and Disease Caused by Free Radicals

Common disease arising due to the harmful action of free-radicals are emphysema, lungs inflammation, atherosclerosis, osteoarthritis, immune-deficiency, high blood-pressure, joint problems, skin-brown patches and cancer. Factors responsible for excessive formation of free-radical are, over-exposure to high energy radiation, e.g. radioactive rays, ultraviolet rays, X-rays, bright sun light, exposure to environment, pollution due to smoke and smog, cigarette smoke, industrial gases and reactive industrial chemicals, effluent and also due to poor nutrition and intense exercise.

Blood cholesterol is also oxidised by free radicals and this oxidised cholesterol is more damageing to arteries than 'native cholesterol'. Certain cooking methods e.g. deep-frying and re-use of oil, charring during roasting, grilling and barbecuing, produce potential precursor of free radicals and cancer. Left unchecked, these unstable and potentially harmful chemicals, create conditions for infection and reduce immunity. Research have shown that a diet naturally rich or supplemented with antioxidants can provide protection against above-mentioned diseases.

### Ageing, Longevity and Antioxidants

*Simon Melov*, a molecular biologist, at the Buck Institute for Age Research, in California, has shown that specifically designed anti-ageing drugs can quadruple the longevity for mice and there is also a significant increase in the life span of human beings, on administering such drugs. According to him, chemical oxidation, which causes rusting of iron metal exposed to air, can also damage DNA in living organisms, which is supposed to be associated with ageing. Relative long life cycle of certain plant spices is due to the fact that they have built-in production of antioxidants. *Melov's* studies, published in the *Journal of Neuroscience*, showed that the antioxidants act by zeroing the toxic molecules and free radicals, which are produced in body cells and are necessary pollutants in burning energy, which contribute to ageing process. Vitamin A, C and E are most effective anti-ageing antioxidants.

### How do Antioxidants Act to Neutralise Harmful Free Radical?

Some of the antioxidants act as sacrificial molecules, i.e. they preferentially react with free radicals, thereby getting destroyed producing harmless by-products, but protecting vital constituents of the body. Such molecules are themselves reducing agents, e.g. vitamin C and the pigments, anthocyanin and flavones, collectively called bioflavonoids. In other cases, particularly the metals acting as antioxidants, are costituents of antioxidant enzymes, which are naturally present in the body. These metals can exist in variable oxidation state. After neutralising free radicals by getting oxidised to a higher oxidation state, these are reconverted, by redox systems of the body into the original lower oxidation-state. Thus, these act as catalyst and react in a continuous cycle. Vitamin C is also regenerated by body redox system and acts as free radical scavenger. Synthetic phenolic antioxidants sometimes react with free radicals to produce resonance stabilised, weakly reactive free radicals, which are not harmful. Moderate amounts of tea, coffee, cocoa and chocolate, which contain antioxidant flavonoids are good for health. Chewing of betel leaf can be harmful due to tobacco, but catechu (katha) used in it contains a bioflavonoid catechin, which is an effective antioxidant.

Thus, the myth about non-vegetarian food being good for health is no more true, because it does not supply required amount of antioxidants. In general, vegetarian food is better source of antioxidants. Antioxidants increase immunity of human body and slow the ageing process.

## INHIBITING RANCIDITY IN SOAPS AND COSMETICS

Population today has increased manifold, with a high rise of urban settlements in different parts of India. Soaps and cosmetic products, produced in one part of our nation are sent to areas located far away that have, varying environmental and climatic differences from the place of its origin. Soaps and cosmetic formulation today, significantly comprise of fats and oils derived from natural plants. Lipids extracted from vegetable sources have limited stability upon storage. Fats start deteriorating from the very moment it is extracted from the vegetable or plant source.

Soaps and cosmetics that are made out of these naturally occurring lipids are susceptible to degradation in the same manner as the original oil or fat and so have to be preserved in the same manner. The preservation of fat or oil based products, has been a challenge to people for a long time now. The fatty acid components of toilet soaps and cosmetics, is such, that can readily undergo auto-oxidation due to atmospheric oxygen. This auto-oxidation is greatly favoured by heat, light and is strongly catalysed by heavy metals, particularly copper or iron, causing product discolouration and formation of objectionable degradation products of unpleasant odour. This oxidised condition is called 'rancidity'.

Our sensory characters of taste and smell readily notice rancidity. Rancidity may be due to oxidation either by bacterial action and enzyme-induced hydrolysis or by atmospheric oxygen (auto-oxidation). The development of rancidity, cause colours and texture changes, reduce shelf life and impair the quality of the product.

Auto-oxidation is difficult to prevent as this chemical reaction has very low activation energy. Oxidation of oil and fats depends on the number of double bonds present and their inherent arrangement. The oxidation also depends on the pro-oxidants or substances that favour oxidation present either inherently or added accidentally and inhibitors (anti-oxidants) i.e. substances that inhibit oxidation and are present naturally in fats and oils.

Naturally occurring fats and oils contain characteristics antioxidants to protect the oil from detoriation during the normal life of the plant seed or fruit. Antioxidants are present in all higher plants in their wood, bark, stems, leaves, flowers, seed, fruit, even pods and pollens. Chemically these natural antioxidants are phenolic or polyphenolic compounds. Some common antioxidants found in nature include tocopherols, flavanoids, cinnamic acid derivatives, gallic acid and phosphotides. It is noteworthy that even at low concentrations, proportional to the content present, these natural antioxidants are effective to keep the oil or fat stable during its lifetime. Rancidity develops in a very characteristic fashion comprising of two stages:

1. First induction stage wherein decomposition is slow and the rate varying. Decomposition is directly dependent on the presence of natural occurring inhibitors of oxidation or on the presence of pro-oxidant substances.
2. The second stage where the reaction rate increases logarithmically.

Enzymes, present in the oxidisable portion of the material or in micro-organisms that have gained entry may cause rancidity. Some of these enzymes attack saturated fatty acids at their β carbon atom to produce ketonic acid and methyl ketones.

$$CH_2CH_2COOH \longrightarrow -\underset{\underset{O}{\|}}{C}CH_2COOH \longrightarrow -\underset{\underset{O}{\|}}{C}CH_3$$

This is referred to as 'ketone rancidity' and is different from 'oxidative rancidity' that is caused due to atmospheric oxygen. Where micro-organisms are involved, sterilisation techniques, aseptic processing, good and proper cold storage, packageing and transportation employed can usually solve the problem largely. If it is practically difficult to follow all the above methods then antiseptic preservatives may also be used at optimum levels to overcome the situation.

The decomposition of oil- or fat-based soaps and cosmetics, by oxidation may be kept in check by use of chemical additives that prevent or diminish oxidation and subsequent degradation of the product. Antioxidants are chemical additives that are effective against oxidation by atmospheric oxygen. Product protecting agents like antioxidants and chelates or sequestering agents (synergists) are usually used in a combination as stabilisers system in cosmetic products to retain the pleasant appearance (e.g. colour, odour, etc.) and to ensure stability and efficacy of incorporated active ingredients (e.g. fragrances). Modern day cosmetic products also contain UV-filters to prevent photo-labile ingredients like fragrances, colourants, etc. which are effected by photolytic deterioration.

The selection of the appropriate product protection system, depends on:

1. The different ingredients used in the product that have to be protected.
2. Type of product, viz., emulsion, gels, non-aqueous formulation, etc.
3. Type of packageing, viz., opaque, transparent, glass or plastic, etc.

To fully understand and appreciate the usage of these additives, let us understand in brief, how an antioxidant functions.

Rancidity is very readily noticed by smell, but detected chemically in lipids by the presence of hydroperoxides. Hydroperoxides further degrade, producing aldehydes and esters that are characteristic of rancidity. Many theories have been postulated to explain the auto-oxidation of oils and fats and the role of antioxidants in its prevention. However, no mechanism is universally accepted, to explain how small amounts of foreign additives are able to control and prevent this natural phenomenon of degradation by auto-oxidation. In fact, if the chain life of auto-oxidation is long then even trace amount of antioxidants is sufficient to prolong the stability of oils and fats by greatly reducing the rate of atmospheric oxidation.

It is believed that antioxidants act at the very beginning or at the starting stage of oxidation, by effectively preventing the formation of peroxides that are the first detectable products of rancidity. Moreover, as only small amounts of antioxidants are required to fulfil this role, it can be assumed that, they are very different from reducing agents that become oxidised and are used up comparatively faster. The most acceptable theory to explain this phenomenon suggests the interference with the chain reaction due to formation of free radical mechanism.

A free radical is formed by the loss of hydrogen atom from the methylene group next to a double bond. The formation of a free radical is triggered by agents such as heat, light, trace metal impurities like copper, nickel, iron or manganese or any other pro-oxidant present in oils and fats.

$$R'CH_2CH = CHCH_2R'' \longrightarrow R'\underset{\overset{|}{*}}{C}HCH = CHCH_2R''$$

Atmospheric oxygen then converts this free radical formed into peroxide, this is also a free radical.

$$R'\underset{\overset{|}{*}}{C}HCH = CHCH_2R'' + O_2 \longrightarrow R'\underset{\overset{|}{OO-*}}{C}HCH = CHCH_2R''$$

The peroxide radical now reacts with another molecule of the original oxidisable substance to produce a hydroperoxide and another free radical.

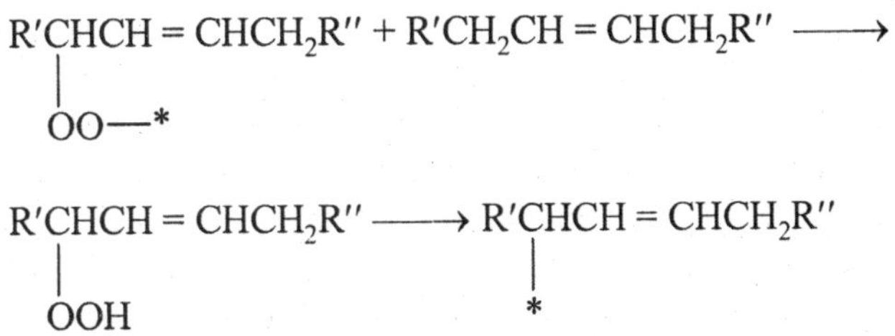

The free radical, thus produced, then takes up oxygen and starts all again as before. Thus, proceeds the continuous chain reaction, with the hydroperoxides decomposing to culminate in the formation of undesirable aldehydes and esters.

Antioxidants it is believed act, by stopping the chain reaction at its very beginning by supplying a hydrogen atom thus destroying the free radical produced and preventing it from being converted to peroxide.

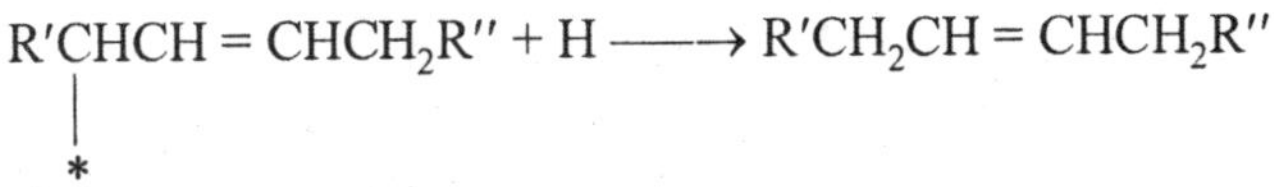

This theory is supported by the fact that *ortho* and *para* derivatives of dihydric and polyhydric phenols are among the most effective antioxidants they being capable to donate a hydrogen atom, by undergoing keto-enol tautomerism.

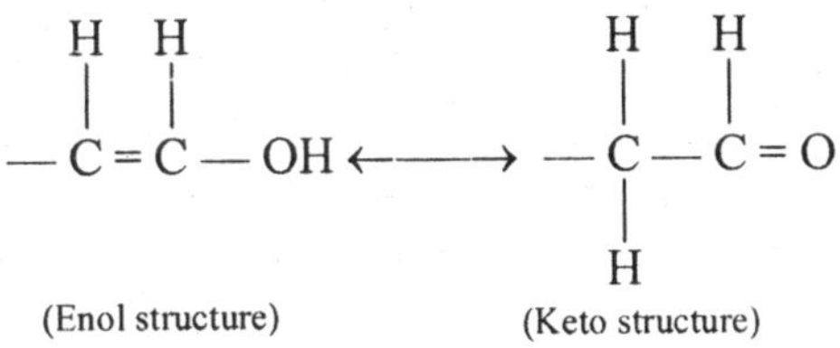


Keto-enol tautomerism

## Basic Requirements for an Ideal Antioxidant in Soaps and Cosmetics

1. It should be effective in providing sufficient degree of protection at low concentration levels.
2. Antioxidants should be compatible with different ingredients or components used in the manufacture of soaps and cosmetics.
3. It should be easily and readily soluble or dispersible in the medium and easy for application.
4. It should be stable to pH variations/changes and under process conditions. The stability should carry throughout the shelf life/storage of the product.
5. It should be free from objectionable odour and colour.
6. It must be non-toxic and non-irritant, forming no harmful oxidation products during its lifetime and should have no physiological activity.
7. It should not constitute a safety risk to consumers and there should be sufficient toxicological support for use in soaps and cosmetics.
8. It should be inexpensive, easily affordable and readily available.

Various classifications have been suggested. However, considerable confusion persists in its classification. The most important classes of antioxidants that are relevant to soap and cosmetics are the primary antioxidants and synergists. Sequestrants and chelating agents are also listed as synergist due to their marked synergism with the primary antioxidants. Synergists are compounds which increase the effectiveness or the activity of primary antioxidants and chelating agents, complex metallic ions, primarily copper, iron and nickel, that act as catalyst in lipid oxidation.

Synergists by themselves have very little or no effect on lipid oxidation, but generally act by enhancing the effect of antioxidant. A significant synergistic effect that occurs between a primary antioxidant and a synergist result in increased hydrogen donation that is greater than one can expect from the sum of two antioxidants. Their method of action is not always clear, but they are said to help in some cases, due to their sequestering of metallic ions that are known pro-oxidants. Synergist may also function as regenerating agents of primary antioxidants, as hydroperoxide decomposers and by other means of sparing primary antioxidants. The use of synergist is of particular value in emulsified fats and oils. Many of the synergists are water soluble like, EDTA, ascorbic acid, citric acid, tartaric acid, phosphoric acid, etc. while some are oil soluble viz., ascorbyl palmitate, etc.

In an emulsified lipid medium, the entrained air present in the medium and the larger surface areas that is exposed to air makes the oil more susceptible to oxidation. The reaction of the aqueous phase is also very important and in this connection, the water-soluble acid synergists are very valuable. Oil soluble synergists are effective in acidic but not in alkaline emulsions.

EDTA salts (ethylenediaminetetraacetic acid as *di* and *tetra* sodium salt), citric acid, tartaric acid, phosphoric acid, EHDP (ethane hydroxy 1,1 diphosphonic acid), magnesium silicate, are perhaps most commonly used as sequestrants. EDTA chelates copper, EHDP, citric acid, tartaric acid, phosphoric acid and magnesium silicate chelate iron. Citric acid, tartaric acid and phosphoric acid are said to prevent metal ion induced degradation that usually occurs at higher pH.

Of the hundreds of compounds, proposed to inhibit oxidation and prevent deterioration of oxidisable substance only a few of them are popular. Primary antioxidants are compounds that eliminate free radicals' chains in a typical auto-oxidation. BHT, BHA, TBHQ and gallic acid esters (propyl gallate) are popular antioxidants commonly used in soaps and cosmetics.

### Butylated Hydroxy Anisole (BHA)

This is a mixture of two isomers (85 per cent of 2-tert-butyl-4-methoxyphenol and 15 per cent of 3-tert-butyl-4-methoxyphenol) and is commercially available as white waxy flakes or tablets with a faint characteristic phenolic odour and a melting point in the range of 48° to 63°C. It has a molecular weight of 180.25 and is highly soluble in oil and alcohol but insoluble in water.

BHA is more effective in inhibiting oxidation in animal fats than vegetable fats. BHA is particularly useful in protecting the colour and flavour of essential oils and short chain fatty acids as those contained in coconut and palm kernel oils typically used in soap, shampoo and cosmetics. Fat solubility is increased by the tertiary butyl side chain in the *ortho* and *meta* position to the hydroxy group ensuring a good carry through protection. BHA is stable under mild basic conditions but has a somewhat objectionable odour at elevated temperatures.

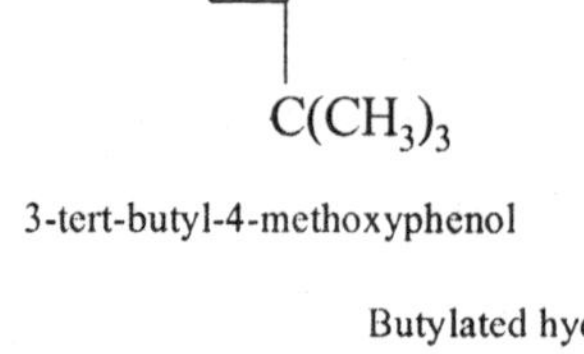


3-tert-butyl-4-methoxyphenol

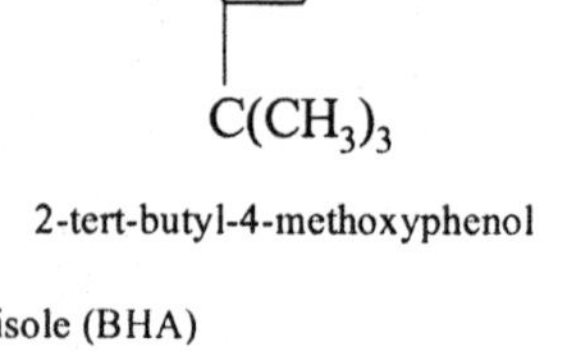


2-tert-butyl-4-methoxyphenol

Butylated hydroxy anisole (BHA)

### Butylated Hydroxy Toluene (BHT)

2,6-di-tert-butyl-4-methylphenol or 2,4-di-tert-butyl-4-methylphenol, like BHA, is a hindered phenol. It is relatively weak when used alone in vegetable oils, but more effective when combined with other antioxidants or synergist.

It is a white crystalline solid with a molecular weight of 220 and a melting point of 69.7°C. Its properties are very similar to BHA in action, but have the advantage, that at raised temperature it is odourless, unlike BHA that has a disagreeable odour at elevated temperatures.

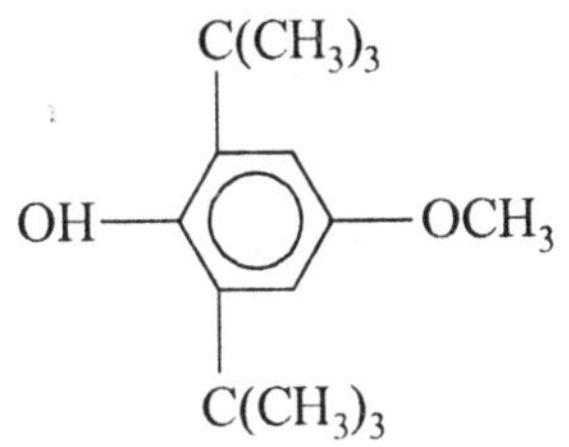


Butylated hydroxy toluene (BHT)

### Tertiary Butyl Hydroquinone (TBHQ)

TBHQ is a white or beige coloured crystal, moderately soluble in oil/fats and slightly soluble in water. TBHQ has a molecular weight of 166.22 and melts at a temperature range of 126.5 to 128.5°C. TBHQ is regarded as the best antioxidant when compared with others due to its several advantages. It is more potent and resistant to heat and does not cause discolouration, as it does not complex with copper or iron.

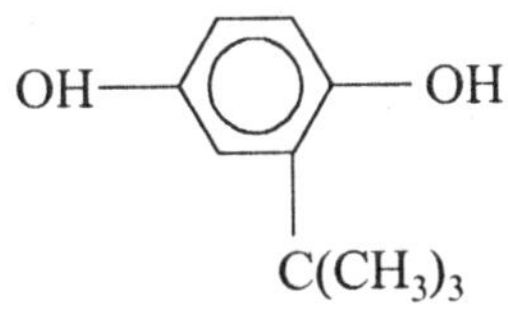


Tertiary butyl hydroquinone (TBHQ)

### Gallic Acid (3,4,5-Trihydroxybenzoic Acid) Esters

Esters like propyl gallate, butyl gallate, octyl and dodecyl gallate that are highly potent antioxidants due to their trihydroxy structures are also used as antioxidants for fats and oils. Propyl gallate of molecular weight 212.2, the more commonly used gallic acid ester, is a white to creamy-white crystalline powder. It has a slight odour with a melting point of 146°C to 150°C. Lower gallates as propyl and butyl gallate are slightly soluble in water and fats, but increased solubility is observed in case of higher gallates like octyl and dodecyl gallates. The various gallates have approximately equal antioxidant activities at equimolar concentrations. Although the gallates are effective, antioxidants with lower volatility than BHT and BHA, their use is relatively less due to their tendency to form coloured complexes with the non-chelated metal ions of iron and copper.

Tinogard™ TT that was recently introduced to the soaps, detergent and cosmetic industry is a sterically hindered phenolic antioxidant. It is claimed to be a highly effective, non-discolouring stabiliser to prevent deterioration of soaps, fatty acids and surfactants, protecting bar/liquid soaps and other rinse-off products efficiently against thermo-oxidative degradation.

Chemically, it is pentaerythritol tetrakis (3-(3,5-di-tert-butyl-4-hydroxyphenyl), having a molecular weight of 1178 and melting point between 110° to 125°C. It is a white free flowing odourless and tasteless powder, compatible in soap-type formulations. Exhibiting solubility characteristics, very similar to BHT, its effectiveness is enhanced by using a combination system along with chelate-type synergists.

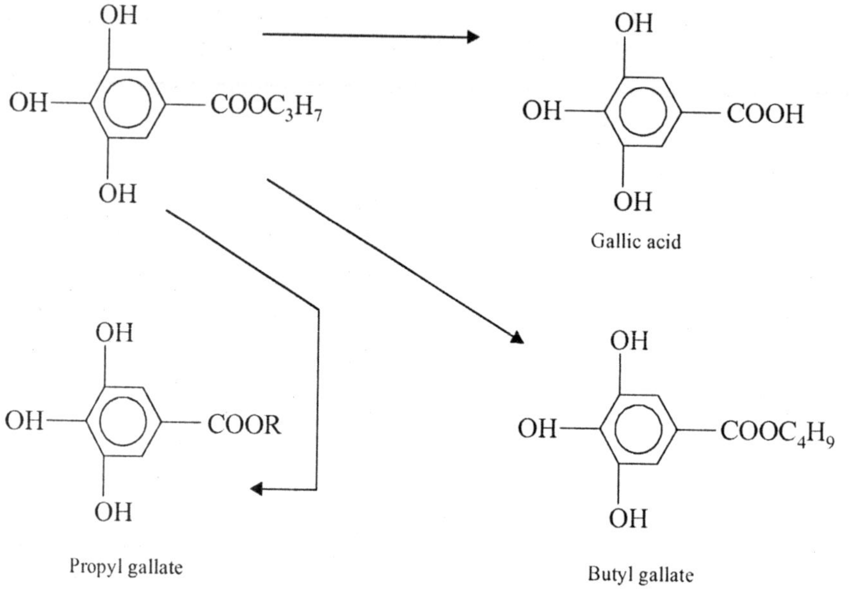


Gallic acid esters

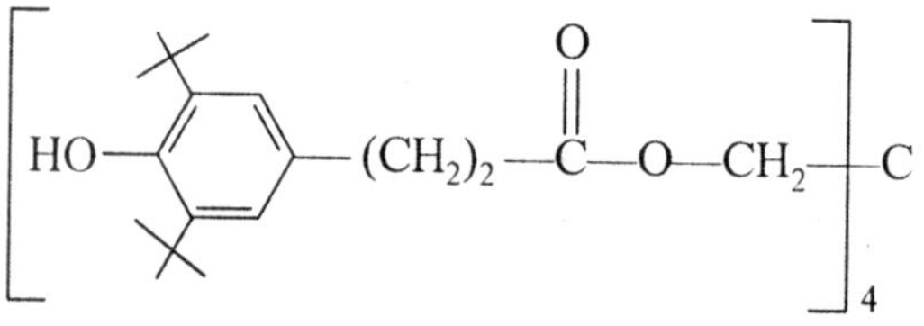


Tinogard™ TT

Antioxidants are usually used as a combination in a liquid formulation system in fats and oils. The different antioxidants are selected to take advantage of the various potent properties of the individual compounds. It is also dependent on the type of product and it needs on requirements. Combining different antioxidants into a liquid formulation system has many advantages. It is convenient and helps one to have a better control/accuracy in applications and ensure a complete distribution of the antioxidant in oils and fats. Antioxidants in formulation system with metal chelating agents help take benefit of the synergistic properties of the antioxidant and acid combinations and minimise the discolouration problems that are associated with it.

In soap cakes the concentration levels for antioxidants range between 0.01 to 0.05 per cent w/w of the formulation. In liquid soaps and other liquid formulations (e.g. shampoos) concentration between 0.005 to 0.03 per cent w/w is sufficient. Actual incorporation levels however, depend on the formulation of its ingredients, processing conditions and long-term stability requirements. Optimum levels have, therefore to be standardised, specific for an application.

It is very important that the antioxidant is thoroughly blended into the oil lipid phase on incorporation in any product. Although the method of addition is influenced on the type of product, the processing equipment's and conditions, it is added directly as a concentrate solution in oil, fat, through a solvent like fragrance or in an emulsified form. It can be sprayed or added along with other ingredients. In case of soaps and soap products, the antioxidant and synergist are added to the oil/fatty acid blends used during manufacture. The incorporation of additives is carried out early in the process. Sometimes it is also incorporated into the formulation through fragrance, as most of the antioxidants used are soluble in aroma chemicals. Product protecting agents like antioxidants and chelates (synergists) are necessary as stabilisers in cosmetic products, to retain a pleasant appearance and to ensure stability and efficacy of incorporated active ingredients. It is also equally important that the absence of yellow discolouration to soap and cosmetics, caused by the antioxidants itself, is confirmed by stability studies before the product is put for sale in the market.

BHA, BHT, TBHQ and propyl gallate are permitted in foods at levels not to exceed FDA- and USDA-permitted levels resulting in a maximum total antioxidant content of 0.02 per cent (200 ppm) of the total fat or oil content including the essential (volatile) oil content of the food. The compounds may be used singly or in combination; except that TBHQ is not lawful for use in combination with propyl gallate (21 CFR 182.3169; 182.3173; 184.1660; 172.185; and 9 CFR 318.7) (Table 15.8). Tocopherols occur in vegetable oils, the prime commercial source being soyabeans. They are extracted from tocopherol-rich deodourised distillates which are by-products from steam stripping of vegetable oils during their final processing step.

**Table 15.8.** Antioxidant usage levels permitted by the FDA.

| *Application* | *Max. usage level (ppm)* | | | | | *Federal* |
|---|---|---|---|---|---|---|
| | *BHA* | *BHT* | *PG* | *TBHQ* | *Tocopherols* | *regulation* |
| General use | 200 | 200 | 200 | 200 | GMP | 21 CFR 182.3169<br>182.3173<br>184.1660<br>172.185<br>182.3890 |
| Dehydrated potato shreds or potato flakes | 50 | 50 | – | – | – | 21 CFR 172.115 |
| Dry breakfast cereals | 50 | 50 | – | – | – | 21 CFR 172.110 |
| Dry diced glazed fruit | 32 | – | – | – | – | 21 CFR 172.110 |
| Chewing gum base | 1000 | 1000 | 1000 | – | – | 21 CFR 172.615 |
| Margarine | 200 | 200 | 200 | 200 | GMP | 21 CFR 166.110 |
| Mixed nuts | 200 | 200 | 200 | 200 | GMP | 21 GMP 164.110 |
| Active dry yeast | 1000 | – | – | – | – | 21 CFR 172.110 |
| Poultry products | 100 | 100 | 100 | 100 | 300 | 9 CFR 381.147 |
| Dry sausage | 30 | 30 | 30 | 30 | – | 9 CFR 318.7 |
| Fresh sausage, pizza topping, beef patties | 100 | 100 | 100 | 100 | – | 9 CFR 318.7 |
| Dried meats | 100 | 100 | 100 | 100 | – | 9 CFR 318.7 |
| Rendered animal fat | 100 | 100 | 100 | 100 | 300 | 9 CFR 318.7 |

Chapter 16

# Analysis of Oils and Fats

## INTRODUCTION

Analysis of natural or technological products is an essential step which specifies the nature of occurring constituents and their relative quantities. In most cases, oils and fats are a complex mixture, and analysis is based on the implementation of physical, chemo-physical, chemical, biochemical, and sensorial methods. The results obtained reflect the nature, purity and freshness of the product.

Very often the first step of analysis is to extract the oil from the matrix, which is an oleaginous raw material (seeds, fruits), a food product (meat, cheese, pastry, etc.) or a household or industrial product (lubricants, cleaning products, cosmetics, etc.). Quantitative information (content) aside, the prepared sample can then be subjected to a more or less detailed examination of its characteristics. Simple determination by physical (density, refractive index) or chemical methods will provide sufficient general information. Frequently, a more detailed knowledge of the product is required and this may be obtained by a combination of different methods using somewhat sophisticated instrumental techniques (e.g. mixture of volatile contaminants such as solvents by polycoupling gas chromatography, mass spectrometry and dynamic headspace technique.

This will make it possible to establish an 'identity' card for the oil or fat. When this operation is performed repeatedly on well-defined samples it produces averaged data and a range of values as shown in the composition charts. Besides this information related to constitution, analysis will also provide data on the alteration state resulting from triglyceride hydrolysis (acidity) or fixation of oxygen on fatty acid hydrocarbon chains with its consequences (auto-oxidation, thermo-oxidation) on the presence of undesirable compounds (water, impurities, contaminants) as well as the taste, smell and texture of products.

The analytical methods for fats and oils are largely standardised: nationally by AFNOR: (i) internationally by ISO; (ii) IUPAC; (iii) EC; and (iv) other organisations. Therefore they will not be reproduced here but their references, main fields of application and validity will be refered to and discussed if necessary.

## LIST OF A FEW STANDARDISATION ORGANISATIONS

| | |
|---|---|
| AFNOR | Association Francaise de Normalisation (French Standards Association) |
| ISO | International Standards Organisation |
| IUPAC | International Union of Pure and Applied Chemistry |
| EC | European Community |

| | |
|---|---|
| FIL | International Dairy Federation |
| AOCS | American Oil Chemists Society |
| AOAC | Association of Official Analytical Chemists |
| FOSFA | Federation of Oilseeds and Fats Associations |
| BSI | British Standards Institution |
| DIN | Deutsches Institute für Normung (German Standards Institute) |

## DETERMINATION OF COMPOSITION

### Traditional Chemical Tests

These tests provide the basis for chemical analysis of fats and in spite of all progress in laboratory techniques, some of them still play a major role for quality control. Let us take a few examples:

#### *Saponification value (AFNOR, NFT 60–206)*

The saponification value (SV) is the quantity (mg) of potassium hydroxide required for saponification of one gram fat. After hot saponification with a titrated ethanolic potassium solution, excess reagent is titrated. The quantity of reagent required by the reaction can be deduced and the value can be calculated. For a given triglyceride weight, the quantity of potassium hydroxide necessary for saponification increases with diminishing fatty acid chain length. The saponification value reflects the average length of the fatty acid chain of the relevant fat.

| | |
|---|---|
| Sunflower oil (linoleic and oleic acids in large proportion) | SV = 190 |
| Coconut oil (lauric acid in large proportion) | SV = 260 |

This test still plays a practical role in the soap industry. When the fat is a fatty acid (palmitic acid, oleic acid, etc.), the saponification value and acid value (AV) are the same, which is why the term ester (EV) value is sometimes used, expressed as EV = SV – AV. Note that the measurement may be defective when the analysis is preformed on wax-containing fats (lanolin, jojoba, spermaceti).

#### *Iodine value (AFNOR, NFT 60–203)*

The iodine value (IV) is the quantity (gram) of iodine fixed per a hundred gram fat under laboratory conditions.

Addition of iodine on the double bonds will only be measurable in the presence of excess reagent (Wijs liquor with iodine trichloride) and provided contact time is sufficiently long (1 hr.). Fixed iodine is measured differentially with the help of sodium thiosulphate.

The method is unreliable for a double bond in an $\alpha$-position of a carbonyl group in presence of conjugated double bonds (steric prevention). The classification of so-called siccative (IV $\geqslant$ 180), semi-siccative (IV = 130), non-siccative (IV $\leqslant$ 90) qualities has practically been abandoned although it directly reflects the degree of unsaturation. Contact time is reduced to 3 minutes by a fast method using mercury salt catalysis, and the method is used in industrial control of fat and oil hydrogenation.

Finally, it must be pointed out that from detailed information such as fatty acid composition, the iodine value of an oil or a fat can easily be re-calculated by taking into account the additive properties of their characteristics.

#### *Hydroxyl value (AFNOR, NFT 60–213)*

The hydroxyl value is the quantity of potassium hydroxide expressed in milligrams necessary for neutralising acetic acid in combination by acetylation of one gram of fat.

In practice, the operation is done in heat by using acetic anhydride as well as pyridine as reagent playing a catalyst role. At the end of the reaction (1 hour) the added water allows forming of acetic acid from excess which is determined by ethanolic potassium hydroxide. A blank test is achieved under the same conditions.

This determination is of importance to industrial lipochemistry for quality control of following matters:

1. Castor oil.
2. Monoacids and dihydroxystearic.
3. Fat alcohol.
4. Partial glycerides, glycerol.

### *Unsaponifiable matter content (AFNOR, NFT 60–205)*

The unsaponifiable fraction of a fat is a group of compounds extractable by lipid solvents (hexane, ethyl ether) after treatment of fat by ethanolic potassium hydroxide solution under fixed conditions.

The fat is saponified by normal ethanolic KOH at high temperature for one hour. The unsaponifiable substances, soluble under these conditions in a reactional medium become less soluble by cooling and by reducing the alcoholic strength by adding water.

The extracts (generally three but sometimes more) are gathered, then washed as per the case with water or alcohol-water. The solvent is evaporated and the dry residue is weighed. The unsaponifiable content is deducted, which as in natural fats, is generally low and lies between 0.3 and 1.5 per cent. An abnormally high content indicates the possibility of pollution by a mineral oil or a substance which is not of glyceridic nature. It is thus a characteristic feature of fat. Preparation of the unsaponifiable fraction, as explained later, is also of great interest in order to isolate and analyse certain minor components, e.g. the sterols which thus constitute one criteria for purity in quality control of fats. Certain fats of vegetable or animal origin (wheat germ oil, shea butter marine oils particularly liver oils) have a very high unsaponifiable content (2–10 per cent and more), this function can even become dominant as far as lipid composition is concerned. This is also the case with waxes analysed under normal conditions.

## Glyceride Analysis

Glyceride analysis is carried out by the application of practical methods using liquid chromatography, gas chromatography, enzymatic hydrolysis, and chromatography with supercritical fluid.

### *Liquid chromatography*

According to the separation principle involved, it is useful to distinguish three modes of chromatography: adsorption, partition, exclusion. Each one of them, but with different degrees, shows applications for analysing glyceride mixtures.

#### *Liquid adsorption chromatography*

It is the simplest method, either by operating on a column (CL) or on a thin layer (TLC). However, it offers only limited possibilities in view of the fact that separation between components in this case, requires a very high polarity difference between each one of them.

We will mention *mono*, *di*, triglyceride and fatty acid separation which can be achieved on a thin layer of silica gel with hexane/ethyl oxide/formic acid mixture as developing solvent. (80:20:2 v/v/v). Under these operating conditions, the following $R_f$ are observed:

1. Triglycerides 0.66.
2. Free fatty acids 0.42.

3. 1,3 diglycerides 0.23.
4. 1,2 diglycerides 0.18
5. Monoglycerides 0.08.

After separation, the different fractions are visualised, extracted from silica and then quantified (weight, colorimetry, gas chromatography with internal standard). Such a mixture can also be analysed with a TLC device with a rod and a flame ionisation detector. By HPLC with a 5 μm Lichrosorb S160 column these components can be separated in 22 minutes with a gradient elution using toluene, hexane, ethyl acetate.

Chromatography of fat on a silica column with 5 per cent water, and with an adjusted polarity eluant (hexane/ethyl oxide 87/13/v/v) makes it possible to separate it in an apolar fraction ('normal' triglycerides) and a polar fraction (partial glycerides, fatty acids but also specially modified triglycerides). This method is used for the quality control of frying oils. An interesting extend of this type of chromatography is the modification of adsorbent characteristics by adding $Ag^+$ ions. These give a weak energy complex with π electrons in ethylene bonds. This property is the basis of the triglyceride separation technique. In practice, the silvery layers are obtained by addition of a silver nitrate solution to the adsorbent before preparing TLC plates. A 5 per cent $AgNO_3$ content is considered as optimum (separation/cost) but sometimes other contents (upto 30 per cent) are advised. The plates have to be activated at a high temperature (160 to 190°C), and then stocked away from light before use. Different solvents are proposed for separating triglycerides such as benzene, chloroform, and mixtures of these two solvents with ethyl ether for the first and alcohol for the second. The visualisation of plates is mainly done with a dichlorofluorescein hydroalioholic solution. The quantification is always a long and delicate operation done generally by scraping the band, destruction of the complex, extraction of components and finally measuring. Triglycerides are separated according to their unsaturation, the most unsaturated components being the most retained and for a same ethylene bond, the *cis* form having an $R_f$ lower than the *trans* form.

Finally, another way of proceeding is the formation of derivatives. Fixing of iodine chloride on glyceride molecules (but only unsaturated) which modifies their polarity and makes it possible to separate them in HPLC adsorption on silica (Lichrosorb S160 5 μ) can be cited as an example.

### *Liquid partition chromatography*

The numerous triglycerides that a natural fat contains differ from one another by the nature of fatty acids which compose them and the position of the latter on glycerol. Generally, this shows itself in differences at the level of molecular weight and unsaturation. If only fat chains are considered, it becomes customary to speak of the number of partition (NP) that is to say, carbon condensation, two times less the number *n* of double bonds of the molecule:

$$NP = NC - 2n$$

ex.triolein $NP = 54 - 2 \times 3 = 48$

For a long time, it is known that liquid chromatography with reverse phase polarity that is an apolar fixed phase and a polar mobile phase makes it possible to separate fatty chains according to their partition number. One of the problems encountered during this analysis is the detection of the compounds taking into account the chemical structure of triglycerides. The UV detector cannot be retained because qualitatively and quantitatively it gives only partial information, and difficult to interpret (response to 210 nm according to unsaturation and no detection of saturated triglyceride chains). The refractive index (RI) detector is a customary use but is not very sensitive and remains limited in these possibilities (gradient of solvents unusable). However, an analytical method of HPLC of triglycerides of vegetable oils, separated according to their partition number using a RI has recently been standardised (IUPAC 2.324).

*Perrin* and *Prevot* have studied the possibilities of the application of a laser light diffusion detector to fat analysis, particularly to that of triglycerides. The detector's response to be added has to be carefully controlled with known concentration solutions (determination of response coefficient, linearity field) before undertaking the analysis. The detection limit is 0.3 μg for triolein, which is quite satisfying. This detector responds to all non-volatile compounds in eluant pulverisation conditions (30 to 80°C), admits technical solvents, presents a weak dead volume and makes it possible to work with the solvent gradient. This last advantage, in connection with improvement in HPLC (microparticular silica) stationary phase specifications, leads to an analysis of good quality. For example, the chromatogram of a soyabean oil obtained is shown in Fig. 16.1.

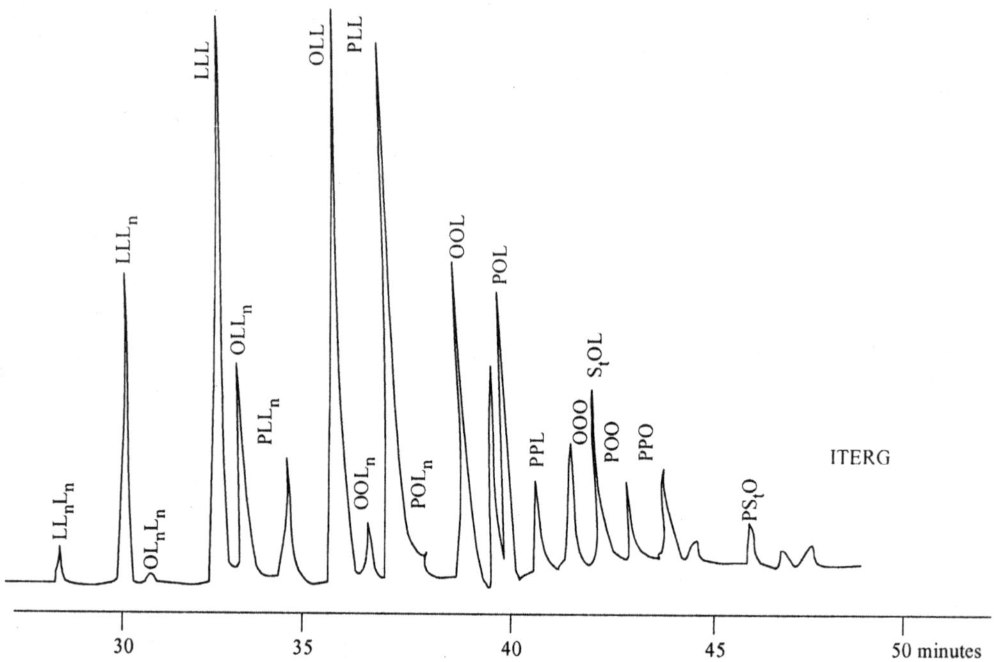


**Fig. 16.1.** Triglycerides of soyabean oil.

In view of an identification of separated components, *Goiffon* has shown that in the case of trisaturated triglycerides, a linear relation exists between the number of carbons and the retention logarithm (or selectivity) a, in relation to triolein. He has applied this results to other triglyceride families which makes it possible to construct a diagram log $\alpha = f$ (delta) from which, knowing the relative retention of an unknown triglyceride in relation to triolein, it is possible to identify it.

*Perrin* and *Naudet* have extended the field of application of the method by underlining the influence of the bonding rate on the separation between components and developing the quantitative aspect from it.

### *Liquid exclusion chromatography*

The main applications of this type of chromatography is the separation of triglyceride polymers and their measurement in heated fats (frying oil). This analysis, in relation to the study of thermoxidation consequences, will be given latter in the chapter.

### *Gas chromatography (GC)*

Medium chain triglycerides such as trilaurine can be eluated in GC on the condition that short columns and apolar or less polar phases (silicon) are used. It is necessary to have devices with a programming oven whose temperature could go up to 350°C while the injection and detection device will be raised to 350°C and sometimes even more. First of all, these are packed columns with a length of 0.5 to 0.7 m, treated supports and stationary phases of the methylpolysiloxane family (SE 30, OV1) with a low impregnation rate (1 per cent). Only in the case of mounting in opposition (two columns) for correcting derivates of base lines, some satisfying layouts are obtained. A demonstrative example is the analysis of coconut oil triglycerides with a separation of 2 in 2 carbon atom components. By using the longer columns (1.8 m) and a medium polarity phase (OV17) *Litchfield* obtained partial separation of homogenous triglycerides having the same number of carbon atoms but whose fatty acids differed from each other by their unsaturation; tristearin, triolein, trilinolein.

A method of determining triglyceride composition grouped by an even number of carbon atoms is standardised (IUPAC 2.323).

The development of glass capillary columns, then of silica, leads to improvements in the study of triglyceride mixtures by this technique. These columns, in comparison to packed columns, show among other things, a great permeability, which constitutes a favourable factor for the analysis of high molecular weight compounds. It is necessary to choose short capillary columns (5 to 6 m) having rather large internal diameters ( $\varnothing \geqslant 0.3$ mm) and a very thin film for stationary phases (0.05 μm to 0.1 μm). The needle injector, which is a simple system, is often mentioned, but *Grob*, in an evaluation of modes of injection for this, shows importance of injector 'on column' in terms of quantitativity and repeatability (CVr = 1 to 3 per cent). A separation example is given in Fig. 16.2.

*Grob* recommends to use longer columns (15–20 m) to separate four components of a mixture resulting from tristearin and triolein interesterification. By combining a high efficiency ($L_c$ = 25 m) and a sufficient selectivity with a thermostable stationary phase of up to 370°C (phenylmethylsiliconegum) *Geeraert* presented, for different vegetable oils, separations between triglycerides, comparable in terms of resolution, to those which the HPLC procures. However, the necessity to standardise has to be underlined to assure oneself of the quantitativity of results, because triglyceride response coefficient increase with their unsaturation degree. This limits the application possibilities of the method.

Finally, for complex mixtures containing triglycerides, partial glycerides and free fatty acids encountered during industrial lipochemistry operations or in biological extracts, a method which consists of forming silyl ethers prior to analysis on short capillary columns impregnated with a polars phase. A method has recently been standardised on this same principle (IUPAC 6.001).

### *Enzymatic hydrolysis*

Certain living organisms possess lipolytic enzymes liable to act on glyceride ester bonds by taking into account the positions occupied on glycerol. Lipase extracted from a pig's pancreas occurs specifically on the level of 1 and 3 positions of triglycerides, and thus allows obtaining one 2-monoglyceride and free fatty acids. An analytical method based on this principle has been developed in the framework of a research carried out by the International Olive Oil Council and has been standardised (IUPAC and AFNOR NF ISO 6 800). After the enzymatic hydrolysis of fats at determined temperature, medium and pH conditions, the monoglyceride is isolated by TLC and the fatty acids are transformed into methyl esters for determining the composition by GC. The method is useful for checking olive oil purity.

Other enzymes like phospholipase $A_2$ can be used and this is specific of the position 2 of glycerophosphatides whereas the lipase of bacterial origin like the My lipase leads to a non-selective hydrolysis.

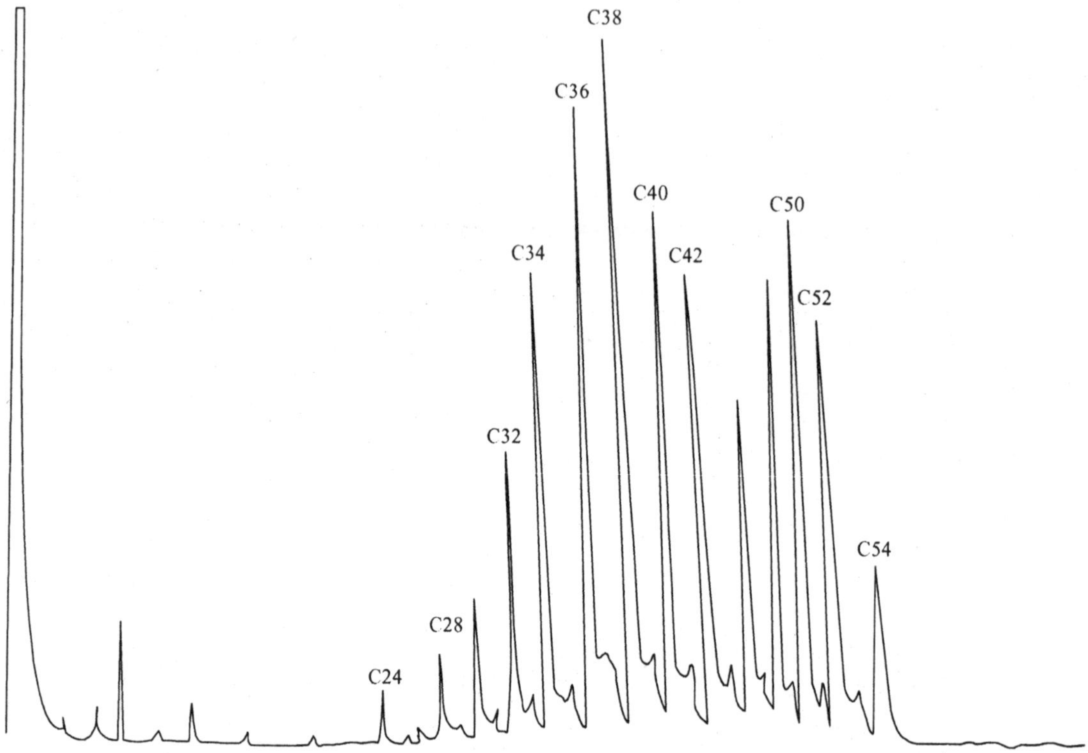


**Fig. 16.2.** Triglycerides of butter oil. (Separations depending on the number of carbon atoms).

### *Mass spectrometry (MS)*

Until 1980s, research on triglyceride identification using this technique was achieved by coupling GC/MS in electronic impact, $M^+$ and $[M - 18]^+$ fragments are very small. The intense peaks correspond to the loss in acyl-oxy grouping or $[M - RCOO]^+$ and $RCO^+$ ions, eventually supplemented by the remaining glycerol of m/e = 74. According to the triglyceride structure in each case, one can obtain two or three fragments. In chemical ionisation with ammonia, the quasi molecular ion $[M + NH_4]^+$ is intense and brings about several triglycerides for one GC peak only.

Improvement in the performance of triglyceride HPLC and the progress in HPLC/MS coupling instrumentation actually offers interesting possibilities for the identification of complex triglyceride mixtures.

### *Chromatography with supercritical fluid*

*Perrin* and *Prevot* showed that in the current analytical cases done by GC or HPLC, the use of a supercritical fluid has not brought about any improvement either in terms of resolution or in reducing the duration of analysis.

## Analysis of Fatty Acids

As fatty acids are the main components of glycerides, their study will occupy a major place in the analytical chemistry of fats. It has been seen, that in the first instance, overall information regarding chain length, unsaturation, presence of secondary groups can be obtained by determining the iodine, saponification, hydroxyl, ester value etc. These informations are, however, inadequate to determine the nature of the diverse fatty acids which are present, their proportions, factors which represent characteristics identifying the fat, thus enabling one to judge its purity and importance on the food processing level as well as commercial level.

Most analytical methods used in that field represent direct applications of chromatographic and spectrometric techniques. Some determinations are, however, based on other principles (for example, enzymatic reactions). Rarely, fatty acids are analysed in the free state. Otherwise they are at first transformed into derivatives and these are, generally, esters.

### *Preparation of esters of fatty acids for analytical purposes*

The reaction mechanisms leading to the formation of esters from triglycerides or from fatty acids alcoholysis esterification may be recalled, that for directing the reaction towards the formation of esters, it will be advantageous to shift the equilibrium in the direction of esterification, by using an excess quantity of alcohol and extracting from it esters by a third non-polar solvent (heptane). Choice of the method to be adopted for preparing esters is influenced by the sample's free fatty acids content and possibly by the presence of particular fatty acids: those with short chains, polyunsaturated, with oxygenated secondary groups.

#### *Methyl esters*

We will first have a look at the alternatives put forward by the AFNOR-standard. Then, we will indicate the various possibilities for preserving, on the chain, less stable groupings (epoxides) or elucidating, from the same, the structure (position of the ethlyene bonds). Lastly we will give details concerning the rapid micro method, of proceeding from refined fats in the laboratory.

1. AFNOR standards NF 60–233.
   It covers four distinct methods:
   (a) A general method: At first the fat is saponified and then the fatty acids are esterified in the presence of boron trifluoride. Esters are concentrated in the heptane acting as a third solvent. This method is rather fast (about 20 minutes) but can not be applied to butyric fats in a quantitative manner.
   (b) A method applicable to neutral fats: This involves a *trans* esterification of triglycerides with the help of hot methanol, and in the presence of potassium hydroxide. After adding water, esters are extracted with heptane.
   (c) A method applicable to acidic fats and fatty acids: This is an operation executed in the hot state with a methanolysis in alkaline medium (methanolic solution of sodium methylate), in the course of which there is a saponification of free fatty acids and methanolysis of triglycerides. This is followed by a methanolysis in acid medium (hydrochloric methanol), transforming the soaps into fatty acids and then into esters. After reducing the alcoholic content (by water addition), the esters are extracted by heptane.
   When the sample is composed solely of fatty acids, only esterification in acid medium is necessary.
   (d) A method applicable to neutral fats or less acidic fats (acid value $< 2$) and possibly containing butyric acid: The *trans* esterification is effected in alkaline medium (methanolic potassium hydroxide, cold), in the presence of a third solvent (heptane) to prevent the loss of the

entire or a part of methylic esters of fatty acids with the shortest chains on account of their high volatility. It is a very rapid process (less than 30 seconds).

The above four methods of the standard cover the majority of the cases encountered in normal practice. However, certain variants are worth mentioning.

2. Other methods: Micro methods relating to the esterification of neutral fats or fats with a low acidity have been discussed. Hence, they have something in common with the standardised method discussed above. The method used in laboratory for neutral fats and reproduced below is inspired by the same. The only difference is represented by an additional stage of neutralisation by methanolic HCl for the purpose of avoiding introduction of alkaline agent into the injector and the column.

Micro method for the preparation of methylic esters of neutral fat is given below:

In a 5 ml tube that can be closed:

1. Introduce 2 drops of fats, weighing about 40 mg.
2. Add 1 ml hexane.
3. Shake for 2 seconds.
4. Introduce with the pipette 0.2 ml of methanolic NaOH 2N. Fix the cork.
5. Shake vigorously for 10 seconds.
6. Hold in the laboratory bath at 50°C for 20 seconds.
7. Shake for 10 seconds.
8. Add 0.2 ml of 2N methanolic HCl.
9. Shake and let decant.
10. Syringe out the necessary quantity of the higher phase and if necessary dilute for injection.

They may have a more specific field of applications, related for example to oils having chains with less stable functional groups (epoxides). With this aim, *Metcalfe* and *Wang* proposed to catalyse the reaction by weak base as tetramethyl ammonium hydroxide whereas *Cecchi* employ methyl-tertio-butyl ether as a third solvent. If the fatty acids exhibit functional groups other than carboxyl (e.g. secondary alcohol group of ricinoleic acid) or if it is designed to determine the form or position of the ethlyene bonds, methylation will be accompanied by other reactions. The alcohol groups will be esterified (acetates, trifluoacetates) or better etherified (silyethers) as can be seen while analysing the fatty alcohols of the unsaponifiable fraction. For the study of structures, it is possible to refer to the work of *Mallet*, who made a comparative study of the significance of forming other derivatives after esterification, namely, hydrogenated, epoxidised, isopropylic derivatives, butylboronate, trimethyl silyether derivatives.

### *Esters of higher alcohols*

The difficulties encountered while analysing methylic esters of fats containing among other things short chains such as those coming from dairy products, from qualitative viewpoint (difficulties in resolving between the first products and the solvent) as well as quantitative one (losses due to volatility) have led to the use of esters of higher alcohols varying from ethanol to pentanol. *Traitler* showed, that ester's response coefficient declines as the alcohol chain becomes longer. It remains practically the same for all the fatty acids, when butylic or pentylic esters are chromatographed. *Wolff* and *Fabien* recently developed a method based on the formation of isopropylic esters after fat extraction with a hexane/isopropanol mixture.

## ***Gas chromatography of fatty acids***

Generally, fatty acids are chromatographied in ester form and only rarely in free state when one is interested in the first terms (from C2 to C5).

The modalities of this analysis are specified by the AFNOR T 60–234 standard as explained below:

1. Equipment: The chromatograph is equipped with a temperature regulated furnace, a flame ionisation detector and a system for injection suitable for the type of column used. It may be mentioned, that in many cases, good results can be obtained using capillary column with split injector, a simple device, provided it is properly adjusted and checked. A good utilisation of such an equipment is a matter of training and experience.
2. Columns: Choice and quality of columns used influence the result of analysis.
   (a) Packed columns: Now they have been practically abandoned. It is, however, pointed out, that best results were obtained (2000 to 2500 plates/m and total resolution between the peaks representing methyl stearate and methyl oleate) in the case of a length of two to three metres, with a support material of granulometry of 80–100 mesh, impregnated with a polar stationary phase like a polyester (DEGS, BDS) at a rate of 5 to 10 per cent.
   (b) Capillary columns: As shown by *Prevot* and *Mordret*, these columns, by their general characteristics (high permeability, high β-coefficient, great length) offer either a high efficiency required for resolving certain components (isomers) or possibilities for rapid analysis, as required for control operations.

   Their manufacturing technology has made rapid strides. The metal constituting the tubes has been given up in favour of glass, which in turn has been replaced by fused silica, offering flexibility and inertia. The processes of deactivation of the walls and *in situ* polymerisation of stationary phases make it possible to eliminate the phenomena of activity (responsible for irreversible adsorptions of whole or part of polar components) and to produce stable films of stationary phase and of controlled thickness. The chromatograms of methyl esters of rapeseed oil and tallow illustrate these possibilities (Figs 16.3 and 16.4).

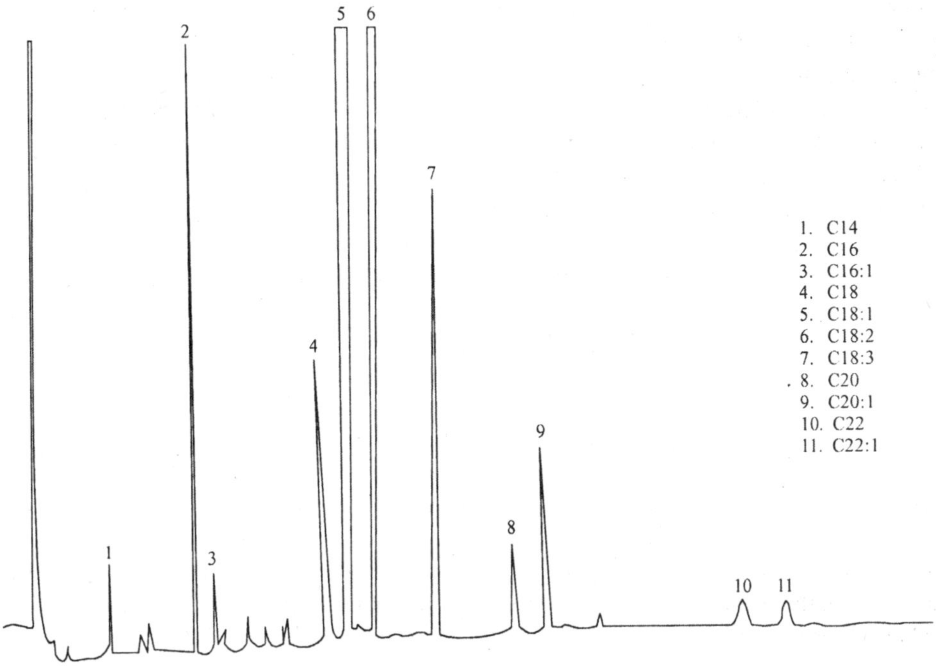


**Fig. 16.3.** Methylic esters of a rapeseed oil.

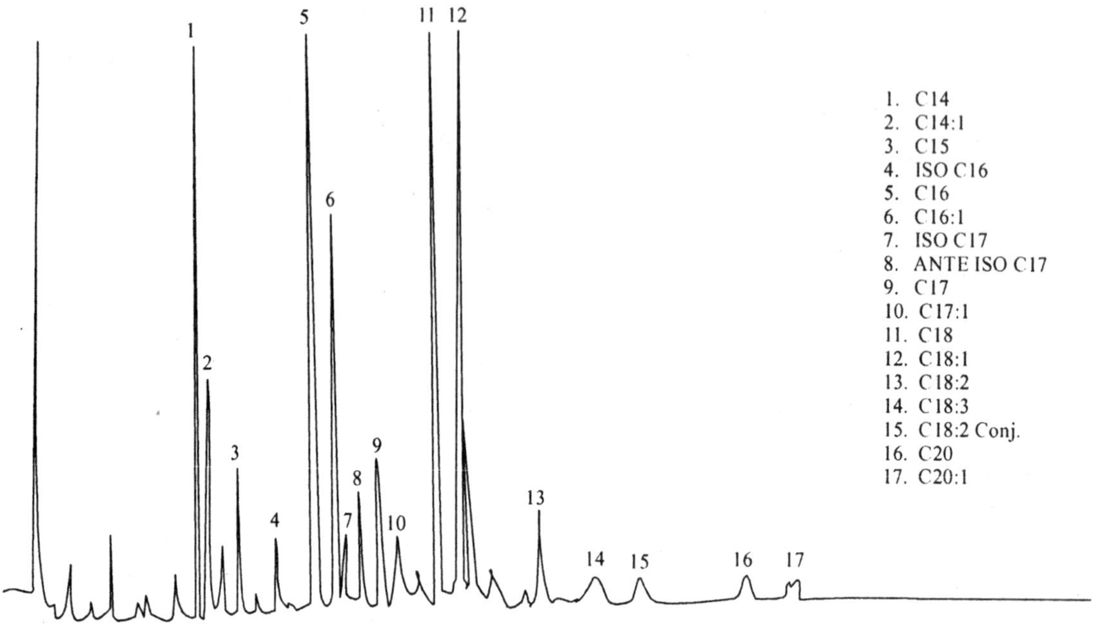


**Fig. 16.4.** Methyl esters of a tallow.

### *Choice of stationary phases*

For obtaining a given separation, the selection of stationary phase is primordial. It should be remembered, that for the same chain length (for example the methyl esters of stearic, oleic, linoleic acids), with a non-polar phase, (SE30, OVI), the components of the mixture will be eluted in the order of their boiling points (linoleic, oleic and stearic). On the other hand, with a polar phase the order will be that of increasing unsaturation of the chains (stearic, oleic, linoleic). In general, when the polarity differences among the components are weaker (in the case of isomers), it will be necessary to increase the polarity of the stationary phase, for getting the required separations.

In the case of fatty acids-mixtures, as found in natural fats containing components of varying chain lengths and also in the case of components with the same chain length of varying unsaturations, it will be preferable to use stationary phases from average to high polarity. It must be mentioned phases constituted by polyglycols (Carbowax 20 M, FFAP, SP1 000, CPWax, etc.), polyesters (BDS), polysiloxanes more or less substituted by high polarity groupings such as cyanopropyls (Silar, 5CP, CPSil 88 ...).

### *Qualitative aspect*

Strictly speaking, gas chromatography is not a method of qualitative analysis. Nevertheless there may be a relationship between the retention of a peak representing a component and its identity.

It is possible to compare the retentions of a mixture's component and those of a reference compound. There is a good probability of identity if the retentions are the same on two columns having phases of opposing polarities. For a series of homologues (for example the methyl esters belonging to the series of saturated fatty acids), there exists a linear relationship between the logarithm of the distance of retention and the number of carbon atoms in the chain belonging to the type:

$$\log.dr = f(nC) + b$$

We can thus plot a curve with nC on the x-axis and log.dr on the y-axis. With the help of two reference points (for example methyl myristate and methyl stearate), a straight line will be traced and the transfer of log.dr values for the different peaks on the ordinate will yield either integral values enabling to identify other homologues (pentadecanoate, hexadecanoate, methyl heptadecanoate) or non-integral values corresponding to methyl esters, which are not saturated homologues.

Non-integral values obtained earlier constitute what is known as equivalent chain length (ECL) which characterise the different fatty acids chromatographied on the column impregnated by the chosen stationary phase.

Example: Carbowax 20 M (immobilised phase)

| Fatty acids | ECL |
|---|---|
| C18:0 | 18.00 |
| C18:1 (n-9) | 18.23 |
| C18:2 (n-6) | 18.69 |
| C18:3 (n-6) | 19.00 |
| C18:3 (n-3) | 19.34 |
| C20:0 | 20.00 |
| C20:1 (n-9) | 20.20 |
| C20:4 (n-6) | 21.16 |
| C20:5 (n-3) | 21.81 |
| C22:0 | 22.00 |
| C22:6 (n-3) | 24.10 |

Many tables have been established and published. Among the latest ones, we will mention the table of *Peyronel* for the phase of Carbowax 20 M and that of *Christie* concerning the phases of Silicon, Carbowax, Silar 5 CP and CPSil 84. They can be used for a given case after verifying their applicability, as per their own working conditions, on particular samples or mixtures. A slight difference between the table values and the experimental results (a few hundredths) is due to an inevitable but slight difference in polarity between the two stationary phases employed. In that case, a correction by increments may be necessary.

This method is simple and of great service in the absence of more powerful means (gas chromatograph coupled with mass spectrometer).

### *Quantitative aspect*

Determination of the fatty acids composition is based on the method of internal standardisation by assuming that the areas of the peaks, represent the totality of components. Content of a given component expressed as a percentage weight of methyl esters can be obtained thus:

$$\%i = \frac{K'iAi}{\sum K'A} \times 100$$

K'i = correction factor of the compound 'i' is relating to $K_{C16}$ (reference = 1.00).

Ai = Surface of the peak corresponding to the compound i.

If it is accepted that Ki = $K_{C16}$, which is generally the case for chain lengths which are not too far removed, the formula is simplified as:

$$\%i = \frac{Ai}{\sum A} \times 100$$

For the purpose of measuring a single component as, for example in the case of butyric acid (IUPAC 2.310) or for evaluating the 'non-chromatographiable' portion of a fat (exhausted frying oil), we add to a weighed quantity of the sample, a known quantity of a reference fatty acid (heptadecanoic, pentanoic acids), constituting the internal standard. In that case:

$$\%i = \frac{ms\ Ai}{m\ As} \times 100$$

i = component to be measured.
m = mass of the sample in mg.
ms = mass of the internal standard in mg.
As = surface of the internal standard.
Ai = surface of the component or components to be measured.

*Measurement of free fatty acids*

The determination is concerned mainly with short fatty acids (C2 to C5). The success of the analysis depends on the column's quality, especially when traces are involved. The material of the packed column should treated by phosphoric acid. As for capillary columns, apart from the choice of a sufficiently polar stationary phase (polyglycol, FFAP) the Grob-test of the column should be satisfactory for the elution of the compound of acidic character (ethyl hexanoic).

### Liquid chromatography of fatty acids

*Adsorption chromatography*

This technique, use in the simplest form (thin silica layer), enables the separation of the esters of fatty acids, only when they differ from one another through the presence of oxygenous groups of diverse polarities such as epoxy, keto, hydroxy, dihydroxy groups on their chains. Its application is limited. Modification of adsorbents by silver nitrate is very important, because it makes possible the separation of components in terms of number, position and configuration of the ethylene bonds from methyl ester mixtures of fatty acids. Thus, it was possible for *Gunstone* to study the *cis* and *trans* isomers of octadecene- 9 methyl oate. *Dudley* and *Anderson* distinguished methyl esters containing 1 to 6 ethylene bonds. *Pelloquin* and *Ucciani* chromatographed on silver impregnated Florosil column, methyl esters of soyabean before and after hydrogenation, with yields and purities of fractions (monoenes, dienes, trienes) exceeding 90 per cent. Cation exchanging resins (Amberlite type) having fixed silver, have been used for isolating the geometric isomers, methyl, octadecenoates and octadienoates. *Vigneron* fractionated the methyl esters of hydrogenated oil by HPLC with the help of an adsorbent impregnated at 10 per cent by silver nitrate. They separated diverse classes of monoenes and dienes as well.

*Partition chromatography*

Methyl esters of fatty acids can be analysed by HPLC on modified octadecyl phases. For separating a mixture of esters, *Scholfield* employed an universal detection of differential refractometry type and worked in the isocratic mode. Critical pairs such as methyl oleate and palmitate were not resolved. *Borsch* prepared phenacyl esters which he chromatographed with an acetonitrile-water mixture, then with pure acetonitrile, determining the diverse esters through UV detection. Resolution between components of the mixture was thus improved very much. In general, this method is of interest for separating without adulteration the fragile components (hydroperoxides of fatty acids, oxidised acids). But it does not offer the performances of analysis by gas chromatography for all the other cases.

### *Fatty acid spectrometry*

In the study of the structure of fats and more particularly that of fatty acids, methods based on spectrometry have been occupying for a long time a very important position, not only for identifications derived from them but also for the measurement made possible by them. We will concentrate mainly on UV and IR spectrophotometries and mass spectrometry. Visible spectrophotometry will be dealt with while studying the colour of fats. There are a certain number of colourimetric methods (phosphorus for example) based on this technique. In absorption spectrophotometry, there is a relationship between the concentration of the substance to be analysed and the quantity of light absorbed. The absorbance A (or optic density) of a solution is defined by:

$$A = \log \frac{I_o}{I}$$

$I_o$ = luminous intensity of incident beam.
$I$ = luminous intensity of emerging beam.

*Beer-Lambert's* law specifies that for a given fat in solution at a wavelength $\lambda$, the absorbance A is proportional to the thickness of the liquid layer 1 crossed by the light beam and to the concentration c of the fat in the solution:

$$A = k.I.c$$

where, k is the extinction coefficient of the fat.

In the special case, where c = 1 per cent and I = 1 cm and assuming that *Beer-Lambert's* law holds good at this concentration, we talk of specific extinction $E^1{}_1$, which is given by:

$$E^1{}_1 = k \text{ (k = specific extinction coefficient)}$$

#### *Ultraviolet spectrophotometry*

Absorption by the UV light is related to the nature, number and the position of the double bond (s) contained in the fatty acid molecule. The individual double bond of *cis* configuration absorbs at 206 nm, but with a relatively low specific extinction coefficient. This determination, apart from practical difficulties (purity of solvents, performance of the instrument) is not of much practical interest. Conjugate dienes, trienes and tetracenes absorb at about 232, 270 and 320 nm respectively. Exact values of wavelengths at maximum absorption depend on the form and position of ethylene bonds. For example 9–11 *cis-trans* linoleic acid exhibits a maximum absorption at 234 nm and a specific extinction of 874, whereas for 9–11 *trans-trans* linoleic acid, these values are at 231 nm and 1170 respectively. Same holds good for conjugate trienes and tetraenes.

At the same time it may be said, that as the number of double bonds increases, the wavelengths of maximum absorption shifts towards the visible range. Proportions of fatty acids with conjugated double bonds in natural fats are very low, but for some exceptions (tall oil, tung oil). Consequent to the isomerisations that can result from oxidation, action of alkaline agents, heat, these compounds can be formed in considerable quantities.

The determination of specific extinctions of fats in UV light (NF T 60–223) is a useful criterion for checking the state of oxidation to establish the bleaching treatment (formation of conjugate trienes) and thus detect the addition of refined oil to virgin oil. It is also helpful for carrying out certain measurements such as the one for polyunsaturated fatty acids having a *cis-cis* dienic configuration, after enzymatical oxidation (NF ISO 7847) or the one for conjugated and polyunsaturated fatty acids.

*Infrared spectrometry*

This technique of functional analysis and quantitative analysis might be very useful for checking the products of industrial lipochemistry. It is necessary to operate according to the general rules of the method. As for the analysis of natural fats, its field of application is very much reduced.

The portion of the spectrum in question extends from 0.8 to 15 μ. Among the numerous absorption bands, we shall mention only a few, which are the most characteristic ones:

| Bonds | Value |
|---|---|
| C — H in — CH = CH — *cis* | 3.31 μ |
| C — H in — CH = CH — *trans* | 10.34 μ |
| C — O in C = O ester | 5.75 μ |
| C — O in C = O free acids | 10 to 11 μ |
| O — H in — OH associated alcohol | 3.0 μ |
| O — H in — OH free alcohol | 2.76 μ |
| O — H in — OH peroxides | 3.0 μ |

For carrying out the measurements from fats, it is preferably to convert the same into esters and thus avoid having to work with free fatty acids (strong absorption of the acid group between 10 and 11 μ). The main application concerns the determination of *trans* fatty acids content through the measurement of the *trans* isolated double bond, which absorbs at 10.34 μ. The spectrophotometer is calibrated by a solution of methyl elaidate. The method has been standardised (IUPAC 2.207).

*Mass spectrometry*

Regarding the analysis of fatty acids by this technique, *Morin* and *Chazan* took stock of its possible applications in the field of fats. It is interesting to avail of two modes of ionisation.

*Electronic impact*

Saturated fatty acids

Molecular ion $M^+$ is apparent. Same is the case with $[M-31]^+$ corresponding to the loss of a methoxy. The base peak corresponds to m/e 74 (fragment originating from the rupture of the bond 2, 3 and migration of a hydrogen: *McLafferty* re-arrangement) and we see peaks of the series m/e (59 + 14 n) $[CH_3OCO(CH_2)_n]^+$ and those of the series m/e 69, 83, 97, ($CnH_{2n}$ +1).

Unsaturated fatty acids

The spectra are distinct from those of the corresponding fatty acids. They differ among themselves according to their degree of unsaturation. On the other hand, for a given unsaturation, it is usually difficult, even impossible to directly distinguish the positional and geometrical isomers, resulting from molecular re-arrangements during ionisation. Molecular ion $M^+$ is abundant and happens to be confirmed by the elimination of a methoxy $[M-31]^+$ or of a methanol $[M-32]^+$ (base peak for oleate). The peak m/e 74 is also encountered. There are also the peaks $[M-74]^+$ and $[M-116]^+$. It may be noted, that the peak m/e 91 has an abundance, which increases with the number of double bonds.

The determination of form and position of the double bonds involves preparation of derivatives, which enable specific fragments to be obtained and hence localised. A large variety of derivatives has been proposed. *Dommes* have compared the advantages of employing isopropylidenes, butyl boronates, methyl ethers, trimethylsilyethers (TMS). *Mallet* in their study of a complex mixture of methyl esters containing *cis* dienes and monoenes, give preference to trimethylsilyethers. These are obtained after many reactions: fixing osmium tetroxide on the double bond, hydrolysis leading to a glycol, formation of TMS. For each double bond, we get two prominent peaks of relative intensity, enabling the localising of the bis-TMS system, hence the double bond.

### Fatty acids with oxygenated groups

In the case of epoxyesters, there is formation of ions, arising from the fragmentation on both sides of the cycle. For esters of ketostearic acids, characteristic peaks result from the break up of β with respect to the functional group. It is the same for the hydroxy acids, which, in addition, can be justified by the formation of TMS, which brings about the appearance of considerable peaks representing the ions of m/e 73 $(CH_3)_3Si^+$ and m/e 75 $(CH_3)_2SiOH^+$.

### Chemical ionisation

We use as reaction gas either ammonia or methane, which give out $NH_4^+$ and $CH_5^+$ ions respectively. These ions will react with the molecules to be analysed.

When compared with the electronic impact technique, we get here the formation of a much more limited number of ions. Nevertheless there is a strengthening of the intensity especially of quasi-molecular ions such as $M + H^+ \longrightarrow M + 1$. This mode of ionisation gives more information about the molecular weight and isomerism. For example, the spectrum of stearic acid contains only one peak of base at m/e 285 and two peaks of low intensity. This variant is useful for identifying polyunsaturated fatty acids, as for example arachidonic acid.

## Spectrometry by Nuclear Magnetic Resonance (NMR)

This technique is based on the magnetic properties of nuclei, which, in the case of certain atoms ($^1H$, $^{13}C$ etc.) when placed in a magnetic field, can enter into resonance and absorb energy.

In actual practice, depending on the value of the magnetic field, we can talk of low resolution NMR (frequencies from 10 to 20 MHz) and high resolution NMR (frequencies beyond 60 MHz), the latter enabling the study of molecular structure. The idea is based on the fact, that, for example in the case of proton, the frequency of resonance is not the same for different protons in the molecule. It is affected by environment. There are characteristic chemical shift (expressed in ppm), with respect to a reference compound. The NMR of carbon ($^{13}C$) can yield more detailed and accurate information. Implementation of this technique is limited to a few very specialised laboratories.

## *Identification of polyunsaturated or fatty acids through chemical or biochemical reactions*

### *Catalytic hydrogenation*

From an unsaturated fatty acid, we can obtain its saturated homologue through a microhydrogenation in the presence of platinum oxide and hence determine the chain length of the acid.

### *Oxidation*

A large variety of reagents has been proposed and depending on the case, there is oxygen fixation, with or without the break-up of the chain.

### *Permanganic oxidation*

Initially used for differentiating between saturated and unsaturated fatty acids, it makes it possible to obtain carboxylated scission products, which will be analysed either directly or after esterification.

### *Ozonolysis*

This method consists of two stages: formation of ozonides and then their decomposition in a reducing medium so as to obtain aldehydes and aldesters.

#### *Epoxidation*

Epoxides are easily formed by the action of peracids (e.g. perbenzoic acid) on *cis* and *trans* bonds. These derivatives can be used as such (separable by thin layer chromatography) or after being converted into glycols and then TMS.

#### *Enzymatic oxidation*

Polyunsaturated fatty acids having a 1–4 dienic *cis-cis* configuration are oxidised in the presence of lipoxydase, leading to the formation of conjugate dienes absorbing at 235 nm. We can thus measure the linolenic and linoleic acids without differentiating between them. The method is standardised (AFNOR NF ISO 7847).

## Analysis of the Minor Components

### *Phospholipids*

Phospholipids of plant or animal origin are complex mixtures which will be analysed by differing techniques depending on the supposed contents in the sample (per cent or ppm) and the information required (overall content, detailed composition). They are usually accompanied by other lipids mainly triglycerides, partial glycerides, fatty acids and sometimes complex lipids.

#### *Phospholipids insolubilisation by acetone*

The low solubility of phospholipids in acetone enables them to be precipitated and be measured by gravimetric analysis. The method is simple but lacks accuracy. It is applicable especially to concentrates such as commercial lecithins. It is standardised (IUPAC 5.433). Based on this property, a nephelometric measurement of phospholipids in soyabean oils and corn oil was proposed by *Sinram*. According to *Morin* and *Prevot*, this method yields only a rough estimate for low content (< 0.4 per cent), which becomes sufficiently satisfactory for higher contents. It is very rapid and could be useful in a first screening of oil batches.

#### *Measurement of phosphorus*

The advantage of this determination lies in determining the phospholipids irrespective of their contents in the sample.

On the basis of average molecular weight, the quantity of phospholipids can be deduced from the measured phosphorus content using the relationship:

$$PL = P \times 26.5$$

PL = Phospholipids
P = Phosphorus
26.5 = Average coefficient varying according to the nature of the phospholipids and their state of 2oxidation.

There are different methods available for phosphorus measurement. They are based on:

#### Colourimetry

Two standardised methods (AFNOR NFT 60–227 and NFT 60–228) are available based on the same principle. They vary only in terms of the sample's size and the nature of the reagents figuring in the formation of complexes. These result in diverse sensitivities and hence in different fields of application. The sample is incinerated in the presence of magnesium oxide and the ashes are treated by a strong acid

($HNO_3$ or $H_2SO_4$). In the former case, a yellow complex of phosphovanadomolybdate is formed. In the so-called 'molybdenium blue' method, the molybdic ions in the presence of a reducing agent (hydrazine) yield, with the phosphate, a blue complex whose absorption is measured at 830 nm.

The first method is applicable for checking crude or semi-refined oils, whereas the second one deals with the refined ones. It must be remembered, that for these determinations homogeneity of sampling is very important. In spite of requiring a rather long execution, these methods serve as reference. *Ames* has proposed a micro-method of colourimetric measurement of phosphorus for reduced quantities of samples, such as those obtained from fractionation through TLC or through HPLC.

### Atomic absorption spectrometry

This method has been recently standardised (IUPAC 2.423). It allows measuring upto 1 ppm and is hence suited for controlling refined oils. Paradoxically, some difficulties persist in the case of crude oils, probably due to the homogenisation or dilution of the samples.

### Emission spectrometry

*Vigneron* utilised this technique with an instrument equipped with an arc plasma source (ICP), placed at maximum emission on one of the lines of phosphorus ($\lambda = 213.6$ nm). The sample was introduced in the plasma under aerosol form. Though being simple, quick and sensitive, the stability of the system should often be verified and the method is better suited for refined oils than for crude oils.

### *Analysis of individual phospholipids*

There is abundant literature about the analysis of these compounds on account of their biological importance, their impact on the preparation of fats, interest shown in them as industrial raw materials.

From a usually complex lipid extract, we have, at first, to isolate the phospholipids, concentrate them, without, however, modifying their structures. For this, we will select fine techniques such as dialysis for which *Ajana* and others have described a simple method, or a chromatographic pre-fractioning on small silica cartridges. An approximately complete separation of different phospholipids can be obtained with liquid chromatography.

Thin layer chromatography on silica gel is still the basis for methods used quite often in laboratories. Here we will mention only two important trends:

Single development of the layer by a ternary system, such as chloroform/methanol/water 25:10:1 (v/v/v) will give the order of migration as follows, but without a complete resolution of the mixture's components: cardiolipids and phosphatidic acid, phosphatidylethanolamine phosphatidyl serine, phosphatidylinositol, phosphatidylcholine, sphingomyeline, lysophosphatidylcholine.

Development in two dimensions improves these separations and often enables the determination of other components such as glycolipids, ceramides. We can employ the system proposed by *Nichols*.

Ist dimension: Chloroform/methanol/ammonia 7N 65:30:4 (v/v/v).

2nd dimension: Chloroform/methanol/acetic acid/water 170:25:25:6 (v/v/v/v).

Quantification is then done by scraping the spots and by microdosage of phosphorus by colourimetry, which is a long and difficult operation.

The method HPLC with micro-special silica column now appears to be preferred for effecting the separation of different classes of phospholipids. For a long time this technique had limited application due to the absence of a universal detection method, available commercially and not affected by the gradient of solvents, an indispensable condition for eluting the components. For separating diverse phospholipids of animal origin, the detection in that case being ensured by a wire and flame ionisation.

By employing light diffusion detector, *Ajana* analysed commercial lecithin's on silica column SI 60 5 μ and on diol modified silica column with graded elution of methanol in chloroform.

A method of measuring phospholipids in soyabean lecithins with UV detection at 206 nm and elution with a mixture of hexane/isopropanol/acetate buffer 8:8:1 (v/v/v) has been standardised (IUPAC 5.302).

### *Unsaponifiable matter*

In natural fats, the unsaponifiable matter usually represents a minor fraction (0.5 to 2 per cent). However, there may be a very large number of components contained in it. After determining its content, it is often necessary to continue its analysis towards a better characterisation of fats and possibly controlling its purity.

However, before undertaking a detailed study of the unsaponifiable matter, in order to simplify the task of the analyst, it is customary to initially fraction it for isolating the diverse groups of compounds contained therein and then study each fraction with a view to separating, identifying and measuring the individual components.

#### *Analysis of the unsaponifiable matter*

At first, we will mention the chromatography's on alumina or silica column with the succession of solvents (or mixtures thereof) of ascending polarities enabling the gradual elution of different classes of components. This method still retains its interest for recovering the fraction of 'aliphatic hydrocarbons' by employing a small column of silica and hexane. The analysis of the unsaponifiable fraction by thin layer chromatography (TLC) has been the theme of many papers and clarifications. This simple method, employing silica layers (adsorption) and mixtures of solvents of adjusted polarities (hexane/ethyl oxide 50:50 v/v, hexane/ethyl acetate 80:20 v/v) allows to separate the majority of the components of the unsaponifiable matter.

As a guide, we give below the $R_f$ observed in the case of an unsaponifiable fraction in fats of plant origin (soyabean oil, sunflower oil, etc.), chromatographed on a layer of silica G and eluted with the mixture hexane/ethyl oxide 50:50 (v/v).

| | | | |
|---|---|---|---|
| Aliphatic hydrocarbons | 0.97 | Triterpenic alcohols | 0.45 |
| Squalene | 0.93 | Aliphatic alcohols | 0.38 |
| | | Δ-Methyl sterols | |
| β-Carotene | 0.87 | Δ5-Sterols | 0.28 |
| α + γ-tocopherol | 0.65 | Δ7-Sterols | 0.24 |
| δ-Tocopherol | 0.55 | Hydroxylated carotenoids | 0.03 |

The advantage of TLC lies also in its utilisation for micro-preparative purposes. A plate of 20 × 20 cm and 0.25 mm thickness can receive upto 20 mg of sample deposited in the form of a wide band. After a non-destructive development, visualisation of the bands, it is possible to isolate the silica by scraping, to extract it and to obtain a small sample for the pursuit of the analysis. We can recover in this manner sterols, alcohols, hydrocarbons. This technique has been covered by standards for every case.

Possibilities have been studied for the direct analysis of the unsaponifiable matter (hence without prior thin layer chromatography) through gas chromatography on capillary column. We can get an 'unsapogram' that could be used for rapid screenings, which, however, require delicate interpretation due to possible interferences among the components.

#### *Hydrocarbon analysis*

Gas chromatography on capillary column allows easily to analyse of this fraction, which, apart from (*n* and *iso*) paraffin's, also contains squalene and possibly some other terpenes (phytenes).

The chain lengths generally vary from 8 to 36 carbon atoms. With columns having less polar or non-polar stationary phases O VI, SE52 or equivalent and temperature programming (80°C to 320°C; 5 to 10°C/mn), we can obtain satisfactory resolutions between components. The appearance of the graph is generally a distribution of the peaks in Gauss curve, which is more or less regular, the maximum being located near C24, often with a very major peak of LEC ≈ C28 corresponding to squalene. Measurement is possible by internal calibration.

*Analysis of fatty alcohols, triterpenic alcohols, 4-methyl sterols*

On account of the low polarity differences existing among these compounds, these three groups of components are badly resolved during TLC of the unsaponifiable matter on silica gel under the conditions indicated earlier. The fractions of triterpenic alcohols and 4-methyl sterols, after formation of the derivatives of acetates or trimethyl silyethers, are analysed by gas chromatography under conditions similar to those chosen for sterol analysis. A method for measuring aliphatic alcohols for the purpose of quality control of olive oils and whose principle is analogous to the one that has just been described, was standardised in Italy. A method for the measurement of a triterpenic diol, erythrodiol, which are found mainly in the oil from pomace olive oil, has also been standardised (IUPAC 2.431).

*Analysis of sterols*

Keeping in mind the importance of this analysis in checking the purity of fats, different method for determining the sterolic composition have been standardised by numerous bodies (AFNOR, ISO, IUPAC, FIL, etc.) over the past 30 years and more. For AFNOR, we will mention the standard NFT 60–232. The underlying principle is always the same. The unsaponifiable fraction of the fat is isolated and fractioned by thin layer chromatography (TCL). Total sterols are recovered, transformed into derivatives (TMS) and then analysed by gas chromatography. While there have not been many modifications in the first stage of the analysis, notable improvements have been brought about in that of separation and identification of individual sterols. The non-polar phase SE 30 in gas chromatography was replaced by a stationary phase of average polarity, OV 17, by *Itoh*. This has enabled the determination of avenasterols and obtain better resolutions among the components.

The chromatograph of sterols of sunflower oil is given as an example (Fig. 16.5). As a result of collaborative joint studies carried out on samples from reliable sources, it was possible to prepare new tables of composition.

The combination of GC/MS is often used for identifying the sterols in the form of acetates of trimethylsilyethers. The NMR of proton is at times necessary for confirming the position and configuration of the ethylene bonds, of the carbon atom in C24, whereas the NMR of C13 gives out a 'finger print' of the product under study.

Finally, the analysis of sterolic mixtures by HPLC can yield two sorts of possibilities. In adsorption mode, *Iatrides* and *Mordret* have isolated, from the unsaponifiable matter on a micro-preparation scale, total sterols, the Δ5 and Δ7 sterols. On bonded column C18, it has been shown, that it is possible to obtain pure compounds (relevant for preparation), though this method, even when optimised, does not have the resolution power as gas chromatography.

*Analysis of vitamin D*

Its main aim is to determine the contents of two forms (Δ2 and Δ3) of this vitamin, which are structurally very close to sterols from which they are derived. Taking into account low contents (less than ppm) that might be tested for in certain fats (e.g. margarines) and the fragility of these compounds, the unsaponifiable fraction will be prepared with care, the vitamin D being isolated and then measured following two HPLC (reverse phase, adsorption).

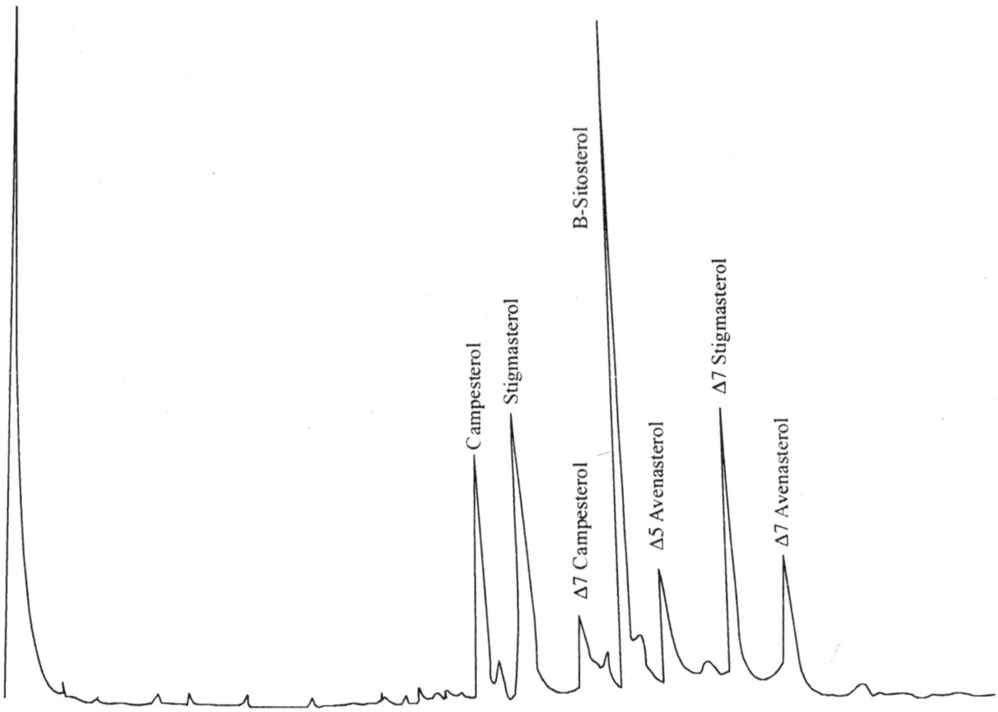


**Fig. 16.5.** Chromatogram of sterols of sunflower oil.

*Analysis of tocopherols*

The tocopherols possess a vitamin power and play the role of natural antioxidant, hence the abundance of papers dealing with their analysis. One of the major difficulties encountered is with regard to their instability (oxidisability), necessitating working precautions if it is desired to isolate them: short term saponification, purification of reagents, presence of inert gas, addition of antioxidant.

At the outset, it is their reducing properties, that have been taken advantage of their characterisation and measurement. We can cite transformation $Fe^{+++} \longrightarrow Fe^{++}$ followed by a colourimetry of ferrous ion (Emmerie Engel reaction) or conversely, oxidation of tocopherols into tocopherylquinones and their determination by pulsating polarography or, in the case of α-tocopherol, formation of a strongly fluorescent phenazine derivative, lending itself to the continuous flow analysis. Total tocopherols can be obtained through the TLC of the unsaponifiable fraction and separated into individual tocopherols through gas chromatography on capillary column impregnated with methyl-phenyl-polysiloxane (OV17). The method standardised by the AFNOR (NFT 60–239) is based on this principle. It is rather long and provides only a centesimal composition of the tocopherolic fraction and not a content (difficulty of selecting an internal standard). The technique which is best suited for analysing tocopherols is HPLC on silica column with a mixture thereof 1.5 per cent of isopropanol in hexane. *Carpentier* separated and measured four tocopherols in different samples of vegetable oil by employing UV detector at 295 nm and recommended a 2 per cent mixture of terbutyl methyl ether in isooctane as well as a detection by UV fluorescence, which would enable one to operate directly on fats. A similar method has been standardised (IUPAC 2.432), which also permits UV detection, at a less specific value, which means more safety in working with the unsaponifiable fraction. This method might be extended to the measurement of the esters of tocopherols (acetates), provided the fluorescent characteristics are better known.

### *Analysis of waxes*

For definite products such as bees wax, carnauba wax, jojoba wax, spermaceti or for minor components of fats, like waxes that may be contained in sunflower, linseed, olive oil meal the purpose of the analysis is generally not the same. In the first case the idea is to establish a composition, in view of, for example, a purity check, whereas in the second case the interest is more towards a measurement keeping in mind the impact that these compounds, even when present in small quantities, might have on the characteristic appearance of refined oils.

### *Determining the composition of waxes*

In a normal state, waxes behave like the unsaponifiable matter, under standardised conditions of analysis. It has been found, that the ester bonds of these compounds were broken only under the special conditions of saponification with an increase in the concentration of alkali agent (N to 2N) arise in the boiling point of the reaction medium (addition of essence E) and lastly an increased duration (1 hr. to 10 hr.). Hence, we are confronted by an analysis of an unsaponifiable fraction (fatty alcohols with the possible presence of hydocarbons sterols, etc.) and a saponifiable fraction, keeping in mind that certain waxes (such as lanoline, beeswax), might also contain special fatty acids (branched chains, secondary groups). This necessitates taking precautions (additional derivatisation) and often causes difficulties in identification).

Present in small quantities in oils, waxes are saponified completely or in part, which explains the presence of fatty alcohols in the unsaponifiable matter. Waxes can be fractionated by TLC by using e.g. carbon tetrachloride as eluant and, if required, slightly heating in a trough (oven temperature at 40 or 50°C). *Valmalle* and *Karleskind* demonstrated, the benefit of gas chromatography at high temperature for directly analysing the waxes and checking their purity by converting jojoba wax into fatty alcohols and ethylic esters, the whole is then analysed by gas chromatography.

### Measurement of waxes in fats

At the outset we can talk of two groups of methods. One is based on the crystallisation of waxes and the other on chromatographical fractionation.

The cold test (AOCS, Method CC 11–53) is a conventional method of checking the dewaxing from sunflower oil. By cooling the oil it aims to bring about the crystallisation of waxes and hence the cloud formation. The time taken for its appearance at a given temperature (0°C) provides information about the presumed wax content. For a concentration exceeding 100 ppm of wax, an oil cloud cover within one hour. *Caupeil* proposed the replacement of visual observation by the use of laser light diffusion system, which enables the detection of even lesser quantities. Separation of waxes by liquid chromatography (CL or TLC) dispenses with the crystallisation stage. In that case we can measure the waxes by gas chromatography. *Henon* described, in 1986 a method based on this principle, which is rapid (2 hr.), repeatable (CVr = 5 per cent) and sensitive (detection limit 10 ppm).

### *Analysis of carotenes and its derivatives*

### Measurement of total carotenoids

This analysis consists in measuring β-carotene, which is for this type of compounds, generally the most abundant and common in fats of vegetable origin. We take as a basis, the fact, that these pigments exhibit adjacent structures and hence all of them absorb light at values between 400 and 460 nm in a rather similar manner. As information it may be mentioned, that the coefficient $E_1^1$, of specific extinction of β-carotene dissolved in hexane is 2500 and that of α-carotene is 2710.

Depending on the abundance in fat, analysis will be carried out directly on the sample itself or after preparing the unsaponifiable matter. Methods based on this principle have been standardised by the BSI and AOAC and are recommended for checking palm oil fractions and oils quality.

#### Determining the composition and measurement of individual carotenoids

Analysis of carotenoids is a delicate operation as these compounds are fragile (isomerisable), photosensitive and oxidisable. Precautions have to be taken while preparing the unsaponifiable matter (obscurity, low temperature, inert gas). The most appropriate separation methods are still liquid chromatographies. Crossing over mixed columns of magnesia-hyflusupersel (1:1), the extracts can be purified, whereas the HPLC with reverse phase polarity results in very good separations between hydrocarbon carotenoids and hydroxyl carotenoids. More recently, *Ng* and *Tan* carried out a complete analysis of palm oil carotenoids by HPLC with a visible UV detector with a block of diodes.

#### Measurement of vitamin A

When compared to the analysis of individual carotenoids, measurement of vitamin A by HPLC does not present any particular difficulty. A method has been standardised (FIL n° 378–09–1989) for the measurement of this vitamin in powdered milk, but it is also suitable for carrying out this determination in a fat.

### *Non-glycerides*

It may be recalled, that this term applies to a set of compounds usually representing a minor fraction of fats, which is complex and contains three types of products:

1. 'Really' unsaponifiable components (hydrocarbons, fatty alcohols, free sterols) for whose analysis we have mentioned are the main methods.
2. Components, which, under the action of an alkali agent are decomposed into an unsaponifiable and a saponifiable matter (esters of sterols, waxes).
3. Components, which, under the action of an alkaline agent and/or by virtue of their structures, polarities, volatilities, are eliminated and are thus lost to analysis (polyphenols, compounds responsible for aroma).

A special methodology has to be adopted for isolating these compounds from fats, unless they can be analysed directly when they exhibit adequately specific properties (absorption of light of characteristic wavelengths, fluorescence, etc.). *Jacini* and *Fedeli* described a method for preparing non-glyceridic concentrates from fats, a non-destructive method, based on a physical process (crystallisation at low temperature in solvent medium). Other possibilities are offered in a similar manner with molecular distillation, liquid extraction, preparative liquid chromatography on column or on thin layer.

#### *Sterol esters*

*Gordon* and *Griffith* proposed a methodology for analysing these compounds. After adding an internal standard (cholesteryl palmitate) to the lipid extract, the TCL on silica enables the isolation of these esters, which can then be analysed either by HPLC with ODS bonded column or through gas chromatography on capillary column type OV 17 at high temperature.

#### *Chlorophylls*

The corresponding pheophytins and chlorophylls can be measured globally by visible spectrometry and directly on the fat (specificity of absorption of these molecules within the range of spectrums wavelength). The AOCS method (Cc 13d-35) recommends the taking of measurements at 630, 670 and 710 nm. Results are expressed as so many ppm of chlorophyll with respect to fat. A method presently under study at the IUPAC (WG 2/89) is based on the same principle. Analysis of individual pigments is cloosely linked with the discovery of chromatography and now rests on the working of the HPLC.

### *Polyphenols*

Diverse substances such as tyrosol or hydroxytyrosol, present in olive oil can be extracted with a mixture of methanol/water (80:20 v/v) from a hexanic solution of the fat. *Cortesi* separated these two polyphenols through HPLC on RP-18 column with the mixture of methanol/acetonitrile/water (46:48:8 v/v/v) whereas *Forcadell* propose their measurement through gas chromatography after forming the TMS derivatives.

### *Components of fragrances*

Virgin oils or prepared fats (butter, margarine) contain small quantities of more or less volatile components, which are responsible for their odours. They can be analysed by gas chromatography (Carbowax column 20 M, temperature from 60 to 180°C at 3°C per minute) provided that they have been isolated at first from their matrix and, possibly, concentrated. For this purpose of the static and dynamic head space techniques are the most suitable.

## Detection of Adulterated Fats and Oils

In the fats and oils industry, purity of edible fats produced should be tested as carefully as their quality.

In fact, it is to be ensured that commercial contracts are respected and that products put up for sale comply with the national and international regulations.

The manufacturer is therefore required to guarantee that information, especially regarding nature and composition of fats, given on labels, is correct.

The analyst has various means to determine the composition of a product and provide evidence of adulteration. These include, among others:

1. Analysis of fatty acids.
2. Analysis of sterols and tocopherols.
3. Analysis of certain triglycerides.

Other analyses may also be carried out: *trans* fatty acids; fatty acids in the 2 position, special values.

### ***Fatty acid composition***

Gas chromatography (GC) determination of fatty acid composition is generally an easy mean of testing the purity of a fat and detecting major adulterations. It suffices to compare readings with those published by laboratories using the latest chromatographic techniques and based on authentic oil samples covering different geographical locations. Absence of specific fatty acids in certain oils especially requires verification. Peanut and sunflower oils for example should be tested for linolenic acid content. In fact, it has been found that C18:3 contents in genuine peanut oils, whatever their geographic origin, should at the most be equal to 0.3 per cent that of sunflower oil 0.2 per cent. Low linolenic content is therefore one of the characteristics of peanut and sunflower oil and should be taken into account to detect an adulteration by soyabean, rapeseed or even linseed oils. The Codex Alimentarius actually indicated the maximum content of C18:3 of both these oils, which is 0.3 per cent.

Checking whether C18:3 contents are low requires the use of efficient chromatography columns coated with a stationary phase which is perfectly suitable for obtaining a narrow peak for linolenic acid, separated from other peaks which could interfere such as C19:0, C19:1 and *trans* isomers of C18:3. Capillary columns coated with butane diol succinate (BDS) or polyethyleneglycol with molecular weight of 20,000 are particularly suitable for these analysis.

However, in some cases, fatty acid analysis alone is insufficient to determine the purity of an oil, although the composition found is within the published values. Thus, peanut oil with a high linoleic content may be from South America or a mixture of African peanut oil and sunflower, corn or grapeseed oils. In order to detect such blends, it is important, therefore, to complete the study of fatty acid composition by that of sterol, tocopherol and triglyceride.

### Analysis of sterols and tocopherols

Some sterols and tocopherols are specific to certain oils and can be used as indicators of adulteration. For example:

1. Brassicasterol (detection of rapeseed oil).
2. Δ7-Stigmasterol (detection of sunflower and safflower oils).
3. Cholesterol (detection of animal fats).
4. δ-Tocopherol (detection of soyabean oil).
5. Tocotrienols (detection of rapeseed and palm oils).

Thus an abnormally high content of brassicasterol or δ-tocopherol in sunflower oil is the sign of an adulteration by respectively rapeseed oil or soyabean oil. An abnormally high content of Δ7-stigmasterol in peanut oil is the sign of adulteration by sunflower oil or safflower oil. The calculation of ratio of certain sterols helps in detecting an adulteration, for example for olive oil the ratio:

$$\frac{\beta\text{-Sitosterol}}{\text{Campesterol} + \text{Stigmasterol}}$$

Other unsaponifiable components like alcohols or hydrocarbons were not studied much for the control of adulteration of oils with the exception of terpenic dialcohols (uvaol-erythrodiol) specific to grapeseed oil and olive pomace oil. Whereas techniques for fatty acid analysis are easy to perform, sterol analysis are more difficult because they require a succession of delicate manipulations, completed by GC. Until recently, an equivalent method was used for analysing tocopherols that increased the risk of loss or deterioration of tocopherols by oxidation.

Now, tocopherols can be directly measured from the oil by HPLC on silica, provided that there is a fluorimetric detector whose excitation wavelength has been fixed at 290 nm and the emission wavelength at 330 nm, these wavelengths being specific for detecting tocopherols. An internal standard allows to calibrate the equipment and to quantify the tocopherols.

However, in some cases (crude oils of corn, soyabean, sesame) some peaks can interfere with those of tocopherols. In that case, a HPLC analysis should be conducted on the unsaponifiable matter of oil and sometimes one must have recourse to analytical procedures used for sterols. Whatever the outcome, these analytical techniques provide results that should be used with caution. In fact:

1. Chromatographic analysis should be done most carefully. It should be ensured that there is no interference with searched peaks. A continuous check of the response of standards added for calibration must be carried out.

   Thus, for sterols it is necessary to use efficient capillary columns coated with OV 17, or better with SE 54, capable of separating the Δ7-stigmasterol peak from a Δ-5,24-stigmastadienol which might lead to error in peanut oils.

For tocopherols, one must constantly check the possibility of detecting traces of δ-tocopherol and separate tocotrienol peaks from those of tocopherols.

2. The specific indicators of adulteration vary according to the geographic origin of seeds. It is important to have as many samples of different origins as possible in order to define the most frequently encountered concentration limits.
3. *Rossell* and *Jawad* have shown that sterol contents can be modified and that the content of Δ7-stigmasterol in particular can be increased by isomerisation of β-sitosterol at high temperatures (8 hrs. at 240°C or 2 hrs. at 260°C or 30 minutes at 280°C).
   However, in normal conditions of refining, the percentage of main sterols is not modified much in spite of new components being created and an appreciable reduction in the global quantity during neutralisation.
   Besides, tables of sterol and tocopherol composition are normally established from crude oils although they are used for calculating the adulteration of refined oils.
4. Some pure oils can have peaks with retention times which are identical to those of the specific indicators being researched. There exists, therefore, a threshold value below which adulteration cannot be calculated.

These observations explain why it is not possible to calculate precisely the adulterating oil content from the only determination of certain sterols or tocopherols. The compositions obtained in this way would not be realistic and would not match with the actual compositions of fatty acids.

In some cases, it is necessary to obtain some additional information such as determination of specific triglycerides in order to make this calculation.

### *Study of triglyceride composition*

Table 16.1 lists the extreme values of linoleic acid. These values are only indicative, but they show that adulteration of oil containing little of C18:2 (peanut-olive) by an oil with high C18:2 content (sunflower, safflower, grape-seed, corn) will increase trilinolein content. In fact, even in different kinds of peanut oils with a high linoleic acid content the percentage of trilinolein is not more than 3.6 per cent of total triglycerides.

**Table 16.1.** Extreme values of linoleic acid.

| *Oil* | *Linoleic acid, %* | *Trilinolein, %* |
|---|---|---|
| Olive | 3.6–12.8 | 0.1–0.7 |
| Peanut | 18.0–38.9 | 1.0–3.6 |
| Rapeseed (*) | 20.5–23.3 | 6.8–10.1 |
| Soyabean (*) | 53.8–56.0 | 18.2–23.1 |
| Corn | 55.3–60.8 | 18.8–25.5 |
| Sunflower | 62.0–71.2 | 23.8–35.7 |
| Grapeseed | 66.3–72.8 | 30.9–38.9 |
| Safflower | 71.1–74.1 | 39.5–43.4 |

(*) Trilinolein + dilinolenostearine + linolenolinoleoolein.

Similarly, adulteration by linseed oil can be detected by the presence of trilinolenin, whereas adulteration by rapeseed or by soyabean oil can be detected by dilinolenolinolein and dilinoleolinolenin.

Consequently, with a study of the triglyceride composition, it is possible to provided evidence or confirm such falsifications. Triglyceride composition is determined by HPLC on columns packed with C18 alkyl-bonded-phase silica. The detection system consists of a differential refractometer or a detector using diffusion of light. With the latter, one can use an elution gradient and thus elute the most saturated triglycerides in the best conditions. This can be useful in the study of the composition of certain fats.

### *Methodology*

With the help of a diagram, we have described the method to be adopted by the analyst to ensure that the oil under analysis is pure. Numbers mentioned in the figures are only indicative because they are given in percentage with respect to fractions studied and the contents of these fractions can also be subjected to major changes after refining especially in case of tocopherols. Generalising the use of internal standards by laboratories for these different analyses will bring a considerable improvement because this will make possible to then define the values in percentage with respect to oil.

Cases of involuntary adulteration due to errors in conditioning or handling of tanks can often be quickly demonstrated by only testing the composition of fatty acids. Logically, cases of deliberate adulteration take place in oils whose consumption is large and therefore, whose prices are high. The falsification is done by adding small quantities of one or more cheaper oils. Procedure to be followed to detect adulterations in olive oil is shown in Fig. 16.6.

The European regulation as well as the International Olive Oil Council prescribe oil purity tests including various physico-chemical parameters such as:

1. The content of saturated acids in the 2-position of triglycerides to verify that there is no re-esterified oil.
2. Specific extinction at 270 nm, limited to a maximum value to ensure absence of refined oil. This determination can eventually be completed by other spectrophotometric measures (extinction after chromatography on alumina, change in extinction around 270 nm). Other criteria to detect refined olive oils in virgin olive oil have been proposed in particular ratios of certain triterpenic alcohols, presence of *trans* isomers and formation of stigmasta-3,5 diene.

Detection of olive-pomace oil can be carried out by the determination of erythrodiol along with sterols but other additional controls have been recommended by the International Olive Oil Council, such as alcanols and waxes, restricting the content of these characteristic components to 250 ppm, in virgin olive oil.

The European regulation imposes limits for certain sterols so as to guarantee that there is not adulteration by seed oils. A study of triglycerides composition should also be done because in case polyunsaturated oils are added, the trilinolein content will exceed the limit normally permitted for virgin olive oils.

In other cases, adulteration will generally modify the sterol composition and especially increase the stigmasterol content. Presence of high oleic sunflower will be detected by an increase in Δ7-stigmasterol content. Peanut and sunflower oil (Figs. 16.7 and 16.8).

Apparently normal contents of C18:2 and C18:3 do not guarantee that these oils are pure although this could be the conclusion that the analyst will arrive at if he restricts himself to fatty acid composition.

Presence of palm olein in peanut oil can be shown by looking for tocotrienols or some carotenoids whereas in case sunflower high oleic can be detected by the presence of Δ7-stigmasterol. Percentages are expressed relatively to the class of products considered.

In case of sunflower oil, as seen in Fig. 16.8, checking the composition of tocopherols determined by direct HPLC of oil besides analysis fatty acids is a practical and fast means to detect various cases of adulteration.

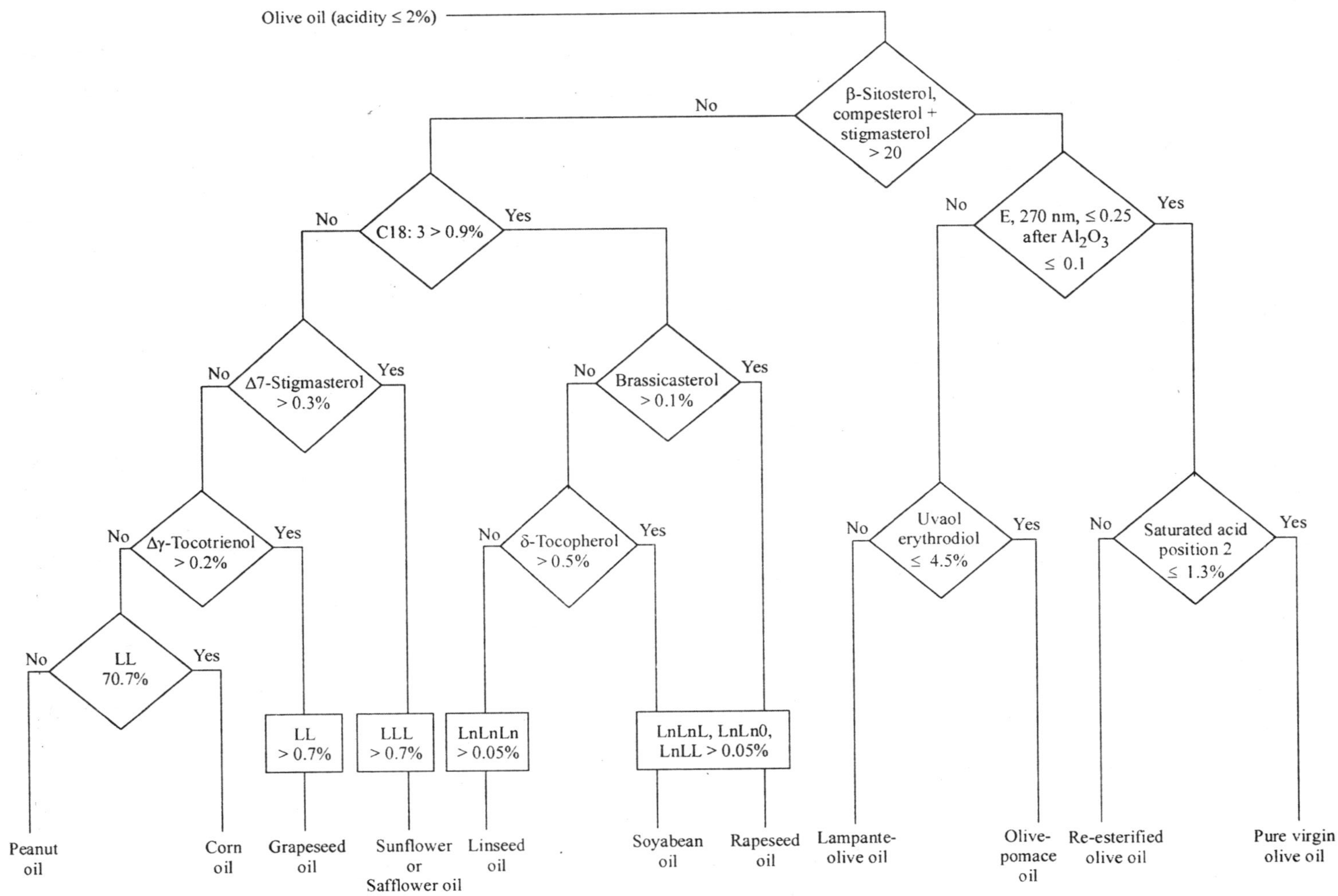


**Fig. 16.6.** Procedure to be followed to detect adulterations in olive oil.

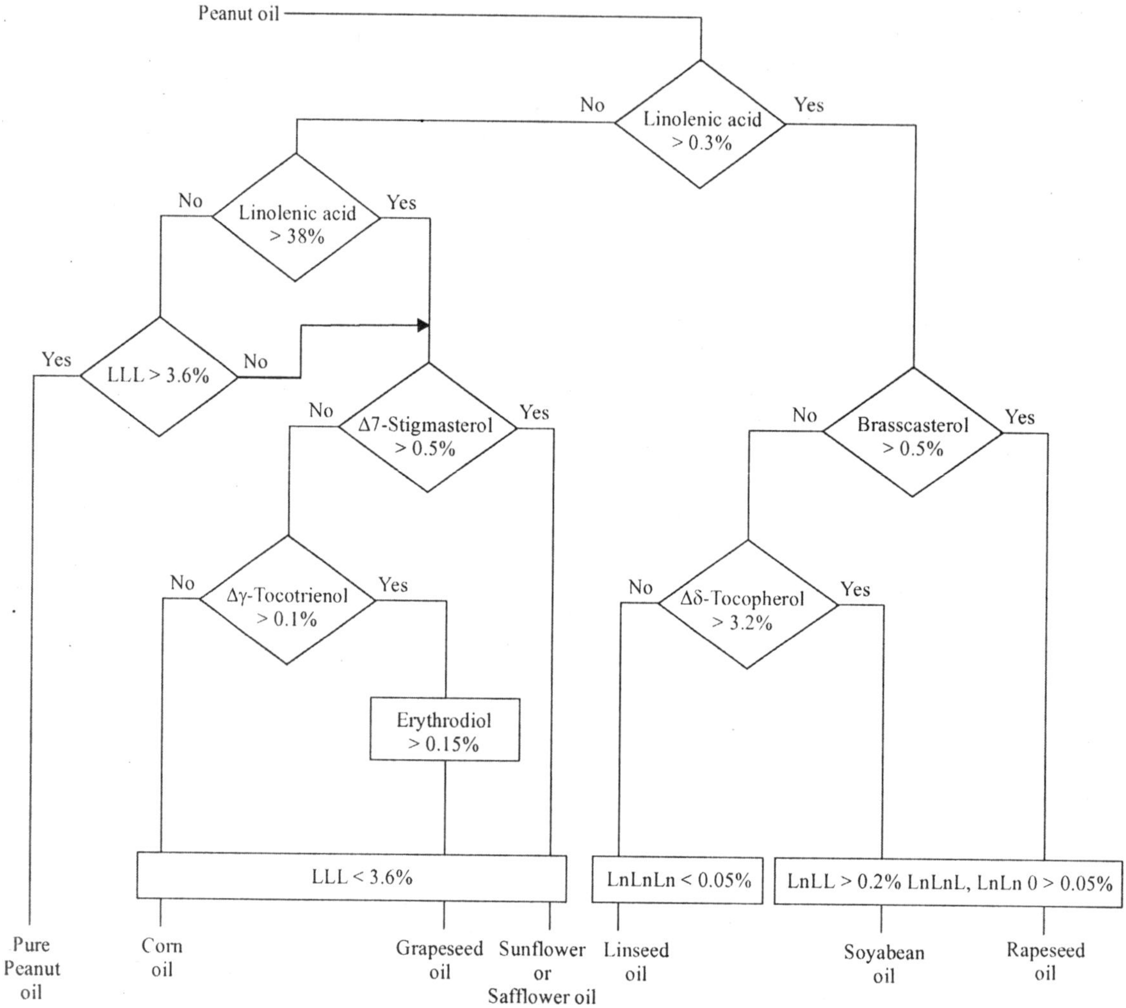


**Fig. 16.7.** Procedure to be followed to detect adulterations in peanut oil.

Let us notice there are varieties of sunflower cultivated in the Mediterranean basin whose C18:2 content lies between 51 and 60 per cent. With the exception of trilinolein, which is well below than of metropolitan varieties, contents of other specific indicators remain practically the same.

### *Fats for margarine industry*

Fluid and solid fats used to manufacture margarines and low-fat spreads can be of vegetable or animal origin. Depending on the performances required for the product, the crystallised fatty phase will be made up of variable quantities of products having a high natural melting point from natural origins or obtained by hydrogenation or by fractionation of certain fats. Possibilities of formulation are therefore very numerous as we can found fatty phases that are completely vegetal or mixed with animal fats, containing or not *trans* fatty acids. Certain formulae are composed of more than five different oils because of performance reasons but also commercialy available cost.

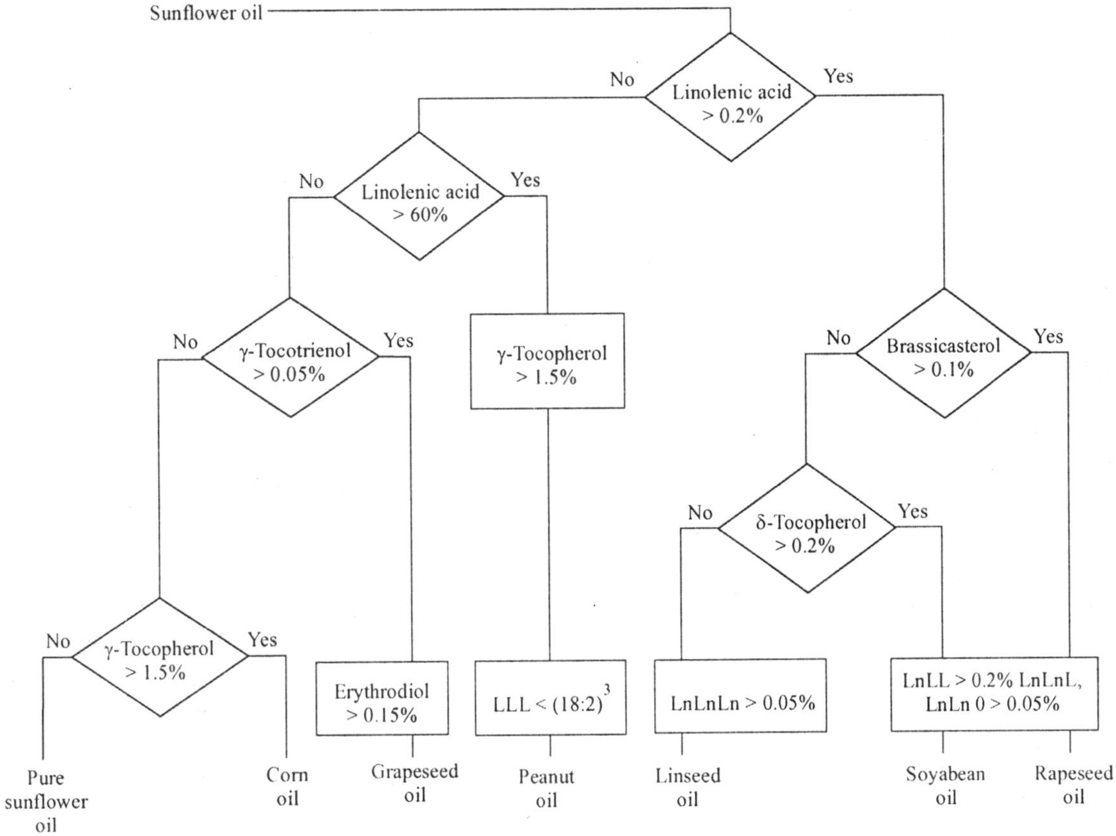


**Fig. 16.8.** Procedure to be followed to detect adulterations in sunflower oil. Percentages are expressed with respect to class of product under consideration.

Calculating the composition of such mixtures precisely may seem illusory considering that processing of certain fats can largely modify the contents of specific indicators required in this kind of study. However, the mention on the label 'vegetable margarine' can be easily tested by checking that the cholesterol content is very low and does not exceed about 2 mg/100 gram fatty phase. Actually, it must be taking into account that the sterol chromatograms of certain oils, at the cholesterol retention time, can have peaks up to 0.2 per cent relative to total sterols and that certain aqueous phases containing butter milk can, by themselves, bring traces of cholesterol.

Besides, more often, the solid fat phase includes palm oil which, on an average, contains 2–3.5 per cent cholesterol with respect to total sterols. In case of margarines not containing any major quantities of palm oil, presence of oil of animal origin can be concluded only if cholesterol content clearly exceeds 0.5 per cent relative to total sterols.

As indicated in Table 16.2 examination of the composition of fatty acid and completed by analysing sterols and tocopherols allow generally to determine the nature and the proportion of principal components of the fatty phase. Triglyceride analysis has limited applications in this case since entire or a part of the fat could be interesterified.

**Table 16.2.** Composition of fatty acid.

| *Specific indications to be searched* | *Indications* |
|---|---|
| C18:3* and brassicasterol | Liquid rapeseed oil |
| Brassicasterol | Hydrogenated rapeseed oil |
| C18:3* and δ-tocopherol | Liquid soyabean oil |
| δ-Tocopherol | Hydrogenated soyabean oil |
| C 18:2* and Δ7-stigmasterol | Liquid sunflower oil |
| Δ7-Stigmasterol | Hydrogenated sunflower oil |
| C4, C6, C8* ... and cholesterol | Milk fats |
| C12*, C14* | Palm kernel/cocount oil |
| C16:1, C 20:1, C 22:1 isomers* | – |
| Odd acids and cholesterol | Fish oils |
| C16* and tocotrienols | Palm oil |

* Corrected according to other oils intake.

Techniques of analysis are practically the same as those described earlier. It is however advised to determine the composition of fatty acids with a capillary column coated with BDS or PEC MW20,000 phase and a gas chromatography equipped with a temperature programming and an 'on-column' injector so as to measure all the fatty acids from C4 to C26 in the same analysis, on the condition that the equipment has been calibrated before hand. As for contents of *cis* and *trans* monoenes, dienes and trienes, they are determined with a capillary column coated with a poly-propylcyanosilicon phase.

In the first approximation, the percentage of liquid oil shall be obtained from the amount of polyunsaturated fatty acids (C18:2 and C18:3) because solid fats are generally devoid of appreciable quantities of this type of acids. By deducting from the total composition the content of fatty acids coming from the fluid oil, we shall get the composition of fatty acids of the high melting point fatty phase. This phase can comprise various oils, even in small quantities, so as fat solid index curve be optimal.

Lastly, from information provided by the composition of fatty acids, sterols, tocopherols, we shall be required to take recourse to a succession of approximations and simulations with the help of propose to data bases a reasonable composition.

Thus some adulterations are easy to detect. This is the case of rapeseed oil, in which contents of total sterols, much higher than in most other common oils (except corn oil) give a brassicasterol peak even if addition of rapeseed is less than 3 per cent.

Others are more difficult to detect, especially those which require the determination of certain tocopherols. Industrial treatment and/or the oxidation state of the sample, can reduce certain specific indicators or show parasite peaks (in case of virgin olive oil). In such a case, one must resort to triglyceride analysis. Some mixtures are difficult or even impossible to detect, especially when several oils have been added in small quantities. Purity of certain oils rich in linoleic acid, such as corn oil, and safflower oil is difficult to control. The introduction of new oil seed cultivars (rapeseed and soyabean with low linoleic acid content, sunflower and safflower with high oleic acid content) complicate the work of the analyst who must regularly update his data banks. It is hazardous to conclude the adulteration of an oil on the basis of one indicator alone. In most of the cases, it is necessary to complete analysis of fatty acids by adding sterols and/or tocopherols and sometimes by measuring the most unsaturated triglycerides.

However, inspite of use of mathematical tools like those proposed by *Van Niekerk*, it is not possible to suggest compositions precisely, because of variability in the contents of specific indicators studied. Appendix I highlights the estimation of adulteration by means of a specific indicator from the unsaponifiable matter.

## APPENDIX I

Estimating adulteration by means of a specific indicator from the unsaponifiable matter is given as:

$$X = \frac{100\, m_2 \left(T - a_2\right)}{m_1 a_1 - m_2 a_2 + T\left(m_2 - m_1\right)}$$

where

$m_1$ = weight of total sterols or tocopherols in g/100 gram adulterating oil A.
$m_2$ = weight of total sterols or tocopherols in g/100 gram of pure oil B studied.
$a_1$ = content in per cent of characteristic indicator of A.
$a_2$ = content in per cent of this indicator in B.
T = content in per cent of indicator to be detected in B adulterated by A; percentages $a_1$, $a_2$, T are calculated with respect to total sterols or tocopherols.
X = percentage of A in B.

Example: Presence of sunflower oil in peanut oil.
Specific indicator: Δ7-stigmasterol.

| | *Lower limit* | | *Upper limit* | |
|---|---|---|---|---|
| | *Peanut* | *Sunflower* | *Peanut* | *Sunflower* |
| Δ7-Stigmasterol (%) | 0.1 | 7.8 | 1.2 | 11.5 |
| Total sterols (g/100 g) | 0.2 | 0.37 | 0.33 | 0.51 |

Mathematically, it is shown that X is maximum if $a_1$, $a_2$, $m_1$ are minimum and $m_2$ is maximum whereas X is minimum if $a_1$, $a_2$, $m_1$ maximum and $m_2$ minimum. We therefore have three possibilities.

| | Percentage of sunflower in the blend | | |
|---|---|---|---|
| | $T < 0.2$ | $0.2 < T < 1.2$ | $T > 1.2$ |
| X maximum | 0 | 12.9 | $\frac{33\,(T - 0.1)}{2.853 - 0.04\,T}$ |
| X minimum | 0 | 0 | $\frac{20\,(T - 1.2)}{5.625 - 0.31\,T}$ |

### Observations

Although in the majority of cases, Δ7-stigmasterol content of peanut oil is ⩽ 0.2 per cent (detection limit of equipment), some samples extracted from peanut seeds contained up to 1.2 per cent.

Consequently, if in a peanut oil, the specific indicator content is found to be ⩽ 1.2 per cent, we should conclude with the benefit of doubt that this oil is pure even though it can contain up to 12.9 per cent sunflower. If this is the case, trilinolein content should be abnormally high (≈ 1.5 per cent brought by peanut oil + ≈ 4 per cent by sunflower oil) and exceed the values normally found in peanut oils with high C18:2 content. Trilinolein measurement will therefore allow detection of such a blend.

## DETERMINATION OF ALTERATION

The main chemical reactions likely to affect fats and oils during their production (extraction of crude oil, refining) or their use, are hydrolysis, oxidation, polymerisation, apart from other minor reactions, such as isomerisation of ethylenic bonds and cyclisation reactions.

Consequences of these reactions are the formation of various decomposition products of fatty chains. These chemical modifications have repercussions on the physical properties of fats. There are many chemical and/or physical methods for measuring alteration:

1. Chemical values and quick tests characterising the most representative modifications.
2. Analytical methods for global measurement of part or all of the alteration products.
3. More complex analytical methods involving various fractionations for a detailed composition of alteration products.

Table 16.3 gives the group of criteria and the methods for analysis that can be applied to assess the quality of fats in terms of alteration. We have differentiated physical, chemical and physico-chemical methods including those techniques which are simple to follow and do not require skilled manpower and laboratory work.

**Table 16.3.** Main methods for determining alterations.

| *Description* | | *Physics* | *Physico-chemical* | *Chemical* |
|---|---|---|---|---|
| Simple tests | Hydrolysis | Smoke | | |
| | Oxidation | Odour | | |
| | Thermo-oxidation | Colour-foam<br>Bath fragrances<br>Fried products | | Fritest-Oxyfritest<br>Veri-Fry TPM test |
| | Hydrolysis | Smoke point | Partial glycerides<br>(IUPAC 2.321)<br>+ TLC, GC | Acid value<br>AFNOR 60.204<br>60.221<br>IUPAC 2.201 |
| | Oxidation | UV<br>AFNOR 60.223<br>IUPAC 2.505 | Volatile products<br>(GC) | Peroxide value<br>AFNOR T 60.220<br>IUPAC 2.501<br>AOCS Cd 8.53 |
| | | Sensorial analyses | | *p*-anisidine value<br>IUPAC 2.504<br>ISO 6885<br>TBA test<br>IUPAC 2.531 |
| | | | | Carboxyl value |
| | | | | Hydroxyl value<br>AFNOR 60–213<br>IUPAC 2.241 |
| | | | | Epoxide value<br>IUPAC 2.251 |

*(Contd...)*

**Table 16.3.** Main methods for determining alterations. *(Contd.)*

| *Description* | | *Physics* | *Physico-chemical* | *Chemical* |
|---|---|---|---|---|
| Laboratory methods | | | | AOCS Cd 9.57 |
| | Thermo-oxidation | Refractive index IUPAC 2.102 AFNOR 60.212 | Polar glycerides IUPAC 2.507 ISO 8420 | |
| | | Viscosity AFNOR 42.011 60.100 51.032 | Triglycerides Polymer AFNOR 60.247 IUPAC 2.508 | |
| | | Dielectric constant Food oil sensor Foam test | Polar acids or esters Polymerised acids or esters | |
| | Thermal alteration | | Cyclic monomers | Iodine value AFNOR 60.203 IUPAC 2.205 |
| | | | Geometrical isomers of fatty acids | |
| | | Accelerated test for oxidative stability: oven test, ASTM test, SWIFT test (IUPAC 2.506, AOCS Cd 12.57) modified SWIFT test (AFNOR 60.219), RANCIMAT. | | |

It may be mentioned that the quick tests indicated under Table 16.3 do not have any strict scientific value. They constitute, at the most, alarm signals which should lead to analysis, using more rigorous laboratory methods.

## Alteration by Hydrolysis

Hydrolysis of fats, whether enzymatic or chemical, leads to the formation of free fatty acids and partial glycerides: monoglycerides and diglycerides. In case of lauric fats, when capric, lauric and myristique acids are liberated, an unpleasant flavour, qualified as soap odour, is released. Measuring free acidity of a fat is one of the good methods to determine its alteration by hydrolysis.

### *Free fatty acids (AFNOR T 60–204 and 60–221; IUPAC 2.201)*

Acidity (A per cent) is the percentage of free fatty acids, conventionally expressed according to the nature of fats in oleic acid, in case of the majority of fats, palmitic acid in case of palm oil or lauric acid in case of lauric fats (coconut, palm kernel).

The acid value (AV) is the number of milligrams of potassium hydroxide required to neutralise the acidity in 1 gram fat.

The number giving oleic acidity A is practically half of what has been found as acid value:

$$A = \frac{1}{2}AV$$

because the ratio between the molecular weight of oleic acid and potassium hydroxide is a factor close to 0.5. There are two titrimetric methods to determine acidity. These differ from each other because of the nature of solvent used to dissolve fat: ethanol/diethyl ether mixture (1/1, v/v), or pure ethanol. In both cases, titration of free fatty acids (FFA) is done with an ethanolic solution of potassium hydroxide.

Both these methods produce practically the same results. The pure alcohol method has an important advantage as far as safety is concerned because of inflammability of diethyl ether. In highly coloured fats, where, because of the colour, one cannot clearly see the change in the colour indication, it is necessary to use the potentiometric method (AFNOR 60.221 or IUPAC 2.201): the neutralisation graph of FFA is arrived at with the help of the pH meter; the equivalent point gives the quantity of soda required by titration. For refined fats, acidity is always below, 0.1 per cent, for virgin oils, acidities are much higher and variable. Determining the composition of FFA can be done by gas chromatography (GC) after fractioning the total fats on a column packed with an aminopropyl-modified silica.

### *Partial glycerides (IUPAC 2.321)*

Separation and measurement of partial glycerides can be achieved by various chromatographic techniques.

#### *Liquid chromatography on open column*

Filled with deactivated silica by adding 5 per cent water (IUPAC 2.321) and elution of triglycerides and then of diglycerides followed by monoglycerides with benzene/ethyl ether mixtures that become more and more rich in ether. After evaporation of organic solvents, the residues are weighed and the percentage of each fraction is calculated.

Benzene, considering its high toxic level, should be, as far as possible, avoided and can be replaced by toluene or preferably by hexane. But, the method should be followed and volumes and proportions especially of solvents constituting the mobile phase should be modified. Filled with deactivated aluminium by adding 5 per cent water and elution of various fractions with mixtures of petroleum ether/ethyl ether with proportions varying from 70/30 to 10/90. For both these methods and in case of highly acidic oils, it is desirable to eliminate FFA by ion exchange chromatography. Thin layer chromatography (TLC) on silica gel as per the IUPAC 2.321 standard. Whereas, separation of various classes of glycerides is easy, on the other hand quantitative operational success is difficult. Evaluation is generally done by densitometry. A more attractive solution is the use of TLC coupled with flame ionisation detection using a latroscan-type device. Generally, all the techniques described above are long and difficult to use.

#### *Gas chromatography (GC)*

Separation of FFA/monoglycerides/diglycerides/triglycerides can be done by GC on short capillary columns not exceeding 10 m and impregnated with a lower polar stationary phase SE 30 type. Silylation of FFA and partial glycerides is necessary. Under optimum temperature conditions and for the analysis period not exceeding 30 minutes, the response coefficients are equal to one. Generally, content of partial glycerides of refined fats is to the order of 1–5 per cent with the exception of palm fats which can contain upto 10–12 per cent.

### *Other tests*

The continuous emanation of smoke by a hot fat means either that the temperature is too high and therefore the smoke point has been reached, in which case, it is advisable to lower the temperature or that the oil has been acidified since the smoke point is lowered by about 30°C while the acidity of an oleic oil increases from 0.01 to 0.1 per cent, and by 85°C if the acidity reaches 1 per cent.

## Thermooxidative Alteration

As there are many different methods of determination of thermooxidative alteration and it is quite difficult to suggest a particular type of analysis because the nature of the products formed which we are trying to measure is so much dependent not only on the type of fat used by also on its application (heating or not, ventilation, light, flat frying, deep frying, etc.).

### *Determination of oxidation*

The first products formed by oxidation are peroxides or hydroperoxides which then evolve towards more stable structures: volatile and non-volatile products.

#### *Specific chemical values*

A certain number of chemical values are used for controlling oxidation, the best known being the peroxide value.

Peroxide value (AFNOR T 60–220, IUPAC 2.501, AOCS Cd 8–53).

The peroxide value (PV) is a very useful criterion with a satisfactory sensitivity for assessing the first stages of an oxidative deterioration.

By definition, the peroxide value is the number of micrograms of active oxygen of the peroxide contained in a gram of fat and capable of oxidising potassium iodide to iodine.

This value is often expressed in millimoles per kilogram or in milliequivalents of active oxygen per kilogram of fat.

$$\text{PV (mmole/kg)} = 1/16 \times \text{PV } (\mu g/g)$$
$$\text{PV (meq/kg)} = 1/8 \times \text{PV } (\mu g/g)$$

The fat is dissolved in a mixture of chloroform and acetic acid and a solution of potassium iodide is added. After reacting for 5 minutes in darkness, the iodine is titrated with a 0.002 N solution (for PV less than 100) of sodium thiosulphate in presence of starch as indicator.

This method does not have any specific difficulty and its repeatability is good.

The main causes of error which can affect the determination are:

1. Addition of the liberated iodine on the ethylene bonds of fatty chains giving then a value by default.
2. Formation of iodine by oxidation of the iodides in presence of the oxygen dissolved in the fat, which would then give an excessive value.

   Other methods have been proposed and can be applied in case we have only very small quantities of samples (biochemical studies, etc.):

   (a) Measurement of the absorption of ions $I_3^-$ at 350 or 290 nm in methanol/acetic acid medium. This method is 1000 times more sensitive than titration with sodium thiosulphate.

   (b) Colourimetric determination at 560 nm of the liberated iodine after addition of starch to hydrochloric acid; the limit of detection being of the order of 0.05 micromole of hydroperoxide.

   (c) Modification of the standardised method of titration by a potentiometric determination with a platinum electrod: this method is particularly suitable for low peroxide values (limit of detection of 20 nanoequivalents of hydroperoxided).

   (d) Other methods, mostly colourimetric, have been used: reaction of the hydroperoxides with ferrous thiocyanate, ferrous chloride, diphenylcarbazide, –2.6 dichlorophenolindophenol, and titanium tetrachloried. A very sensitive method (limit of detection of 25 pmoles) has been developed; the principle is the oxidation by the active oxygen of the peroxides, of the dichlorofluorescein which gives a fluorescent intermediate. It should be noted that none of these methods has been standardised and that neither their repeatibility nor their reproducibility are known. Their significance at these levels in fats is not obvious.

As already mentioned, the determination of the PV is a sure method for predicting subsequent deterioration of organoleptic qualities of the oil. However, and in no case, can the peroxide value account for the past oxidative of the oil; a fat may be rancid or have a mediocre organoleptic quality without having a high peroxide value. If the decomposition of the hydroperoxides is relatively slow at room temperature, the rate of decomposition on the other hand, becomes very high as soon as the temperature reaches 100°C. We should not forget that the purpose of the last stage of refining, namely the deodourisation at 180–220°C, is, besides elimination of the volatile odours products, total destruction of the hydroperoxides.

In fact, the peroxide value is a useful determination to foresee the future behaviour of a fat stored at a temperature which is not very high. In no case can this value be used for heated fats because under these conditions, the decomposition of peroxides is faster than their formation.

A freshly refined fat has a peroxide value of less than 1 meq/kg. In trade, taking into account the best before date of one year, it is not rare to come across oils having peroxide values which can reach 10 meq/kg and without any defect from the sensory point of view.

### Para-anisidine value (IUPAC 2.504, ISO 6885)

The *p*-anisidine value (pAV) enables the determination of mainly α-unsaturated aldehyde compounds.

The principle is based on the reaction, in acetic solution, between the aldehydic compounds present in the sample and the para-anisidine to give a yellow coloured complex compound whose absorbance is measured at 350 nm. The pAV is by convention 100 times the value of the optical density, measured in a 1 cm cell, of a solution of 1 gram fat in 100 ml mixture of solvent and reagent. The class of the aldehydes is the most important from the sensory point of view. A compound such as the *trans*-2, *trans*-4 decadienal, whose typical odour is described as that of frying and having linoleic acid as precursor, leads under the conditions of the reaction to molar absorbance 4–5 times higher than those of the saturated aldehydes. The pAV enables therefore to assess the present state of rancidity of fats. It is, however, obvious that other components aldehydic or not, and not responding to this test, contribute in a significant manner to the alteration of the flavour of fats. The pAV is often associated with the PV to give the Totox value (total oxidation value):

$$\text{Totox} = 2\text{PV} + \text{pAV}$$

The Totox value has the advantage of combining the oxidative part of the oil (pAV) and its potentiality to give secondary oxidation products (PV). It is considered that good oils generally have a Totox value of less than 10.

### Thiobarbituric acid (TBA) test (IUPAC 2.531)

Certain oxidation products of fat chains react with TBA by giving a red pigment one of the absorbance maximum of which is at 530 nm.

The TBA value is defined as the increase in absorbance at 530 nm due to the reaction of 1 mg fat in 1 ml butanol containing a specific quantity of thiobarbituric acid.

If this test is commonly used, the nature of the products reacting with thiobarbituric acid is, however, not well known. If we know that malondialdehyde (MDA) forms a chromagene with TBA by condensation of two molecules of TBA and one of MDA, the presence of MDA in the oxidised fats is not obvious: for instance, it has been shown that the products of auto-oxidation of methyl linoleate, even for peroxide values reaching 2000 meq/kg, do not give any coloured pigment with the TBA. On the other hand, for fatty acids with three or more than three ethylene bonds, the colour intensity increases with the degree of unsaturation of the fat chains. This therefore seems to indicate that only hydroperoxides possessing unsaturation in the β-position of the —OOH group and therefore capable of giving cyclic peroxides (mono- or bicyclic) are likely to lead to the MDA.

We know also that certain chemical compounds such as sugars and oxidised proteins give a positive response with TBA.

To conclude, TBA test is particularly suitable in case of fats having at least triunsaturated fat chains and it is widely used for testing meat products.

Recent work by *Tatum* deserves attention, which involves measurement of MDA in biological media and the liver: MDA-TBA pigment is separated by reverse phase paired ion, and detection is carried out by fluorescence. The detection limit is a few picomoles of MDA.

### Carbonyl compounds

The most frequently used method is that of *Henick*: the 2,4-dinitrophenylhydrazine reacts in acid medium with the carbonyl compounds to give -2,4-dinitrophenylhydrazones which are red pigments in alkaline medium. The measurement of absorbances at 430 nm (saturated carbonyl compounds) and at 460 nm (unsaturated carbonyl compounds) enables the assessment of the total carbonyl compounds content. Another colourimetric method uses benzidine acetate which gives, with the carbonyl compounds, a complex with absorption at 350 nm.

One of the disadvantages of these methods is the fact that the hydroperoxides get decomposed under the given operating conditions.

One of the advantages of these two methods is that they measure components which contribute to the alteration of the flavour of fats.

### Other chemical values

Other less usual chemical values can account for the presence of certain oxidation products:

1. Hydroxyl value (AFNOR 60–213; IUPAC 2.241) is the number of milligrams of potassium required to neutralise acetic acid which combines by acetylation with 1 gram fat.
2. Oxygen of the epoxide group (IUPAC 2.251; AOCS Cd 9–57) is determined by titration with a solution of hydrobromic acid in acetic acid in the presence of violet crystal as coloured indicator.
3. Physico-chemical methods.

### Measurement of oxidised acid (IUPAC 2.502 and 2.503)

Two methods have been standardised by IUPAC. The first is based on rendering oxidised acids insoluble in hexane: after saponification of the fat, extraction of the unsaponifiable fraction and decomposition of soaps with hydrochloric acid, non-oxidised FA are extracted with hexane and the oxidised FA are dissolved in ethanol. The weight of the residue, after evaporation of the ethanol enables the calculation of the percentage of oxidised FA. This method, widely used in Germany, is long and delicate; moreover, it cannot be easily applied to edible fats which contain very little oxidised acids.

The second method is based on the separation of the oxidised and non-oxidised acids by thin layer chromatography. After quantitative detection by a sulphochromic mixture, intensity of the various stains is evaluated with a photodensimeter. The results are expressed in grams of ricinoleic acid per 100 gram FA cleared of the unsaponifiable fraction. This method requires experienced personnel and it is time-consuming. However, it is very well suited for the determination of small quantities of oxidised acids, which is the case of oils which are not heated.

### UV spectrophotometry (AFNOR 60–223; IUPAC 2.505)

Auto-oxidation of fats containing polyunsaturated fatty acid (PUFA) chains is accompanied by conjugation of the ethylene bonds: hydroperoxides of the linoleic acid and conjugated diene systems resulting from their decomposition absorb in the vicinity of 232 nm. Secondary oxidation products and particularly α-diketones or α-unsaturated ketones show a maximum absorption around 268 nm.

Therefore, the stronger the absorbance at 232 nm, the more peroxidised the fat; the stronger the absorbance at 268 nm, the richer the fat in secondary oxidation products. In case of fats containing conjugated triene systems existing naturally (tallows) or resulting from industrial processing (bleaching on bleaching earth), absorbance at 268 nm is no more specific of the secondary oxidation products, because the conjugated trienes have a triple absorption band at 262, 270 and 278 nm.

This method of oxidation evaluation is therefore closely linked with fatty acid composition of the oil; it can therefore be applied only in a comparative manner to oils of similar chemical compositions. In case of refined fats, the sum of the conjugated dienes and trienes present in the oil can give an idea of the state of oxidation of the oil before the passage on bleaching earths; the higher this value, the richer was the oil in peroxides at the time of its bleaching.

### Analysis of volatile products

The physico-chemical method most suitable for their study is GC.

Several methods are followed:

1. Without prior concentration: by direct injection or by static headspace; however, these methods lack sensitivity.
2. After extraction and concentration of the volatile compounds: in order to obtain concentrates, the procedure followed is either by distillation and concentration of the volatile products on a cold finger in contact with liquid nitrogen, or by a carrier gas (dynamic headspace) in closed or open circuit.

An interesting report by *Mordret* compares the advantages and disadvantages of the various methods: use, rapidity, memory effect, sensitivity, suitability for serial sample testing, and cost.

#### *Dynamic headspace method*

The dynamic headspace methods are generally used for the analysis of volatile products. If there is no specific difficulty for using these methods the interpretation of the results is still delicate for several reasons:

1. According to the extraction or desorption temperature and that of the chromatographic column, the components analysed correspond more or less to the volatile compounds actually present in the sample at the time of analysis. If, at a low temperature (up to 70°C), the problem does not arise, it is no longer so at higher temperatures: components recovered correspond not only to the volatile compounds actually present but also to the volatile compounds formed by thermal degradation of their precursors.
2. It is not rare to obtain chromatographs having a hundred or even several hundreds of peaks: the problem is then to link the physico-chemical analysis responses with those of the sensory analysis.

There is a duality between the results of the physico-chemical analysis and those of the sensory analysis. Even if the sensory analysis is not tampered with, in many cases the physico-chemical analysis provides valuable information on the quality of fats (comparison of fats of the same type, perfecting and checking the efficiency of deodourisation, etc.).

### Chemiluminescence

We speak of chemiluminescence when light is emitted during a chemical reaction. A certain number of works seem to indicate that this method may explain the state of peroxidation or oxidation of fats. Extremely simple equipments are required for chemiluminescence method and the results are obtained very fast. However, the interpretation of the result is very delicate and several questions remain unanswered. This method cannot be used at present to assess the quality of fats in terms of oxidative alteration.

## ***Methods of determination of thermo-oxidative alteration***

The methods described earlier are preferentially applicable to the verification of the evolution of fats at standard temperatures.

When heating at temperatures higher than 100°C, oxidation phenomena are amplified and we then avail in addition to the above methods, overall methods of appreciation of the thermo-oxidative products of alteration.

### *Physico-chemical methods*

They generally resort to chromatography on triglycerides or on the latter's fat chains (FA, methyl esters of FA). In the second case, actually degraded FA are measured.

### Polar glycerides (IUPAC 2.507-ISO 8420)

A French decree No. 86–857 stipulates that 'fats and oils whose polar compound content is higher than 25 per cent are unfit for human consumption'.

The principle is based on the adsorption, on slightly hydrated silica column (5 per cent), of chemical species more polar than normal triglycerides.

The non-modified triglycerides fraction is eluted with an hexane/$Et_2O$ 87/13 mixture, polar glycerides content is assessed by working out the ratio between the difference in the quantity charged on the column minus the weight of the fraction of normal triglycerides collected and the quantity charged. This method, which does not involve any sophisticated equipment, has a duration of execution of about three hours. It is obvious, however, that a trained experimenter can simultaneously carry out about eight measurements.

With this method, only polar species are measured. We have moreover some reservations as regards the retention of the epoxide chains under these conditions.

Not only the products resulting from the oxidative alteration (oxidative monomers and polymers) will be determined but also the components resulting from alteration by hydrolysis such as diglycerides, monoglycerides and FFA. Purely thermal polymers are not measured.

### Limitations of the method

This method has the following limitations:

Since we are operating by difference for calculating the polar glycerides content, certain precautions should be taken with regard to the recovery of the normal triglycerides fraction. We are of the view that the difference between the charge and the normal triglycerides content should always be at least equal to 200 mg. If the method specifies a charge of 1 gram, this means that the determination becomes accurate only from 20 per cent of polar glycerides. With 5 per cent polar glycerides, an error of 20 per cent is easily committed. It is not rare to find in case of unheated refined oils, polar glycerides contents close to 5 per cent and even to 10 per cent for palm fats. Under these conditions, and taking into account the legal limit fixed at 25 per cent, it is obvious that such fats start with a strong handicap even if their life period is longer. This is, of course, due to the presence of partial glycerides. Acidity of the fat, when it exceeds a certain level, leads to a systematic underestimation of polar compounds: silica is totally deactivated by adsorption of FFA. Besides, it has been shown that certain very polar compounds are not eluted and that their quantity increases with the degree of alteration.

There are more modern or faster methods of determination of the polar glycerides:

*Prevot* proposed the determination of polar glycerides and polymers of frying oils by GC on filled column of stationary phase at 3 per cent SE 30 on Chromosorb W.

The second method, developed by *Sebedio* is miniaturisation of the standard method of determination of polar glycerides. It uses Sep Pack silica cartridge and involves a shorter time of analysis and lower consumption of solvent.

### Polymers of triglycerides (AFNOR 60–247; IUPAC 2.508)

This measurement is carried out by exclusion chromatography, the principle of which is very simple: the large molecules, excluded from the totality or only from a part of the pores of a rigorously calibrated stationary phase, migrate faster than the small molecules which can enter a larger number of pores. The separation is therefore based not on the physico-chemical interactions with the stationary phase but on the size of the molecules in solution. By this method all polymers of oxidative and thermal nature are measured. Normal triglycerides and monomer oxidised triglycerides are co-eluted under the peak of the monomer triglycerides. This method is fast since the time of analysis is of order of 10 minutes; however, it requires rather costly equipment.

The advantages it have as compared to the determination of polar glycerides. And in the particular case of animal fats meant for animal consumption, for which the polar glyceride method is not suitable, the determination of the polymers enables one to say whether there has been heating or not.

Several circular analyses carried out at AFNOR and IUPAC enabled one to come out with the following points: repeatability is very good: for polymer contents of about 5 per cent and more, the coefficient of variation of reproducibility is 5 per cent. This is a good result. On the other hand, for contents close to 2 per cent, coefficient of variation of reproductibility is 25 per cent.

AFNOR has adopted this method as an experimental standard by fixing its area of validity from 3 per cent onwards. IUPAC has adopted this method under the same conditions.

If in most of the cases detection has been carried out by differential refractometry, UV detection at 234 nm has also been used: only species responding at 234 nm, i.e. the conjugated diene systems are detected: normal triglycerides are therefore not detected and so is the case of purely thermal polymers. One disadvantage is that it is quite difficult to obtain a quantitative measurement with this type of detection.

There is a rather good corelation between the triglyceride polymers content and partial glycerides content: triglyceride polymers represent generally 40 to 70 per cent of the polar glycerides. It may therefore be advantageous, for a screening, to determine the triglyceride polymers since this method take only about 10 minutes. If the content is close to 10 per cent, the analysis of the polar glycerides should be done to check whether the value is more or less than 25 per cent.

### Alterated fatty acids or esters

As we have already pointed out, the passage to the fatty chains enables the measurement of what is really alterated.

Several methods have been standardised some of them are given below:

1. Adsorption chromatography of methyl esters on alumina: new chemical species (NCS).
2. Adsorption chromatography of FA on silicic acid: minor products.

3. Adsorption chromatography of esters on silica by adaptation of the polar glycerides method.
4. GC of FA methyl esters with internal standard (generally C17 acid) so as to be able to determine, by difference, that which cannot be chromatographed.

There is in fact, a relationship between the polar glycerides (PG) and the polar esters (PE):
if PG/PE = 3 means that there is one chain alterated by PG.
if PG/PE = 1.5 means that there are two chains alterated by PG.
if PG/PE = 1 means that there are three chains alterated by PG.

We generally find a ratio ranging between 2.2 and 2.8: a simple calculation shows that in PG there are about four PG with one alterated chain for every PG with two alterated chains. In other words, for 100 fatty chains present in the PG, only 40 are oxidised. A PG/PE ratio > 3 is obviously not possible: it simply indicates the presence of partial glycerides in PG.

### Polymerised esters or acids

The column used for triglycerides polymers enables also the measurement of polymerised esters or FA.

#### *Complex physico-chemical methods*

There are several fractionation procedures by combination of various chromatographic methods. *Ottaviani* working on methyl esters, first carried out a chromatography with reverse phase chromatography on polyethylene column; and obtained a fraction which was hardly polar and which they sub-fractionated by adsorption chromatography on alumina into a polar sub-fractions which was analysed by exclusion chromatography to obtain the oxymonomer and oxypolymer contents. This rather complex method was particularly useful for the detailed analysis of heated oils and especially for the structural identification of the products formed. *Gere*, carried out two fractionations on triglycerides of heated oils: one using exclusion chromatography followed by a thin layer chromatography and the other using adsorption chromatography followed by exclusion chromatography.

The procedure followed by *Dobarganes,* is the most interesting because it is the most practical: the polar fraction of the glycerides obtained by the IUPAC 2.507 method is analysed by exclusion chromatography, which enables to obtain the polymerised and dimerised oxidised triglycerides and the monomer oxidised triglycerides contents as well as the diglycerides, monoglycerides and FFA.

*Dobarganes* and *Perez Camino*, isolated the polar glycerides fraction of oil and prepared methyl esters which were afterwards fractionated by adsorption chromatography: non-polar methyl esters fraction eluted with a Hex/$Et_2O$ 88:12 mixture contains the non-alterated methyl esters of PG fraction and apolar dimers (thermal dimers).

### Oxidised sterols (oxysterols)

The mechanism of formation of oxysterols is one of the ways of formation is heating of fats. The qualitative and quantitative analysis of oxysterols is very important because these products are considered as atherogenes. Generally, the fat is saponified and the unsaponifiable fraction is subjected to analysis by TLC with detection by flame ionisation, by GPC on SE 54 capillary column after derivation or by HPLC on silica column.

### Cyclic monomers (purely thermal alteration)

The investigation and determination of cyclic monomers have been the subject of a considerable number of works. Generally, all these methods follow the same procedure with the exception of a few details: preparation of methyl esters, hydrogenation of methyl esters, inclusion with urea, and GC with internal calibration. In any event, these methods remain difficult and time-consuming and require experienced personnel. Another method has been very recently developed by *Ribot:* methyl esters of the heated oil are hydrogenated in presence of an internal standard (palmitate or ethyl margarate) and the cyclic monomers are isolated by reverse phase HPLC; the fraction containing the cyclic monomers and the internal standard is then analysed by GC. According to *Ribot*, this method is 20 times more sensitive than the methods followed till now. The detection limit is approximately 50 ppm cyclic monomers.

#### Isomers of natural FA

The consequence of heat treatments undergone by fats is the geometrical isomerisation of ethylene bonds (*trans* fatty acids) in, however, rather small proportions. The delicate determinations of geometrical isomer contents necessitate a complex procedure: the heated oil is saponified and methyl esters are prepared and then fractionated according to their polarity on silica column. In the apolar fraction, fractionation with urea is carried out in such a way as to eliminate the cyclic monomers that might interfere. A semi-preparative HPLC on octadecyl modified silica enables isolation of FA according to their unsaturation. Each fraction of methyl esters is then analysed by TLC on silver nitrate impregnated silica. After extraction of the chosen areas, the fractions are analysed by gas chromatography on polar column. In all cases it is a question of sophisticated methods requiring experienced personnel.

#### *Physical methods*

Among various measurements that can be done, following should be particularly mentioned:

#### *Refractive index (IUPAC 2.102; AFNOR 60.212)*

It increases with the degree of alteration of the oil. When overheated, the increase may be in the range of 3 to 5 per cent. It is quite difficult, however, to use such a measurement to assess the degree of oxidation because the parameters influencing this index are so varied.

#### Viscosity

While frying, the increase in viscosity may reach 20–70 per cent of the initial value depending upon the type of oil examined. Several viscosimeters are used: viscosimeters with fall of the ball (AFNOR 42–011), viscosimeters with *Ubbelohde* and *Cannon-Fenske* capillary tube, *Oswald* type viscosimeters (NFT 51032), and rotating body viscosimeters. Measurement of viscosity may be a good test to assess the degree of alteration of fats, provided the initial oil is available. The dielectric constant also increases with the oxidative deterioration of fats and therefore with the increase in polar compound content. *Croon* have shown that there is good correlation between the FOS reading and the PG content: a polar compound content of 25–27 per cent corresponds to a FOS reading of 3–3.2. Some precautions should be taken, however: stabilisation of the apparatus, absence of water in the fat, low PG and FFA contents. These three determinations are therefore more suitable for testing fats in use because they lay down the condition that we should be able to have the initial sample.

#### Foam test

The presence in heated oils of polar glycerides which are to a large extent polymers of triglycerides, modifies in a rather significant manner the surface properties. The principle of the test is as follows: a specific quantity of oil (15 ml) is placed in a graded tube. This tube is dipped in a bath thermo-regulated at a suitable temperature (175°C). After 20 minutes, a cellulose stopper impregnated with a specific quality of water (1.5 ml) is introduced in the oil. The level reached by the foam is read every 15 seconds for 3 minutes. This level increases during the first minute, passes through a maximum and then decreases slowly. The average and maximum volumes of foam are afterwards calculated. This foam test enables to judge whether an oil which is being used has reached a sufficient degree of alteration to cause formation of a fine and persistent foam.

### *Simple tests*

For heated oils, two Merck colourimetric tests can easily be used.

#### Fritest

It is sensitive to carbonyl compounds.

#### Oxyfritest

It contains oxydo-reduction indicators which react with all the oxidation products.

For both these tests, some hot oil and a few drops of check solutions are poured in a tube. After agitation, the colour obtained is compared with that of the diagnostic colours. There is often a good concordance between the values of these tests and those of the polar compounds. However, they cannot be sufficient by themselves to arrive at a certain conclusion about the state of oxidation of the heated oils. The reaction is sometimes difficult to interpret in case where fat has its own colour because of its nature or the food products where they are used.

Another test, used in the United States, is sold in the name of veri-fry TPM diagnostic test by the company Libra Laboratories Inc. (44, Stelton Road, Piscataway, NJ, USA): the product reacting with certain compounds of the oil is already contained, in the form of gel, in the tube. A table indicates the probable polar glyceride contents based on the colours obtained: 0–5, 5–11, 12–16, 17–22 and 23–27 per cent PG.

## ***Methods of evaluation of oxidative stability***

The methods described so far enable to know the physical or chemical characteristics of a fat at a given time. Another question that may claim is that of knowing for how long it can be preserved and therefore its resistance to oxidation. Theoretically, the later can be really measured only by an actual preservation test under conditions close to natural conditions. However, and on account of their duration, these practical ageing tests under normal conditions are not very much used and we often resort to accelerated oxidation tests, which provide a quick answer. Several accelerated methods have been developed: some of these tests use heating with or without light.

### *Oven test*

It is carried out in a thermo-regulated oven set at 60°C and lit by a 60 W lamp. The oil (20 gram) is placed in a glass chiller of diameter 70 mm. The peroxide value is determined at regular intervals, after homogenisation of the fat by agitation. An alternative is the Schaal oven test which uses 50–100 gram fat maintained at 63°C in an oven. The sample is examined at regular intervals till a 'rancid' odour is detected. The peroxide value is also used to follow the progress of the fat. If the peroxide value is measured, the induction period can be easily determined. Generally, no concordance of time between the induction period, determined by the peroxide value and the development of 'rancid' odours and taste is obtained.

Under the conditions of these tests, a non antioxidised fat withstands for about thirty hours. These tests certainly provide results which are close to reality than some other determinations. However, they require several determinations of peroxide and it is recommended to have qualified personnel, operating under standard conditions, determine the oxidative stability by sensory analysis.

These tests are not commonly used and none of them is standardised.

*ASTM test or Sylvester test*

The fat sample (6 g) is placed in a closed vessel, heated to 100°C and under oxygen pressure of 50 psi. The oxygen reacts slowly with the oil and this causes a reduction in pressure. This decrease in pressure against time can be followed up by means of a mercury gauge connected to a recorder.

An automated apparatus is marketed under the name of Fira-Astell and it enables simultaneous study of six samples. The curves obtained enable to determine the induction period: about 20 min for a lard, 24–48 hrs. for vegetable oil and up to 250 hrs. for a cocoa butter.

The disadvantages linked with this system are problems of tightness as well as an increases in pressure when hydroperoxides get decomposed to give rise to volatile products. This test is not very much used.

*Swift test or AOM test (IUPAC 2.506; AOCS cd 12–57) modified Swift test (AFNOR T 60–219) and Rancimat*

The Swift test, also known as AOM (active oxygen method) consists in bubbling a current of air at a speed of about 8 litres/hour in 20 grams of fat maintained at 98°C. Through samples taken periodically, the variation in the peroxide value is observed.

The period of induction under the conditions of the accelerated test is thus determined i.e. practically the time required for the fat to reach a peroxide value close to 20 meq/kg.

In France, the modified and AFNOR-standardised Swift test is preferred: if the principle remains unchanged, the oxidation is not more detected by the measurement of the peroxide value, but more simply by the change in colour of a solution of a coloured indicator in which the air which has served for the oxidation of the fat get mixed. In fact, when the oxidation becomes appreciable, by-products of oxidation are formed: among these we find acid volatile products (formic, acetic, propionic, butyric, valeric acids, etc.) which cause the coloured indicator to change into a slightly alkaline solution. The coloured indicator chosen is cresol; its colour changes at pH 7.1–7.2 from violet to yellow whereas the fat shows a peroxide value close to 20 meq/kg. The Swift test and the modified Swift test give therefore results which are very similar. The latter, however, eliminates the numerous determinations of the peroxide value.

As compared to the test described earlier, the Swift test and its modified version have a certain advantage: the secondary oxidation products, which are themselves oxidation catalyst, are continuously eliminated by the passage of air. We therefore come closer to the normal conditions of use of fats.

The duration of resistance with the Swift test of fats of various types is as follows:

| | |
|---|---|
| Peanut oil | 20–30 hours |
| Sunflower oil | 9–12 hours |
| Rapeseed oil | 8 hours |
| Non-antioxidised tallow | 3 hours |
| Crude tallow | Up to 6–8 hours |
| Non-antioxidised lard | 2–8 hours |
| Olive oil | 20–60 hours |

The repeatability of the method is fairly good: less than 10 per cent deviation between two determinations for fats resisting for less than 10 hours.

An automatic version of the Swift test known as Rancimat the acid volatile products are trapped in distilled water and this induces a change in conductivity which is continuously measured by means of a platinum double-conductivity cell. The impression of the analysis curves is automatically done right from the beginning of the analysis. Six samples can be tested simultaneously.

The results obtained by this method are repeatable: for instance, for six tests carried out on a sunflower oil we obtained an average of 10.6 hr. and a standard deviation of 0.29 which leads to a coefficient of variation of 2.6 per cent.

By way of information, we give below a few results obtained for various samples of commercial oils of the same type: hours (T = 98°C).

| | | |
|---|---|---|
| Peanut oils | 1 | 22.5 |
| | 2 | 17.2 |
| | 3 | 33.4 |
| Sunflower oils | 1 | 10.8 |
| | 2 | 11.0 |
| | 3 | 11.8 |
| | 4 | 8.7 |
| Rapeseed oils | 1 | 17.6 |
| | 2 | 17.5 |

Generally, AOM or Swift test values are rather correlated with those obtained with *Rancimat*.

$Y = 1.003\,x - 0.093$, $r = 0.996$ determined for 26 olive oils.

$Y = 0.982\,x + 1.2$, $r = 0.987$ determined on peanut, sunflower, rapeseed oils and on tallow fats.

Y = AOM or SWIFT value, × = *Rancimat* value

### Important observation regarding all accelerated tests

All methods require that the glassware be properly washed. Generally, the procedures and cleaning agents are indicated for each method.

#### *Other accelerated tests*

Other tests have been proposed:

Differential calourimetric analysis (DCA) by detection of an exothermic reaction corresponding to the quick formation of hydroperoxides. The method is well correlated with the AOM test and it has the advantage of being very fast. However, it has not significantly developed.

Thermo-gravimetric analysis: the variation in the weight of the heated sample is continuously followed up. GC determination of total volatile products by accelerated oxidation of the oil at 170°C for 20 minutes by a gas current (nitrogen containing 250 ppm oxygen). The results are well correlated with the AOM values.

### *Conclusion regarding all the accelerated tests of resistance to oxidation*

If as we saw AOM or Swift tests as well as the automatic test with the *Rancimat* are widely used and are mentioned in the specifications of the companies, it is on the other hand quite difficult and delicate to extrapolate the results of an oxidation in the neighbourhood of 100°C with that which takes place simultaneously at ordinary temperatures: the methods of hydroperoxide formation not being the same, secondary oxidation products resulting therefore are different and do not necessarily have the same impact on the organoleptic qualities of fats. On the other hand, these tests are quite suitable for the determination of the comparative efficiency of antioxidants.

## DETERMINATION OF FOREIGN SUBSTANCES

### Water and Impurities

Edible fats and oils are constituted mostly of glycerides. Nevertheless, especially when crude, they might contain water and other non-fat matters commonly known as 'foreign' substances.

The latter compounds are mainly remaining fragments of the oilseeds or of the adipose tissues (except for various cases of pollutants), partially soluble or dispersed in crude fats. These substances are likely to impair the quality and preservation of fats. Refined edible oils and fats are practically devoid of such compounds. Some foreign substances are voluntarily introduced during the manufacturing of certain emulsified fats in order to obtain a specific texture or quality.

### *Water and volatile substances*

Oils and fats can contain water and volatile compounds as a result of the extraction method (water from vegetable or animal tissues released during crushing or melting, extraction solvent), the transformation method (fractionation in solvent) or the transport (salted water ...). Often, liquid oils and fats that are not clear contain water which is soluble in lipids at about 0.05 to 0.1 per cent. Greater amounts of water can be found in crude oils and fats, generally emulsified because of partial glycerides and phospholipids present in tank bottoms. Besides being a foreign matter, water is likely to impair the quality of oils and fats. For instance, water is a favourable medium for bacterial growth or for the activity of enzymes catalysing hydrolysis or oxidation. As water concentration widely varies, different measuring methods with various scopes have been determined.

#### *Water determination by physical methods*

#### Methods for determining moisture and volatile matter

Two methods are available based on the drying up of a known weight of oil or fat, on hot plate or in a drying oven.

They cannot be applied to oils and fats high in water or in acidity. This is especially true for lauric fats.

These methods are detailed in the French standard T 60.201 corresponding to the ISO 662 standard and IUPAC 2601 method. These methods are described also in the BS 684 1.10 1982 standards. AOCS describes two methods: Ca 2b 38 for the naked lights method and Ca 2c 25 (2) for the oven method.

#### *Hot plate (or sand bath) method*

This method consists in heating over a hot plate or over a sand bath, a known amount of oil or fat enclosed in a calibrated porcelain capsule, continuously and gently stirred with a small thermometer or else with an agitator.

Bubbles of steam appear at the bottom of the capsule. Water is released as the bubbles surface.

It is important to:

1. Avoid any overheating that could result in errors by excess, due for instance to an entrainment of free fatty acids or simply to projections.
2. Avoid any prolonged heating, particularly when working with fats highly susceptible to oxidation, that could result in an error by default due to fat oxidation.

For a skilled operator (particularly needed if water content exceeds a few per cents), this method is satisfactory in terms of repeatability and reproducibility. Moreover, it is fast.

#### *Drying oven method*

This method consists in causing the release of water and volatile compounds from a known amount of oil placed in a crystalliser into a oven kept at 103°C (235°F).

Duration of stay in the drying oven should:

1. Ensure the complete release of water and volatile compounds.
2. Cause the least oxidation possible (especially when studing a drying oil).

Therefore, it is important to take the oil out of the oven periodically, cooling it down in a humidity-proof housing, in order to record the weight loss regularly. If kept too long in a drying oven, siccative or highly susceptible to oxidation fats and oils (linseed oil) can gain weight as a result of oxidation. This method can not be applied when water content exceeds a few per cents. It is as satisfactory as the previous method in terms of repeatability and reproducbility.

### Method for determining moisture by azeotropic entrainment

Water forms an azeotropic mixture with toluene or xylene-type aromatic hydrocarbons. When this azeotrope is distilled, moisture is entrained and separated once condensed.

This method, known as the Dean-Stark method, allows the measure of moisture contents exceeding 0.5 per cent. This technique is specific to moisture determination in the absence of water-soluble solvent. Normally, it is not altered by the presence of a small amount of glycerol, in particular if the distillation is cautiously undertaken. It should be noted that the determination of a refractive index on isolated water avoids any parasitic entrainment. This method is covered in the NF T 60.218 French standard of November 1984 corresponding to the ISO 934 standard and to the IUPAC 2602 method of 1980. It corresponds also to the Ca 2a 45 AOCS method.

### *Moisture determination by a chemical method*

Specific to the determination of moisture, this general method, known as the *Karl Fisher* method, can be used to measure moisture in oils and fats. Its implementation is nevertheless delicate due to the low solubility of fats in the *Fisher* reagent.

*Fisher* reagent is made of a blend of iodine and sulphur dioxide in solution in methanol and pyridine.
In the presence of water, the sulphuric anhydride reduces the iodine.
Determining the reduced iodine enables to calculate the water concentration.
This method applies to the determination of:

1. Low water concentrations in fats (< 1 per cent).
2. Water in fats susceptible to oxidation, or particularly in acidic fats containing short chain fatty acids (lauric fats).
   This method corresponds to the application of the French NF T 60–225 standard corresponding to the IUPAC 2603 method.
   This method is equivalent to the Ca 2c 84 AOCS method.
   The use of an automated device or of specific reagents overcomes the difficulties linked to.
3. The use of a toxic solvent.
4. The use of very hygroscopic solvents, requiring, in the manual application of the method, some preventive measures against humidity (rinsing and drying of the glassware).

### *Choosing a method to implement*

By knowing the nature and rapidly observing the oils and fats, one can chose the methods to implement.

| *Nature of the fat* | *Observations* | *Method to implement* |
|---|---|---|
| Oils and fats (except for lauric fats) | Low susceptibility to oxidation<br>Low in acidity | Hot plate or drying oven |
| | Low water concentration | |
| | Susceptible to oxidation | Hot plate |

*(Contd...)*

| *Nature of the fat* | *Observations* | *Method to implement* |
|---|---|---|
| | Acidic | Karl Fisher |
| | High water concentration | Azeotropy |
| Lauric fats | – | Hot plate |
| | | Karl Fisher |

Repeatability and reproducibility of these methods depend on the moisture concentration. They can be defined as follows:

| *Methods* | *Moisture and volatile matter concentration (Designating the method as optimum)* | *Repeatability* |
|---|---|---|
| Hot plate or drying oven | 0–2.0% | 0.05–0.10% |
| Azeotropy | > 2.0% | 0.50% |
| Karl Fisher | < 0.5% | 0.01% |

### *Volatile compounds*

Volatile compounds are generally minor components in oils and fats, eliminated during the deodourisation step of the refining process. Accidentally, volatile substances can be found in minor quantities in oils and fats, for example as a result of a pollution during transport or following migration from the packaging. Then, they are subjected to a particular investigation. Volatile compounds can be found in crude oils when the solvent used for the extraction step is not completely removed. This can be worrisome in terms of security during transport or storage of these oils.

Concentration in residual solvent in soya oil is subject to regulations specifying that these oils must:

1. Either have a flash point below 250°F, according to the cc 9b 55 AOCS method in retorts.
2. Or an hexane concentration below 0.1 per cent.

Hexane content in oils is determined by gas chromatography.

The 607 IUPAC method uses the head space technique corresponding to AFNOR and ISO 9832 methods.

The direct injection method developed by *Prevot* has been adopted by *Fosfa*. The oil to be examined is injected directly into a chromatograph and collected onto a glass wool micro-column. This method is particularly well suited for serial testing.

### *Foreign substances*

Foreign substances or impurities consist of all compounds (but water and solvents) that are not glycerides, fatty acids or unsaponifiable matter proper to oils and fats.

Thus, impurities include:

1. Visible macroscopic matters.
2. Substances dispersed or even solubilised in oil, produced for instance by fatty acids derivatives such as metallic soaps.

Impurities are important for several reasons:

1. They are prejudicial to the aspect and presentation of marketed refined oils and fats.
2. They can impair the quality of the oil. For instance, metallic soaps can induce oxidation and water can favour bacterial growth (in case of seed meal in vegetable oils or in case of meat in animal fats).

3. They are not included in the price of the oils and fats. Their presence in excess can justify a non-negligible rebate, or even jeopardise the label of 'safe, loyal and of marketable value'.

By convention, impurities are defined as any substance insoluble in solvents, under specific experimental conditions, and separated by filtrating the fat dissolved in the chosen solvent. The conventional nature of this definition presents no disadvantage for the determination of impurities of mineral (sand, earth, metallic particles, metal oxides from tanks), vegetable (seed meal) or animal (meat meal) origins. This conventional nature of the definition is less well-suited to the determination of oil-soluble organic compounds and of solvents such as metallic soaps, in particular lime soaps. The same is true of oxidation products, which have a variable solubility coefficient depending on the nature and temperature of the solvent, and of phospholipids the solubility of which depends on their hydration and even on their degradation rate.

Therefore, it is important that procedures for the determination of impurities be properly followed.

The type of filter used (paper or fritted glass with varying porosity levels, used for instance in the AOCS Ca 3a–46 method), the quality of its manufacturing, and in the case of fritted glaze, the number of use can affect the results. The solvent used for the determination of impurities must be the one mentioned in the standards. In general, diethyl oxide, petroleum ether, or, in few specific cases, trichloroethylene (olive-pomace oil, for instance). Knowing the impact of the solubility coefficient of impurities, the volume of solvent used, as well as the way it is used, should be controlled also. In fact, the use of large volumes of solvent can render soluble some impurities that are usually hardly soluble (i.e. metallic soaps).

In this regard, determining impurities becomes a complex problem in the case of oil mills residues susceptible to oxidation such as 'acidic oils'. Indeed, the filters usually get clogged during determination of impurities in such products due to the fatty acid oxidation by-products. As a result, the operator tends to increase the amount of solvent recommended in the standards which yields poorly repeatable results.

In such conditions, one available technique consists in determining impurities (sand, earth...) by extraction in a Soxhlet starting from the cold insoluble matter in the solvent issued from the application of a standard method for determining impurities. Other impurities such as metallic soaps, oxidised fatty acids, are determined by the application of such techniques as determining cations by atomic absorption, determining ashes alkalinity for sodium soaps.

In case of impurities consisting of metallic soaps in fats meant for the soap industry, it is possible to consider only their mineral part as impurities, for instance to consider only the calcium in calcium soaps.

Two main methods of determination of impurities, used particularly in commercial transactions, are described:

1. In the AFNOR (NF T 60202), ISO (663) and IUPAC (2604) methods using solvents with a low distillation point (hexane, diethyl oxide, carbon disulphide, cold trichloroethylene).
2. In the AOCS (Ca 3–46) method using hot kerosene and petroleum ether (under such operating conditions, lime soaps for instance are not considered as impurities).

Carbon disulphide and diethyl oxide are dangerous solvents. Their handling, and in particular, their distillation should be done by following closely the recommendations mentioned in the standard. Diethyl oxide should never be distilled twice. The solubility of major impurities are shown in Table 16.4.

Separated impurities are dried at 103°C (217°F). Their concentration is expressed as a percentage of crude product. The mineral or organic nature of the separated impurities can be identified by the determination of ashes. In such case, it is necessary to use an ashless filter. Impurity concentration usually encountered in crude oils are low: generally 0.05 per cent and very rarely 0.1 per cent.

**Table 16.4.** Solubility of the major impurities.

| *Impurities* | *Solvents* | | | | |
|---|---|---|---|---|---|
| | *N Hexane* | *Diethyl oxide* | *Carbon disulphide* | *Toluene* | *Trichloroethylene* |
| Earth, sand various impurities (bone) | Insoluble | Insoluble | Insoluble | Insoluble | Insoluble |
| Carbohydrates | Insoluble | Insoluble | Insoluble | Insoluble | Insoluble |
| Protein | Insoluble | Insoluble | Insoluble | Insoluble | Insoluble |
| Phospholipids (Except for hydrated or transformed phospholipids) | Soluble | Soluble | Soluble | Soluble | Soluble |
| Oxidised fatty acids | Partially soluble | Soluble | Soluble | Partially soluble | Soluble |
| Metallic soaps (calcium, iron) | Partially soluble | Partially soluble | Partially soluble | Soluble | Soluble |

Last, it should be noted that certain trade contracts express together impurity and moisture contents. This method should be banned as it does not allow to differentiate between an elevated but acceptable moisture level and a high concentration in impurities which would deny the 'sound, loyal and of marketable value' character to the oil or fat considered. In spite of international standards (IUPAC, ISO), international trade does not always specify the use of these methods. Thus, if FOSFA, bought back by IASC, specifies the ISO method for the determination of moisture and volatile matter, it specifies the BS method for the determination of impurities.

### *The specific case of sodium determination in fats*

Sodium soaps constitute a particular case in the determination of impurities.

The presence of sodium soaps reveals an imperfect refining. In this case, sodium soaps are generally found as traces and the methods related to the determination of sodium (which could correspond to the determination of other forms of sodium) or dischargeable fatty acids (applicable to a concentration from 0.2 per cent up) are not well-adapted.

Direct determination of soaps by volumetry in acetone medium may be carried out according to the procedure described in AFNOR NF T 60217 standard. It is facilitated by the low dielectric constant of this solvent. This determination is facilitated also when undertaken into two steps: using acetone at 3 or 4 per cent water in which oils are not soluble and soaps are soluble. The alkalinity due to sodium soaps is titrated with aqueous solution of hydrochloric acid very dilute (N/100) in presence of bromophenol blue.

Detection of soaps is particularly sensitive as a concentration in combined soda of 0.0001 per cent triggers a frank greenish blue colouring of the titration solution (yellow in the absence of soap).

This well-adapted method for refined oils and fats fails in presence of:

1. An acidic oil or fat (acidity > 1 per cent).
2. Oils and fats rich in phosphatids or in amino acids which are alkaline as they are amphoteric.

This method, which enables the determination of an alkalinity, does not only determine sodium soaps. It determines for instance traces of lime soaps in crude bone fats.

### *The specific case of emulsified edible fats*

Water, or even milk (at concentrations ranging from 16 to 50 per cent), non-glyceridic compounds, at a concentration of a few per cent (dairy protein, sugar, salt) can be introduced in fats or blended in emulsified edible fats (margarines, spreads or low-fat butter). They can be detected through various methods.

### *General methods*

Water found in margarines can be determined by the hot plate method. Implementing this method is delicate but yield satisfactory results when heating is moderate, stirring is regular and, eventually, with a last evaporation of water in a drying oven. Another method is preferred: weight loss in a drying oven at 103°C (217°F) in presence of sand or pumice-stones according to a standard procedure (AFNOR-ISTV drying oven method VI-2-1986).

The non-fat content can be determined by a classic method for the determination of impurities on the dry fat obtained following moisture determination by the naked lights method. Sodium chloride content is determined by classic argentometry on an aliquot of the separated water following the melting, decantation, or centrifugation of fat.

### *Specific method*

Moisture and non-lipid parts of emulsified fats can be determined globally by deduction thanks to a method whose main purpose is to determine the total fat content. This method is based upon the absorption of moisture and non-lipid matter on a anhydrous sodium sulphate column. Fat is eluted by an organic solvent. The solvent is distilled and the collected fat is weighed. This weight deducted from a hundred is compared to the standard. It indicates the total moisture and non-fat content.

## Antioxidants

Antioxidants (or antioxygens) are employed for protecting the organic molecules from oxidation. Thus they are used in various industries (rubber, plastics, pharmaceutical products).

Nevertheless, their role is the most important in food processing industries, where they are mostly employed for preventing the oxidative adulteration of oils and fats. This phenomenon of spontaneous oxidative adulteration proceeds in two stages: in the first stage, unsaturated fatty chains fix oxygen from air and form peroxides. These peroxides are unstable and evolve spontaneously, with or without oxygen consumption and lead to the formation of oxidation, adulteration and polymerisation products.

The formation of these products have harmful technological consequences (for example, during refining process). Further, these products are liable to bring about modifications in taste and smell of fresh products. Finally, we should not neglect the physiological effects brought about by the ingestion of oxidised fats. Antioxidants are hence oxidisable molecules, which by consuming the oxygen from air, slow down the rate of peroxide formation. There are also compounds known as synergists. These products, which by themselves exhibit only little, even no antioxidising activity, strengthen the action of actual antioxidants. This phenomenon of synergy can, elsewhere be produced also with two (or more) actual antioxidants.

Some antioxidants, such as tocopherols ($\alpha$, $\beta$, $\gamma$, $\delta$) are present in the natural state in vegetable oils but, the majority of them are synthetic and generally contain one or more phenolic nuclei liable to fix free radicals by taking up a hemiquinonoid structure. In practice, all the antioxidants can act as a catalyst to the peroxide formation, if their concentration in fats is too high.

### ***Principal antioxidants used***

#### *Hydroquinone*

It is the simplest phenolic derivatives. Its antioxidising activity is excellent, but on account of the toxicity of the oxidation products formed, it has no use in food processing industry.

*Pyrogallol*

It has a better antioxidising activity than hydroquinone, but it presents the same problems of toxicity.

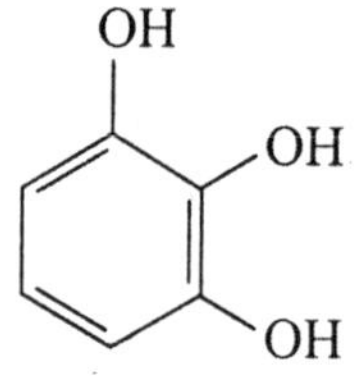


*Gallic acid and gallates*

Propyl, butyl, octyl and dodecyl gallates are employed more frequently than gallic acid, whose antioxidising activity is, however, more important than that of its derivatives. Gallates are all the more soluble in oils and fats as their aliphatic chain is long. Gallates are effective in concentrations of 50 to 100 mg/kg.

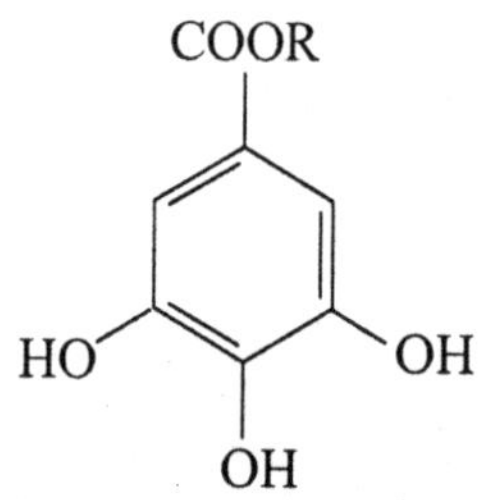


*Nordihydroguiaiaretic acid (NDGA)*

It is a diphenol, with a considerable antioxidising activity. It is effective at concentrations of a few hundred milligrams per kilogram.

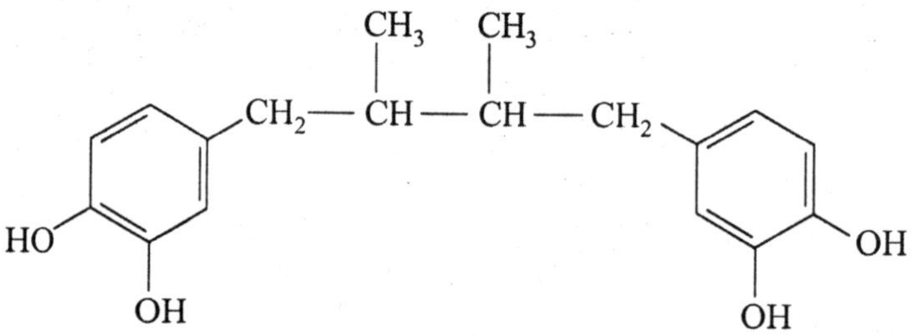

*Butylhydroxyanisol (BHA)*

The commercial product is a mixture of tert-butyl-2 methoxy-4 phenol (or tert-butyl-3 hydroxy-4 anisol) called 3-BHA and of tert-butyl-3 methoxy-4 phenol (or tert-butyl-2 hydroxy-4 phenol) called 2-BHA.

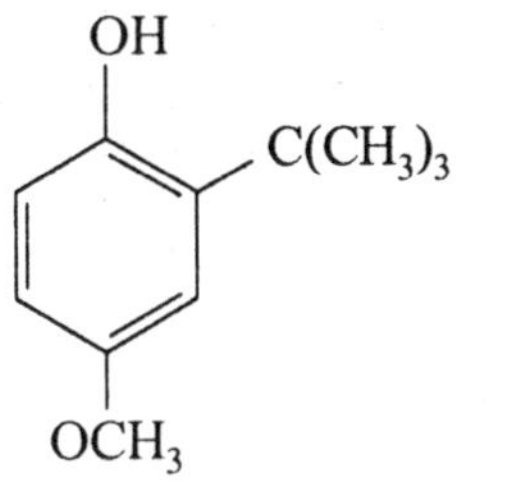


3-BHA

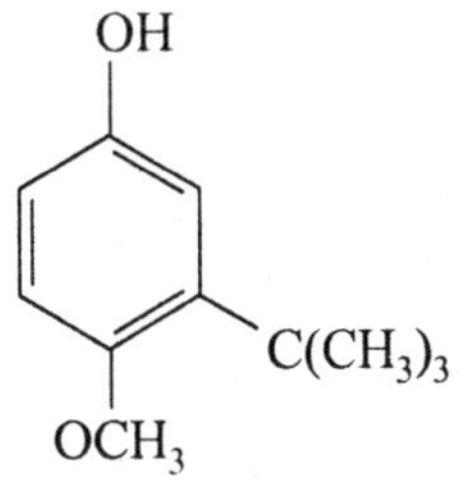


2-BHA

Though less active than 3-BHA, 2-BHA also has a synergist role towards 3-BHA. The main advantage of BHA is its high thermal stability, hence its use in frying oils, e.g. for potato chips. If it is a good antioxidant for animal fats, such as lard, its sole use is not, on the other hand suitable for vegetable oils, which already contain natural antioxidants. The activity of BHA can be augmented by effects of synergy, when used in mixture together with gallates or NDGA or even but more slightly with BHT. With BHA too one runs the same risk of oxidation catalysis, as obtained with gallates and NDGA.

*Butylhydroxytoluene (BHT)*

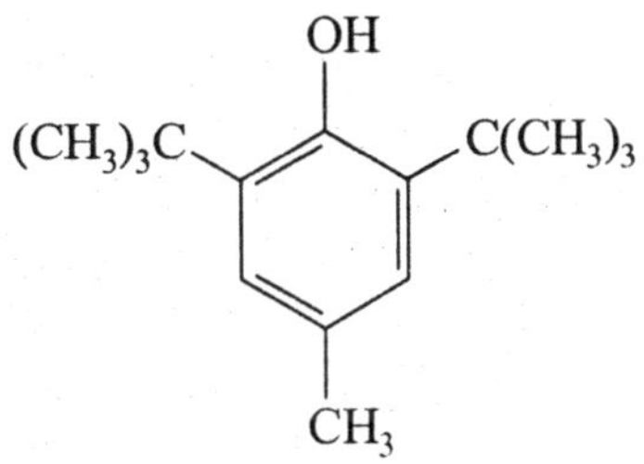


Its antioxidant properties are slightly higher than those of BHA with which it is often blended. These last two antioxidants are permitted in most of the countries in concentrations of 200 mg/kg. Their use is permitted for oils and fats but prohibited for milk fats.

*Tert-butyl hydroquinone (TBHQ)*

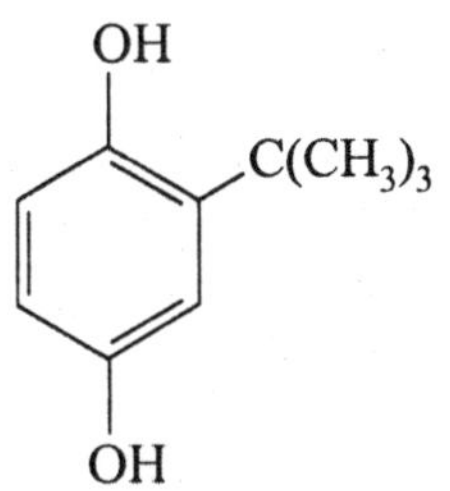


The use of this antioxidant is forbidden in all European countries. It is permitted in USA, Australia, New Zealand and South Africa. Its main disadvantage lies in the fact of its rapid oxidation in the presence of air, to give out tert-butylquinone.

*Trihydroxybutyrophenone (TBHP)*

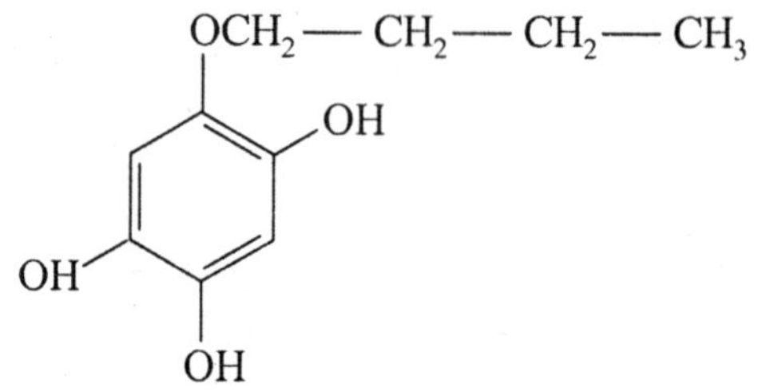

2,4,5-Trihydroxybutyrophenone

*Ethoxyquine*

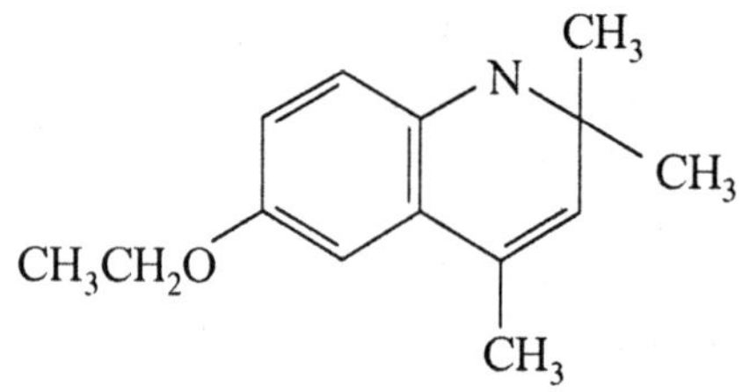

1,2-Dihydro, 6-ethoxy, 2,2,4-trimethylquinolein

Also known under the name of santhoquin, its activity is very strong and close to that of hydroquinone. Its use is permitted for animal feeds.

*Tocopherols*

Among natural tocopherols present in oils, it is δ-tocopherol which is the most effective antioxidant.

The antioxidant activity of tocopherols can be magnified by natural synergists existing in certain oils.

*Natural antioxidants*

Following demands from consumer associations, oils and fats industry is now going in for natural sources of antioxidants, such as clove extract, rosemary extract, sage extract or lecithin of soyabean, the latter acting rather on account of its sequestering capacity.

*Some synergists*

Ascorbic acid

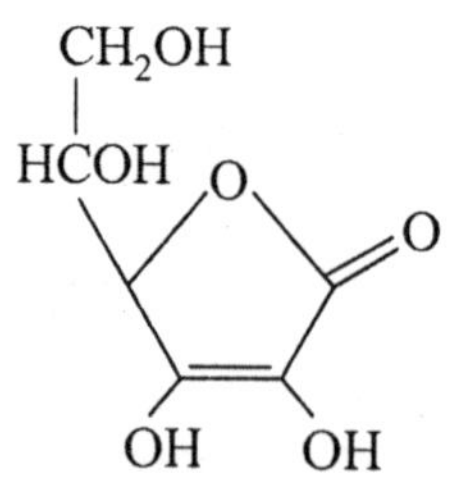

It acts by being transformed into dihydro-ascorbic acid, in other words by hydrogenising phenolic antioxidants oxidised by the peroxides and sequestering oxidation-enhancing metals such as iron.

Ascorbyl palmitate

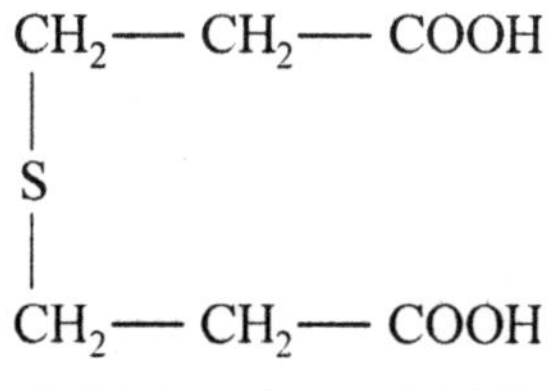

3,3′-Thiodipropionic acid (TDPA)

It exhibits synergic characteristics towards BHA.

Citric acid

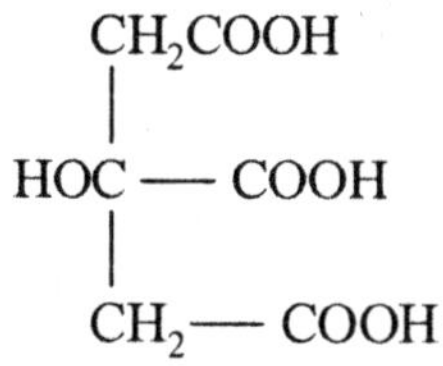

Phosphoric acid

$$PO_4H_3$$

### *Methods of antioxidant analysis in fats*

The principal techniques used are as follows:

1. Colourimetry.
2. UV spectrophotometry.
3. Fluorimetry.
4. Electrochemistry (polarography, voltametry, potentiometry).
5. Chromatography (on paper, on thin layer or on column).
6. Gas chromatography (GC).
7. High performance liquid chromatography (HPLC).

#### *Colourimetric methods*

These are generally old methods, employed mainly for qualitative analysis or for measurements of relatively high contents in the case of a single known antioxidant being present.

The advantages of these methods lie in their rapidity and relatively low working cost. However, they are of low sensitivity and of low specificity.

Further, it often becomes necessary to add on stages of solvent extraction or purification (by steam extraction or by thin layer chromatography for example). *Komaitis* proposed a method for measuring BHA through reaction in alkaline medium with NN-dimethyl-*p*-phenylenediamine in the presence of light oxidising agent such as potassium ferricyanide. A purple colouring is formed, which is extracted by carbon tetrachloride and whose absorbance is measured at 550 nm. In order to reduce the interferences from other phenolic products, *Komaitis* recommend a method of extraction by steaming. BHA also reacts with *p*-*N*-methyl-aminophenol. This reaction enables the measurement of gallates as well. The maximum absorptance is 550 nm for the gallates and 510 nm for the BHA. The antioxidants are extracted by methanol after dissolving of the oil in carbon tetrachloride. But this method has the disadvantage of necessitating a purification stage when the two antioxidants are simultaneously present.

*Komaitis* proposed also the measurement of TBHQ, BHA and gallic acid with 3-methyl-2 benzothiazalinone hydrazone hydrochloride (MBTH) in the presence of ceric ammonium sulphate. This method exhibits low detection limits; 1 ppm for TBHQ and 0.5 ppm for BHA. Two other methods of measuring TBHQ were compared by *Viplava*. They involve the reaction in alkaline medium with 4-aminophenazone in the presence of sodium periodate (maximum absorbance at 490 nm) and the reaction with sodium nitroprussiate in the presence of hydroxylamine hydrochloride (maximum absorbance 650 nm).

These two methods are sensitive and selective. But there is an interference by propyl-gallate in each of these methods and a method of separation by thin layer chromatography is proposed. *Guttierez* introduced improvements in these two methods of BHA measurement by employing reagents, which have been in use since the fifties (*Gibbs* reagent: 2,6-dichloro-*p*-benzoquinone-4-chlorimine; and Ehrlich reagent: sulphanilic acid and sodium nitrite). With this system we can have very rapid measurements, taking just a few seconds instead of 15 minutes. But it does not eliminate the main disadvantage of these two measurement methods, namely the low selectivity on account of the interferences from the gallates.

*UV spectrophotometric methods*

These methods are employed for measuring antioxidants exhibiting a phenolic nucleus. Due to the presence of interfering natural components, it is necessary to calculate the concentration in each antioxidant at many wavelengths or to proceed with a selective extraction. Hence, these measurement methods are at present employed exclusively after a separation by HPLC.

*Fluorimetric methods*

These methods have the same disadvantages as the earlier ones and, just as in the case of UV spectrophotometric methods, they are used only in combination with a system of HPLC.

*Electrochemical methods*

These methods (polarography, voltametry and potentiometry) make use of the faculty of the antioxidants which can be easily oxidised. Polarography is a rapid, very precise and very sensitive method for measuring of antioxidants. Some of its results have been considered to be better than those obtained by chromatographic techniques. It enables the direct measurement of antioxidants present in oil. Nevertheless, the risks of interferences might necessitate purification stages. As a result, the future of these electrochemical methods lies more in their employment as a system of detection combined with high performance liquid chromatography.

A comparative study was carried out with an amperometric detector on one hand and spectrophotometric and spectrofluometric detectors on the other for measuring BHA, BHT and TBHQ after separation by high performance liquid chromatography. It appears that the amperometric detection is as specific as the other two detection methods and also much more sensitive.

*Thin layer chromatographic (TLC) methods*

These methods are preceded by a solvent extraction of antioxidants. They have been the subject of by numerous publications, which define a mode of extraction, an adsorbant, a mobile phase and a development reagent. For quantitative analyses, it is necessary to proceed with an extraction of spots followed by a measurement with colourimetry or with UV spectrophotometry. It is also possible to have direct measurements through densitometry. For complex mixtures, it may be necessary to use two-dimensional chromatography. Nevertheless, TLC has mainly become a method for extracting antioxidants, with a view to measuring them precisely by means of chromatographic techniques such as GC or HPLC.

### *HPLC methods*

Measurement of antioxidants by HPLC has formed the subject of numerous publications during the past few years. It seems to be an analytical technique, which is the easiest to be implemented as well as the most effective. It has the advantages of specificity, sensitivity and precision. There are also possibilities of automation. In addition, it is also possible to measure a large number of antioxidants. In fact, by resorting to techniques employing an elution gradient it is possible to separate antioxidants with widely varying polarities ranging from propyl gallate, a polar compound, to BHT, a non-polar compound. The polar nature of the majority of antioxidants renders them particularly well suited for reverse phase chromatography and many methods using this technique have been developed, most of them in spectro-photometric detection.

However, these methods necessitate extraction stages, generally those using solvent, with or without a succeeding stage of pre-concentration.

These additional manipulations lead to problems of yield and might cause losses due to oxidation of antioxidants among other things. Such losses are considerable especially in the case of TBHQ or BHT.

Further, it will be necessary to eliminate the lipid residues, which have been extracted along with, before proceeding with the reverse phase injection. That is why, certain authors developed direct-phase chromatographic techniques enabling direct injection. But *Indyk* points out that the NDGA, TBHP and PG antioxidants are not eluted satisfactorily in normal phase. Detection by spectrophotometry is the most commonly applied technique, most of the time at a wavelength around 280 nm. A few authors exploit the reducing properties of phenolic antioxidants for proposing an electrochemical detection. In facts, this detection method enables measurements without the traditional interference problems of spectrophotometric detection. Spectrofluorometric detection too has been the subject of many papers with a view to improving the sensitivity and selectivity of measurement of antioxidants. But it is not possible to measure BHT in this manner as it is almost non-fluorescent.

### *Gas chromatographic methods*

These methods are applicable to most of the antioxidants.

Two techniques can be used: measurement after prior extraction stage or measurement by direct injection of fat. The second technique has the advantage of dispensing with cumbersome manipulations and increases the precision of the results. It may even be indispensable in case of TBHQ measurement so as to avoid losses due to its oxidation in the presence of air. Direct injection of fat can be carried out by using glass wool pre-column or by proceeding with a derivatisation of silylation type or trimethylsilylation or methylation or acetylation. We can analyse BHA, BHT and TBHQ antioxidants with the help of polar or non-polar stationary phases. On the other hand, for PG and NDGA, their highly polar character necessitates a pre-column derivatisation so as to reduce the retention times (Table 16.5).

Derivatisation offers the advantage of realising acceptable retention times for most of the antioxidants. There is also an improvement in the selectivity of the methods of measurement.

Further, it is also possible to increase the sensitivity by proceeding with a trifluoro-acetylation, which enables the use of an electron capturing detector, thus dividing the detection limit by a factor of 10. Tocopherols are measured by GC of their trimethylsilylether derivatives after extracting the unsaponifiable fraction and fractionation by thin layer chromatography (AFNOR NF T 60–239). A new technique was proposed by *Wyatt* making it possible not to use the preliminary extraction phase because, by exploiting the volatile character of certain antioxidants, he carried out TBHQ measurement in oils with the help of dynamic head space technique.

The working conditions of these methods are re-capitulated in Table 16.6.

**Table 16.5.** Measuring conditions of different antioxidants by HPLC.

| *Antioxidant* | *Sample* | *Stationary phase* | *Mobile phase* | *Detection* |
|---|---|---|---|---|
| BHT | Animal feed | Lichrosorb RP 18 (250 × 4.6 mm) | $H_2O$/MeOH (5/95) | Spectrophotometric (280 nm) |
| BHT | Composite products | Kieselgel C18 (250 × 4 mm) | MeOH/$H_2O$ (90/10) | Spectrophotometric (283 nm) |
| PG, TBHP, TBHQ NDGA, BHA, BHT | Dry products | Supelcosil C18 (150 × 4.6 mm) B: $H_2O$ + 5% acet. ac. | Gradient: A: $CH_3$ CN + 5% acet. ac. B: $H_2O$ + 5% acet. ac. | Spectrophotometric (280 nm) |
| Ascorbic acid | Beer | Ubondapak C18 (300 × 3.9 mm) | Citrate buffer pH 4.4 electrochemical + EDTA $0.5.10^{-3}$ M + amine $1.10^{-3}$ M | |
| BHA, BHT | Dried fish | TSK gel $SiO_2$-60 (250 × 4.6mm) | Butyl acetate/heptane (10/90) | Spectrophotometric (285 nm) |
| BHA, BHT, DG, PG, OG | – | Waters C18 (150 × 3.9 mm) | $H_2O$ + $H_3PO_4$ pH 5.5/Me OH (35/65) | Spectrophotometric (280 nm) and electrochemical |
| BHT, BHA | Plasma | Lichrosorb RP 18 | $CH_3$ CN/$H_2O$/ acetic acid (45/54/1) | Fluorescence 273/320 nm |
| BHT, BHA | Plasma | Lichrosorb RP 18 (150 × 4.6 mm) | MeOH/$H_2O$ + $KH_2PO_4$ (73/27) | Fluorescence |
| Ascrobyl palmitate | Oils and fats | Chromegabond diamine (250 × 4.6 mm) | MeOH/$H_2O$ + $KH_2PO_4$ (70/30) | Spectrophotometric (255 nm) |
| HA, BHT, TBHQ | Oils | Diasil CN and unisil QCP (150 × 5 mm) | $CH_3$ CN/$KH_2$ $PO_4$ 0.05M (30/70) $CH_3$ CN/$KH_2$ $PO_4$ 0.1 M (50/50) | Spectrophotometric (285 nm) Fluorescence (285/315 nm) amperometric |
| BHA, BHT, TBHQ | Oils and fats | Porasil (300 × 3.9 mm) or Rad-pak cyano or 2 Rad-pak $SiO_2$ | Hexane/$CH_2$ $Cl_2$/$CH_3$ CN (85.0/9.5/5.5) (88.1/9.8/2.1 | Spectrophotometric (280 nm) |
| BHA, PG, DG OG, TBHQ, NDGA | Oils and fats | Lichrosorb diol (250 × 4.6 mm) | Hexane/dioxane/$CH_3$ CN (62/28/10) | Spectrophotometric (280 nm) |
| BHA, BHT, TBHQ QG, DC, PG, NDGA | | Lichrosorb 100 CH C18 (250 × 4.6 mm) | Gradient A: $H_2O$ + 1% acetic acid B: $CH_3$ CN + 1% acetic acid | Spectrophotometric (280 nm) |
| BHA, BHT, PG, DG DG, DGA, TBHQ, | Fats | Lichrosorb RP 18 (250 × 4.6 mm) | Gradient A: $CH_3$ CN | Spectrophotometric (280 nm) |

*(Contd...)*

**Table 16.5**. Measuring conditions of different antioxidants by HPLC. *(Contd .)*

| *Antioxidant* | *Sample* | *Stationary phase* | *Mobile phase* | *Detection* |
|---|---|---|---|---|
| TBHP | | | B: $H_2O/pH_3$ | Fluorescence (290/325 nm) |
| NDGA, BHA, BHT Ascorbyle palmitate, Tocopherols | | Ultrasphere (250 × 4.6 mm) | $CH_3$ $CN/H_3$ $PO_4$ 0.05 M (80/20) or gradient A: Me OH B: $H_3PO_4 + H_2O$ pH 3 | Spectrophotometric (280 nm) electrochemical |

**Table 16.6.** Conditions of measuring diverse antioxidants by GC.

| *Antioxidants* | *Products* | *Filling materials* | *Column temp.* | *Detector* |
|---|---|---|---|---|
| BHT derivatives (cyclohexanedione) | | 2% OV-1 on Chromosorb W | 150°C | FID |
| BHA, BHT, TBHQ (Sylilated derivatives) | Vegetable oils | 10% GE-versilube F50 on gas chrom Q 100–120 | 150–210°C | FID |
| BHA, BHT Ionox 100 | Oils | Glass wool Pre-columns + SE 30 10% on chromosorb W-LA 80–100 mesh | 160°C | FID |
| BHA, BHT, TBHQ | Soyabean oil | Tenax GC + 10% 1-polymetaphenoxylene | 140–250°C | FID |
| BHT, BHA | Oils and fats | SP2250–3% on supelcoport 80–100 mesh | 160°C | FID |
| BHA, BHT, TBHQ (Silylated derivatives) | Milk products | 7% Se30 + 30% Xe60 on Chromosorb W-HP 100–120 mesh | 135°C | FID |

### *Extraction of antioxidants*

An extraction which is as complete as possible is the first condition to be met for effecting quantitative analyses and in case of antioxidants, there is not such thing as a universal method.

In fact, it is necessary to cover a wide range of concentrations and also to keep in mind the different properties of each antioxidant. PG is much more polar than BHT and it is not possible to find a single method which could give good extraction yields for both these products at the same time. In certain cases, it might be necessary to take the precaution of effecting vacuum extraction or of extracting in a nitrogen atmosphere: this is true for BHT, which is volatile or with TBHQ, which oxidises rapidly in contact with air to form tert-butyl-quinone.

The most frequently used extraction methods are solvent extraction and distillation extraction.

There are two types of approaches employed for solvent extraction:

1. Dissolving the fat phase in a non-polar solvent and successively extracting with a polar solvent, followed by a concentration of the polar phase. Solvents which are used most often are

acetonitrile or water/alcohol mixtures. The main disadvantage of this method is the heavy consumption of solvent.

2. Direct extraction of fat phase by a polar solvent. This method is not very precise.

It is also possible to dissolve the fat phase directly in an appropriate solvent. This method has the advantage of eliminating with the extraction stage, however, in case of, say a measurement by gas chromatography, it might be necessary to protect the column with a glass-wool pre-column.

Solvent extraction (or by directly dissolving the fat phase in a solvent) leads to an extract, which necessitates purification on account of co-extraction of certain compounds from the matrix.

This stage of purification can be effected by liquid-liquid partition or by passing through a column or even by TLC. Steam extraction results in satisfactory yields, except for gallates which are destroyed at high temperature.

### Standardised methods

AFNOR has standardised certain number of working methods for identifying and measuring the antioxidants. The NFT 60–35 norm enables the qualitative measurement of antioxidants by TLC after extraction with acetonitrile. The mobile phase is made up of a mixture of benzene/petroleum ether/acetic acid and the development is assured by 2,6-dichloroquinone-4-chlorimide. Gallates can be measured by ultraviolet spectrophotometry after extraction with ammonium acetate/acetonitrile mixtures. Specific absorbance are measured at 272 and 320 nm. This method is applicable only in the absence of BHA or BHT. GC for measurement of BHA and BHT has been standardised. Oils or fats are injected directly after dissolving in carbon sulphide on a column lined with 10 per cent DC 200 and provided with a pre-column for trapping triglycerides. Tocopherols too are quantified in the form of trimethylsilylethers by GC after extracting the unsaponifiable fraction and fractionation of the same by TLC.

An HPLC method has also been standardised by IUPAC. This method enables one to resolve a mixture of gallates, BHA, BHT, TBHQ, NDGA after extraction with acetonitrile from fat dissolved in hexane through reverse phase chromatography followed by a spectrophotometric detection at 270 nm.

Eluant is a mixture of water/acetonitrile/acetic acid whose proportions vary along a gradient.

Table 16.7 summarises the measurement of seven antioxidants by IUPAC method (HPLC).

**Table 16.7.** Measurements of seven antioxidants by the IUPAC method (HPLC).

| *Sample* | *Oils* | | *Lard* | |
|---|---|---|---|---|
| | *CV Repeatability %* | *CV Repeatability %* | *CV Repeatability %* | *CV Repeatability %* |
| BHA | 3.8 | 8.1 | | |
| BHT | 7.2 | 13.8 | | |
| Propyl gallate | 4.9 | 9.7 | 5.4 | 5.9 |
| Octyl gallate | 2.5 | 6.9 | 2.0 | 3.6 |
| Dodecyl gallate | 7.9 | 9.5 | 2.2 | 8.0 |
| TBHQ | 3.9 | 12.2 | 5.2 | 20.4 |
| NDGA | 0.4 | 9.3 | 4.7 | 14.7 |

With the possibility of analysing a mixture of at least seven antioxidants, as well as the good results obtained during these circuits, this method appears to be one of the most promising one.

## Metallic Traces

### *Importance of quantifying metals in fats and oils*

Quality control of crude, refined or developed vegetable or animal fats and suggests the control of oils, very low levels of metallic traces. In fact, some metals, though present in traces, exercise a great influence on the stability and toxicity of the products.

The origin of metals is sometimes difficult to determine. Some metals belong to the biosynthetic cycle: they are trace elements whose presence is necessary for the life and growth of animals and vegetation. For example, copper or manganese can be found in crude oil at a level of several ppm. Metallic traces are not always harmful, nor does their presence always imply pollution.

Technology is sometimes responsible for the increase of metallic contents, on account of usage or wear and tear of the apparatus (Fe, Cu, Cr) or the incomplete elimination of certain catalysts (Ni, Cu). For example, it is possible to find a residue of nickel of a few ppm in a shortening or a hydrogenated vegetable oil in an accident case of bad elimination of catalyst. The traditional content in refined fats is of the order of some ppb. It is useful to mention here, that the legislation imposes a general residual limit rate of 0.2 ppm for catalysts permitted in hydrogenation, with the exception of chromium, where it should be less than 0.05 ppm. Meanwhile, we observe that refining eliminates a lot of metals present in a natural state in crude oils. For example, the iron content can exceed several tens of ppm in crude soyabean oil, less than 1 ppm after neutralisation and only several tens of ppb after deodourisation. Still, it is not uncommon to find several hundreds of ppb of copper in crude oil, while the traditional content of copper in refined fat is several units or a few tens of ppb.

Finally, contamination can be responsible for the abnormal content of metals such as arsenic, lead, cadmium, chromium and tin. That, the level of toxicity of these elements has generally an effect on the nervous and hepatic systems, does not need to be demonstrated. These contaminants can originate from industrial waste, atmospheric pollutants, fertiliser and pesticide residues, and from packaging (i.e. soldering of metal boxes, stabilisers, and pigments of plastic materials).

Thus, it is possible to find several hundreds of ppb of lead in certain crude vegetable oils or several tenfolds of ppb of tin in conditioned fats, packed in metallic boxes or in PVC bottles. But, in almost totality of the cases, the content of heavy metals in refined fats and oils reckoned as the quantities consumed is lower than the permitted levels in foods. Furthermore these permissible levels are not clearly defined and the data available are inconsistent. We may point out that WHO has fixed the cadmium level for daily consumption by an adult at 55 μg, the permitted daily intake of arsenic mineral is 50 μg/kg or 3750 μg/adult (the maximum content permitted in fats is of the order of 1 ppb) and the maximum rate of lead intake is 350 μg/person/day.

#### *Quantify metallic traces and their control*

The control of metallic traces is necessary at two levels: one, at the nutrition and legislative level, because though the presence of heavy metals is hardly unlikely, it is better to be sure that their levels are well below permissible limits; two at the technological level. In fact, certain metals, like copper and iron in particular, are called pro-oxidisers because they are catalysts in the formation of hydroperoxides, and hence interfere in the acceleration of rancidity. Copper, has a catalystic power for concentration of the order of 5 ppb and the iron content around 30 ppb is necessary. Chromium, nickel, zinc and cobalt reduce to a lesser extent the resistance level of oil during oxidation. A common characteristics of different metallic traces liable to be present in fats, is their low content which is of the order of only a few ppb in some cases. It is hence necessary to avail of an analysing instrument which is sufficiently sensitive and specific for their measurement. Theoretically, different techniques can be used: X-fluorescence, spectrometry with a plasma torch, neutronic activation, colourimetry and atomic absorption. Some techniques are not sensitive enough (colourimetry, fluorescence) while others require costly materials (plasma torch, neutronic activation).

### Spectrometry with emission of plasma

Spectrometry with emission of plasma has been tested in the case of fats either as induction current plasma (ICP) or as direct current plasma (DCP). In both cases, the ionised gas is argon. The temperatures attained with a plasma source (6000–10,000°K), allows the dissociation of the most stable chemical bonds. The measures can be carried out in a solvent medium (e.g. dilution to 50 per cent (m/v) in petroleum ether), so as to homogenise the nebulosity of the sample in plasma. The method has the advantage of being quick and of leading to large linearity areas: we can hence measure different elements and their relative concentrations, in a sample solution, simultaneously or sequentially. This method is equally sensitive, but detection limits are higher in some cases such as those obtained with an atomic absorption furnace: 50 ppb for iron, 100 ppb for lead, 30 ppb for cadmium.

Users of these method mention the difficulty in regulation and the multiple risks of destabilisation of plasma which cause frequent re-calibrations. Moreover, the noise issuing from the presence of complex organic matrix like oil, can be disturbing. Spectrometry with emission of plasma, although performing efficiently, is not perfectly suitable for fats.

### Spectrophotometry by atomic absorption with flame

Spectrophotometry by atomic absorption with flame was historically the first application of atomic absorption to the direct measurement of elements in fats. It was based on old researches carried out for determining of alkalines sodium, potassium, calcium and magnesium. Its principle is simple: oil is diluted in an organic solvent, sucked into a nebulisation chamber and then atomised in a flame generated by a mixture of air and acetylene.

Oil is diluted to 10 or 20 per cent (m/v) in methylisobutylketone or in a mixture of isoamyl acetate/ methanol 1:1 (v/v) for example. In these conditions, the effect of the matrix on viscosity is negligible, and the calibration curves can be obtained in a solvent medium. Although detection limits can reach several tens or hundreds of ppb, the elements measured are generally present at the level of ppm.

The measurement by flame is seldom used now-a-days in the case of fats, with the exception of the control of soap elimination after neutralisation. This method is not sensitive enough to measure the pro-oxidising or heavy metals.

### Spectrophotometry by atomic absorption with electrothermal atomisation

Spectrophotometry by atomic absorption with electrothermal atomisation (without flame) is chosen as the best method for the measurement of metallic traces in fats and oils because it is sufficiently sensitive and specific and investment is affordable by testing laboratories. Moreover, progress in the sphere of atomic absorption, has lead to considerable improvement in methods for direct measurement and for low detection limits in the case of fats.

We will not discuss the theory of electrothermal atomisation. It is possible to refer to the recent works of *Hoenig* and *Kersabiec* which set out the current state of knowledge regarding instrumentation and methodology. Let us recall here the general principle of the method: the sample is introduced into a graphite tube called the atomiser. After thermal dissociation of the product obtained by heating the inner walls of the tube or furnace (Joule effect), the analyte is atomised at a very high temperature

(2700°C for iron). The atomic vapour formed absorbs radiation emitted by the luminous sources which crosses the atomiser in the optical axis of the spectrophotometer. The quantity of luminous energy absorbed is directly proportional to the atomic concentration. Choice of a specific luminous source and characteristic wavelength of an element permits selective measurement of one metal amongst all atoms formed.

Contrary to the traditional process of atomisation by flame, where the sample is brought in the form of an aerosol which leads to a heavy dispersion of atoms, the use of an electrothermal atomiser allows atomic vapours to concentrate, and hence improves the yield of atomisation; consequently sensitivity limits are considerably improved.

Depending on the elements, the lamps are hollow cathodes (iron, copper or nickel) or, electrodeless discharge lamps (tin, lead). Atomisers are furnaces made from of very pure graphite, cylindrical in shape, and which can be recovered with pyrolytic carbon. This improves resistance and chemical inertia of graphite. Furnaces can be equally equipped with an interior platform. Regulation of temperature in furnaces is ensured by photodiodes in the contact-electrodes: it is thus possible to obtain more than 2800°C in approximately one second. Currently, transversal heating tubes with integrated platforms are being perfected which perform better in terms of homogeneity of temperature, efficiency of atomisation and increasing the life of the furnace. The furnaces are swept by an inactive gas (argon or nitrogen). To improve the yield of atomisation, this step can be carried out after interruption of the sweeping of the inert gas (stop-flow system). Non-specific absorptions, due to matrix or parasitic elements, are corrected either by a source of continuous radiation (deuterium lamp), or by Zeeman effect, generally with a transversally indirect pulsating magnetic field.

Before measuring by atomic absorption, it is necessary to foresee the primordial and often delicate step of sample preparation. For most of food matrix, mineralisation of samples was possible preliminary to measurement by a wet method (digestion in a suphonitric or a sulphochromic medium) or by a dry method (calcination on flame or in a furnace or in microwave digester). This step is often long and difficult because it is a source of pollution and losses of elements to be determined. Measuring determination is then possible on the mineralised sample in an aqueous medium, generally acid medium.

In the case of a fat or a lipid extract, although this method can be used, it is far more interesting to obtain direct determination after dissolving the sample in an organic solvent. The mixture of oil and solvent is directly injected into the atomic absorption furnace. The solvent is evaporated there and the sample is mineralised *in situ* by increasing the temperature progressively.

This type of measurement has the advantage of minimising the risks of contamination and of reducing considerably the duration of analysis. However, it requires perfect operating conditions, especially in the programming of the furnace temperature, so as to ensure the complete elimination of the organic matrix without loss of the element to be determined before the atomisation step. Moreover, it is not possible to carry out any concentration of the element prior to its measurement.

The principle parameters which allow the optimisation of methods of measurement metals in an oil medium by atomic absorption, are as follows:

1. Treatment of the graphite furnace by a refractory carbide (niobium or tantalum) to reduce volatility of certain metals and to accelerate mineralisation of the organic matrix.
2. Use of air to improve destruction of the matrix by oxidation, and to allow formation of less volatile metal oxides.
3. Use of secondary surface of atomisation (L'Vov platform) to obtain more isothermic conditions in the furnace (this surface is heated by radiation and not by conduction) and to improve the yield of atomisation.

4. Use of a matrix or analyte modifier (palladium or ammonium dithiocarbamate) to stabilise the element to be determined under a less volatile chemical form.

After the description of operating instructions common to the measurement of all metals in fats, we will give the conditions particular to the measurement of principal pro-oxidising and toxic elements.

## General methodology applicable to fats and oils

### Preparation of glassware

Great care in the cleaning of glassware used during the determination of metallic traces is indispensable. After the usual washing and rinsing with distilled water, glassware is treated vigorously either with cold acid using sulpho-chromic mixture, or hot acid using a solution of 10 per cent diluted nitric acid. After this treatment, glassware is rinsed with distilled water and then dried.

### Preparation of samples

Oils, fats, or lipid extracts should be melted if necessary and homogenised before analysis. Certain authors recommend a hot agitation (50–60°C) others, a cold one. The basic difficulty is to obtain representative and homogeneous samples from crude oils e.g. if the sample contains fines it is better to agitate it till a homogeneous suspension is obtained instead to eliminate them by filtration.

#### Choice of dilution solvent

It is necessary to use a good organic solvent of fats. Depending on the authors and the matrix, the following solvents are recommended: chloroform, cyclohexane, heptane, methylisobutylketone or a mixture of methylisobutylketone methanol 80:20 (v/v). These solvents should be analytical quality grade so that they do not become a source of pollution. The fact that increasingly, auto-samplers are being used, it is preferable to choice less volatile solvents to minimise the risk of evaporation and concentration of samples placed in advance in the injection cups. In this aspect, methylisobutylketone, (boiling point = 116.5°C) has an advantage over other solvents. *Prevot* advises moreover the addition of methyl alcohol to improve dispersion and homogenisation of crude oils which can contain water, fines and phospholipids.

Besides, determination in a solvent medium requires a teflon injection capillary and a storage and pumping system, compatible with the organic solvent.

#### Choice of dilution ratio

For a manual injection with volumetric syringes or micro-pipettes, it is possible, but difficult in practice, to use oil or pure fat. With auto-samplers, which allow rapid determination and a standardisation of injected volumes, it is necessary to dilute oil in a solvent for reasons of viscosity and wettability. Depending on the element to be determined, its concentration and the matrix, the ratio of dilution varies between 10 and 50 per cent w/w or w/v, oil/solvent.

In the case of hydrogenated or solid fats, it is preferable to achieve the highest dilution, compatible with great sensitivity and to inject hot samples. Without an automatic heating sampler, it is impossible to obtain serial measurements. The volume injected is generally fixed at 20 microlitres.

### Calibration

Analytical points for calibration are traditionally prepared by successively diluting a first solution. Different standard products can be used: mineral standards such as iron chloride, (available as normadose or Titrisol) are soluble in water and therefore, recommended less for mixture of oil and solvent. Organo-metallic powder standards such as copper cyclohexabutyrate or iron tris(1-phenyl-1,3-butanedione)

(Eastman products) are soluble in a solvent medium by adding hexanoic acid. Lastly, organo-metallic standards in mineral oil (Conostan products) with a concentration of 5000 ppm, have the advantage of being easy to use and preserve.

Dilute solutions can be made with a refined vegetable oil, eventually freed of metals by passing it on an alumina column.

Working standard solutions (having concentration between 5 and 500 ppb, depending on elements) are prepared daily and should rigorously contain the same weight or volume of oil and solvent as the samples so that viscosity and quantities injected are identical.

*Furnace preparation*

*Hocquellet* has developed a technique for treating tube surfaces by a tantalum or niobium refractory carbide. The method consists of preparing a concentrate solution of niobium or tantalum nitrate or oxide, spreading it on the inner and outer walls of the tube and making tantalum or niobium carbide *in situ* by heat treatment. This technique is adopted by IUPAC standard for measurement of iron.

Presence of refractory carbide confers to the furnace a reducing power which favours carbo- reduction of metal oxides and therefore, reduces their volatility. It also accelerates mineralisation of the organic matrix and increases the life-time of the tube. Use of niobium or tantalum is almost systematic for analysis of heavy metals in fats.

*Thermal programming*

Thermal programming is hardly the most important factor of a method. It controls the accuracy of the measurement. In fact, the matrix is to be totally destroyed so as to avoid interferences during atomisation due to mineralisation smokes (case of non-specific absorption) and which would not be compensated by the correction system, thus entailing errors in measuring specific absorption. We must also not entail losses of elements to be measured by volatilisation before atomisation.

While preparing a temperature programme, it is useful to monitor evolution of specific absorbance and non-specific absorbance depending on the mineralisation temperature. Graphs obtained then help us to define the 'optimal point' corresponding to the maximum absorbance (sensitivity) and the minimum background (specificity).

Generally, increase in temperatures should be progressive and it is advisable to separate them by successive stages. The rate of increase is important as the duration for which a given temperature is maintained.

A thermal programme includes the following four steps:

1. Drying step: between 70 and 150°C eliminates solvent.
2. Charring step (mineralisation): if the element to be measured is not volatile (iron or copper), we can select temperature ranges between 800 and 1200°C. For volatile elements (lead, cadmium) we can use lower temperatures (450 and 650°C) but in the presence of air. Mineralisation times varies between 30 and 100 seconds.
3. Atomisation step: depending on the elements, temperature to be attained very quickly varies between 1700 and 2700°C with very brief period (3–5 seconds).
4. Cleaning step: the tube is depolluted by maintaining for a few seconds a temperature of 2,700–2800°C.

*Measurements*

Results are expressed in absorbance corrected or not by the background (minimum signal due to the optical system and the presence of atomiser) and compensated by the non-specific absorbance. Depending on the case, measurements can be calculated with peak area or peak height (the latter being specially useful for measurement near the detection level).

In case of an autocalibrated equipment, the result is directly expressed as an analyte content by using calibration curves. It should however, be mentioned here that considering the matrix and the fact that diluting oil is never fully exempt from metals, calibration corresponds to a 'doping' method. Consequently, the 'zero' value should be subtracted from the value of each sample.

*Validation of a method*

Some criteria, general or specific to atomic absorption help in validating a method:

1. Linearity area: it corresponds to the analyte concentrations for which the Beer-Lambert rule is verified.
2. Characteristic mass: this is a value which evaluates the sensitivity of the method. It is defined as the mass of an element expressed in pg (= $10^{-12}$ g) likely to produce a signal, of 0.0044 absorbance unit (absorption of 1 per cent).
3. Detection limit: it corresponds to the concentration or to the mass of the analyte which produces a signal equal to three times the value of the background. The value of the background is in practice based on the average of ten successive readings. Detection limit is generally expressed in ppb of element in the sample.
4. Determination or quantification limit: it corresponds to the lowest concentration or mass of the analyte which can be determined with a reasonable statistical certainty. Depending on the authors, it corresponds to 2 to 5 times the detection limit.
5. Exactitude: the correctness of the method can be checked by applying the procedure on a certified reference. Unfortunately, there is, till date, no oil with certified metal contents.
6. Precision: it is represented by the variation coefficient or ratio of the standard deviation and the arithmetic mean of n values.
7. Repeatability and reproducibility: these values are usually calculated by intra and inter-laboratories tests.

### Specific methods for fats and oils

Table 16.8 gives the practical conditions for direct measurement of principle metals in fats. These conditions are standardised or only published. Certain parameters remain practically unchanged irrespective of the method: injected volume being 20 µl, vector gas argon with a flow of 300 ml/minute drying and tube cleaning steps, system of 'stop flow' during atomisation.

**Table 16.8.** Operating conditions for determination by atomic absorption. 1. Pro-oxidising metals.

| *Metal* | *Fe* | *Cu* | *Ni* | *Co* | *Zn* |
|---|---|---|---|---|---|
| Wavelength | 302.1 nm | 324.7 nm | 232.0 nm | 240.7 nm | 213.9 nm |
| Slit-width | – | – | – | 0.2 nm | 0.5 nm |
| Furnace | Graphite + niobium | Graphite | Graphite | Graphite | Graphite |
| Standard | Organo-metallic (Conostan) | Organo-metallic (Conostan) | Organo-metallic (Conostan) | Cobalt cyclohexabutyrate | Zinc cyclohexabutyrate |
| Solvent | Heptane | Heptane | Heptane | MIBC | MIBC |

*(Contd...)*

**Table 16.8.** Operating conditions for determination by atomic absorption. 1. Pro-oxidising metals. *(Contd .)*

| *Metal* | *Fe* | *Cu* | *Ni* | *Co* | *Zn* |
|---|---|---|---|---|---|
| Rate of dilution | 50% | 50% | 50% | 50% | 50% |
| Calibration range | 0–1000 ppb | 0–200 ppb | 0–1000 ppb | 0–200 ppb | 0–100 ppb |
| *Thermal programme* | | | | | |
| Drying | – | – | – | 200°C–25 s | 150°C–25 s |
| Mineralisation | 1200°–80 s | 900°C–80 s | 1200°–80 s | 800°C–25 s | 650°C–60 s |
| Oxygen-rich atmosphere | – | – | – | – | + |
| Atomisation | 2700°C–6 s | 2700°C–6 s | 2700°C–6 s | 2680°C–4 s | 2200°C–3 s |
| Detection limit | Estimated: 25 ppb | Estimated: 5 ppb | Estimated: 10 ppb | 10–15 ppb | 5 ppb |
| Reference method | IUPAC–2.631 (18) | IUPAC–2.631 (18) | IUPAC–2.631 (18) | Wolff (34) | Wolff (34) |

Operating conditions for determination by atomic absorption. 2. Toxic elements

| *Metal* | *Pb* | *Cd* | *As* | *Sn* | *Cr* |
|---|---|---|---|---|---|
| Wavelength | 283.3 nm | 228.8 nm | 193.7 nm | 224.6 nm | 357.9 nm |
| Slit-width | – | 0.7 nm | 2.0 nm | 0.7 nm | – |
| EDL lamp | + | + | + | + | – |
| Furnace | Graphite + platform | Graphite + platform + tantalum | Graphite + platform + niobium | Graphite + tantalum | Graphite |
| Standard | Organo-metallic (Conostan) | $CdCO_3$ | $AS_2O_3$ | Dibutyl-tin-bis (ethyl-2-hexanoate) | Chromium tris (hydroxy-2-acetophenone |
| Matrix modifier | Lecithins (2%) | DOTC (2%) | DOTC (2%) Pd | – | – |
| Solvent | Cyclohexane | Chloroform | Chloroform | MIBC | MIBC |
| Rate of dilution | 50% | 10% | 25% | 33% | 10% |
| Calibration range | 0–100 ppb | 0–10 ppb | 0–100 ppb | 0–250 ppb | – |
| Thermal program | | | | | |
| Drying | 200°C– 30 s | 70°C/120°C–62 s | 70°C–51 s | 120°C–50 s | 150°C –30 s |
| Mineralisation | 650°C–100 s | 450°C–51 s 600°C–3 s | 800°C–6 s 900°C–25 s | 500°C–35 s 950°C–20 s | 400°C–40 s 1000°C –62 s |
| Oxygen-rich atmosphere | – | + | – | – | + |
| Atomisation | 1700°C–5 s | 2400°C–8 s | 2500°C–5 s | 2600°C–5 s | 2700°C–5 s |
| Detection limit | Estimated: 5 ppb | 0.5 ppb | 3 ppb | <1 ppb | 12 ppb |
| Reference method | IUPAC–2.632 (19) | P. Hocquellet (12) | P. Hocquellet (12) | P. Hocquellet (12, 13) | Di Giovacchino (9) |

Index: MIBK = Methylisobutylketone; EDL = Electrodeless discharge lamp; DDTC = N, N Diethyldithiocarbamate.

### *Pro-oxidising metals*

Iron, copper and nickel are covered by the IUPAC standard method which was proposed by *Hendrikse*. These non-volatile elements do not cause any special difficulty regarding temperature programme and choice of furnace. For measurement of copper, *Kundu* and *Prevot*, however, advise the use of an oxygen-rich atmosphere during charring. Measurements of copper and nickel generally do not show any problem of repeatability.

Iron is probably the most delicate element to be measured in fats and oils. One reason is the presence of different chemical forms in oil. The IUPAC method recommends the treatment of the tube with niobium. In spite of this, sensitivity and repeatability of the method remain very low: for a content of 150 ppb, the average absorption expressed as peak area is 0.170 with a variation coefficient of 21 per cent. Recent works have proposed a more sensitive method for concentrations below 150 ppb (refined oils) by using a more sensitive wavelength (248.3 nm) and a L'Vov platform.

In case of cobalt, *Kundu* indicates the influence of the nature of oil on sensitivity of the method. This result can, in fact, be seen in case of different elements. Calibration curves do not have the same linearity nor the same sensitivity with soyabean or peanut oil, for example. This matrix effect can be explained by the presence of components or metals doing interferences.

In case of zinc, which is volatile, oil has to be mineralised at a low temperature in the presence of air.

### *Toxic elements*

Like iron and copper, lead measurement has been described in a standardised method. Because of relative volatility of this element, losses before atomisation can occur and it is necessary to use a matrix modifier. The standard method provides the addition of lecithins in the solvent. *Hocquellet* suggest use of a chelatant, which is N, N diethyldithiocarbamate diethylammonium (DDTC). To improve sensitivity of measurement, he suggests an additional treatment of the oven using tantalium carbide and a charring step in an oxygen-rich atmosphere.

In case of cadmium and arsenic, use of a matrix modifier, a platform and a refractory carbide enable very low detection levels to be achieved. Partially volatilised cadmium from 400°C is stabilised as sulphur in the presence of DDTC and mineralisation is carried out under oxygen. Arsenic is generally measured by hydrure ($H_3As$) method because of its high volatility. However, direct measurement is possible in presence of palladium which gives a more stable palladium arsenide.

Measurements of tin and chromium give no specific difficulty. The case of mercury is not described in Table because its determination, by generating vapour method, is very special.

Thus measurement of metallic traces in fats and oils benefits the performances, sensitivity and specificity of the atomic absorption spectrometry with electrothermal atomisation. It is almost always possible to attain by this technique the level of ppb and to measure most of the traces of metallic elements present in fats.

As an indication, here are some examples of contents in commonly used vegetable oils: refined sunflower oil contains about 20 to 30 ppb of iron, 10 ppb copper, 10 ppb nickel, 3 ppb chromium , 5 ppb manganese, 3 ppb mercury, less than 3 ppb arsenic, less than 0.5 ppb cadmium and less than 10 ppb lead. Virgin olive oil can contain 70 to 3600 ppb of iron, a few tens of ppb of copper, 2 to 50 ppb of nickel, 2 to 500 ppb chromium, 2 ppb mercury, less than 10 ppb cadmium, 40 ppb lead and 3 to 15 ppb tin (in metallic packaging). For more details, refer to Table 16.9.

The type of matrix enables the direct determination of the element content in a solvent medium. If for certain elements (iron, copper, nickel, lead) there are recognised and validated methods, for others (arsenic, cadmium, chromium, tin), optimisation and/or application of the proposed methods can still be performed. The special case of toxic heavy metals has today become all important, especially, with the development of virgin oils for dietary use, or any fatty product other than refined oil.

**Table 16.9.** Some contents of metallic traces (ppb) in differents fats oils and fatty products.

| *Reference samples* | *As* | *Cd* | *Cr* | *Co* | *Cu* | *Fe* | *Hg* | *Mn* | *Ni* | *Pb* | *Sn* | *Zn* |
|---|---|---|---|---|---|---|---|---|---|---|---|---|
| Vegetable fats | | | | | | | | | | | | |
| Refined peanut oil | | <2 | 3 | 10 | | | 4 | 10 | | 5–10 | 22* | |
| Crude repeseed oil | | | | | 10 to 1100 | 1650 to $10^6$ | | 130 to 960 | | | | |
| Neutralised rapeseed oil | | | | | | 20 to 1550 | | | | | | |
| Bleached rapeseed oil | | | | | | <5 to 1400 | | | | | | |
| Refined rapeseed oil | <3 | | 3 | | 3 to 8 | <5 to110 | | <2 to 10 | | 5 | 10 | |
| Crude cottonseed oil | | | 200 | | | | | | | 1200 | | |
| Refined corn oil | | | | | | | | | 7 | | | |
| Virgin olive oil | | <10 | 2 to 500 | | | 70 to 3600 | 2.5 | | 2 to 48 | 40 | 3 to 15** | |
| Crude palm oil | | | | | | 6000 | | 1105 | | | | |
| Refined palm oil | | | | | | 1100 | | | | | | |
| Crude soyabean oil | | | 3 | ⩽ 15 | 5 to 35 | 550 to 9500 | | 230 to 670 | | | | |
| Refined soyabean oil | <3 | <0.5 | 3 | ⩽ 15 | <5 to 30 | <10 to 560 | 2 | 2 to 190 | | 21 | 1 | 20 to 460 |
| Hydrogenated soyabean oil | | | | | | | | | 1200 | | | |
| Crude sunflower oil | | | | | 40 | 2000 | | | 20 | | | |
| Neutralised sunflower oil | | | | | | 25 to 40 | | | | | | |
| Bleached sunflower oil | | | | | 10 | | | | | | | |
| Refined sunflower oil | <3 | <0.5 | 3 | | 10 | 20 to 30 | 3 | 5 | 10 | 5–10 | | |
| Animal fats | | | | | | | | | | | | |
| Cod liver oil | 2300 | | | | | | | | | 10 | 19 | |
| Goose fat | | | | | | | | | | | 15** | |
| Crude menhaden oil | 10400 | 40 | | | | | | | | 320 | | |
| Refined menhaden oil | <700 | | | | | | | | | | | |
| Lard | | <10 | | | | | | | | | | |
| Others | | | | | | | | | | | | |
| Butter | | 15 | | | 25 | 165 | | | | 101 | | |
| Margarine | 3 to 22 | 20 | | | | | 4–10 | | | | | |
| Fatty phase margarine | | 3–4 | | | | | | | 40 to 135 | <5 to 35 | | |
| Aqueous phase margarine | | 1 to 8 | | | | | | | | 5–7 | | |
| Shortening | | | 6 to 40 | | | | | | 70 to 2700 | | | |

*** PVC bottle; and ** Metal box**

To conclude, it is useful to mention to the analyst who wishes to elaborate a new atomic absorption method or to adopt an already existing method to a new equipment, that, considering the high number of parameters (oven, temperature, modifier), a method of organising tests such as the experimental multifactorial theory, can be selected.

## Analysis of Polycyclic Aromatic Hydrocarbons (PAH)

Polycyclic aromatic hydrocarbons (PAH) are produced during combustion of organic matter and thus are universal contaminants of our environment and, in particular, pertaining to man's food intake. Parallely, certain technological processes applied to food can, under certain conditions, add to this pollution: thus smoking of food and application of severe thermal treatments. Toxicity of PAH, or at least some among them is now very well known: by themselves PAH are not toxic, on the contrary, some of their metabolites have a pronounced mutagenic and cancerogenic activity. The PAH with strong cancerogenic activity are dibenzo (a, h), anthracene, benzo (a) pyrene, and dibenzo (a, h) pyrene. The activity is lower for the benzo (b) and benzo (j) fluoranthene metabolites.

Except in exceptional cases, PAH contents in fats are in general lower than 1 µg/kg (1 ppb) for every PAH. However, it seems that, depending on their origin and method of drying used, the coconut fats are more or less polluted. Refining and treatment with activated charcoal reduces them to a great extent.

Analysis of PAH in food products is mainly an analysis of traces. The inherent difficulties of this type of titration are of two types:

1. Necessity to extract, isolate and purify the PAH which may be present in a very complex mixture. The difficulty is further increased by the fact that oils present an apolar character very much like those of PAH.
2. If the separation of PAH between them is relatively easy, on the other hand, one should be able to detect them at very low level of concentration (ppb) and identify them.

### *Extraction and concentration of PAH*

If one considers an average content of PAH in a total of 10 µg/kg and one treats 10 gram sample, one should extract quantitatively 0.1 µg PAH. Methods actually proposed for fats are many and varied and well reflect the difficulties encountered during isolation of PAH.

We can distinguish three types of methods:

1. Formation of complexes between coffein and PAH.
2. Direct extraction of PAH from oil with, mixtures dimethylformamide/water.
3. Saponification of fat and extraction of PAH with isooctane or cyclohexane.

These methods are in fact more often a succession of liquid-liquid extraction than single extraction.

#### *Drawbacks of operating modes*

This operating mode has various drawbacks:

1. Need to use large volumes of sample and organic solvents, which are sometimes volatile and toxic. Moreover, determinations are very costly, considering the price of very pure solvents.
2. Alteration and/or loss of compounds, for at the end of extraction the organic phases have to be evaporated.
3. It is hardly possible that extraction is quantitative. Generally, the recovery rate varies from 30 to 200 per cent.
4. Pollution due to solvent (s).

5. Methods are time-consuming and laborious and can be carried out only by a competent analytical assistant capable of performing each technique.

After extraction of PAH, chromatographic purification is carried out with silica, alumina, florsil, calflo E, resin XAD 2, or gel biobeads.

A rapid technique which does not require handling was perfected at ITERG. PAH is isolated by donor acceptor complex chromatography: 1 gram (4 mm column diameter) or 4 gram (8 mm column diameter) oil are injected into an HPLC column packed with a 10 μm tetrachloropthalimidopropyl modified silica. Neutral lipids and minor compounds of oils are eluted with a mixture of hexane/methyltertisobutylether 75:25. PAH electron donors are re-taken and eluted with pure dichloromethane. PAH fraction recovered is then analysed by reverse phase chromatography. The recovery of different PAH is nearly 100 per cent, even at the level of 1 ppb. The method's detection limit is 0.1 ppb.

### *Analysis and quantification of PAH*

Separation of PAH is performed either by GC on a polar phase (SE54, DB5) or, most frequently, by HPLC on octadecyl-bonded silica with fluorimetric detection and time-programming wavelengths.

Figure 16.9 shows a chromatogram of a mixture of 16 PAH separated by HPLC on column Vydac 201 TP 54 (grafted silica C18) with a mobile phase composed of acetonitrile and water and whose proportions vary from 55:45 (v/v) to 100:0 in 40 minutes.

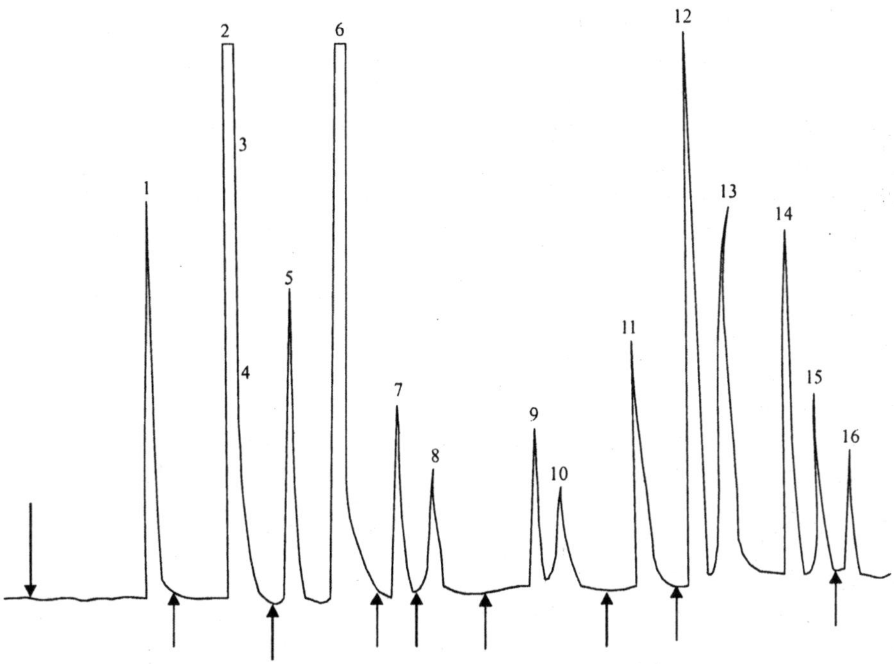


Fluorimetric detection with programming of pairs γ exc./γ em.
Injection of 0.2 ng of each PAH

1. Naphthalene
3. Acenaphtene
4. Fluorene
5. Phenanthrene
6. Anthracene
7. Fluoranthene
8. Pyrene
9. Benzo (a) anthracene
10. Chrysene
11. Benzo (b) fluoranthene
12. Benzo (k) fluoranthene
13. Benzo (a) pyrene
14. Dibenzo (a, h) anthracene
15. Benzo (g, h, i) perylene
16. Indeno (1,2,3-c,d) pyrene

**Fig. 16.9.** HPLC chromatogram of a mixture of 16 PAH.

Quantitative analysis is performed either with the help of an internal standard or by measured addition of PAH in oil. There is no standardised method for testing of entire PAH; however, a method has been adopted at IUPAC for testing of benzo (a) pyrene which is a good tracer for PAH at four or more aromatic rings; a 400 mg fat sample is loaded on a column containing 22 gram alumina with 10 per cent water. Benzo (a) pyrene is eluted by 85 ml hexane, the evaporation residue dissolved in 25 μl tetrahydrofurane and then analysed by HPLC on C18 phase. Benzo (a) pyrene is detected by fluorescence (excitation 381 nm/emission 403 nm). The main advantages of this method are its simplicity and speed.

### Phytosanitary Residues (Pesticides) in Oils and Fats

The word 'pesticide' refers to a general definition of all chemical products used to fight:

1. Vectors of human and animal diseases.
2. Enemies of plants and crops, standing or harvested.

'Pesticide' therefore defines all phytosanitary products such as insecticides, fungicides (meant to fight mildew and fungus), weedkillers. All these products correspond to two or three hundred pure or mixed active matters, in two to three thousand commercial specialities worldwide.

Past the introduction of intensive techniques for animal husbandry and agriculture, the use of these phytosanitary products once considered particularly advantageous for improving the production, is now subject to:

1. A limit in use, in terms of the protection of the environment (in case of organochlorinated insecticides and some weedkillers: Atrazine.
2. The setting of a maximum admissible concentration in food, to reduce their eventual level of contamination, in terms of public health.

Food contamination is linked notably to chemical perenniality of certain insecticides (such as organochlorinated insecticides due to their insolubility in water, their solubility in lipids and their resistance to hydrolysis). In addition, contamination can have various origins:

1. In vegetable production, from a bad, not adapted, overdosed or late use.
2. In animal production, from the ingestion of polluted food or from an excessive treatment of the breeding premises.

The importance of these risks of contamination thus leads to an intensive research for the possible presence of residues of phytosanitary products in food.

Research for such products in oils and fats is often limited because of:

1. The specificity of use of phytosanitary products (disinfecting of soil, weeding of fields, protection against insects or fungus) or to the timing of their use.
2. The specificity of their use leading to observe, for instance, that the growing of rapeseed, independently from general problems of seed storage is the one considering the number of animal parasites susceptible to harm it (small altica, large altica, meligethe, weevils of siliqua, stems, terminal buds) for which the greatest number of insecticides are likely to be used (Tables 16.10 and 16.11).
3. The toxicity (or the variable perenniality) of phytosanitary products which is particularly high for insecticides and more precisely for organophosphorated systemic insecticides (Table 16.12). As a result, within prevailing laws and regulations, the search is limited to:
4. The most persistent as for example organochlorinated insecticides, even though their scope of use is limited.

5. To insecticide residues whose use is authorised generally and in particular.
    (a) The main organophosphorated insecticides in non-refined oils and fats.
    (b) Eventually, insecticide residues made with synthetic pyrethrinoid and carbamate residues constituting either insecticides or fungicides.

Generally, the search for residues of the latter three types of insecticides are limited considering the thermal and chemical instability of organophosphorated insecticides, the usually low toxicity of pyrethrinoids for man, an above all, in this last case, the extremely low amount of active matter used in agricultural treatments as these are exceptionally efficient products.

### *Searching and determining residues of the main insecticides*

The search and determination of residues of any insecticide is based upon four operations: extraction, concentration, purification and determination (Table 16.12).

First of all, good quality solvents and specialised laboratory glassware are required. Their cleanliness should be checked as impurities could interfere with or degrade the residues of insecticides eventually present.

#### *Extracting insecticides residues*

In this case, extraction applies to oil seeds and oil meals. Several techniques are available; the most frequently used consist in:

1. Superficial washing: These techniques only extract contact insecticides and not systemic insecticides.
2. Extraction following prior crushing in Soxhlet-type devices.

Extraction can also be done by maceration. In any case, the extracting solvent is cautiously concentrated, in general in a rotating evaporator. It should be noted that these solvents are fat solvents (except for methanol); therefore, the extracted residues of insecticides are mixed with lipids or lipo-soluble compounds.

#### *Concentration and purification of insecticides*

This is an important step as it conditions the determination. It can be carried out by a liquid/liquid extraction or by distillation.

#### *Liquid/liquid extraction*

This old and traditional technique consists in separating two non-miscible solvents with differing polarities.

This technique was implemented specifically for the determination of halogen insecticide residues and also for analysing pyrethrinoids-based insecticide residues. In general, the concentrated solution of insecticide is extracted with acetonitrile saturated with petroleum ether. Insecticides preferentially pass into the acetonitrile while lipo-soluble compounds remain in the petroleum ether.

The acetonitrile solution is saturated with water and insecticides are once again extracted in petroleum ether. The solvent solution is concentrated then purified on an absorption or a filtration gel chromatographic column. The lining of the absorption columns can be made of various absorbents: florisil, celite, magnesia, charcoal. In case of a search for organochlorinated or phosphorated insecticide residues, this technique unfortunately requires important amounts of solvents (500 to 600 ml). In addition, its implementation is delicate and requires a lot of time.

#### *Extraction by distillation*

This very general technique also known as 'sweep co-distillation' was made official by AOAC in the US but with a limited application to the following organophosphorated insecticed residues:

carbophenotion, diazinon, ethion, malathion, and parathions. It is based upon the separation of insecticides by entrainment in a current of nitrogen and solvent vapours at high temperature. Numerous studies have confirmed its efficiency and polyvalence in the search for organophosphorated and organochlorinated insecticides residues.

In such conditions, triglycerides and waxes are not entrained while insecticides are in totality, and fatty acids and some oxidative by-products are only partially. This operation is undertaken in a column lined with inert filling, crossed by a gas current and placed into an isothermal oven.

## Material

The material to be used corresponds to the diagram of principle shown in Fig. 16.10.

It comprises:

1. A vertical oven with thermal regulation.
2. A column lined with glass wool, sand, glass marbles in defined proportions.

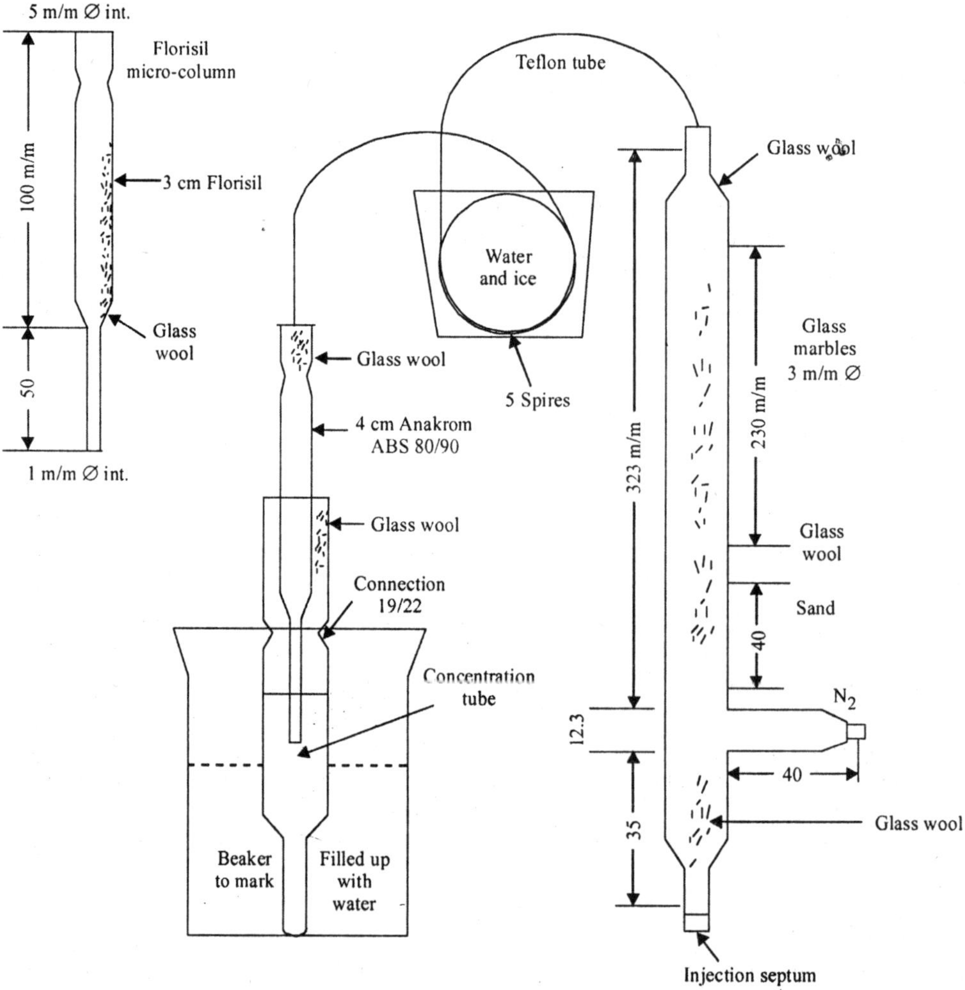


Fig. 16.10. Insecticides extraction.

This column has an entry at its base for the vectorial gas and an opening sealed with an elastomer plug for the injection of the extracted sample containing the insecticides. A condensation system completes the device.

### Procedure

The operating mode consists in concentration insecticides at a temperature of 248°C (478°F), with a vector gas flow of 600 ml/minute, while injecting directly into the system 1 ml of extract containing the insecticide or a solution with 50 per cent oil or fat to be examined. The solvent is injected regularly every 2 minutes for 30 minutes. The choice of solvent depends on the nature of the insecticides to be concentrated and purified:

Organochlorinated insecticides: Hexane

Organophosphorated insecticides: Ethyl acetate

It should be noted that, especially for the analysis of organophosphorated insecticide residues, material should be designed so that insecticides never come in contact with any metallic part.

Obtained solutions of insecticides in solvent are concentrated to a millilitre volume.

They can be examined as such when searching and determining organophosphorated insecticides residues. They must be purified of a florisil micro-column when searching for organochlorinated insecticides residues.

### *Detection and determination of insecticides residues*

Even though the detection and determination of insecticides residues can be achieved by high performance liquid chromatography, it is practically always done by gas/liquid chromatography.

### *Columns*

More than for any other analysis by gas/liquid chromatography and in the absence of a specific detector, and even with columns filled with efficient phases, the detection and absolute identification of insecticides residues can be done only by using many columns or by using a highly specific column. The best performing columns are the capillary columns filled for example with 5 per cent the phenyl methyl silicones (organophosphorated insecticides residues) and with CP SIL 8 BC (organochlorinated insecticides residues).

### *Detectors*

Detectors are either specific detectors for the identification and determination of all insecticides residues or polyvalent detectors used mainly for their identification.

### Mass spectrometry detector

This complex and costly type of detector allows an absolute identification of the residues of every insecticides (provided they have been entered in its memory space).

### Electron capturing detector

This type of detector is specific to organochlorinated insecticides and is also used for determining synthetic pyrethrinoid-based insecticides.

They comprise:

1. Tritium source detectors: These detectors, whose highest working temperature is 230°C (446°F), are susceptible to rapid pollution and require the use of highly purified insecticides solution which, in turn, induces loss.
2. Nickel 63 source detectors: These detectors are more polyvalent as they work up to a temperature of 320°C (608°F).

Electron capturing detectors are not specific to organochlorinated insecticides which allows them to be used for the determination of synthetic pyrethrinoids. However, minor components in oils rich in conjugated ethylene bonds can interfere in the determination of insecticides, in particular some oxidative by-products.

*Thermoionic detectors*

These are flame ionisation detectors (FID), modified by the adjunction of a bead of potassium chloride at the detector's head.

In this type of detector, combustion of an ingredient containing phosphorus considerably increases the sensitivity of detection, reduces by as much the amount of impurities and is very useful in the testing of organophosphorated substances.

They can be replaced with flame photometry detectors (FPD) or nitrogen phosphorus detector (NPD) often allowing to reduce the purification steps thanks to their high sensitivity. Other, more specific detectors are adapted to the determination of specific insecticides residues.

*Determination*

The determination of insecticides residues is carried out by comparing the test results with insecticides solutions with a known titer. These standard solutions must be renewed frequently. The determination consists in the comparison between the peaks height of the standard solution and that of the insecticides solution, diluted as required.

A qualitative confirmation of the results can be done by thin layer chromatography; however, reliability improves when using mass spectrometry detectors.

In all cases, it is recommended to test a sample enriched with the researched insecticide residues, as the search for some insecticides residues, in particular organophosphorated ones, (diflubenzuron for instance) can require the use of specific methods adapted to detect low concentrations.

*The specific case of carbamate residues*

These phytosanitary products are mainly fungicides, some of which also possess insecticides properties.

Often, the active agents of these products are hydrolysed, e.g. photochemically.

Their determination therefore requires specific methods.

Two active substances of the carbamate type are used in rapeseed culture.

1. Pyrimicarb: Its detection and determination are done on seeds by chloroform extraction and on oils by their dilution in this solvent. The chloroform solution is washed successively with a soda solution, water and a phosphoric buffer. Following the elimination of chloroform, the residue is submitted to a steam entrainment in presence of a normal soda solution. It frees the diethylamine which is collected in a normal solution of hydrochloric acid.
   Diethylamine is determined by colourimetry in the form of a yellow complex in presence of copper (diethyldithiocarbamate).
2. Mercaptodimethyl: Its detection and determination are done by extracting the insecticide residue with methanol from the oil in an hexane solution. The alcohol solution of insecticide is purified by liquid extraction. Determination is done by high performance liquid chromatography with fluorescene spectrometry following a hydrolysis at the column outlet ($3_3$).

***Insecticide residue contents in oils and fats***

Phytosanitary products are made up of a large number of different chemical classes.

Thus, if differences in toxicity appear, there are also differences in chemical behaviour susceptible to influence the insecticides becoming.

In the annual index of phytosanitary products, beetroot is the most frequently cited plant among extensively cultured crops likely to be treated with phytosanitary products. Among oil seeds, rapeseed is the most frequently mentioned (Tables 16.11 and 16.12).

Rapeseed can be affected by about ten diseases and infested with about twenty different animal parasites. Whereas, sunflower is threatened only by four or five illnesses, mainly cryptogamic, and by two or three predators.

Therefore, it is not surprising that rapeseed is generally the most contaminated of crude oils. Also, the adipose tissues of cattle grown in breeding stables is the most contaminated crude material of the animal fat industry. The life span of these animals influence the level of pollution of their fatty tissues. The elimination of insecticides residues is slow, particularly for organochlorinated insecticides residues. The nature of the insecticide used, some of which are systemic (and therefore enter the plant), is likely to influence further the oil pollution, as seen in olive oil with the use of such insecticides to protect the harvest from the olive tree fly (Tables 16.10 to 16.12).

**Table 16.10.** List of the main phytosanitary products (insecticides) used in France for the protection of the main oilseeds crops.

| *Active matters* | *Rapeseed* | *Sunflower* | *Soya* | *Olive* | *Corn* |
|---|---|---|---|---|---|
| *Synthetic pyrethrinoids* | | | | | |
| Alpha methrine[3] | + | | | | + |
| Bifenthrine | + | | | | |
| Cyfluthrine | + | | | | |
| Cypermethrine | + | | | | + |
| Delta methrine | + | | | + | + |
| Esfenvalerate | + | | | | |
| Fenvalerate | + | | | | + |
| Fluvalinate | + | | | | |
| Lambda cyhalothrine | + | | | | |
| Permethrine | + | | | | + |
| Tralomethrine | + | | | | |
| *Organochlorinated* | | | | | |
| Endosulphan | + | | + | | |
| Lindane[4] | + | | | | + |
| *Carbamates* | | | | | |
| Aldicarbe | | | | | + |
| Bendiocarbe | | | | | + |
| Benfuracarbe | + | | + | | + |
| Carbaryl | | | | + | |
| Carbofuran | + | | + | | |
| Carbosulphan | + | | | | |

*(Contd...)*

**Table 16.10.** List of the main phytosanitary products (insecticides) used in France for the protection of the main oilseeds crops. *(Contd.)*

| *Active matters* | *Rapeseed* | *Sunflower* | *Soya* | *Olive* | *Corn* |
|---|---|---|---|---|---|
| Furathiocarbe | + | | | | + |
| Mercaptodimetur[5] | + | | | | |
| Methomyl | | | | | + |
| Pyrimicarbe | + | | | | |
| *Organophosphorated* | | | | | |
| Bromophos | + | | | | |
| Chlormephos | | | | | + |
| Chlor pyriphos ethyl | | | | | + |
| Diazinon | | | | + | |
| Dicofol | | | + | | |
| Diflubenzuron | | | | | + |
| Dimethoate | | | | + | |
| Finithrothion | | | | | + |
| Fenthion | | | | + | |
| Formothion | | | | + | |
| Isophenphos | + | | | | |
| Malathion | + | | | | |
| Methidathion | + | | | | |
| Parathion ethyl/methyl | + | | | | |
| Phorate | + | | | | + |
| Phosalonne | + | | | | |
| Phosphamidon | | | | + | |
| Phoxime | | | | | + |
| Thiometon | | | + | | |
| Terbuphos | + | | | | + |
| *Miscellaneous* | | | | | |
| Hexythiazox | | | | | + |

3. Cyper methrine isomer; 4. Gamma HCH; and 5. Methiocarb.

**Table 16.11.** List of the main phytosanitary products (insecticides) allowed in France for protecting seeds (warehouse or seeds).

| |
|---|
| Bio resmethrine |
| Delta methrine |
| Permethrine |
| Chlor pyrimiphos methyl |
| Dichlorvos |
| Malathion |
| Methoprene |

*(Contd...)*

**Table 16.11.** List of the main phytosanitary products (insecticides) allowed in France for protecting seeds (warehouse or seeds). *(Contd .)*

| |
|---|
| Pyrimiphos methyl |
| Pyrethrine |
| Aluminium phosphide |
| Magnesium phosphide |

**Table 16.12.** General information regarding insecticides.

Organic insecticides used on crops

| | *Solubility* | | | *Sensitivity* | | *Determination method* | *LD50 ingestion in rats mg/kJ (3)* |
|---|---|---|---|---|---|---|---|
| *Active matters* | *Water* | *Organic solvent* | *To temp. in °C* | *To acids* | *To alkalies* | | |
| *Pyrethrinoids* | | | | | | | |
| Alpha methrine (Cypermethrine isomer) | No | Yes | >220 | No | Yes | GC/ECD | 251–4123 |
| Bifenthrine | No | Yes | >165 | ? | ? | GC/ECD | 54.5 |
| Cyfluthrine | No | Yes | ? | ? | ? | ? | 500–800 |
| Cyhalothrine | No | Yes | No | No | Yes | GC/ECD | 243 |
| Cypermethrine | No | Yes | >220 | No | Yes | GC/ECD | 251–4123 |
| Deltamethrine | No | Yes | >220 | No | No | GC/ECD | 251–4123 |
| Esfenvalerate | No | Yes | >220 | ? | ? | GC/ECD | 436 |
| Fenvalerate | No | Yes | No | No | No | GC/ECD | 450 |
| Fluvalinate | No | Yes | No | Yes | Yes | GC/ECD | 260–280 |
| Permethrine | No | Yes | No | No | No | GC/ECD | 430–4000 |
| Tralomethrine | No | Yes | No | Yes | No | GC/ECD | 99–157 |
| *Organochlorinated* | | | | | | | |
| Endosulphan | No | Yes | No | No | Yes | GC/ECD | 80–100 (1) |
| Lindane (HCH) | No | Yes | <180 | No | Yes | GC/ECD | 88–270 |
| *Carbamates* | | | | | | | |
| Aldicarbe (E) | Yes | Yes | No | Yes | Low | HPLC/FID | 1 |
| Bendiocarbe (E) | No | Yes | No | Low | Yes | GC/ECD | 40–156 |
| Benfuracarbe (E) | No | Yes | ? | ? | ? | ? | 138 |
| Carbaryl (E) | No | Yes | No | ? | Yes | GC | 850 |
| Carbofuran (S) | Low | Yes | No | ? | Yes | GC | 8–14 |
| Carbosulphan (S) | Low | ? | ? | ? | ? | ? | 250 |
| Furathiocarbe (S) | Low | Yes | No | ? | ? | GC | 137 |
| Mercaptodimetur | No | Yes | ? | No | Yes | HPLC or specific method | 100 |
| Methomyl (E) | Yes | Yes | Yes | Yes | Yes | GLC/FDP | 17 |
| Pyrimicarbe (E) | Yes | Yes | No | Yes | Yes | GC/FID | 147 |

*(Contd...)*

**Table 16.12.** General information regarding insecticides. *(Contd .)*

Organic insecticides used on crops

| | Solubility | | | Sensitivity | | Determination method | LD50 ingestion in rats mg/kJ (3) |
|---|---|---|---|---|---|---|---|
| *Active matters* | *Water* | *Organic solvent* | *To temp. in °C* | *To acids* | *To alkalies* | | |
| *Organophosphorated (2)* | | | | | | | |
| Bromophos (E) | No | Yes | No | Yes | No | GC | 3750–8000 |
| Chlormephos (E) | No | Yes | No | Yes | Yes | GC/FID | 7 |
| Chlor pyriphos ethyl (E) | No | Yes | No | No | Yes | GC | 135–163 |
| Diazinon (E) | No | Yes | No | Yes | Yes | GC | 300–400 |
| Dicofol (E) | No | Yes | No | No | Yes | GC | 668–842 |
| Diflubenzuron (E) | No | Yes | >230 | Low | Yes | HPLC or GC | >4640 |
| Dimethoate (E and S) | Yes | Yes | ? | No | Yes | GC | 500–600 |
| Fenitrothion (E) | No | Yes | <145 | No | Yes | GC | 800–890 |
| Fenthion (E) | No | Yes | <210 | ? | No | GC | 190–315 |
| Formothion (E and S) | Yes | Yes | Yes | Yes | Yes | GC | 365–500 |
| Isophenphos (E) | Yes | Yes | ? | ? | ? | GC/FID | 28–38 |
| Malathion (E) | No | Yes | <157 | ? | ? | GC | 2800 |
| Methidation (E) | No | Yes | Yes | Yes | Yes | GC | 25–54 |
| Parathon/ethyl methyl (E) | No | Yes | <162 | No | Yes | GC | 3.5–13 |
| Phorate (S) | No | Yes | No | Yes | Yes | GC | 17.4 |
| Phosalone (E) | No | Yes | ? | ? | ? | GC | 120–170 |
| Phosphamidon (S) | Yes | Yes | No | Yes | Yes | GC | 17.4 |
| Phoxime (E) | No | Yes | Yes | No | Yes | GC | 2500–5000 |
| Terbuphos (E) | No | Yes | No | Yes | Yes | ? | 1.5–5.4 |
| Thiometon (s) | No | Yes | No | Yes | Yes | GC | 120–130 |
| 2. Organic insecticides used in warehousing. | | | | | | | |
| Bioresmethrine | No | Yes | ? | No | Yes | GC/FID | 7070–8000 |
| Chlor pyrifos methyl | No | Yes | ? | Yes | Yes | GC/ECD | 1630–2140 |
| Diehorvos | Yes | Yes | No | Yes | Yes | GC | 56–108 |
| Permethrine | No | Yes | No | No | No | GC/ECD | 430–4000 |
| Pyrimiphosmethyl | No | Yes | ? | Yes | Yes | GC | 2050 |

(1) Varies according to aqueous or oily presentation.
(2) Insecticide with external or internal impact (systemic, entering the plant and making it phytotoxic).
(3) Level of toxicity related to the LD50.
– ⩽ 100: extremely dangerous.
– ⩾ 1000: slightly dangerous.
GC = gas chromatography.
HPLC = high performance liquid chromatography
ECD = electron captor detector
FID = flame ionising detector

The evolution of the organochlorinated insecticide content (lindane gamma isomer of HCH) in rapeseed seeds, meals and oils have been studied as shown below:

| *Product* | *Oil content in %* | *HCH content in ppm* |
|---|---|---|
| Seeds | 43.0 | 0.065 |
| Oil meal, pressed | 13.5 | 0.019 |
| Oil meal, extracted | 1.3 | 0.002 |
| Oil | – | 0.165 |

Results point out that insecticide residues, especially the organochlorinated type, due to their fat solubility, spread evenly in oil meal and in oil, in an almost linear way. It was noted also that extraction oils were more polluted than press oils. The influence of refining on the becoming of organochlorinated insecticides residues was studied. It appears that the elimination of such products takes place during the deodourisation step of refining and that its efficiency increases with temperature. These results were seen with different types of vegetable oils and with animal fats. There is no similar work done on other residues of phytosanitary products such as organophosphorated insecticides, carbamates or synthetic pyrethrinoids. However, it is reasonable to believe that the refining process leads to the total elimination of insecticides residues for several reasons:

1. The volatility of some insecticides, such as dichlorvos.
2. The water solubility of others, such as phosphamidon or dimethoate.
3. The instability in alkaline medium of many of them, such as most of the organophosphorated and some pyrethrinoids.
4. The sensitivity to temperature of some others, such as the formothion.
5. The capacity for steam at high temperatures under high vacuum to entrain the majority of them, including the most residual types such as the organochlorinated or some synthetic pyrethrinoids.
6. Degumming in phosphoric or citric acid medium.
7. Soda neutralisation.
8. Various steps of water washings during degumming and neutralisation.
9. Deodourisation in the eventual presence of citric acid, lead to the elimination of most phytosanitary residues eventually present.

It is worth noting, regarding animal feeding, that oil meals are cleared, during oil extraction, of the quasi totality of physosanitary residues. Some regulations, such as the Codex Alimentarius, indicate the highest admissible concentrations is food items, including oils and fats, as shown in Table 16.13. However, these limits are significant only for crude oils and fats as the refining process almost guarantees the elimination of all phytosanitary products.

## Pollutants Due to Packaging or Processing

Research on pollutants eventually introduced by packings or by processes in food products is of great concern to the analysts in order to ensure harmlessness of the products made available to the consumers.

Pollution of foods can lead to several consequences:

1. Sensory: Through adulteration of food not necessarily toxic, but rendering it in edible (appearance, odour, taste).
2. Toxicological: Through presence of toxic undesirable chemical products in food.

**Table 16.13.** Maximal value for main phytosanitary products (insecticides) in oils and fats.

| *Active matters* | *Product (seeds, oils or fats)* | *Maximum content* |
|---|---|---|
| | Insecticides for the protection of crops | |
| *Synthetic pyrethrinoids* | | |
| Alphamethrine | 1–7 | E |
| Bifenthrine | – | – |
| Cyflutrine | – | – |
| Cypermethrine | 8–19 | D |
| Delta methrine | 1–6 | B |
| | 7 | A |
| Esfenvalerate | – | – |
| Fenvalerate | – | – |
| Fluvalinate | – | – |
| Lambda cyalothrine | – | – |
| Permetharine | – | – |
| Tralomethrine | – | – |
| *Organochlorinated* | | |
| Endosulphan | 1 | E |
| | 8 | A |
| Lindane | 5 | A |
| | 7 | E |
| | 13 | E |
| *Carbamates* | | |
| Aldicarbe | – | – |
| Bendiocarbe | 15–18 | A |
| Benfuracarbe | – | – |
| Carbaryl | 1 | E |
| | 3 | F |
| | 15, 16, 18 | C |
| Carbofuran | 1–7 | B |
| | 15–18 | A |
| Carbosulphan | – | – |
| Furathiocarbe | – | – |
| Mercaptodimetur | – | – |
| Methomyl | – | – |
| Pyrimicarb | 3 | D |
| | 5 | C |
| *Organophosphorated* | | |
| Bromophos | 5 | C |

*(Contd...)*

**Table 16.13.** Maximal value for main phytosanitary products (insecticides) in oils and fats. *(Contd.)*

| *Active matters* | *Product (seeds, oils or fats)* | *Maximum content* |
|---|---|---|
| | 9 | G |
| | 12 | C |
| | 15 | F |
| | 16 | D |
| Chlormephos | – | – |
| Chlor phyriphos ethyl | 1 | A |
| | 8 | A |
| Diazinon | 2–3–4 | A |
| | 9 | F |
| Dicofol | – | – |
| Diflubenzuron | – | – |
| Dimethoate | – | – |
| Fenithrothion | 7 | A |
| Fenthion | 9 | E |
| Formothion | – | – |
| Malathion | – | – |
| Methidathion | 3 | C |
| | 8 | E |
| | 15–18 | <A |
| Parathion ethyl/methyl | 8 | A |
| Phorate | – | – |
| Phosalone | 5 | B |
| | 16 | D |
| Phosphamidon | – | – |
| Phoxime | – | – |
| Tiometon | 3′ | D |
| | 8 | B |
| | 12 | A |
| Terbuphos | – | – |
| Bioresmethrine | – | – |
| Permethrine | 1 | D |
| | 4 | E |
| | 5 | A |
| | 15 to 17 | E |
| | 18 | A |
| Chlor pyriphos ethyl/methyl | 18 | A |
| Dichlorvos | 5 | G |

*(Contd...)*

**Table 16.13.** Maximal value for main phytosanitary products (insecticides) in oils and fats. *(Contd.)*

| *Active matters* | *Product (seeds, oils or fats)* | *Maximum content* |
|---|---|---|
| | 5 | G |
| | 6 | F |
| | 15 to 18 | A |
| Pyrimiphos methyl | 14 | G |
| | 3 | > G |
| Aluminium phosphide | – | – |

Legend:

1. Identification of oilseeds and fats:

| *Seed* | *Oil or Fat* | *Meat* |
|---|---|---|
| Cotton 1 | Cotton 8 | Bovine 15 |
| Safflower 2 | Olive 9 | Caprine-ovine 16 |
| Peanut 3 | Sunflower 10 | Porcine 17 |
| Sunflower 4 | Soya 11 | Poultry 18 |
| Rapeseed 5 | Rapeseed 12 | |
| Soya 6 | Cocoa butter 13 | |

2. Identification of maximum values (in mg/kg): expressed in mg/kg of total fat, for meats

A 0.05
B 0.1
C 0.2
D 0.5
E 1.0
F 2.0
G 5.0

Food pollution problems are common. In regard to fats or foods containing fats (oils, butters, margarines, pastes for spreading on bread with reduced liquid content), which are very sensitive to contaminations by their ability to absorb volatile odours from the surroundings.

Unfortunately, the example of painters working on a production floor and unvoluntarily contaminating food products produced several dozen metres away is not uncommon.

French and European regulations regarding the packaging of food items are complex and of the utmost importance. Analysis of pollutants introduced by packings or by processes was mainly developed for materials and packaging made of plastic material. For packagings of other kinds, for which there is few or no official texts (papers used for food products, for example), techniques described for packings and materials made of plastics, are often generalised.

This section deals with:

1. Overalls migrations.
2. Specific migrations.

### *Overall migrations*

The principle of overall migration consists in determining the mass of products migrating from one plastic material packing towards a liquid which suppose to simulate a food product, this is determined conditions of time and temperature. The simulant suppose to represent fats is refined olive oil or, for technical reasons connected with method of analysis, a mixture of synthetic triglycerides or of sunflower oil. Depending on whether the materials is meant for general or more specific applications, test conditions are different. But usually it is: 10 days at 40°C. Tests should be carried out minimum in triplicate and it is necessary at times to adapt the conditions when the material undergoes physical or other modifications during tests. At the end of the tests, demonstration of eventual migration is done by iodometric or chromatographic techniques.

#### *Principle of migrations*

The principle is as follows:

The sample subjected to tests, of known mass and/or surface, is wiped and weighed after contact with the simulant. Liquid simulant eventually absorbed is extracted on flowing back by trichlorotrifluroethane. Mass of simulant liquid (refined olive oil generally) is determined, either by measuring iodine number by Wijs' method, or by chromatography in gaseous phase or methyl esters of fatty acids in the presence of an internal standards of margaric acid (C17). This allows the determination of quantity of simulant absorbed by the sample and to correct variation in weight of the material of the quantity of simulant eventually retained by the material. With this correction the sample cannot help but gain weight. The method, which is hardly precise, can lead to total negative migrations.

#### *Triple tests*

Henceforth, triple tests should be done as below:

1. In normal conditions, for example, 10 days at 10°C which furnishes the result of total migration M1.
2. In normal conditions, but with contact duration doubled, which provides the results of total migration M2.
3. In normal conditions, but with contact duration tripled, which provides the result of total migration M3.

Tests are considered as conforming if M1 (10 days) and (M3-M2) (30–20 days) do not exceed the limit of overall migrations.

Limit of total migration in fatty simulant is: 60 mg/kg or 10 mg/dm$^2$ with an analytical tolerance of– – 20 mg/kg or 3 mg/dm$^2$.

A number of justified criticisms have been directed towards the tests of total migration:

1. Blind tests, hardly precise, difficult to interpret. Possible toxic character of products released in the simulant is not stated with precision.
2. Disappearance of volatile compounds.
3. Very insufficient precision for low values of migration.

Tests of total migration, specially tests carried out with a fatty simulant have been the subject of a number of works for example at CGDUMA (General Commission of the Methods of Analysis of the DGCCRF), RNE (National French Test Network), CEN (European Standardisation Centre).

Interlaboratory co-operation is envisaged to study the conditions:

1. Of test (total immersion, filling, use of cell for multilayer products, for example).

2. Of choice of simulant (pentane, heptane).
3. Of simulant extraction time (48 hrs.).

In certain countries, for example in the USA and in the Netherlands, fatty simulant is replaced by hexane or isooctane with reflux extraction lasting a few hours. Results obtained with these simulants are generally superior to results obtained by using a simulant triglyceride.

### *Specific migrations*

The principle of specific migration consists in determining the quantity of a known and identified component of packing released out in food item which is as under:

1. Monomers and other original substances which can be used for making materials and objects of plastics material (Section A).
2. Monomers and other substances which can continue to be used while waiting for a decision on their inclusion in the section A (Section B).
   It gives indispensable information to an analyst:
3. EEC reference number in the field of packing materials.
4. Registration number of Chemical Abstracts Service.
5. Chemical denomination.
6. Restrictions.
   (a) Limit of specific migration.
   (b) Maximum quantity permitted of residual substance in material or object.
   (c) Other restrictions.
   The definition of substances sought after helps in selection of method of analysis.
7. Static or dynamic headspace.
8. Gas chromatography (GC).
9. Detection in flame ionisation of any other kind (nitrogen, phosphorus) or use of mass spectrometer.
   Very few official texts or any other text deal with specific migrations, nevertheless, those that should be mentioned are:
10. Vinyl chloride monomer which should not be found more than 1 mg/kg in the material or object made of plastic and 0.01 mg/kg in the food product.
11. Vinylidene chloride.
12. Vinyl acetate.
13. 1,3 Butadiene.
14. Acrylonitrile.
15. Monoethylene glycol and diethylene glycol.
16. Monomerous styrene.
17. Isophtalic and terephthalic acids high performing chromatography liquid.

It should be noted that research on such substances is at times very difficult, in spite of improved analytical techniques (capillary columns). Thus researchal vinyl chloride monomer in a vegetable oil is difficult, as a number of parasite peaks are liable to interfere.

For this reasons, detection by mass spectrometry (MS) is strongly advised for identification of peaks. Then one passes over to traditional detection by ionisation of flame, which permits to make a satisfactory quantitative determination. For this determination the GC/MS coupling is at present the simplest method.

### *Other points concerning packings*

1. Sensorial analysis of fats which are in contact with packaging is recommended. Styrene monomer can easily be located, its sensitivity being clearly higher than that of instrumental techniques. One means to verify food adaptability of packing at least under the organoleptic aspect consists in carrying out the Robinson test described by the ICCA (International Cocoa and Chocolate Association). Passage of one or more volatile compounds of packing into a fat can thus be often shown.
2. Adaptation of very resistant plastic materials for use in microwave ovens or freezers leads to the questioning of existing methods. It is suggested to carry out simulation tests that are as close as possible to working conditions.
3. Fats can contain traces of water (anhydrous lactic fat, for example) or can be emulsified with water (butter, margarine). This may lead to:
   (a) Migrations of mineral elements, catalysts of oxidation (iron, copper) or toxic (heavy metals).
   (b) Microbiological contamination through development of pathogenic germs.
4. Fats can render soluble certain compounds or packaging pollutants, which compels one to verify the absence of:
   (a) Packaging components: (i) coating waxes; (ii) waxes for removal of mildew; and (iii) colouring agents.
   (b) Pollutants which may emanate from packaging: (i) dioxines from papers contaminated by bleaching treatment; and (ii) polychlorobiphenyl (PCB).
   (c) Various solvents (halogenated) emanating from inks, glues and varnishes deposited on packing.

For these volatile pollutants, as for monomers, these are headspace techniques coupled with GC/MS which give the most satisfying qualitative and quantitative results.

## Pollutants Caused by Transport

There are three main causes of contamination of crude or refined oils and fats during transport and especially during bulk maritime transport:

1. Nature and state of metal coating of containers (vats, tanks, shafts).
2. Nature and state of plastic coating of containers (vats, tanks, shafts).
3. Presence of residues of prior loadings.

### *Metal coatings*

Use of non-coated metal packings in the interior constitutes an important source of contamination of oils and fats by metal traces during transportation. This contamination can be of purely mechanical origin on account of the fats that metal or rust particles can get detached from side walls and can contaminate oil or fat. The greater the deterioration of side walls of the containers due to corrosion (vats, tanks, shafts), the greater the risk of contamination is.

In the case of storage or transportation of acid oils, this contamination can, moreover, be due to action of free fatty acids which are going to interact with metals to form metallic salts (soaps).

This contamination of oils and fats by metal traces has two major disadvantages:

On the one hand, it can lead to faster adulteration of fat as a result of presence of certain metals like iron or copper which are well known oxidising catalysts. On the other hand, presence of certain heavy toxic metals (arsenic, lead, cadmium) can pose a serious problem for public health. It is to be noted that this risk of contamination by metal traces has far less serious consequences in case of oils and coarse fats than in case of refined products ready to be consumed. As a matter of fact, as several studies have shown, refining eliminates large majority of metal traces likely to be found in crude oil. For transportation of oils and refined fats, it is however, imperative to use materials limiting to the maximum, this risk of contamination by metal traces (inox, varnish or plastic film adapted to food).

Measuring of metal elements in oils and fats is carried out, in most cases, through spectrometry of atomic absorption.

### *Plastic coatings*

Plastic coatings like those used to protect side walls of tank of ships (polyurethane and epoxy resins), if they enable avoiding contamination of oils by metal traces, can also be a source of oil pollution. The components of plastic polymers and particularly certain additives (catalysts, stabilisers, pigments, etc.) and monomer residues can be found in oil as a result of migration phenomenon.

### *Residues of prior loadings*

Because of development of commercial exchanges between countries, ships used for transportation of oils and edible fats can also be used to transport products from chemical or petrochemical industry with the objective of reducing costs of transportation.

This situation constitutes the most important risk of contamination of oils and fats during transportation by sea and can have several sources:

1. It can, first of all be due to bad cleaning of side walls of tanks, pipe works and pumping systems. It is to be noted that this cleaning operation, which should be done regularly between each loading, is rendered difficult because certain chemical products which are particularly difficult to dissolve and because of the difficulty of reaching certain parts of vats, pipe network and pumping system.
2. The second cause of pollution of oil loading may be due to error in handling. In the face of the complexity of networks of pipes and systems of valves and pumps serving different tanks of some ships, one cannot exclude human error of handling during operations of loading, unloading or siphoning. This error can lead to dispatch a part of load of one tank to another. And the realisation that on the same ship certain tanks containing edible oils have by their side other tanks containing chemical products, some of which are particularly toxic, one can imagine the danger represented by such a situation.

Number of chemical contaminants likely to pollute a cargo of crude oil is very high, thus, from reading the books about other ships, which give the nature of all products transported earlier, one could identify more than 150 potential chemical contaminants. These different contaminations do not present the same risk to human health. Four important factors are going to define the real extent of danger represented by different pollutant chemical molecules, namely:

1. Solubility of pollutant in oil.
2. Toxicity of polluting molecule.
3. Facility of elimination of pollutant during refining.
4. Reactivity of polluting molecule towards constituents of oil.

From these four criteria, it is possible to establish a classification of different chemical contaminants in terms of the danger which they really represent.

Federation of Oils, Seeds and Fats Associations (FOSFA) has, in this way, made out and published the first list of chemical products prohibited as cargo, transported as previous cargoes of edible oil (Table 16.14).

*Methods of measuring chemical contaminants*

Analytical means to be employed for detection and determination of chemical contaminants in oils depend, first of all, on degree of contamination and on the molecule shown.

### High contaminations

Large quantities of chemical contaminants can be relatively easily shown following modification in principal physico-chemical characteristics (smoke point, flash point) of oils and fats and in particular, their unsaponifiable fraction. An abnormally high content of unsaponifiable matter or certain of its components, like hydrocarbons, enables to show the high contamination by non-glyceride components.

It is to be noted that IUPAC has standardised a detection method for mineral oils in oils and fats of animal and vegetable origin. This method (IUPAC 2611), which is operated by thin layer chromatography, allows detection ceiling of 0.01 per cent (100 mg/kg) to be attained.

However, this analytical technique does not help to differentiate hydrocarbons naturally present in oils from those arising out of external contamination. For vegetable oils with natural hydrocarbon content comprising 100 mg and 300 mg/kg, IUPAC method will indicate with certainty that there really is pollution only if rate of pollution is greater than 500 mg/kg (0.05 per cent).

Another disadvantage of IUPAC method, which may be pointed out is that this technique is not applicable in research and is measure of volatile hydrocarbons (gas oil, fuel).

**Table 16.14.** List of chemical products prohibited.

| | |
|---|---|
| Monomers vinyl acetate | Glutaraldehyde |
| | Hexamethylene diamine |
| Cyanohydrine acetone | Oil for transformer (PCB) |
| Acrylic acid | Isocyanates |
| Nitric acid | Methyl methacrylate monomers |
| Butyl acrylate | Methylstyrene monomers |
| Ethyl acrylate | α-Methylstyrene monomers |
| Methyl acrylate | *p*-Methylstyrene monomers |
| Acrylonitrile | Morpholine ethanol |
| Additives for lubricating oils | Phthalates such as: |
| Adiponitrile | Di-octyl phthalate (DOP) |
| Furfurylic alcohol | Di-iso octyl phthalate (DIOP) |
| Aniline | Di-iso nonyl phthalate (DINP) |
| Cardura E | Di-iso decyl phthalate (DIDP) |
| Dibutylamine | Di-allyl phthalate (DAP) |
| Diethanolamine | |
| Diethalanetramine | Products containing lead |
| Di-isopropylamine | *n*-Prolylamine |
| Di-propylamine | Propylene oxide |
| *m*-Divinylbenzene | Pyridine |
| Epichlorhydrine | Epoxy resins (untreated) |
| Ethanolamine | Styrene monomers |
| Ethylenediamine | Carbon tetrachloride |

### Low contaminations (traces)

For very low rates of contamination, starting from a few mg/kg (ppm) to a few hundred μg/kg (ppb) which are most commonly observed at the level of accidental pollutions, the best technic adapted in the majority of cases is gas chromatography (GC). The only constraint imposed by this technique of measuring through GC is the desired pollutants be volatile.

Several variants of this chromatographic technique can be used in terms of choice of injection system and type of detector retained.

### Direct injection

This is by far the simplest technique to be carry out. It consists of injecting a small quantity of oil (100 μl) directly into the injector chamber of chromatograph. This injection technique gives, in a number of cases, good results, however, the sensitivity is not always appreciable taking into account low volume of oil injected. One should also take into account the fact that in order to preserve the life span of columns and to avoid effects of matrix which may appear with time it is preferable to intercalate either an insert (glass wool) or a pre-column between the injector and the column.

### Static headspace

This technique which is presently the one mostly used in finding of volatile pollutants, is based on thermodynamic equilibrium existing in closed vessel between a liquid phase (oil) and an overfloating gaseous phase (volatile products). To do this, one introduces a few grams of oil in an air-tight flask (penicillin type), the whole thing is then kept in a drying oven at a temperature and for a period of time fixed in such a manner so as to have the best desorption of desired pollutant, one part of gaseous phase being then taken out by means of a gas syringe before being injected into the chromatograph. It should be noted that the static headspace was retained as analysis technique in the method of measuring hexane traces in extraction oils (IUPA method) or in the method of measuring of tetrachlorethylene (perchlorethylene) in olive oils (EEC method).

### Dynamic headspace

Principal disadvantage of static headspace is that this technique lacks sensitivity in finding very small quantities of pollutant on account of the effect of dilution of volatile products in gaseous phase. Dynamic headspace corrects this disadvantage by achieving, before injection in column of the chromatograph, a pre-concentration of these compounds. In the market there are several systems of which one named DCI (desorption-concentration-injection) plate permits carrying out easily and rapidly different operations of desorption, concentration and injection.

In these devices, the volatile compounds are desorbed by heating of oil in an oven having a regulation system for temperature.

Cleaning of oven by an inert gas (helium) enables displacement of equilibriums of concentration into complete desorption of volatile components of their matrix. They are then concentrated on a trap filled with an absorbant (tenax) which can, moreover, be cooled at low temperatures (liquid nitrogen) so as to have better concentration for very volatile products. The trap is then heated again suddenly so as to enable instantaneous restoration to gaseous state of volatile components which are then injected directly in the column of the chromatograph. At the level of detection, several detectors can be used whose choice will depend on the type of pollutants to be identified and on the ceiling of proportioning that one wants to obtain.

### Flame ionisation detector (FID)

This is the detector most commonly used in GC because of its good detection sensitivity. Thus one easily obtains ceilings of 1 ppm. Moreover, it meets a great majority of organic compounds with a large range of linearity making it largely polyvalent.

### Electron capturing detector (CED)

This type of detector has the advantage, with respect to FID, of being much more sensitive which permits it to attain extremely low ceilings of measurement, particularly for chlorinated compounds (ppp).

This is the type of detector which was accepted by in the EEC method of measurement of tetrachlorethylene in olive oils, the limit of detection obtained being 0.01 ppm (10 ppb).

### Mass spectrometry (MS)

Mass spectrometry used as detector in fragmentography enables to attain a selectivity clearly greater than that given by any other system of detection and this, while retaining good sensitivity during measurement.

The choice of this type of detector is found particularly judicious in case of doubt on the identification of a peak. Therefore by setting the spectrometer on one or several characteristic ions of the polluting molecule to be identified, one can identify it with certainty. DCI/GPC/SM coupling appears at present to be the only technique enabling to show and to proportionate different volatile pollutants of oils and fats taking into account its good selectivity and polyvalence. It should be noted that this type of coupling has been accepted in the project method of FOSFA which is the subject of a collective study in different laboratories.

## DETERMINATION OF PHYSICAL CHARACTERISTICS

### General Characteristics

While physical characteristics are hardly ever used to identify fat or to verify its purity, they have nonetheless some importance on the commercial (i.e. to associate a weight to a given bulk volume of oil transported) and on the technological level (i.e. use of the refractive index in hydrogenation).

#### *Specific gravity—weight of a litre of oil*

Specific gravity (often known as density) depends on the chemical composition of the oil and its temperature.

1. For a given chain length, the greater the unsaturation, the higher is the density.
2. For a given unsaturation, density increases as length of the fatty chain decreases.
3. The presence on fatty chains of secondary functions (alcohol, ketone, epoxy, etc.) increases the density.
4. In a narrow range of temperature (about 10°C), it is assumed that density is a decreasing linear function of temperature. For most oils, this variation is of 0.00068/°C. The International Association of Seed Crushers (IASC) gives slightly different values for certain oils:
   0.00071 for coconut
   0.00070 for palm kernel
   0.000675 for linseed
   0.00066 for rapeseed (canola)

The density of fats is precisely measured with a pycnometer (specific gravity bottle) as per usual methods.

The method has been standardised. However, as the known density is almost always used for the conversion from volume to weight of oil, the absolute specific gravity (that is under vacuum) is of no interest. The standardised determination of the 'weight of a litre of oil in air' is preferred.

The difference, due to the air pressure, is of about 0.0002. Nonetheless, both methods refer to water density values at different temperatures. Given values in IUPAC and ISO standards differ by 0.0010. Commercial consequences of this difference are important. Therefore, both standards should hopefully be harmonised.

This determination being used very often for crude oils containing gums in suspension, the laboratory sample should be taken in conditions such that it is representative of the batch (dynamic sample taken during unloading or distinct samples taken from oil and feet). Also, in the laboratory, the content of the pycnometer should tally with the laboratory sample. In particular, during the operations of level adjustment, it should be checked that there is no loss of the clear oil floating above the gums. Repeatability of the pycnometric methods is defined in standards; it can be estimated at 0.0002. Repeatability is improved when using a method based on a completely different principle: the frequency variation of a non-damped oscillator, comparable to a liquid-filled diapason, is related to this liquid's density.

Therefore, it is possible to use devices built according to this principle in order to determine the density of oils with a theoretical repeatability of $10^{-6}$. In practice, repeatability is only of $10^{-5}$ as it is difficult to rigorously install a thermostat into the measuring cell.

Table 16.15 gives the range of specific gravity at 20°C (84°F) of some animal and vegetable fats.

**Table 16.15.** Specific gravity of oils at 20°C.

| *Oleic vegetable oils* | | *Linoleic vegetable oils* | |
|---|---|---|---|
| Almond | 0.911–0.917 | Safflower | 0.923–0.926 |
| Peanut (Africa) | 0.914–0.917 | Cotton | 0.917–0.925 |
| Peanut (South America) | 0.917–0.920 | Corn | 0.919–0.923 |
| High oleic safflower | 0.912–0.917 | Grape seeds | 0.923–0.926 |
| Erucic rapeseed | 0.910–0.916 | Walnut | 0.924–0.926 |
| Low-erucic rapeseed (canola) | 0.914–0.917 | Sesame | 0.921–0.923 |
| Mustard | 0.912–0.920 | Soya | 0.921–0.924 |
| Olive | 0.910–0.916 | Sunflower | 0.920–0.925 |
| *Industrial vegetable oils* | | *Animal oils* | |
| China wood | 0.938–0.943 | Whale | 0.916–0.927 |
| Linseed | 0.927–0.936 | Herring | 0.918–0.923 |
| Bud | 0.924–0.926 | Menhaden | 0.925–0.930 |
| Oiticica | 0.970–0.980 | Egg | 0.939–0.940 |
| Ongokea | 0.975–0.988 | Neat's foot | 0.911–0.914 |
| Castor | 0.957–0.968 | Sardine | 0.923–0.932 |
| Dehydrated castor | 0.931–0.955 | | |

At 40°C (104°F), 'hard' fats (animal fats, coconut, palm kernel oils) have a density lying between 0.900 and 0.910 with the exception of palm oil whose density lies between 0.895 and 0.900.

### Refractive index

The refractive index depends, as the specific gravity, on the chemical composition of the fat and on its temperature:

1. It increases with unsaturation with the presence of secondary functions on the fatty chains.
2. In a narrow range of temperature (about 10°C), the refractive index is a decreasing linear function of temperature.

For most oils and fats, this variation is of 0.00035/°C.

Though the measurement of the refractive index of fats does not require any specific precaution, the method is standardised by IUPAC, ISO and AFNOR. The three identical standards do not specify the repeatability or reproducibility. Reproducibility can be estimated to be of about $10^{-4}$. The use of differential refractometers improves the repeatability and allows continuous measurements. Determining the refractive index can be a rapid and simple method to monitor hydrogenation or fractionation.

Table 16.16 shows the refractive indices at 20°C (68°F) of most common fluid fats.

**Table 16.16.** Refractive index at 20°C of fluid fats.

| *Oleic vegetable oils* | | *Linoleic vegetable oils* | |
|---|---|---|---|
| Almond | 1.471–1.472 | Safflower | 1.473–1.476 |
| Peanut (Africa) | 1.470–1.472 | Cotton | 1.471–1.473 |
| Peanut (South America) | 1.472–1.474 | Corn | 1.472–1.476 |
| High oleic safflower | 1.468–1.471 | Grape seeds | 1.473–1.476 |
| Erucic rapeseed | 1.472–1.473 | Walnut | 1.475–1.476 |
| Low-erucic rapeseed (canola) | 1.472–1.473 | Sesame | 1.472–1.476 |
| Mustard | 1.473–1.474 | Soya | 1.473–1.477 |
| Olive | 1.468–1.471 | Sunflower | 1.474–1.476 |
| *Industrial vegetable oils* | | *Animal oils* | |
| China wood | 1.518–1.525 | Whale | 1.471–1.475 |
| Linseed | 1.479–1.481 | Herring | 1.471–1.476 |
| Bud | 1.475–1.476 | Menhaden | 1.472–1.480 |
| Oiticica | 1.511–1.517 | Egg | 1.471–1.472 |
| Ongokea | 1.510–1.515 | Neat's foot | 1.470–1.471 |
| Castor | 1.477–1.482 | Sardine | 1.476–1.485 |
| Dehydrated castor | 1.482–1.490 | | |

The refractive index of animal fats at 40°C (104°F) lies between 1.457 and 1.463, that of lauric fats (coconut, palm kernel oils) between 1.448 and 1.452 and that of palm oil between 1.449 and 1.458.

### Colour

The colour of fat can be determined by a trichromatic process. However, this method is so complex that it is practically never used. The same is true of the IUPAC spectrometric method which is, however, easy to implement with recording spectrophotometers. It consists in determining the absorption spectrum of the fat between 400 and 700 nm, on the fat as is or, if absorption is too high, on the fat diluted in carbon tetrachloride.

Methods mostly used in industrial and commercial sectors are:

1. The *Lovibond-AOCS* or BSI method which can be applied to all fats.
2. The *FAC* method essentially used for animal fats.
3. The *Gardner* method more especially used in the paint and varnish industry.

These three methods compare, visually in theory, the colour of the sample studied (cleared, if necessary, by melting at a temperature exceeding by 10°C its melting point and by filtration) to conventional standards.

An approximate correspondence is found between these three methods (Fig. 16.11).

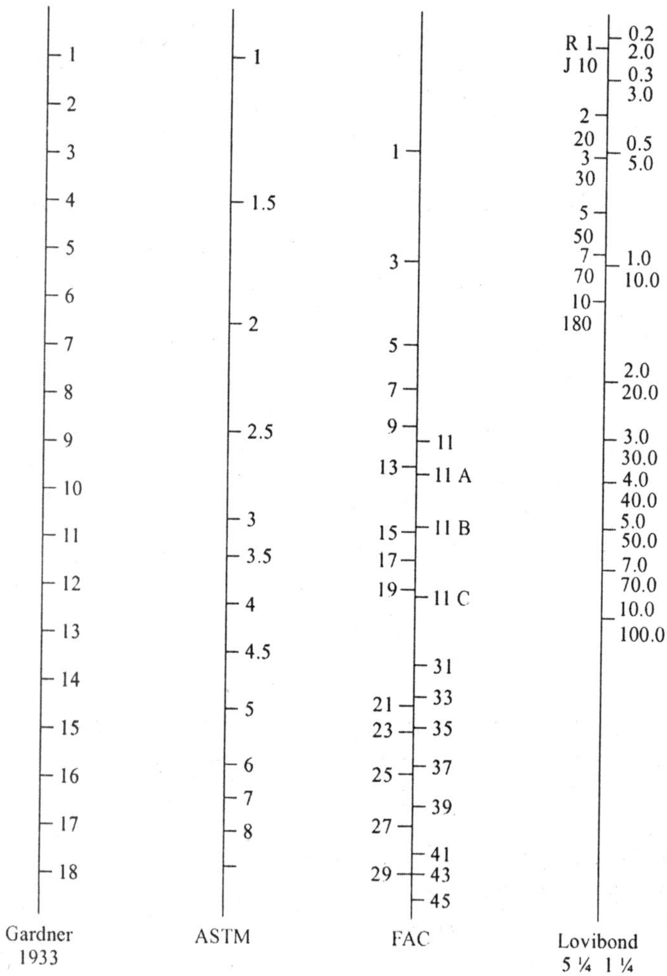


**Fig. 16.11.** Equivalence between different empirical methods of colour determination.

## *Lovibond method*

Lovibond method consists in comparing the colour of the sample with a set of yellow and red coloured glasses whose superimposition yields a colour identical to the colour of the sample. Each glass carries its number of 'Lovibond units' which add up. For instance, n°5R glass gives the same reading as superimposition of glasses 3R + 2R.

The values ratio of the yellow and red glasses is specified in the standard for each type of fat. This ratio should be strictly respected considering the conventional character of the measurement and the low sensitivity of the human eye to yellow light. In general, the ratio lies between 6 and 10 yellow units for each red unit.

### *FAC method*

FAC method consists in visually 'framing' the colour of the sample between two out of 29 standard glasses numbered oddly, by convention, from to 45, and distributed in four groups according to the prevailing colour:

| | |
|---|---|
| Slightly coloured | 1 to 9 |
| Yellow | 11 to 11C |
| Dark | 13 to 19 |
| Very dark (green prevailing) | 21 to 29 |
| Very dark (red prevailing) | 31 to 45 |

### *Gardner method*

Gardner method consists in visually framing the colour of the sample between two out of 18 standard glasses numbered from one to 18. *Gardner* method used tubes filled with standard solutions instead of coloured glasses.

Thus, repeatability and reproducibility of standard methods of colour determination are not specified. Nevertheless, it is obvious that if repeatability is hopefully correct, reproducibility, which includes the direct observation of the colour by two operators, can not eliminate the individual dimension and, therefore, can not be very precise. This is why Lovibond has adopted a spectrometer with interferential filters equipped with a calculator that transforms visible spectrum data lying between 400 and 700 nm into Lovibond units or into *FAC* and Gardner colours.

Thus, it allows to use the IUPAC-AFNOR method in a way that is conventional but still familiar to international trading. The colour of hardened fats (margarine, soaps) can be measured by reflectrometry using devices generally set with interferential filters.

## *Viscosity*

Fats viscosity depends on their chemical structure and temperature. An increase in molecular weight results in an increased viscosity. At a given chain length, the greater the saturation, the more viscous is the fat. The presence of secondary functions on the fatty chains increases significantly the viscosity. For all fats, viscosity decreases as temperature rises. However, (except for castor oil), the viscosity of most usual fats varies much less with temperature than that of mineral oils. Table 16.17 shows the viscosity of the major fats and oils.

These indications are only approximate for the viscosity of crude oils differs from that of the corresponding refined oils, as lecithins found in crude oils greatly increase the viscosity. There is not specific method for measuring the viscosity of oils and fats. General methods are applied, particularly for fluids, such as the methods by flow used for petroleum products or shearing methods used for oils and fats such as butters and margarines.

**Table 16.17.** Viscosity of fats (in centipoises).

| *Name of oil* | *Viscosity temperatures* | | |
|---|---|---|---|
| | *20°C (68°F)* | *50°C (122°F)* | *100°C (212°F)* |
| Almond | 68–76 | 22–24 | 7–8 |
| Castor | 815–1055 | 124–148 | 18–20 |
| Coconut | – | 17–20 | 5–6 |
| Corn | 55–66 | 20–24 | 6–7 |
| Cotton | 65–69 | 21–23 | 6–7 |
| Erucic rapeseed | 86–97 | 26–31 | 8–9 |
| Grape seeds | 53–58 | 16–18 | 6–7 |
| Linseed | 42–47 | 18–20 | 5–6 |
| Low erucic rapeseed (canola) | 72–82 | 23–28 | 8–9 |
| Mustard | 80–85 | 25–27 | 8–9 |
| Olive | 75–79 | 24–26 | 7–8 |
| Palm | – | 25–31 | 7–8 |
| Palm kernel | – | 17–20 | 5–6 |
| Peanut | 77–82 | 24–27 | 7–8 |
| Soya | 53–58 | 19–23 | 6–7 |
| Sunflower | 51–62 | 18–22 | 6–7 |
| Lard | – | 25–31 | 7–8 |
| Neat's foot | 81–84 | 25–27 | 7–8 |
| Tallow | – | 26–31 | 7–8 |

### *Smoke point—flash point—fire point*

The smoke point is the temperature at which a sample heated progressively and in well-defined conditions smokes continuously. In fact, this parameter whose determination is very imprecise is applied only to frying oils. The flash point is the temperature at which, in the presence of a flame, a clear inflammation of the vapours from the heated sample is produced. The fire point is the temperature at which, when continuously heating the sample beyond its flash point, the oil itself catches fire and burns for at least five seconds following the withdrawal of the flame. Fatty acids have a smoke point, a flash point and a fire point very much lower than that of corresponding glycerides.

As a result, the smoke point, the flash point and the fire point of an oil vary considerably according to its free acidity. The oxidative state of an oil or fat has only little effect on these characteristics. Table 16.18 indicate the range of these variations.

**Table 16.18.** Lowering of smoke, flash and fire points according to the acidity of the oil.

| *Acidity of oil* | *Smoke point* | *Flash point* | *Fire point* |
|---|---|---|---|
| 0.10% | –30°C | –10°C | 0°C |
| 1.0% | –85°C | –20°C | –50°C |
| 10% | –100°C | –85°C | –100°C |
| 100% | – | –120°C | –130°C |

Free fatty acids, but also all the inflammable components of an extraction oil having a high vapour pressure (hexane, in particular), can lower the flash point of an oil. Determination of the flash point is used to evaluate the content in residual hexane of an oil. For soya oil, commercial contracts require a minimum value of 250°F (121°C).

Devices of the 'Cleveland' type allow the determination of the smoke point, flash point and fire point, whereas those of the 'Pensky martens' type only give the flash point in retort.

Repeatability and reproducibility of the measure of flash and fire points given by ASTM are, for petroleum products:

| | Repeatability | Reproducibility |
|---|---|---|
| Flash point | 8°C | 16°C |
| Fire point | 6°C | 12°C |

These measures, when carried out on saponifiable fats, are less reproducible due to their elevated flash and fire point [about 200 and 350°C (392 and 662°F) in refined oils] and to the resulting difficulties in respect to the heating rate recommended in standards.

### *Thermophysical data*

There are no determination methods specific to fats for such data. We will focus on giving few values regarding various fats, taken from *Solomons*' work.

The specific heat (pertaining to mass) at a constant pressure is the quantity of heat needed to raise the temperature of 1 gram of fat by 1°C (Table 16.19).

**Table 16.19.** Specific heat (in cal/g/°C).

| *Name of oil* | *0* | *20* | *50* | *100* | *150* |
|---|---|---|---|---|---|
| Cotton | 0.395 | 0.415 | 0.445 | 0.495 | 0.545 |
| Corn | 0.390 | 0.414 | 0.432 | 0.469 | 0.507 |
| Grape seed | 0.376 | 0.394 | 0.411 | 0.447 | 0.486 |
| Soya | 0.420 | 0.433 | 0.455 | 0.488 | 0.523 |
| Sunflower | 0.403 | 0.424 | 0.455 | 0.505 | 0.555 |
| Lard | – | 0.31 | 0.48 | – | – |
| Tallow | – | 0.31 | 0.55 | – | – |

The thermal conductivity is the quantity of heat which passes during one time unit across a volume unit of fat whose temperatures on opposite faces differ by 1°C (Table 16.20).

**Table 16.20.** Thermal conductivity (in cal/g/°C).

| *Name of oil* | *0* | *20* | *50* | *100* | *150* |
|---|---|---|---|---|---|
| Cotton | 0.146 | 0.144 | 0.140 | 0.134 | 0.128 |
| Corn | 0.155 | 0.149 | 0.145 | 0.134 | 0.123 |
| Grape seed | 0.148 | 0.145 | 0.139 | 0.128 | 0.118 |
| Soya | 0.156 | 0.151 | 0.146 | 0.135 | 0.125 |
| Sunflower | 0.146 | 0.144 | 0.140 | 0.134 | 0.128 |
| Lard | – | 0.20 | – | – | – |
| Tallow | – | 0.30 | 0.16 | 0.14 | – |

The heat of fusion is defined as the number of calories absorbed by 1 gram of fat when passing from a hardened to a liquid state without any change in temperature (Table 16.21).

**Table 16.21.** Average melting heat (in cal/g).

| | | | |
|---|---|---|---|
| Peanut | 19.0 | Palm | 25.8 |
| Cocoa | 35.0 | Palm kernel | 32.0 |
| Rapeseed | 19.8 | Lard | 29.2 |
| Coconut | 30.6 | Soya | 17.9 |
| Cotton | 16.8 | Tallow | 27.3 |
| Linseed | 18.2 | Sunflower | 19.0 |
| Olive | 20.3 | | |

The calorific power of most fats is close to 9.4 calories per gram. Only fats rich in short chain fatty acids or in fatty acids having secondary functions have a significantly lower calorific power. For example, the calorific power of castor oil is of 8.9 cal/gram.

### *Melting point and solidification*

If fatty acids and pure glycerides have true melting points, oils and fats do not have any defined melting point, but a more or less wide melting interval. Considering the polymorphism of glycerides, the phenomenon of reversible solubilising, the phenomenon of surfusion, and of eutectic constitution, the melting interval depends on the conditions under which the fat was solidified (maturation) before determining its melting temperature.

The study of fats behaviour in heat and in cold is therefore extremely difficult, which explains the multiplicity of methods proposed for its study.

#### *Melting point*

Although the 'melting point' is a characteristic appearing in many commercial contracts, it is very difficult to define a reproducible method to measure it. All the proposed methods use capillary tubes. To reach an acceptable reproducibility, the operators should use tubes that are as similar as possible. AOCS recommends using tubes from one single supplier; IUPAC proposes to select capillary tubes by verifying the consistency of their diameter with a calibrated metal gauge.

AOCS standard plans only one type of fat conditioning and observes its complete melting in closed capillary. IUPAC standard tables on two modes of fat maturation depending on whether the studied fat is known to have a marked polymorphism. Operating in open capillary tubes immersed in water, the melting point is the temperature at which the fat slides in the tube. IUPAC standard gives indications on the repeatability and reproducibility of the method. AOCS standard for the determination of the melting point in open tube (softening point) is close to IUPAC standard and considers a particular maturation for palm oil.

#### *Slip point*

Contrarily to the methods for determining the melting point which requests a rigorous maturation of the sample, the slip point is meant to study the behaviour of hardened fat (margarine type of fats or emulsions) as is, without melting nor prior maturation. This method was standardised by AOCS. Instead of capillary tubes, it introduces fat as is, in 1 cm-internal-diameter cylinders, immersed in salted water.

*Flow test*

This method standardised by AOCS measures in a very conventional way the temperature at which is stirred plastic-malleable fat can slide from one to the other end of a tube when simply reversed.

*Cloud test*

The cloud test is the temperature at which a cloud appears during progressive cooling of the sample heated at 130°C (266°F), a cloud caused by the start of crystallisation. The method is described by AOCS.

*Cold test*

The objective of this test is to verify that an oil 'resists' to cold, i.e. there is no crystallisation when maintained for 5.5 hours at 0°C (32°F). This test, standardised by AOCS, is applied to control the 'winterisation' steps.

*Titer*

This is the oldest method for studying the solidification of fats or more exactly the solidification of fatty acids which constitute them: during the cooling of fatty acids, at the end of a period of supercooling, crystallisation is accompanied by a more or less clear-cut rise in temperature. The 'titer' is the highest temperature then attained (also known as 'conditioned solidification point of fatty acids').

ISO-AFNOR and AOCS standards are practically identical and proceed to the cooling down of liquid fatty acids under continuous agitation. IUPAC standard saponifies fat with a potassium hydroxide solution in ethanol instead of glycerol and does not require continuous stirring, which can lead to different results. The repeatability of the method is of 0.2°C.

The titer is mostly used in the control of specifications for tallow and stearines.

*Solidification curve*

Instead of measuring the temperature, as in the preceding method, this method determines the cooling curve of a sample of melted fat in strictly defined conditions. Cooling is achieved without stirring, in a flask with a double jacket, under vacuum. Some fats, in particular cocoa butter and related fats, posses a characteristic cooling curve with a lowest temperature followed by a highest temperature. The temperature and time needed to attain the lowest and highest temperatures are relatively specific. The method was standardised by IUPAC.

*Differential heat analysis*

When a constant heat per unit of time is applied to a fat and when this fat does not show any modification in state, in the analysed temperature interval, its temperature is an increasing linear function of time. If, on the contrary, the fat studied melts or changes its crystalline state, its temperature is no longer a linear function of time.

Rather than proceeding to absolute measure of temperature, one measures the difference in temperature $\Delta T$ between the sample studied and a reference sample not showing thermal phenomena in the temperature interval studied. If one takes $\Delta T$ with reference to temperature of a standard sample, a characteristic thermograph of the fat studied is obtained. $\Delta T$ is positive when the observed thermal phenomenon is exothermic and negative when it is endothermic.

The thermograph allows a definition of the melting range (which is the temperature interval where thermal phenomena take place) and the temperatures at which each of the major thermal phenomena take place (Fig. 16.12). As the melting point, the melting range depends on the maturing conditions of the sample. The method allows the obtention of characteristic thermographs for each fat. Also, it shows that transformations still take place at temperatures below –30°C for very unsaturated oils, e.g. linseed oil. Nevertheless, attempts to standardised and quantify obtained results have not been sufficiently reproducible to be used in controlling commercial transactions.

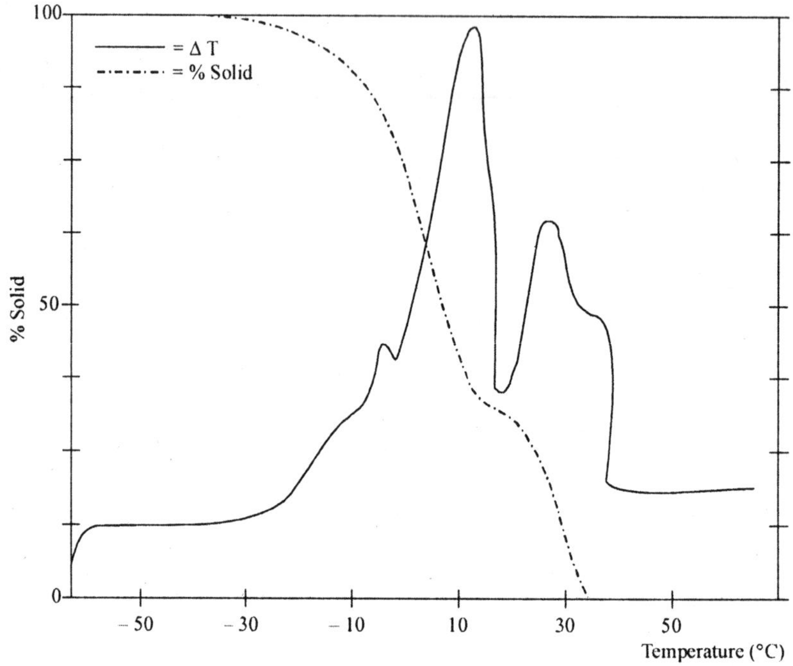


**Fig. 16.12.** Melting thermographs and solid content of butter fat.

The multiplicity of methods for studying the melting and solidification underlines the complexity of this study, but also the need for users to have differing information according to the use these methods are meant for.

For examples:

1. ATD, which shows all melting or solidification phenomena taking place in a complex fat such as certain hydrogenated fats, is a technique particularly well-adapted to control the texture of raw material, the identity of thermographs practically assuring the identity of the products.
2. The melting point in closed capillary tubes, which indicates the temperature at which all crystals disappear from a fat, can inform on the taste of this fat. If the melting point is greater than 37°C (98.6°F), the consumer will find it unpleasant.

Information provided by the determination of solid content according to temperature is obviously more interesting as they give, in addition, indications on fats plasticity and spreading capacity.

## ANALYSIS OF OILSEEDS AND OILS MEALS

### Analysis of Oilseeds

Oilseeds, like any other plant products, are subject to changes in their composition, which depend mainly on:

1. Mode of cultivation.

2. Geographic origin.
3. Varieties cultivated.
4. Climatic factors.

The market value of oilseeds depends on a number of parameters. The main analytical parameters are:

1. Water content.
2. Impurity content.
3. Oil content.
4. Acidity of oil.

To these four parameters, normally included in commercial contracts, can be added a number of other criteria given below to better judge the nutritional quality of different oilseeds. These additional criteria are:

1. Composition of fatty acids of oil.
2. Protein content.
3. Cellulose content (fibre).
4. Content of antinutritional components (erucic acid, glucosinolates, gossypol, aflatoxins)

Analytical parameters are determined in the laboratory, using measurement techniques of which the majority have been standardised both at the national as well as the international level (Table 16.22).

**Table 16.22.** Principal standardised methods for analysing oilseeds.

| *Determinations* | *AFNOR* | *ISO* | *IUPAC* | *FOSFA* | *AOCS* |
|---|---|---|---|---|---|
| Reduction for analysing a sample for laboratory | NF V 03–902 | ISO 664 | 1.001 | | |
| Water and volatile matter content | NF ISO 665<br>NF V 03–909 | ISO 665 | 1.121 | ISO 665 | Aa 3–38 Cotton<br>Ab 2–49 Peanut<br>Ac 2–41 Soyabean<br>Ai 2–75 Sunflower |
| Impurity content | NF ISO 658 | ISO 658 | 1.101 | ISO 658 | Aa 2–38 |
| Hexane extraction (oil content) | NF ISO 659 | ISO 659 | 1.122 | FOSFA official method | Aa 4–38 Cotton<br>Ab 3–49 Peanut<br>Ac 3–44 Soyabean<br>Ai 3–75 Sunflower |
| Oil content | | | | | |
| Low resolution nuclear magnetic resonance spectrometry | NF V 03–907 | ISO 5511 | 1.123 | FOSFA official method | – |
| Acidity of oil | NF V 03–906 | ISO 729 | 1.151 | BS 684 | Aa 6–38 Cotton<br>Ab 5–49 Peanut<br>Ac 5–41 Soyabean |
| Protein | NF V 18–100 | ISO 5983 | – | ISO 5983 | Aa 5–38 Cotton<br>Ab 4–50 Peanut<br>Ac 4–41 Soyabean<br>Ai 4–75 Sunflower |
| Fibre | NF V 03–040 | – | – | AOCS Ba 6–84 | Ba 6–84 |

NF = Norme Francaise (French standard).

### *Reduction of samples for laboratory*

This reduction should lead to a reduced sample called 'sample for analysis' or 'sample of test' on which various analytical tests are to be carried out in reality. This operation of reducing samples for the laboratory into samples for analysis has been normalised at the national level (NF V 03–902 norm) as well as the international level (ISO 664 norm). Both these standards, which are equivalent, define the minimum mass that the samples should attain depending on the seeds (Table 16.23). For practical reduction, it is recommended to use the two types of dividers available on the market, that operate by dichotomic division.

**Table 16.23.** Mass of samples for analysis (test) as per standard ISO 664.

| *Type of seeds* | *Mass (g)* |
|---|---|
| Copra | 1000 |
| Average and large sized seeds including: | 500 |
| Castor | |
| Palm kernel | |
| Peanut | |
| Shea nut | |
| Pumpkin | |
| Sunflower | |
| Soyabean | |
| Cotton | |
| Small sized seeds including: | 200 |
| Hemp | |
| Lin | |
| Rapeseed | |
| Rape | |
| Oil poppy | |
| White mustard | |
| Black mustard | |
| Sesame | |

#### *Conical divider*

Seeds are fed into the divider with the help of a feeding hopper. They fall on a metallic cone at the base of which are located a series of openings. Two collectors alternatively re-group the even and odd openings and thus divide the sample in two identical halves, practically identical. By re-introducing in each division one of the two fractions in the divider, we get a representative reduced sample. The number of times division which is to be carried out depends on the importance of the initial sample and the size of the sample we want to arrive at for the purpose of analysis. Conical dividers which exist in various sizes can be used for treating large samples (many kilograms) and are particularly well adapted for dividing small and average sized seeds (rapeseed, lin, sesame, soyabean) (Fig. 16.13).

#### *Slit dividers*

This equipment which operates on the same principle as the conical divider by dichotomic division is made of a series of parallel grooves (rifles) which alternatively direct the seeds to the right and to the left in two parallelepipedic recipients.

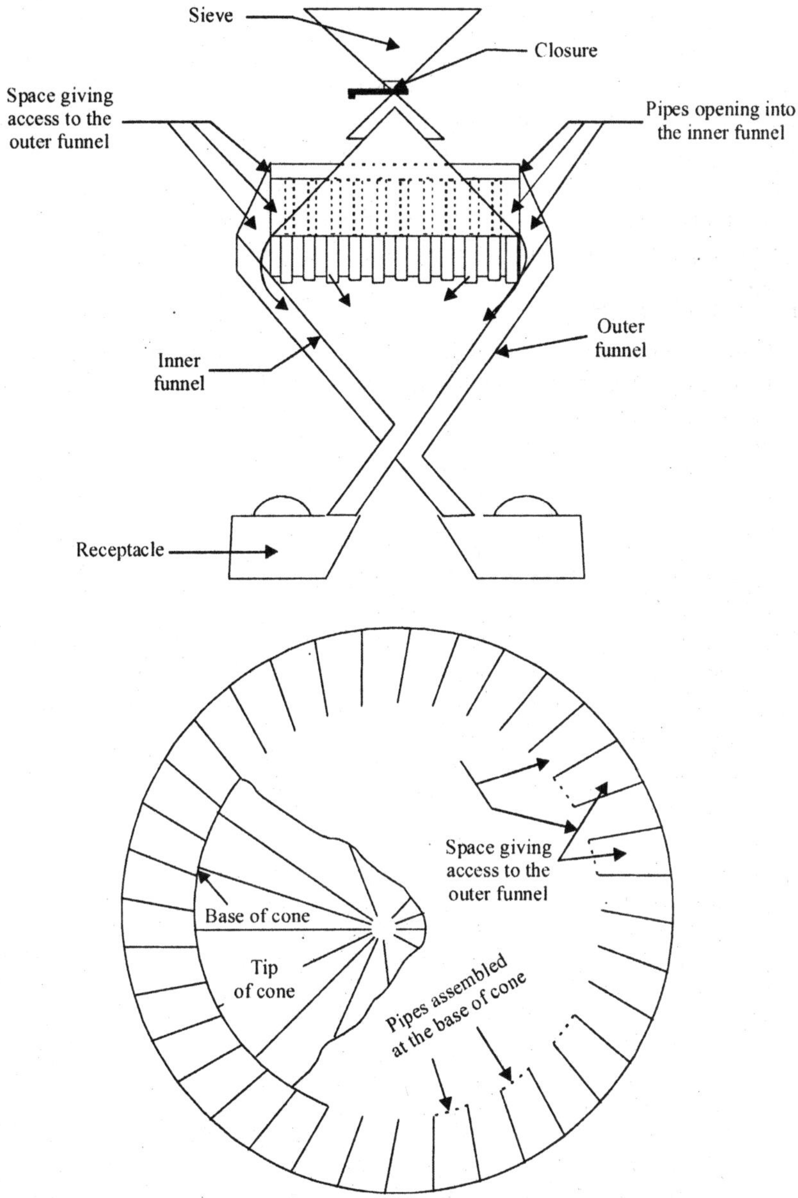


**Fig. 16.13.** Conical divider.

The main advantage of this divider is that it is entirely open and any obstruction of one or more rifles can be immediately detected, which facilitates cleaning. As far as inconveniences are concerned, it must be noted that slit dividers presently available on the market, because of their size, can only treat up to a maximum of 1 kg. To obtain with the help of these dividers (conical or slit), a reduced and most representative sample possible, it is necessary to take certain precautions for their use, which are:

1. Place the divider on a perfectly plain surface.

2. Check that an opening or slit is not blocked before use.
3. Before division, homogenise the initial sample by passing it through the divider several times.

In case of samples made up of large size seeds (karite, palm kernel) or other seeds that do not flow easily (cotton), use of dividers, whether conical or rifle type is impossible. In that case, concise technique of division by quarters has to be adopted. It consists of spreading all the seeds of the sample on a plane surface and dividing them in four quarters. Three are eliminated. The fourth quarter is again divided in the same manner until a reduced sample of the desired size is obtained.

In order to divide in quarters, we may use the cross.

### *Determination of water content*

Determination of water content is a difficult method in spite of its seemingly simple procedure.

The different methods proposed for determination of water content in oilseeds has a conventional character and can be divided into two following categories:

1. Reference method (standardised).
2. Quick method.

#### *Reference methods*

In these methods, water as well as volatile matter content in oilseeds correspond to a loss in mass of oilseeds by desiccation in an oven under given conditions. NF ISO 665 norm, which is presently recognised as a reference method, both nationally as well as internationally, suggests drying in an isothermal oven at a temperature of 103 ± 2°C.

Determination is made either with:

1. Whole seeds in case of small seeds (rapeseed, lin), and sunflower, safflower, and cotton seeds.
2. Crushed seeds in case of large and average sized seeds.

In practice, determination is made on a sample of 5 mg of grounded seed, or on 5 to 10 mg of whole seeds, placed in a metal or glass receptacle fitted with a cover. The vase containing the seeds is put in the oven for three hours. Subsequently, after cooling in a drier, they are weighed. Seeds are then put back in the oven every hour until a constant weight is achieved. (The difference between two weights should be less than 5 mg).

Water content (H per cent) is obtained by the formula:

$$\text{H per cent} = \frac{\text{Loss of mass by desiccation}}{\text{Weight of the sample}} \times 100$$

The repeatability of this method is 0.2 per cent in absolute terms. The main inconvenience of this reference method is that it is relatively long because of repeated weight taking operations. Also, in some cases (routine analysis) it can be advantageously replaced by a method which is more practical and which, at the national level, has been standardised. (Standard NF V 0.3–909 norm).

In this practical method, drying is done in one oven only over a total duration of 15 hrs. (one night). Results obtained are almost identical to those obtained by the reference method. Determination of water content by desiccation requires that certain precautions regarding liability of water are to be taken. One must always remember that water content in an oilseed sample is likely to change with time due to surrounding atmospheric changes. A humid sample will have a tendency to dry whereas a dry sample will get hydrated until the water content levels with the hydride characteristics of air.

Taking into account these observations, the following conditions should be respected:

1. Determining humidity only on a sample which has been conditioned in a perfectly air tight container.

2. Preserve the sample at the temperature ambient in the laboratory. Avoid it to be exposed to all heating source (sun, radiator).
3. As far as possible, leave the sample for 24 hours in the containers, before transfer the sample in an air tight receptacle to corresponding volume (filled upto 2/3).
4. Before taking the test sample, homogenise the sample by turning the recipient upside down.
5. In case crushed seeds are used, use a crushing device which restricts heating to the maximum extent possible (refrigerated crusher).
6. While adding water (crushing, weighing) work as fast as possible.
7. Use an oven having good heat distribution (homogeneous temperature inside) and a good thermal inertia.

Multicellular ovens 'chopin' perfectly assures these conditions and are therefore specifically recommended for this determination.

*Quick methods*

In certain cases, especially at the time of collecting seeds by stock agencies (SA), it is not possible to use the method of slow drying because of their reaction time. In this case, we normally use humidimeters more or less automated which operate by the dielectric capacity and which gives result in a few seconds.

In these apparatus, a given volume of seeds is placed between the plates of a condenser. The dielectric capacity (Cs) or of the system is linked to the dielectric coefficient (εs) of the seed, to the surface (S) and to the distance.(d) of the condenser plates as per formula:

$$Cs = \varepsilon S\ (S/d)$$

The electric coefficient (εs) of seeds being dependent on their humidity and the nature of seeds, it is thus possible to determine the humidity of unknown samples by calibrating the humidimeter with seeds whose humidity is known.

Use of humidimeters is regulated in France. These devices should comply with certain technical specifications and should respect certain rules as far as their use is concerned.

To be recognised, a humidimeter, in the first place, should give results with a tolerance of ± 8 per cent with respect to the reference value obtained by slow, drying and a minimum of ± 0.4 per cent tolerance in case of values less than 5 per cent.

Thus, for a humidity value of 10 per cent determined by the reference method, the error tolerated with humidimeter should be:

$$\frac{10 \times 8}{10} = \pm 0.8 \text{ per cent}$$

Any value between 9.2 and 10.8 per cent will therefore be considered correct; which also shows the limitations of these devices and the necessity to test them regularly.

To be recognised, humidimeter should also undergo an annual checking by an approved body by following the procedure defined in NF V 03–915 norm.

Besides humidimeters operating by measuring the dielectric capacity, there is a series of devices operating on a different principle which also give results in a very short time and which are:

1. Thermidimeters which quickly dry by desiccation at high temperature in a high capacity drier (many hundreds of samples).
2. Infrared lamps which dry the samples until constant weight is obtained. This type of device usually has an incorporated weighing scale.
3. Devices operating by reflection spectrometry in the nearest infrared.

### *Determination of impurity content*

Besides their reducing oil content, impurities present in a lot of oilseeds are likely to:

1. Change the nature of oil.
2. Change the nature of oil meal.
3. Bring about a faster deterioration of the pressing material.

NF ISO 658 norm defines impurities in oilseeds as any foreign, organic and inorganic bodies other than seeds of the base species. These impurities can be sub-divided into three distinct categories:

1. Dusts which pass through a sieve having round holes of a diameter depending on the type of seeds to be analysed (Table 16.24).

**Table 16.24.** Diameter of openings of sieve as per NF ISO 658 norm.

| *Nature of product* | *Diameter of holes mm* |
|---|---|
| Copra | 2.0 |
| Small seeds (rapeseed, oil poppy, lin) | 0.5 |
| All other seeds (sunflower, soyabean) | 1.0 |

These dusts can themselves be classified as:

Sterile dusts ⟶ Sand-earth

Fatty dusts ⟶ Debris of seeds (flour)

It may be noted that in case of peanut flours are not considered as impurities.

2. Impurities other than oilseeds which correspond to foreign bodies in the seeds under analysis (stones, vegetable remains, metal pieces, seeds of non-oleaginous plants) and all inherent elements of the oilseeds under examination (fragments of branches and leaves, remains of hull). It may be noted that in case of seeds sold in husks, especially for sunflower, loosing husk is considered as impurity only if its proportion is more than the corresponding kernels in the sample, as a result of adding husk intentionally.
3. Oil seed impurities are oilseeds other than the species under analysis. They especially include wild mustard seeds (*Sinapis arvensis*) present in the rapeseeds (*Brassica napus*).

The principle of measuring consists of isolating each of the three categories of impurities by making the sample go through a succession of sieving operations and manual sorting (spring tweezers) as per one of the three diagrams shown in Figs. 16.14, 16.15 and 16.16 and the type (size) of oilseeds under analysis.

Determining impurities whether they are of oilseeds or not is done as per standards only by visual observation without any help of an enlarging device (magnifying glass).

Sometimes, this can lead to difficulties in identifying certain impurities. One of the most difficult cases is the identification of wild mustard (*Sinapis arvensis*) in a sample of rapeseed (*Brassica napus*). Macroscopic differences existing between these two types of seeds being very little, a magnifying lens can be used. In case of doubt, we may use the fluorescence test described in the EEC regulation (regulation no. 3025/75 of 17/11/1975 which appeared at JOCE on 20.11.1975) to differentiate these two types of crucibles.

In this test, doubtful seeds are deposited in a Petri box on a filter paper impregnated with a 5 per cent aqueous solution of KOH at 5 per cent. In these conditions wild mustard seeds emit, under ultraviolet rays (365 nm), a fluorescent halo ranging from yellow to geenish blue whereas rapeseeds, do not emit any fluorescent light.

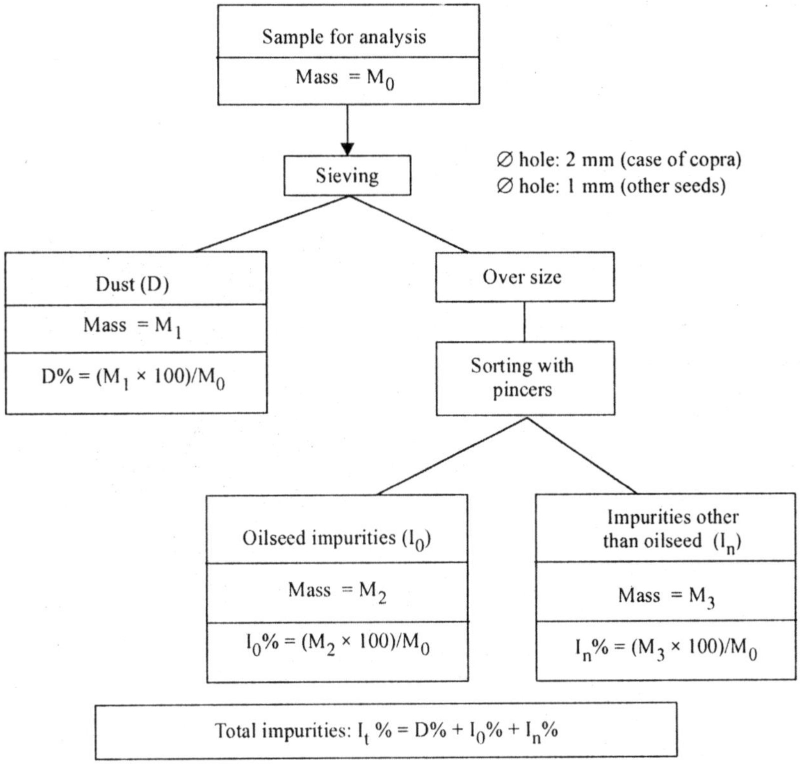


**Fig. 16.14.** Determination of impurities in oilseeds.

The result of total impurities, which are expressed in percentage of mass of seeds, is equal to the sum of impurities of all three categories (dust + non-oleaginous impurities + oleaginous impurities).

The result being the arithmetic mean of two readings taken simultaneously on the same sample, the difference between these two values will not be more than the limits given in Table 16.25. In the opposite case, two new readings will have to be taken.

**Table 16.25.** Repeatability of determination of impurities as per NF ISO 658.

| *Impurity content* % | *Maximum difference admissible* % |
|---|---|
| Upto 0.5 included | 0.2 |
| From 0.5 excluded to 1.0 included | 0.4 |
| From 1.0 excluded to 2.0 included | 0.6 |
| From 2.0 excluded to 3.0 included | 0.8 |
| From 3.0 excluded to 4.0 included | 1.0 |
| From 4.0 excluded to 5.0 included | 1.2 |
| From 5.0 excluded to 6.0 included | 1.4 |
| Above 6 | 1.6 |

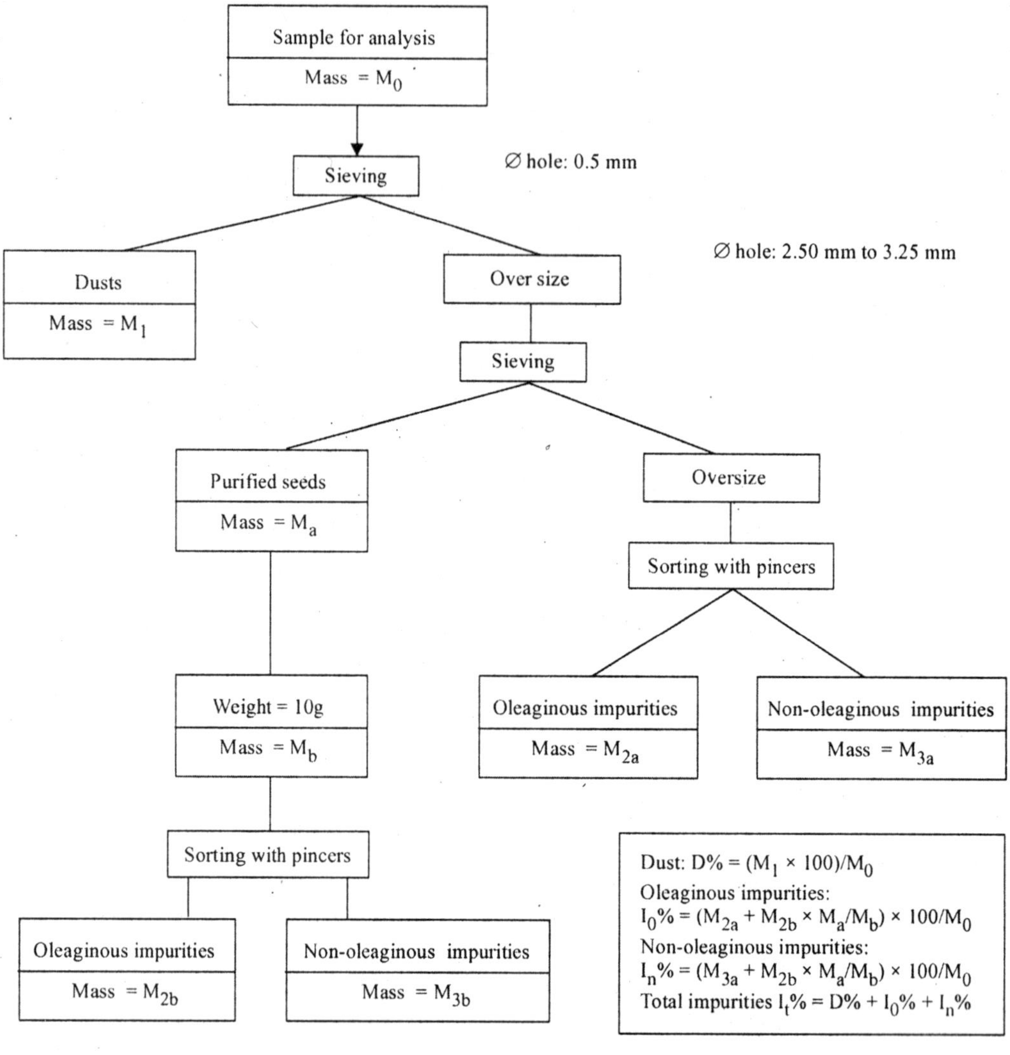


**Fig. 16.15.** Determination of impurities of small oilseeds.

This standardised method, the only one recognised for commercial transactions of oilseeds is a long and tedious operation. For certain unofficial analysis (receiving and experimenting in culture) where it is not necessary to precisely know the impurity contents, one can advantageously use a laboratory type cleaner-separator (Westruff/NSL 1000/Labofix) which along with combining the sieving and densimetric separation gives approximate result in a few minutes.

### *Determination of oil content*

Oil content of oilseeds is the essential quality factor and the one which has most repercussion on the market value of oilseeds. Most methods that have been standardised nationally as well as internationally have, by convention, defined oil as being all the components extracted from the seed by a given solvent (hexane or petrol ether) and under operating conditions specified by each standard. NF ISO 659 norm, which in most national and international transactions is recognised as the official method, operates on a sample of 10 gram of crushed matter which is subjected to a series of extraction and crushing operations until the entire seed is fully used up. This is done as per diagram reproduced in Fig. 16.17.

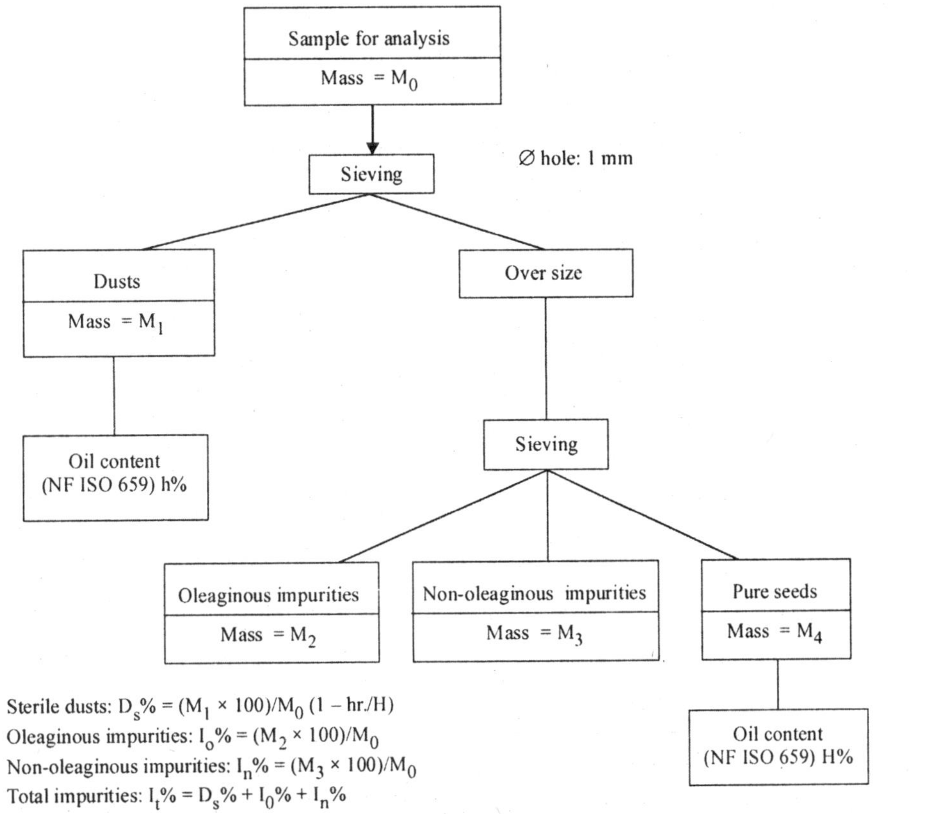


**Fig. 16.16.** Determination of impurities in peanut seeds.

This standard, however, precise, gives a certain latitude to the operator on three important points:

*Extraction apparatus*

The standard does not define the type of extractor to be used. In practice two apparatus are generally used to determine oil content in oilseeds, Soxhlet and Bolton-Williams (Fig. 16.18). The difference between these two extractors is that Soxhlet besides solvent extraction vapours, also allows a delipidation of the product by maceration, results obtained by any of the two apparatus being however, similar. It should be noted that Bolton-Williams scores over Soxhlet as far as robustness is concerned.

*Extraction solvents*

For the extraction solvent the norm recommends use of *n*-hexane technique but they also tolerate use of petrol ether. The main inconvenience of petrol ether is that it is not properly defined chemically and because of this, it gives results that are less reproducible than in cases where hexane is used as extraction solvent. The differences in result seen between the use of these two solvents, however, remain restricted. **Petrol** ether, on an average gives results less than 0.05 to 0.20 per cent in absolute terms with respect to **hexane.**

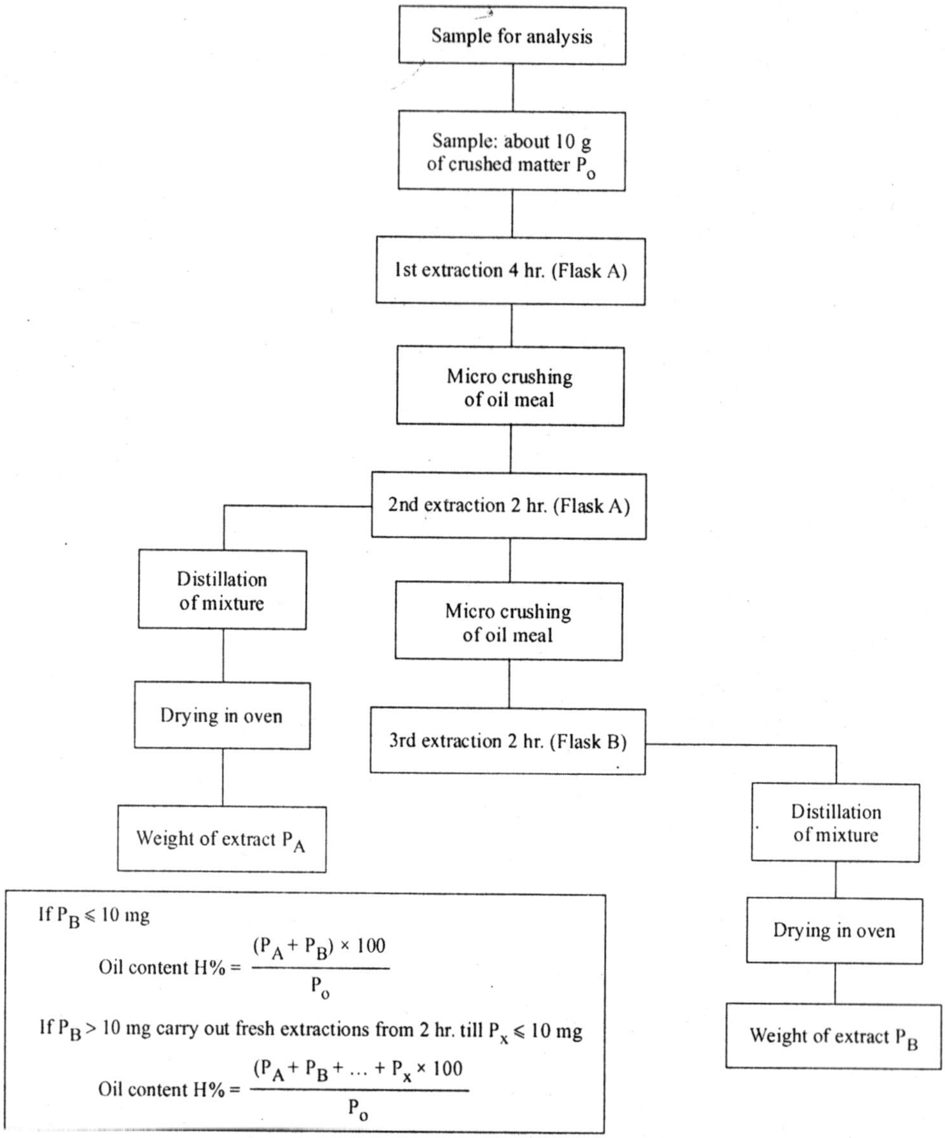


**Fig. 16.17.** Determination of oil content as per NF ISO 659 norm.

### *Crushing conditions*

For different crushing operations, the standard does not precisely indicate the type of crusher to be used, it only indicates the fineness of the crushed substance to be obtained keeping in mind the type of seed. The first crushing which is done on entire seeds before extraction operation can be done with different crushers, mill or grater depending on the hardness of seeds to be treated. In many cases, use of a simple coffee mill enables to correctly carry out this operation. The only precautions to be taken regarding the choice of the crusher are that it should cause minimum heating and pressure on the seeds so as to restrict loss of water by drying and loss of oil by exudation.

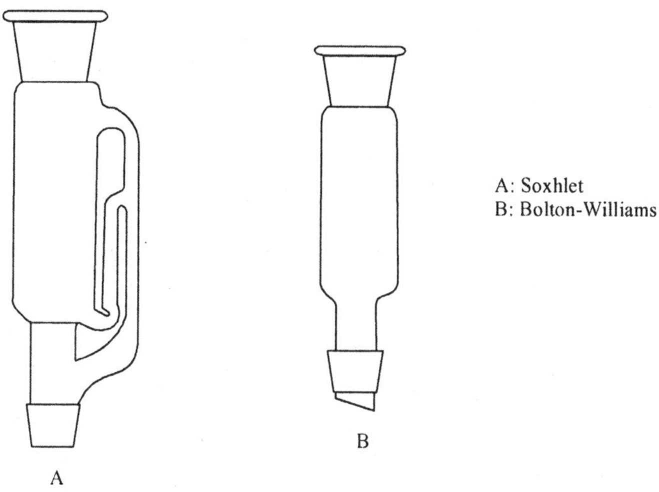


**Fig. 16.18.** Extractors.

Other operations of crushing, carried out after the first extraction have more incidence on the result of quantity determination because these operations break the walls of fat cells thereby facilitating diffusion of solvent in the product and ensure complete delipidation.

NF ISO 659 norm recommends for these crushing operations, the use of a mechanical mirco-crusher, capable of producing a paste of the seeds, with a thickness less than 160 mm. The type of apparatus which gives the best fineness is the steel ball type micro-crusher, 'Dangoumeau'. Its use in the laboratory has been standardised.

In case, such a micro-crusher is not available, the norms provides for replacement of mechanical crushing by manual crushing with the help of a pestle in a mortar along with sand. It is to be noted that if this crushing technique can give satisfactory results for isolated measurement, it cannot be applied only in case of analysis in series, as often, the fatigue of the operator hinders in getting an efficient crushing.

Besides NF ISO 659 norm, there are several other methods to determine oil content by extraction using solvent which also have been standardised or recognised nationally and internationally.

### *FOSFA method*

The protocol of the FOSFA method is the same as that of NF ISO 659. The number of extractions has been precisely fixed at 3, which range over a total duration of 8 hours ( 4 + 2 + 2 hr.). The rate of oil extracted during the third crushing, in this case, doesn't need to be checked whether it is less than or more than 0.1 per cent like the NF-ISO norm, thus saving enough on time. Comparative studies done with both these methods (NF/ISO and FOSFA) have not shown significant differences in the results obtained.

### *AOCS method*

In America, AOCS has not standardised a single protocol for determining the oil content of oleaginous fats. They have standardised an individual method for each type of seeds as given below:

| *Oilseeds* | *AOCS standards* |
|---|---|
| Soyabean | Ac 3–44 |
| Sunflower | Ai 3–75 |

*(Contd...)*

| *Oilseeds* | *AOCS standards* |
|---|---|
| Peanut | Ab 3–49 |
| Cotton | Aa 4–38 |
| Corn (germ of) | Aj 4–89 |

All the AOCS methods for determining oil content differ significantly in their principle with respect to NF-ISO and FOSFA methods.

The main differences seen in the operating protocols of American Standards, with respect to the ones used in the European community are:

1. Single use of petrol ether as extraction solvent.
2. Single sampling of 2 to 5 gram depending on the type of seeds.
3. Single extraction without micro-crusher during 4 to 6 hours depending on the type of seed to be analysed, except in case of AOCS Ab 3–49 norm dealing with peanuts (2 hr. + 2 hr.).

Comparative tests carried out between AOCS and ISO methods have shown that the American methods usually give lesser oil rates than the oil rates obtained by methods used in Europe, the differences being from 0.2 to more than 1 per cent in absolute terms, depending on the type of seeds analysed.

The differences in the result are explained mainly by the fact that the majority of AOCS Standards suggest single extraction without use of micro-crusher.

Reference methods (ISO-FOSFA-AOCS) that we have just described are long to use (8 to 10 hrs.) because of repeated crushings and extractions involved. It is often necessary to get result in a short span of time, hence use of methods called 'quick'.

### *Quick methods*

AFNOR has recently standardised a quick oil extraction method using hexane (NF V 03–908 norm) which brings about that is approximately identical to that obtained by the Reference method (NF, ISO 659) and that too in a time limit of 3–4 hrs. In this method, a sample of 10 gram of seeds is introduced in the bowl of the Dangoumeau crusher with 30 to 40 ml of hexane and a steel ball having a 20 mm diameter. After 5 minutes of crushing, all the crushed material and the miscella is transferred into a Soxhlet type extractor to complete delipidation under hot conditions (extraction time 2 hrs. 30 minutes).

If this method reduces considerably the extraction time with respect to reference method, it forces to take precaution to avoid loss of oils especially during crushing in solvent (bowl should be air-tight) and while transferring the crushed material and miscella, from the bowl of the crusher into the extractor.

In the field of methods called 'quick' for determining oil by extraction, it may be mentioned that semi-automatic extracting machines available on the European market are of two types:

1. Soxtec of Tecator.
2. Automatic Soxhlet of Buchi.

In these machines, which simultaneously allow treatment of 2 to 6 samples with the least handling, extraction is generally done in two phases:

1. The crushed seed placed in an extraction thimble is first macerated directly in boiling solvent which produces quick delipidation of the product in a short span of time (about 30 minutes).
2. In the second phase, thimble tube is placed on top of solvent vapour so that last traces of oil can be washed away.

If this method gives a result in less than two hours, trials that have been carried out show that this method quite often gives wrong results (0.2 to 0.8 per cent) with respect to those obtained by the reference method

(NF-ISO 659). The efficiency of this method, to a large extent, depends on the crushing of the seed before extraction is done. If both the techniques that we have just described significantly reduce the analysis time with respect to reference methods, they still require 1 to 3 hours to obtain a result.

Besides, there are other techniques to determine oil content in oilseeds. These techniques give results in a few minutes but operate on completely different techniques of extraction by hot solvent.

### *Measurement by densimetry*

This technique consists in crushing a given weight of seeds in the presence of a given volume of solvent having a known high density.

After a fixed crushing time, the density of the miscella (mixture of oil/solvent) recuperated by filtration is determined with the help of an appropriate densitometer.

The decrease in the density of the miscella, being proportional to the quantity of oil present in the mixture, is possible, by taking help of already prepared table for each oilseed, in order to determine oil content in seeds under analysis.

There are two principal devices in the market:

1. Oleometer D10 commercialised by 'Normandie Labo' which uses orthodichlorobenzene as solvent (density = 1.306 at 20°C).
2. 'Fosslet', commercialised by 'Foss Electric' which uses perchlorethylene as solvent (density = 1.620 at 20°C).

This technique, involving a densitometer, requiring only little investment is mainly used in laboratories of stockists (cooperatives—agricultural negotiators). It gives result in 10 to 15 minutes with a good reliability with respect to reference method. It is however, prudent that users of this technique regularly check whether their equipment is in good working condition by analysing samples of seeds with standard oil content. As far as inconveniences of the densimetry technique are concerned, it may be stated that use of solvents is relatively toxic for the operator. Hence, the recommendation of operating under an aspiration unit.

### *Measurement by nuclear magnetic resonance (NMR) spectrometry at low resolution*

This method by NMR is based on contents of hydrogen atoms of liquid glycerides which are the main components of oils extracted from oilseeds. Two types of low resolution devices can be used for this:

#### Continuous wave device

This apparatus does not differentiate hydrogen atoms of glycerides from those of water, which implies that seeds should be dried before analysis. This technique has been standardised both by AFNOR (NF V 03–907) and by ISO (5511). Texts of both these standards which are identical give the testing conditions of the device and the precautions to be taken regarding readings.

#### Pulsated wave device

This apparatus helps in differentiating between hydrogen atoms constituting glycerides from those of water. This enables to get the measurements of water and oil without resorting to drying.

However, it may be noted that simultaneous measurement of these two parameters (water and oil) is possible only in a specific humidity range of seeds to be analysed:

| | |
|---|---|
| Rapeseed-sunflower | maximum 10 per cent |
| Soyabean | maximum 14 per cent |

This method of simultaneously measuring the contents of oil and water by using pulsated NMR spectrometry is presently undergoing standardisation by ISO.

The principal advantages of these methods of determining oil by low resolution NMR are rapidity of the measuring (1 to 3 minutes). This enables laboratories equipped with this device to deal with a large number of samples with good reliability with respect to reference method (repeatability: 0.6 per cent).

Another advantage of this analytic technique is that it operates directly on seeds in a non-destructive manner without using any solvent or a chemical reagent.

The only factor limiting the development of this method in laboratories is the high cost of the equipment.

### *Measurement by reflection spectrometry in the near infra-red (IR)*

This recent technique of analysis by spectrometry in the near IR is based on the fact that main components of organic matter absorb radiations in the field closer to infra-red which goes from 800 to 2500 nm.

These absorption bands are mainly due to C—H, O—H, N—H and C—O chemical bonds that we find in all organic compounds and particularly in fats. Absorption, as per the Beer-Lambert law, is proportional to the concentration. It is theoretically possible, after identifying characteristic bands and calibration, to determine the contents of different organic components. If this technique of analysis is attractive because of its large polyvalence (simultaneous determination of various analytical parameters of various products) and its rapidity, it is today facing a number of obstacles which are restricting its development in laboratory analysis. The main difficulty is calibration on which reliability of results would depend. Calibration will integrate all the variability parameters of analysed products (content, origin, date) so as to determine for each product and for each parameter reliable calibrating factors which would have to be regularly checked. Among other difficulties of this technique, preparation of samples also needs to be mentioned as it will strongly influence the results. It is to be noted that a number of working groups formed by organisations like AFNOR and BIPEA (Inter Professional Bureau of Analytical Studies) to study these methods determining oil content by reflection in the closer infra-red, with the aim of preparing precise operating protocol for preparing samples and calibrating equipments. In the more precise domain of determining oil content of oilseeds, only FOSFA has till now, standardised a method by infra-red reflectance for only analysing soyabean seeds.

### ***Determining acidity of oil***

Determining acidity of oil extracted from oilseeds is also very important as it gives the degree of adulteration (hydrolysis of glycerides) of fats and is used to evaluate the losses to be achieved during refining (neutralisation) of crude oil obtained from these oilseeds. Most commercial contracts indicate a fixed limit for the acidity (2 per cent in case of sunflower). Beyond this value, cuts in the sale price are made, which depend on the table showing the acidity value in the lot of seeds examined. This calculation has now been standardised, both by AFNOR (NF V 03–906) as well as by ISO (ISO 729). In both these standards which are equivalents, acidity is calculated from oil that is obtained when extraction by hexane as per NF-ISO 659 norm is carried out. Once the extracted oil is mixed in a mixture of diethylic oxide/ethanol (1 + 1), free fatty acids present in the oil are determined by titration using an ethanolic solution of potassium hydroxide (KOH).

Results of acidity can be expressed in two different manners i.e. (i) by acid value (AV); and (ii) by acidity (FFA).

At the commercial level, generally it is the second method—expressing acidity results (FFA) which is used in contracts. Interlaboratory test results made at the international level have shown the mediocre accuracy of this method for determining acidity in oilseeds, the coefficient of change in the reproducibility of this method ranging from 18 to 31 per cent depending on the type of seeds analysed. Poor reliability of this method is not due to titrimetric measurement of free fatty acids in oil.

Reproducibility values obtained during direct determination of acidity on oils and fats are totally satisfactory.

Therefore causes of differences in the method for determining acidity in oilseeds should be found out in the oil extraction conditions of these seeds (sampling—extraction under heat-drying).

For this reason, presently, there is an effort at ISO to better define these oil extraction parameters. The main proposal for the amendment of the present standard are:

1. Operating on a larger sample.
2. Extracting either under heat for a fixed time, or under cold (crushing in solvent medium or percolation on columns).
3. Drying the oil after extraction at a temperature and for a well-defined period.

From these recommendations, two new methods have been arrived at extractions (under heat and under cold) and are presently being tried in laboratories at international level before being adopted as a standard of ISO as replacement of the present test. For certain oilseeds, the standard for determining acidity (NF or ISO) is not directly applicable. This is especially the case for measuring acidity in cotton seeds for which volumetric determination of free fatty acids becomes very difficult because of high colouration level which, in turn, is due to gossypol. In this case, it is necessary to resort to potentiometric determination of acidity as explained in the AOCS norm (Aa 6–38). Another special case is that of the determination of acidity of lauric seeds (palm kernel or copra) whose oil contain fatty acids of short chains, which, because of their volatile nature may get partly eliminated during drying of oil. In this case the solution is not calculated directly the acidity on dried oil but on the miscella (oil-solvent mixture).

#### *Determining the composition of fatty acids of oil*

In certain cases, determining the composition and content of certain fatty acids can be necessary to judge the nutritional quality or to define the variety of analysed seeds.

It is in this manner that we may determine the fatty acid content of certain oilseeds like:

1. Erucic acid in rapeseed.
2. Oleic acid in sunflower seeds called 'oleic'.
3. Gamma-linolenic acid in seeds of borage of onager or in seeds of black currant.

#### *Determining the content of proteinic and cellulosic matter*

These two determinations which are used to evaluate the nutritional quality of oil meals obtained from analysed seeds are generally carried out on seeds from which oil has been extracted as per standardised methods for animal foods.

### Analysis of Oleaginous Oil Meal

Oleaginous oil meal used as the main protein source in animal food are generally sold on the basis of its protein and fat content (Profat) to which three other analytical criteria can be added, they are:

1. Water content (humidity).
2. Cellulose content (fibre).
3. Mineral content (ash).

To these five analytic parameters which are generally considered in case of commercial contracts, certain other criteria can be added in order to be able to judge well the nutritional quality of any oil meal. They include the following:

1. Amino acid composition.
2. Protein solubility (digestibility).

3. Phosphorus and calcium content.
4. Starch and sugar content.
5. Microscopic analysis (purities).
6. Residual hexane content.
7. Antinutritional component content (urease, Aflatoxins, glucosinolates).

As a result of having determined all these analytic parameters, well defined modes of operation have to be employed to achieve standardisation. Most of these methods achieve standardisation on a national (AFNOR) or European (EEC) level rather than on an international (ISO) level; this is more valuable in a specific domain or oleaginous oil meal (oil-mill) than in the more general field pertaining to animal food (Table 16.26).

**Table 16.26.** Main standardised methods for analysing oleaginous oil meal.

| *Determinations* | *AFNOR* | *ISO* |
|---|---|---|
| Sample preparation for trial | NF V 03–926 | ISO 5502 |
| *Water and volatile matter content* | | |
| Oil meals from oilseeds | NF V 03–921 | ISO 771 |
| Animal food | NF V 18–109 | ISO 6496 |
| *Fat content* | | |
| Oil meals from oilseeds | NF V 03–924 | ISO 734 |
| | NF V 03-925 | ISO 736 |
| Animal food | NF V 18–104 | ISO 5986 |
| Protein content | NF V 18–100 | ISO 5983 |
| Cellulose content | NF V 03–040 | ISO 5498 |
| (Weende method) | | |
| *Ash content* | | |
| Oil meal from oilseeds | NF V 03–922 | ISO 749 |
| Animal food | NF V 18–101 | ISO 5984 |
| *Insoluble ash content in hydrochloric acid* | | |
| Oil meal from oilseeds | NF V 03–923 | ISO 735 |
| Animal food | NF V 18–102 | ISO 5985 |
| *Calcium content* | | |
| By titrimety | NF V 18–107 | ISO 6490/1 |
| By atomic absorption | NF V 18–108 | ISO 6490/2 |
| Phosphorus content | NF V 18–106 | ISO 6491 |
| Sugar content | – | – |
| Starch content | – | – |
| Available lysine content | NF V 18–103 | ISO 5510 |
| Residual hexane content | NF ISO 8892 | ISO 8892 |

### *Sample reduction of laboratory*

As in the case of oil seed analysis, sample reduction for laboratory constitutes a very important preliminary phase of true analysis. In the case of oil meals, this operation has brought in standardisation at national

(NF V 03–926) as well as international (ISO 5502) level. These two standards, which are equivalent, recommend the use of divisors among other things which operate by dichotomy (conical or rifle shaped) like the ones already described for oil seed sample division and reduction.

### *Water content (humidity)*

As in the case of oil seed analysis, two types of methods can be used at this juncture:

1. Reference methods (standardised).
2. Quick methods.

#### *Reference methods*

In these methods, water content of an oil meal is defined by convention as the weight loss it undergoes in conditions determined by dessication. In the case of NF V 03–921 and ISO 771 standards which are equivalent and which are applicable only to oleaginous oil meals, dessication is carried out on a trial basis on 5 grams in an isothermic oven with a regulated temperature of 103°C ± 2°C which goes on until constant weight (difference between the two weights which are lower than 5 mg) is obtained. In the case of NF V 18–109 and the EEC directive whose field of application covers all animal food products (raw materials and final products), the product is kept in an oven for 4 hours, fixed arbitrarily. Other stoving parameters are similar to those defined for oil meal analysis alone (sample trial = 5 mg/ temperature = 103°C). It is to be observed that the difference in results obtained by using different methods remains negligible (a maximum of 0.2 per cent in absolute value).

#### *Quick methods*

Use of humidimeter, thermidimeters, reflectance apparatus to the close infra-red such as the ones described for oil seed analysis can be recommended for finding out quickly the oil meal humidity.

### *Fat content*

Fat content of oil meals can be determined by reflux extraction in extractors (Soxhlet, Bolton-Williams) of the type described for determining oil content of oilseeds.

Various methods standardised for this determination differ basically in the choice of extraction solvent chosen (diethylic oxide—hexane). Since a long time, animal food production industry has recommended the use of diethylic oxide (ether) as extraction solvent as defined in the AFNOR standards (V 03–925 and V 18–104) and ISO (736 and 5986).

Today, we see an increase in the method using hexane as it was decided by the (NF V 03–924/ISO 734) standards and the second EEC directive (71/393 mod. 84/4).

This modification in the choice of the extraction solvent is on the one hand due to security reasons pertaining to work in the laboratory and on the other hand due to the fact that hexane extraction corresponds better to what we call fat fraction (lipid). Diethylic oxide has a disadvantage that, in addition to lipids, it extracted a certain quantity of non-fat substance which as a result led to an overdose of fat and calculation mistakes pertaining to energy value of food products. Moreover, the nature of extraction solvent of standardised methods meant for such a determination varies according to the time of extraction chosen. In norms relating to oil meal analysis (oil mills) extraction operation is the same as in the case of oilseeds. In the latter, extraction has many phases such as a first extraction for 4 hours, followed by a micro-grinding of the oil meal (Dangoumeau) and a new extraction for 2 hours. Methods relating to the animal food sector include a single 6 hour extraction without any intermediary grinding phase. For quickly calculating the fat content in oil meals, reflection spectrometry near IR is most commonly used in industry.

### *Protein content*

Protein content of an oil meal is calculated by the Kjeldahl method on the basis of nitrogen content. It has been the object of standardisation at AFNOR (NF V 18–100), ISO (5983) as well as EEC (3rd Directive 72/199). In this method organic nitrogen is transformed into ammoniacal nitrogen (ammonium sulphate) by mineralisation by concentrated sulphuric acid in the presence of a catalyst (mercury oxide, selenium, copper oxide or sulphate). Ammonical nitrogen is replaced by an alkali (sodium) and ammonium liberated is titrimetrically measured by a sulphuric acid solution. Protein rate is obtained by multiplying the total nitrogen content found out by the conventional factor 6.25. It is to be noted that laboratories obtain this measurement more often with the help of apparatus (autoanalyser) and the various stages of the Kjeldahl method (mineralisation, flow, titration) have become more or less automated. These apparatus help in increasing considerably the analysis capacity of a laboratory when compared to the application of the manual method prescribed in the standards. The repeatability of the Kjeldahl method is 1 per cent of the relative value. The reproducibility of this method as it is seen in the analysis chains of IBPAS is between 0.4 and 0.6 per cent of the absolute value. In the case of quick unofficial methods, it is better to talk about reflection spectrometry in infra-red and also about the Dumas method.

### *Cellulose content (fibre)*

The only method which has been standardised for the time being by AFNOR (V 03–040) and ISO (5498) is derived from the one proposed more than a century ago by the 'Weende' agronomic station in Germany.

In this method, cellulose matter corresponds to the insoluble organic residue obtained after acid ($H_2SO_4$ 0.26 N) and alkali (NaOH 0.31 N) reaction on finely ground oil meal. For fatty oil meals (expellers), it is better to subject it to delipidation by hexane as a preliminary step to these two reactions.

It is to be noted that the two hydrolysing agents used in the 'Weende' method dissolve hemicellulose strongly. They also dissolve a non-negligible portion of cellulose and lignin. Thus, the residue obtained by this method represents part of the vegetal wall. Hence the name insoluble cellulose index is given to this method. Keeping in mind the limits of the 'Weende' method, many researchers have proposed new analytic techniques which help in evaluating better the different components of the vegetal wall.

Right now, it is the method proposed by an American *Van Soest* which carries weight at the international level. This analytical technique whose principle is reproduced on Fig. 16.19 helps in separating successfully three residues. They are:

1. NDF (neutral detergent fibre) which constitutes the main part of the vegetal wall.
2. ADF (acid detergent fibre) which corresponds to the lignocellulose.
3. ADL (acid detergent lignin) which corresponds to lignin.

From the measures of these three residues, we can estimate the content in:

Hemicellulose = NDF – ADF

Cellulose = ADF – ADL

Due to the complexity and length of completion of the Van Soest method, it is less used in laboratories. 'Weende' method is the most commonly used method these days. It is to be noted that to achieve these measurement methods, laboratories use to a greater extent semi-automatic apparatus (Fibertec) which allow 2 to 6 samples to be treated simultaneously. These apparatus are used basically because of filtrate crucibles which help in finishing all operations such as reactions, filtration, calcination and weighing without any decantation. In turn this helps considerable gain of time and limits losses.

As for measuring precision in the 'Weende' method, AFNOR and ISO standards show a 0.4 per cent repeatability in the absolute value for cellulose level less than 10 per cent and 4 per cent relative value for cellulose levels higher than 10 per cent.

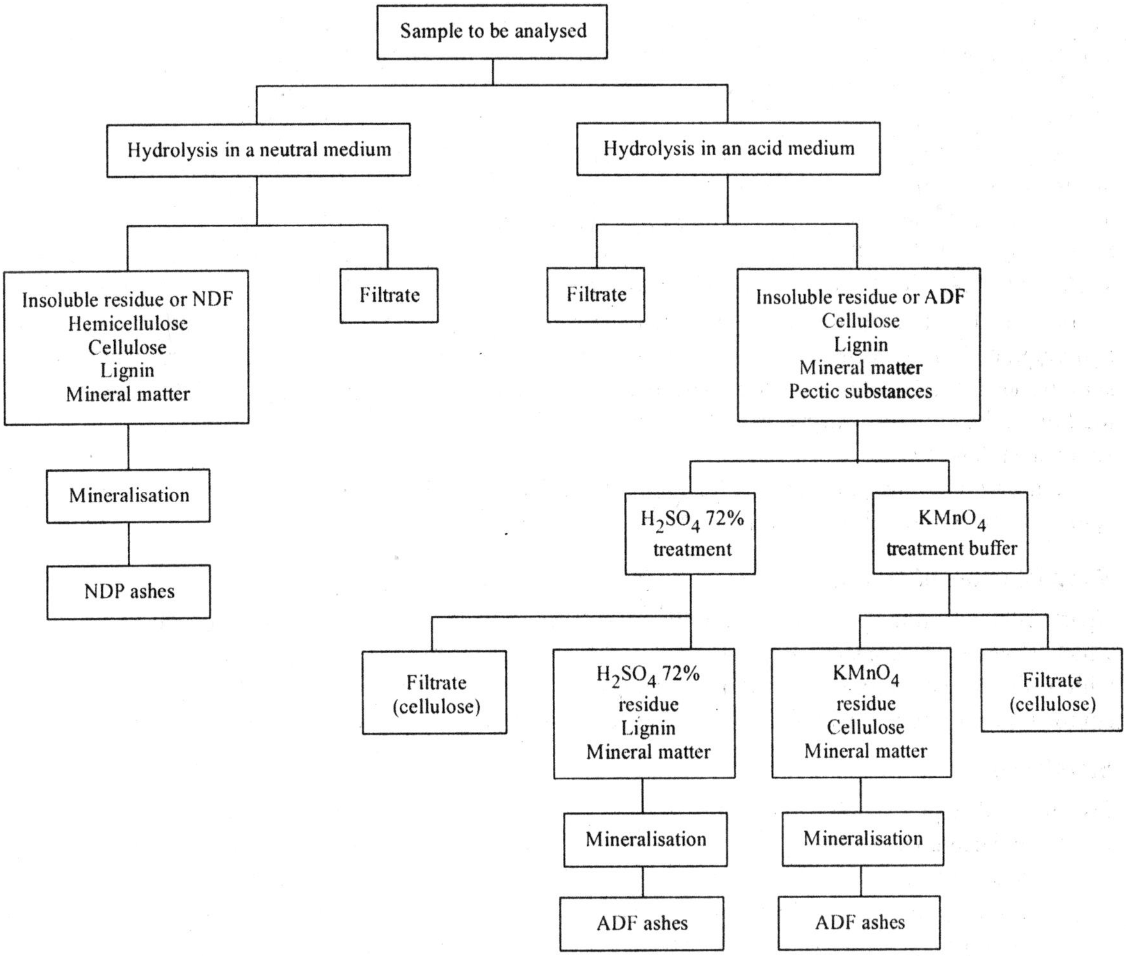


**Fig. 16.19.** Diagram of the principle of Van Soest method.

As for measuring, reproducibility, differences as shown in the IBPAS (Inter-professional Bureau of Analytical Studies) circuits range between 0.3 and 1.3 per cent absolute value with a 0.7 per cent average.

As in the case of quick determination of water content, fat and proteins, reflection spectrometry in near IR can be planned for cellulose determination after rigorous sampling.

### *Mineral matter content (ashes)*

According to AFNOR (V 03–992/V 18–101), ISO (749/5984) standards and the EEC directive (71/250), crude ash is defined as the residue obtained after calcination at 550°C in a pulley block oven for a 5 gram oil meal sample. This incineration temperature has been retained to a certain extent because it shows compromise, certain carbon particles can remain in the residue whereas mineral substances (chlorides) can evaporate at 550°C. Measuring reproducibility of ash obtained by these methods can be evaluated on the basis of IBPAS circuit at 0.2 per cent absolute value for ash rates less than 6 per cent. For higher mineral matter content, inter-laboratory differences can reach 0.5 per cent. Crude ash determination can sometimes be completed by a specific research of silica substances (sand, earth) which can pollute the oil meal. This determination which has been standardised in France (NF V 03–923/V 18–102), in the EEC (1st directive 71/250) and also internationally (ISO 735 and 5985) consists of treating crude ash with warm hydrochloric acid 3N. Insoluble residue in the acid consists of silica.

Complementary to crude ash measures, specific determination of calcium and phosphorus will be necessary in certain cases. Phosphorus measurement is generally obtained by colourimetric measure at 430 nm of the phosphovanadomolybdic complex compound (yellow). This method has been standardised by AFNOR (NF V 18–106) and ISO (6491). Calcium can be measured either titrimetrically (NF standards V 18–107 or ISO 6490/1) or by atomic absorption spectrometry (NF V 18–108 or ISO 6490/2).

In titrimetric method, crude ash is treated with a concentrated hydrochloric acid solution and calcium is precipitated in the form of calcium oxalate. After dissolving the precipitate in sulphuric acid, oxalic acid formed is titrated with a 0.1 N potassium permanganate solution. The atomic-absorption spectrometry method such as the one standardised for animal food products, operates at a wave length of 422.7 nm in an air-acetylene flame.

Measurement is obtained after calcination of the sample at 550°C, collection of ash with hydrochloric acid and appropriate dilution in the presence of lanthan.

### *Specific determinations*

Apart from traditional determinations that we have studied (humidity, proteins, cellulose, ash) and which are generally considered for commercial contracts, there are certain other specific analytic criteria which aim at estimating better than nutritional quality of oil meal (protein solubility, amino acid composition) or at exposing the presence of toxic or antinutritional substances.

#### *Starch content*

Two methods can be used for this purpose:

1. Polarimetric method (Erwers method) which is right now the only one to be known and EEC standardised method for animal food products. In this analytic technique, starch present in the sample is hydrolysed by a hot diluted hydrochloric acid solution. After decantation and filtration, rotatory power of the solution obtained is measured by polarimetry. Starch sample content is obtained by multiplying the rotatory power by a conventional factor depending on the type of product analysed.
2. In the enzymatic method which is more recent than the polarimetric method, starch is dispersed and then undergoes amyloglucosidase action. Glucose liberated is measured by spectrophotometry by the glucose-oxidase enzymatic method.

It is to be noted that this method, due to its super specificity, systematically gives results which are inferior to the polarimetric method results.

#### *Sugar content*

The method standardised by the European standard involves measuring the reduction number of sugars before and after inversion. Apart from this generally accepted method, specific measures of certain sugars (glucose-saccharose) can be obtained enzymatically.

### *Amino acid composition and content*

Amino acid composition and content which are elements which constitute proteins and in certain specific amino acids called 'essential' acid (lysine, methionine, arginine) in certain cases can be necessary to better evaluate the nutritional quality of the oil meal. This determination standardised by AOAC (42.258–264) has been obtained after protein hydrolysis, separation of the cation ($Na^+$) exchange resin and spectrometric detection in visible range after post-column determination of different amino acid by ninhydrine. Protein hydrolysis is generally carried out by 6 N hydrochloric acid action in a sealed tube, under vacuum, at a temperature of 110°C during a time period which varies between 24 and 72 hours (AOAC 42.263 method). For amino sulphated acids oxidation has to be carried out prior to hydrochloric hydrolysis. As for the special case of tryptophane, an alkaline hydrolysis has to be carried out. There are a lot of apparatus (LKB, Biotronick) which help in conducting all operations such as separation, derivation, detection and qualification of amino acids.

However, it is to be noted that in the last few years, a number of amino acid measuring methods have been proposed by high performance liquid chromatography (HPLC) and this offers the advantage of asking for a less expensive but more polyvalent equipment compared to the amino acid measurer.

It can be explained here that detection in the visible range (ninhydrine) can be replaced advantageously by fluorescence measure. The aim of this replacement will be to lower detection sensitivity, which is particularly interesting for exploring the presence of minute quantities of amino acids. Since amino acids are not naturally fluorescent, it become necessary in this technique to fix a chromogeneous fluorescent group for them such as OPA (ortho-pthalidaldehyde) of FMOC (9-fluorenylmethoxycarbonyl chloride, 9-fluroenylmethylchloro-formate).

Finally, in this field of amino acid analysis, it is always better to mention the specific measuring method of lysine available which can be assimulated and this has been standardised by AFNOR (NF V 18–103) and also by ISO (5510).

### *Protein solubility*

In addition to determining total nitrogen content by Kjeldahl method, a certain number of complementary, analytical techniques are proposed to evaluate the availability (digestibility/solubility) of these proteins for different animal species.

Many methods have been proposed:

1. Protein solubility in soda at 0.2 per cent.
2. Protein solubility in pepsin and HCl.
3. Soluble nitrogen in artificial saliva (ruminant).

### *Residual hexane content*

Measuring residual hexane in oil meals is usually done on the oil mills for two important reasons. On the one hand, it helps in keeping a count on hexane loss during extraction; on the other hand, it ensures that oil meal coming out after desolvanisation contains residual amounts of solvent prescribed by security standards. This is besides the fact that this solvent present potential dangers because of its inflammability and property to form explosive mixtures in the atmosphere. In order to carry this out, a method of measuring the total residuary hexane content has been standardised by AFNOR and ISO (NF ISO 8892). In this method a 5 mg sample of the oil meal is introduced in a water tight 50 to 60 ml flask (penicillin type). This is placed in an oven for 90 minutes at 110°C for complete desorption of hexane that may be present in the sample analysed.

A part of air found on top (0.5 ml) is brought out with the help of a gas syringe and analysed by gas chromatography on a filled or capillary column (SE 30 types). From a standardisation curve obtained previously, we can determine hexane quantity in milligrams per kilogram contained in the analysed sample. As for the accuracy of this method, the coefficient of reproducibility variation that we obtain is between 20 to 30 per cent of the relative value.

### *Microscopic examination*

Microscopic examination helps in revealing foreign matter from vegetal or mineral sources in oleaginous oil meals, particularly those which have an exotic origin and which are often adulterated on purpose (soyabean from Brazil, linseed from Argentina, rapeseed from China, peanut from Sudan).

This examination also helps in revealing the presence of toxic or undesirable elements in certain oil meals (mustard in rapeseed, datura from soyabean, castor in linseed). Microscopic examination also gives the complementary traditional physico-chemical analysis a good means of verifying the purity of different oil meals.

Observation technique is mainly based on identification of the visual aspect of vegetable fragments and botanical species. This calls for a good visual memory from the researcher and for it is a science where experience acquired is fundamental.

## Antinutritional Components

An extrapolation of the definition of nutrition given by the French dictionary 'Le Petit Larousse' leads us to define antinutritional components as 'compounds capable to affecting digestive, respiratory, circulatory, excretory and endocrine functions which bring to cells elements favouring their growth, maintenance of their forms, their function, and elimination of waste matter'.

### *Nature of the antinutritional components*

The main antinutritional constituents of oilseeds and of their oil meals are summarised in Table 16.27. Their effects are also listed.

**Table 16.27.** Antinutritional components of oilseeds.

| *Seeds* | *Antinutritional components* | *Effects induced* |
|---|---|---|
| Rapeseed (and other crucifers) | Glycosinolates degraded by myrosinase in | No direct effect |
| | Isothiocyanates | Inappetance |
| | Vinyl-5 oxazolidine 1,3, thione-2 (VTO) | Goitrigenous |
| | | Inhibited growth |
| | | Hepatic hypertrophy |
| | thiocyanate ions | Goitrigenous |
| | nitriles | Teratogenic effect |
| | | Embryo mortality |
| | Sinapin | Inappetence |
| | | Fish taste in coloured eggs |
| Soyabean | Ureasic activity | Digestibility |
| | Antitrypsic activity | Digestibility |
| Sunflower, rapeseed, soyabean | Phytic acid | Sequestering agents of bivalent cations |
| | Tanins | Reduce digestibility |
| Castor | Ricine | Toxic protein |
| | | Agglutination of red blood corpuscles |
| Cotton | Free gossypol | Toxic product |
| | Bond gossypol | Decreases the availability of lysine |

### Glucosinolates (GLS)

GLS are a uniform class of more than 90 compound which are found in a certain number of dicotyledons. But a majority of cultivated plants containing GLS belongs to the crucifer family and are more often of the Brassica genera. GLS are thioglucosides (Fig. 16.20) which can be classified into four main families according to the chemical nature of the lateral chain (R). In rapeseed, there are about ten main GLS (Table 16.27).

**Table 16.27.** Main GLS of rapeseed.

| *Class* | *Nature of lateral chain R* | *GLS* |
|---|---|---|
| Alkenyl | Butene-3-yl | Gluconapin |
| | Pentene 4-yl | Glucobrassicanapin |
| Hydroxy alkenyl | 2-Hydroxybutene-3-yl | Progoitrin |
| | 2-Hydroxypentene-4-yl | Gluconapoleiferin |
| Arylalkyl | 2-Phenylethyl | Gluconasturin |
| | 4-Hydroxybenzyl | Sinalbin |
| Indolyl | Indo-3-ylmethyl | Glucobrassicin |
| | 4-Hydroxyindol-3-ylmethyl | 4-OH-glucobrassicin |
| | 1-Methoxyindol-3-ylmethyl | Neoglucobrassicin |

Intact GLS do not seem to present direct physiological effect. Their degradation, either by myrosinase (thioglucoside glucohydrolase, an enzyme which co-exists with GLS in seeds, in different cellular compartments, or by microflore intestinal enzymes of animals which have consumed rapeseed meal, leads to the formation of products with well-known physiological effects. In practice, GLS are measured only as 'potential precursors' of various degradation products. According to the nature of GLS and the surrounding conditions, different products are formed (Fig. 16.20).

There are numerous and varied measuring methods for GLS:

1. Traditional physico-chemical methods (measuring intact and desulphated GLS by chromatography).
2. Reactions leading to the formation of products measurable by photometry (palladium test, thymol test).
3. Analysis of degradation products (glucose, sulphate, etc.).
4. Physical measure (X-ray) fluorescence spectrometry or reflection spectrometry in near infra-red (RSNIR).
5. Immuno-chemical techniques.

The choice of method depends on time, precision and cost constraints.

#### Methods for measuring individual GLS

All methods given under this heading allow individual identification and quantifying of each GLS. The content of each glucosinolate is expressed as μ moles per gram of seeds. Total sample content is obtained by calculating the sum of individual glucosinolate content.

During commercial transactions, GLS content is expressed conventionally in relation to the seeds with 9 per cent water content. A parallel humidity determination permits in carrying out this mathematical correction.

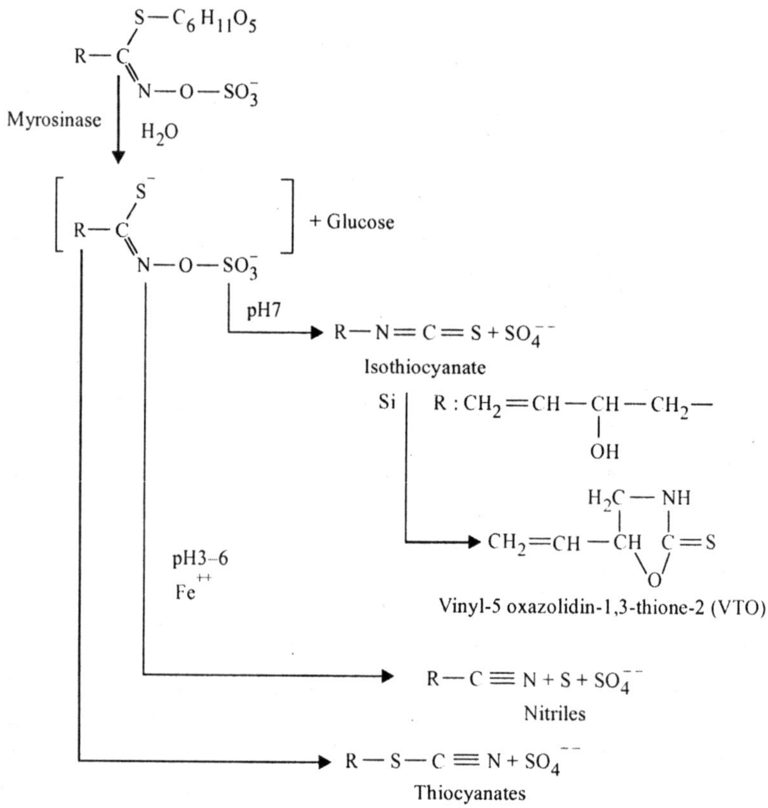


**Fig. 16.20.** Glucosinolate hydrolysis products.

## Measuring intact GLS

A priori, this is the best method because it analyses non-derivated GLS. GLS are extracted with boiling methanol solution which inactivates myrosinase at the same time. The extract is then purified by ion exchange (DEAE Ecteola cellulose) and analysed by high-performance liquid chromatography (HPLC) with polarity of inversed phases with ion. Elution is obtained by iso means and GLS are detected in UV at 235 nm.

Recently, *Clossais-Besnard* have improved the initial 'Helboe' technique concerning GLS extraction and choice of columns.

However, this method is now less frequently used because it provides insufficient separations between internal commercialised standards (sinigrine, glucotropaeoline) and GLS found in extracts.

### *Measuring desulphated GLS*

This can be measured either directly by HPLC or by GC of silylated derivates.

## HPLC

This method will soon be standardised by ISO.

It consists first to inactivate myrosinase from ground seeds by adding a 70 per cent boiling methanol solution and then extracting GLS with the same solution by agitation. GLS from the extract are then desulphated enzymatically on an ion exchange column (DEAE Sephadex A25). Desulphated GLS

(DSGLS) are then eluted by water. This preparation method allows the obtention of pure GLS derivatives. DSGLS are then analysed by HPLC with polarity of reversed phases using *n*-alkyl grafted stationary phase (C8 or C18) transplantation and an elution gradient (variation of acetonitrile rate in water from 0 to 25 per cent). Detection is carried out in UV at 229 nm. Response factors have been determined experimentally. The effective performance of separation of this method allows the use of both internal and commercial standards (sinigrine or glucotropaeoline) and is thus highly reliable.

#### Gas chromatography

This technique used the same DSGLS extracts as the previous method. It is necessary to sililate before subjecting them to chromatography on a OV 7 column with a temperature gradient for indolglucosinolate elution. Detection is carried out with flame ionisation.

This method has a tendency to produce irregular results for unstable indolglucosinolates at experimental temperatures. Moreover, response factors have been fixed arbitrarily and this could explain the difference between the results of this method and those of the HPLC reference method. However, it produces satisfactory results for alkenyl GLS. This method allows the use of sinigrine and glucotropaeoline as internal reference solutions and allows very efficient separations with a capillary column.

### *Total GLS measuring methods*

Different techniques suggested give a general value of GLS content, expressed in $\mu$ moles/g seeds.

#### Enzymatic measurement of released glucose

The principle is to measure the glucose released by myrosinase action on GLS. Myrosinase can be prepared in the laboratory, or it can be purchased.

In order to obtain reliable results, it is better to degrade previously fixed GLS on an ion exchanger with the help of myrosinase. Released glucose is then measured by using reactive boxes.

1. Either by the glucose-oxidase/peroxidase system with a measurement at 610 nm.
2. Or by the hexokinase/G-6-P dehydrogenase system producing a coloured derivative measured at 520 nm or NADP reduction measured at 340 nm.

Glucose measurement is largely used in quick GLS estimation tests when seeds are received for storage. They use narrow reactive bands or tablets which are generally used for measuring glucose in urine or blood. These tests can be done quickly but their accuracy is mediocre.

#### Thymol reaction

Glucose of GLS is dehydrated into 5-hydroxymethylfurfural in the presence of sulphuric acid. This then reacts with thymol and gives a coloured compound. *Brzezinski* and *Mendelewski* used this principle for measuring total GLS in rapeseed. This is a very sensitive method, but it is less selective for impurities.

#### Formation of a complex compound with palladium chloride

GLS react with palladium chloride, producing a coloured complex compound whose intensity has been measured at 425 nm. This technique, widely used by breeders because of its rapidity and automation potential, was made more precise by *Mooller* who introduced purification of extracts.

#### Measuring released sulphate

GLS are extracted from ground rapeseed with water and broken up by myrosinase. Sulphur released from the molecule is measured by ionic chromatography with conductimetry or inverse absorptiometry detection in UV at 265 nm (sometimes it is necessary to adjust the wave length according to eluant and detector performances).

First, a blank test should be carried out by inactivating seed myrosinase by boiling methanol, so as to measure free sulphate that does not result from glucosinolate degradation.

#### Measuring sulphur by X-fluorescence

This very fast method of analysis is probably the most used in commercial GLS analysis. It is based on the fact that GLS contain sulphur which can be measured by X-fluorescence and that there is a close relation between total sulphur content of the seed measured by X-fluorescence and GLS content (r2 = 96–99 per cent).

In rapeseed, sulphur is basically linked to proteins and GLS. Other compounds containing a sulphur do not play an important role. However, high variations in protein content influence GLS determination. one per cent variation in protein content in a sample causes a 2 μ mole g-1 error on the value of the GLS content by X-fluorescence. Analysis is carried out on approximately 20 gram of previously dried (if water content exceeds 10 per cent) and ground seeds. The ground seeds are deposited in a cupel suited to the measuring apparatus and pressed. Sample preparation before analysis (grinding and settling) can strongly influence the results.

Total sulphur determination is carried out in a X-fluorescence spectrometer, under vacuum in a helium atmosphere to avoid sulphur fluorescence absorption by air. The major barrier in vulgarising this technique is the apparatus cost for, it is necessary to use spectrometers, energy dispersers, which are sufficiently selective to measure sulphur. Prior to any measurement, the apparatus must be standardised using three reference rapeseed samples with known sulphur content. GLS content of the analysed sample is determined by comparison of the rays emitted and standardised results (ISO, 1990). This is a quick method (around 3 minutes) which can be automated. Results obtained can be correlated with those obtained by conventional reference method (r2 = 96–99 per cent). However, a certain number of undefined factors can influence the results and hence its use has to be justified.

#### Reflection spectrometry in near infra-red

*Lilas* and *Furtoss* have found out that GLS give specific absorption bands in near infra-red (NIR) at 1674, 1660, 1618, 1650, 1632 nm. This has allowed us to use this technique for the analysis of rapeseeds and rapeseed oil meals after previous standardising of the apparatus with the help of previously analysed samples. Measurement, which is fast (20–30 seconds), can be done on the whole seeds. The moisture content of seeds at the time of harvest affects the granulation of ground seeds and the surface condition of the sample in the measuring capsule, which in turn affects the result. For this reason, NIR can currently only be used for sorting seeds when they come in, but not for commercial purposes.

#### Immunoenzymology

This is a new method, developed in Australia. Recent results (CETIOM, 1990) allows us to predict a use of it:

1. Either in the form of reactive-caskets which allows total GLS to be determined quickly and precisely when the seeds come in during harvest.
2. Or by using measuring micro-titration sheets in the laboratory for determining total GLS as well as a specific GLS.

### *Isothiocyanates (ITC)*

Isothiocyanates are produced by degradation of GLS under myrosinase action. They could be partly responsible for the inappetence attributed to rapeseed oil meal. ITC are measured by a standardised method (AFNOR, 1985; ISO, 1983) in which the potential ITC quantity is measured in the sample. GLS are broken up at 7.0 pH by adding myrosinase prepared by the *Appelqvist* and *Josefsson* method. The ITC produced are extracted by dichloromethane and analysed by GC. Total ITC content is expressed in mg/gram dry, defatted sample.

#### *Vinyl-5 oxazolidine-1,3 thione-2 (VTO)*

This product comes from breaking up of progoitrin by myrosinase. Measuring is done by a standardised method (AFNOR, 1985, ISO, 1983). VTO is extracted by diethylic oxide and extract absorption is measured in UV, the VTO having maximum absorption of around 250 nm. A blank test is necessary to determine absorbance due to enzymatic support.

VTO is expressed as mg/gram dry, defatted sample.

#### *Thiocyanate ions*

They come mainly from breaking up arylalkyl and indolglucosinolates. Measuring is basically done by the Jurges and Thies method. This colourimetric method is based on the fact that thiocyanate ions, transferred into BrCN, react with pyridine and barbituric acid, producing a polymethic red colour which absorbs at 586 nm.

#### *Nitriles*

Recently, *Gardrat* have developed a measuring method by head space technique in GC to replace older methods, in response to technical progress of the apparatus (columns, detectors, etc.).

The stationary phase of GC permits a desorption of nitriles in the presence of water at 105°C. These nitriles are injected in a chromatograph on a Carbovax 20 M column at 60°C. The apparatus is equipped with a thermoionic detector. A better nitrile extraction has been obtained by a dynamic method using a desorption-concentration-injection (DCI platinum) coupled with chromatography.

#### *Sinapine*

Sinapine is a choline ester of sinapic acid. It has a bitter taste and contributes an unappetising effect to rapeseed oil meal. It could be the cause of the fishy taste in eggs laid by hens that have consumed such a meal. There are many measuring methods, but none of them is standardised. Absorption determination at 585 nm of the coloured compound extraction formed with titanium tetrachloride is a method which seems to be being replaced by HP, thin layer chromatography sinapine absorption is then measured at 326 nm and the quantity is estimated by a standard curve. For identifying the different esters of choline, use of many conjugated separation techniques (electrophoresis, thin layer or liquid chromatography) gives more reliable results.

#### *Phytic acid, phytates and tanins*

These compounds which affects the nutritional value and digestibility of oil meals will be mentioned only as a reminder in this review of different methods for analysing antinutritional factors. They are generally found in oil meals of the main oilseeds (rapeseed, sunflower and soyabean). Methods used in their analysis are numerous and none of them are standardised.

Phytic acid can combine strongly with four main mineral elements (Zn, Cd, Fe, Mg), producing insoluble compounds. During analysis, phytates are generally hydrolysed and phytic acid or liberated phosphorus are measured by a coloured reaction. Phytic acid can also be measured by HPLC. However, the method which seems to give the most reliable results and the most in use combines a phytate concentration by ion exchange and then a colourimetric measurement.

### Tanins

Tanins represent a very large class of compounds. They can be divided into two groups: condensed and hydrolysable tanins. In rapeseed oil meal, condensed tanins are quantitatively more important. Research relating to these products is numerous and old. Research carried out in other fields (sorghum, tobacco) have shown that chromatographic methods will be best adapted for such an analysis.

### Antinutritional soyabean components

#### Ureasic activity

This technique is intended to measure residual ureasic activity of products prepared from soyabean oil meals. It has been standardised (AFNOR, 1980; ISO, 1978; AOCS, 1973). Ureasic activity is defined as the ammonia nitrogen quantity released per minute in operational conditions. It is expressed in mg/gram. The ground test sample is mixed with a urea solution, after 30 minutes at 30°C, liberated ammonia is neutralised by an excess of hydrochloric acid treated by a sodium hydroxide solution.

#### Antitrypsic activity

Inhibitors of trypsin in soyabean oil meals (raw and roasted) and in concentrates or isolates are determined by this method (AOCS, 1983). Antitrypsic activity is measured by inhibitory breaking-up action in benzoyl-DL-argine-*p*-nitroanalide hydrochloride by trypsin. The unit of activity is defined in this method.

### Castor bean: measurement of ricine activity

Castor seeds contain a toxic protein called ricine which is present in all varieties. This phytotoxine is one of the most toxic proteins; equivalent to tetanus, botulism and diptheria toxins (lethal dose 0.5 mg/kg-1 in rabbits). It destroys capillary circulation in organs and lymphatic tissues and can cause agglutination of an erythrocyte suspension. Activity of this compound is generally measured by a photometric method based on erythrocyte agglutination. A castor agglutinine solution is prepared from castor oil meal and is tested on an erythrocyte preparation (guinea-pig blood). One unit of erythrocyte agglutination, has been defined. It corresponds to the quantity of test solution which causes 50 per cent agglutination of suspended erythrocytes in 60 minutes under test conditions.

### Cotton: gossypol determination

Gossypol is a polyphenolic compound substituted by two aldehyde groups. It is concentrated on pigment glands present in cotton seed cotyledons. It exists in free or combined state.

In free state, it can be toxic in pig, rabbits and poultry products. During cottonseed oil meal production, aldehyde groups can react on terminal amino groups of lysine. In the combined state it can reduce the biological availability of amino acid. Measuring this compound under different forms has been standardised (AFNOR, 1986; ISO, 1985). It consists of gossypol extraction in the presence of amino-3 propanol, either by mixing propanol-2 and hexane for determining free gossypol or by dimethylformamide for measuring total gossypol. Gossypol is then transformed into gossypol-dianiline whose absorption is measured by spectrophotometry at maximum absorption (between 435 and 445 nm).

Gossypol content, free or total, is expressed in milligrams per kilogram of the product.

Linked gossypol content is for a given sample equal to:

Total gossypol content—free gossypol content.

AOCS has standardised very close methods for measuring free gossypol as well as total gossypol (AOCS, 1987). Presently a HPLC technique is mostly used for measuring gossypol, there being no standardised method.

## MYCOTOXINS-AFLATOXINS

The colonisation by certain toxigeneous mildew of oil-seeds having the power to secrete mycotoxins is an important factor which alters the sanitary quality of these seeds and derived products (oils and oil meals). The aflatoxins (Fig. 16.21) have been the first mycotoxins revealed in the 1960's following a severe and massive poisoning which took place in Great Britain in the rearing of turkey, by using contaminated peanut oil meals as food for poultry.

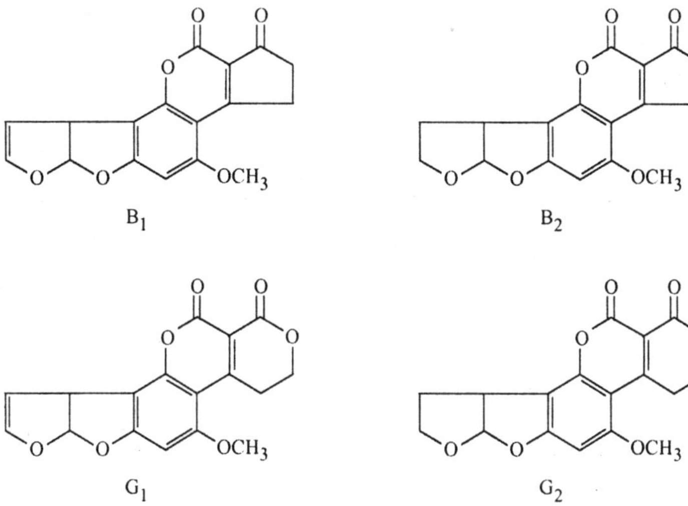


**Fig. 16.21.** Formulae for aflatoxins of group B and G.

Since, then a great number of other mycotoxins have been identified (Zearalenon, Trichotecens, Patulin, Ochratoxin, Sterigmatocystin) and these are present in a number of food stuffs (oleaginous substances, cereals, tubercles).

In this section we will limit ourselves to the studies of only measuring methods of aflatoxins of group B and G, for two main reasons:

1. Aflatoxins, are mycotoxins which most frequently contaminate tropical oleaginous substances (peanut, cotton) and the derived products (oil meals, oils).
2. Among all the known mycotoxins, $B_1$and $G_1$ are highly risky for human and animal health due to presence of carcinogenic, mutagenic and teratogenic molecules.

Moreover this high toxicity explains why the basic regulations and recommendations are so elaborated in many countries. In France aflatoxins are main concerned (Tables 16.29 and 16.30).

Here we will discuss the main principles of these different analytical techniques which could be classified in three distinct groups:

1. Biological methods.
2. Physico-chemical methods.
3. Immunological methods.

**Table 16.29.** Content of aflatoxin, $B_1$ in animal foods (raw materials and finished products).

| *Raw materials (oleaginous oil meals)* | *Aflatoxin $B_1$ in mg/kg (ppm)* |
|---|---|
| Peanut, copra, palm kernel, cotton | 0.2 |
| *Finished products (animal foods)* | *Aflatoxin $B_1$ in mg/kg (ppm)* |
| Simple foods | 0.05 |
| Complete foods for cattles, sheeps and goats, except for milk cattle, calves and lambs | 0.05 |
| Complete foods for pigs and poultry, except for young animals | 0.02 |
| Other complete foods | 0.01 |
| Supplementary foods for cattles, sheeps and goats except for supplementary foods for milk cattle, calves and lambs | 0.05 |
| Supplementary foods for pigs and poultry except for young animals | 0.03 |
| Other supplementary foods, in particular for milk cattle | 0.01 |

**Table 16.30.** Maximum concentration in aflatoxin $B_1$ in food products (raw materials and finished products) meant for human food fixed by the Superior Council of Public Hygiene in France.

| *Products* | *Aflatoxin $B_1$ in microgram/kg (ppb)* |
|---|---|
| Oils and fats (virgin oils or mixed with refined oils) | 5 |
| Products consumed by children and adults | 1 |
| All other food products (raw materials and finished products) | 10 |

Before examining the different measuring methods, we cannot overlook the delicate sampling problem which directly conditions the analysis results in laboratory, keeping in mind the great heterogeneity of contamination observed, mainly on seeds.

This is how various studies have shown that more than two-thirds of the variability of the results on aflatoxin measurings on peanut lot have been due to single sampling operation.

In view of its complexity, the sampling operation has not yet obtained totally satisfactory solution.

There are many parameters on which one can intervene to be able to improve the sample representativeness.

1. Multiplying the number of sampling tests.
2. Increasing the size of tests and sampling tests.
3. Increasing the number of measurements obtained from the same sample.

### Biological methods

These methods which have been the first ones to be used measure the lethal dose (DL 50) of aflatoxin for a living human being chosen according to its great sensibility to this mycotoxin.

Several biological materials can be used for these tests among which we can mention: (i) one day old duckling; (ii) embryo of a chicken; (iii) bacterias; (iv) insects; and (v) mollusk eggs.

These biological methods of aflatoxin measurement are now-a-days more and more supplanted by physico-chemical methods for three main reasons:

1. First of all, these tests are relatively long to settle, the delay in response varying from 4 hours to 1 month which makes it more difficult for the industrial control where results should be communicated as fast as possible.

2. On the other hand, some of these biological tests require the availability of animal breeding in laboratories and special equipments, which limits their use to a very small number in specialised laboratories, and in fact a relatively expensive method.
3. Finally and above all, these methods lack accuracy and specification and cannot be used as semi-quantitative methods.

On the other hand, they can represent a certain kind of interest as confirmation test of effective presence of aflatoxin in an extract to be studied in case of doubt.

### *Physico-chemical methods*

It is the field of physico-chemical methods of aflatoxin measurement that the highest number of analytical techniques have been published and few amongst them are subject to standardisation (Table 16.31).

**Table 16.31.** Standardised principle methods of aflatoxin measurement in oilseeds and oil meals.

| *Measurement method of aflatoxin as per type of oleaginous product* | *AFNOR* | *EEC Directive* | *AOAC* | *IUPAC* | *AOCS* |
|---|---|---|---|---|---|
| Animal food (raw materials-finished products) | NF V 18–200 | 7° 76/372 (JOCE 15/4/76) | – | – | – |
| Peanuts and derived products | | | | | |
| CB method | – | – | 968–22 | 18–31 | – |
| BF method | – | – | 970–45 | – | Ab 6–68 |
| Coconut-copra oil meal | – | – | 971–24 | 18–9 | Ah 1–72 |
| Cotton seeds and derived products | | | | | |
| TLC method | – | – | 980–20 | – | Aa 8–83 |
| ELISA method | – | – | 989–06 | – | – |
| Soyabean seeds | – | – | 972–27 | – | – |
| Pistachio | – | – | 974–16 | – | – |

Whatever be the obtained analytical technique, we always find the same analytical steps (Fig. 16.22) which include four main stages:

1. Extraction of aflatoxins of substratum.
2. Purification of extract.
3. Separation and quantification of mycotoxin.
4. Confirmation of identity of toxin.

#### *Extraction*

The extraction stage is very important as it directly conditions the results of this measurement. The extraction solvent should be extracted on the basis of its affinity for aflatoxin molecules (solubility) and on the other hand for its compatibility towards the substratum on which the measurement takes place. From the number of surveys carried out, it appears that the chloroform extraction method has the best rates of aflatoxin recovery in most of the substratums, which explains why this solvent extraction has been used in numerous methods and also standardised by the EEC, AFNOR, AOAC and the IUPAC. This chloroform extraction is generally carried out in the presence of water in a low percentage (10 per cent). The role of water is to facilitate the penetration of the polar solvent in the substratum of vegetal nature.

It should be noted that chloroform extraction can be carried out only in substratums which do not contain much lipids (less than 10 per cent). In fat products such as oilseeds, it is absolutely necessary to carry out a hexane dilipidation beforehand.

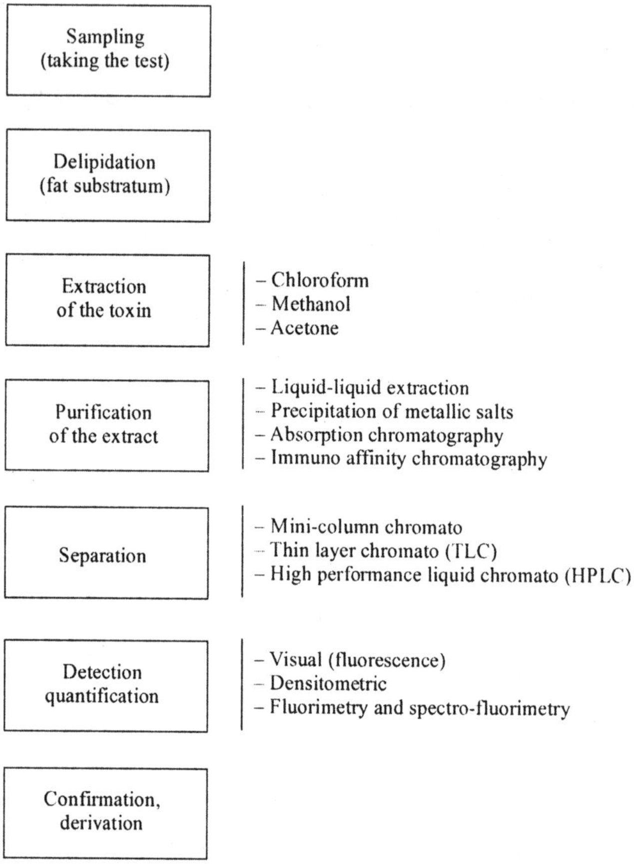


**Fig. 16.22.** Diagram of aflatoxin measurement.

In order to avoid this extra dilipidation stage, some authors have proposed to replace chloroform by solvent or solvent couples in which lipids are a little soluble or insoluble. This is particularly the case for the solvent extraction proposed in the BF method which operates with a methanol water phase (55 + 45) and a hexane phase, method mentioned in the AOAC (970–45) and AOCS (Ab 6–68) standards for extracting aflatoxins from seeds, butters and peanut oils. The acetone/water (85 + 15) couple as proposed in the AOAC (980–20) and AOCS (Aa 8–83) standards for cottonseeds and derived products also helps in carrying out the process without any prior dilipidation of the substratum.

The extraction, rightly speaking, of the aflatoxin is obtained by mechanical agitation of the sample which is finely ground with the retained solvent.

These are various methods employed for carrying out this operation (Table 16.32).

1. Magnetic agitator (slow agitation, 30 minutes).
2. Shaker agitation (slow agitation, 30 minutes).
3. Mixer (fast agitation, 1 to 4 minutes).

**Table 16.32.** Principle methods of aflatoxin on extract.

| *Standard methods* | *Substrat* | *Extraction solvent* | *Time and system of agitation* |
|---|---|---|---|
| NF V 18–200 | Food for animals (raw material and finished product) | | |
| AOAC 968–22 | Peanut oil meal | Chloroform/water (90 + 10) | Slow agitation (magnetic agitator or shaker) |
| AOAC 971–24 | Copra oil meal | | |
| AOAC 972–27 | Soyabean seeds | | 30 minutes |
| AOAC 974–16 | Pistachio | | |
| IUPAC 18–31 | Peanut oil meal | | |
| AOAC 970–45 | Peanut seeds and oil meal | Methanol/water (55 + 45) | Fast agitation (mixer) |
| AOAC Ab 6–68 | Peanut butter | + hexane | 1 minute |
| AOAC 980–20 | Cottonseeds and derivatives | Acetone/water (85 + 15) | Fast agitation (mixer) |
| AOCS Aa 8–83 | | | 1 + 3 minutes |

### *Purification*

Inspite of using very selective extraction techniques, the extract obtained contains, apart from toxin, various impurities (liquid, oxidation products, pigments, proteins) which can simulate the toxin or interfere with it during measurement. To avoid this, purification is a necessary step.

Numerous purification techniques were proposed, some of which are:

1. Purification by precipitating the impurities through purifying agents such as lead acetate as recommended by the AOAC 980–20 method for cleaning cotton extracts.
2. Purification through liquid-liquid partition in a separating funnel which operates through a simple transfer of the toxin molecule from the extraction solvent (acetonic or methanolic) to another solvent (chloroform).
3. Purification through absorption chromatography in a silica column is the most developed method and is found in the principle standardised methods (AFNOR, EEC, AOAC, AOCS).
   In this technique the extract to be purified is deposited on the top of the column and eluated by solvents with increasing eluent forces.
   The first two elutions are generally carried out by passing hexane and diethyl oxide (ether) which helps in eliminating, mainly lipids and pigments contained in the extract.
   The toxin is then recovered by passing it through a column containing a mixture of chloroform/methanol (97 + 3).
4. Finally the process of immuno-affinity column chromatography which involves antigen-antibody reactions was recently proposed for purifying extracts.

### *Separation and quantification*

Three different chromatographic techniques can be used for separating and quantifying aflatoxins, namely:

1. Mini-column chromatography.
2. Thin-layer chromatography (TLC).
3. High performance liquid chromatography (HPLC).

### Mini-column chromatography

These methods were basically developed as quick semi-quantitative techniques for detecting aflatoxins and are particularly interesting as a screening method. In these analytical techniques, aflatoxins are detected when under ultraviolet light (365 nm) a blue fluorescent ray of light appears in the body of the mini column packed with different absorbent supports (silica, florisil).

### Thin layer chromatography

Thin layer chromatography over silica (silica gel) is most commonly used as separation technique for the measurement of aflatoxins. This technique is applied in half the standardised methods. Numerous solvent systems were proposed for developing these plates, normally they are mixtures with a variable proportion of chloroform, methanol, acetone and benzene. In order to improve the quality of separations, we can use silica gel plates of very fine granulometry (HPTLC) or carry out a double development (bidimensional chromatography).

Detection and quantification of the toxin is done under ultraviolet light (365 nm), by comparing the fluorescence of a spot of the extract of known concentration. This evaluation can be done either visually (with the naked eye), or with the help of fluorodensitometer. This separation and quantification helps in attaining a detection level of 0.2 to 0.4 mg for $B_1$ aflatoxin, which in turn enables a measurement of 0.5 to 2.0 mg/kg (ppb) of aflatoxin in a sample. If the detection sensitivity of this method is relatively satisfying, it is not the same when it comes to the precision of the measurement.

Thus, after numerous results of circular analyses carried out by the French (AFNOR) as well as by the Americans (Smalley Test AOCS) we have variation coefficients (VC) of reproductibility between 25 and 50 per cent with a relative value. Accuracy of the measurement is as unsatisfactory as the low rate of contamination of the sample.

#### *High performance liquid chromatography (HPLC)*

Numerous methods of aflatoxin measurement by HPLC have been published during the last few years; but till now only a few have been standardised or officially accepted. For aflatoxin measurement, two separation techniques by HPLC can be used, i.e. absorption chromatography on a silica column and inverse phase partition chromatography (grafted non-polar phase), aflatoxins are detected at the exit of the column by fluorimetric measurement. It should be noted that $B_1$ and $G_1$ aflatoxins have a very low fluorescence in a liquid medium, hence it is necessary to carry out derivation of these molecules in order to increase their fluorescence and thus the detection level. This derivation is generally done before the sample is injected into the column by the action of trifluoroacetic acid (TFA) which transforms the $B_1$ and $G_1$ aflatoxin molecules into their hemiacetals which are more fluorescent ($B_2a$ and $G_2a$). Iodine was also proposed as a reagent for post-column derivation of aflatoxins.

The HPLC technique for aflatoxin measurement is preferred over the TLC due to the former's detection sensitivity (0.05 to 0.2 mg) and its great specificity as well as its automatisation facility which helps in increasing the productivity of a laboratory. As far as fidelity is concerned the inter-laboratory studies carried out by this technique have, unfortunately not shown any significant improvement compared to TLC technique used for determining the reproducibility coefficient.

The main disadvantage of this technique is that it requires a very heavy and sophisticated equipment which unlike the TLC equipment cannot be installed in any laboratory.

#### *Confirmation*

The comparison of Rf and retention time of spots or peaks of the extract with that of the aflatoxin standards is not always sufficient and only presumes the presence of a mycotoxin.

In fact, however, carefully the purification and separation of the extraction is carried out, they can always contain a certain number of components possessing the same chromatographic and fluorescent characteristics as the aflatoxins. The identity of the toxin is generally confirmed by derivation which changes the chromatographic and fluorescent characteristics of the aflatoxin molecule. The simplest test consists in pulverising the plate after development with a 50 per cent sulphuric acid solution which transforms the blue fluorescence of the group B aflatoxin into yellow fluorescence. But this is a presumptive test and it actually only helps in confirming the absence of the toxin. Another test which is more specific than the preceding one consists in forming a hemiacetal derivative of aflatoxin through hydratation on the molecule by the action of trifluoroacetic acid (TFA).

A comparison of Rf of the derivatives formed from suspected toxins in the extract with those obtained under the same conditions from the standard helps in confirming the presence or absence of the toxin in the extract.

### *Immunological methods*

Immunological techniques initially used for human and animal health diagnostics are being increasingly proposed for aflatoxin measurement in food products and right now numerous ready to use immunological detection kits are being developed. Immunological techniques are based on the principle of antigen-antibody. Here the antigen being the aflatoxin to be detected and the antibody is the anti-mycotoxin immunoglobulin (monoclonal antibody) obtained from an animal following the injection of the substance to be measured (immune reaction). Among the different immunological measurement techniques only two methods have been developed in the domain of aflatoxin analysis.

#### *Enzymo-immunological measurements (EIA and ELISA tests)*

In these tests, the antigen or the antibody is marked with an enzyme which acts as a tracer as it gets coloured in the presence of a specific chromogen subtrat. The colour obtained can be photometrically measured or visually seen. It is in the field of immuno-enzymatic reactions that have helped in the commercial manufacture of a large number of aflatoxin measurement kits.

These kits, in the form of ready to use measurement cases can be divided into two groups depending on their use.

1. Detection kits which help in quickly determining whether the sample contains or not an aflatoxin in quantities higher than the level fixed beforehand (semi-quantitative evaluation). It can be visually measured (with the naked eye). These kits which are generally very simple to assemble, give very quick results and constitute very good screening test and can be used beyond the laboratory walls (field test).
2. Measurement kits which help in quantitative determination of the aflatoxin content in a sample. These measurement kits generally use polystyrene microtitration plates, the reading is done photometrically after the determination of a standard range.

The advantages of enzyme-immunological dosages compared to physico-chemical techniques (TLC-HPLC) are due to the great specificity of monoclonal antibodies which help in working directly on non-purified extracts thus saving a lot of time at the measurement stage. But actually, in the laboratories, the development of these new measurement techniques is subject to their recognition or officialisation by the competent authorities.

#### *Immuno-affinity chromatography (IAC)*

The IAC is mainly used as a technique for extract purification by using anti-bodies possessing a certain specificity (affinity) towards aflatoxin molecules. Purification is done by passing the extract through mini-columns in which anti-bodies are fixed over an inert support. Aflatoxin molecules are linked to anti-bodies whereas the various impurities contained in the extract are eluted by passing through various buffers. The toxin is then eluted through an appropriate organic solution. The toxin contained in this eluent can then be quantified by traditional physico-chemical measurement techniques (mini-column, TLC, HPLC). The advantage of this purification technique, apart from the fact that it can be quickly assembled, is its great selectivity which helps in obtaining very clean extracts thus restricting the risks of interference at the measurement stage.

## ANALYSIS OF FAT DERIVATIVES

### Food Additives (Monoglycerides and Derivatives)

Food additives, like emulsifiers (monoglycerides and derivatives), are generally, related to a class of chemical compounds, known as 'surfactants'. These compounds can modify the interfacial tension between fluids of different polarities, or between solids and fluids or solids and solids, by absorbing itself at the intefaces. They are characterised by a molecule revealing the joint existence of a polar group (generally alcohol or acid groups) and hydrophobic groups (more or less branched carbon chains). This peculiarity helps them in playing a pivotal role at the polarity limits and as a texture stabiliser. The list of emulsifiers permitted in food products laid down by the European Economic Community consists of products of varied nature.

However, if we exclude alkali metal salts of edible fatty acids, the chemical structure of food emulsifiers can be pictured as a polyalcoholic nucleus esterified by a variable number of acid groups. Table 16.33 indicates these structures. Sucroglycerides have the peculiarity of corresponding to a mixture of sucrose, glycerol and interesterified fatty acids resulting in a real mixture of mono and diglyceride sucroesters. Lecithins, not mentioned in the Table 16.33, are a more complex mixture, since some acid groups of phosphoric acid can be esterified by amino acids with an alcohol group: choline or ethanolamine.

**Table 16.33.** Esters and esterified polyols.

| *Esters* | *Esterified polyols* | *Acid groups* |
|---|---|---|
| Mono and diglycerides of fatty acids | Glycerol | Fatty acids |
| Acetic ester of mono and diglycerides of fatty acids | Glycerol | Acetic acid<br>Fatty acids |
| Lactic ester of mono and diglycerides of fatty acids | Glycerol | Lactic acid<br>Fatty acids |
| Citric esters of mono and diglycerides of fatty acids | Glycerol | Citric acid<br>Fatty acids |
| Tartaric ester of mono and diglycerides of fatty acids | Glycerol | Tartaric acid<br>Fatty acids |
| Monoacetyltartatric and diacetyltartatic ester of mono and diglycerides of fatty acids | Glycerol | Tartaric acid<br>Acetic acid<br>Fatty acids |
| Mixed acetic and tartaric esters of mono and diglycerides of fatty acids | Glycerol | Tartaric acid<br>Acetic acid<br>Fatty acids |
| Sucroesters | Sucrose | Fatty acids |

*(Contd...)*

**Table 16.33.** Esters and esterified polyols. *(Contd .)*

| *Esters* | *Esterified polyols* | *Acid groups* |
|---|---|---|
| Sucroglycerides | Sucrose<br>Glycerol | Fatty acids |
| Polyglyceridic esters of fatty acids | Polyglycerols | Fatty acids |
| Esters of propylene-glycol and fatty acids | Propylene glycol | Fatty acids |
| Stearoyl-2-lactylate of sodium or calcium | Lactic acid and its dimer (Na or Ca salt) | Stearic acid |
| Stearyl tartarate | Stearylic alcohol | Tartaric acid |

All these characteristics together results in the multiple uses of emulsifiers greatly overflowing the simple property of emulsification. Table 16.34 indicates, though not exhaustively, some physical properties of the above mentioned compounds:

**Table 16.34.** Physical properties of esters.

| *Functions* | *Examples of application* | *Emulsifiers* |
|---|---|---|
| Emulsification | Water-in-oil phases (like margarine or lightened fat) | Monoglycerides |
| Complexation of bread starch | Agent for preventing from going stale | Monoglycerides of saturated fatty acids |
| Interaction with proteins | Improving elasticity and viscosity of wheat gluten | Diacetyltartaric esters |
| Proliferation | Dispersion of fat phases in biscuit making | Monoglycerides, lactic esters |
| Stabilising foaming | Creams, topping, etc. | Acetic esters, lactic esters |
| Crystalline modification | Changing stable crystalline form $\alpha$ to form $\beta$ of fats | Acetic esters, lactic esters |
| Covering | A protective film preventing the surface from drying up in pork shop, fruits and vegetables | Acetic esters |
| Lubrication | Agents for removing moulds from sweets | Acetic esters |
| Anti-oxidation activators | Complexation of metals (such as iron, copper etc.), oxidation factors of fats. | Citric esters |

Analytical techniques use these particularities either for carrying out an identification (qualitative analysis for detecting the presence or absence), or for a measurement.

In order to restrict ourselves to 'derivatives of fats' in the strict sense of the term, derivatives, which is used in food products, we shall only talk about the following compounds:

1. Mono and diglycerides of fatty acids.
2. Acetic esters of mono and diglycerides of fatty acids.
3. Lactic esters of mono and diglycerides of fatty acids.
4. Citric esters of mono and diglycerides of fatty acids.
5. Tartaric esters of mono and diglycerides of fatty acids.
6. Monoacetyltartaric and diacetyltartaric esters of mono and diglycerides of fatty acids.
7. Mixed acetic and tartaric esters of mono and diglycerides of fatty acids.
8. Polyglyceride esters of fatty acids.

In other words we are only talking about of emulsifiers whose hydrophillic group is made up of a single or polymerised glycerol nucleus.

### *Physico-chemical characteristics of additives*

Criteria of physico-chemical characteristics of additives is given below:

| *Mono and di glyceride content* | *Free fatty acid content* | *Total glycerol* | *Free glycerol* | *Sulphated ash* | *Water content* | *Polyglyceride content* |
|---|---|---|---|---|---|---|
| > 70% | < 3% (*) | > 16% and < 33% | – 7% | < 0.5% (**) | < 2% | < 4% Dimer < 1% other polymers |

(*) Content of free fatty acids is expressed as oleic acid.
(**) Content of sulphated ash is determined by calcination at 800 ± 25°C.

| *Content of esters and fatty acids* | *Content of free fatty acids* | *Total glycerol polyglycerides* | *Free glycerol polyglycerides* | *Sulphated ash* | *Characteristics of polyglycerol fraction* |
|---|---|---|---|---|---|
| > 90% | < 6% (*) | > 18% < 60% | < 7% | < 0.5 % (**) | Basically dimers and tetramers, heptamers < 10% |

(*) Content of free fatty acids expressed as oleic acid.
(**) Content of sulphated ashes determined by calcination at 800 ± 25°C.

After extraction from water, free glycerol is measured as under.

Mono and diglycerides remaining in the chloroform solution after extraction of free glycerol (a monoglyceride) can be specifically measured by periodical oxidation of a diols according to the reaction:

$$\begin{array}{lcl} CH_2\text{—}OH & & \\ | & & \\ CH\text{—}OH + IO_4H^- & \longrightarrow & HCHO + CHO + H_2O + IO_3^- \\ | & & \qquad\quad | \\ CH_2\text{—}OOCR & & \qquad\quad CH_2\text{—}OOCR \end{array}$$

Quantity of periodate transformed into iodate is determined by a titrated solution of thiosulphate through difference between the test and a black test (IUPAC No 3.121). Characteristics of polyglycerols are determined after saponification and extraction of fatty acids by gas chromatography.

| *Names of esters* | *Total carboxylic* | *Total free acidity (***)* | *Free glycerol* | *Total glycerol* | *Sulphated ash (*)* |
|---|---|---|---|---|---|
| Acetic E472a | > 9% and < 32% | < 3% | < 2% | > 14% and < 31% | < 0.5% |
| Lactic E472b | > 13% and < 45% | < 3% | < 2% | > 13% and < 30% | < 0.5% |
| Citric E472c | > 13% and < 50% | < 3% | < 2% | > 11% and < 29% | < 0.5% |
| Tartaric E472d | > 15% and < 50% | < 3% | < 2% | > 12% and < 29% | < 0.5% (**) |

(*) Content of sulphated ash is determined by calcination at 800 ± 25°C.
(**) This content is carried to maximum of 10 per cent for partially or totally neutralised product.
(***) Total free acidity is expressed in oleic acid.

The analysis methods necessary for controlling the established characteristics are those which have been defined by the UN Food and Agricultural Organisation (FAQ), mainly in the monographs concerning emulsifiers. They also determine the acceptable limits of arsenic and heavy metal content in the food product.

| Name of esters | Total tartaric acid | Total acetic acid | Free tartaric acid | Free acetic acid | Free fatty acid | Total glycerol | Free glycerol | Sulphated ash |
|---|---|---|---|---|---|---|---|---|
| Mono Diac tartaric E472e | > 10% and < 40% | > 8% and < 32% | — | — | < 3% (*) | > 11% and < 28% | < 2% | < 0.5% (**) |
| Mixed E472f | > 20% and < 40% | > 10% and < 20% | < 1% | > 5.5% and < 48.5% | < 3% (*) | > 12% and < 27% | < 2% | < 0.5% (**) |

(*) Content of free fatty acids expressed in oleic acid.
(**) Content of sulphated ash is determined by calcination 800 ± 25°C.

### *Analysis of emulsifiers in finished products*

The American Oil Chemists Society (AOCS) standardised a method which can only be applied to the form of monoglycerides.

The principle is based on a periodical oxidation of alcohol groups situated at the alpha position of one another. It cannot be applied in the presence of free glycerol.

Generally, the analysis of a food product containing emulsifier takes place in three stages:

1. Extraction
2. Identification.
3. Measurement or evaluation.

#### *Extraction*

Emulsifiers are concentrated in the fat phase of the extract. This section deals with these extraction techniques of fat matter present in polar compounds (extraction through a mixture of solvents such as chloroform/ethanol or methanol).

*Dieffenbacher's* studies describe precisely the conditions of extraction and enrichment of numerous types of food products mainly by using techniques of eliminating non-polar fats through chromatography over a column of silica gel.

#### *Identification*

The International Union of Pure and Applied Chemistry (IUPAC) developed various identification techniques for isolated emulsifiers, or in a mixture, through chromatographic methods (thin layer chromatography, gas phase chromatography after derivation in silyle's esters or high pressure liquid chromatography by using a detection UV at 220 nm of carboxyl groups).

These methods basically help in identifying single emulsifiers or even those present in a mixture.

#### *Measurement or evaluation*

Chromatographic techniques were quickly used by solving the problem of measurement, especially by using of gas chromatography method.

A method for a quick evaluation of compounds in mono, di and triglycerides coupling the TLC techniques with enzymatic methods of hydrolysis of glycerides (tri or partial) and the measurement of glycerol is used. After separation over a thin layer of silica, the different glycerides are extracted from the support by ether, then concentrated. They are then hydrolysed by a lipase and an esterase into glycerol, along with the formation of free fatty acids. This hydrolysis requires a perfect emulsion between the fat phase containing partial glycerides and the aqueous solution buffered with enzymes.

Liberated glycerol is determined through UV spectrometry according to the following sequence of enzymatic reactions:

$$\text{Glycerol} + \text{ATP} \xrightarrow[\text{Glycerokinase}]{} \text{Glycerol 3 phosphate} + \text{ADP}$$

$$\text{ADP} + \text{Phospho-Enol pyruvate} \xrightarrow[\text{Pyruvate kinase}]{} \text{ATP} + \text{Pyruvate}$$

$$\text{Pyruvate} + \text{NaDH} + H^+ \xrightarrow[\text{Lactate dehydrogenase}]{} \text{Lactate} + \text{NaD}^+$$

Quantity of oxidised nicotinamide adenine dinucleotide ($NAD^+$) is proportional to the quantity of glycerol present. Development of enzymatic techniques in a food medium led us to include the diagram not only mono-glycerdies but also other categories of emulsifiers. Thus it is possible to dose lactic, citric and acetic esters of mono and diglycerides. Each partial glyceride isolated by TLC is hydrolysed chemically, then the glycerol and the carboxylic acid (acetic, lactic, citric) are enzymatically measured according to the operating protocol.

## Glycerines

The hygroscopicity of glycerol requires special precautions during sampling and preparation in order to avoid humidification. The sampling regulations and preparation of laboratory sample are given in detail in ISO and AFNOR standards, with complete technical concordance.

### *Glycerol measurement*

Methods standardised by IUPAC, ISO and AFNOR are identical in their principle:

Cold oxidation of glycerol by periodate in a strong acid medium and potentiometric titration of formic acid produced by the reaction:

$$CH_2OH—CHOH—CH_2OH + 2IO_4^- \longrightarrow HCOOH + 2HCOH + 2IO_3^- + H_2O$$

It is not possible to titrate formic acid by soda in a presence of excess periodate, because the periodate, neutral salt, gets transformed into pH7 in paraperiodate according to the reaction:

$$IO_4^- + H_2O \longrightarrow IO_5H_2^-$$

Paraperiodate has properties of a weak acid and interferes in the test of formic acid. Therefore, before titrating formic acid, it is necessary to destroy excess periodate by glycol whose oxidation only produces formol according to the reaction:

$$CH_2OH—CH_2OH + IO_4^- \longrightarrow 2HCOH + IO_3^- + H_2O$$

The reproducibility of the method is not specified in any of the three specifications. The repeatability of the method is to the tune of 0.1 per cent according to 80 per cent glycerine analysis. This method which is very specific in the absence of sugar, is well adapted to the test of combined glycerol in fats and free glycerol in soaps or more complex products such as face creams.

However, its sensitivity is weak and does not allow a measurement of glycerol of test if its content in the research product is less than 0.5 per cent. Therefore, it is necessary to refer to formic aldehyde colourimetric test by chromotropic acid, which is a sensitive but unspecific method.

### *Ash measurement*

The glycerine from soap factory contains a high percentage of sodium chloride, whose risk of loss during calcination under unregulated temperature is not negligible. That is why the standardised methods have described, since a long time, two different techniques of ash measurement as per the analysis concerning 'saponification' glycerine (glycerine obtained by hydrolysis of fats) 'lye' glycerine (glycerine from soap factory). By imposing IUPAC, ISO and AFNOR, standards 1972, calcination under regulated temperature (750°C ± 10) has made it possible to adopt a simple and unique mode of operation for all glycerines.

It is to be pointed out that ISO and AFNOR standards have elaborated a method for determining sulphated ashes for glycerines containing less than 0.5 per cent of ash. None of these standards give indication of the reproducibility of the method.

### *Water measurement by Fischer's reagent*

The methods standardised by IUPAC, ISO and AFNOR (S) are identical in their principle:

Direct water titration on a sample for analysis put in a solution in methanol.

ISO and AFNOR also elaborate an operating mode based on back titration by a solution titrated with water in methanol, of an excess of *Fischer* reactive used in a known volume for dissolving the sample for analysis. If the glycerine is alkaline a millimole of free base taken as one millimole of water, it is necessary to introduce a corrective factor, taking into account alkalinity.

These standards do not give indications of the reproductibility of the method which contrary to the general rule, has an absolute value as satisfying as the low water content.

### *Calculation of 'matter (organic) non-glycerol' content*

By convention the term 'matter (organic) non-glycerol' (*Mong*) designates, the difference between 100 and the sum of glycerol, ash and water contents:

'*Mong*' per cent = 100 – (Glycerol per cent + ash per cent + water per cent)

This definition has been standardised by IUPAC, ISO and AFNOR.

Organic matters other than glycerol are essentially made up of polyglycerols (resulting from thermic polymerisation of glycerol) and eventually by fatty acids, fatty acid salts, amino acids and thermic decomposition products of glycerol (particularly oxalic acid), decomposition catalysed by iron salts and particularly aluminium. This determination replaces the ancient method of determining the 'non-volatile residue at 160°C and organic matters' (old French standard AFNOR T60 354 1962) for measuring of all the non-volatile impurities at 160°C and by deducting from ash content, non-volatile organic impurities content. In order to avoid possible thermic decomposition of glycerol, it was necessary to eliminate aluminium before determining the non-volatile residue and taking into account the ponderal influence of aluminium and therefore the need for measurements.

The tedious method was imprecise when the analysed glycerines were rich in ash (glycerine from lye). It did not have volatile glycols in 'organic matters'.

*Mong* calculation is simpler than the evaluation of old method. Values obtained are more repeatable but unfortunately IUPAC, ISO and AFNOR standards do not indicate the reproductibility value.

### *Determination of acidity and alkalinity*

The standardised methods of IUPAC, ISO and AFNOR are practically identical: by definition acidity (alkalinity) is the quantity expressed in milli-equivalents per 100 gram strong base (or strong acid), necessary for neutralising the acidity (or alkalinity) of the sample in presence of phenolphthalein. The measurement of 'saponification equivalent' should be connected to these determinations which is the number of KOH millimoles necessary for saponifying the esters contained in 100 gram glycerine. The method has been standardised by IUPAC.

### *Arsenic measurement*

This method is an application to glycerines of the traditional photometric test method is silver diethyldithio-carbamate; arsenic is transformed into arsine by nascent hydrogen, the arsine is led into a solution (DDTC) Ag where it precipitates silver according to the reaction:

$$AsH_3 + 6Ag\ (DDTC) \longrightarrow 6Ag + 3H\ (DDTC) + 3As\ (DDTC)$$

The colloidal silver precipitate gives a purple colour to the solution whose absorbance is determined at 540 nm. The methods standardised by IUPAC, ISO and AFNOR are practically identical with the exception of the field of application limited to arsenic content greater than 0.5 mg/kg in IUPAC standards instead of 0.1 mg/kg in the two others. These methods are very difficult, as it becomes necessary to ensure that the recovery of arsenic is higher than 80 per cent while proceeding with a second test of glycerine to which a known quantity of arsenic is added. If this is not the case, it is necessary to undertake a new measurement after calcination of glycerine in the presence of magnesium oxide.

Use of this method is only justified if the presence of arsenic leads to major inconveniences. Paradoxically, this measurement is not retained for checking of pharmaceutical quality glycerines.

### *Density determination*

The volumic mass of purified glycerines, e.g. 'dynamite glycerine', is often used for checking their purity. This determination is subject to special precautions owing to the very high hygroscopicity of concentrated glycerines. ISO and AFNOR have standardised this determination by recommending the use of either *Gay-Lussac*'s traditional pycnometer which is not particularly well adapted or *Reischauer's* pycnometer, which has a narrow graduated neck and which requires the use of a vacuum filling and under pressure device. IUPAC combines density determination, with the relative density tables of water-glycerol mixtures.

#### *Determination of colour*

This determination only concerns in practice, glycerines that are distilled and are chemical and pharmaceutically pure. Thus they have very little colour and the usual methods of measuring the colour of fats cannot be applied to them. IUPAC has standardised a method for measuring the colour of glycerine by visual comparison of samples with a standard solution of potassium chloroplatinate and cobalt chloride (Hazen scale) or glasses coloured with colour that is identical to the standards.

### *Characteristics of main type of glycerines*

Characteristics of raw glycerine and purified glycerine are shown in Tables 16.35 and 16.36.

**Table 16.35.** Characteristics of raw glycerines.

| *Description* | *'Lye' glycerine (soaplyse crude glycerol)* | *'Saponification' glycerine (crude glycerol by hydrolysis)* |
|---|---|---|
| Glycerol(%) | > 80 | > 88 |
| Ashes (%) | < 10 | < 1.0 |
| Water (%) | < 10 | – |
| *Mong* (%) | < 2.5 | < 1.5 |
| Arsenic (ppm) | < 2.0 | < 2.0 |
| Propane 1,3 diol (%) | < 0.5 | < 0.5 |
| Sugars | Non-detectable | Non-detectable |

**Table 16.36.** Characteristics of purified glycerines.

| *Description* | *Technical glycerine* | *Dynamite glycerine* | *Chemically pure glycerol* |
|---|---|---|---|
| Glycerol | > 99.0 | > 99.0 | > 99.0 |
| D 20/20 | 1.261–1.264 | 1.261–1.264 | 1.261–1.264 |
| Alkalinity-acidity (mill eq/100 g) | 0.32 | 0.32 | 0.064 |
| Sapo. equivalent (mill eq/ 100 g) | – | 0.64 | 0.64 |
| Colour (113 mm cell) | J < 5.0<br>R < 1.2 | J < 5.0<br>– | –<br>– |
| Hazen colour | – | – | – |
| Sulphated ashes (%) | < 0.010 | < 0.010 | < 0.010 |
| Chlorides (%) | < 0.010 | – | Non-detectable |
| Iron (ppm) | < 2 | – | – |
| Sugars | – | – | Non-detectable |
| Reducing substances | – | Negative | Conforming to BS test |
| Organic chlorine | – | – | Conforming to BS test |
| Arsenic (ppm) | – | – | < 2.0 |
| Lead (ppm) | – | – | < 1.0 |
| Heavy metals (ppm) | – | – | < 5.0 |

The corresponding methods of checking are described in BS 571 standard (1979 and 1983 for 20 heavy metal parts). These corresponding methods and specifications are analogous, but not identical to that of the European Pharmacopoeia.

## Soaps

The properties of soap are very different depending on the nature of cations that they contains. The analytical checks to which they are subjected also differ according to the group to which they belong. Various types of soap dealt in this section are:

1. 'Alkaline soaps' (sodium and potassium soaps).
2. 'Weak base soap' (ammonia, ethanolamine, amines, etc.).
3. 'Metallic soaps' (cation is a metal other than sodium or potassium).

### *Alkaline soaps*

Alkaline soaps are commercialised under diverse forms: at least 14 varieties. Besides two main components (salts of fatty acids and water) they include certain secondary components:

1. Glycerides having escaped saponification or voluntarily added to 'high-fat soaps'.
2. Residual glycerol or voluntarily added to 'glycerine soaps'.
3. Free fatty acids or free base acids, carbonate.
4. Sodium chloride or potassium chloride.
5. Mineral (carbonates, silicates, talc, titanium oxide) or organic matter (amidon, caseine, saw dust etc.).
6. Special additives (EDTA, solvents, medicinal substances, etc.).

Alkaline soaps are soluble in water and ethanol where they are completely ionised. Therefore, it is possible to have recourse to traditional acidimetry techniques to analyse them. In order to facilitate the interpretation of main methods used in analytical checking of soaps, we think it necessary to recall, at the outset, their essential acidimetric properties.

### *Acidimetric properties*

#### In ethanol

Soaps and fatty acids are soluble in ethanol and their acido-basic properties are quite comparable to those of acetates in concentrations that are generally used for analytical chemistry.

Therefore, one can determine the alkalinity of $RCOO^-$ ions of a soap by a solution titrated with strong acid according to the reaction:

$$RCOO^- + H^+ \longrightarrow RCOOH$$

by potentiometry or by using bromophenol which turns yellow from blue thus indicating the end of titration. Free fatty acids can be titrated by alcoholic potash according to the reaction:

$$RCOOH + OH^- \longrightarrow RCOO^- + H_2O$$

by potentiometry or by using pheonolphthalein as indicator which turns pink.

#### In water

In water, the solubility of fatty acids is very low (it is between $10^{-4}$ and $10^{-6}$ moles/litres approximately). Acidimetric properties of soaps in aqueous solution are therefore totally different from those of acetates though the dissociation constants of fatty acids are close to that of acetic acid:

In M/10 concentration, sodium salts of fatty acid having more than 12 carbon atoms have a pH greater than 9.5. The higher the pH, the less soluble the fatty acid, that is to say, for saturated acids, the chain is longer. Lauric acid soaps thus give less alkaline solutions than palmitic and stearic acid soaps. If they are more 'aggressive' towards epidermis, it is not because they are more alkaline (or more 'hydrolysed') than palmitic or stearic soaps but because of their specific properties of short chain vis-a-vis the epidermis.

Like acetates, the soaps have vis-a-vis acids buffer properties but the buffer effect concerns a pH zone much more alkaline than that of acetates (Fig. 16.23). As a result of this we can easily titrate the $RCOO^-$ ions by a strong acid, using helianthine, the methyl red or bromophenol blue as indicator.

### *Determination of water and volatile matters*

Certain soaps, particularly the domestic hard soaps, called 72 per cent, dry up gradually in the air. The general samples and the making of specific samples requires the taking of special precautions to ensure the representativity of the analysed sample. These precautions are described in detail in French Standard T 60.302.

IUPAC, ISO and AFNOR specifications are practically identical and propose two methods:

1. Measurement of water and volatile substances eliminated by heating at 103° ± 2°C.
2. Measurement of water and volatile matters soluble in water and insoluble in xylene (or toluene) by azeotropic distillation. This second method is less sensitive and less reproductible than the stoving method therefore, it is advised only in particular cases.
3. Analysis of a soap containing a volatile solvent insoluble in water.
4. Analysis of a 'potash soap' of linseed oil whose fatty acids would oxidise, become heavy at the time of heating at 103°C and consequently will lead to a wrong water content by defect. In case

of 72 per cent type domestic hard soaps, 72 per cent type dehydrated gradually in the air, the results of analysis, so far as total water content and fats are concerned, is brought by mass calculation to a fresh state which is the invoiced mass and is generally marked on the soap. Therefore, it is necessary to know the mass in fresh state $M_0$ of all soap pieces, their mass $M_1$, during analyses and to bring back by calculation the humidity found experimentally, to the initial humidity of soap with the help of the formula as per French specification NFT 60 302:

$$\text{Per cent water in fresh state} = 100 - \text{DE} \times \frac{M_1}{M_0}$$

where DR is the percentage of dry residue found at the time of analysis.
Reproducibility of this method by using stoving is 0.25 per cent without, however, the standards indicating the specific water content for which this value was established (we known by experience what is valid for water contents about of 30 per cent). Reproducibility by azeotropic method is 1 per cent.

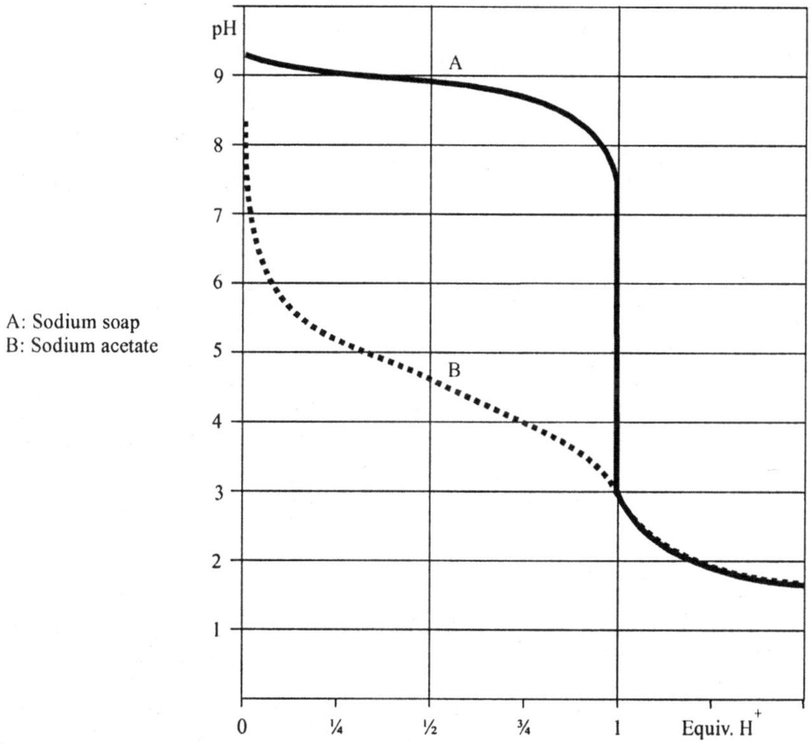


**Fig. 16.23.** Neutralisation graph of acetate and soap in water.

*Determination of 'total fats'*

The term 'total fats' means the fats obtained by decomposition of soap by a strong acid. It is made up of fatty acids of soap, glycerides (or fatty acids) which have escaped saponification, unsaponificable fraction of fats being used for making the soap. If resinic acids are incorporated into them (abietic, neoabietic acid of colophane) they are found again in the total fats which are sometimes called 'total raw fatty acids'.

IUPAC, ISO and AFNOR specifications identical in their principle, differ only on certain details:

After acid decomposition of soap, the liberated fats are extracted by a solvent (diethylic oxide in IUPAC and AFNOR specifications and petroleum ether in ISO specification). After the evaporation of the solvent, fatty acids which constitute the essential of extracted fats are neutralised in alcoholic medium by a strong base titrated solution (soda in IUPAC specification, potash in ISO and AFNOR specifications). The soap, thus formed is dried and weighed. From this mass and volume of the strong base solution used for neutralisation, we calculate the 'total fat content'. In case of soaps which get dehydrated in the air, total fat content thus obtained has to be brought back by calculation to the fat content of the soap in a fresh state by using the formula:

$$\text{Fat per cent in fresh state} = \text{Fat per cent in analysis} \times \frac{M_1}{M_0}$$

where $M_1$ is the mass of soap at the time of analysis and $M_0$ is its mass in fresh state.

Solvent diethylic oxide which is more polar than petroleum ether allows a better extraction of polyunsaturated acids, hydroxylated acids (castor oil soap) and oxidised acids.

*Determination of total alkali content*

This determination is governed by IUPAC specifications. It is included in ISO and AFNOR specifications for determining total fats. The principle of method is identical: soap, in aqueous solution is decomposed by a known quantity, in excess, of a solution titrated with hydrochloric acid (or sulphuric). After separation of fatty acids liberated in this way, the excess of acid is determined by a soda titrated solution.

The determination of total alkali is of little interest because this amount is the sum of three constituents present in most of the cases in very different concentrations: alkalinity combined with fatty acids which for a sodium soap represents 13 per cent of the mass of dry soap, free caustic alkali and carbonates each of which constitutes less than 0.1 per cent of dry soap and can more often be individually determined by specific methods.

*Determination of free caustic alkali or free acidity*

'Free caustic alkali' is the excess of strong base (soda or potash) present in the soap.

By air carbonation excess of strong base transforms progressively into carbonate. In order to determine the caustic alkali in a soap, it is essential to differentiate the strong base from its carbonate.

1. Either by using, for dissolution of sample, a soap and strong base solvent which leaves the carbonate insoluble.
2. Or by precipitating carbonate and soaps under the form of insoluble barium salts in water.

The first method which is 'ethanol' method is applicable only to sodium soaps owing to solubility of potassium carbonate in ethanol.

The second only applies to soaps rich in free caustic alkali due to its non-negligible absorption by very high precipitate of soaps and barium carbonate. With the exception of special cases, it is only used for analysis of potassium soaps.

Commercial soaps which often contain simultaneously a small quantity of unsaponified fats and an excess of strong base, could give rise to, during making of soap solution, a strong base reaction on 'unsaponified', this risk is all the more notable when the quantity of the excess strong base is greater. Therefore, there is no precise method for free caustic alkali titration and all the methods have a conventional character which makes it necessary to follow rigorously the operating methods described in the specifications. IUPAC, ISO and AFNOR specifications are practically identical.

The 'ethanol' method determines the free soda in methanolic soap solution (in the presence of phenolphthalein or by potentiometry). Barium chloride method determines, by a solution titrated with strong acid, the free potassium in ethanol soap solution at 60 per cent after having precipitated soap and carbonate by barium chloride. The ethanol method allows, if need be, measurement of free acidity of soap which will not contain free soda. The reproducibility of the ethanol method is 0.02 per cent for standard commercial soaps and of barium chloride method is 0.5 per cent.

*Determination of total free alkali and carbonates*

The 'total free alkali' is composed of free caustic alkali (soda or potash) and free carbonated alkali (sodium or potassium carbonate). If total free alkali has been determined by the 'ethanol' method, sodium carbonate can be titrated by the following method which is not standardised but is quite fast:

After having measured free soda, add to the alcoholic solution an equal volume of distilled water, neutralised in the presence of phenolphthalein. Sodium carbonate is dissolved in 50–50 water-alcohol mixture which becomes red in the presence of carbonate. Titrate the carbonate by chlorohydric acid 0.1 N upto bleaching of phenolphthalein.

$$1 \text{ ml solution } 0.1 \text{ N} = 10.6 \text{ mg of carbonate}$$

The method standardised by IUPAC, ISO and AFNOR consists in titrating the sum of free caustic alkali and free carbonated alkali by the reaction of a soap solution in 95 per cent ethanol with an excess solution of sulphuric acid, the excess being titrated on return after boiling in order to drive out the released of $CO_2$.

Repeatability is 0.05 per cent. This value concerns the total free alkali of a commercial soap. It does not concern the determination of free carbonated alkali which can be deducted from it.

*Determination of unsaponified and unsaponifiable fraction*

The unsaponified matter is made up of free acids, *mono*, *di* and triglycerides present in the soap.

The unsaponifiable fraction, normally made up of unsaponifiable fat components for making soap can also contain certain additives incorporated in soap (fat alcohols, sterols, etc).

Methods standardised by IUPAC, ISO and AFNOR are equivalent.

In order to avoid that the free strong base does not saponify the 'unsaponified matter' during ethanol soap solution is made, it is done in the presence of an equal volume of sodium bicarbonate solution which immediately reacts on the strong base and transforms it into carbonate which in diluted solution, cannot saponify the unsaponified triglycerides. When soap solution is made, 'unsaponified + unsaponifiable fraction' are together extracted from hydroalcoholic solution at 50 per cent by washed, dried and weighed by hexane. It is then put in alcohol solution to titrate free fatty acids which occur by hydrolysis of soap during washing and then titrated with the unsaponifiable fraction according to a technique derived from the standardised unsaponifiable measurement method in the fats.

In soaps containing unsaponifiable additives such as fatty alcohol, sterols, sulphur, etc. these additives get together in 'unsaponified + unsaponifiable matter' and in unsaponifiable matter. That is why the specifications indicate that the method is not applicable to this type of soap. We believe that the method remains applicable (except if the additive is a perfume) but its interpretation must take into account the presence of additive whose nature has to be known beforehand.

The complex and relatively long method of implementation has poor reproducibility because the unsaponifiable matter is not determined directly but by the difference between the results of three experimental determinations, 'unsaponified + unsaponifiable matter,' together, free fatty acids and unsaponifiable matter. The last experimental determination, generally done on substantially weak quantities after a succession of operation is itself known to be difficult to repeat.

Besides, IUPAC, ISO and AFNOR specifications do not give any indication of the repeatability and reproducibility of the method.

### *Determination of chlorides*

The principle of IUPAC, ISO and AFNOR is identical: argentimetric measurement (volumetric or potentiometric) of chlorides. The measurement being done in a highly acidic medium (pH less than 1), soaps are decomposed and fatty acids insoluble in water, disturb the measurement. Therefore, they have to be eliminated before the measurement either by filtration (IUPAC and AFNOR) or before acidification and adjunction of silver nitrate by precipitation in the form of calcium soaps (ISO). This second technique runs the risk of leading to chloride loss owing to its partial absorption by a very important precipitate of calcium soaps.

### *Determination of free glycerol*

The glycerol content of soaps is very different when it comes to 'salted out' soaps (toilet, domestic hard soaps), 'glycerine' toilet soaps or saponification soaps (potash soaps, industrial soaps).

IUPAC, ISO and AFNOR specifications describe two different methods according to the presumed glycerol content of the analysed soap.

For glycerol contents greater than 0.5 per cent the three specifications prescribe the measurement of formic acid obtained by periodic oxidation of glycerol remaining in water phases after acidic decomposition and extraction of fatty acids by hexane or diethylic oxide.

IUPAC and AFNOR specifications use diethylic oxide to extract fatty acids while ISO specification advise petroleum ether to avoid any risk of glycerol solubilisation in the extraction solvent. But this solvent can, in certain cases, extract the totality of fatty acids with difficulty (e.g. castor oil soap).

For glycerol contents less or equal to 0.5 per cent, three practically identical specifications determine, after acid decomposition of soap and extraction of fatty acids, the glycerol by formaldehyde colourimetry (resulting from oxidation of glycerol by periodate) with the help of chromotropic acid. The periodic oxidation of polyols possess two closely related 'primary alcohol' functions giving rise to aldehydes, the method is defective if the soap contains this type of polyol, particularly sugars or glycols.

Although more specific to glycerol, formic acid measurement after periodic oxidation is also faulty if the soap contains sugar. Reproducibility of glycerol measurement in quantity greater than 0.5 per cent established by a circular analysis re-grouping 23 laboratories is 0.2 per cent for a free glycerol content of 2.7 per cent. Reproducibility for contents less than 0.5 per cent is not specified in any of the three specifications. Other methods have been proposed particularly the gas chromatography with capillar columns applied directly to water phase after the decomposition of soap more than the glycerol measurement in soaps, they are used for analysis products more complex than the soap of—cosmetic milk or face cream type.

### *Determination of ethanol insoluble matter*

Besides carbonates which are present in a more or less important quantity in all soaps, the insoluble matters in ethanol are essentially composed of additives voluntarily incorporated in soaps: mineral additives such as titanium oxide or zinc oxide, organic additives such as casein, starch, saw dust and certain silicates in industrial soaps. Methods standardised by IUPAC, ISO and AFNOR specifications are practically identical: the soap is put in a 95 per cent boiling ethanol solution. The insoluble residue is filtered, washed and dried at 103°C and weighed. The reproducibility is 0.05 per cent when the insoluble content is less than or equal to 1 per cent and 0.1 per cent for insoluble contents greater than 1 per cent.

### *Determination of fatty acid composition*

'Raw' fatty acids extracted from diethylic oxide can be used easily for analysis by GLC. The copra or palm oil content of a soap often constitutes an important commercial characteristic. The content is evaluated from lauric acid content determined by GLC. Therefore, it is important that the preparation of methylic esters and analysis be done according to techniques which do not bring about a loss of lauric acid. Resinic acids of colophane (abietic, neoabietic acids) can be used in certain countries at certain times to increase the foaming power of soaps.

There is also a method standardised by IUPAC for resinic acid measurement in soaps. It is based on the esterification speed difference between the fatty acids (easily esterified) and resinic acids (esterified with difficulty). In specific operating conditions for a given concentration, the esterification of fatty acids by methanol by using sulphuric acid as catalyst can be complete while that of resinic acid is practically zero. Resinic acids can therefore be titrated by acidimetry (titrimetry or potentiometry). A blank test has to be carried out with fatty acids without resins to take into account very low quantity of fatty acids which remain non-esterified. The method can be used to determine the resinic acids in 'tall-oil'.

### *Study of additives*

Additives which can be incorporated in various types of soaps are many and there is no general method of examination or a particular standardised method for EDTA (tetrasodic salt of aceticdinitrilotetraethylene acid) which can be tested directly after acid decomposition of the soap by complexometric titration with a solution of copper sulphate titrated in the presence of an appropriate coloured indicator: mono sodic salt of (pyridyl-2-azo) 4-resorcinol which turns from yellow to pale pink at the end of titration.

The method is specific except in the presence of more than 10 per cent of sodium salt of another chelating agent of calcium, nitrilo triacetic acid (NTA).

Reproducibility established by a circular analysis re-grouping 12 laboratories is 0.06 per cent for a soap containing 0.3 per cent of EDTA tetrasodic salt.

Very often additives are found to be insoluble in alcohol or in unsaponified + unsaponifiable matter.

In an insoluble alcohol one looks for:

1. The metallic pigments (zinc oxide, titanium oxide, lithopone) by mineral chemistry techniques after elimination of eventually present organic substances by calcination (zinc sulphur contained in lithopone is thus transformed into sulphate and zinc oxide).
2. Organic additives as casein and starch which will be identified and titrated by certain specific methods (for example determination of nitrogen content by the Kjeldahl method for 'measuring' casein).

On the unsaponified + unsaponifiable fraction, one looks for lanolin 'extra-fats' such as almond oil which can generally be identified and evaluated by gas chromatographic analysis, methylic esters, alcohol and sterols of the unsaponifiable fraction. It is to be recalled that sulphur is again present in this fraction but to recuperate it quantitatively, it is preferable to carry out extraction with diethylic oxide.

### *Colour iodine (colour on saponification)*

This method standardised by AFNOR is in reality a conventional measuring method of oxidisation of fats. But since it is only used for testing tallow meant for soap industry, it is logical to deal with it here.

It consists of saponifying 5 gram tallow by about 5 N potash in well specified conditions and then comparing visually the colour of the alkaline soap solution to an iodine solution obtained by pouring an aqueous solution of 5 gram/l iodine in a known volume of water to match the colour. The result is expressed in 1/10 ml of iodine solution used.

### *Bleaching test*

The soap making industry, for checking its raw materials, particularly tallows, uses a test meant for determining their aptitude at bleaching. The melted fat is shaken in well defined conditions (specified in AOCS standards) with a bleaching earth before measuring its colour. The choice of earth, respective weight of earth and fats and the importance of sample for analysis have a considerable influence on the results.

Results obtained by methods identical in their principle, but consisting of variants, cannot be compared between them.

## *Weak base soaps*

Weak base soaps are generally prepared extemporaneously by a mixture of fatty acids with a weak base.

Contrary to soaps of strong base, they very often contains, a high excess of free fatty acids or free base.

The main analytical determinations to be discussed are:

1. Water content.
2. Free acidity (or basicity).
3. Total fatty acids or amine content.

The possible analysis of fatty acids used for preparing the soap is done by GLC as in case of strong base soaps.

### *Water measurement*

The weight loss method at 103°C is not used because the ammoniac and the aminos, the amino salts and free fatty acids, at this temperature have a high vapour pressure. The azeotropic drive is usable only if the amino is not driven by water-toluene or water-xylene azeotrope, which for example is not the case with monoethanolamine. The best method is the titrimetric test of water by Fischer reactive. In presence of free amino, it is necessary to acidify the soap before titration by anhydrous acetic acid, (or by a correctly measured acetic acid with known water content), the free bases interfering in titrimetric measurement of water.

### *Free fatty acids (or bases) measurement*

In 95 per cent ethanol solution, weak base soaps are completely ionised, for example in case of an ammonium soap in:

$$RCOO^- + NH_4^+$$

If the soap is stoichiometrically neutral (a mole of fatty acid for a basic mole) the solution has as much $RCOO^-$ ions as $NH_4^+$ ion.

In 95 per cent alcohol solution we can:

1. Volumetrically measure $RCOO^-$ ions by a strong 0.1 N acid solution, titrating potentiometrically (or in presence of bromophenol blue), according to which

   $$RCOO^- + H^+ \longrightarrow RCOOH$$

   Or VA the volume of strong acid solution used per gram of soap.
2. Volumetrically determining $NH_4^+$ ions by an alcoholic potash solution 0.1 N, potentiometrically or till phenolphthalein turns to pink, according to which:

   $$NH_4^+ + OH^- \longrightarrow NH_3 + H_2O$$

   Or VB the volume of strong base solution used per gram of soap.
3. If VA = VB the soap is made up of a fatty acid mole and a basic mole.

4. If VA < VB the soap contains less of $RCOO^-$ ions, than $NH_4^+$ ions and free fatty acids. As $NH_4^+$ ions which come from ionisation of soap are in a quantity equal to that of $RCOO^-$ ions, the soap contains free fatty acids whose molar content per gram of soap is given as:

$$VB - VA$$

If VA > VB, the soap contains, besides $RCOO^-$ ions, from the free base whose molar concentration per gram of soap is given as:

$$VA - VB$$

If the average molecular weights of fatty acids and base are known, it is possible, thanks to these two volumetric titrations, to calculate the centesimal composition of the soap.

*Measurement of total fatty acid content*

If the average molecular weight of fatty acids is unknown, extraction and weighing of fatty acids should be carried on according to the method proposed for strong base soaps.

If their average molecular weight is known, their content in soap is directly calculable from VA.

*Identification of amines*

Aminos, separated from soap by steam in a sufficient alkaline medium in order to transform the whole of amino soap into strong base soap can be identified by TLC, GLC or HPLC.

***Metallic soaps***

Properties of these soaps essentially used in lubrification (Ca, Ba, Al, Li, Zn soaps), as oxidation catalysts (Pb, Co, Mn soaps) and as impermeabilisation agents (Cu, Fe soaps) are linked to their metallic cation content. But very often metallic soaps contain an excess of metal, coming from preparation, either in the form of oxide or in the form of mineral salt (chloride, sulphate, nitrate). Some of them may contain, as is case with wire works soaps, mineral loads being voluntarily added to modify their physical properties. Therefore, it is not sufficient to use quantity determination by analytical mineral chemistry method, total metal after calcination or by hydrolysis in a concentrated acidic medium for the soap. It is necessary to distinguish metal combined with fatty acids from metal in the form of mineral salt or oxide.

This distinction is much more important in certain cases, e.g. for lithium hydroxysteatates, mineral impurities changing the properties of the soap.

In general, the analysis of a metallic soap thus consists of:

1. The determination of mineral 'impurities' separated from soap by dissolution of the latter in an organic solvent (chloroform, toluene).
2. The test, on purified soap, of total fatty acids and metallic cation after acid hydrolysis in a concentrated acid medium, which decomposes the fatty acid and metallic cation soap soluble in acid water.

The extraction of fatty acids after acid hydrolysis has to be carried out preferably by diethylic oxide, hexane being a poor solvent of certain fatty acids used for preparing metallic soaps (hydroxy and dihydroxystearic acids).

## SAMPLING

### General Conditions—Definitions

'Sampling' may be defined as the entire set of operations which when applied to a sizable quantity of materials (ranging from a few quintals to several thousand tonnes) lead to the compounding of a sample (scaled down) to a convenient size (from some hundred grams to a few kilos). It is this scaled down sample (sample for testing) which is intended for the laboratory so as to determine its analytic properties and thereby find out the market value of the lot from which it has been selected. The sample intended for the laboratory should be 'representative' i.e. it should represent as accurately as possible, the stock or loading from which it has been selected out. In any case, the concept of a representative sample is inseparable from the concept of analysis, the final error of the results of measurement is the sum of errors committed during sampling, sub-sampling and analysis.

It must be noted that very often the error made during the operation of sampling is much greater than the error arising from analysis because correct sampling is a very difficult operation demanding tremendous care specially for seeds because of their very heterogeneous composition (Fig. 16.24).

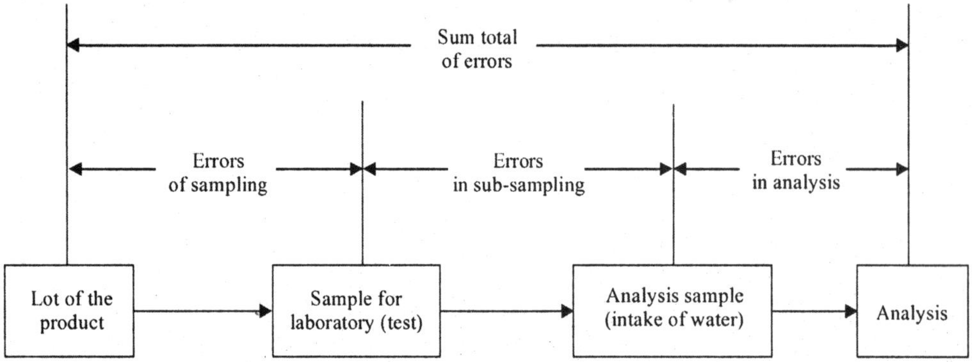


**Fig. 16.24.** Errors of sampling.

Moreover, it is unfortunate that though laboratory analysis has always been regarded as a delicate science to be conducted only by a very qualified and skilled staff, this is not so for sampling. This task is entrusted too often to under qualified people who are hardly aware of how difficult and important this operation is.

Sampling projects and procedures for each type of item would be in the form of recommendations or standards (Table 16.37).

**Table 16.37.** Standards and rules for sampling from seeds, oleaginous oil meals, oils, and fats.

| *Description* | *Oleaginous oilseeds* | *Oil meals* | *Oils and fats* |
|---|---|---|---|
| AFNOR | NF ISO 542 | NF ISO 5500 | T 60–280 |
| ISO | ISO 542 | ISO 5500 | ISO 5555 |
| EEC rule | N° 1470/68 Appendix | – | – |
| FOSFA | ISO 542 | – | ISO 5555 |
| IASC (1) | IASC-2nd section | – | IASC -3rd section |
| GAFTA (2) | – | Rule 121 | – |

(1) IASC: International Association of Seed Crushers.

(2) GAFTA: The Seed and Feed Trade Association

In the section, we will deal with each case of sampling separately oilseeds, oleaginous oils meal, oils and fats.

## Sampling of Oilseeds

Sampling of oilseeds has come under national as well as international standardisation (NF Standard ISO 542). In addition, this standard has been included in the EC regulations (N° 1470/68) and FOSFA commercial contracts. This standard recommends that samples be compounded for each batch of maximum 500 T as per a chart depending on the nature of the seed (size) and on the type of consignment (packed or loose) (Fig. 16.25).

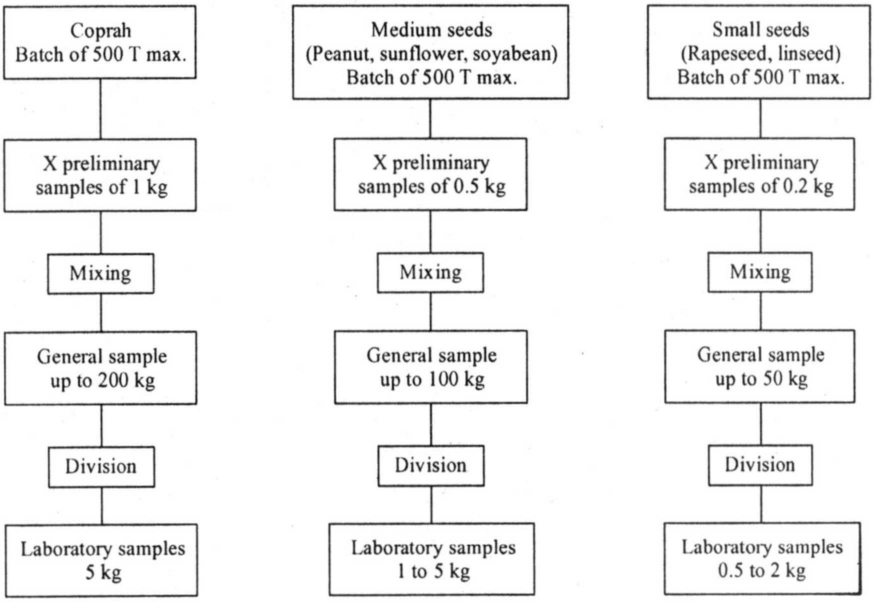


**Fig. 16.25.** Diagram of oilseed sampling (NF ISO 542).

### *Preliminary sampling*

For this stage, two cases must be considered i.e. whether the item is packed in sacks or is loose.

#### *Seeds in sacks*

Preliminary samplings must be carried out on 2 per cent of the sacks selected at random from the entire lot. The operation of selection out is usually done with slim seed samplers which allow samples to be picked out at different levels directly through the sides of the (jute) sacks without opening them beforehand.

#### *Seeds in bulk*

Initial samples may be selected from a static mass (seeds at rest), or from a mass in motion.

#### Sampling from a static mass

A 'lot' of seeds at rest is characterised by a heterogeneity in three dimensional spatial compositions (longitudinal-transversal-vertical).

To resolve the problem of heterogeneity, specially on the vertical and longitudinal dimensions, which are the most important, samplings must be done from different points and at different levels in the mass of seeds (height of the layer).

Thus, NF ISO 542 standard recommends that sample should be selected out from a minimum of 3 points from a truckload of oilseeds, and from 5 points, from a wagon load. Equipments used to select samples usually consists of samplers that are akin to the core barrels used for geological exploration. They make it possible to make a forced penetration at a given point in the lot of seeds, and to pick out an increment which represents a column stretching along the entire height of the layer.

Samplers which are generally of the round section type have several orifices spaced out over the entire length at uniform intervals. Openings are obturated by lids when the samplers are placed into the mass of seeds (Fig. 16.26). Then the lids are opened gradually, beginning from the lower orifices so as to ensure that all sections of the orifices are filled equally. After the filling, the lids are closed again and the sampler is removed from the mass.

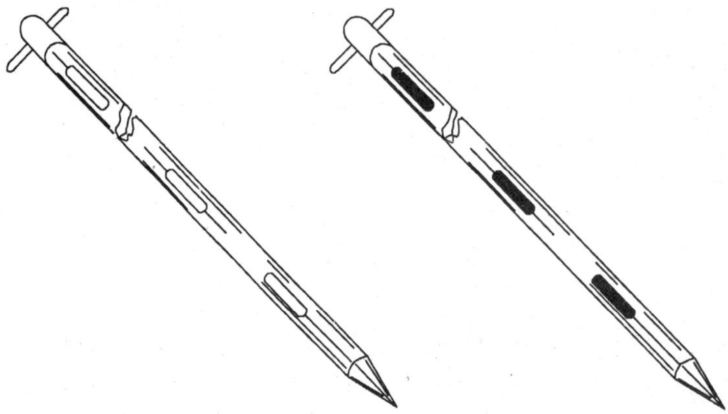

**Fig. 16.26.** Samplers used for sampling.

The size of the orifices of these samplers depend on the granulometry of the seeds to be selected as samples. Usually it is recommended that the size of the orifices should be 3 to 5 times more than that of the seeds to be sampled.

These cylindrical samplers with alveolus may be used manually or mechanically.

1. Manual use of these samplers requires a lot of physical energy from the operator and has to be restricted to taking samplers from limited layers of seeds because of their useful length (maximum 2–3 minutes).
2. To remove the difficulties which arise when manual samplers are used, automatic sampling—selecting systems have been developed. In this type of equipment, the alveolar sampler is usually replaced by the suction sampler which has at its tip one or many openings which suck in the seeds during the entire operation of penetration and withdrawal from the mass of seeds.

The quantity of seeds picked out depends mainly on the length of time for which the sampler stays in the mass of seeds, on the size of the suction openings and the air flow. Even though, the use of samplers (manual or automatic) is the only way of selecting preliminary samples representing the entire lot of seeds at rest, it is unfortunate that this type of instrument is not used very often.

Therefore, selection of samples with a bucket and a spade only from the top of the load of seeds is still a very common sight today.

### Sampling from a mass in moving

All studies have shown that it is always preferable to carry out the initial sampling when the seeds are in moving i.e. during the process of loading, unloading or carrying from one side to another.

In this case, initial samples are taken at regular intervals from the entire flow of seeds in motion. Taking samples of seeds in motion can be done manually or with the help of automatic devices (selectors).

1. Manual sampling is the method most widely used for sampling truck loads. A cup is fixed on to a pole of suitable length and samples are selected from the entire transversal section of the flow of seeds while they are being dumped.
   The main disadvantage of this method of selecting samples manually is that samples selected can easily be falsified, purposely, apart from the disadvantage that it requries tremendous energy and concentration from the operator.
2. If automatic selectors (samplers) are used, the role of human factors is considerably reduced. Thus, they have proven to be more worthwhile. There are several types of devices for this operation which are discussed below.

#### Cross selector

This type of apparatus is most widely used and ensures that the entire flow of seeds in motion is cut at regular intervals with the help of a cutter with a slit varies according to the size of the seeds to be sampled. The cutter works in rectilinear motion (Fig. 16.27a) or in rotary motion (Fig. 16.27b). These sample selectors are fitted with timers which accurately regulate the duration and frequency of the sample picking operation.

#### Screw sample selectors

This kind of sample selectors which equip some dumping trenches, are made of an endless screw thread rotating in a tube. The tube has a horizontal slit over a certain length. The slit is obsturated by a cover which opens at regular time interals. The main disadvantage of this type of sample selectors is that they can select samples only from a portion of the flow of seeds pouring into the trench.

### *Compounding samples for the laboratory*

Different preliminary samplings from a lot are thoroughly mixed in order to compound the combined sample. Then this will be divided into several smaller scaled down samples intended for the laboratory.

For the operation of dividing the combined sample into laboratory samples, it is advisable to use conical or multi-slit dividers like those described in earlier section which deals with scaling down of laboratory samples. At this stage, it is unfortunate that this type of separator is not always used in this deduction operation. Some operators are satisfied with mixing manually the diffrent elementary samples in a vessel of suitable volume (bucket, boiler), the general sample thus compounded serves in compounding the different samples for the laboratory without any particular precaution being taken. The analysis of oilseeds consists in the majority of cases of humidity determination and the results of other analytical parameters being formed in the material as it is (humid), it is important that the container of the laboratory samples be perfectly air-tight.

Therefore, for packaging of samples, plastic or glass jars or failing which plastic pouches must be given preference in the choice of materials. Paper or canvas bags are to be avoided.

Since the laboratory samples are intended for arbitrary analyses in a great number of cases, they must be stamped and labelled properly to guarantee the authenticity and origin of the samples.

## Sampling of Oilseed Meals

Just like in case of oilseeds, the procedure of sampling of oil meals has been standardised jointly under AFNOR and ISO (Standard NF ISO 5500).

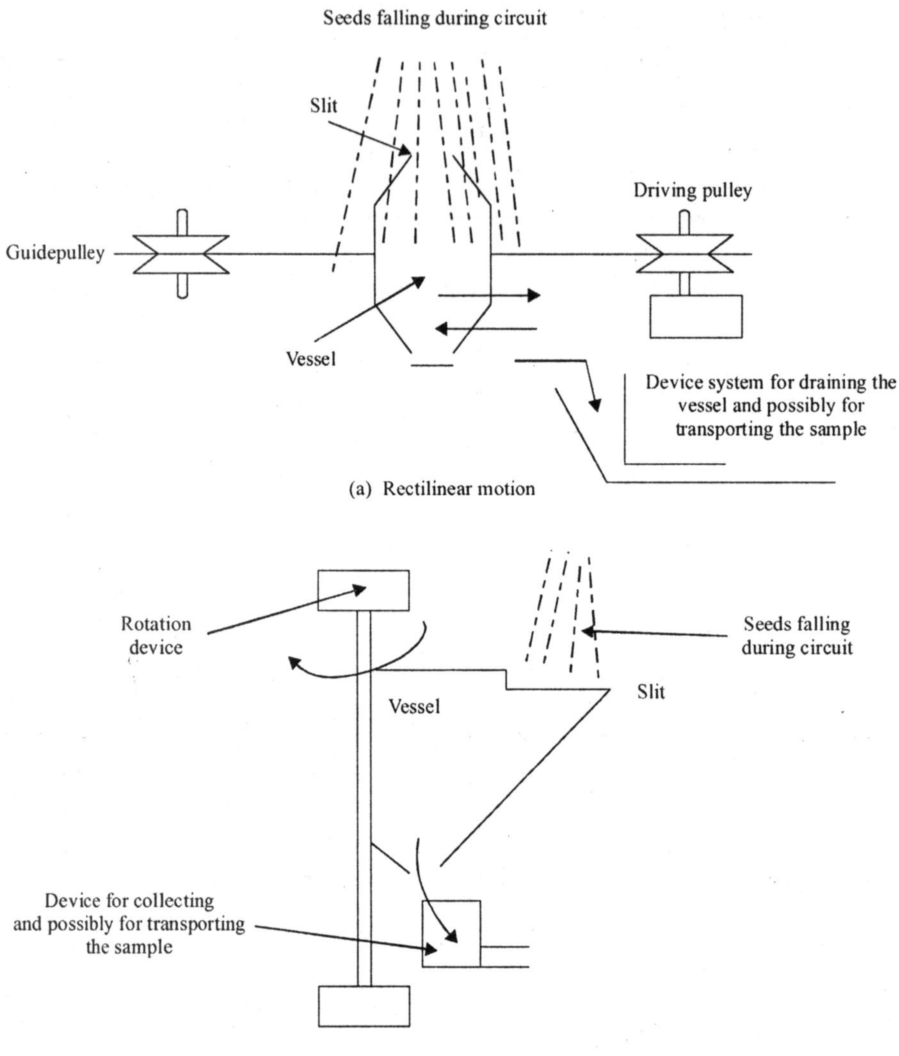


**Fig. 16.27.** Cross sample selectors.

This standard defines the conditions of sampling of oil meals according to their presentation (flours, conglomerates or slabs) and the type of packaging (sacks or loose). The procedure of sampling defined in this standard for oil meals in the form of flour or in conglomerated form (pellets) is very akin to the one which has been previously dealt with oilseeds. Only the cases of sampling of oil meals in the form of slabs and the sampling carried out for measurement of certain undesirable substances (mycotoxins, castor oil) have come under a particular protocol in operation).

### *Sampling of loose and packed oil meals*

Just like for oilseeds, NF ISO 5500 standard recommends compounding of samples for each 500 T batch or segment (Fig. 16.28).

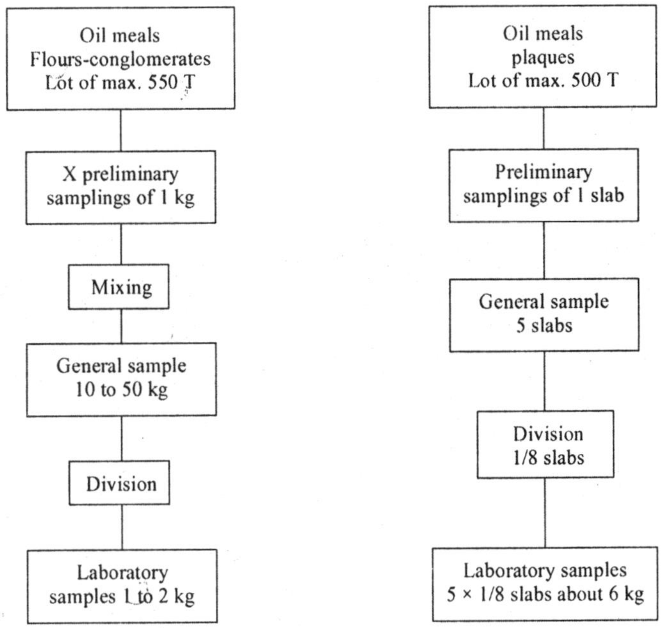


**Fig. 16.28.** Diagram of oilseed meal sampling (ISO 5500).

The method to be used for preliminary sampling for compounding will depend upon the kind of packaging of the oil meals (sacks or bulk).

*Oil meals in sacks*

The preliminary samplings are selected with the help of the same kind of slim seed samplers that are used for oilseeds from as many sacks as are required according to the size of the lot to be sampled (Table 16.38).

**Table 16.38.** Number of sacks to be sampled according to ISO 5500 standard.

| *Description* | *Number of sacks* |
|---|---|
| From batch | To be sampled |
| Up to 10 | Each sack |
| 11 to 100 | 10 picked at random |
| More than 100 | Approximately the square root of the total number selected at random according to a proper sampling plan |

*Oil meals in bulk*

The preliminary sampling can be done when the oil meal is at rest (static), or preferably when it is in motion during the process of loading, unloading or transfer.

The operative protocols to be applied and the instruments for selecting out samples that are to be used (manual seed samplers or automatic pickers) are in both cases identical. The different preliminary samples are then thoroughly mixed as for oilseeds to make combined samples which in their turn are divided to give small samples for the laboratory. This operation of dividing should be done preferably with the help of dividers (conical or with slits) like those described earlier. It should be noted that in case this type of dividers are not available or cannot be used, one can resort to division by quarters.

### *Sampling of slab oil meals*

In the particular case of slab oil meals (expellers), it would be advisable to apply a sampling plan which is little different from the one previously described for oil meals in the form of flour or pellets (Fig. 16.28).

When the product is packed in sacks, a slab must be selected from a particular number of packages (sacks). This number must represent at least 2 per cent of the total number of sacks representing the entire lot. As for the items in bulk, five slabs of oil meals per lot of 500 T must be picked at random. A small scaled down sample will be made from these different samples by taking only a portion of each slab (1/8). It would be advisable to take care that the laboratory sample compounded in this manner is made of pieces where all the parts of the slab are represented (corners-central parts).

### *Sampling of oil meals for detecting and measurement of undesirable substances*

Since contamination of the oil meals by some undesirable substances like mycotoxins or some toxic seeds (castor) is particularly heterogenous and problematical, it would be advisable to apply a sampling project that is much more rigorous than the previous ones, for detecting and measuring them.

Thus, specific protocols for sampling were proposed by several authors chiefly in the area of research and measuring mycotoxins (aflatoxins).

However, it is unfortunate that they are rarely applied in practice, on account of the time required for executing them and the cost of the operations.

In Tables 16.39 and 16.40, ISO 5500 Standard takes into account this particular case of sampling of oil meal in view of research and measuring undesirable substances and toxic seeds.

**Table 16.39.** Number of total samples to be compounded for oil meals in sacks as per ISO 5500 standard.

| *Number of sacks or containers in the lot* | *Minimum number of general samples separated* |
|---|---|
| 1 to 20 | 1 |
| 21 to 200 | 2 |
| 201 to 800 | 3 |
| More than 800 | 4 |

**Table 16.40.** Number of total samples to be compounded for oil meals in bulk as per the NF ISO 5500.

| *Lot* | *Minimum number of general samples separated* |
|---|---|
| Up to 1 T inclusive | 1 |
| 1 to 10 T inclusive | 2 |
| 10 to 40 T inclusive | 3 |
| 40 to 500 T inclusive | 4 |

To resolve the heterogeneity of the contamination, the standard recommends increasing the number of initial samples to be selected, general samples and laboratory samples per lot examined (Tables 16.39 and 16.40). Thus for a lot of 500 T, not one laboratory sample (as is generally done) but four samples should be compounded and each one should be analysed separately.

### *Laboratory samples*

Laboratory samples for determining water content and other parameters of analysis the result of which can be influenced by change of moisture of the product must compulsorily be packed in containers that are completely air tight (glass or plastic jar, or plastic bag). However, it should be noted that it is not advisable to use plastic containers for samples intended for measuring volatile hydrocarbons (hexane).

Since, these samples are generally subjected to arbitrary analyses, like oilseeds, they must be properly stamped and labelled to guarantee the authenticity and origin of the sample.

## Sampling of Animal and Vegetable Fats and Oils

Samplig of fats has been standardised in France (NF T 60–280) and internationally (ISO 5555).

Sampling procedures laid down by these standards are identical and have moreover been included in the recommendations of FOSFA and IASC.

The protocol of sampling to be applied will depend on the type of container other than the nature of the fat to be sampled (fluid or solid).

### *Fats packaged in small units*

For consignments consisting of a large number of separate packages in small quantities (barrels, crates, drums, boxes, bottles) samples should be selected from a suitable number of packages chosen at random from the entire consignment. AFNOR and ISO standards have laid down the precise number of packages which should be sampled depending upon the number of pieces in the consignment and size of the individual packing (Table 16.41).

When initial samples are being selected for compounding the general samples, a distinction should be made between solid fats (solid or semi-liquids) and fluid fats (liquids).

1. For solids fats, an open sampler with semi-circular cross-section or a C-shaped sampler must be used for core-drilling through the entire thickness of the matter in several directions and at different points. This type of sampler can also be used for selecting samples of emulsified fats (margarine, butter, etc.).
2. For packages of liquid fats, it would be advisable to homogenise the contents of the container before picking out the sample.

This homogenisation could be done mechanically (agitator) either manually by rolling or turning the barrels upside down. A cylinder-shaped sampler of similar design as that of an oilseed sampler would be used. This type of sampler is made of two concentric tubes with longitudinal openings spread over the entire length. The two tubes are closely fitted so as to move freely inside one another. In one position the tube is open and allows the oil to come in freely; it becomes an air tight container when the inner tube is turned. Another more rudimentary type of sampler can also be used for selecting samples of fluid oil. It consists of a glass or plastic tube of suitable length which is inserted in the container to be sampled. Then all that is required to be done is to close the end of the tube with a finger and to pull it out of the mass to get a sample from a particular point.

**Table 16.41.** Number of packages to be sampled according to NFT 60–280 and ISO 5555 standards.

| *Size of package* | *Number of packages in consignment* | *Number of packages to be sampled* |
|---|---|---|
| More than 20 kg, up to 5 T max. | 1 to 5 | All |
| | 6 to 50 | 6 |
| | 51 to 75 | 8 |
| | 76 to 100 | 10 |
| More than 5 kg, up to 20 kg incl. | 1 to 20 | All |
| | 21 to 200 | 20 |
| | 201 to 800 | 25 |
| | 801 to 1600 | 35 |
| | 1601 to 3200 | 45 |
| | 3201 to 8000 | 60 |
| | 8001 to 16,000 | 72 |
| | 16,001 to 24,000 | 84 |
| | 24,001 to 32,000 | 96 |
| | More than 32,000 | 108 |
| Up to 5 kg incl. | 1 to 20 | All |
| | 21 to 1500 | 20 |
| | 1501 to 5000 | 25 |
| | 5001 to 15,000 | 35 |
| | 15,001 to 35,000 | 45 |
| | 35,001 to 60,000 | 60 |
| | 60,001 to 90,000 | 72 |
| | 90,001 to 130,000 | 84 |
| | 130,001 to 170,000 | 96 |
| | More than 170,000 | 108 |

### *Bulk fats*

Initial samples may be selected either when the fats are at rest (static mass) or in motion (decanting) as the case may be.

#### *Samples selected from fats and oils at rest*

Before proceeding with the sampling operation, it is important to ensure the homogeneity and liquefaction of the fat as far as possible. Thus, as far as possible for solid fats (tallow, lard, copra, palm kernel, palm etc.), it is necessary to heat up the product so as to bring it to a particular temperature range depending on the product being sampled (Table 16.42). For preliminary sampling, many sampling devices of different sizes suitable for different types of containers (ship tanks, tank wagons, tank trucks, earthen vat) are available in the market. The simplest device consists of a bottle or a metal vessel that is generally weighed and which is fixed to a rod (pole) long enough to reach the lowermost part of the mass to be sampled. A detachable cap closes the opening of the container when the equipment is immersed in the fat.

**Table 16.42.** Temperature range under the fat has to be maintained during the sampling operation.

| *Product* | *Temperature °C* | |
|---|---|---|
| | *min.* | *max.* |
| Safflower, rapeseed, linseed, corn, olive, sesame, soyabean, sunflower | 15 | 25 |
| Peanut, tung | 20 | 25 |
| Cottonseed | 20 | 30 |
| Castor, palmolein, fish | 30 | 35 |
| Coconut | 35 | 45 |
| Palm kernel | 40 | 45 |
| Fats | 45 | 55 |
| Lard, palm, illipe, karite (shea) | 50 | 55 |
| Tallow | 55 | 60 |
| Palm stearin | 55 | 65 |

When the required depth is reached, all that is required to be done is to remove the cup by pulling on the chain or rope so that the container is filled.

Other sampling devices used for fats are:

1. The tilting sampler-plunger consists of a metal cylinder of about 150 mm length and 50 mm diameter fitted with a device which allows it to tilt. The instrument is immerged upside down (with its opening towards the bottom) then turned up when it reaches the required depth with the help of a trigger thus allowing the cylinder to fill up.
2. The plunging type sampler with valve consists of a cylinder opened on the upper side and the bottom of which is fitted with a valve. The valve remains open while the instrument is lowered. When the lowering stops, the valve closes and a sample is taken from the depth while the instrument reaches.

Other kinds to samplers are also available and they are called 'bottom' or 'zone' which can tell whether the contents of the tank are homogeneous or not. Whatever sampling device is used, it is important that many samples are taken out from the entire depth of the tank or cistern, the minimum being three samples from the upper, medium and lower level. Next the different samples are mixed in appropriate proportions to obtain a representative general sample from the lot. This operation becomes difficult sometimes, because certain parts of the loading are not accessible, owing to the geometrical shape or the placement of some tanks which usually happens on ships.

### *Samples taken from oils and fats during decantation*

This sampling process commonly known as 'drop by drop' consists in selecting a small quantity of oil regularly throughout the transfer of the product (loading, unloading). The device consists of a tube of 9.5 mm diameter fixed in the centre of the main outlet line, facing the flow of oil. To obtain good results, with this method of taking samples, it is important that the flow rate of oil be sufficient to produce a turbulence resulting in proper and thorough mixing and stirring of the product.

In addition, it should be noted that this method can be used only for sampling fluid oils which contain very few impurities capable of blocking the channel used for taking out samples.

### *Compounding general and scaled-down samples for the laboratory*

AFNOR and ISO standards recommend that one general sample be compounded for each lot or part of 500 tonnes, by mixing the various initial samples obtained by following the methods discussed in the above paragraphs. The minimum size of the general sample to be obtained depends upon the size of the lot to be sampled (Table 16.43).

**Table 16.43.** Minimum size of universal sample.

| *Contents, tonnes* | *Minimum size of general sample, l* |
|---|---|
| Up to 20 incl. | 1 |
| From 20 to 50 incl. | 5 |
| From 50 to 500 incl. | 10 |

The general sample is then divided into a minimum of 4 scaled down samples (laboratory samples) of a minimum size of 250 gram. It should be noted that for conducting certain kinds of analyses, larger samples of 500–1000 gram have to be compounded.

Concerning packaging laboratory samples, a certain number of precautions have to be taken which are given below:

1. Use clean and dry containers made from materials which is inert towards fats. Thus, all containers with iron or copper are to be avoided because of the pro-oxidising effect of these metals in oils and fats. Containers made of glass are highly recommended: however, some materials such as plastic or metals with internal varnish may be used if some precautions are taken.
2. Fill the containers but leave a minimum volume of air to allow the fats to expand due to temperature variations.
3. Store the samples away from light and heat.

## DETERMINATION OF FAT CONTENT OF COMPLEX PRODUCTS

Generally fat analysis is often applicable to fats which are components of biological mediums, raw materials or compound foods whose fat content and characteristics should be determined.

As the methods described above cannot be applied directly on complex products, fat should be extracted in advance. This extraction should be total without changing the characteristics of the extracted fat.

Both in the case of foods and biological medium, it is difficult to have extraction methods which may respond to these two characteristics, the partial extractions being generally selective, the polar lipids being more difficult to extract by the solvents of low polarity then the 'neutral lipids'.

### Food

The term 'fat' is ambiguous. In compound foods, rich in eggs, for example, does one have all the phospholipids, or on the contrary, only triglycerides. The whole or part of phosphatides is extracted depending on the analytical technique used.

As a matter of fact, what is measured is defined by the method of analysis and varies from one method to another. This finding for example, has led ISO not to standardise the measurement of oil in oil meals, but a method of measuring of hexane extract and a method of measuring of diethyl oxide extract.

Conventional character of the methods obliges one to take recourse to standardised methods whenever they exist.

They can be classified into four principal groups based on:

1. The extraction of fat by a solvent (or a mixture of solvents) without prior treatment other than eventual desiccation of food.
2. The extraction of fat by a solvent (or a mixture of solvents) after a prior chemical treatment (in general an acid hydrolysis).
3. The direct measurement of fat by spectrometry (IR or NMR).
4. For the foods containing at least 80 per cent of fat, the determination of fat by measuring other components and difference at 100.

Table 16.44 gives the principal method used and the foods to which they are applied. All these methods are sensitive to the water content of the foods studied and they can be applied only to the foods whose water content is in conformity with the indications given in the methods.

**Table 16.44.** Principal methods used and foods to which they are applied.

| *Principle of the method* | *Foods* | *Preliminary treatment* | *Extraction solvent* | *Reference* |
|---|---|---|---|---|
| *Direct extraction* | Feedstuffs | Drying on $SO_4$, $Na_2$ | DO | |
| | Oilseeds | – | Hexane | NF V 03 905 |
| | Lacto replacements (free fats) | – | Hexane | |
| | Oilseeds residues | – | Hexane | NF V 03 924 |
| | Oilseeds residues | – | DO | NF V 03 925 |
| | Meats (free fats) | – | Hexane | NF V 04 403 |
| | Miscellaneous | – | Isopropanol + hexane | NF V 83.30 *Wolff* |
| | Miscellaneous | – | Dischloromethane Methanol | *Dieffenbacher* |
| Extraction after prior treatment | Feedstuffs | CIH 3N 100°C | DO | |
| | Starch-Fecula | CIH 4N 100°C | Hexane | NF V 03 610 |
| | Cocoa Chocolate | CIH 8N 100°C | Hexane | Ar. 16–10–75 |
| | Casein | CIH 8N 98°C | Ethanol-DO | NF V 04 382 |
| | Cereals | CIH 8.4N + formic acid 75°C | Ethanol-hexane | NF V 03 713 |
| | Cream | $NH_3$ + Ethanol | DO-hexane | NF V 04 261 |
| | Ice creams | $NH_3$ + Ethanol | DO + PE | NF V 04 261 |
| | Milks | $NH_3$ + Ethanol | DO + PE | NF V 04 214 |
| | Milk (formol added) | CIH 12 N 80–85°C | DO | NF V 04 215 |
| | Margarines | CIH N | Hexane | IUPAC |
| | Milk products | CIH 4N 100°C | Hexane | NF ISO 82 623 |
| | Meats | CIH 4N 100°C | Hexane | NF V 04 402 |
| Physical methods | Oilseeds and oilseeds residues | Desiccation | – | RMN NF V 03 907 |

*(Contd...)*

**Table 16.44.** Principal methods used and foods to which they are applied. *(Contd .)*

| *Principle of the method* | *Foods* | *Preliminary treatment* | *Extraction solvent* | *Reference* |
|---|---|---|---|---|
| | Oilseeds and oilseeds residues | – | – | Reflectivity IR |
| Methods by difference | Butter | | | |
| | Margarines | | | |
| | Oils | | | |

DO = Diethyl oxide; and PE = Petroleum ether.

### *Methods by direct extraction*

In order to obtain reproducible results the methods by direct extraction imply that the extraction is done upto its 'completion'. In other words, they impose, as in the case of analysis of seeds, several successive extractions including re-homogenisation of test sample between each extraction.

However, in spite of precautions taken, they do not always permit complete extraction of fat. This, in particular, is the case for meats and still more for milk and milk based foods or containing milk proteins: a typical case is that of lactoreplacers for feeding calves for whom the direct extraction can extract only 8 to 10 per cent of fat from the products which in fact contain 30 per cent of fat. This leads to the use of the notion 'free fat' even if the physico-chemical meaning of the term is not precise. It has on the other hand, an undeniable practical importance. The 'free' fat of lactoreplacers gets, for example, oxidised more easily than the total fat. The choice of extraction solvent can lead, in case of same food, to remarkably different results: diethyl oxide, more polar than hexane, extracts more easily the oxidised fats, the phosphatides and even certain amino acids. It, therefore, gives higher fat contents than hexane or petroleum ether. Two methods which use polar and apolar solvent mixtures permit the extraction, at room temperature, the totality of lipids and seem to have a very extended field of application, including, in particular, milk products. Both are based on the use of a 'dry column' constituted of sodium sulphate and celite, the food being either extracted beforehand by a couple of solvents, or incorporated in powdered form in the components of the column. The fat being eluted by a mixture of isopropanol and hexane. (*Wolff* and *Castera*) or by a mixture of dichloromethane and methanol (*Dieffenbacher*).

*Wolff* and *Castera* method was adopted, after three circular analyses organised by CGDUMA, was adopted by AFNOR as experimental standard.

Besides the standardised methods, other methods were proposed, for example:

1. Doing several successive extractions of 30 seconds in microwave oven. 4 or 5 extractions permit, a total extraction in case of fats and cereals to be obtained. But it appears to be less than that obtained by traditional methods.
2. Use $CO_2$ as supercritical extraction solvent which permits:
   (a) A rapid, selective or total extraction of the fat depending on the operating conditions to be obtained.
   (b) An easy elimination without inconvenience to the operators of extraction solvent.

This method is likely to undergo a significant development in near future.

### *Methods by extraction after preliminary processing*

The food analysed is subjected, before extraction with a chemical treatment, most often to acid hydrolysis, meant to destroy the lipoproteins and to modify the structures of peptides and carbohydrates which resist extraction of lipids. Conditions of chemical treatment are quite variable depending on the nature of foods, and even depending on the treatments undergone by the food.

Thus, milk will not be analysed in the same manner depending on whether the milk is raw, concentrated or powder:

1. Fat of raw milk is measured with etheroammoniacal extraction (Rose-Gottlieb method).
2. If formol was added in the milk (in order to conserve the sample), the preceding method is no more applicable. In that case one has to do hydrochloric ether extraction.
3. For concentrated or dried milk products, or for milk base foods containing cereal products which are no more completely soluble in ammonia medium, one has to resort to an acid hydrolysis, but the hydrolysis conditions and specially the extraction solvents are not those of the preceding method.

Fat extracted by a method comprising a prior chemical treatment is no more identical to the fat contained in food: glycerides and still more phospholipids are partially or totally hydrolysed depending on the concentration of the acid used, duration and temperature of the treatment. This phenomenon is particularly important for foods rich in eggs or lecithins. On egg yolk for example, the rate of extraction of the fat after hydrolysis can be 60 per cent of extractable fat without prior hydrolysis. On the qualitative plane, hydrolysis prevents analysing correctly the extracted fat. This is particularly important in the case of flour and egg mixture (like biscuits and noodles with eggs): not only the egg content can not be calculated but the extraction of cholesterol is incomplete and its measurement is more meaningful if it is done on the fat extracted directly without prior hydrolysis. One can generally say that the methods using a chemical treatment before extraction do not permit correct study of composition and 'quality' of the extracted fat. They are therefore particularly unsuitable to the analysis of compound foods whose composition is desired to be known. They are, in general, reserved for the determination of the fat content of well defined simple foods.

### *Physical methods*

Low resolution NMR measures the protons of the liquid components of the foods. After elimination of water, the remaining liquid components are fats.

The methods is therefore usable in principle in the measurement of fats in foods. With respect to the methods described earlier, it offers the advantage of speed and the simplicity of execution. In addition, it does not destroy food. But to be able to apply it, one should known exactly the nature of the fat present in food, as the apparatus should be standardised with a fat identical to that one wishes to measure. Measurement and standardisation should be done strictly at the same temperature, the latter being sufficiently high so that the fat is entirely liquid. The method, in spite of its advantages, is therefore inapplicable to the foods whose fat is not known or whose composition, essentially known, can vary in large limits. In practice, it is used only for oilseeds and oilseeds residues. It is the same with reflectivity in the near IR which permits, in principle, to measure simultaneously the fat, proteins and the cellulose, but can be applied only on products whose different components are perfectly known.

### *Methods by difference*

These methods are applied only for the foods containing at least 80 per cent of fat. In this case, it can be more simple and more precise to measure components other than the fat content, this is particularly evident for a crude oil, whose fat content is generally close to 99 per cent is determined much more strictly by measuring water and the impurities or which directly determines the fat content. But this is also the case for butter and margarine whose content of fat is traditionally measured by difference of 100 of the sum water + non-fat.

### *Fat content in some foods*

Table 16.45 highlights the average contents of 'total fats' of a certain number of foods.

**Table 16.45.** Average contents of total fats.

| *Food* | *Fats, %* |
|---|---|
| Powder cocoa | 20–25 |
| Chocolate | 32–35 |
| Wheat flour | 1.0–1.5 |
| Fecula | Traces |
| Bread | 0.8–1.0 |
| Rice | 0.4–0.6 |
| Noodles | 1–3 |
| Almonds | 50–55 |
| Nuts | 60–65 |
| Hazel nuts | 50–60 |
| Brazil nuts | 60–68 |
| Avocado | 16–24 |
| Black olives | 34–38 |
| Green olives | 10–15 |
| Mushrooms | 0.1–0.5 |
| Cow milk | 1.5–4.0 |
| Unsugared condensed milk | 9–10 |
| Butter | 81–83 |
| 1/2 Butter | 41–42 |
| Butter oil | 99–100 |
| Cream | 20–30 |
| Soft paste cheese | 20–30 |
| Hard pasté cheese | 25–35 |
| Fresh white cheese 30% FM | 6–7 |
| Yoghurts | 1.5–4.0 |
| Margarines | 81–84 |
| Minarines | 41–42 |
| Corn | 2–3 |
| French fries | 7–10 |
| Potato chips | 40–45 |
| Large sausage | 27–30 |
| Sausage | 20–45 |
| Salami | 40–50 |
| Full eggs | 10–12 |
| Egg yolk | 32–38 |

When the fat content was regulated in France, the indications given in Table 16.45 do not necessarily correspond to the French legislation, but to the usual values encountered in European countries. Meats are not mentioned in the Table 16.45, as the usual data concerning the entire carcasses are without significance from the point of view of food. On the other hand, it is difficult to evaluate from the nutritional point of view, the content of fat in the morcels of meat which depends, to a considerable measure on the manner in which one prepares them and cooks them. In general, fat content of muscles is lower by 3 per cent and can even, in certain cases (for example the white of chicken), be lower than 1 per cent. The conditions of breeding animals have, in addition, a big influence on the fat content of meats. At present, they have tendency, in France, to favour breeding of animals having lean meat.

## Biological Materials

In biological materials (tissues, digestive juice, blood, lymph, feces, etc.), the fat is made of various lipid classes: glycerides, free and esterified sterols, phospholipids, glycolipids, etc. Moreover, these lipids are present with other non-lipid components, like proteins and carbohydrates. In a very general manner, the measurement of total fat, that is total lipids, necessitates their prior extraction. On the other hand, certain classes of lipids such as free fatty acids, triglycerides or even cholesterol can at times be specifically measured without prior extraction, directly in some biological medium (blood, for example).

### *Extraction of total lipids*

#### *General conditions of extraction*

These conditions should be suitably chosen to guarantee a quantitative, selective and undergrading extraction of lipids.

#### Quantitative extraction of lipids

Inside biological materials, the lipids contact with other components, specially proteins, hydrophobic bonds, Van der Walls bonds, hydrogen or ionic bonds. Moreover, these biological materials usually contain lipids of very different polarities: for example, triglycerides and esters of cholesterol having very long apolar changes ($\geqslant$ C20), besides polar lysophospholipids and hydrosoluble short-chain fatty acids or gangliosides. It results from these findings that the quantitative extraction of lipids requires the use of a mixture of two miscible solvents: one apolar, one polar.

The latter, generally an alcohol, permits to break the interactions between lipids and proteins, specially by causing denaturation of these last. Solubility and therefore extraction of lipids are then ensured by the co-existence, in the mixture, of solvents of different polarities. Relative proportions of the polar and apolar solvents and volume of the mixture of solvents with respect to weights of the biological material should be appropriate in such a manner that water present in this biological material does not entail dephasing of two solvents during extraction.

However, when an anhydrous biological sample, for example a lyophilised sample, is to be extracted, its partial re-hydration may be necessary, before it comes in contact with the mixture of solvents, in order to have a quantitative extraction of lipids.

Sequential addition of polar solvent, then of apolar solvent, in suitable proportions, is preferable to the direct use of the mixture of two solvents; the biological material is left in contact (or crushed in presence, if it is a tissue) with the polar solvent for about five minutes before adding the necessary volume of apolar solvent and carrying out the extraction.

## Selective extraction of lipids

Non-lipid apolar components (hydrophobic proteins, for example) and polar components (carbohydrates, peptides, etc.) of the biological materials can be carried away by apolar and polar solvents, respectively, of the mixture. If the presence of such contaminants is detrimental to subsequent dosages (for example, weighing of lipid extract) their elimination should be done by a process (washing of lipid extract, partition between solvents, various chromatographies) not entailing losses of lipids. That excludes washing procedures, operations of evaporation of solvents and of heating when, for example, the extracted lipid contains hydro soluble or volatile lipids such as short-chain fatty acids. The determination of one or several given lipids (cholesterol, triglyceride, etc.) in the lipid extract by specific methods (enzymatic, for example) does not necessitate elimination of non-lipid contaminants.

## Non-alterating extraction of lipids

Extraction conditions should entail neither hydrolysis, nor oxidation of lipids. For these reasons, the methods of hot extraction or extraction in acid or alkaline medium, which are at times used for foods, are not advisable for biological materials and will not therefore be considered here. Solvents should be distilled before use and added with an antioxidising agent such as 2,6-di-tert-butyl-*p*-cresol (BHT; 0.02 g/100 ml). The lipid extract obtained is freed of its solvents at room temperature, under vacuum in a rotative evaporator. Preferably, the lipid extract should not be dried, if it has to be subsequently subjected to analyses, in order to prevent any oxidation. If a weighed amount of lipids has to be used to known the total lipid content of the onset biological material, only the aliquot part of the extracted lipid will be totally freed of solvents up to constant weight, the rest being only concentrated and used for analyses. The lipid extract should be stored not dry but in solution in a solvent like chloroform at –20°C and under nitrogen atmosphere. If BHT present in the lipid extract interferes with a measurement, it can be eliminated by thin layer chromatography, extemporaneously to this measurement.

### *Recommended methods of extraction*

These methods should be applied taking into account the remarks or recommendations made earlier.

The procedure of *Folch* consists in treating 1 gram biological material by 20 ml chloroform/methanol mixture (2/1– vol/vol), at room temperature for about 1 hr. After separation of eventual solid tissue, the solution is added with 0.2 volumes of an aqueous 0.88 per cent KCl solution in order to cause separation of hydroalcoholic and chloroformic phases. This latter is washed three times according to the conditions described in the method. The three aqueous washing phases and initial hydroalcoholic phase together contain gangliosides, whereas the washed chloroformic phase encloses all the other lipids. This last phase is practically made dry under nitrogen, the residue is immediately taken up by chloroform and the solution obtained is rapidly filtered in order to eliminate the non-lipid compounds which would have been carried away.

A variant of this method developed by *Folch* which uses a volume of solvent 5 times weaker and takes into account the volume of water carried by the material to be extracted.

On account of the risks of toxicity of chloroform and methanol, other binary mixtures of solvents have been proposed: dichloromethane/methanol or hexane/isopropanol. Extraction performances with these mixtures are comparable to those of the method of *Folch*. Extraction by the hexane/isopropanol mixture permits, moreover, to prepare by acid catalysis directly done on the hexane/isopropanol lipid solution, the isopropylic esters of fatty acids (of $C_4$ to $C_{20}$) in view of their analysis by gas chromatography.

### *Quantification of lipids*

General analytical methods of fats are applicable to the extracted lipids from biological materials. Therefore, we will consider only in brief here certain biochemical methods commonly used to measure specifically a given type of lipids in the presence of others, either in the lipid extract of a biological medium, or directly in the medium. Every time it is a question of measurement by enzymatic means, various companies commercialise kits ready for use containing the mixtures of chemical and enzymatic reagents, it is enough to dissolve in the buffer solution in order to obtain the 'reagent solutions'. These solutions are generally stable for a sufficiently long time, between +2 and +8°C, allowing several measurements to be carried. Detailed modes of operations are furnished with each kit. Here we will give only the general principles of measurement.

#### *Enzymatic measurement of triglycerides*

The following is the measurement principle. A lipase hydrolyses totally the triglycerides in free fatty acids and *sn*-glycerol. This is then transformed as below:

$$sn\text{-glycerol} + \text{ATP} \xrightarrow{\text{Glycerokinase}} sn\text{-glycerol-3-phosphate} + \text{ADP} \quad \text{... (16.1)}$$

$$\text{ADP} + \text{Phosphoenolpyruvate} \xrightarrow{\text{Pyruvate kinase}} \text{ATP} + \text{Pyruvate} \quad \text{... (16.2)}$$

$$\text{Pyruvate} + \text{NADH} + \text{H}^+ \xrightarrow{\text{Lactate dehydrogenase}} \text{Lactate} + \text{NAD}^+ \quad \text{... (16.3)}$$

With ATP = adenosine triphosphate; ADP: adenosine diphosphate; NADH and $NAD^+$ = reduced and oxidised forms, respectively, of nicotinamid adenin dinucleotide. Transformation of reduced NAD into oxidised NAD is accompanied by a decrease in absorbance at 340 nm which is proportional to the quantity of glycerol, therefore of triglycerides. From this decrease in absorbance is deduced the quantity of glycerol (triglycerides), either by reading on a reference curve traced with a standard range of glycerol, or by calculation from the molar extinction coefficient of NAD and the convention-factor of glycerol into the triglycerides furnished with the kit. The total duration of enzymatic measurement (reactions 16.1 + 16.2 + 16.3) is of 10 minutes at 25°C. The measurement can be done directly on certain biological medium such as blood serum.

#### *Free and esterified cholesterol determinations*

The following sequence of enzymatic reactions takes place in these measurements:

$$\text{Cholesterol ester} + H_2O \xrightarrow[\text{esterase}]{\text{Cholesterol}} \text{Cholesterol} + \text{Free fatty acids} \quad \text{... (16.4)}$$

$$\text{Cholesterol} + H_2O + O_2 \xrightarrow[\text{oxidase}]{\text{Cholesterol}} \Delta 4\text{-Cholesterol} + H_2O_2 \quad \text{... (16.5)}$$

$$2H_2O_2 + \text{Phenol} + \text{Amino} + \text{4-phenazone} \xrightarrow{\text{Peroxidase}} (\text{Monoimino-}p\text{-benzoquinone})\text{-4-phenazone} + 4H_2O^+ \quad \text{... (16.6)}$$

The colour intensity due to formation of (monoimino-*p*-beznoquinone)-4-phenazone is measured at 500 nm. From the absorbance, the corresponding quantity of cholesterol is calculated with the help, either of a reference curve traced with a cholesterol standard range treated in the same conditions, or of a conversion coefficient given in the directions for use in the measurement kit. For each studied sample, two tests are done in parallel, in the presence of chemical and enzymatic reagents corresponding to reactions 16.4, 16.5 and 16.6 (1st test) or 16.5 and 16.6 (2nd test). The absorbance measured with the 1st test and 2nd test correspond, respectively, to total cholesterol (free and esterified) and to free cholesterol, present in the studied sample. The difference between these two absorbances corresponds to the esterified cholesterol present in the sample.

These measurements are done directly on the biological medium, when the latter is favourable (blood serum, for example), if not on lipid extract. If the latter contains an antioxidant, BHT for example, it should be freed of it by thin layer chromatography (TLC). A pre-fractionating by TLC may also be necessary to eliminate the phospholipids: if they are too abundant with respect to the cholesterol in the extracted lipid, their presence can influence the reading of absorbance. The lipid extract or the TLC fractions are dried, and then taken up again in 50 µl of acetone or isopropanol. In all cases, 1 ml of suitable solution of enzymatic reagents is added to the sample studied (serum, lipids, etc.). After 10 minutes at 37°C the absorbance is measured against a corresponding 'white', without sample. The range of vary between 6 and 40 nmoles of total cholesterol.

### *Quantification of phospholipids (PL)*

#### Chemical quantification of total PL

Total lipid extract (1 to 30 atoms of phosphorus) is placed in a pyrex glass tube (120 × 10 mm) and solvents were evaporated. After addition of 100 µl of twice distilled $H_2O$ (or of 100 µl – $x$ µl if the measurement is done directly on $x$ µl of an aqueous biological liquid) and of 60 µl of $MgNO_3$ at 10 per cent in the 96°C ethanol, the tube is heated (flame of a Bunsen burner or bath of a digestion apparatus) in order to proceed to phosphorus lipid mineralisation. Then 300 µl of an aqueous HCl 0.5 N solution are added and the tube heated for 15 minutes in a boiling bath, in order to decompose the pyrophosphates eventually formed during mineralisation.

After addition of ascorbic acid (100 µl of a 10 per cent solution in bidistilled water) and of ammonium molybdate (600 µl of a 3.4 mM solution in $H_2SO_4$ N), the colouration is developed at 45°C during 20 minutes. The absorbance is read at 820 nm against a 'white'. A reference curve (traced in the same conditions from a standard range of potassium phosphate) permits to know the number of phosphorus nano-atoms present in the sample. The corresponding weight of phospholipids is then calculated by taking an average molecular weight equal to 775.

Measurement can be done on the fractions of PL obtained by TLC, in the presence of silica gel. In this case, the volumes of twice distilled water and of $MgNO_3$ solution are respectively carried to 200 µl and 100 µl, the other volumes being unchanged. The absorbance is read on the supernatant after centrifugation.

#### Enzymatic quantification

This method permits the determination of all the choline-phospholipids (phosphatidylchloine + lysophosphatidylcholine + sphingomyelin) present in a medium without their prior isolation:

$$\text{Phospholipids} + H_2O \xrightarrow{\text{Phospholipase D}} \begin{array}{c}\text{Choline + phosphatidic acid}\\ \text{(or lysophosphatidic acid, or}\\ \text{N-acyl sphingosyl-phosphate)}\end{array} \quad \ldots (16.7)$$

$$\text{Choline} + 2O_2 \xrightarrow{\text{Choline oxydase}} \text{Betain} + 2H_2O_2 \quad \ldots (16.8)$$

Then, the measurement proceeds according to the reactions 16.5 and 16.6 of the cholesterol determination. All the enzymatic reactions demand only 10 minutes at 37°C. But this measurement requires prior elimination of BHT, if the latter is present in the analysed sample. The response is linear from 0 to 100 n moles, that is upto $77.5 \times 10^{-6}$ gram of phosphatidylcholine.

$$\text{Phosphatidylcholine} + H_2O \xrightarrow{\text{Phospholipase C}} \text{1,2-diglyceride} + \text{Choline phosphate} \quad \text{...(16.9)}$$

$$\text{Choline phosphate} + H_2O \xrightarrow[\text{phosphatase}]{\text{Alkaline}} \text{Choline} + PO_4^{3-} \quad \text{...(16.10)}$$

$$\text{Choline} + \text{ATP} \xrightarrow[\text{kinase}]{\text{Choline}} \text{Phosphorylcholine} + \text{ADP} \quad \text{...(16.11)}$$

Then, the measurement proceeds by the reactions 16.2 and 16.3 of the triglycerides determination. Incubation time is higher (70 minutes) than that (10 minutes) of the preceding measurement and it involves more handlings.

Chapter 17

# Environmental Aspects of Animal and Vegetable Oil Processing

## INTRODUCTION

This chapter is an assemblage of information on the impacts that processing of animal and vegetable oils have on water quality, air quality and land pollution through manufacturing process waste-waters, gaseous and particulate emissions to the air and waste materials. The manufacturing processes covered are oil-seed extraction, edible oil refining and soap and fatty acid manufacture.

One intent is to provide information sufficient for an inexperienced engineer or technician to be informed of the environmental control needs of the industry. A second is to give a new member of production management a basic understanding of the environmental control aspects of the processes.

For each phase of the environment, the sections discuss the sources of the potential environmental insult, pollution impacts, regulations and control methods.

Edible oil is extracted from the seeds by means of expellers and solvent extractors. In rural areas ghannies/kolhus are used for extracting oil from the seeds. The oil expelled from the seed is either filtered and packed or sent to the refinery for further purification. The cake/meal discharged from the mills/expellers which contain about 15–20 per cent oil are sent or sold to the solvent extraction units to extract the residual oil.

After preparatory treatment like cleaning, decortication/delinting the seeds are sent to the expellers. If ground nut is used, it is sent to the decorticators where the shell is separated from the kernel and blown out by using a blower. Seeds are steam cooked and taken to the expellers. During crushing, oil is expelled out and falls in the trough below the chamber and the residue comes out from the expeller as oilcake. The oil is then filtered through filter press and stored in tanks. The oil expelled is either sent to the refineries or sent to the filtration units for further purification. The cake removed from the filter press is sent or sold to the solvent extraction plant for recovering the residual oil. Typical process flow diagram is presented in Fig. 17.1.

This industry whether small, medium or large, cause pollution in one way or the other and create occupational hazards knowingly or unknowingly. The types of pollution and hazards are associated with the following: (i) air; (ii) noise; (iii) effluent; (iv) fire and explosion; (v) toxicity and allergy; and (vi) bacteria and enzymes.

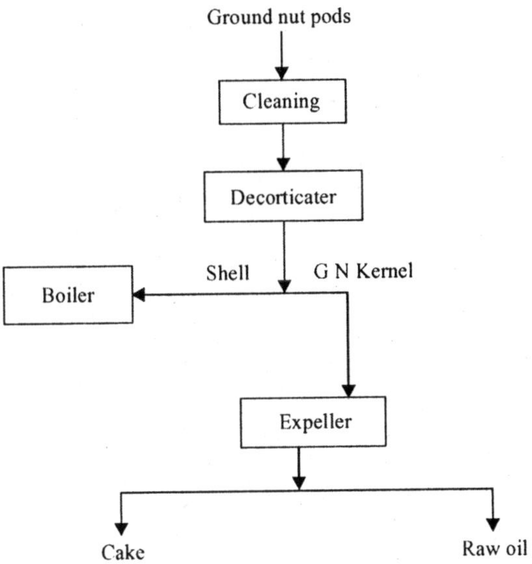


**Fig. 17.1.** Typical flow diagram for oil-expellers.

## AIR POLLUTION

Table 17.1 gives percentage incidence of specific complaints of the human body created due to air pollution vis-a-vis other weather conditions. Air pollution occurs during seed cleaning and decortication when dust particles float. Cottonseed delinting or defibration give rise to cotton fibres and dust. In solvent extraction plants, fine meal particles and traces of normal hexane vapour pervade the atmosphere. Objectionable acrid odour is generated when non-edible oilseeds like neem and pongam are cooked and crushed in expellers.

**Table 17.1.** Effect of air pollution on the human body.

| *Chronic condition* | *Created due to* | |
|---|---|---|
| | *Air Pollution* % | *Other weather conditions* % |
| Eye complaints | 54 | 3 |
| Asthma | 34 | 22 |
| Nose complaints | 29 | 18 |
| Throat complaints | 28 | 12 |
| Allergies | 15 | 11 |
| Coughing | 12 | 10 |
| Bad headaches | 11 | 5 |
| Shortness of breath | 11 | 10 |
| Chest pains | 10 | 4 |
| Feeling tired and low | 5 | 5 |
| Stomach complaints | 2 | Less than 1 |

When rubberseed and kusum (*Schelecheira trijuga*) are being cooked and crushed, nauseating odour spreads in view of the presence of cyanogen compounds in the seeds. A pervading odour of paint is a feature of stand oil and boiled linseed oil preparation plants. Sulphurous fumes are liable to leak during sulphonation of alkyl-benzene in detergent industry and in Turkey Red Oil preparation. In open kettle soap boiling, open steam goes out into the surroundings. Glycerine vapours are lachrymating. Bromination of oils is a hazardous laboratory operation because bromine vapours attack the mucous membrane. Fine chemical particles are ejected during spray-drying of synthetic detergents and soap powder.

Air pollution common to all units having boilers is the product of soot and colloidal carbon particles which surround the vicinity. An atmosphere saturated with the vapours of perfumed products brings on headaches which vary in intensity. In more serious cases, migraine, vertigo and even a state of nausea are experienced.

Inhalation of irritant vapours of certain essential oils causes rhinitis. It is possible that the joint action of essential oils, reinforced or modified by the association of chemicals, brings about nervous and psychological states taking the form of general fatigue, sick headache, decreased olfactory acuity, irritability and unforeseeable psychosomatic manifestations linked with the individual's condition. Table 17.2 gives a summary of air polluting agents.

**Table 17.2.** Air pollutants and industries.

| *Pollutant* | *Industry where occurring* |
|---|---|
| Dust, fine vegetable matter | Seed cleaning, decortication, dehulling |
| Cotton fibres | Cottonseed delinting |
| Meal particles | Solvent extraction industry |
| *n*-hexane vapour | Solvent extraction industry |
| Bad odour | Inedible oilseeds crushing |
| Paint odour | Paint and varnish industry |
| Sulphurous vapour | Detergents and TRO industry |
| Steam | Soap industry |
| Fine chemical particles | Synthetic detergent and soap industry |
| Colloidal carbon, soot and smoke | Boiler, furnaces and burners |
| Vapours of chemicals and essential oils | Perfumery and essential oil |

## Air Pollution Impacts

As already discussed air emission sources, animal–vegetable oil processors need have limited concerns. The single source of attention to date has been the solvent, hexane, used in extraction of the oilseed. The volatile organics are controlled in air quality programmes because they are involved in the formation of photochemical smog. The smog is a result mainly of the interaction of hydrocarbons from automobile emissions and other sources with nitrogen oxides under the influence of sunlight to form ozone. Ozone then reacts further with the substances in polluted air to form a variety of substances. One of the products of reaction between ozone and hydrocarbons is free radicals that accelerate smog formations. Photochemical smog is linked to eye and bronchial irritation, plant and crop damage and reduced visibility. To date, the other air pollutants for which control requirements have been or are being set include particulate matter, oxides of nitrogen, sulphur dioxide, photochemical oxidants and carbon monoxide. These, plus volatile hydrocarbons, are specific pollutants identified by the US EPA as pollutants that have an adverse effect on public health or welfare.

## Odour Sources

Odour appears to be a logical subject to associate with air pollution discussions but on close examination is quite different. Odour is not subject to mass limitations because in most cases no specific substance can be quantitated. Odour is also not directly health related, so the basis for regulation is absent. However, both odour and air pollution involve the atmosphere and control methods may be similar.

Animal and vegetable oils as they are separated from fresh animal and oilseed sources do not have a significant or strong odour. However, on storage, with time and the growth of bacteria and on auto-oxidation the triglycerides will be broken down to short-chain fatty acids and other species that have an odour that most people find highly objectionable. Those materials are also fairly volatile. Therefore, on storage and transport of oil care is taken to keep bacterial degradation and autoxidation from occurring.

Odourous materials naturally present in the oil or developed during processing are removed from the finished products. In soapmaking processes the odourous materials are transferred to water, mainly to the condenser water used in sprays or in the steam condensate itself if surface condensers are used. Specifically, the fatty acid still condensate may contain some objectionable odourous material. If a direct condensing system is used, the odour will be in the water. If this water is recirculated over a cooling tower, odours will be present in the air emissions of the cooling tower.

Similarly, in edible-oil refining, odourous materials are contained in the vapours from the deodouriser operation. These vapours will be contained in the condensate from surface condensers or in the condensing water of direct condensers. These odours will be transferred to the exhaust of the cooling tower if the direct condensing water is cooled and recycled.

Miscellaneous odour sources include any vents of steam to the atmosphere. Also, if a direct-to-atmosphere steam eductor is used for removing non-condensibles from the distillation and deodourising operations, the process may be objectionable.

The acidulation operation for recovering triglycerides or fatty acids from soapstock, soapy water, or fatty acid still bottoms may also be a significant odour source. The source of these streams is discussed in the preceding section on water pollution. The acidulation process, if done batchwise in open tanks, usually includes boiling the tank contents. The vapours arising from this can be highly odourous.

In addition to odours entering condensing water of recirculated systems, the oil and fatty acids and other material in the recirculated water can under certain conditions be degraded by bacterial action and autoxidation and thus odours can be created in the system. The odourous material are most likely produced in the more static sections of the basin under the cooling tower or in the hot well in the skimming basin.

## Control of Air Emissions and Odour

As already discussed the significant air emission from industrial plants involved in animal–vegetable oil processing is primarily from some solvent extraction operations. Only a small segment of the industry utilises solvent extraction, namely, those plants utilising solvent extraction for separating the raw oil from oilseeds. The losses of the solvent from oilseed extraction has been the subject of investigation by the US EPA. It would be inappropriate at this time to speculate on what control technology is practical and economically feasible. Managers of these plants are apprehensive about the costs and the energy needs likely to be involved. All plants have solvent recovery and recycle systems; the degree of recovery is a question of economics.

The industry has faced isolated instances of odour control problems. Air emissions and odour are discussed together in this control section because it is estimated that 50 per cent or more of the complaints about air pollution deal with odours. Also, the control of odours in most of the cases involves air emissions released with the odours.

As discussed in a preceding subsection on odour sources, the major source is from condensing water recirculating systems. Odours from such recirculated systems of oil-bearing water can originate in materials condensed from the process vapour streams or from anaerobic conditions of bacterial growth in portions of the water system.

The correction of problems from bacterial growth is relatively easy. The addition of a bactericidal agent, such as chlorine or chlorine dioxide, has been reported to be successful. The dose is low and the residual chlorine dissipates readily in the discard stream. Manufacturers of chlorine dioxide and chlorine will advise potential customers including dose requirements. Many recirculating contaminated cooling water systems do not experience odour problems. An ample discard stream helps to maintain low concentrations of contaminants.

Reduction of odours in recirculated water streams is being accomplished by treatment of the recirculated water and by treatment of the exhaust air from the cooling tower, or both. The treatment of the water stream with chlorine dioxide has been reported to achieve a significant odour reduction as reflected in panel observations of the plant atmosphere and the number of complaints. Further reduction in odour has been reported to be achieved by mist sprays of dilute hypochlorite solutions into the offgases of the cooling tower. In assessing potential odour problems from recirculating contaminated water in cooling towers and the effectiveness of control, it has been found that threshold odour measurements on the recirculating contaminated water in cooling towers and the effectiveness of control, it has been found that threshold odour measurements on the recirculating water are an effective tool and are easier to turn than are odour threshold measurements on the air emissions.

One such measure is the addition of powdered activated carbon to the recirculating water. This carbon is discarded as a component of the discarded water stream from the process. It seems improbable that this measure is effective because the high concentration of other organics in the waste-water would take up much of the adsorptive capacity of the carbon.

In several cases the control measure taken was to abandon the direct vapour condensing systems and substitute surface condensers. The condensed water vapours and organic vapours from the surface condensers are discharged to gravity-settling systems. After removal of the flotable matter, the wastes are discharged to the waste-water treatment system. The condensing surfaces become coated with congealed material. For this reason the surface condensers must be installed in multiple units.

## WATER POLLUTION

### Sources of Waste-Water

Water and chemicals diluted in water are important process materials in refining raw oil and fat and production of soap and fatty acids. The water used becomes a waste-water stream conveying away the chemicals, impurities and heat. Undesired constituents of the oil are eliminated from the products in this manner. Some valued oil is inevitably lost in these waste-waters. In addition to the process waste-waters themselves, the waste-water include oil and chemicals lost to flows and surfaces through leaks and spills. The latter become part of housekeeping wastes from such operations as washing floors and paved areas. Liquid wastes also originate from sanitation and cleaning measures in which water supplemented with cleaning agents is used in sanitising pipes, tanks and equipment.

Water is used for temperature control in cooling and condensing operations. Waste-waters are generated from indirect heat exchangers. Such water, used in a once-through-and-to-waste fashion, will not have any changes in its characteristics other than increased temperature, provided, of course, that there are no leaks of process streams into the cooling water. Indirect cooling water is commonly passed over cooling towers and recycled. In such cases some discharge to waste must be practiced to avoid an intolerable increase in the mineral content of the recirculating water.

Large volumes of waste-waters also originate in systems for creating low pressures for evaporation, distillation, or drying. Water vapour in these systems is condensed by sprays or streams of cold water. The condensing water is discharged from the system through a pipe that terminates below a water surface and thus provides a seal to maintain a vacuum. Volatile matter from the materials being processed will be part of the vapours reaching the condensing water sprays. Also, some droplets of the material being processed may carry over or be entrained in the stream of vapour to the condensing water. Thus, waste-water from such condensing systems will contain condensed and entrained volatile materials. Such condensing water may be passed over cooling towers and recycled. Some discharge to waste must be practiced from such recirculating systems to avoid an intolerable increase in the minerals in the water supply and an excessive concentration of contaminants from the process.

Steam is used in heating tanks of oil to maintain fluidity and for many process steps. This steam condenses and thus yields its heat. The condensate is returned to the boiler feed where it is economic to do so and where the possibilities of contamination are acceptable. At many locations the condensate enters the sewer at the point of generation. This condensate is unpolluted except in those instances where there is a leak into the steam line from the process stream.

### *Sources of waste-waters from oil extraction*

Mechanical or solvent extraction processes for separating oil from the oilseed or other source do not cause the discharge of major amounts of waste-waters or water-borne pollutional materials. Oil extraction plants are commonly at different geographical locations than are oil refining plants. Consequently oil extraction plants are considered separately in this discussion.

Plants that use solvents to extract the oil use large volumes of cooling water in indirect condensing systems. The water is used to condense the solvent in the solvent recovery system. This water is recirculated over cooling towers and only that necessary to limit the buildup of dissolved solids enters the sewer. In preparation for extraction of soyabean and other seeds, the beans are pressed into thin sections. This operation creates appreciable dust. Usually exhaust ducts convey the dust to scrubbers or filters to control the emissions of particulate matter. Waste-waters from such wet scrubbers constitute a considerable part of the routine wastes from extraction operations. In addition to the bean preparation, the remaining seed after extraction is usually processed in some way for marketing as animal feed. Such operations may include drying and packaging and other steps from which the dust must be controlled.

Mechanical extraction plants have minor quantities of waste-waters. Both methods share waste-waters from cleaning and housekeeping operations. Small volumes of steam condensate used in heating the oils to maintain them in liquid state will be discarded.

A small fraction of soyabean oil is subjected to a prerefining process referred to as degumming. This can be a part of the extraction plant or a preliminary step at an oil refining plant. The degumming process is employed to remove lecithin, a mixture of several phosphatides. This commercial product, including the chemical lecithin, is used as a food additive and for industrial purposes. Degumming consists of contacting the oil with a small volume of water, a process referred to as hydration. Lecithin and other water-soluble materials are subsequently removed in the water stream by centrifugation. The lecithin is then dried. The oil is filtered using cloth or cloth and filter aid before lecithin removal. The market for lecithin is limited; therefore, the degumming process is carried out on only a fraction of soyabean oil produced, about 30 per cent. The market value of the raw oil is the same regardless of whether the oil is degummed. Waste-water production from these operations is low and of little importance. Degumming is important to the quantity of pollutional matter discharged because if lecithin is not recovered, it becomes part of the waste-waters from subsequent processing.

### *Sources of waste-waters from oil refining*

The objective of the process of refining edible oils is to separate undesired non-triglycerides from the triglycerides. Some vegetable oils contains about 2–4 per cent of organic compounds that are not triglycerides or free fatty acids. The raw oil will usually contain a few percent of fatty acids, mono- and diglycerides, glycerine, lecithin, phosphatides, sterols, sugar, a variety of colour and flavour-producing compounds, particles of the oil seed and low concentrations of other nontriglycerides constituents. The oil supply to which the refining process is applied is either raw fat or oil or soyabean oil that has been degummed.

The refining process used practically exclusively is the extraction of the oil with a solution of caustic or a combination of caustic and soda ash. The alkaline materials convert the free fatty acids to sodium soaps and dissolve other materials from the raw oil. The alkaline extracts, referred to as soapstock, is separated from the oil by centrifugation. The centrifugation produces two streams, the extracted oil steam and the alkaline extract. The alkaline extract contains practically all the fatty acids and water-soluble non-triglycerides.

A few plants in the United States use steam distillation for refining instead of the process just described. The alkaline extract from alkaline refining is acidified (acidulated) for recovery of the fatty acids and any emulsified oil. The agent used to neutralise the alkaline material is sulphuric acid. The soap stock may be heated to help release the emulsified oil. Separation of the fatty acid and oil from the extracting solution may be accomplished either by centrifugation or gravity sedimentation or by the two operations in series. The water and the materials it contains are discarded. The fatty acids and oil separated in the acidulation plant are used in animal feed or for raw materials for a fatty acid processing plant. The soap stock may be shipped to a central or independent plant for the acidulation process.

Immediately following the extraction of the oil with caustic, the oil stream is extracted with water to remove any fatty acids soaps remaining with the oil stream or any water-soluble materials, including any traces of caustic. The water stream, referred to as soapy water, is separated from the oil by centrifugation. In the majority of plants the water extract (the soapy water) is added to the alkaline extract and passed through the acidulation process just described. In some plants, particularly those which ship the soap stock to other locations for acidulation, the soapy water is discarded.

The alkaline-extracted oil must be further processed to remove trace components that contribute flavour and odour, to create the bland, colourless product desired by edible oil consumers.

The introduction of this discussion of refining mentioned plants that do not use alkaline extraction for refining but instead use steam stripping at low pressures for the essential refining step. This process is similar to the deodourisation process described later in this section, under 'Sources of waste-water from deodourising'. Statements have been made that the waste-waters are reduced in magnitude compared to alkaline refining. This could refer to the absence of the neutralised alkaline minerals in the waste.

### *Bleaching*

The removal of colour is accomplished by an adsorption process using natural clays whose adsorptive properties are suitable. A typical clay is a diatomaceous earth with the desired adsorptive property. The clay is added continuously to the oil steam, the slurry is mixed in a contact chamber and finally the clay is separated from the oil by filtration. Filtration equipment is typically plate and frame or septum type filters. Filtration is continued until pressure limits are reached, or the flow rate is reduced to a limit, or the filter has a full load of clay. The oil remaining in the lines, the filter spaces and the filter clay is purged into the product flow by use of air or nitrogen. The routine followed at different plants in further separation of oil from the clay differs from plant to plant. A hot water wash may follow the air purge. This oil-laden water flush is discharged to a gravity oil separator. The oil is skimmed from this separator and is marketed for animal feed or returned to process. The small volume of water enters the sewer.

These purging operations produce a spent filter clay containing at least 35 per cent oil. This spent clay from decolourising constitutes the major solid waste material of the industry and is discussed in a subsequent section on waste materials. However, in addition to the waste-water from the purging operation previously discussed, other waste-waters enter the sewer from the spent clay processing operation at some plants. The oil-laden spent clay must be cleaned from the filtration equipment. At many plants this is done by manual labour. The cake is knocked or scraped from the leaves and falls to a trough below. A screw conveyor in the trough moves the plastic clay mass to a bin or truck tank. Labour time in such operations plus problems with the pyrophoric property of such spent clay has led some plants to the practice of hosing the cake from the filters and receiving trough with water. The resultant slurry then flows to a truck tank. There are at least four ways in which such a slurry may be further processed. Three of these involve the sewering of liquid wastes. In former years this entire slurry was discharged to the sewers. In circumstances where the slurry was transported and treated satisfactorily, this was a technically feasible practice. Sewage treatment has become more expensive and the practice may not be economical today.

Another alternative is to dewater the slurry with a rotary drum filter. The relatively dry cake is discharged to a bin or a truck. The water is discharged to the sewer. This water is referred to as 'spent-filter-clay-transport waste-water'.

Still another alternative is to process the slurry for oil separation and recovery. The operation consists of adding limited quantities of caustic, which helps to separate some oil from the clay. The process is carried out batchwise; each caustic treated batch is heated and then settled. The oil in representative operations is reduced from an original content of 35 per cent by weight of solids to 20 per cent or lower. The batch is settled, the oil layer skimmed, the water layer drawn to the sewer and the clay slurry either hauled as a slurry or dewatered as described in the previous paragraph. The decanted water or filtrate is another form of the spent filter cake transport water.

### *Sources of waste-waters from deodourising*

Volatile organics not removed by previous processes are removed by steam stripping at low pressure and high temperature. A significant percent of the oil and of volatile oil-soluble constituents are removed from the oil. Much of the material is the more volatile, short chain length, fatty acids. Modern deodourisers are continuous, multistage systems. The quantity of steam used requires tremendous quantities of cooling water for its condensation. A number of systems may be employed to remove material from the water vapour before it is condensed. Initially, these may include an operation for removing entrained material in the form of droplets. This is a gravity separation system in a chamber providing reduced velocities of the vapour, typically followed by a partial condensation system to capture the majority of fatty acids. One such partial condensation system utilises a vegetable oil or a fatty acids that is not volatile at temperature maintained in the partial condenser. This oil at low temperature is circulated to a vapour contactors, where it cools the vapour stream to just below the condensation temperature of the volatile organics that are to be removed from the vapour stream. Not all the organics can be condensed without condensing water vapour also, which is not desired in such a partial condenser. It is obvious that the vapour temperature, the condensing oil temperature and the pressures must be delicately balanced to achieve the optimum removal.

Following the partial condenser, the vapours pass to a system for condensation of the water vapour as well as any remaining volatile organics. Such condensation may be accomplished with either direct or indirect condensers. The direct condensers use large flows of water, thousands of gallons per minute. This water is typically recirculated over cooling towers. The organics carrying over to the condensing water build up to high concentrations. Some insoluble organics such as fatty acids can be removed by gravity settling from this stream. A significant blowdown stream is typically discarded to the sewer.

Some systems operate without a partial condenser. The steam and the vapourised material are condensed together in a spray of water. The cooling water is recirculated over a cooling tower. A settling and surge tank is included in the system. The organic material which floats to the surface in the surge tank is periodically removed. Makeup to these systems is minimal, if any is necessary at all, because of the addition of condensed steam to the systems. A significant diversion of the recirculating water to the sewer is practiced to control the buildup of soluble matter.

In the winter the amount of heat dissipated by convection rather than evaporation is so great that more condensate enters such systems than evaporates in the cooling tower. When such conditions prevail, there is a need to continuously or periodically discharge the excess.

The production and characteristics of the waste-waters from these vapour condensing systems was materially changed at a number of plants in recent years because of local odour problems with the recirculating systems. Some odourous materials in the recirculated cooling water with direct condensing systems are released to the air. One solution is to use surface condensers so that the condensate is not mixed with the cooling water. The cooling water is uncontaminated and may be recirculated without significant odour production. The indirect condensers condense the water vapour as well as any material condensable at the temperatures and pressures involved. The stream of condensed steam, fatty acids and other organics is relatively small, making effective gravity separation of floatable material possible. The skimmed water is discharged to waste. A second waste comes from cleaning the indirect condenser of oil and fat sticking on the surfaces. This oil is removed by isolating the condenser and circulating hot water through the vapour side. This flush water is also passed through the gravity settling tank to waste.

### *Sources of waste-waters from hardening*

Following processes previously described, a portion or all of the oil processed may be subject to another process referred to as hardening if the objective is to increase the congealing temperature of the product. The addition of hydrogen to the unsaturated linkages lowers the temperature of congealing. The property of being semisolid or plastic at ambient temperatures or slightly higher is important for products such as margarine and the solid form of shortening. The addition of hydrogen is accomplished by reacting the oil with hydrogen in the presence of a nickel catalyst. The nickel catalyst is a compound of nickel deposited on diatomaceous earth. The reactants and the catalyst are intimately mixed at the desired temperature and pressure and when the reaction is complete, the catalyst is removed by filtration. The filtration equipment employed is of the plate and frame type. The catalyst has a limited active life and must eventually be discarded or regenerated. This problem is discussed in the section of this chapter on solid wastes.

The generation of waste-water from hardening has not been significant in the past. Any waste-water that result consist of water overflow from hydrogen gas tank seals and housekeeping wastes. Since, the mid 1970s, information on the significance of nickel in the environment has caused much concern, as summarised in a subsequent section on pollutional properties, toxic pollutants. At times of repair of pumps, tanks and piping systems, small losses of the oil and nickel slurry to floor drains may occur. Also, the cleanup of filters and filter cloths where used results in waste-water. In such practice the material is adsorbed in granular commercial cleaning material and disposed of in the manner of a waste material and not discarded in the sewer.

### *Sources of waste-water from winterising*

For the production of an edible oil end product that will remain clear and unclouded at temperatures of household storage and use, a process is employed for removing those triglycerides whose solidification temperature is in the range of storage and use. This is accomplished by cooling the oil to low temperature and removal of the solidified material. No routine waste-water are produced from the oil processes, so there is no need to discuss the operation further. Auxiliary to this process is usually one of a variety of refrigeration systems for cooling the oil. These involve indirect heat exchanges with which are associated recirculated cooling water systems. The usual discharge from such recirculation systems to control buildup of minerals is employed. An alternative is the use of large volumes of once-through cooling water. A concern of such recirculation systems can be ammonia leaks or spills if such material is involved in the refrigerant system.

### *Sources of waste-water from drying*

For some products it is desirable to remove all traces of water from the oil so that shelf life is not affected by trace amounts of water. This is done by drying the oil under vacuum. Direct condensation of the vapourised water or other volatile material by water sprays is practiced. Ordinarily the quantity of entrained oil or other volatile material is insufficient to produce a significant concentration in the condensing water. However, misoperations do occur and oil will appear in significant concentrations in condensing water recirculating systems. The waste-water stream from such recirculating systems may occasionally have a low concentration of dispersed oil.

### *Sources of waste-water from filtration*

To ensure that no sediment is present in a final product and to assure clarity, the refined and purified oil may be subjected to a final filtration prior to transfer to its container or as a part of such a packaging operation. Filtration is done by plate and frame systems using as filter media diatomaceous earth similar to or the same as is used in decolourising. There are no routine liquid waste-water from this system. Filtration system of earlier years utilised filter cloths that were periodically washed similarly to a clothing laundering operation. Filters using cloths have been replaced by metallic or plastic materials. The removal of oil from such cake may involve similar waste-water to those discussed in decolourising.

### *Sources of waste-water from packing*

The filling, packaging and warehousing of edible oil products does not involve the release of routine process waste-water. During processing, the oil is maintained at temperatures that destroy bacteria or other life forms. The oil is handled in closed systems and blanketed with nitrogen so that atmospheric sources of any organisms are minimal and oxidation is minor. The system is essentially sterile at all times so that the usual periodic cleanouts for sanitising purposes, such as those found in a milk-bottling plant, are not needed. At times of maintenance or weekend shutdowns, steam flushing of pipes, tanks and equipment may result in some condensed steam and flushing water being added to the sewer. Water with detergents or caustic may be used occasionally in cleanup.

The packing line and even the warehouse may experience breaking of bottles. Clean-up procedures may include flushing to the sewers. A peculiar waste-water may be associated with warehousing. The waste-water comes from the service area for the carts. These carts, if of the internal combustion engine type, may be washed down and the washwater discarded in the sewer. In one instance, drainage from this area, which likely included some used crankcase oil, contributed a small amount of mineral oil.

### *Sources of waste-water from storage and handling*

Numerous sources of wastes originate in the unloading and storing operations. One is leaks and spills from tank car and tank truck unloading and loading. Small losses can occur at each uncoupling of a line and loss of the line contents will take place unless precautions are taken. Glands of pumps used to transfer the oil may leak.

A major sources of oil-bearing waste-water can take place from the washouts of tanks, trucks and railway cars. At one plant this loss of vegetable oil was the greatest single source of oil to waste. Furthermore, during winter months this loss was twice the quantity of such losses in the summer months. This likely reflected the effect of cold tank walls retaining more oil or slowing drainage.

Major quantities of waste materials may be in waste-water from cleaning storage tanks. Storage tanks of raw oil can accumulate major amounts of sediment, pieces of beans or nuts, or in the case of animal fats, of hair, bone and flesh. Some plants practice removal of much of such accumulations and mere disposal in the manner of solid waste material. Smaller residues are washed from the tanks in the waste-water from cleaning.

Much of the waste-water that enters the sewers serving the storage tanks and unloading areas that are not under roof is rainfall runoff. The drainage from such areas may be slightly polluted from leaks and drips and constitutes a significant volume of waste-water.

### *Sources of waste-water from soapmaking*

Waste-water volume and characteristics from soapmaking are markedly affected by which of two processes are employed. The manufacture of soap by the batch kettle process has a very low discharge of waste-water. In the kettle, a purified fat and sodium hydroxide are reacted to form soap and glycerine. The aqueous caustic stream is used countercurrent to the oil in a series of batch reactors so that the caustic is essentially exhausted. After the final contact, the soap is drawn off the top. The lower layer is water containing the colour bodies and other impurities from the fat or the oil. The material, called *niger*, is recirculated to fresh fat reactors until the alkali content is low. The niger can be processed to recover a low grade dark soap. The water residual after separation of the dark soap is eventually neutralised and discarded to the sewer when it becomes particularly loaded with colour bodies. Salt and sulphuric acid may be used in the process of recovering the dark soap and these will be present in the waste-water.

To prepare a high-quality soap by the batch process, the fat must be refined before use. One of the most frequently used methods employs activated clay as the extraction agent. No liquid wastes are produced except for washouts, provided that the clay is handled dry or as a waste material and not discarded to the sewer. Other ways in which fats are refined include caustic extraction, steam stripping and proprietary treatments. Waste-water from caustic extraction have been discussed in the section on edible oil refining. Steam stripping produces a waste of condensate and condensing water that is discarded to the sewer after separation of the fatty acids by gravity settling.

The alternative process of making soap by fatty acid neutralisation produces a small quantity of waste-water. The fat splitting operation preliminary to soap making has significant volumes of waste-water. The equipment for mixing of the fatty acid and caustic and other materials that may be used is only occasionally washed out. A small volume of water containing unreacted alkali may be separated from the product and discarded to the sewer. For some products it is necessary to remove water from the product of the neutralisation step. This may be accomplished by subjecting a thin layer of the product to a vacuum. The vacuum is created by use of steam and a barometric condenser. Very small amounts of volatile fatty acids may be drawn over to the condensing water. The waste-water from this operation will be the once-through condensing water or the discard stream from a recirculating cooling water system.

Fatty acid manufacture by fat splitting produces several liquid waste streams. The hot fat at high pressure is contacted with water counter-currently and hydrolysis takes place. The acids are released to atmospheric pressure and the vapours are condensed by water sprays and discarded to the sewer following gravity separation of the floatable fatty acids. The fatty acids are purified by distillation at low pressures. The low pressure is achieved commonly by a barometric condensing system using steam ejectors and direct water sprays for condensing. The condensing water may be recirculated over cooling towers. Waste-water consist of a discard stream from this system. The recirculating water is passed through a gravity settling system for recovery of floatable fatty acids and oil. The unreacted oil and impurities will accumulate in the bottoms of the fatty acids still. These still bottoms are treated with acid to break emulsions, floatable oil recovered and they are then neutralised and discarded to the sewer. Typically, this is a batch operation.

The other product stream from acid splitting is that containing the glycerine. Glycerine streams from the kettle processes are neutralised by the addition of mineral acid. The glycerine is evaporated to a low water content usually by a two-stage process. Barometric condensers are used in the two stills and the vapour are condensed by sprays of cold water. Although mist eliminators are commonly employed, there will be some carryover of crude glycerine to the condensing water. The major volume of waste-water is the condensing water or the discard stream from a recirculating condensing water system. The still bottoms are typically dark brown amorphous solids when cooled. This material is commonly disposed of in the manner of a solid waste.

The process of forming bars or other form of soap seldom involves a process waste, although reworking of scrap can produce a waste drawoff. There may also be waste from washout of equipment.

## ECOLOGICAL EFFECTS

Very recently, control of the environment and the effects of chemicals on the ecology have become prominent and well-publicised subjects. As far as the detergent industry is concerned, the control of foam and syndets in water systems as an indication of pollution and its effect on the ecology was the first step in adequate control.

### Biodegradable Detergents

Soap use is historical and it is exceptional to find reference to water contaminated by it. The fact that water hardness reduces soap to a non-foaming state is one reason why it has not been considered a major pollutant and the further fact that it is almost without exception a straight-chain compound is another more pertinent reason. The latter characteristic is important to river water and sewage disposal as was evident when replacement of soap by syndets became more universal. Sewage plants began to display beds of foam at their effluent ends, some of them covering entire streams for several hundred feet. Then too, foaming of drinking water from ground-water supplies where septic fields were in use was a further indication of pollution, both by syndets and viable organisms. Presence of foam was a sure indication that biodegradation had not occurred.

It became evident that something must be done about 'hard' syndets. 'Hard' because they were not straight-chain compounds and resisted sewage treatment and river-water degradation. 'Soft' or easily degraded surfactants were to become a necessity. The surfactant chosen for study because of its wide usage was alkyl-benzene sulphonate. The alkyl portion of the compound was the focus of attention: its source may be chlorinated kerosene or a suitable olefin. Extensive research demonstrated that a tertiary carbon atom in the alkyl chain was the point at which biodegradation ceased. The organisms were unable to utilise the chain any further when such an atom was encountered, whereas with straight-chain compounds such as soap no difficulty was found. The answer to this was a straight-chain kerosene or olefin, and suitable synthesis control to prevent chain rearrangement, or more suitable use of a preferred straight-chain olefin in the synthesis.

The straight-chain product finally used was termed LAS, linear alkylbenzene sulphonate (as contrasted with branched chain ABS). Today any surfactant to be used successfully must be biodegradable, as shown in Fig. 17.2. A third generation of agents is beginning to draw attention, and includes nonionics and nonionic sulphates.

## Detergent Additives

### *Phosphorus compounds*

Immediately following the development of biodegradable syndets to control pollution by surfactants, the overall pollution of rivers and lakes became the next focus of attention. Lakes in particular, as recipients of inflow from agricultural lands and either the effluent from inadequate sewage treatment plants or raw sewage were becoming so tainted that in effect it might be said the lakes were dying. The apparent effect was a change in the aquatic life with more frequent and extensive algal blooms. These blooms reduced the oxygen content of the water, killing much of the aquatic life, and in effect stagnating the water body by depositing organic debris which consumed still more oxygen. Reasons for this change were attributed principally to high phosphorus contributions from run-off and sewage effluent, and secondarily to nitrogen and carbon dioxide. Ordinarily this process, called eutrophication, would take thousands of years, but it has been greatly accelerated by overnourishment.

Since the phosphorus content of sewage is not greatly reduced by treatment, syndets which contain from 5 to 50 per cent phosphates became the immediate target for control. The *pros* and *cons* continue, while congressional and state actions would ban phosphates in detergents. The problem of phosphate control is being attacked through more effective removal during sewage treatment and by substitution of agents capable of replacing phosphorus as a builder.

### *Phosphorus substitutes*

The usual builders which might partially substitute for phosphates are sodium or other water-soluble carbonates, and sodium or potassium silicates. Both are more effective in improving syndet soil removal than are essentially inert electrolytes such as sodium sulphate or sodium chloride. For liquid detergents perhaps a greater degree of dilution is possible, i.e. more water to replace phosphates, but this is not a good solution to the problem. What is needed is a product which is completely biodegradable, is not a prime nutrient either as used or when degraded and has the properties contributed by phosphates. That is, the replacement should soften water, preferably by sequestration and be as effective in promoting detergency as are the polyphosphates. It should be noted here that both soap and surfactants require builders for effective detergency, and that reversion to soap with the aforementioned remaining builders will result in soil removal levels which are no longer considered satisfactory.

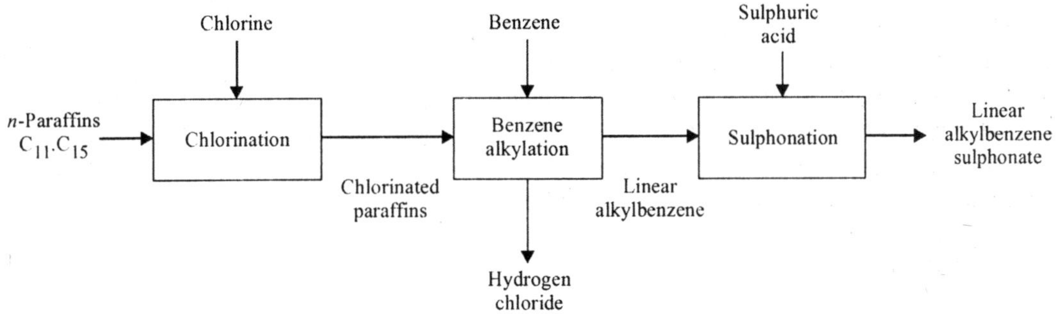


Process and products like these are begining to attract more attention.

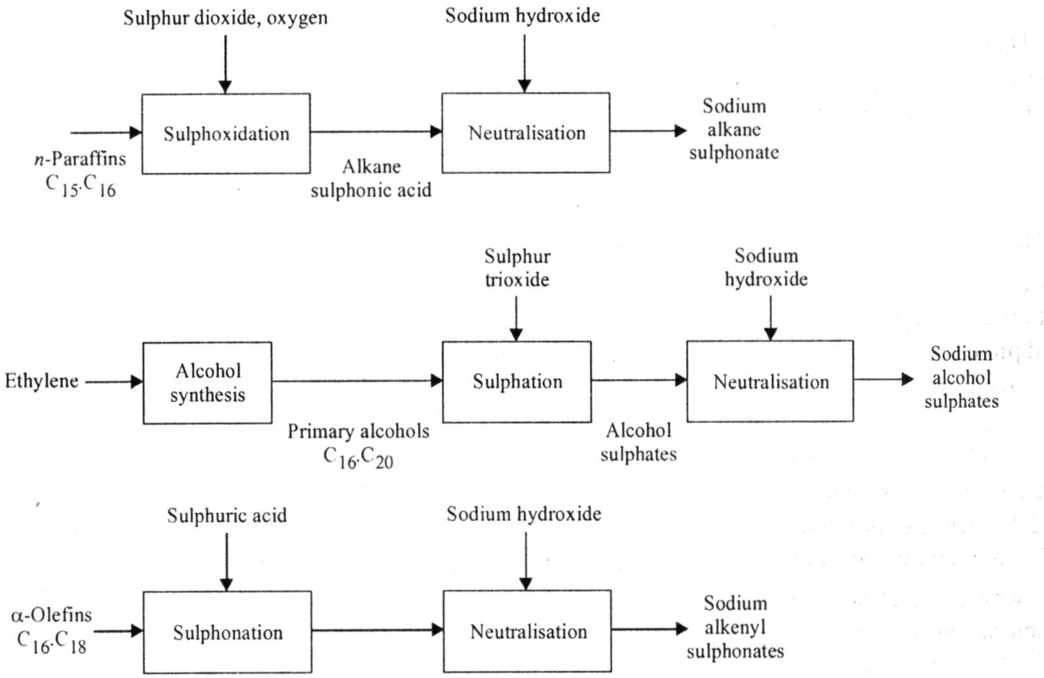


**Fig. 17.2.** Flow diagrams for manufacture of some biodegradable surface-active agents.

A variety of polymeric products have been tested as sequestrants or chelating agents, but none has been found to be as effective as the original agents.

Possibly the first compound to be fully investigated for detergency was ethylenediaminetetracetic acid, EDTA:

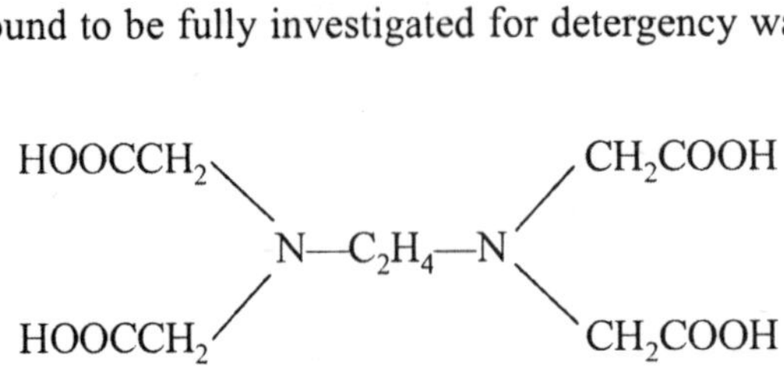


Each of the four carboxy groups is capable of reacting with metallic ions to form the tetrasodium salts, for example, or to sequester or chelate two calcium ions, forming soluble rather than insoluble salts. This compound and its variants have been extensively investigated as substitutes for STP. Although it is effective, its cost is excessive.

Another compound which received attention as a substitute for phosphates is also a chelating or sequestering agent. Nitrolotriacetate, NTA, when used in adequate amounts proved essentially as effective as EDTA but at lower cost.

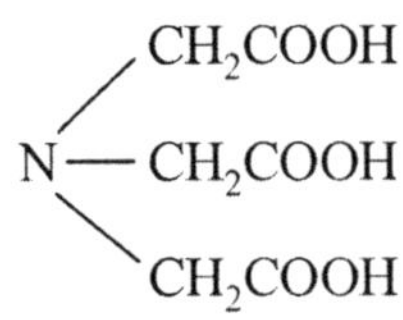


NTA was the compound which the detergent manufacturers started to use to replace STP. Its use was terminated, however, when it was disclosed that birth defects in rats and mice occurred when they ingested NTA complexed cadmium or methyl mercury, chemicals which are potentially present in ground water. This left no suitable candidate for replacement, and for the time being the only replacements are the conventional alkaline builders and water.

## LAUNDRY WASTES

One of the most common liquid wastes discharged is laundry waste. As a rule the waste is discharged into municipal sewers. Larger volumes of soapy wastes from power laundries, scouring establishments and plants employing soap solutions may require pre-treatment before discharge into municipal sewers or must be treated separately.

Soapy solutions have high oxygen demands; pure soaps require almost twice their weight in oxygen for satisfaction of the 5-day BOD and commercial varieties of soap about 1½ times. The volume of laundry waste normally is 5 to 10 per cent of the average daily flow of sewage, but the waste is from 10 to 20 times stronger. Average analyses of domestic and commercial laundry waste are shown in Table 17.3, but laundry practice varies and both stronger and weaker wastes are frequently observed.

Separate treatments for clarification by means of chemicals have included lime, alumino-ferric, calcium chloride, iron salts, alum and salt (sea) water. Treatment with alum and acid has been claimed to be less costly than ferric sulphate-lime or acid-lime treatment, but purification results comparable to acid-alum treatment with ferric sulphate and lime at about equal cost have been reported. It appears that adjustment of pH to 6.4 to 6.6 with acid prior to coagulant addition produces optimum purification at lowest cost. The use of carbon dioxide to lower alkalinity and the pH value prior to coagulation has produced excellent results.

**Table 17.3.** Average analyses of commercial and domestic laundry wastes.

| *Description* | *Commercial* | *Domestic* |
|---|---|---|
| pH | 10.3 | 8.1 |
| Total alkalinity (ppm) | 511 | 678 |
| Total solids (ppm) | 2114 | 3314 |
| Volatile solids (ppm) | 1538 | 2515 |
| BOD, 5-day (ppm) | 1860 | 3813 |
| Oxygen consumed (ppm) | 868 | 1045 |
| Grease (ppm) | 554 | 1406 |

The oxidation of laundry waste can be successfully accomplished on trickling filters, handling rates of application of 1.5 mgad despite grease loadings up to 490 pounds per acre foot per day, pH values of 11.0 and alkalinities of 1,000 ppm in the raw waste.

Activated sludge oxidation of laundry waste is feasible, but longer periods of aeration than commonly used for sewage are required.

One difficulty seldom mentioned, but frequently encountered in treating the waste with chemicals, is settling. The release of carbon dioxide on treatment causes partial flotation of the sludge. Experiments on treatment with flotation by vacuation produced better results than settling. Treatment of waste-water after equalisation reduces the chemical demand and aids in clarification. Pre-aeration or flocculation prior to settling aids removal of impurities. Laundry waste treatment can be done as under:

1. To remove about 75 per cent of oxygen consumed, solids, and grease, laundry waste can be treated most economically by acidification with $H_2SO_4$, $CO_2$, or $SO_2$, followed by coagulation with alum or ferric sulphate. It can also be coagulated partly and completely by many salts and acids and by lime, but in most cases such treatment is too costly.
2. Laundry waste can be purified to a high degree by means of conventional and probably high-rate trickling filter, or by the activated sludge process if long aeration periods are used.
3. The sludge obtained can be dried on sand beds directly or very probably digested anaerobically, or filter pressed. Final disposition of undigested sludge can be accomplished by a soap recovery process or by incineration.
4. Waste treated by chemical coagulation can be further purified by passage through a biological filter or by the activated sludge process.
5. Sewage containing any percentage of laundry waste can be treated on a biological filter of adequate capacity and correctly designed for this purpose.
6. The activated sludge process employing normal aeration periods can handle sewage containing laundry waste in concentrations approaching 20 per cent.

## ENVIRONMENTAL PROBLEMS

As in other industries, some problems of an environmental nature can arise in and in the environs of a factory, due to noise, gaseous and liquid effluents, and solid wastes. This section is concerned, however, with two problems—lack of adequate biodegradability and eutrophication, which can occur when waste-water containing certain detergents are discharged from the homes and factories in which they have been used. The two problems are quite distinct.

## SYNTHETIC SURFACTANT WASTES

In the past, the increased use of synthetic surfactants, especially detergents reached to the point of creating problems in municipal sewage plants due to excessive foaming and inability to reduce the organic content of the sewage effluent in the developed countries. These compounds which are not degradable are called bio-hard. The need for biodegradable surfactants has become very prominent due to the emerging lobbies of environmental activists. The biodegradable types can be oxidised to simple end-products such as methane is known as biologically soft syndets. This has presented the industry with both challenges and opportunities. The molecular structure of chains is important in determining whether a compound is biologically soft or not. As the straight chain-normal paraffin structure gives a soft compound and a branched or iso-paraffin structure resists biodegradation, the biological-persistent branched alkylbenzene chains are being replaced by linear chains made by new technologies.

### Biodegradability

When waste soapy water is discharged into a sewage system, or directly into a river or lake, no real problems normally arise. Excess soap is precipitated by the hardness in the receiving water and it is also readily broken down by micro-organisms such as those which play a major role in sewage treatment. Soap is, therefore, described as biodegradable. This is not the case with certain non-soapy surfactants. Problems arose with foam on small rivers due to the discharge of waste liquors from textile factories. NSD products had serious adverse effects on sewage treatment plants and on the rivers, etc. into which the effluents, treated or untreated, were discharged. Even water from wells near places where these effluents were discharged was sometimes found to contain foam. At that time the trouble was traced to alkyl benzene sulphonates with highly branched alkyl chains, in particular those derived from propylene tetramer (the so-called PT alkylbenzenes, or alkylates). These surfactants do not break down sufficiently readily in sewage treatment plants and they are described as biologically 'hard', or as not biodegradable. The micro-organisms attack side chains two carbon atoms at a time and the process is stopped if a tertiary carbon (one to which only alkyl groups and no hydrogen atoms are attached) is reached. This led to the linear alkylbenzene sulphonates, often called LABS or LAS, in which the side chains are derived from suitable, substantially straight-chain, hydrocarbons. In fact, this was a reversion to improved forms of the 'keryl' benzenes which were used before the PT type was developed and came to be the dominant anionic surfactant. LABS surfactants are accepted as adequately biodegradable, although they are not broken down as readily and completely as soaps and other surfactants derived from fats, or synthesised to contain a completely unbranched chain with an even number of carbon atoms and no benzene ring.

Non-ionic surfactants can also lack adequate biodegradability, but problems of analysis of low concentrations in effluent water make regulation more difficult than for anionics. The alkyl phenol ethylene oxide types, once widely used, are being, or have been, phased out by voluntary agreement. There is still some argument, but they are probably not sufficiently degradable to be acceptable. The octyl and nonyl side chains in these compounds are highly branched. Other non-ionics based on straight-chain, or relatively straight-chain, alcohols are currently acceptable, but there is some uncertainty regarding the fate of the polygylcol chains. Surfactants with long polyethylene oxide and/or polypropylene oxide, chains are particularly suspect.

The biodegradability of cationic surfactants has been little discussed, but it is probably not important, because of the mutual precipitation when most anionic and cationic surfactants are mixed.

### Eutrophication

It is a natural process for lakes and ponds to silt up gradually. When little nutrient is present, and hence there is little plant life, the condition is called oligotrophic. When rivers and rains bring in nutrients, algae and other plants grow and the condition becomes eutrophic. 'Cultural' eutrophication can occur if human activities, usually the discharge of large quantities of sewage into relatively small lakes, or into lakes from which little, or no water is discharged to the sea, result in the presence in the water of excessive quantities of nutrients. This abundance of nutrients leads to a massive explosion of the population of algae, with production of oxygen. At the end of the growing season most of the algae die and bacteria feed on the dead plant material. This process requires oxygen, that is the destruction of the dead algae imposes a biological oxygen demand (BOD), similar to that caused by the discharge of milk, or any other destructible organic matter, including normal sewage. In extreme cases, this can eventually turn the lake water from an aerobic to an anaerobic condition (one lacking dissolved oxygen), which kills fish and other animals. Under anaerobic conditions, anaerobic bacteria flourish and may produce toxins and bad odours.

## LINEAR ALKYL BENZENE (LAB)

### Environmental Issues

LAB sulphonate (LAS) emerged in detergent market as an alternative to sodium dodecyl benzene sulphonate (DDBS) which has been phased out due to problems of persistent foam in rivers and sewages. In view of growing concern about impact of LAS on environment, several studies have been undertaken on biodegradability and possible hazards of this product. Some inferences drawn are as follows :

1. LAS is mineralised in the environment to form $CO_2$, water and inorganic sulphates.
2. Its biodegradability as per OECD method is over 98 per cent.
3. The half life of the product in natural water is in the range of 4 to 9 days.

Though LAS is biodegradable, efforts are being made to develop detergents with more positive biodegradability profile. These include *n*-paraffin sulphonates, alpha-sulphomethyl ester of detergent alcohols and alkyl polyglycosides.

## METHODS OF SAFETY EVALUATION OF SYNTHETIC DETERGENTS

Detergents are used almost universally on regular basis. Their safety aspect has been the subject of numerous investigations in the past. The committee responsible for the formulation of this standard felt the need for specifying test methods which can evaluate the safety aspects of synthetic detergents used by the consumers. These test methods have been specified in two parts of this standard. Part 1 of this standard prescribes the test methods for irritant potential of synthetic detergents. These tests are designed to investigate the irritancy potential of detergent products which, during use, involve repeated immersion of hands/forearms and skin contact with the detergent for extended periods of time. The test for skin sensitisation is being covered in this second part.

Detergent products which come in contact with skin should be free from skin sensitising/skin allergy producing components. Such components even when present in minute levels (ppm — part per million concentration) in the formulation, would cause itching, redness, swelling and blister formation on the skin. The onset of such episodes is usually a delayed response from the time of first contact with the product which makes it difficult to identify the incriminating product/ingredient, even by an experienced medical expert. The detection involves cumbersome procedure of patch testing on skin with all suspect materials on people who are susceptible to such skin allergy. The intractable nature of this skin condition may cause unnecessary hardships to the consumers. Hence, it is essential that manufacturers of detergent products make sure that their products are free from such hazard, by subjecting their products/ingredients to a predictive skin sensitisation test using a sensitive animal model.

The allergic skin reaction described above is not caused by toxic or irritant properties of the substance (which need to be tested separately) but is due to the increased sensitivity and harmful immunological response to these substances caused by repeated contact. Identification of such allergenic ingredients/ products is possible through the guinea pig maximisation test (GPMT). This standard describes the principle and method of test. Presently, the facilities for these test methods are limited. However, there is scope and imperative need for rapidly expanding and providing these facilities.

These tests are designed to investigate the irritancy potential of detergent products which, during use, involve repeated immersion of hands/ forearms and skin contact with the detergent for extended periods of time. The test for allergenicity (skin sensitisation) will be covered in the subsequent part of the standard, when the test procedure and other details have been standardised. Till such time, if it is reported that a particular detergent is producing very frequent allergic reactions, the matter should be referred to a competent dermatological laboratory.

Presently, the facilities for these test methods are limited. These facilities are available at organisations like the Haffkins Institute, Mumbai; Industrial Toxicological Research Centre, Lucknow; and the All India Institute of Medical Sciences, New Delhi. However, more and more organisations are acquiring these facilities.

## POLLUTIONAL PROPERTIES

### Oxygen Depression

Much organic matter in waste-water will be degraded by bacteria ever present in the environment. An initial population of bacteria is omnipresent in rivers and lakes. In degrading the material, the bacteria utilise dissolved oxygen present in the water. This oxygen is also used by gill-breathing fish in their life processes. Therefore, biodegradable organic matter in a stream or lake may result in a lowering of the dissolved oxygen content so that the fish are harmed or killed. The common measurement of biodegradable organic matter is the biochemical oxygen demand (BOD) test. This test measures the amount of oxygen used in degradation of the organic matter under test conditions simulating to a degree those in a stream.

The standard BOD test period is 5 days and the temperature 20°C. The BOD is the amount of oxygen used expressed as oxygen and varies from about one-half to two or more times the weight of the organic matter oxidised. During this 5-days test, about 68 per cent of the carbon and hydrogen in the waste-water compounds is oxidised. All the organic matter in waste-water from animal and vegetable oil processing are readily biodegradable, so the waste-water have a significant BOD, that is, use of oxygen in degradation. The ratio of BOD to biodegradable organic matter for triglycerides is about 1.8:1. Therefore, the oil is eliminated in biological treatment.

### Suspended Solids

Another common parameter used in measuring pollutional characteristics is the suspended solids test. This measures materials that can be filtered using specific filter media. In the waste-water under discussion there will be some suspended solids consisting of fat particles and perhaps filtering clay. Suspended solids are of interest because of their effect on appearance of the water as well as their potential of settling out on the river or lake bottom.

## Acidity–Basicity

The presence of acids producing too low a pH or of alkaline materials producing too high a pH will adversely affect aquatic life. A pH in the range of 6–8 is usually specified in discharges to public water. Both acids and alkaline materials are used in these industrial processes and there is a potential of either high or low pH.

## Oil

Unfortunately for the processors of animal and vegetable oils, in early water pollution control activity, oil of all kinds is generally lumped together as one measurement and controlled as one parameter. The standard analytical procedure measures any material extractable by a specified solvent under acidic conditions. The substances measured include petroleum oil, animal–vegetable fat or oil, fatty acids, soaps and any other compound that is extracted. A water quality standard based on this 'total' oil measurement has been adopted by two or three states and limitations on total oil have been specified in effluent standards for a few industry categories. The most recent technical policy document for water quality standards of the US Environmental Protection Agency (EPA) discusses differences in the pollutional properties of two major kinds of oil–petroleum and animal–vegetable. Both kinds of oil have a wide variety of compounds of differing properties; nevertheless, broad categorical differences exist between the two kinds mentioned. The potential pollutional properties of petroleum oil include:

1. Low rate of aerobic biodegradation in the environment and thus ability to persist for extensive periods in the water.
2. Complete lack or biodegradability in the absence of oxygen; therefore indefinite persistence when agglomerated with silt and settled in bottom muds.
3. Impairment of the flavour of fish flesh.
4. Resultant taste and odour in drinking water.
5. Creation of fire hazards or explosive atmospherics in sewers and treatment plants.
6. The pollutional properties of animal–vegetable oils include the consumption of oxygen in biological degradation as described for the BOD test previously at rates comparable to most biodegradable organics. It is more dramatic to list the pollutional properties that animal–vegetable oil do not exhibit that petroleum oil does:
   (a) Animal–vegetable oil triglycerides are not toxic to fish or aquatic life; this oil is a part of commercial fish food.
   (b) Animal–vegetable oil does not persist for great lengths of time in a river flow.
   (c) Animal–vegetable oil does not persist in bottom deposits where oxygen is absent.

The two kinds share some properties, such as that of spreading to thin layers on water surfaces, for example, causing oil slicks and other unaesthetic effects. Therefore, there is logic in common restrictions on oil spills or discharge of free floating oil of any kind. However, with the universal application of biological treatment, the animal–vegetable oil remaining in treated effluents will be in solution, finely dispersed or associated with suspended particulates; therefore, such effluents are not involved in slicks or aesthetic problems.

The most meaningful pollutional measurement for animal–vegetable oil is the BOD test. Biochemical oxygen demand effluent standards are adequate for control. One state, Illinois, has recognised in its effluent standards for oil the two different kinds of oil and permits separate measurement and reporting. A standard (approved) analytical measurement for petroleum oil (hydrocarbons) is available.

### Toxic Pollutants: Zinc, Nickel and Phenols

Three substances currently listed as toxic pollutants in federal water pollution control law have the potential of being present at detectable levels in waste-water of some plants of the industry. The three substances are nickel, zinc and phenols.

One or more of three properties form the basis of listing these compounds, namely, direct adverse effects at relatively low concentrations on saltwater aquatic life, on freshwater aquatic life and on human health. The criteria (standards) for the protection of the aquatic environment for these pollutants are in the process of being established by the US EPA as provided under federal law. Technology-based limitations on these substances in industrial waste effluents containing significant concentrations are in the process of being established by the US EPA. The current limitations for the waste-water of the soap manufacturing industry do not contain limitations on any of the compounds listed as toxic because the compounds have not been found to be present in significant concentrations. Effluent limitations for the edible-oil-refining industry have not been proposed at this time. The preliminary document for the effluent limitations for the edible-oil-refining industry does not propose limitations on any of the compounds.

Nickel has the potential of being present in waste-water of edible-oil plants that carry out hydrogenation of oils with a nickel catalyst. Not all edible oil refining plants perform this operation. Zinc may potentially be present in waste-water from fat splitting (the hydrolyser). Many plants do not use zinc in the hydrolysis process. Substances that analyse as phenols have been found in low concentrations in the effluents of two edible-oil-refining plants. Phenols are defined as materials that react with the reagent, 4AAP, in a specified analytical procedure.

The incentives of loss prevention and water quality protection, coupled with the treatment necessary to remove the biodegradable organic matter, will probably produce waste-water from these industries that do not contain concentrations of phenols of significance to the aquatic environment or to human health.

## WASTE-WATER TREATMENT PROCESSES AND TECHNOLOGIES

This section briefly describes the various treatment systems encountered in fats and oils and oleochemical applications along with their capabilities relative to the parameters of interest. The processes listed are generally consistent with the order in which they might appear in a treatment facility. Table 17.4 presents the potential effluent quality for oil and grease for various physical–chemical processes.

**Table 17.4.** Treatment process efficiencies.

| *Process* | *Freon extractable oil and grease (mg/L)* |
|---|---|
| Well designed gravity separation, neutral pH range | 100–500 |
| Well designed gravity separation, low pH, elevated temperature | 50–150 |
| Dissolved air flotation, chemically assisted | 30–100 |
| Pressure diatomaceous | 50–150 |

### Gravity Oil Separation

This processes can be considered as in-plant process control or as a waste-water treatment process. In either case, it is essential that adequate gravity separation be provided at every fats and oils handling facility. In one instance associated with the process design of waste treatment facilities at a new, grassroots edible oil refining and processing plant, the waste collection and in-plant controls were being designed by others on the refinery engineering team. When questioned as to why no gravity oil separation facilities were provided or desired, the new plant manager noted that this was to be a state-of-the art refinery and would experience no oil spills, due to excellent process operations controls; therefore, no gravity separation facilities were required. Needless to say, some months after the facility started up, some very expensive retrofitting was required to provide in-plant gravity separation capabilities.

Where is the best place to locate gravity separation facilities in an oilseeds operation? The answer is in as many places as possible and as close to the sources as possible. This is not always practical or efficient. The main considerations are listed below:

1. Prevent cross-contamination of a higher quality recoverable product with a lower quality material whenever possible; i.e. do not combine packaging area drainage with deodouriser cooling water blowdown which may contain pesticide residue.
2. Place gravity separators wherever waste-water gravity flow ends and pumping begins. If there is any kind of a wet pit involved in the pumping, the oil will float to the surface despite whatever attempts are made to keep it in suspension. Therefore, one should take advantages of this and construct a gravity separation section as part of or adjacent to the wet pit.
3. All waste streams must receive oil separation of some sort prior to any further treatment. Every oilseeds processing plant requires facilities for capturing of floatable fats and oils resulting from both day-to-day losses and larger spill situations.

Gravity oil separators for these applications should be designed for rise rate loadings to accommodate peak flows, if practical. Experience dictates rise rates in the range of 10 to 20 L/minute per $m^2$ of effective separators surface area.

The rise rate concept is applicable to the design of continuous flow separation devices. Many designers use retention time for separator sizing. This is acceptable if desirable separator geometry dimensions such as length to width ratios are maintained. The concept of retention time may be applied directly for design of batch settling vessels.

Skimming of oil grease from the surface of separation vessels can be performed on a manual or continuous basis. The viscosity of many oils used in vegetable oil processing at ambient temperature is such that surface contact skimming devices (belts, ropes) are not particularly effective. Chain and scraper type skimmers are effective, but not physically adaptable to every skimming situation. Shop fabricated, manually operated decant devices have been quite effective. In summary, few, if any, commercially available skimming devices are useful for skimming higher melting point oils.

A portion of the flow to a separator often originates in rainfall from oil contaminated runoff areas. This may result in flow rates too high to practically handle at desirable rise rates. This often dictates a need to handle process waste flows and contaminated runoff in separate systems.

## Equalisation

The variance in process conditions, rainfall contributions and related considerations make equalisation facilities of stabilisation of waste flow and strength essential.

The volume of equalisation required will depend on the variation of incoming flows and strength over a reasonable time duration. Experience indicates that the volume of equalisation required for flow and strength for the typical oilseeds operation will be about the same. The only waste-water stream not requiring equalisation is the effluent from the milling and extraction operation. This stream is composed primarily of the sewer water evaporator steam condensate and is usually without much variation.

The volume of equalisation required can be determined using a volumetric influent hydrograph approach. The required volume can be determined graphically to the level of accuracy required. A check of influent and resulting effluent strength variation can be made to confirm the suitability of the volumetric design for a specific application. In general, the equalisation surge volume required will vary between 30–50 per cent of the daily average flow, but this can vary from each application.

### Dissolved Air Flotation

Dissolved air flotation (DAF) is a technique that utilises minute air bubbles to enhance the flotation of oils and suspended materials not removable in reasonably sized gravity separators. The bubbles are formed by pressurising all or part of the waste flow and introducing air to a pressurisation cell. Subsequent depressurisation allows the bubbles to form, according to Henry's law.

This process is enhanced by the addition of polymers, alum, or other flocculants. This chemical addition provides for coagulation and flocculation of colloidal oils and solids into larger particles that can be easily removed.

In fats and oils applications, this techniques has been shown to be very effective in removing FOG and associated insoluble BOD; however, significant quantities of waste sludges are generated. Table 17.4 shows the removal rates that can be expected for various processes discussed in this presentation.

The selection of a chemical flocculent must consider pH conditions as well as the removals desired. In the applications of DAF in a fully integrated oilseeds plant, alum and polymers are effective. An alternate approach is to use sodium hydroxide, the most commonly used refinery waste neutralisation agent, to bring pH up to a range of about 5.0, then lime can be used as a final pH adjustment chemical. The resulting insoluble calcium sludge will perform the same function as alum at less cost. The lime sludge may settle better than it floats, however, and flotation should be evaluated against settling.

The use of chemical flotation in this application may be less desirable considering other processes. DAF has been applied effectively as a trimmer process following other treatment facilities for FOG removal requiring a 100 mg/L FOG limit. This can sometimes be accomplished using food grade approved polymers. In that event, the recovered materials can be sold as a low grade feed fat rather than disposed of as a waste sludge.

### Filtration

In-line filtration using diatomaceous earth is an effective device for reducing FOG and associated insoluble BOD to very low levels. It will not remove soluble BOD components due to acidulation wastes.

Filtration has not found extensive use in fats and oils waste-water treatment applications; however, it does provide an effective treatment process, particularly as an alternative to DAF as a trimmer device. Another advantage of filtration is its familiarity to refinery personnel as it is used in other in-plant processes. The principal disadvantage of filtration is that it also produces another waste sludge to be disposed of.

### Acid Hydrolysis

An alternative approach used successfully at several fats and oils handling facilities employs a low pH, high temperature hydrolysis effect to recover oils and greases. Figure 17.3 shows this process.

If the plant acidulates soapstock and/or washwater, the resulting acidulation waste will contains sufficient residual heat and acidity to affect the entire waste stream pH and temperature. This obviously assumes an alkali refining process rather than a physical refinery. This will produce an acidulation effect on residual oils. These oils can then be readily removed in a gravity separation process.

The process can be used at plants that do not acidulate soapstock; however, site specific studies should be conducted to determine pollutant removals to be expected. In the overall scheme of things, acidulation of washwater as part of the treatment process will greatly improve effectiveness.

This process offers a number of advantages.

1. Process control is relatively easy. If the waste stream pH stays at 3.5 or below and a temperature of 37.7°C (100°F) or higher and soapstock and/or washwater is acidulated, relatively constant effluent results can be obtained virtually independent of the raw influent waste characteristics (high FOG and BOD levels). The exception to this is carryover of excess middle phase from an inefficiently operated acidulation system. This material will not gravity phase separate and will actually inhibit the process effectiveness for other plant waste streams. Other sources of protein, such as extraction and milling meal fine carryover into the crude oil, will also have an adverse effect on this (and other) oil separator processes.
   Under proper conditions the effluent BOD strength in the separator underflow will be equal to the soluble BOD load from acidulation plus the additional contribution from the residual oil and grease in the separator effluent.
2. The materials recovered in the separator will contain a relatively low moisture and high total fatty acids (TFA).
3. If the facilities can be constructed near acidulation, existing acid handling, steam and recovered oil storage tanks can be utilised for both purposes. The primary disadvantage is the need to utilise material of construction that will withstand the high temperature and low pH conditions. A wide variety of tankage, process controls and related equipment have been utilised in designing these treatment facilities. This is dependent on refinery size and process, client preferences, existing equipment to be utilised and other related factors.

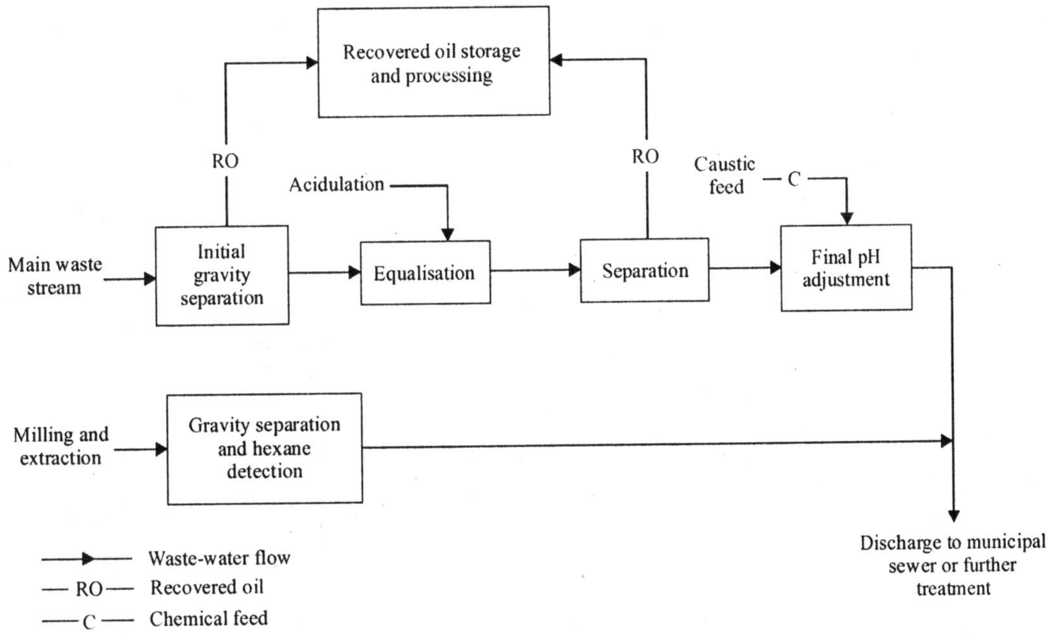


**Fig. 17.3.** Acid/hydrolysis process.

Neutralisation of the separator effluent is usually accomplished with sodium hydroxide; however, lime or ammonia are occasionally used in special situations, as noted previously. A two-stage system using well mixed tankage of suitable size is recommended. The neutralisation invariably occurs on the steep part of the titration curve in these applications. Any inappropriate design, such as an excessively remote pH controllers, will result in a non-functional system. As a result, it is highly recommended that an individual experienced in designing pH control systems reviews the final design.

## Other Processes

Comments are made here relative to processes that have had only marginal success in oilseed processing. Various membrane technologies have been applied to these waste streams. While pollutant removals are good in some applications, membrane life and regeneration have presented problems. As a result, operating costs for periodic membrane replacement are quite high.

At least one attempt has been made to utilise activated carbon to remove residual organics. The nature of the soluble components in the acidulation stream (the primary sources of soluble) is such that they cannot be effectively removed by activated carbon.

Several processors attempted to use a mixed media (sand, anthracite) filter to remove residual oils. This technology was adopted from petroleum waste applications; however, it did not function well in oilseeds applications. This was due primarily to an inability to backwash these units caused by the highly viscous nature of these oils at ambient temperature. Attempts have been made to apply inclined plate and other packaged media separator systems. These have not functioned well in edible oil applications. This is also due to the viscous oil coating over the separator media.

## Biological Treatment

Wastes from oilseeds operations have been successfully treated using biological treatment in any number of circumstances. Traditionally, pretreated effluent has been further treated in combination with domestic wastes in publicly owned pretreatment works (POTWs). Also, a number of biological systems are in place at oilseeds plants either as pretreatment or for direct discharge. Figure 17.4 is a diagram showing relative rates of biodegradation for a variety of oils. Clearly, it can be seen that oils of animal and vegetable nature are highly biodegradable and compatible with municipal domestic wastes.

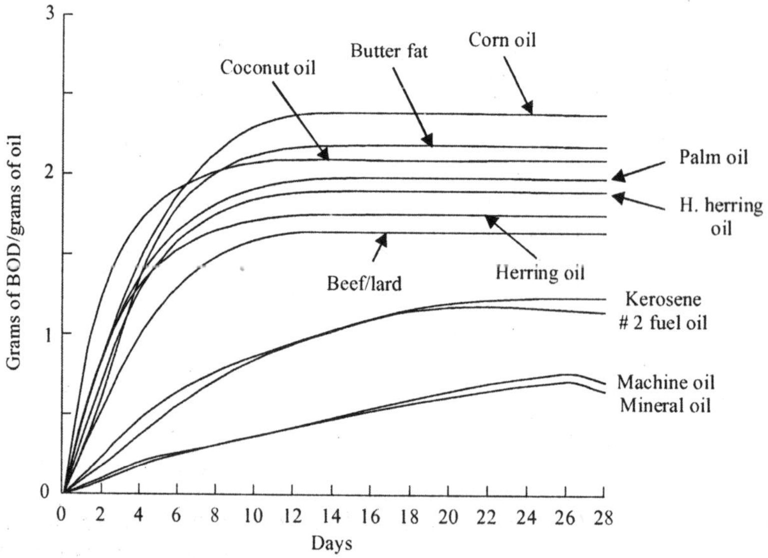


**Fig. 17.4.** Biological uptake rates.

Good gravity separation of floatable oils is mandatory prior to further treatment in a biological facility. Additional oil removal pretreatment may be required for oil value, prevention of sewer line blockage problems, or merely to reduce the size of the biological facilities.

Various biological processes have been used with good results, including extended aeration/activated sludge, sequential batch reactors and biofilm reactors. Anaerobic treatment has not been effective due to high sulphate levels.

## Treatment for Direct Discharge

The first step in treatment of waste-water from this industry is the removal of floatable oil by gravity settling. This process is, of course, applied to wastes that have been minimised by in-plant control as previously discussed. The oil remaining after gravity settling is in a dispersed state or in solution. The removal of the dispersed oil and other biodegradable organics can be accomplished by either of two treatment systems: (i) biological treatment; and (ii) chemically enhanced air flotation followed by biological treatment. The choice is an economical one between chemically enhanced air flotation and the additional biological treatment capacity. Practically all the plants in the industry have chosen to use chemically enhanced air flotation followed by biological treatment. Some managers have utilised air flotation because it offered a backup process to gravity settling when a slug or an emulsifying agent loss would cause abnormal passthrough of oil through gravity settlers. Troublesome massive overloads of biological treatment with oil are thus avoided.

The dissolved air flotation process consists of creating minute air bubbles in the waste-water introduced to a settling tank; these air bubbles tend to become attached to surfaces such as the surfaces of oil particles in the waste-water (Fig. 17.5). A few air bubbles on a particle of oil will rapidly buoy it to the surface. The air bubbles seldom become attached in sufficient numbers to smaller oil particles to permit their removal, so air flotation alone will not be very effective. A necessary step is formation of the small oil droplets into larger agglomerates by use of coagulating chemicals, hence the name chemically enhanced air flotation. Chemicals, such as aluminium hydroxide (added as aluminium sulphate) or an organic flocculant, form bridges between particles so that an agglomerate of oil particles is formed that is large enough for the air bubbles to buoy them to the surface.

Small air bubbles may form by mechanically creating them or by a process that causes them to be created by air coming out of solution. The latter process, called dissolved air flotation, accomplishes the solution of air by pressurising the flow at about 50 psi and contacting the water with air. In the flotation tank the pressurised water is released to atmospheric pressure and the air comes out of solution in the form of fine bubbles. Various designs are marketed embodying the air flotation process. In principles, they resemble gravity flotation. The surface overflow rates may be four times or more than those used in gravity flotation because the rise rate of the particles is greater with attached air bubbles.

The flotation process effectiveness is very dependent on the chemical coagulation step. Coagulation will not take place optimally outside a certain pH range nor in the presence of higher concentrations of dispersing agents, emulsifiers and surface-active materials. The latter are likely to be present in the waste in variable concentrations. Dispersing and surface active agents such as caustic, detergents and phosphates are likely to be present from cleanup operations. Emulsifiers such as mono- and diglycerides are naturally present and such compounds are used in salad dressings and margarine for their emulsifying properties. Therefore, the process is not 100 per cent reliable. For good results in a majority of cases, the waste-water must be rendered more uniform in composition by provision of a large equalisation basin. Close control of the pH so that it remains in the desired range is also a prerequisite. Even with reasonable application of these precautions, chemically enhanced air flotation cannot be depended on under the technical control and level of attention practical in the field to consistently and continuously produce an effluent of low oil content. Occasionally, the emulsion simply cannot be broken for a variety of reasons, such as variations in acid and alkaline conditions and in the concentration and kind of emulsifiers. The science or art of coagulation is not at the state that perfect performance can be assured.

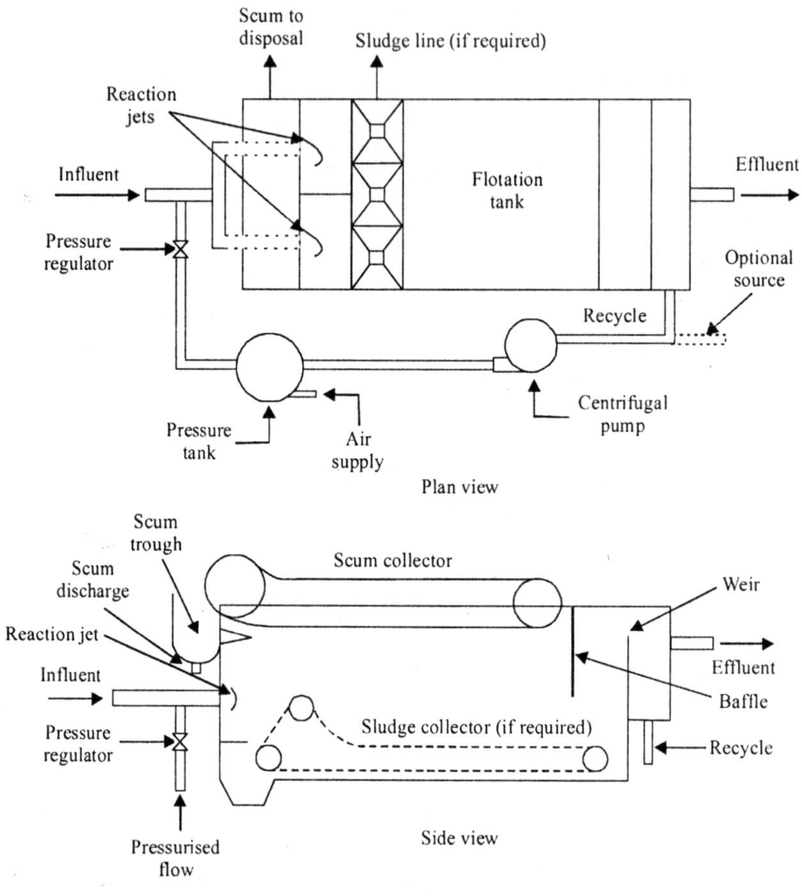


**Fig. 17.5.** Dissolved air flotation process equipment. This model utilises rectangular tanks and pressurisation of a recycle stream.

The effluent from the gravity-settling step or from chemically enhanced air flotation must be given extensive biological treatment to meet the categorical limitations for direct discharge to be imposed by the US EPA. Biological treatment is described briefly in the subsection on municipal treatment. Two biological treatment processes are presently utilised for biological treatment of waste-water from animal and vegetable oil processing–the activated sludge process (Fig. 17.6) and the aerated lagoon process. Operating data from existing plants show that the loading parameters used in designing such processes for domestic sewage or other biodegradable industrial waste-water apply. The biodegradable material to micro-organism ratio (pounds of BOD per pound of biological mass) is in the same range as used in such designs for municipal sewage. One plant is using a combination of activated sludge and aerated lagoon treatment.

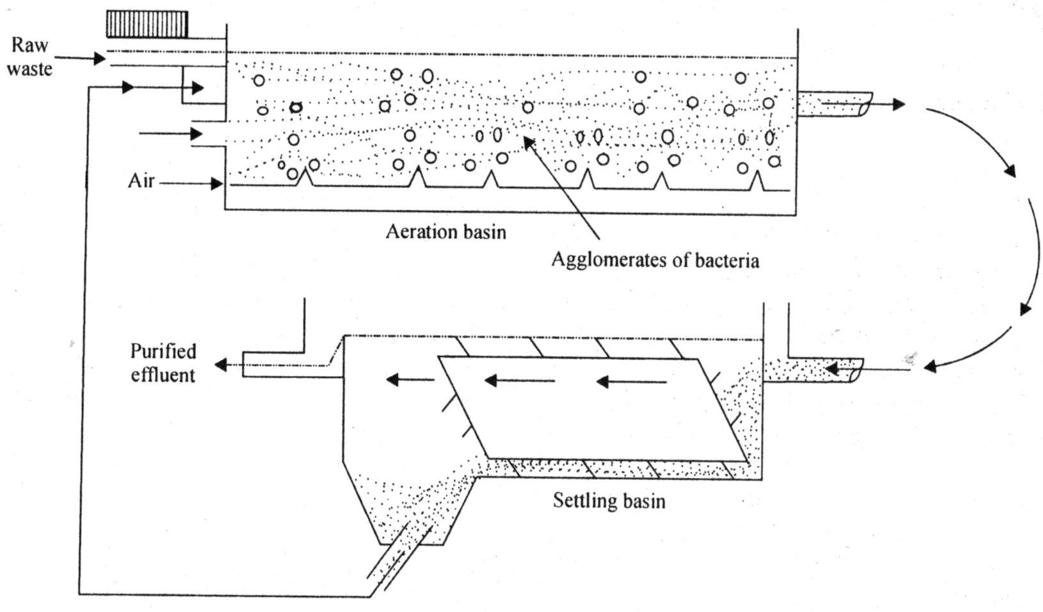


**Fig. 17.6.** Activated sludge waste-water. Treatment process consists of contacting wastes with bacteria culture in aerated basin and settling to remove and recycle settleable bacterial agglomerates.

At plant locations on small streams an additional treatment requirement has been applied. This consists of filtration through beds of fine material, including sand. The processes used in one of the more complex treatment systems is shown in Fig. 17.7.

As discussed in the section on pollutional effects, one or possibly a few animal and vegetable processing plants practicing joint treatment are or will be subject to limitations on phosphorus. The practice and methodology for removing phosphorus from waste-water is limited to the precipitation of phosphorus occurring in the form of phosphates. Phosphorus in these industries waste-water are in the form of organic phosphorus compounds and it is doubtful that the present precipitation technology for phosphate will apply. The treatment technology has yet to be worked out.

## In-plant Control

Water pollution control by operator attention to preventing losses to sewers and the use of equipment that inherently has less loss of materials to waste-water are major elements in a control programme. A quantity of process material that is negligible in terms of production is very important and significant as a percent of materials in waste-water.

A complete water pollution control programme includes:

1. Minimisation of materials entering the waste-waster by keeping valued materials in the product stream.
2. Separation of materials for marketing as by-products that otherwise would be associated with waste-water.
3. Recovery of materials from waste-water for reuse or recycle.
4. Treatment of the minimised waste-water.

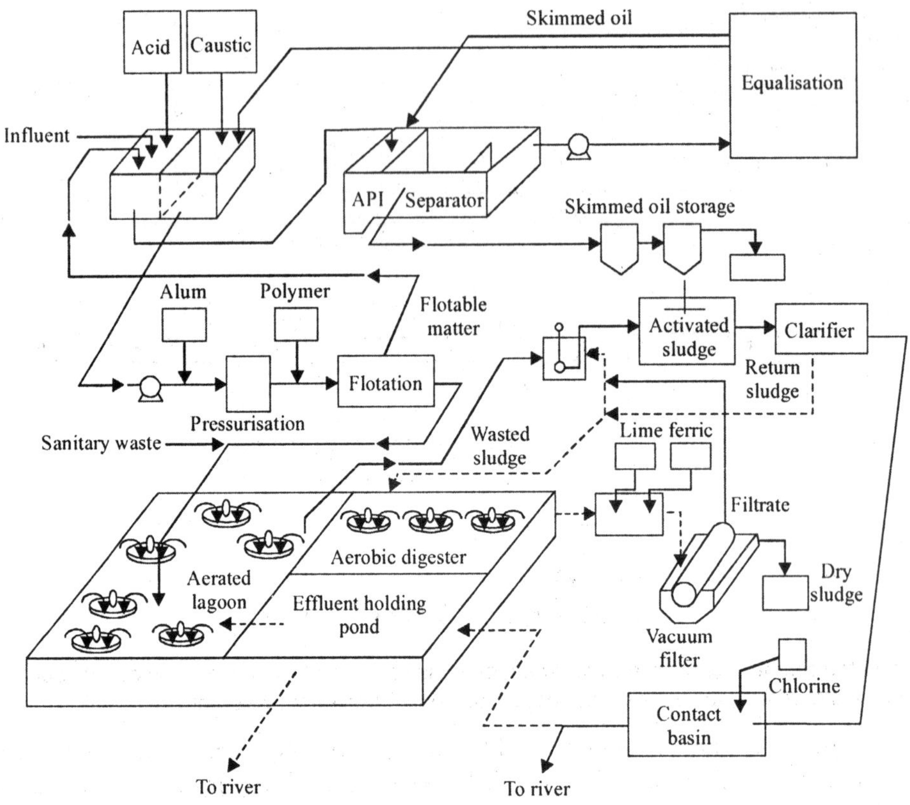


**Fig. 17.7.** Complete complex treatment train.

The cost of waste-water treatment to meet the more severe limitations is such that prevention and recovery incentives are greatly supplemented by costs of treatment.

Pollution prevention measures, referred to as 'good housekeeping', include such maintenance steps as never allowing a leaky pump or stirrer gland to continue leaking. 'Pollution prevention' means using seals or equipment that minimise leaks and the immediate repair of those that leak. It means never draining oil or process material from a line or tank or pump to the sewer. The material should be returned to process or disposed of in the manner of a solid waste. 'Good housekeeping' means scooping up spilled material and picking up remnants on absorbent material rather than flushing the surfaces to the sewer.

Pollution prevention means not allowing tanks to overflow and by using alarms, alert operators and catchment systems. Most importantly, it involves not operating equipment at over its rated capacity if it is at the cost of producing a high quantity of materials in waste-water. It also means avoiding misoperations and correcting those that occur without delay.

## WASTE MATERIALS/SOLID WASTES

The words 'waste materials' or 'solid wastes' in environmental control vernacular generally means those materials that are discarded other than as pollutants in the waste-water or in emissions to the atmosphere. Waste materials may be solids, liquids, semisolids, slurries, or even contained gases. Much of the solid waste from industry is similar to household garbage and refuse as far as environmental significance is concerned. Some waste materials from industrial plants could be hazardous, that is, harmful to humans or other organisms if not properly disposed of. The disposal facilities must be approved and meet certain criteria.

### Sources of Waste Materials in the Oil Processing Industries

Waste materials from industrial operations include:

1. Spent bleaching earth.
2. Spent metal catalysts.
3. Oilseed residues accumulating in raw-oil storage tanks.
4. Animal hide scrap, bone, hair and other residue accumulating in raw animal fat storage tanks.
5. Diatomaceous filter earth and powdered activated carbon from filtering and decolourising processes.
6. Bottom residues from some distilling operations.
7. Oil and coagulating chemical slurry from chemically enhanced air flotation of waste-water.
8. Sludges of biological matrix created in biological treatment.
9. Impurities separated from the raw oil for which there is no by-product use.
10. Sludge from gravity settling tanks used to recover flotable oil.
11. Off-quality or contaminated products or raw materials.
12. Residues of unreacted raw materials and process chemicals from batch soapmaking operations.

### Solid Waste Impacts

Solid wastes of the industry, if not properly disposed of, could have significant environmental impact. For the most part, the solid wastes of the industry are similar to household garbage and refuse with respect to environmental significance. One interest, of course, is simply aesthetic. Also, if left open and unprotected, such material will breed flies, rats and odours. In a dump or a landfill, a certain amount of liquid will ordinarily drain from the material. This drainage, termed *leachate*, can seriously degrade the quality of groundwater or surface water it might reach. Degradation of the organic matter may produce slightly acidic conditions in the leachate that will tend to dissolve metals that may be present.

Improper incineration of solid waste can cause excessive emissions of particulate matter. Volatile organic materials might enter the air in damaging amounts.

### Solid Waste Regulation

For regulatory purposes, solid wastes are divided into two categories: 'hazardous' waste and waste similar to household garbage and refuse in disposal requirements.

A second way wastes are classified as hazardous is by naming specific wastes as identified by industry category and process source. Still a third way is by certain characteristics. One such is corrosivity, which is simply pH limits. Another is ignitability; the test for liquids is a flash point of 140°F and for solids is a judgement based on whether or not when ignited the material 'burns so vigorously and persistently as to create a hazard'. The stated intent is to control thermally unstable solids liable to spontaneous chemical changes. An other properly is reactivity, which covers explosiveness and/or generation of toxic fumes. Another test is an extraction procedure and analysis for 13 specific elements or compounds. The compounds are those named in the primary drinking water standards of the US EPA. Concentration limits are specified.

A few waste materials of the fat and oil industry are possible candidates for eventual designation as hazardous wastes under present definitions. Spent nickel catalysts used in hydrogenation of vegetable oils and containing such oil may be considered hazardous based on their testing for reactivity or in interpretation of the language of 'burning vigorously and persistently'. Much of the spent catalyst will be processed for regeneration. Nickel and zinc catalysts may also be designated as hazardous wastes based on their toxicity to aquatic life if the proposed water quality criteria for nickel and zinc are established.

Another possible candidate for designation as a hazardous waste is spent bleaching earth of high oil content from purification of edible oils. Bleaching earth with high (≥ 20 per cent) oil content, when exposed to the atmosphere at warm ambient temperatures, will spontaneously reach ignition temperatures. Spent bleaching earth of lower oil content, which may be achieved by oil recovery or by mixing with earth or other matter, is not pyrogenic. Likewise, spent bleaching earth if wet or in slurry form is not pyrogenic. Even the dry oily earth when spread in a thin layer will not ignite spontaneously.

Other materials that the fat and oil industry uses are potentially hazardous because they are in the range of pH defined to make it a hazardous waste, namely, pH less than 2.5 or greater then 12.5. However, the sodium hydroxide and sulphuric acid streams commonly used in the industry are part of the process and not wastes. These materials are neutralised before discharge in waste-waters. However, if there is any off-quality acid or caustic that is discarded and hauled off site for treatment and disposal, the plant involved would be a hazardous waste generator.

Whether or not a waste material is in the hazardous category makes a great difference in the regulations that apply and also in the costs of disposal. If a plant stores, processes, or disposes of hazardous wastes, a permit will be required from the agency with jurisdiction. Shipments must be accompanied by a manifest containing specified information.

Regulation of the non hazardous wastes are limited to the hauler and the disposer. As a minimum, the hauling regulations will require that the wastes be covered to prevent windblown loss and to control odours. The disposal must be to approved facilities. The types of approved facility include landfills, incinerators, composting plants, or land farming. 'Land farming' refers to the spreading of appropriate organic wastes, such as organic sludges, on land. The organic material is stabilised by bacterial action and later tilled into the soil. Slurries of spent bleaching earth have been disposed of in this manner. The use of land spreading is subject to limitations based on the amount of toxic metals such as cadmium in the sludge and the crop to be grown on the land. The productivity of poor soil may be enhanced by the material added. Not all metals are objectionable in controlled amounts; for example, zinc is a necessary trace element for plant growth.

Because of the contribution to the volume of leachate that might be produced in a landfill, the disposal of water slurries of waste materials of any kind to a garbage–refuse-type landfill may be prohibited.

A special waste material control is that required for PCBs. These compounds are used for heat transfer purposes in transformers, capacitors and heat-exchange systems.

This material has been found to be very persistent in the environment and to harm beneficial organisms. Regulations have been proposed to eliminate its use in food and food package manufacturing plants. Disposal of the liquid must be to incineration facilities approved by the EPA. Even the electrical transformer bodies or other equipment in which it was contained, if disposed of, have to be delivered to approved facilities. Other fluids used in transformers may have become contaminated with PCB and any such fluid with 50 ppm or more of PCB is regulated as to disposal and handling in the same manner as the pure material.

The EPA intends to issue separate regulations governing the disposal of waste oil. Waste petroleum oil is persistent in the ground, so disposal to landfills will be prohibited. (Animal and vegetable oils are degradable in landfills). Hopefully, the differences in the kinds of oil will be recognised in the regulations and the regulation should be appropriate to each.

## Waste Material Recovery and Disposal Methods

Recovery of the heavy nickel catalyst used in hardening is being practiced by many manufacturers of catalyst. The continuous hydrolysis of fats and oils, which is the cornerstone process of the soap and fatty acid industry, is accomplished at some plants by using medium pressures and catalysts, such as zinc oxides, calcium, or magnesium oxides. The spent zinc catalysts may be recovered and converted to the oxide form for reuse. Thus, this recovered material is not involved in waste material disposal.

Spent bleaching earth has two components with recovery potential, the oil and the earth. Unfortunately, the earth has special properties such as its bleaching (adsorptive) property, which is not easily restored. The deoiled earth was processed in multihearth furnaces and the organic-free product was used for various purposes. The recovery of the oil or portions of the oil is likewise possible. One highly effective method is extraction with solvent.

Individual companies practice partial recovery of the oil. The dewatered bleaching earth with oil content less than the pyrophoric level can be disposed of in a common landfill approved for household garbage and refuse.

Trials have been carried out for the use of the fuel value of the oil in spent bleaching earth. The quantity of material does not justify a facility solely for this purpose. The trial was to test the addition to pulverised coal being fed the boilers of a public utility. The object was more for disposal than use of the fuel value; the test was successful. However, a number of problems arose: (i) there was the pyrophoric problem during storing and transfer, if the oil content was high; (ii) the cold material conceivably could foul up the pulveriser and conveyors by clinging to them; and (iii) there was also the potential of corrosion or scale problems in the boiler and even erosion by the clay particles. The utility decided that there was not enough spent clay to justify further investigation. One concern was that the clay would add to the particulate in the stack emissions. It was reported that the clay was removed in the fly ash settling chamber. Particulate emission was below normal during the run.

Another clever secondary use method that accomplishes disposal is the utilisation of the value of the oil content of the spent bleaching earth as an animal food additive. Fat and oil are normally added to animal and chicken feed. Because of the content of clay and other materials, this source can constitute only a small percent of the feed content. This use is being practiced to a limited extent.

As to actual disposal practice, the disposal of slurries of spent bleaching earth will quite likely become unfeasible because the disposal of liquid wastes at landfills for garbage and household refuse will be prohibited. Ease of filter cleaning and transport of the spent earth is enhanced by making it into a water slurry. Disposal of a waste slurry will probably be permitted only at facilities approved for liquid hazardous material disposal. Use of these will be expensive. Thus, the spent bleaching earth will have to be dewatered or sufficient dry material added to take up the water so that no water will drain from the cake. Dewatering can reduce hauling and disposal costs.

Special arrangements with a commercial landfill have enabled one plant to dispose of the spent bleaching earth in a special landfill separate from the general facility. The clay is covered daily soon after deposit at the site. Because of its isolation from other combustible materials and coverage soon after deposit, there is not concern over the pyrophoric nature of the material. Leachate from the special fill is monitored; so far there has been none.

Recently some plants have begun disposal of the spent bleaching earth slurry by land farming. The slurry- is spread from a truck as a thin layer on land. After spreading, the slurry-laden area is disked, using common farm equipment to incorporate the material in the soil. No problems have been encountered and the agricultural use properties of the soil are expected to be enhanced.

## NOISE POLLUTION

Any industry gives rise to some noise during working. But there is a limiting factor beyond which the workers should not be exposed to noise. Noise causes constriction of blood vessels, dilation of pupils, tension in voluntary and involuntary muscles and reduces gastric secretion. On the other hand, it may be said that a certain amount of minimal noise is necessary because absolute silence is frightening. Noise is generated in seed cleaning, delinting, dehulling, decorticating, flaking and expelling operations. Cake grinders also give rise to noise. Air-cyclone separator generates roaring noise. Some types of conveyors, for example drag elevator, give out noise. Electrode boilers generate car-splitting noises at frequent intervals. Air compressors, vacuum pumps, buzzing motors, etc. are responsible for infra-sound which causes headaches among the persons working in their vicinity. Vibrations are felt near centrifuges, pumps, high speed stirring equipment, mixers with high pressure steam etc. Stamping noise is a familiar experience in soap industry. The standard adopted is a permitted noise exposure on an eight-hour weighted average sound level of 90dBA.

### Reducing Noise Pollution

Noise can be treated at the source or reduced by an enclosure. If these measures are insufficient, ear defenders can be provided for personnel. Fans, compressors, pumps, etc. produce both noise and vibration. These can be suppressed by having resilient mountings, flexible pipe connections, pipe hangers and massive foundations. Aerodynamic noise can be suppressed by fitting line silencers at the air or gas inlet and discharge points where noise arises. Absorptive or reactive designs are selected according to the frequency content of the noise and the flow rate. Suppression of noise is often accomplished by means of silencers. Individual sources radiating noise can be effectively suppressed by the use of damping compounds and acoustic lagging. These curtains form flexible barriers. Lead-loaded clothes and loaded reinforced rubber are among the materials used. These curtains are used around feed pumps, power units, compressors and bottling and canning machinery. Substituting nylon bearings for metal on a conveyor chain belt reduces noise. Another method of noise reduction is by having complete enclosures. There should be no noise leaks. Easy access to maintenance and adequate ventilation are essential. Within the enclosure, exposure limits must be observed and ear defenders worn.

## FIRE AND EXPLOSION HAZARDS

Fire hazards are prevalent in solvent extraction industry. Fine particles of cake powder cause explosion. During stand oil and blown oil preparation also, fire hazard is present. Auto-ignition takes place when certain varieties of oil cakes like kardi, linseed, nigerseed, tobaccoseed and rice bran meal are stored. Filter clothes used to filterpress some drying oils like kardi, linseed higherseed, tobaccoseed and rice bran meal are stored. Terylene filter clothes in oil industry undergo spontaneous combustion.

### Reducing Fire and Explosion Hazards

Thermocouples can be permanently installed throughout the structure. They record the temperature prevailing in the storage area. These thermocouples are not subject to mechanical injury and require very little maintenance. Some of the other devices are capable of providing a digital reading of the temperature detected as well as a typed printout of the temperature. There are continuous fire detector systems which utilise heat sensitive cables that will sound an alarm when a fixed temperature is detected at any point along the entire length of the cable. A hand-held infrared viewer device presents a detailed thermal picture of the entire scene within the field of view. The device clearly reveals temperature differences between closely adjacent objects and between objects and backgrounds, thereby quickly and accurately locating sources of fire or heat build-up. In solvent extraction industry, a continuous sampling combustible vapour detection system can be installed so that vapours escaping from the extraction process equipment can be detected immediately. Ultraviolet light fire detection systems detect flames and initiate responsive action in milli-seconds of time. Explosion protection systems detect an incipient explosion and discharge a suppressant to extinguish the explosion during the milliseconds time lapse between ignition and the build-up of pressure to explosive levels.

## TOXICITY OF OILSEEDS AND ALLERGY

Some people are highly allergic towards certain oilseeds and oils. For example, by mere exposure to the atmosphere of castor seed, meal or hulls, some people get sneezing and cold and others swelling of face and hands. By application of coconut oil to the body or head, some people get headache and rashes on the body. Certain oilseeds like tung, kusum, rubberseed when taken in minute quantities are toxic. Violent purgings, vomitting, stomach-ache and sickness are some of the symptoms. Some people are allergic to essential oils and perfumes which cause skin eruptions, eczema, pruritis, oedema and breathing disorders of the asthma and hay fever type. In extreme case, disorders of nervous type affecting the eyes and the rhinopharyngeal mucous membrane also occur. Groundnuts copra afflicted with aflatoxin can be very risky for eating.

## OCCUPATIONAL HAZARDS AND PROTECTIVE MEASURES

The above types of pollution should be avoided as far as possible at each and every stage to keep the environment safe and healthy. At the same time, the workers' safety and well-being should be maintained. Some of the remedial measures to be considered are given below.

### Air Pollution

Some of the commonly used equipment for controlling air pollution, including particulates and gaseous pollutants, are listed in Table 17.5.

Some of the air depollution equipments in cottonseed mills are dry mechanical separators (cyclones), fabric filters, wet scrubbers, electrostatic precipitators etc. To increase the efficiency of cyclones, a fine spray of water or suitable wetting agent added to the air flow entering a cyclone would be suitable. This will cause greater agglomeration of the dust particles. The efficiency of a cyclone may also be improved by placing a skimmer on the cyclone discharge. The skimmer removes an air layer near the wall of the cyclone containing the bulk of the escaping particles. Fabric filters have an efficiency of 98 per cent. Wet scrubbers have the disadvantage of sludge disposal and possibility of freezing. Another means of reducing the particulate emission from an oil mill is by recycling the air flow. The air instead of passing through a cyclone and discharging directly into the atmosphere is recycled back to the point of lint pickup. This system will eliminate direct discharges of dust particles. In the case of essential oils and perfumery, the handling of these materials should be done under a fume hood at a balanced temperature (20–24°C) free from residual laboratory odours, tobacco smoke, dust and contamination.

**Table 17.5.** Equipment to control air pollution.

| *Basic type* | *Specific type* | *Operating principle of force* | *Collecting efficiency (% by wt.)* |
|---|---|---|---|
| Dry mechanical collectors | (a) Settling chamber | Gravity | Below 50 |
| | (b) Baffle chamber | Inertia | Below 50 |
| | (c) Multi diameter cyclones | Centrifugal | 80–95 |
| | (d) Multicyclones | Centrifugal | 90–95 |
| Cleanable fabric filters | (a) Manual cleaning | Filter cake filtration | 90–95 |
| | (b) Automatic shaker cleaning | Filter cake filtration | 95 |
| | (c) Automatic reverse jet cleaning | Filter cake filtration | 99.9 |
| Air filters | (a) Dry fibre | Fibre filtering | 95 |
| | (b) Absolute paper filters | Paper filtration and dust retaining | 100 |
| Washable viscous coated filters | Automatic viscous filters | Viscous film filtration | 80–85 |
| Air washers | Spray chamber | Water spray filtration | 30 |
| Scrubbers | (a) Cyclone | Centrifugal | 80 |
| | (b) Impingement baffle | Liquid capture | 80 |
| | (c) Venturi | Liquid capture | 99 |
| | (d) Jet | Jet force | 90 |
| | (e) Packed tower | Liquid capture | 90–95 |
| Electrostatic precipitators | (a) Single stage plate/filter | Electric ionisation | 95–99 |
| | (b) Multistage plate/filter | Electric ionisation | 90 |
| Odour absorbers | Shallow bed | Particles retaining force | 95 |
| Gas absorbers | (a) Spray tower | Absorption | 95 |
| | (b) Packed column | Absorption | 95 |
| | (c) Fibre cell | Absorption | 95 |
| Incinerators | (a) Direct | Burning disposal | 95 |
| | (b) Catalytic | Burning disposal | 95 |

## Suggestions

1. Gas masks and hearing protection equipment should be provided for workers in seed cleaning, decortication and delinting units. The former should also be used in detergent and soap units.
2. Gloves should be worn by workers handling acids, alkalies, detergents, soaps, edible fatty protein material and essential oils and perfumery. Footwear should be insisted on.
3. Aseptic handling is necessary in the case of edible flour, protein concentrate, margarine and peanut butter industries to avoid bacteria, micro-organisms, dirt, insect and rodent remnants.
4. Dust collectors, cyclones, electrostatic precipitators and absorbers should be effectively installed in all the units where dust and fumes emanate.
5. Suitable noise and vibration absorbing devices should be built into the processing equipment.
6. Insurance coverage should be provided for workers in solvent extraction industry (in view of the dangerous petroleum, namely hexane, used), vanaspati industry (in view of hydrogen gas which is highly explosive), cake-grinding units (because of hazard of power explosion) etc.

7. Explosimeters or hydrocarbon solvents detectors should be used to give an automatic warning in the event the hexane vapour concentration in the atmosphere exceeds the safe level in solvent extraction plants and paint and varnish units.
8. Formation of slippery or sticky layers on floors, tanks, pipes etc. should be prevented by timely cleaning, washing and scraping.
9. Workers should be educated in handling strong toxic chemicals like oleum, sulphuric acid, caustic soda, soda ash, sodium silicate, sodium tripolyphosphate, amines, alkylbenzene and synthetic optical whiteners (the latter are said to be carcinogenic).
10. A voluntary code of conduct has to be adopted by oil and oilbased industries to reduce and control pollution and to undertake environmental protection from the irreducible pollution.

It may be stated that of all the agro-industries, oil and oilbased industries are the cleanest and the least polluting.

## FLOTATION OF EFFLUENTS FROM SUNFLOWER OIL REFINERIES

Sunflower oil is one of the world's widely used vegetable edible oil. During processing of raw sunflower for obtaining the refined oil, significant amount of alkalines waste-water containing appreciable quantity of organic mass on a vegetable basis are released.

A proposed flow sheet for handling such an effluent has been carried out, employing flotation technique for separation. A flow sheet has been designed on the basis of research results obtained from flotation studies concerning vegetable acids and esters. A flotation machine has been chosen as a separation device. The proposed flow sheet has been designed in significantly simplified manner, so that installation set-up requires limited space. The flow sheet has been tested at the plant site of the largest Bulgarian sunflower refining industry. The organic material rejected during effluent treatment can be used for the manufacturing of soap, detergent and washing powder. The capital investment is low, so that it can be attached with flow sheet of oil refining industries, which has relatively low production capacity.

The recent technology of sunflower processing includes the operation of sunflower oil refineries. The paraffin and hard oils are rejected by vacuum filter in the operation. For improvement of filtration efficiency, oil perlite is used as coating material on the surface of filter cloth, so that hand oil and paraffin can be separated even at ultra fine sizes. It is required number of times of cleaning after dismantling. It is washed with soap liquids at high concentration of NaOH. The waste-water after washing having high oil concentration effluents are rejected in two stages. Normally sunflower oil refineries have acid waste-water effluents.

### Experimental Works in Laboratory Scale

The laboratory experiments work carried out using three kinds of waste-water rejected by one of the sunflower refinery at Sofia. The waste-water are as follows.

1. Water without soap (after first stage washing).
2. Wastes water with soap.
3. Acid waste-water.

The two kinds of waste-water contain oil phase, perlite and paraffin. The third kind of waste-water is acid without perlite and paraffin. When the pH of water is lower then 2, it contains less than 0.8 g/l of oil phase materials. When the pH of the water is around 9–9.5 during the first and second stage washing, the water contains around 2.29 g/l of oil phase materials. The objective of this investigation can be summarised in this manner: (i) separation of perlite from oil phase for recycling; and (ii) removal of oil phase material from water.

The concentration of oil phase material should be less than 100 mg/l, the main technological problems in this investigation are to disperse the oil-perlite aggregates and rejection of the oil phase materials from water. The first operation was carried out in diemulsificator with special desire and fabrication. The diemulsificator was build in laboratory-scale and scaled up to pilot scale. The second operation of the oil phase from waste-water was separated by using flotation technique. Flotation study was carried out using Denver flotation column having 1.2 litres capacity.

In design of the pilot scale, mechanical flotation column M-1 and M-3 were used. The detail material balance of the flow sheet is given in Fig. 17.8. The details of layout of machine installation of the flow sheet is given in Fig. 17.9.

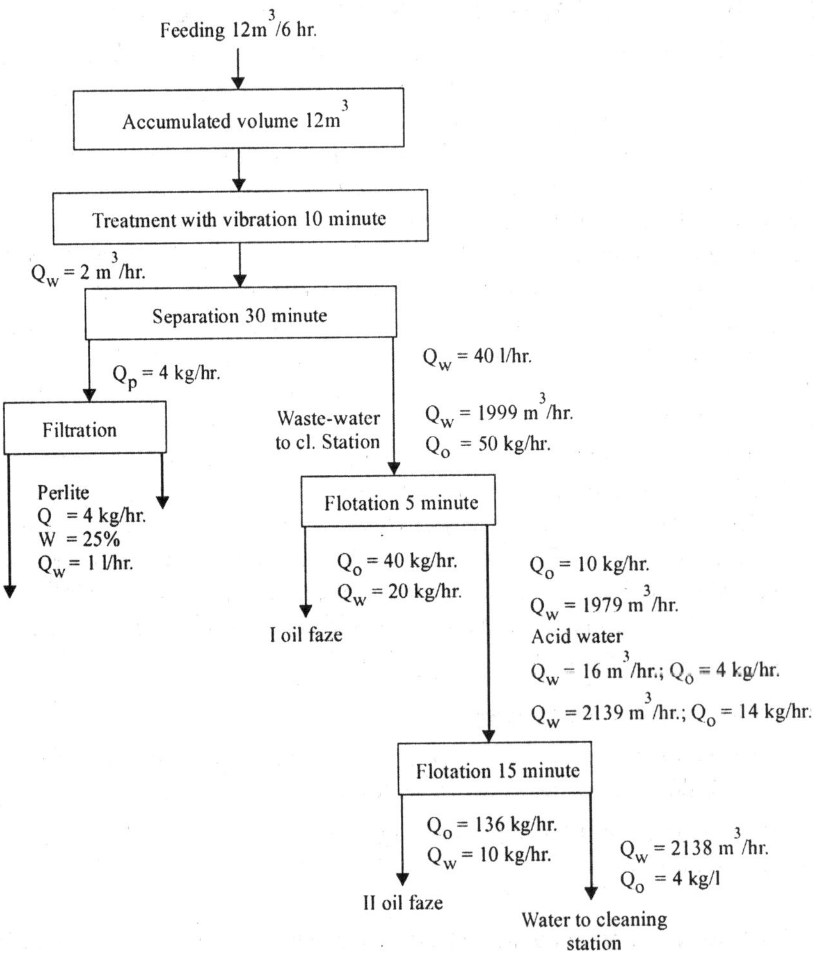


**Fig. 17.8.** Flowsheet for waste-water cleaning .

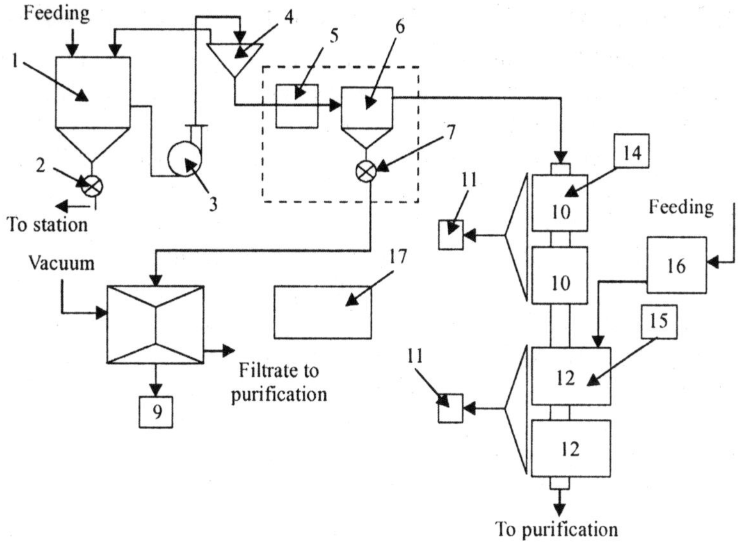


1. Buffer reservoir 12m[3]
2. Tap
3. Pump M-1F
4. Fixing the dose
5. Vibrator
6. Emulsion destroing 2500 l
7. Tap
8. Filter
9. Drum for perlite
10. Flotation machine M-1
11. I oil drum
12. Flotation machine M-3
13. II-oil drum
14. Reagent feeder
15. Reagent feeder
16. Buffer reservoir and acid water feeder
17. Drying

**Fig. 17.9.** Layout of machine installation.

The design of the pilot plant has been done, based on the following principles:

1. Rejection of the impurities from waste-water have to be made in two stages. In the first stage, waste-water was generated with soap mixed and also without soap. In the second stage the waste-water is mixed with acid waste-water.
2. The correlation between waste-water in mixture is settled from quantity of the different kinds of water and their pH. The pH of the waste-water is usually in between 8.5–9.5. After mixing of two kinds of waste-water, it is treated for dispersion of emulsions; of oily material and perlite. Then perlite is separated from waste-water by filtration. After filtration the moisture content in the cake is 15–20 per cent. The paraffin material was separated from the cake by simply drying at 105°C. Even then the perlite contains moisture around 3.5–4.0 per cent in loose state.
3. The waste-water with and without soap after perlite rejection is subjected to flotation for removal of the oil phase material. The oil phase mostly contains carboxylic acid and paraffins.
4. Waste-waters after first stage separation have been mixed with acid water at pH 3, so that soap material changes to carboxylic acids. By flotation, these impurities are separated in form of froth.

Thus, by this technology it is not only cleaning the waste-water, but also, three supplementary products can be obtained. The oil concentrate from the first stage cleaning can be used for production of food. The oil concentrate from the second stage can be used for the production of detergents. The rejected perlite can be used for the fertile of the soil.

## EDIBLE OIL PROCESSING INDUSTRY PROBLEMS OF ENERGY CONSERVATION

Energy saving operations can be planned in a number of ways. The management can install more energy efficient equipments, insulate tanks and pipelines and stop wastages. The batch process could be converted into more effective and efficient continuous ones. There can be recovery and re-usage of energy and optimisation of fuel and electrical energy. The detailed energy audit plan will identify the areas where opportunities exist for improvement.

The subject of energy conservation is receiving attention in all sectors of our economy. The oil processing industry can be no exception. The processing costs are increasing year after year while the prices of the finished products of oils are not matched in the same order resulting in pressure on the profits of corporate sector. Heat recovery or heat interchange has taken increased importance as a result of the energy crisis and sky-rocketing fuel cost. Therefore, not only for existence of the oil industry but also for its future growth it has now become more necessary than ever to conserve energy and reduce the cost of production. The plant manager can plan his energy saving operations in a number of ways. He can turn off energy loads, he can turn down energy intensive operations depending on his priorities.

The oil processing industry can be broadly divided into—oil refining, bleaching, hydrogenation, deodourisation and packing.

The oil refining, bleaching, hydrogenation and deodourisation industry consumes energy to heat and cool the oil as well as to generate the various vacuum levels that are required. Besides, the production of $H_2$ for use in hydrogenation, it is highly energy intensive. Heat recovery has now become an indispensible factor which needs to be studied more closely and thoroughly than has been the case until recently. The use of continuous bleaching and deodourisation equipments reduces the steam consumption by 50 per cent as compared to batch system mostly used in our country. Moreover, the heat generated in exothermic reaction of hydrogenation is 900 kcal/kg oil. This is for one drop of iodine value. The proper heat exchange could save enormous quantity of heat energy in a vanaspati plant. Suitable spiral heat exchangers are available now. The production of hydrogen by water electrolysis is another energy intensive operation. There are various other economical processes for the production of hydrogen and even the electrolysis plants now in operation in several parts of the world consume much less energy than KNOWLES cells being used in Indian industries. The major sub-sections of oil processing industry which consume bulk of the energy are given in Table 17.6.

**Table 17.6.** Energy intensive units of oil processing industry.

| *Refining* | *Bleaching* | *Hydrogenation* | *Deodourisation* |
|---|---|---|---|
| Storage | Storage | Storage | Storage |
| Pre-heating | Slurry | Pre-heating | Per-heating |
| Refining | Heat bleaching | Hydrogenation | Final heating |
| Water-cooling | Cooling | Cooling | Cooling |
| Drying | Filtration | Filtration | – |
| Storage | Storage | Storage | Storage |

In Tables 17.7 and 17.8 the steam consumption with or without the heat recovery systems in neutralising bleaching and deodourising units of oil processing industry is given. The heat generated in the exothermic reaction of hydrogenation is 900 kcal/kg oil for one drop of iodine value. The proper heat exchange could save enormous quantity of heat energy and thus, reduce the total cost of thermal energy by at least 50 per cent.

**Table 17.7.** Neutralising-bleaching line (1 per cent bleaching earth).

| *Neutralising washing, drying-bleaching* | *Without heat recovery* | *With heat recovery* |
|---|---|---|
| Heating of crude oil from 20 to110°C | 88 kg | 38 kg |
| Hot water production | 30 kg | 18 kg |
| Set (s) vacuum equipment | 50 kg | 35 kg |
| Steam blowing of filters | 15 kg | 15 kg |
| Total | 183 kg | 106 kg |

**Table 17.8.** Steam, water and heating energy consumption per tonne processed oil, semi-continuous deodourising plant.

| *Description* | *Without heat recovery* | *With heat recovery* |
|---|---|---|
| Live steam for vacuum production (water at 28°C) | 125 kg | 125 kg |
| Live steam | 17 kg | – |
| Heating energy (oil at 255°C) | 500 Mj | 150 Mj |
| Cooling water for vacuum production (water at 28°C) | 13 $m^3$ | 13 $m^3$ |
| Water for oil cooling (water at 28°C) | 25 $m^3$ | 2.75 $m^3$ |

Energy is used to provide heat, motive power and cooling. Heat is provided by burning fuel to generate a medium which transfers the heat to a point of use. The medium is steam heat transfer fluid, like dowther/mineral oil and hydrogen gas. Exothermic heat is a by-product of the hydrogenation reaction. Motive power of cooling is provided by the steam but generally by electricity.

By analysing the use of energy it is necessary to divide between energy variable with production and operation versus energy that does not vary with production. The latter can be called 'fixed'. The other interesting factor which has to be taken into account for steam production and usage, cooling water availability; the product melting point, and the weather. Therefore, the use of energy in an oil processing plant also has to be qualified with temperature.

## Steps for Energy Conservation

There are five logical steps to be followed in making an energy audit. These are as follows:

1. Draw a process flow diagram.
2. Determine operating conditions, flow rate, temperature, pressure.
3. Establish the audit system–energy input to energy output.
4. Material balance and flow–solids, liquids, gases, etc.
5. Convert material balance to energy balance.

The data obtained for each process step in the energy audit can be listed in the flow diagram to ensure that all energy is accounted for.

### *Heat energy*

The opportunities for pursuing conservation effort in heat energy can be further summarised as follows:

1. Optimise steam boiler efficiency.
2. Insulate tanks and equipment.
3. Establish a good maintenance programme, repair, steam leaks, steam traps, etc.
4. Modify processes as to salvage heat for re-use (heat exchangers and economisers).

### *Electrical energy*

Similarly, opportunities for conserving electrical energy are summarised as follows:

1. Evaluate the energy benefit in the existing hydrogen plant.
2. Energy conservation by designing proper chilling systems.
3. Use of efficiently designed electric motors.
4. Check on idle running of equipment.

## New Researches for Energy Savings

In the oil seed processing plants, energy and solvent cost account for 50 per cent of total operating cost. Energy conservation will continue to be a primary factor influencing change. Fluid-bed system is being tried for drying, cooling and conditioning operations. This process is said to eliminate a number of process steps and save about 50 per cent thermal energy used in conventional process of dehulling and flaking. Similarly, another system for drying with microwaves and infrared is being demonstrated with advantages of energy savings. For savings in solvent, research is being done for commercial use of other solvents like water and supercritical carbon dioxide for oil extraction. The use of enzymes for oil extraction would reduce the consumption of hexane by 70 per cent as 90 per cent of the extraction would be possible by the use of enzyme.

From the above we can conclude that heat recovery has now become an indispensable factor which needs to be studied carefully. Economy in the electrical energy consumption can be achieved in the operations by taking appropriate steps suggested above. The energy conservation programmes may need updating of the age old batch plants with continuous plants, fitted with heat recovery and energy economiser systems.

# References


**Brydson, J.A.,** *Encyclopedia of Food Science and Technology,* Elsevier Applied Science, London.

**Brydson, J.A.,** *Chemistry and Technology of Edible Oils,* Applied Science, London.

**Bunn, C.W.,** *Bailey's Industrial Oil and Fat Products,* Butterworth Scientific, London

**Treloar, L.R.,** *Fatty Acids in Industry,* Clarendon Press, London.

**Gardon, J.L.,** *Introduction to Fats and Oil Technology,* Interscience, New York.

**Bateman, L.,** *Encyclopedia of Chemical Technology,* Applied Science, London.

**Stannett, V.,** *Vitamins and Minerals,* Academic Press, London.

**Wall, L.A.,** *Antioxidants,* Interscience, New York.

**Noll, W.,** *Chemistry and Technology of Fats and Oil,* Academic Press, New York.

**Blow, C.M.,** *Soyabean Utilisation Alternatives,* Butterworth Scientific, London.

**Hofmann, W.,** *Essential Oils and Aromatic Chemicals,* Applied Science Publishers, London.

**Smith, C.K.,** *Spices as Antioxidants,* John Wiley and Sons, New York.

**Holden, G.,** *Chemistry and Technology of Lipids,* Reinhold, New York.

**Lewis, F.W.,** *Soap Manufacture,* John Wiley and Sons, New York.

**Brown, R.P.,** *Food Sciences,* Applied Science Publishers, London

**Morton, M.,** *Dairy Technology and Engineering,* Van Nostrand Reinhold, New York.

**Craig, A.S.,** *Fatty Acids and their Derivatives,* John Wiley and Sons, New York.

**Blackley, D.C.,** *Basic Concepts of Food Technology,* Applied Science Publishers, London.

**Saunders, J.H.,** *Chemistry of Animal and Vegetable Oils,* Interscience, New York.

**Wright, P.,** *Edible Fats and Oil Processing,* Maclaren, London.

**Alliger, G.,** *Chemistry and Biochemistry of Natural Waxes,* Reinhold, New York.

**Scott, G.,** *Cosmetics and Toiletries,* Elsevier, Amsterdam.

**Bruins, P.F.,** *Sunflower Science and Technology,* Reinhold, New York.

**Kraus, G.,** *Olive Oil Technology,* Interscience, New York.

**Donnet, J.B.,** *Elements of Food Engineering,* Marcel Dekker, New York.

**Nourry, A.,** *Chemistry of Toxicity*, Maclaren, London.

**Allen, P.W.,** *Cottonseed Oil,* Maclaren, London.

**Anon, W.,** *Frying Technology,* Applied Science, London.

**Collyer, H.J.,** *Dairy Technology,* Interscience, New York.

**Khromov, M.K.,** *Nutrition and Dietics,* Maclaren, London.

**Mooney, M.,** *Chemistry of Fatty Acids,* Academic Press, New York.

**Norman, R.H.,** *Coconut Oil,* Maclaren, London.

**Penn, W.S.,** *Castor Oil,* Maclaren, London.

**Scott, J.R.,** *Chemistry of Emulsifiers,* Maclaren, London.

**Stamberger, P.,** *Rendering, the Invisible Industry,* Oxford University Press, Oxford.

**Stern, H.J.,** *Flavour and Fragrances,* Maclaren, London.

**Thompson, D.C.,** *Encyclopaedia of Chemical Technology*, John Wiley and Sons, New York.

**Sprague, G.R.,** *Handbook of Lipids,* Vanderbilt, New York.

**Wheelans, M.A.,** *Bakery Products,* Maclaren, London.

**White, J.L.,** *Science and Technology of Surfactants,* Academic Press, New York.

**Nauntron, W.J.,** *Handbook of Vegetables Oils,* Arnold, London.

**Kuryla, W.C.,** *Soyabean Utilisation Alternatives,* Maclaren, London.

**Killeffer, D.H.,** *Sunflower Science and Technology,* Palmerton, New York.

**Cooper, W.,** *Oleochemicals and Allied Products,* Applied Science Publishers, London.

**Brennan, J.J.,** *Toxicity and Safe Handling of Oils and Fats,* Academic Press, New York.

**Allen, N.S.,** *Chemistry of Lipids,* Elsevier, USA.

**Bark, L.S.,** and **Allen, N.S.,** *Analysis of Oils and Fats*, Elsevier, USA.

**Alfrey, T.,** *Chemistry of Oil and Fats,* Interscience, New York, USA.

**Bernhardt, E.C.,** *Chemistry of Lipids,* Pergamon, UK.

**Bernhardt, E.C.,** *Vegetable Protein Utilisation in Human Foods and Animals Feedstuffs,* Van Nostrand Reinhold, USA.

**Bruins, P.F.,** *Canola Meal Feed Industry Guide,* Wiley Interscience, New York.

**Ceresa, R.J.,** *Corn Chemistry and Technology,* Wiley Interscience, New York.

**Ferry, J.D.,** *Handbook of Soap Technology,* Wiley Interscience, New York.

**Harper, C.A.,** *Degumming, Refining, Washing and Drying,* McGraw-Hill, New York.

**Iwen, E.D.,** *Antioxidants: The Ultimate Answer of Oxidation,* Elsevier, USA.

**MacDermott, C.,** *Linseed Refining,* Marcle Dekker, Reinhold, USA.

**Matthews, F.G.,** *Deodourisation—A Basic Understanding,* Van Nostrand Reinhold, USA.

**McKelvey, J.M.,** *Spices, Herbs and Edible Fungi,* Wiley Interscience, New York.

**Cherry, J.P.,** *Lecithins: Sources, Manufacture and Uses,* American Oil Chemists Society.

**Alliger, G.,** *Chemistry and Technology of Lips,* Arnold, London.

**Saunders, J.H.,** *Food Sciences,* Pergamon, UK.

**Campbell, R.,** *Nutrition and Dietics,* Arnold, London.

# Index